NONPARAMETRIC TECHNIQUES IN STATISTICAL INFERENCE

T0297859

NONPARAMETRIC TECHNIQUES IN STATISTICAL INFERENCE

EDITED BY

MADAN LAL PURI

Professor of Mathematics, Indiana University
Bloomington

CAMBRIDGE

AT THE UNIVERSITY PRESS

1970

CAMBRIDGE UNIVERSITY PRESS
Cambridge, New York, Melbourne, Madrid, Cape Town, Singapore, São Paulo, Delhi

Cambridge University Press
The Edinburgh Building, Cambridge CB2 8RU, UK

Published in the United States of America by Cambridge University Press, New York

www.cambridge.org
Information on this title: www.cambridge.org/9780521078177

© Cambridge University Press 1970

First published 1970
This digitally printed version 2008

A catalogue record for this publication is available from the British Library

Library of Congress Catalogue Card Number: 74–116750

ISBN 978-0-521-07817-7 hardback
ISBN 978-0-521-09305-7 paperback

CONTENTS

Part 2: Testing and Estimation 2

Part 3: Order Statistics and Allied Problems

CONTENTS   vii

Part 7: Teaching of Nonparametric Statistics

PREFACE

For the last two decades, the field of nonparametric statistics has been occupying an increasingly successful role in statistical theory as well as in the methods of application. The importance of the developments in this area is now very well recognized not only by its elegant theory, but also because of the significantly improved performances, the nonparametric methods are yielding in various applications in the fields of Biometrics, Agriculture, Economics and industry among others.

Many of the research workers who have been concerned with nonparametric statistics felt the need for a meeting which by providing an opportunity for personal contacts among scholars from different intellectual centres of the world, could survey the present state of the subject, disseminate the recent research, discuss the major open problems, and provide stimulation for further research and scholarship through the lectures of carefully selected speakers. To meet this purpose, the first International Symposium on Nonparametric Techniques in Statistical Inference was held at Indiana University, Bloomington, Indiana, 1–6 June 1969. Professor L. L. Merrit, Jr, Vice-President for Research and Dean of Advanced Studies at Indiana University, was kind enough to make the opening remarks, and Colonel W. R. Trott, Air Force Office of Scientific Research, welcomed the guests. Professors Z. W. Birnbaum and Jerzy Neyman were the main speakers at the final banquet.

The program committee of the symposium consisted of Professor H. A. David, Professor Jack Kiefer, Colonel W. R. Trott and myself. I take this opportunity to express my sincere thanks to the members of this committee for the overall help they gave me in making the symposium a great success. I also benefited greatly by the advice I received from Mr B. R. Agins in the planning of the symposium. To Professors H. A. David and Jack Kiefer, my thanks are due also for their help in selecting the contributors. In this task I was also profited greatly by the advice I received from Professors Z. W. Birnbaum, Jaroslav Hájek, Wassily Hoeffding, Ronald Pyke and Milton Sobel. It is a pleasure to express my deep appreciation to all of them.

The proceedings of the symposium are divided into seven parts. Parts 1 and 2 are devoted to the problems of testing and estimation. Part 3 deals with order statistics and allied problems. Part 4 deals with general theory. Part 5 deals with Ranking and Selection procedures. Part 6 deals with Decision Theoretic and Empirical Bayes procedures. Part 7 is devoted to the teaching of nonparametric statistics at the elementary level.

Professors R. E. Bechhofer, Patrick Billingsley, Julius Blum, H. A. David, H. T. David, S. S. Gupta, Jaroslav Hájek, Wassily Hoeffding, G. Kallianpur, Leo Katz, Oscar Kempthorne, Jerzy Neyman, G. E. Noether, J. W. Pratt, Ronald Pyke, Murray Rosenblatt, S. Sherman, Milton Sobel and Drs A. W. Marshall, William H. Pell and R. G. Pohrer very kindly presided over different sessions.

For acting as formal discussants and/or referees of the papers, my sincere thanks are due to Professors R. E. Bechhofer, H. A. David, H. T. David, Václav Dupač, J. Gastwirth, J. Hájek, Wassily Hoeffding, Oscar Kempthorne, Jack Kiefer, D. S. Moore, J. W. Pratt, P. S. Puri, Ronald Pyke, P. K. Sen, Milton Sobel, Lajos F. Takács, M. Woodroofe, and Drs J. E. Jackson, Gary C. McDonald and A. W. Marshall.

It is also a pleasure to express my appreciation to Mr Paul Bigelow, Professors J. Chadam, and J. Thompson, and Mr Norman Wykoff for taking care of the problems connected with local arrangements. For transportation and other logistic problems very valuable assistance was received from my students, Irvin Bossler, David Donley, Michael Ehrsam, John Gaisser, Carl T. Russel and Norman Wykoff.

The contributors are to be thanked for the fine spirit of co-operation and the prompt handling of the correspondence.

The symposium would not have been possible without the very generous support from the Air Force Office of Scientific Research. My heartiest thanks are due to Dr R. G. Pohrer and Colonel W. R. Trott who took personal interest in the project, and gave me all the help I needed in making this special event successful. It is also a pleasure to express my deep sense of appreciation to Indiana University for providing excellent facilities to hold this conference.

Special thanks are due to the personnel of the Cambridge University Press for taking care of the many complexities of publishing the manuscripts with the utmost care and patience.

Last but not the least my thanks go to the Department of Mathematics and my colleagues therein for their helpful attitude towards the success of the symposium. Special thanks are due to Professor George Springer, Chairman of the Department of Mathematics, who in spite of his multifarious activities, spent a considerable amount of time from the initiation to the completion of this project and gave me very valuable assistance and advice. It is a pleasure to express here my deep appreciation.

MADAN LAL PURI

CONTRIBUTORS

Numbers in parentheses indicate the pages on which the authors' contributions begin.

T. V. Avadhani, Department of Statistics, University of Michigan, Ann Arbor, Michigan (215).[1]

Richard E. Barlow, Department of Engineering, Operation Research Center, University of California, Berkeley, California (159).

R. E. Bechhofer, Department of Operations Research, Cornell University, Ithaca, New York (545).

C. B. Bell, Department of Statistics, University of Michigan, Ann Arbor, Michigan (215).

Z. W. Birnbaum, Department of Mathematics, University of Washington, Seattle, Washington (427).

A. A. Borovkov, Institute of Mathematics, Siberian Division, Akademiia SSSR, Novosibirsk, U.S.S.R. (259).

Ralph A. Bradley, Department of Statistics, Florida State University, Tallahassee, Florida (111).

H. D. Brunk, Department of Statistics, University of Missouri, Columbia, Missouri (177).[2]

S. K. Chatterjee, Department of Statistics, Calcutta University, Calcutta, India (563).

H. A. David, Department of Biostatistics, University of North Carolina, Chapel Hill, North Carolina (402).

H. T. David, Department of Statistics, Iowa State University, Ames, Iowa (154).

Roger R. Davidson, Department of Mathematics, University of Victoria, British Columbia, Canada (111).

V. Dupač, Department of Statistics, Charles University, Prague, Czechoslovakia (73, 75).

J. Durbin, London School of Economics and Political Science, London, England (435).

F. Eicker, Institute for Mathematical Statistics, The Albert-Ludwigs University, Freiburg, West Germany (321).

---

[1] Present address: Department of Statistics, Andhara University, Waltair, India.
[2] Present address: Department of Statistics, Oregon State University, Corvallis, Oregon.

Joseph L. Gastwirth, Department of Statistics, Johns Hopkins University, Baltimore, Maryland (89).

Shanti S. Gupta, Department of Statistics, Purdue University, Lafayette, Indiana (491).

Jaroslav Hájek, Department of Statistics, Charles University, Prague, Czechoslovakia (3).

Wassily Hoeffding, University of North Carolina, Chapel Hill, North Carolina (18, 50).

Peter J. Huber, Department of Mathematics, Eidg. Technishe Hochschule, Zürich, Switzerland (453).

J. Edward Jackson, Eastman Kodak Company, Rochester, New York (126).

Oscar Kempthorne, Department of Statistics, Iowa State University, Ames, Iowa (450).

J. Kiefer, Department of Mathematics, Cornell University, Ithaca, New York (299, 349).

Charles H. Kraft, Department of Mathematics, University of Montreal, Montreal, Canada (267).

B. P. Lientz, System Development Corporation, Santa Monica, California (275).

A. W. Marshall, Boeing Scientific Research Laboratories, Seattle, Washington (174, 465).

Gary C. McDonald, General Motors Corporation, Warren, Michigan (491).

Kenneth S. Mount, Department of Statistics, Iowa State University Ames, Iowa (479).

G. E. Noether, Department of Statistics, University of Connecticut, Storrs, Connecticut (607).

Ingram Olkin, Department of Statistics, Stanford University, Stanford, California (465).

J. W. Pratt, Graduate School of Business Administration, Harvard University, Boston, Massachusetts (602).

Madan L. Puri, Department of Mathematics, Indiana University, Bloomington, Indiana (131).

Ronald Pyke, Department of Mathematics, University of Washington, Seattle, Washington (21, 196).

J. S. Rao, Indian Statistical Institute, Calcutta, India (405).[1]

M. Rosenblatt, Department of Mathematics, University of California, La Jolla, California (199).

Herman Rubin, Department of Statistics, Purdue University, Lafayette, Indiana (579).

Pranab Kumar Sen, Department of Biostatistics, University of North Carolina, Chapel Hill, North Carolina (53).

J. Sethuraman, Department of Statistics, Florida State University, Tallahassee, Florida (405).

Harold D. Shane, Department of Mathematics, Baruch College, City University of New York, New York (131).

M. M. Siddiqui, Department of Mathematics and Statistics, Colorado State University, Fort Collin, Colorado (417).

Milton Sobel, Imperial College of Science and Technology, London, England, and Department of Statistics, University of Minnesota, Minneapolis, Minnesota (515).[2]

M. Sycheva, Institute of Mathematics, Siberian Division, Akademiia SSSR, Novosibirsk, U.S.S.R. (259).

Lajos F. Takács, Department of Mathematics, Case Western Reserve University, Cleveland, Ohio (359).

K. Takeuchi, University of Tokyo, Japan, and Courant Institute of Mathematical Sciences, New York University, New York (283).

Constance van Eeden, Department of Mathematics, University of Montreal, Montreal, Canada (267).

J. van Ryzin, Department of Statistics, University of Wisconsin, Madison, Wisconsin (584).

Willem R. van Zwet, Department of Mathematics, University of Leiden, Netherlands (159).

I. Vincze, Institute of Mathematics, The Hungarian Academy of Sciences, Budapest, Hungary (385).

George H. Weiss, National Institute of Health, Bethesda, Maryland (515).

[1] Present address: Department of Mathematics, Indiana University, Bloomington, Indiana.

[2] Present address: Department of Statistics, University of Minnesota, Minneapolis, Minnesota.

Lionell Weiss, Department of Operations Research, Cornell University, Ithaca, New York (343).

H. Witting, Institute for Mathematical Statistics, University of Münster, Münster, West Germany (41).

M. Woodroofe, Department of Statistics, University of Michigan, Ann Arbor, Michigan (211).

PART 1

TESTING AND ESTIMATION 1

MISCELLANEOUS PROBLEMS OF RANK TEST THEORY

JAROSLAV HÁJEK

1 INTRODUCTION AND SUMMARY

The paper deals with diverse problems, partly unpublished, partly published without proofs in Hájek (1969). Section 3 proposes three methods of density estimation which may be used in choosing proper scores for rank tests. The second of these methods is based on a theorem by Jurečková (1969) on asymptotic linearity of simple linear rank statistics, which is presented in a modified form in § 2. The same result by Jurečková is once again used in § 4 to prove that some statistics for testing scale (the symmetric case) may be adapted for the case when both medians are unknown without inflicting their asymptotic distribution. For a less general result of this kind, derived by a different method, *see* Raghavachari (1965). Section 5 provides a theorem to the effect that one partial ordering of permutations is finer than another. This theorem may be utilized in proving Jurečková's theorem, and also in proving unbiasedness of some nonparametric tests (see Lehmann (1966), and Hájek (1969)). The last section provides an explicit form for asymptotic normality of the conditional distribution of averaged rank statistics, when ties are present.

2 THE BEHAVIOUR OF LINEAR RANK STATISTICS UNDER A SHIFT

For each $N \geqslant 1$ let us consider a simple linear rank statistics.

$$S_N = \sum_{i=1}^{N} c_{Ni} a_N(R_{Ni}), \qquad (2.1)$$

where the ranks $R_{N1}, ..., R_{NN}$ are derived from a sequence of observations $Y_{N1}, ..., Y_{NN}$. Introducing

$$u(y) = 1, \quad x \geqslant 0, \qquad (2.2)$$
$$= 0, \quad x < 0,$$

and putting $\quad s_N(y_1, ..., y_N) = \sum_{i=1}^{N} c_{Ni} a_N \left(\sum_{j=1}^{N} u(y_i - y_j) \right) \qquad (2.3)$

we may also write $\qquad S_N = s_N(Y_{N1}, ..., Y_{NN}). \qquad (2.4)$

[3]

We shall study the increment of S_N, if the Y_{Ni}'s are shifted by a random multiple of some numbers d_{Ni}. That is, considering simultaneously the statistic

$$S_N^* = s_N(Y_{N1} + \Delta_N d_{N1}, ..., Y_{NN} + \Delta_N d_{NN}) \tag{2.5}$$

we shall try to find an asymptotic expression for the difference $S_N^* - S_N$.

Assumptions A:

(A1) The regression constants satisfy the Noether condition:

$$\lim_{N\to\infty} \frac{\max_{1\leqslant i\leqslant N}(c_{Ni}-\bar{c}_N)^2}{\sum_{i=1}^{N}(c_{Ni}-\bar{c}_N)^2} = 0 \quad \left(\bar{c}_N = \frac{1}{N}\sum_{i=1}^{N}c_{Ni}\right). \tag{2.6}$$

(A2) The scores $a_N(i)$ are generated by a function $\phi(t)$, $0 < t < 1$, by either of the following two ways:

$$a_N(i) = \phi\left(\frac{i}{N+1}\right) \quad (1\leqslant i\leqslant N), \tag{2.7}$$

$$a_N(i) = E\phi(U_N^{(i)}) \quad (1\leqslant i\leqslant N), \tag{2.8}$$

where $U_N^{(i)}$ denotes the ith order statistic in a sample of size N from the uniform distribution on $(0,1)$.

(A3) The score-generating function $\phi(t)$ is nonconstant and expressible as a difference of two *nondecreasing* and *square integrable* functions.

(A4) The constants d_{Ni} are such that

$$\sup_{N}\sum_{i=1}^{N}(d_{Ni}-\bar{d}_N)^2 < \infty, \tag{2.9}$$

and $$\lim_{N\to\infty}\max_{1\leqslant i\leqslant N}(d_{Ni}-\bar{d}_N)^2 = 0 \quad \left(\bar{d}_N = \frac{1}{N}\sum_{i=1}^{N}d_{Ni}\right). \tag{2.10}$$

(A5) The constants d_{Ni} and c_{Ni} are concordant in the following sense:

$$(c_{Ni}-c_{Nj})(d_{Ni}-d_{Nj}) \geqslant 0 \quad (1\leqslant i,j\leqslant N). \tag{2.11}$$

(A6) The random variables $\Delta_N = \Delta_N(Y_{N1},...,Y_{NN})$ are *bounded in probability*.

(A7) $Y_{N1},...,Y_{NN}$ is a random sample from a distribution with a density f possessing finite Fisher's information.

Put $$\phi(t,f) = -\frac{f'(F^{-1}(t))}{f(F^{-1}(t))} \quad (0<t<1). \tag{2.12}$$

and $$b_N = \sum_{i=1}^{N}d_{Ni}(c_{Ni}-\bar{c}_N)\int_0^1\phi(t)\,\phi(t,f)\,dt. \tag{2.13}$$

Theorem 2.1. Under assumptions (A 1) *through* (A 7),

$$\frac{S_N^* - S_N - \Delta_N b_N}{\sqrt{(\operatorname{var} S_N)}} \to 0 \tag{2.14}$$

holds in probability.

Proof. The theorem is an easy corollary of a result by J. Jurečková (1969). She showed that for nonrandom Δ_N (2.14) holds uniformly for $|\Delta_N| \leqslant C$. Consequently (2.14) must also hold for Δ_N random and bounded in probability.

Remark 2.1. Instead of (A 4) and (A 5) we could also assume that $d_{Ni} = d'_{Ni} - d''_{Ni}$, where both the constants d'_{Ni} and d''_{Ni} satisfy (A 4) and (A 5).

Remark 2.2. Assumption (A 2) may be replaced by (A 2*): The scores $a_N(i)$ satisfy

$$\frac{1}{N} \sum_{i=1}^{N} [a_N(i) - E\phi(U_N^{(i)})]^2 \to 0. \tag{2.15}$$

3   THREE METHODS OF DENSITY TYPE ESTIMATION

Let $\mathscr{F}_1, \dots, \mathscr{F}_k$ be k distinct density types generated by some one-dimensional densities $f_1, \dots, f_k$:

$$\mathscr{F}_j = \{f : f(x) = \lambda f_j(\lambda x - u), \, -\infty < u < \infty, \, \lambda > 0\} \quad (1 \leqslant j \leqslant k). \tag{3.1}$$

Given a sample $(X_1, \dots, X_N)$ governed by a density of the form

$$p(x_1, \dots, x_N) = \prod_{i=1}^{N} f(x_i), \tag{3.2}$$

where f belongs to $\bigcup_{j=1}^{k} \mathscr{F}_j$ but otherwise is unknown, let us try to locate the type of f, i.e. find j such that $f \in \mathscr{F}_j$. Thus, a decision procedure for this problem will be a function $\delta(x_1, \dots, x_N)$ taking its values in the set $\{1, 2, \dots, k\}$. Assume that the loss incurred by $\delta(x) = d$ under $f \in \mathscr{F}_j$ is given by

$$L(j, d) = 0 \quad \text{if} \quad j = d,$$

$$= 1 \quad \text{if} \quad j = d, \, 1 \leqslant j, d \leqslant k. \tag{3.3}$$

Restricting ourselves to procedures that are invariant relative to the group of positive linear transforms

$$((x_1, \dots, x_N) \to (\lambda x_1 + u, \dots, \lambda x_N + u), \lambda > 0)$$

the risk of a procedure given $f \in \mathscr{F}_j$ will equal

$$R(j, \delta) = 1 - P(\delta(X_1, \dots, X_N) = j \,|\, f \in \mathscr{F}_j). \tag{3.4}$$

Invariant Bayes solutions may be obtained from 'marginal' densities relative to the sub σ-field consisting of events that are invariant with respect to the group of positive linear transforms. These densities can be established for our types as follows:

$$\bar{p}_j(x_1, \ldots, x_N) = \int_0^\infty \int_{-\infty}^\infty \left[\prod_{i=1}^N f_j(\lambda x_i - u) \right] \lambda^{N-2} du \, d\lambda \quad (1 \leqslant j \leqslant k) \quad (3.5)$$

(see Hájek and Šidák (1967), § II.2.2). If all types are considered a priori equiprobable, then the expected risk $(1/k) \sum_{j=1}^k R(j, \delta)$ is minimized by the procedure δ_0 such that

$$[\delta_0(x_1, \ldots, x_N) = j] \Rightarrow [\bar{p}_j(x_1, \ldots, x_N) = \max_{1 \leqslant h \leqslant k} \bar{p}_h(x_1, \ldots, x_N)]. \quad (3.6)$$

Any such procedure satisfies

$$\sum_{j=1}^k R(j, \delta_0) \leqslant \sum_{j=1}^k R(j, \delta) \quad (\delta \text{ invariant}). \quad (3.7)$$

One should expect that

$$\sum_{j=1}^k R(j, \delta_0) \to 0 \quad \text{as} \quad N \to \infty$$

under fairly general conditions concerning the types $\mathscr{F}_1, \ldots, \mathscr{F}_k$. If the ϕ-functions $\phi(t, f_j)$ given by (2.12) satisfy (A 3), we shall be able to infer

$$\sum_{j=1}^N R(j, \delta_0) \to 0$$

from (3.7) and from the fact that

$$\sum_{j=1}^N R(j, \delta_1) \to 0$$

for some other invariant decision method δ_1 to be described next. We conclude the discussion of the procedure (3.6) by observing that the complexity of the formula for $\bar{p}_j$ makes its applicability doubtful.

Another selection procedure is yielded by Theorem 2.1. First introduce the functions

$$s_{Nj}(y_1, \ldots, y_N) = \sum_{i=1}^{[\frac{1}{2}N]} a_{Nj} \left(\sum_{h=1}^N u(y_i - y_h) \right) \quad (1 \leqslant j \leqslant k), \quad (3.8)$$

where the scores $a_{Nj}(i)$ are generated either by (2.7) or by (2.8) with $\phi(t)$ replaced by $\phi(t, f_j) = -f_j'(F_j^{-1}(t))/f_j(F_j^{-1}(t))$. Further put

$$d_{Ni} = 1/\sqrt{N} \quad (1 \leqslant i \leqslant [\frac{1}{2}N]), \quad (3.9)$$

$$= 0 \quad (\tfrac{1}{2}N \leqslant i < N),$$

and denote by $\Delta_N = \Delta_N(y_1, ..., y_N)$ any statistic such that

$$\Delta_N(\lambda y_1 + u, ..., \lambda y_N + u) = \lambda \Delta_N(y_1, ..., y_N) \quad (-\infty < u < \infty, \, \lambda > 0) \quad (3.10)$$

and that

$$0 < p \lim_{f=f_j} \Delta_N(X_1, ..., X_N) = b_j < \infty \quad (1 \leqslant j \leqslant k). \quad (3.11)$$

For example, if $X_N^{(i)}$ denotes the ith order statistic we may put

$$\Delta_N = M[X_N^{(N+1-j_N)} - X_N^{(j_N)}] \quad \left(\frac{j_N}{N} \to \alpha, \, 0 < \alpha \leqslant \tfrac{1}{2}\right). \quad (3.12)$$

Then obviously,

$$b_j = M[F_j^{-1}(1-\alpha) - F_j^{-1}(\alpha)] \quad \left(F_j(\alpha) = \int_0^\alpha f_j(x)\,dx\right). \quad (3.13)$$

Finally put

$$S_{Nj} = s_{Nj}(X_1, ..., X_N) \quad (1 \leqslant j \leqslant k) \quad (3.14)$$

and

$$S_{Nj}^* = s_{Nj}(X_1 + \Delta_N d_{N1}, ..., X_N + \Delta_N d_{NN}) \quad (3.15)$$

and compute the ratios

$$l_{Nj} = \frac{S_{Nj}^* - S_{Nj}}{\sqrt{(\mathrm{var}\, S_{Nj})}}. \quad (3.16)$$

The invariance of l_{Nj} under positive linear transforms and a direct application of Theorem 2.1 provide

$$p \lim_{f \in \mathscr{F}_h} l_{Nj} = p \lim_{f=f_h} l_{Nj} = p \lim_{f=f_h} \frac{\Delta_N \tfrac{1}{4}\sqrt{N} \int_0^1 \phi(t,f_h)\,\phi(t,f_j)\,dt}{\left[\tfrac{1}{4}N \int_0^1 \phi^2(t,f_j)\,dt\right]^{\frac{1}{2}}}$$

$$= \tfrac{1}{2} b_h \rho_{jh} \sqrt{I_h}, \quad (3.17)$$

where

$$I_h = \int_0^1 \phi^2(t,f_h)\,dt$$

and

$$\rho_{jh} = [I_j I_h]^{-\frac{1}{2}} \int_0^1 \phi(t,f_h)\,\phi(t,f_j)\,dt. \quad (3.18)$$

If the types $\mathscr{F}_1, ..., \mathscr{F}_k$ are distinct, we have

$$\rho_{jh} < 1 \quad (1 \leqslant j \neq h \leqslant k), \quad (3.19)$$

while $\rho_{jj} = 1$, $1 \leqslant j \leqslant k$. Consequently the decision procedure δ_1 such that

$$[\delta_1(x_1, ..., x_N) = j] \Rightarrow [l_{Nj}(x_1, ..., x_N) = \max_{1 \leqslant h \leqslant k} l_{Nh}(x_1, ..., x_N)] \quad (3.20)$$

is consistent in the sense that $R(j, \delta_1) \to 0$ if $N \to \infty$, $1 \leqslant j \leqslant k$. This entails

$$\sum_{j=1}^k R(j, \delta_1) \to 0,$$

and, in view of (3.7), also

$$\sum_{j=1}^k R(j, \delta_0) \to 0.$$

In order to apply the selection procedure δ_1 to the choice of proper scores in a rank statistic of form (2.1), we need δ_1 to be a function of the order statistic $X^{(\cdot)} = (X^{(1)}, \dots, X^{(N)})$ only.

This goal may be easily achieved by introducing random variables

$$Y_i = X^{(Q_i)} \quad (1 \leqslant i \leqslant N), \tag{3.21}$$

where $Q = (Q_1, \dots, Q_N)$ is independent of $(X_1, \dots, X_N)$ and assumes each of $N!$ permutations of $(1, 2, \dots, N)$ with probability $1/N!$. Then $(Y_1, \dots, Y_N)$ has the same distribution as $(X_1, \dots, X_N)$ under all densities of the form (3.2) and the selection procedure may equally well be applied to $Y_1, \dots, Y_N$ instead of to $X_1, \dots, X_N$. Thus we have

Theorem 3.1. *Under the above notations, if the functions $\phi(t, f_j)$ are expressible as differences of two nondecreasing and square integrable functions, $1 \leqslant j \leqslant k$, then*

$$\lim_{N \to \infty} P(\delta_1(Y_1, \dots, Y_N) = j \mid q_{Nj}) = 1 \tag{3.22}$$

holds for any sequence of densities $q_{Nj}(x_1, \dots, x_N)$ that are contiguous to densities of the form

$$p_{Nj}(x_1, \dots, x_N) = \lambda_N^N \prod_{i=1}^{N} f_j(\lambda_N x_i + u_N) \quad (-\infty < u_N < \infty, \, \lambda_N > 0). \tag{3.23}$$

Proof. The above considerations yielded the proof for $q_{Nj} = p_{Nj}$, i.e. if f in (3.2) satisfied $f \in \mathscr{F}_j$ for every $N \geqslant 1$. However, if q_{Nj} are contiguous relative to p_{Nj} in the X-spaces, the induced distributions of $(Y_1, \dots, Y_N)$ are also contiguous to p_{Nj} in the Y-spaces, as may be easily seen. Then it suffices to note that convergence to 1 under p_{Nj} entails the same under any contiguous alternative.   Q.E.D.

Let us resume the discussion of how δ_1 may be applied in rank testing: Let us test the hypothesis of randomness against the alternative of two samples differing in location or against a general location shift alternative. Then the pertinent statistic is of form (2.1), where the scores $a_N(i)$ should correspond to the underlying density f in the well-known way. If the type of f is not known, we put forward a certain number of density types and compute for them the quantities l_{Nj} applied to $Y_1, \dots, Y_N$ generated as a random permutation of order statistics $X^{(1)}, \dots, X^{(N)}$. The density type providing the largest l_{Nj} is then chosen to generate the scores and the test is performed as if these scores were decided upon before knowing $X_1, \dots, X_N$. Since under the hypothesis of randomness the vector of ranks $(R_1, \dots, R_N)$ is independent of the vector of order statistics, the above selection of scores does not invalidate the significance level. For moderate

N we can be deciding between three types corresponding to the following ϕ-functions, for example:

$$\phi_1(t) = \Phi^{-1}(t) \quad \text{(normal type)},$$

$$\phi_2(t) = 2t - 1 \quad \text{(logistic type)},$$

$$\phi_3(t) = -1 \quad (0 \leqslant t \leqslant \tfrac{1}{4}),$$

$$= 4t - 2 \quad (\tfrac{1}{4} \leqslant t \leqslant \tfrac{3}{4}),$$

$$= 1 \quad (\tfrac{3}{4} \leqslant t \leqslant 1). \tag{3.24}$$

(Φ^{-1} denotes the inverse normal distribution function.) It is advisable to use $\phi_3(t)$ instead of $\phi_4(t) = \text{sign}\,(2t - 1)$, which corresponds to the double-exponential type, since the discontinuity of ϕ_4 at $t = \tfrac{1}{2}$ increases unduly the variability of the corresponding ratio l_{Nj}. Thus the method would consist in testing sensitiveness of the van der Waerden test, the Wilcoxon test, and a test lying 'between' the Wilcoxon test and the median test, to the shift of location of the sample $Y_1, ..., Y_{[\frac{1}{2}N]}$. The test which is most sensitive is then applied to the original problem, as far as the scores are concerned. If ϕ_3 is selected, we can apply the median test as well.

The third selection procedure is based on a partial ordering of ϕ-function and of corresponding types. Within the family of skew symmetric ϕ-functions, we shall say, that ϕ increases more rapidly than ψ, if

$$\phi(t) = b(t)\,\psi(t) \quad (\tfrac{1}{2} < t < 1), \tag{3.25}$$

where $b(t)$ is nondecreasing. Inspecting (3.24), it is easy to see that ϕ_1 increases more rapidly than ϕ_2, and ϕ_2 increases more rapidly than ϕ_3, and ϕ_3 increases more rapidly than $\phi_5(t) = \sqrt{2}\sin\,[\pi(2t - 1)]$, which corresponds to the Cauchy distribution. Further, given two types $\mathscr{F}$ and $\mathscr{G}$ of symmetric densities, we shall say that the densities from $\mathscr{F}$ have shorter (lighter) tails than the densities from $\mathscr{G}$ if for any $f \in \mathscr{F}$ and $g \in \mathscr{G}$, such that their medians are zero, the corresponding quantile function functions $F^{-1}(t)$ and $G^{-1}(t)$ satisfy

$$F^{-1}(t) = a(t)\,G^{-1}(t) \quad (\tfrac{1}{2} < t < 1), \tag{3.26}$$

with $a(t)$ nonincreasing.

In Hájek (1969), § 34, there is given a theorem according to which (3.25) with $b(t)$ nondecreasing entail (3.26) with $a(t)$ nonincreasing, provided $\phi(t) = \phi(t, f)$ and $\psi(t) = \phi(t, g)$, using notation (2.12). Consequently, the normal type has shorter tails than the logistic type, the latter type has shorter tails than the type corresponding to ϕ_3, and the last type has shorter tails than the Cauchy type.

Since (3.26) with $a(t)$ nonincreasing entails that

$$F^{-1}(G(x)) \quad (x>0), \quad \text{is concave,}$$

$$G^{-1}(F(x)) \quad (x>0), \quad \text{is convex,}$$

we can test whether the true ϕ-function increases more rapidly than ϕ_0 as follows: We plot $[\tfrac{1}{2}N]$ points with co-ordinates

$$\left[x^{(N+1-i)} - x^{(i)}, \; F_0^{-1}\left(-\frac{N+1-i}{N+1}\right) \right] \quad (1 \leqslant i \leqslant [\tfrac{1}{2}N]),$$

where F_0^{-1} is the quantile function corresponding to ϕ_0.

If the curve suggested by these $[\tfrac{1}{2}N]$ points is convex, than we may prefer a ϕ-function which is increasing more slowly than ϕ_0; if it is concave, we may prefer a ϕ which increases more rapidly than ϕ_0; if it is straight, we keep ϕ_0. The effectiveness of this method and the possibility of its formalization may be in doubt. On the other hand, it surely is the quickest of all three methods explained in this section.

4   TESTING SCALE WHEN BOTH MEDIANS ARE UNKNOWN

Jurečková's theorem may also be applied to an asymptotic treatment of nuissance medians in testing scale alternatives. First observe that the increment $(S_N^* - S_N)/\sqrt{(\mathrm{var}\, S_N)}$ will be asymptotically zero if

$$\int_0^1 \phi(t)\, \phi(t,f)\, dt = 0. \tag{4.1}$$

Such a situation occurs if ϕ is symmetric,

$$\phi(t) = \phi(1-t) \quad (0<t<1), \tag{4.2}$$

and $\phi(t,f)$ is skew symmetric,

$$\phi(t,f) = -\phi(1-t,f) \quad (0<t<1). \tag{4.3}$$

Obviously, the skew symmetry of $\phi(t,f)$ is equivalent to the symmetry of f about its median.

Consider two samples $X_1, \ldots, X_m$ and $X_{m+1}, \ldots, X_{m+n}$ and assume that the respective densities are of the form $\sigma f[\sigma(x-\mu)]$ and $\lambda f[\lambda(x-\nu)]$. If $\mu = \nu$, proper rank tests for testing $\sigma = \lambda$ against $\sigma \neq \lambda$ (or $\sigma > \lambda$, or $\sigma < \lambda$) are based on statistics

$$S_N = s_N(X_1, \ldots, X_N), \tag{4.4}$$

where $$s_N(x_1, \ldots, x_N) = \sum_{i=1}^{m_N} a_N\left(\sum_{j=1}^{N} u(x_i - x_j) \right) \tag{4.5}$$

and the scores $a_N(i)$ are generated by (2.7) or by (2.8). If $\mu \neq \nu$, but we have some estimates $\hat{\mu}_N$ and $\hat{\nu}_N$ for μ and ν_1 respectively, we may try to prove that the statistic

$$S_N^* = s_N(X_1 - \hat{\mu}_N, \ldots, X_m - \hat{\mu}_N, X_{m+1} - \hat{\nu}_N, \ldots, X_N - \hat{\nu}_N) \quad (4.6)$$

possesses the same limiting distribution as S_N.

Assumptions B:

(B1) The scores $a_N(i)$ are generated by (2.7) or (2.8), or satisfy (2.15), where $\phi(t)$ is square integrable, fulfils (4.2), and is nondecreasing for $t \in (\frac{1}{2}, 1)$.

(B2) Random vector **X** is governed by the density

$$p_N(x_1, \ldots, x_N) = \prod_{i=1}^{m_N} f(x_i - \mu) \prod_{j=1}^{n_N} f(x_{m+j} - \nu) \quad (m_N + n_N = N), \quad (4.7)$$

where μ and ν are arbitrary and f is a fixed density which is symmetric and possesses finite Fisher's information.

(B3) The estimates $\hat{\mu}_m$ and $\hat{\nu}_n$ are square-root consistent, that is, random variables $\sqrt{(m)}\,(\hat{\mu}_m - \mu)$ and $\sqrt{(n)}\,(\hat{\nu}_n - \nu)$ are bounded in probability.

(B4) $\min(m_N, n_N) \to \infty$.

Theorem 4.1. *Let assumptions B be satisfied and denote by ES_N and var S_N the expectation and variance of S_N under the hypothesis of randomness.*

Then S_N^* is asymptotically normal with parameter $(ES_N, \text{var } S_N)$.

Moreover, if densities q_N are contiguous to densities p_N of (4.7), *and if S_N is asymptotically normal with parameters $(a_N, \text{var } S_N)$ under q_N, then S_N^* is also asymptotically normal $(a_N, \text{var } S_N)$ under q_N.*

Proof. Since S_N is a rank statistic, we may write equivalently

$$S_N = s_N(X_1 - \mu + \Delta_N d_{N1}, \ldots, X_m - \mu + \Delta_N d_{Nm}, X_{m+1} - \nu$$
$$+ \Delta_N d_{N,m+1}, \ldots, X_N - \nu + \Delta_N d_{NN}), \quad (4.8)$$

where

$$\Delta_N = (\mu - \hat{\mu}_N - \nu + \hat{\nu}_N)\left(\frac{1}{m_N} + \frac{1}{n_N}\right)^{-\frac{1}{2}} \quad (n_N = N - m_N, \hat{\mu}_N = \hat{\mu}_{m_N}, \hat{\nu}_N = \hat{\nu}_{n_N})$$
$$(4.9)$$

and
$$d_{Ni} = \left(\frac{1}{m_N} + \frac{1}{n_N}\right)^{\frac{1}{2}} \quad (1 \leqslant i \leqslant m_N), \quad (4.10)$$

$$= 0 \quad (m_N < i \leqslant N).$$

Now we shall check if Assumptions A of Theorem 2.1 are satisfied. As for (A 1), we have $c_{Ni} = 1$, $1 \leqslant i \leqslant m_N$, and $c_{Ni} = 0$, $m_N < i \leqslant N$, so that (B 4) entails (2.6). Further, (B 1) implies (A 2), (A 2*) and (A 3). Next, definition (4.10) of the d_{Ni}'s entails (2.9), and (B 4) entails (2.10); consequently (A 4) holds. (A 5) is obvious, and (A 6) follows from (B 3). Finally, (B 2) entails the satisfaction of (A 7) for

$$(Y_{N1}, ..., Y_{NN}) = (X_1 - \mu, ..., X_m - \mu, X_{m+1} - \nu, ..., X_N - \nu).$$

Now under (B 1) and (B 2) we have (4.2) and (4.3), and, consequently, also (4.1). Thus $(S_N^* - S_N)/\sqrt{(\text{var } S_N)} \to 0$ under p_N. From this all conclusions of the theorem easily follow.   Q.E.D.

Examples. The theorem is applicable among others to the following statistics: The Klotz statistic, the quartile statistic, the Mood statistic, the Capon statistics, the Ansari–Bradley statistic; these statistics are described in Hájek and Šidák (1967), §III. 2. 1. Further it also applies to the adapted Wilcoxon test (the Siegel–Tukey adaptation, e.g.) and the adapted van der Waerden test; see Hájek (1969), §16.

5 PARTIAL ORDERINGS OF PERMUTATIONS

As is described in Lehmann (1966), §7, one employs two distinct partial orderings of permutations. We shall call them Ordering 1 and Ordering 2, and instead of writing 'better ordered in the sense of Ordering 1' we shall write 'better ordered[1]'; 'better ordered[2]' will have a similar meaning.

Ordering 1. We say that a permutation $q = (q_1. ..., q_N)$ is better ordered[1] than $r = (r_1, ..., r_N)$ if q may be obtained from r by a number of steps, each of which consists in correcting an inversion. (To correct an inversion means to interchange two elements r_i and r_j such that $i < j$ and $r_i > r_j$.)

Ordering 2. We say that a permutation $q = (q_1, ..., q_N)$ is better ordered[2] than $r = (r_1, ..., r_N)$ if

$$[i < j, r_i < r_j] \Rightarrow [q_i < q_j]. \tag{5.1}$$

Ordering 1 does not entail Ordering 2, as is shown by example (*a*) in Lehmann (1966, p. 1149). On the other hand, we present below a theorem to the effect that Ordering 2 entails Ordering 1. This might seem to be in contradiction with example (*b*) in Lehmann (1966, p. 1149). However, it is not so, since 'correcting an inversion' in Lehmann's paper is understood as exchanging the positions of two *neighbouring* elements r_i and r_{i+1}, $r_i > r_{i+1}$.

Theorem 5.1. *If q is better ordered[2] than r in the sense of* (5.1), *then q may be obtained from r by a number of steps, each consisting in correcting an inversion such that the involved numbers differ by* 1.

Proof. If $r = q$ the proposition is trivial. Assume $r \neq q$ and denote by $d = (d_1, \ldots, d_N)$ the inverse permutation to r, that is, $r(d_i) = i$, $1 \leqslant i \leqslant N$. Certainly, we do not have $q(d_1) < q(d_2) < \ldots < q(d_N)$, since that would entail $q(d_i) = i$, and in turn, $q = r$. Consequently, there is a k, $1 \leqslant k \leqslant N$, such that

$$q(d_k) > q(d_{k+1}). \tag{5.2}$$

Since r is better ordered[2] than q,

$$[d_k < d_{k+1}, r(d_k) = k < k+1 = r(d_{k+1})] \Rightarrow [q(d_k) < q(d_{k+1})]. \tag{5.3}$$

Since $q(d_k) < q(d_{k+1})$ is false, in view of (5.2), and $r(d_k) < r(d_{k+1})$ is true, we cannot have $d_k < d_{k+1}$. Consequently we have $d_{k+1} < d_k$, that is, $k+1$ precedes k in r. If we correct this inversion, the state of all other pairs will remain unchanged: the inversions will occur exactly where they were before. Thus the new permutation r' will be better ordered[2] than r, and q will be better ordered[2] than r', since for the pair where the inversion was corrected we have (5.2). Obviously, after a finite number of steps of this kind we must obtain a permutation which coincides with q.

Example. Obviously, $q = (2, 1, 3, 4, 5)$ is better ordered[2] than

$$r = (3, 2, 5, 1, 4).$$

Correcting the inversion $(2, 1)$ in r, we obtain $r^{(1)} = (3, 1, 5, 2, 4)$; correcting $(3, 2)$ in $r^{(1)}$, we obtain $r^{(2)} = (2, 1, 5, 3, 4)$; correcting $(5, 4)$ in $r^{(2)}$, we obtain $r^{(3)} = (2, 1, 4, 3, 5)$; finally, correcting $(4, 3)$ in $r^{(3)}$, we obtain

$$r^{(4)} = (2, 1, 3, 4, 5) = q.$$

Theorem 5.1 may be used to the proof that the

$$s_N(y_1, \ldots, y_N, \Delta) = \sum_{t=1}^{N} c_i a\left(\sum_{j=1}^{N} u(y_i + \Delta d_i - y_j - \Delta d_j) \right)$$

is nondecreasing in Δ, if $a(1) \leqslant \ldots \leqslant a(N)$ and (2.11) hold, (*see* Hájek (1969), §7). A different proof is provided in Lehmann (1966), and still another in Jurečková (1969).

6 ASYMPTOTIC NORMALITY UNDER TIES

Let $X_1, \ldots, X_N$ be a random sample from an arbitrary distribution, possibly discrete. Then the set of order statistics $X^{(i)}$ decomposes into g groups of tied values:

$$X^{(1)} = \ldots = X^{(\tau_1)} < X^{(\tau_1+1)} = \ldots = X^{(\tau_1+\tau_2)} < \ldots$$

$$< X^{(\tau_1+\ldots+\tau_{g-1}+1)} = \ldots = X^{(N)}.$$

The occurrence of ties is characterized by the vector of tie-sizes $\tau = (\tau_1, \ldots, \tau_g)$, where g as well as τ_i's are random. Introducing averaged scores

$$a_N(i, \tau) = \frac{1}{\tau_k} \sum_{j=T_{k-1}+1}^{T_k} a(j), \qquad (6.1)$$

if $T_{k-1} = \tau_1 + \ldots + \tau_{k-1} < i \leqslant \tau_1 + \ldots + \tau_k = T_k,$

let us investigate the asymptotic distribution of the statistic

$$\bar{S}_N = \sum_{i=1}^{N} c_{Ni} a_N(R_{Ni}, \tau), \qquad (6.2)$$

where $R_{Ni} = \sum_{j=1}^{N} u(X_i - X_j)$ as before. If the distribution of X_i's is not continuous, then it is not true that $R = (R_{N1}, \ldots, R_{NN})$ is a permutation of $(1, 2, \ldots, N)$ with probability 1. However, the conditional distribution of $\bar{S}_N$ given τ is the same as if R were distributed uniformly over the space of permutations, in view of the special form of $a_N(i, \tau)$ (see Hájek (1969), § 29).

The statements concerning the asymptotic distribution may have three different forms:

Form 1. We assert that $\bar{S}_n$ is asymptotically normal with parameters $(E S_N, \sigma_N^2)$, where σ_N^2 is independent of τ, but depends on the distribution function F of the X_i's, if F is not continuous. (*See* Vorlíčková (1969), Theorem 3 of § 3.)

Form 2. We assert that the *conditional distribution of $\bar{S}_N$ given τ* differs from $\mathcal{N}(E\bar{S}_N, \text{var}(\bar{S}_N \mid \tau))$ in the Kolmogorov distance by less than ϵ with probability greater than $1 - \epsilon$ if $N \geqslant N(\epsilon)$. (*See* Vorlíčková (1969), Theorem 4 of § 3.)

Form 3. We explicitly establish regions W_N such that the Kolmogorov distance between $\mathscr{L}(\bar{S}_N \mid \tau)$ and $\mathcal{N}(E S_N, \text{var}(\bar{S}_N \mid \tau))$ is smaller than ϵ if $\tau \in W_N$ and $N \geqslant N(\epsilon, \{W_N\})$.

Form 2 provides us with the knowledge that $\mathscr{L}(\bar{S}_N \,|\, \tau)$ should have been close to $\mathscr{N}(E\bar{S}_N, \mathrm{var}\,(S_N\,|\,\tau))$ with great probability. Form 3 enables us to check whether this really happened, if τ is known.

Let us now prove a theorem in which the assertion takes on Form 3.

Theorem 6.1. *Let the scores $a_N(i)$ satisfy (2.15), where ϕ is nonconstant and square integrable. Let $X_1, ..., X_N$ be independent and equidistributed.*

Then for every $\epsilon > 0$ and $\eta > 0$ there exists a $\delta = \delta(\epsilon, \eta)$ such that

$$\max_{1 \leqslant i \leqslant N} (c_{Ni} - \bar{c}_N)^2 < \delta \sum_{i=1}^{N} (c_{Ni} - \bar{c}_N)^2 \qquad (6.3)$$

together with
$$\sum_{i=1}^{N} (a_N(i) - \bar{a}_N)^2 > \eta \sum_{i=1}^{N} (a_N(i, \tau) - \bar{a}_N)^2 \qquad (6.4)$$

entails
$$\sup_{-\infty < x < \infty} |P(\bar{S}_N \leqslant E\bar{S}_N + x\sqrt{\mathrm{var}\,(\bar{S}_N\,|\,\tau)}\,|\,\tau) - \Phi(x)| < \epsilon. \qquad (6.5)$$

Proof. Recall that we may assume that $P(R = r) = 1/N!$ for all permutations r, without changing the conditional distribution of $\bar{S}_N$. Then keep η fixed and give the assertion the equivalent limiting form, which reads as follows: If (2.6) holds, then $\mathscr{L}[(\bar{S}_N - E\bar{S}_N)/\sqrt{\mathrm{var}\,(\bar{S}_N\,|\,\tau)}\,|\,\tau] \to \mathscr{N}(0, 1)$. According to Theorem 4.2 of Hájek (1962), it suffices to show that

$$\left[\frac{k_N}{N} \to 0\right] \Rightarrow \left[\dfrac{\displaystyle\max_{1 \leqslant i_1 < ... < i_{k_N} \leqslant N} \sum_{\alpha=1}^{k_N} [a_N(i_\alpha, \tau_N) - \bar{a}_N]^2}{\displaystyle\sum_{i=1}^{N} [a_N(i, \tau_N) - \bar{a}_N]^2} \to 0\right] \qquad (6.6)$$

holds for any choice of $\tau = \tau_N$ such that (6.4) holds. Square integrability of ϕ and (2.15) entail that

$$\left[\frac{k_N}{N} \to 0\right] = \left[\dfrac{\displaystyle\max_{1 \leqslant i_1 < ... < i_{k_N} \leqslant N} \sum_{\alpha=1}^{k_N} [a_N(i_\alpha) - \bar{a}_N]^2}{\displaystyle\sum_{i=1}^{N} [a_N(i) - \bar{a}_N]^2} \to 0\right]. \qquad (6.7)$$

On the other hand the averaging process (6.1) implies

$$\max_{1 \leqslant i_1 < ... < i_{k_N} \leqslant N} \sum_{\alpha=1}^{k_N} [a_N(i_\alpha, \tau_N) - \bar{a}_N]^2 \leqslant \max_{1 \leqslant i_1 < ... < i_{k_N} \leqslant N} \sum_{\alpha=1}^{k_N} [a_N(i_\alpha) - \bar{a}_N]^2, \qquad (6.8)$$

which in conjunction with (6.4) yields (6.6).   Q.E.D.

Form 3 may be given also to the limiting theorem concerning the chi-square-type statistics. Let $X_1, ..., X_N$ split into k samples and denote by $\bar{S}_j$ the sum of scores $a_N(i, \tau)$ over the jth sample.

Theorem 6.2. *Under conditions of Theorem* 6.1 *put*

$$\bar{Q}_N = (N-1)\left[\sum_{i=1}^{N}(a_N(i,\tau)-\bar{a}_N)^2\right]^{-1}\left[\sum_{j=1}^{k}\frac{\bar{S}_j^2}{n_j}-N\bar{a}_N^2\right], \qquad (6.9)$$

where n_j denotes the size of the jth sample.

Then for every $\epsilon > 0$ and $\eta > 0$ there exist an $M = M(\epsilon,\eta)$ such that (6.4) and

$$\min(n_1,\dots,n_k) > M \qquad (6.10)$$

entail
$$\sup_{-\infty<x<\infty}|P(\bar{Q}_N \leqslant x\,|\,\tau)-P(\chi^2_{k-1}\leqslant x)| < \epsilon, \qquad (6.11)$$

where χ^2_{k-1} possesses the chi-square distribution with $k-1$ degrees of freedom.

Proof. Theorem 6.2 is an easy consequence of Theorem 6.1.

If testing the hypothesis of independence, we have a sequence of pairs $(X_1,Y_1),\dots,(X_N,Y_N)$. Let τ_x and $(R_{N1},\dots,R_{NN})$ be the tie-sizes and ranks associated with $(X_1,\dots,X_N)$; let τ_y and $(Q_{N1},\dots,Q_{NN})$ be the same quantities for $(Y_1,\dots,Y_N)$. Consider the statistic

$$\bar{S}_N = \sum_{i=1}^{N} a_N(R_{Ni},\tau_x)\,a_N(Q_{Ni},\tau_y). \qquad (6.12)$$

Theorem 6.3. *Let the scores $a_N(i)$ satisfy* (2.15), *where ϕ is nonconstant and square integrable. Let the samples $(X_1,\dots,X_N)$ and $(Y_1,\dots,Y_N)$ be independent and each equidistributed.*

Then for every $\epsilon > 0$ and $\eta > 0$ there exists an $N(\epsilon,\eta)$ such that $N > N(\epsilon,\eta)$ together with

$$\min\left\{\sum_{i=1}^{N}[a_N(i,\tau_x)-\bar{a}_N]^2,\ \sum_{i=1}^{N}[a_N(i,\tau_y)-\bar{a}_N]^2 > \eta\sum_{i=1}^{N}[a_N(i)-\bar{a}_N]^2\right\} \qquad (6.13)$$

entails (6.5) *with $\bar{S}_N$ given by* (6.12) *and $\tau = (\tau_x,\tau_y)$.*

Proof. The proof may be based on Theorem 6.1, if we identify

$$c_{Ni} = a_N(i,\tau_y).$$

Since, under (6.13),

$$\frac{\max\limits_{1\leqslant i\leqslant N}[a_N(i,\tau_y)-\bar{a}_N]^2}{\sum\limits_{i=1}^{N}[a_N(i,\tau_y)-\bar{a}_N]^2} \leqslant \frac{1}{\eta}\frac{\max\limits_{1\leqslant i\leqslant N}[a_N(i)-\bar{a}_N]^2}{\sum\limits_{i=1}^{N}[a_N(i)-\bar{a}_N]^2}\to 0, \qquad (6.14)$$

we may apply Theorem 6.1.   Q.E.D.

REFERENCES

Hájek, J. (1961). Some extensions of the Wald–Wolfowitz–Noether theorem. *Ann. Math. Statist.* **32**, 506–23.

Hájek, J. (1969). *A course in nonparametric statistics.* San Francisco: Holden–Day.

Hájek, J. and Šidák, A. (1967). *Theory of rank test.* Prague: Academia and London: Academic Press.

Jurečková, J. (1969). Asymptotic linearity of a rank statistic. *Ann. Math. Statist.* **40**, 1889–1900.

Lehmann, E. (1966). Some concepts of dependence. *Ann. Math. Statist.* **37**, 1137–53.

Raghavachari, M. (1965). The two-sample scale problem when locations are unknown. *Ann. Math. Statist.* **36**, 1236–42.

Vorlíčková, D. (1970). Asymptotic properties of rank tests under discrete distributions. *Wahrscheinlichkeitstheorie verw. Geb.* **14**, 275–89.

DISCUSSION ON HÁJEK'S PAPER

W. HOEFFDING

Instead of elaborating on Professor Hájek's interesting and valuable paper, I will, in the spirit of the title of his paper, make a few methodological remarks on the problem of the asymptotic distribution of a linear rank statistic.

Let $X_1, \ldots, X_n$ be independent random variables with respective continuous distribution functions $F_1, \ldots, F_n$, and let $R_1, \ldots, R_n$ be the corresponding ranks. Consider the linear rank statistic

$$T_n = \sum_1^n c_i \phi\left(\frac{R_i}{n+1}\right),$$

where the c_i are constants and ϕ a function finite on $(0, 1)$.

In his paper in *Ann. Math. Statist.* (1968), Professor Hájek proved the asymptotic normality of T_n under mild assumptions. He used his projection lemma to approximate T_n by a sum of independent random variables. With this approach the mean of the asymptotic distribution comes out as ET_n. The following alternative approach is akin to von Mises' (1947) treatment of what he called differentiable statistical functions and also to the approach of Chernoff and Savage (1958).

Let $u(x) = 0$ or 1 according as $x < 0$ or $x \geqslant 0$,

$$S(x) = (n+1)^{-1} \sum_1^n u(x - X_j), \quad S^*(x) = \sum_1^n c_j u(x - X_j).$$

Then we can write $\qquad T_n = \int_{-\infty}^{\infty} \phi(S(x))\, dS^*(x).$

Note that $ES(x) = \bar{F}(x)$ and $ES^*(x) = F^*(x)$, where

$$\bar{F}(x) = (n+1)^{-1} \sum_1^n F_j(x), \quad F^*(x) = \Sigma c_j F_j(x).$$

If ϕ is absolutely continuous inside $(0, 1)$ and the integrals in what follows exist, we have $\qquad T_n - \int \phi(\bar{F})\, dF^* = L_n + R_{1n} + R_{2n},$

where $\qquad L_n = \int \phi(\bar{F})\, d(S^* - F^*) + \int \phi'(\bar{F})\, (S - \bar{F})\, dF^*,$

$$R_{1n} = \int \{\phi(S) - \phi(\bar{F})\}\, d(S^* - F^*),$$

$$R_{2n} = \int \{\phi(S) - \phi(\bar{F}) - \phi'(\bar{F})\, (S - \bar{F})\}\, dF^*.$$

[18]

Here L_n is a sum of independent random variables with zero means and is almost identical with Hájek's projection approximation for T_n. Under the conditions of Theorem 2.1 of Hájek's (1968) paper, which include the assumption that ϕ has a bounded second derivative, it is rather straightforward to show that $ER_{i,n}^2/\sigma_n^2 \to 0$ for $i = 1, 2$, where $\sigma_n^2 = \operatorname{var} L_n$. (Assume $\Sigma c_i = 0$; otherwise only a trivial modification is needed.) This provides an alternative proof of Hájek's Theorem 2.1. It is easy to show in this case that the centering constant ET_n of Hájek's theorem may be replaced by $\int \phi(\bar{F}) \, dF^*$.

However, under the conditions of Theorem 2.3 of Hájek's (1968) paper, which assumes much less about ϕ and somewhat more about the c_i, a direct proof of the asymptotic negligibility of $R_{1n} + R_{2n}$ seems to be much more difficult. Hájek's main tool is his powerful variance inequality.

REFERENCES

Chernoff, H. and Savage, I. R. (1958). Asymptotic normality and efficiency of certain nonparametric test statistics. *Ann. Math. Statist.* **29**, 972–94.

Hájek, J. (1968). Asymptotic normality of simple linear rank statistics under alternatives. *Ann. Math. Statist.* **39**, 325–46.

von Mises, R. (1947). On the asymptotic distribution of differentiable statistical functions. *Ann. Math. Statist.* **18**, 309–48.

ASYMPTOTIC RESULTS FOR
RANK STATISTICS†

RONALD PYKE‡

1 INTRODUCTION

I appreciate the opportunity to speak at this Symposium. When I accepted
Professor Puri's invitation to give an expository paper, I did so because
I believe that an expository paper is a *good thing*, in the spirit of Sellar
and Yeatman's dichotomization of British history, 1960. It is clear
however that a complete exposition of the enormous literature which
exists today on limit theorems for rank statistics should not be attempted
within a short paper. Moreover, the book by Hájek and Sidak (1967)
already provides an excellent coverage of most of this literature. I shall
instead simply attempt to impart some personal comments on the
philosophy of limit theory and to illustrate by means of two examples
one particular approach to asymptotic results for rank statistics.

The significant role of limit theory in Nonparametric Inference, and
hence the importance of a limit theorist or 'limitor', is empirically verified
by the following statement:

'64% of all papers in Nonparametric Inference
are concerned primarily with asymptotic results.' (1)

This finding is based on the personal assignment of weights 0, $\frac{1}{2}$ and 1 to
the random [*sic*] sample of 33 papers listed in the program of this Symposium. The weights were assigned according to a paper's proportional
concern with asymptotic results. The sample average of these weights
was 21/33. Although the announced finding (1) is based on a very quick
and subjective analysis, I am sure it truly reflects the high concentration
of limit theory in our subject. I suspect if one made a more careful and
objective evaluation of this concentration, based say on the printed
proceedings of this Symposium, a higher estimate would be obtained.
This emphasis on asymptotic results is of course a quality of statistical
literature as a whole, and not just of Nonparametric Inference.

'Practical problems are finite; tractable problems are infinite.'

† An invited paper presented at the First International Symposium on Nonparametric
Techniques in Statistical Inference, Indiana University, Bloomington, Ind., 1–6 June
1969.
‡ The research described herein was supported in part by the National Science Foundation under G–5719.

[21]

Throughout the history of our subject, Statisticians have been forced to solve most of their problems in the relative paradise at infinity, and then to determine empirically (usually) or theoretically (seldom) the ease (or rate) with which this paradise is achieved. In this connection, reference is made to Kiefer (1967) who identifies the determination of rates of convergence as one of three basic problems in statistics.

Limit theory is a collection of tools which Statisticians use to determine which problem if any in the paradise at infinity is the best fit to his problem in the finite world. These tools are continually being sharpened and modernized, and occasionally replaced by more powerful ones. If one observes the history of Limit Theory from the time of Bernoulli to the present, this continual evolvement is clearly evident. Also clear is the way in which the practical problems encountered by Statisticians, or Probabilists, have so often determined the new directions for Limit Theory. Perhaps one could picture the relationship between the practicing Statistician, the applier of existing limit theorems and the creator of new theory in somewhat the following way.

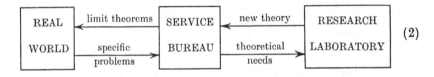

$$(2)$$

The Statistician in the real world often meets problems which require limiting results. Usually when he approaches the Statistician in the Service Bureau he finds that an existing tool (limit theorem) is available and ideally suited for his needs; for example, Lindeberg's Central Limit Theorem, or the Normal approximation to the Binomial or Poisson, or the weak convergence of the empirical process to a Brownian Bridge. Occasionally however no tool is available; as for example when the Wilcoxon–Mann–Whitney statistic was first proposed (1945, 1947) or when the Kolmogorov statistic D_n was considered (1933). In this case the Service Bureau Statistician sends a description of the type of tool needed on (up?) to the Research Laboratory where hopefully 'Limitors' will be able to produce the desired new result. Of course many Statisticians work in all three areas just by changing hats at appropriate times. (Undoubtedly a Limitor's hat is bell-shaped!) Certainly it is very common for identifiers of problems and dispensers of tools to be the same individuals. It is unfortunately less common for Limitors in the Research Laboratory to play one of the other roles as well. Consequently their output may not be as applicable as it should be, nor as important as the methods they developed in the process of creating their output. These methods usually require much more time to communicate, as the following example indicates.

In the 1930s and 1940s separate limit theorems were obtained by Kolmogorov, Smirnov and others for specific functions of the empirical process. (E.g. D_n, D_n^+, W_n^2; cf. Darling (1957).) In the 1940s, asymptotic results were derived by Kac, Erdos, Seigert, Mark and others for specific functions of the partial sum process. (E.g. maximum, sums of squares, number of positive summands, etc.) Although these specific results were immediately available, a Statistician requiring a limit theorem for a slightly different function of either of these processes would be out of luck. The methods that had been used (e.g. the invariance principle) were however widely applicable and by 1951 and 1952 Donsker had formulated general theorems for partial-sum and empirical processes respectively. It was not however until the 1956 papers of Prokhorov, Skorokhod, Chentsov and Billingsley that the general method might be said to have been completely presented. Although a few additional basic papers were published in 1957 (e.g. Kimme, LeCam and Skorokhod; the reader is referred to Billingsley (1968) for complete references), it is safe to say that by 1956 the tool of *weak convergence* was available to Service Bureaus for dispensing to Statisticians with problems involving either partial sum processes or empirical processes. However, for a variety of reasons (for example, not every branch of the Service Bureau stocked this tool, not every customer of the Service Bureau was qualified to operate this tool, not every problem on which the tool could be used was properly identified as such) the tool of weak convergence was slow to be utilized in our statistical literature. Perhaps it is fair to say that it took 12 years, from 1956 to 1968, for the necessary communication of methods to be completed. The year 1968 is chosen since that was the year of circulation of two texts on the subject, by Billingsley (1968) and by Parthasarathy (1967), as well as the incorporation of the subject into recent texts such as that of Brieman (1968). Numerous seminars on weak convergence have been held this year throughout our universities, motivated primarily by the existence of these books.

Perhaps this 12 year lag from research journal to text book is a relatively short one compared to other examples in the evolution of knowledge. Nevertheless, I think it indicates the need of more direct communication between the Real World and the Research Laboratory. If for example those who consider themselves workers in the latter could offer their services directly to Statisticians, who wish to know for example the asymptotic distribution of specific statistics, probably both time and space in journals would be saved. This circumvention of the Service Bureau, which can conveniently be identified with the collection of theorems or tools in the literature, can often save time and effort simply because it is easier for the Limitor to use his methods to give a direct proof for a specific statistic than it is for the statistician to verify that one of the Literature's general theorems apply to this special case. This is especially true for rank statistics and linear combinations of order statistics whose general theorems all have somewhat unwieldy assumptions. Methods often apply even when the general theorems derived by these methods do not.

2   LINEAR RANK STATISTICS: NOTATION

The main concern of this paper is with the limit theory of rank statistics. Some notation is therefore necessary. Let $X = (X_1, X_2, ..., X_n)$ and $Y = (Y_1, Y_2, ..., Y_n)$ be r.v.'s associated with d.f.'s F and G respectively. Set $N = m+n$ and $\lambda_N = m/N$. N is also used to stand for the pair (m, n). Let F_m and G_n be the empirical d.f.'s for X and Y, respectively, and define empirical processes for $0 \leqslant t \leqslant 1$ by

$$U_m(t) = m^{\frac{1}{2}}[F_m \circ F^{-1}(t) - t], \quad V_n(t) = n^{\frac{1}{2}}[G_n \circ G^{-1}(t) - t]. \tag{3}$$

For the pooled sample $(X_1, ..., X_m, Y_1, ..., Y_n)$ let $H_N = \lambda_N F_m + (1 - \lambda_N) G_N$ be the empirical d.f. and set

$$H_\lambda = \lambda F + (1-\lambda) G \quad (0 \leqslant \lambda \leqslant 1; H = H_\lambda). \tag{4}$$

If R_{Ni} $(1 \leqslant i \leqslant N)$ is the number among $X_1, ..., X_m$ which do not exceed the ith order statistic among the pooled sample $X_1, ..., X_m, Y_1, ..., Y_n$ then $\{R_{Ni}\}$ are the relative ranks of the X-sample and could be expressed as $R_{Ni} = mF_m \circ H_N^{-1}(i/N)$. Write $Z_{Ni} = R_{Ni} - R_{Ni-1}$ for $1 < i \leqslant N$ and $Z_{N1} = R_{N1}$. Linear rank statistics are those of the form

$$T_N = m^{-1} \sum_{i=1}^{N} c_{Ni} R_{Ni} = \sum_{i=1}^{N} c_{Ni} F_m \circ H_N^{-1}(i/N), \tag{5}$$

where $\{c_{Ni}: 1 \leqslant i \leqslant N\}$ is a set of constants. A summation by parts in (5) yields the alternative form

$$T_N = m^{-1} \sum_{i=1}^{N} c_{Ni}^* Z_{Ni}, \tag{6}$$

where $c_{Ni}^* = c_{Ni} + ... + c_{NN}$. This latter form is the one used by Chernoff and Savage (1958) and that of (5) was used by Pyke and Shorack (1968).

Each form may be written more conveniently. Let ν_N denote a signed measure which puts mass c_{Ni} on the point i/N and let J_N be a function defined on $(0, 1)$ by $J_N(t) = c_{Ni}^*$ for $i - 1 < Nt \leqslant i$. Then (5) and (6) may be written respectively as

$$T_N = \int_0^1 F_m \circ H_N^{-1} d\nu_N, \tag{5'}$$

$$T_N = \int_{-\infty}^{\infty} J_N \circ H_N dF_m. \tag{6'}$$

The relationship between ν_N and J_N is clear. The advantage of representation (6') is that after a natural Taylor's expansion of the integrand to replace H_N with H, the main term becomes in view of dF_m a sum of iid r.v.'s whenever the X_i's are; all of the difficulties from this representation are involved with proving that remainder terms in the expansion may be neglected. An advantage of representation (5') is its adaptibility to situations in which the limit of $\{\nu_N\}$ has a discrete part, or equivalently, the case of discontinuous limits to $\{J_N\}$.

A capsule review of the development of Limit Theory for statistics of this type is as follows. The first general limit theorem for a broad class of statistics of the form (5) or (6) was obtained by Dwass (1956). His theorem applied to the subclass for which $c_{Ni}^* = p(i/N)$ and p is a polynomial. Dwass' approach was to extend the concept of U-statistics to two samples of unequal size, utilize the methods of Hoeffding (1948) to derive a central limit theorem for these U-statistics and then to show that his class of rank statistics were special cases of these.

Two years later, Chernoff and Savage (1958) derived exceptionally general and applicable limit theorems using representation (6'). Their approach was to postulate the existence of a sufficiently smooth limiting weight function J satisfying

$$T_N - T_N' = o_p(N^{-\frac{1}{2}}), \quad \text{where } T_N' = \int J \circ H_N dF_m.$$

(Define $J_N(1) = J(1) = 0$.) It thus suffices to study T_N'. Upon making a Taylor's expansion of J, $J \circ H_N = J \circ H + (H_N - H)J'(\theta)$, the statistic $T_N'' = \int J \circ H dF_m$ suggests itself. Since T_N'' is a sum of independent and identically distributed r.v.'s, its asymptotic normality is easily studied. However considerable work is necessary to show that the difference $T_N'' - T_N' = o_p(N^{-\frac{1}{2}})$. Chernoff and Savage state that their result 'is not given in the most general form possible', but in fact it and its corollaries have been sufficiently general for most applications. Relatively slight extensions of their results are given, with somewhat shorter proofs, in Govindarajulu et al. (1967) and in Pyke and Shorack (1968) by weakening the smoothness and tail-behaviour conditions on J. In the latter the smoothness is dropped to the point of allowing jumps in J.

The method used in Pyke and Shorack (1968), referred to hereafter as PS, was to study primarily the weak convergence of a two-sample empirical process suggested by the representation (5'). These authors then consider conditions under which $\{\int . d\nu_N\}$ is a sufficiently regular sequence of functionals.

The theorems referred to in the above paragraphs all deal with fixed alternatives, albeit most authors did consider questions of uniformity over large classes of alternatives. A different type of problem is treated by Hájek in several papers since 1960, namely the problem of *contiguous alternatives*. For these alternatives it is possible to generalize from the two-sample case to quite general problems in which each observation has a possibly different df. In particular, regression models under contiguous alternatives are then included. The results obtained by Hájek are widely applicable, and to a great extent involve assumptions that are more easily verified. In the course of his studies Hájek introduces several methods; in particular, methods utilizing moment inequalities, comparison methods, the method of contiguity, and most recently a projection method. (See for example Hájek (1968), the text by Hájek and Sidak (1967) and other references cited therein for discussions of these methods.) These are very useful methods, they having already been applied to several other problems in the recent literature.

The main motivation for the Pyke–Shorack investigation was twofold: (i) to illustrate the applicability of weak convergence to these limit theorems and (ii) to indicate the usefulness of the alternative representation (5) in handling cases of discontinuous J-functions. (Concerning the first aspect, the applicability of weak convergence, it should be noted that back in 1957 there were discussions between Chernoff and Savage and other dwellers of Sequoia Hall, Stanford University, about the possibility of the then new ideas in Prokhorov (1956) being applied to rank statistics.) Our use of weak convergence was indirect through a method of using a.s. convergent versions of weakly convergent processes. The meaning of this is clarified in the following paragraphs and then illustrated for a one-sample problem in § 3. We return again to the two-sample problem in § 4.

Consider the false statement:

'Convergence in law implies a.s. convergence'. (!)

Perhaps this should be stated as: 'Weak convergence implies strong convergence.' In any event, taken literally, (!) is false. However, slightly clarified it becomes a convenient slogan, and represents a useful method. The clarification is given by the following 'true' result.

Theorem 1. *Suppose* (S, d) *is a separable metric space and* $\{X_n: n \geqslant 0\}$ *are S-valued r.v.'s defined on a probability space* $(\Omega, \mathscr{A}, P)$. *Then* $X_n \xrightarrow{L} X_0$ *implies the existence of a probability space* $(\Omega^*, \mathscr{A}^*, P^*)$ *and r.v.'s* $\{X_n^*: n \geqslant 0\}$ *defined thereon for which* $\mathscr{L}(X_n) = \mathscr{L}(X_n^*)$ *for all* $n \geqslant 0$ *and* $d(X_n^*, X_0^*) \to 0$ *a.s.* $- P^*$.

This result is due to Skorokhod (1956) in the case of a complete and separable metric space. Without completeness, the theorem is due to Dudley (1968). Many interesting cases of weak convergence involve non-separable metric spaces, as for example the case of empirical processes under the supremum metric. For these cases a modified definition of weak convergence is needed under which an analogue of Theorem 1 is proved by Wichura (1969).

The import of Theorem 1 is that for any problem involving only convergence in law, one can without loss of generality use a sequence of processes which not only converges in law but in fact converges in the appropriate metric almost surely. (One might as well of course assume it converges everywhere.) If for example one seeks the limiting df of a statistic $K_n(X_n)$ for given functionals $\{K_n\}$ it suffices to check the pointwise limit of $K_n(X_n^*)$ under the assumption that $d(X_n^*, X_0^*) \to 0$. This method is described in greater detail and with several applications in Pyke (1969). It is illustrated further by the example of the following section.

3 AN APPLICATION OF A.S. CONVERGENT PROCESSES

In 1955, Blackman proposed an estimator based on distance methods for a translation parameter. Specifically, suppose one observes a sample $X_1, X_2, ..., X_m$ of independent r.v.'s each from the same continuous distribution function which is known to be some translate of the specified df F. Blackman considered

$$M_m(\theta) = \int_{-\infty}^{\infty} [F(x - \theta) - F_m(x)]^2 \, dF(x - \theta) = \int_{-\infty}^{\infty} [F(y) - F_m(y + \theta)]^2 \, dF(y)$$

for $-\infty < \theta < \infty$, as a measure of fit of the empirical df F_m to the θ-translate of F. His estimate $\hat{\theta}_m$ of θ is defined as any value (including $\pm \infty$) at which M_m is a minimum. Since M_m is continuous on $[-\infty, \infty]$

(it has a limit of $1/3$ as θ converges to $\pm\infty$), the estimator exists. Blackman (1955) proved the following result.

Theorem 3.1. *If F has a continuous and bounded third derivative,* $m^{\frac{1}{2}}\theta_m \overset{L}{\to} a\, N(0, \sigma^2)$ *r.v. with*

$$\sigma^2 = \nu^2/\mu^2, \quad \mu = \int_{-\infty}^{\infty} f^2 dF,$$

$$\nu^2 = \int_0^1 \left[\int_0^y f \circ F^{-1}(t)\, dt - \int_0^y t^{-2} dt \int_0^t zf \circ F^{-1}(z)\, dz \right]^2 dy, \qquad (7)$$

where $f = F'$.

The proof of this result given by Blackman requires 11 pages. Given below is a shorter proof under weaker assumptions based on an application of uniformly convergent empirical processes. This approach also yields as a by-product a simpler form for the limiting variance σ^2. The fact that after 14 years it is possible to prove a result more simply is not surprising nor in any way is it critical of the original proof. It would to the contrary be surprising if the 'Limit Theory Research Laboratory' mentioned earlier had failed in 14 years of active operation to obtain any simplifications. The original proof of Blackman is extremely clever and interesting. Based as it is on a Taylor's quadratic expansion of H_m, his approach is quite natural and easily motivated to the reader.

Consider now the following approach. As in (3) let

$$U_m(t) = m^{\frac{1}{2}}[F_m \circ F^{-1}(t) - t]$$

define the empirical process for $X_1, ..., X_m$ when transformed to $[0, 1]$. It is known that $U_m \overset{L}{\to} U_0$ with respect to Prokhorov's metric d on $D[0, 1]$ where U_0 is a Brownian Bridge (cf. Billingsley (1958)). Since (D, d) is a complete separable metric space one obtains from Theorem 1 the existence of versions $\{U_m^*: m \geq 0\}$ of these processes for which $d(U_m^*, U_0^*) \to 0$ a.s. However, the a.s. continuity of U_0^* and the equivalence of d-convergence and uniform convergence for sequences converging to continuous functions allows one to conclude that $\rho(U_m^*, U_0^*) \to 0$ a.s. for these versions, where ρ is the uniform 'sup.' metric. (This suffices to prove that $U_m \overset{L}{\to} U_0$ in (D, ρ) according to the Dudley–Wichura definition.) We henceforth drop the * from the notation and assume simply that U_m *is a fixed function* (it suffices to assume that it actually has the appearance of a realization of the empirical processes; m jumps of magnitude $m^{-\frac{1}{2}}$, piecewise linear of slope $-m^{\frac{1}{2}}$ and equal to zero at 0 and 1) *which converges uniformly to a continuous function U_0.*

Define $K_m(b) = mM_m(bm^{-\frac{1}{2}})$ so that

$$K_m(b) = m\int_{-\infty}^{\infty} [F(y) - F_m(y + bm^{-\frac{1}{2}})]^2\, dF(y)$$

$$= \int_{-\infty}^{\infty} [m^{\frac{1}{2}}\{F(y + bm^{-\frac{1}{2}}) - F(y)\} + m^{\frac{1}{2}}\{F_m(y + bm^{-\frac{1}{2}}) - F(y + bm^{-\frac{1}{2}})\}]^2\, dF(y). \qquad (8)$$

In terms of U_m this gives

$$K_m(b) = \int_{-\infty}^{\infty} [bf(x_{y,m,b}) + U_m \circ F(y + bm^{-\frac{1}{2}})]^2\, dF(y), \qquad (9)$$

where $x_{y,m,b}$ is the appropriate point between y and $y + bm^{-\frac{1}{2}}$ which exists by the mean value theorem. If we assume f is continuous and bounded it follows by the dominated convergence theorem that for each b

$$K_m(b) \to K(b) \equiv \int_{-\infty}^{\infty} [bf + U_0 \circ F]^2\, dF.$$

Since K is a quadratic, its minimum occurs at

$$\hat{b} = -\int_{-\infty}^{\infty} fU_0 \circ F\, dF \Big/ \int_{-\infty}^{\infty} f^2\, dF = -\mu^{-1}\int_{-\infty}^{\infty} fU_0 \circ F\, dF.$$

But this is a linear function of U_0. Since the law of U_0 is that of a Brownian Bridge and since the variance of $\hat{b}$ with respect to that law is finite, then $\hat{b}$ is a Normal r.v. In particular, $\hat{b}$ is a $N(0, \sigma_0^2)$ r.v. with

$$\sigma_0^2 = 2\mu^{-2}\iint_{x<y} f^2(x)f^2(y)\, F(x)\,[1 - F(y)]\, dx\, dy. \qquad (10)$$

In order to prove that the limiting distribution of $\hat{b}_m = m^{\frac{1}{2}}\hat{\theta}_m$ is that of $\hat{b}$, it remains to verify that the 'location of the minimum of K_m' converges to the 'location of the minimum of K'. It would then follow that the variances σ_0^2 and σ^2 must be the same, an identity which can be checked directly by routine but somewhat lengthy manipulations. To verify this 'interchange of limits', we must show that K_m is sufficiently 'U-shaped'.

By Minkowski's inequality applied to (9)

$$[K_m(b)]^{\frac{1}{2}} \geqslant \left\{\int_{-\infty}^{\infty} [bf + U_m \circ F]^2\, dF\right\}^{\frac{1}{2}} - \left\{\int_{-\infty}^{\infty} [\Delta_m(y, b)]^2\, dF(y)\right\}^{\frac{1}{2}}, \qquad (11)$$

where

$$\Delta_m(y, b) = b[f(y) - f(x_{y,m,b})] + U_m \circ F(y) - U_m \circ F(y + bm^{-\frac{1}{2}}). \qquad (12)$$

Since $U_m \to U_0$ uniformly it follows that

$$|U_m \circ F(y) - U_m \circ F(y + bm^{-\frac{1}{2}}) - U_0 \circ F(y) - U_0 \circ F(y + bm^{-\frac{1}{2}})| \to 0$$

uniformly in y and b. Moreover, since F has a bounded derivative f, bounded by C say, so that $|F(y+bm^{-\frac{1}{2}})-F(y)| \leqslant Cbm^{-\frac{1}{2}}$, it follows from the uniform continuity of U_0 that

$$|U_0 \circ F(y) - U_0 \circ F(y+bm^{-\frac{1}{2}})| \leqslant \omega_{U_0}(\delta)$$

whenever $Cbm^{-\frac{1}{2}} \leqslant \delta$, where ω is the modulus of continuity. Thus if f is assumed to satisfy a Lipschitz condition: $|f(x)-f(y)| \leqslant C_1|x-y|$ for all x, y and some constant C_1, it follows easily from (12) that for any $B > 0$, there exist constants $\{C_m\}$ such that $C_m \to 0$ and

$$|\Delta_m(y,b)| \leqslant C_m \quad \text{for all } y \text{ and all } b \in (-B, B).$$

This is sufficient to show using (4) that the location of the minimum *over* $(-B, B)$ of K_m conveys to the location of the minimum of K.

By Minkowski's inequality applied to (8)

$$[K_m(b)]^{\frac{1}{2}} \geqslant \left\{m \int_{-\infty}^{\infty} [F(y+bm^{-\frac{1}{2}})-F(y)]^2 \, dF(y)\right\}^{\frac{1}{2}}$$

$$- \left\{\int_{-\infty}^{\infty} [U_m \circ F(y+bm^{-\frac{1}{2}})]^2 \, dF(y)\right\}^{\frac{1}{2}}. \quad (13)$$

Since $U_m \to U_0$ uniformly the last term in (13) is bounded uniformly in b and n. Also, for $b > B$

$$m \int_{-\infty}^{\infty} [F(y+bm^{-\frac{1}{2}})-F(y)]^2 \, dF(y) \geqslant m \int_{-\infty}^{\infty} [F(y+Bm^{-\frac{1}{2}})$$

$$- F(y)]^2 \, dF(y) \xrightarrow[m \to \infty]{} B^2 \int_{-\infty}^{\infty} f^2 \, dF$$

by the dominated convergence theorem. Since a similar statement holds for $b < -B$, it follows that for any large constant N one can choose B sufficiently large, so that $K_m(b) > N$ for all $|b| > B$ and all sufficiently large m. This together with the statement at the end of the preceding paragraph suffices to establish the desired interchange in limits. This proves

Theorem 3.2. *If F has a density f which satisfies a (first order) Lipschitz condition, then $m^{\frac{1}{2}}\hat{\theta}_m \xrightarrow{L}$ a $N(0, \sigma^2)$ r.v. with $\sigma^2 = \sigma_0^2$ given by (7) or (10).*

It might also be pointed out that the limiting df of the minimum $mM_m(\hat{\theta}_m) = K_m(m^{\frac{1}{2}}\hat{\theta}_m)$ is that of

$$K(b) = \int_{-\infty}^{\infty} (U_0 \circ F)^2 \, dF - \frac{\left\{\int_{-\infty}^{\infty} fU_0 \circ F \, dF\right\}^2}{\int_{-\infty}^{\infty} f^2 \, dF}$$

$$= \int_0^1 U_0^2 \, dt - \mu^{-1}\left\{\int_0^1 gU_0 \, dt\right\}^2,$$

where $g = f \circ F^{-1}$.

4   LIMIT THEORY FOR LINEAR RANK STATISTICS

In limit theorems for two-sample statistics as in (5′) or (6′) one has a certain amount of freedom in the choice of a sequence of centering constants. For convenience I am going to choose $\mu_N = \int_0^1 F \circ H^{-1} dv_N$ which is suggested by (5′). (Recall that μ_N depends on N through H as well as v_N.) One is then interested in the limiting behaviour of the centered statistic,

$$T_N^* = N^{\frac{1}{2}}(T_N - \mu_N) = N^{\frac{1}{2}} \int_0^1 [F_m \circ H_N^{-1} - F \circ H^{-1}] dv_N. \qquad (14)$$

In this form one is led naturally to a study of the integrand, namely the *two-sample empirical process* defined by

$$L_N(t) = N^{\frac{1}{2}}[F_m \circ H_N^{-1}(t) - F \circ H^{-1}] \qquad (0 \leqslant t \leqslant 1). \qquad (15)$$

In terms of L_N, (14) may be rewritten simply as

$$T_N^* = \int_0^1 L_N dv_N. \qquad (14')$$

If all quantities in this expression were deterministic, several results from analysis would then be available for describing the limiting behaviour of T_N^*. The most general result might be the following combination of Lebesgue's dominated convergence theorem and the Helly–Bray theorem (cf. Lamperti (1964)): If $\{f_n : n \geqslant 0\}$ is a sequence of functions such that $f_n \to f_0$ uniformly, if $\{\mu_n : n \geqslant 0\}$ is a sequence of probability measures such $\mu_n \overset{w}{\to} \mu_0$ and if f_0 is continuous a.e. $-\mu_0$, then $\int f_n d\mu_n \to \int f d\mu$. A related theorem is given by Doubrovsky (1945).

To see that we can in fact consider L_N to be deterministic, we first express it in terms of the two 1-sample empirical processes U_m and V_n defined in (3):

$$L_N = (1 - \lambda_N)\{\lambda_N^{-\frac{1}{2}} B_N U_m(F \circ H_N^{-1}) - (1 - \lambda_N)^{-\frac{1}{2}} A_N V_n(G \circ H_N^{-1})\} + \delta_N \qquad (16)$$

for all $0 < t \leqslant 1$, where $\delta_N = A_N(H_N \circ H_N^{-1} - H \circ H^{-1}) N^{\frac{1}{2}}$ and

$$A_N(t) = [F \circ H^{-1}(u_t) - F \circ H^{-1}(t)]/(u_t - t) \qquad (u_t = H \circ H_N^{-1}(t)),$$

while B_N is defined by $\lambda_N A_N + (1 - \lambda_N) B_N \equiv 1$. Thus L_N is essentially a linear combination of U_m and V_n under random transformations of 'time', in which the 'coefficients' A_N and B_N are differential quotients of $F \circ H^{-1}$ and $G \circ H_N^{-1}$ respectively. These 'coefficients' are therefore bounded (by λ_N^{-1} and $(1 - \lambda_N)^{-1}$, respectively) and under reasonable assumptions should converge to certain derivatives. Furthermore, the random transformations $F \circ H_N^{-1}$ and $G \circ H_N^{-1}$ satisfy $\rho(F \circ H_N^{-1}, F \circ H^{-1}) \to 0$ a.s. and $\rho(G \circ H_N^{-1}, G \circ H^{-1}) \to 0$ a.s. (Lemma 2.3 of PS). Define as a natural limit for L_N the function

$$L_{ON} = (1 - \lambda_N)\{\lambda_N^{-\frac{1}{2}} b_N U_0(F \circ H^{-1}) - (1 - \lambda_N)^{-\frac{1}{2}} a_N V_0(G \circ H^{-1})\}, \quad (17)$$

where a_N and b_N are the derivatives of $F \circ H^{-1}$ and $G \circ H^{-1}$ respectively.

A comparison between (16) and (17) suggests a study of the convergence of $U_m(F \circ H_N^{-1})$. Since $U_m \xrightarrow{L} U_0$ in (D, ρ) we can apply (!) as in the previous section and assume that $\rho(U_m, U_0) \to 0$. However this uniform metric is not good enough for many applications since it does not utilize the fact that empirical processes are small near 0 and 1. To remedy this one can show in the iid case for example that $U_m/q \xrightarrow{L} U_0/q$ in (D, ρ) for certain weight functions q. Alternatively one can write $U_m \xrightarrow{L} U_0$ in (D, ρ_q) where $\rho_q(f, g) = (f/q, g/q)$. (Cf. Chibisov (1964) and PS.) If Q represents the class of all such weight functions q, then Q includes in particular all non-negative concave functions whose reciprocals are square integrable over $[0, 1]$. (For example $q(t) = [t(1-t)]^{\frac{1}{2}-\epsilon}$ is in Q.) It is relatively straightforward to establish from this in the iid case that

$$U_m(F \circ H_N^{-1}) - U_0(F \circ H^{-1}) \to 0 \quad \text{in } (D, \rho_q). \quad (18)$$

(Cf. Theorem 2.2 of PS.) By (!) and its subsequent clarification we may henceforth assume that $\{U_m: m \geqslant 0\}$ is a sequence of functions for which $U_m(F \circ H_N^{-1})/q - U_0(F \circ H^{-1})/q \to 0$ uniformly along with a similar assumption for $\{V_n: n \geqslant 0\}$. Under this construction it follows from (16) and (17) that one need only study the convergence of $A_N - a_N$ in order to complete a study of the difference between the processes L_N and L_{ON}.

In view of the representation of T_N^* as an integral of L_N given in (14'), the following 'limit' of T_N^* suggests itself, namely.

$$T_{ON}^* = \int_0^1 L_{ON} \, d\nu_N. \quad (19)$$

Then for any $q \in Q$, their difference satisfies

$$|T_N^* - T_{ON}^*| \leqslant \int_0^1 |(L_N - L_{ON})/q| \, q \, d\nu_N.$$

It follows directly from this and the discussion of the preceding paragraph that $T_N^* - T_{ON}^* \to 0$ if

$$(a) \quad \int_0^1 q \, d|\nu_N| = o(1),$$

and

$$(b) \quad \int_0^1 |A_N - a_n| \, q \, d|\nu_N| \to 0.$$

It is relatively straightforward to put conditions on F, G and λ_N to insure that (b) holds; see Sections 4 and 5 of PS. For example, the following result may be deduced.

Theorem 4.1. *If* (i) $\lambda_N = \lambda + O(N^{-\frac{1}{2}})$ *for some* $\lambda \in (0, 1)$, *and* (ii) *there exists a limiting measure ν for which*

$$\int_0^1 q\,d(\nu_N - \nu) \to 0, \quad \int_0^1 q\,d|\nu| < \infty$$

and $F \circ H_\lambda^{-1}$ *is differentiable a.e.* $- |\nu|$, *then* (b) *holds.*

The main assumption in this theorem from a practical point of view is the condition on the differentiability of $F \circ H_\lambda^{-1}$. Thus one may allow atoms for ν (or equivalently, discontinuities in the J-function of Chernoff and Savage) provided only that F is sufficiently smooth with respect to G at these atoms. Specifically, let C_λ denote the set of points $y \in [0, 1]$ for which

$$\lim_{x \to z} \frac{F(x) - F(z)}{G(x) - G(z)},$$

where $z = H_\lambda^{-1}(y)$, does *not* exist. Then one must have $|\nu|\,(C_\lambda) = 0$.

Since T_{ON}^* is a linear combination of normal r.v.'s with mean zero, it itself is normal with mean zero. It will have a limiting $N(0, \sigma^2)$ df if and only if $\sigma_N^2 \equiv \operatorname{var}(T_{ON}^*) \to \sigma^2$. In any case, it suffices to compute

$$\sigma_N^2 = 2(1 - \lambda_N)^2 \int_0^1 \int_u^1 \{\lambda_N^{-1} b_N(u)\,b_N(v)\,F \circ H^{-1}(u)\,[1 - F \circ H^{-1}(v)]$$

$$+ (1 - \lambda_N)^{-1} a_N(u)\,a_N(v)\,G \circ H^{-1}(u)\,[1 - G \circ H^{-1}(v)]\}\,d\nu_N(v)\,d\nu_N(u). \quad (20)$$

An interesting aspect of the method outlined in this section is that it applies verbatim to any two-sample problem for which the empirical processes U_m and V_n behave in the same way as those for independent and identically distributed r.v.'s. An attempt to stress this in the above discussion has been made by not assuming at the outset in §2 that the X's and Y's are iid but instead speaking of 'r.v.'s *associated* with df's F and G'. For example, one could obtain a Chernoff–Savage theorem for interchangeable r.v.'s without additional work once it were known that for such r.v.'s the empirical processes also satisfy (18). For $q \equiv 1$, this has been studied by Billingsley (1968) but for general q the problem is open. A second application would be to the case of iid pairs (X_i, Y_i) of dependent r.v.'s. Another aspect of the above approach to linear rank statistics is its applicability to the non-linear case. Since the emphasis has been upon a study of the two-sample empirical process L_N, one is able to proceed rather straightforwardly to functions of L_N other than the simple integrals $T_N^* = \int L_N dv_N$ which arose in this section. An example of this extended applicability is outlined in the next section.

5   AN ILLUSTRATION INVOLVING TESTS FOR INDEPENDENCE

In the notation of § 2, consider the statistic

$$S_N = m^{-2} \sum_{i,j=1}^{N} \alpha_{ij}^{N} R_{Ni} R_{Nj}, \tag{21}$$

where $R_{N1}, \ldots, R_{NN}$ are the ranks arising from two independent samples $(X_1, \ldots, X_m)$ and $(Y_1, \ldots, Y_n)$ associated with F and G respectively, and $\{\alpha_{ij}^N\}$ are given constants. Alternatively by a summation by parts as in § 2 one may express S_N in terms of the indicator r.v.'s $Z_{N1}, \ldots, Z_{NN}$, the only difference being the increased notational complexity since (21) involves a double summation. To do this, define constants

$$\{a_{ij}^N \colon 1 \leqslant i, j \leqslant N\}$$

by

$$a_{ij}^N = \sum_{r=i}^{N} \sum_{s=j}^{N} \alpha_{rs}^N.$$

Deleting all unnecessary N's from the notation, the appropriate interchanges of summation yield

$$m^2 S_N = \sum_{i=1}^{N} \sum_{j=1}^{N} \sum_{r=1}^{i} \sum_{s=1}^{j} \alpha_{ij} Z_r Z_s = \sum_{r,s=1}^{N} a_{rs} Z_r Z_s. \tag{22}$$

By the definition of the two-sample empirical process L_N given in (15),

$$R_i = m F_m \circ H_N^{-1}(i/N) = m[N^{-\frac{1}{2}} L_N(i/N) + K(i/N)],$$

where $K = F \circ H^{-1}$. Thus if v_N is defined to be the (signed) measure over the unit square I^2 which puts mass α_{rs} at the point $(r/N, s/N)$ it follows that

$$S_N = \iint F_m \circ H_N^{-1}(u) F_m \circ H_N^{-1}(v) \, dv_N(u, v)$$

$$= \iint [N^{-\frac{1}{2}} L_N(u) + K(u)] [N^{-\frac{1}{2}} L_N(v) + K(v)] \, dv_N(u, v),$$

where all double (single) integrals are over $I^2(I)$ unless otherwise indicated. Let v_N^* denote the measure on I defined by

$$v_N^*(A) = \iint_{A \times I} K(v) \, d[v_N(u, v) + v_N(v, u)] \tag{23}$$

for Borel sets A. Then for

$$\mu_N = \iint K(u) K(v) \, dv_N(u, v),$$

$$N^{\frac{1}{2}}(S_N - \mu_N) = N^{-\frac{1}{2}} \iint L_N(u) L_N(v) \, dv_N(u, v) + \int L_N dv_N^*. \tag{24}$$

It is clear from this representation that the limiting distribution of S_N will depend on which of the two terms on the right-hand side of (3) is dominant. The second term $\int L_N d\nu_N^*$ is a linear rank statistic of the form given in (14'). Consequently, as reviewed in § 2, several methods are available for studying its asymptotic behaviour. In particular conditions can be given under which it has a limiting Normal df. When this is the case one could investigate what additional conditions are necessary, if any, to insure that the first term on the right of (23) is $O_p(N^{\frac{1}{2}})$. Then under these conditions one would have established that asymptotically $N^{\frac{1}{2}}(S_N - \mu_N)$ behaves like $\int L_{ON} d\nu_N^*$.

On the other hand it could be that the measures $\{\nu_N^*\}$ are converging to the zero measure in such a way that $\int L_N d\nu_N = O_p(N^{-\frac{1}{2}})$. In this case the limiting distribution of S_N would be determined by that of the first term on the right of (3), so that under appropriate conditions one would have that asymptotically $N(S_N - \mu_N)$ behaves like $\int\int L_{ON}(u) L_{ON}(v) d\nu_N(u, v)$. In the first case the limit law would be Normal while in the latter case the non-Normal limiting distribution's characteristic function could be given by existing methods for such integrals of Gaussian processes.

The class of statistics S_N given by (21) or (22) has been studied recently by Schach (1969) for the case of (a_{ij}) forming a symmetric matrix and $F = G(= H)$. In this case, $K(v) = v$ and the measure ν_N^* of (23) is determined by the row sums of (a_{ij}). If these row sums equal zero, the second case of the previous paragraph obtains.

6 CONCLUSION

There exists in the current literature a large store of limit theorems for rank statistics under both contiguous and non-contiguous alternatives. However, of possibly greater importance than the theorems themselves is the rich collection of methods that have been used in their derivations. An analogous variety of methods is also present in the recent literature on limit theory for linear combinations of order statistics. The intent of this paper has been to draw attention to the importance of statisticians knowing methods as well as specific theorems, and for the need of improved communication thereof. The discussion centers mainly around the approach to two-sample rank statistics used by Pyke and Shorack (1968), and is illustrated by two applications; namely, an alternate proof of a result by Blackman (1955) and an outline of possible limit results for non-linear rank statistics, as studied in the null-hypothesis case by Schach (1969).

REFERENCES

Billingsley, P. (1968). *Weak Convergence of Probability Measures*. New York: John Wiley and Sons.

Blackman, J. (1955). On the approximation of a distribution function by an empiric distribution. *Ann. Math. Statist.* **26**, 256–67.

Brieman, L. (1968). *Probability*. Reading: Addison–Wesley.

Chernoff, H. and Savage, I. R. (1958). Asymptotic normality and efficiency of certain nonparametric test statistics. *Ann. Math Statist.* **29**, 972–94.

Chibisov, D. M. (1964). Some theorems on the limiting behaviour of empirical distribution functions. *Trudy Matem. Inst. in V. A. Steklova.* **71**, 104–12.

Darling, D. A. (1957). The Kolmogorov–Smirnov, Cramér–von Mises Tests. *Ann. Math. Statist.* **28**, 823–38.

Doubrovsky, V. M. (1945). On some properties of completely additive set functions and passing to the limit under the integral sign. *Isv. Akad. Nauk SSSR*, **9**, 311–20.

Dudley, R. M. (1968). Distances of probability measures and random variables. *Ann. Math. Statist.* **39**, 1563–72.

Dwass, M. (1956). The large-sample power of rank order tests in the two-sample problem. *Ann. Math. Statist.* **27**, 352–74.

Govindarajulu, Z., LeCam, L. and Raghavachari, M. (1967). Generalizations of theorems of Chernoff–Savage on asymptotic normality of nonparametric test statistics. *Proc. of Fifth Berkeley Symp. on Math. Statist. and Prob.* **1**, 609–38. University of California Press.

Hájek, J. (1968). Asymptotic normality of simple linear rank statistics under alternatives. *Ann. Math. Statist.* **39**, 325–46.

Hájek, J. and Šidak, Z. (1967). *Theory of Rank Tests*. Prague: Academia.

Hoeffding, W. (1948). A class of statistics with asymptotically normal distributions. *Ann. Math. Statist.* **19**, 293–325.

Kiefer, J. (1967). Statistical Inference: Panel discussion by J. C. Kiefer, L. M. LeCam and L. J. Savage. *The Future of Statistics.* Edited by D. G. Watts. New York: Academic Press.

Lamperti, J. (1946). On extreme order statistics. *Ann. Math. Statist.* **35**, 1726–37.

Parthasarathy, K. R. (1967). *Probability Measures on Metric Spaces*. New York: Academic Press.

Prokhorov, Yu. V. (1956). Convergence of random processes and limit theorems in probability theory. *Th. Prob. and Applic.* (translated by SIAM), **1**, 157–214.

Pyke, R. and Shorack, G. (1968). Weak convergence of a two-sample empirical process and a new approach to Chernoff–Savage theorems. *Ann. Math. Statist.* **39**, 755–71.

Pyke, R. (1969). Applications of almost surely convergent constructions of weakly convergent processes. *Proc. of Int'l. Symp. on Prob. and Inform. Th. Lecture Notes in Mathematics*, **89**, 187–200. Berlin: Springer–Verlag.

Schach, S. (1969). The asymptotic distribution of some non-linear functions of the two-sample rank vector. *Ann. Math. Statist.* **40**, 1011–20

Sellar, W. C. and Yeatman, R. J. (1960). *1066 and all that*. Harmendsworth: Penguin Books.

Skorokhod, A. V. (1956). Limit theorems for stochastic processes. *Th. Prob. and Applic.* (translated by SIAM), 1, 261–90.

Wichura, M. J. (1969). On the construction of almost uniformly convergent random variables with given weakly convergent image laws. To appear in *Ann. Math. Statist.*

DISCUSSION ON PYKE'S PAPER

J. HÁJEK

I share with Professor Pyke his opinion that the method based on the concept of weak convergence in metric spaces is becoming a standard and unavoidable tool in asymptotic problems of statistics. Another compelling example of this fact has been provided recently by LeCam (1969), who showed that this method is essential in reducing the conditions for asymptotic theory of maximum likelihood estimates close to the level of necessity.

In the application of the weak convergence principle to particular cases, the success depends on our ability to derive uniform bounds for fluctuations of trajectories of the processes under study. For processes with independent increments the result is provided by the Kolmogorov inequality or its Hájek–Rényi modification. For processes whose increments are dependent in the same manner as successive samples without replacement from a finite population, we may still receive adopted inequalities of Kolmogorov and Hájek–Rényi type. The empirical distribution functions belong to such processes. Pyke and Shorack (1968) provide for nondecreasing and non-negative function q the following inequality for $U_m(t)$:

$$P[|U_m(t)| \leq q(t),\ 0 \leq t \leq \theta] \geq 1 - C_\theta \int_0^\theta [q(t)]^{-2}\, dt. \qquad (1)$$

By using the above mentioned inequalities for sampling from finite population [see Rosén (1964) or Hájek–Šidák V. 3.4] we can show that $C_\theta = 1$. In the LeCam paper the inequalities are obtained by utilizing strong dependence of the increments of the likelihood processes. The treatment of simple linear rank statistics in Hájek (1968) is based on the inequality which for the Pyke's statistics T_N writes as follows:

$$\operatorname{var} T_N \leq 21\, m^{-2} \sum_{i=1}^{N} (C_{Ni}^*)^2 \quad \text{if } C_{N1}^* \leq \ldots \leq C_{NN}^*. \qquad (2)$$

This inequality is powerful enough to make the use of (1) unnecessary. In problems where no proper inequalities are available the development stagnates, as is exemplified by asymptotic theory of rank statistics for testing independence.

In the theory of asymptotic distributions of linear rank order statistics four basic approaches have been used, as exemplified by the papers by

Motoo (1957), Pyke and Shorack (1968), Hájek (1962) and Hájek (1968). Motoo used a central theorem for dependent summands published in Loève's probability textbook (Theorem 28.2.C) to obtain Lindeberg type conditions for asymptotic normality of linear rank statistics under the hypothesis of randomness. This result cannot be obtained by any other known method. On the other hand, Motoo's method has not provided any result for distributions under alternatives. Pyke and Shorack (1968) (as well as previously Govindarajulu–LeCam–Raghavachari (1967)) utilized the idea to approximate a given two-sample linear rank statistic by a linear functional on a Gaussian process. The drawback of this approach is that at present it does not extend to cases when each observation may have a different distribution, and in the two-sample case to situations when the sample sizes are not of the same order. There is no sign, however, that by further development the method could not be extended to such cases, too. The approach in Hájek (1962), utilized also in Hájek–Šidák (1967), is based on LeCam's lemmas on contiguity. For noncontiguous alternatives Hájek (1968) presents a method consisting of approximating rank statistic by a sum of independent random variables, which is an idea already present in the theory of U-statistics as well as in the original Chernoff–Savage (1958) paper. This method has been extended in the paper Dupač and Hájek (1969) to cover discontinuous scores-generating functions successfully treated before in Pyke and Shorack (1968). Each of the above four methods has its merits and deserves attention.

To avoid semantic confusion, let me conclude by pointing out the differences in using the term 'linear rank statistic' and in the definition of ranks. In concordance with Hájek and Šidák (1967) I call S a linear rank statistic if

$$S = \sum_{i=1}^{N} a(i, R_i),$$

where $\{a(i,j)\}$ is a square matrix and R_i denote the number of observations smaller than or equal to X_i, $1 \leqslant i \leqslant N$. Pyke uses the same term for a 'two-sample simple linear statistic' for which $a(i,j) = C_j^*$ or 0, if $1 \leqslant i \leqslant m$ or $m < i \leqslant N$, respectively. The ranks R_{Ni} in Pyke's paper under discussion denote the number of observations in the first sample which do not exceed the ith order statistic among the pooled sample.

REFERENCES

Dupač, V. and Hájek, J. (1969). Asymptotic normality of simple linear rank statistics under alternatives II. *Ann. Math. Statist.* **40**, 1992–2017.

LeCam, L. (1969) On the assumptions used to prove asymptotic normality of maximum likelihood estimates. Preprint.

Motoo, M. (1957). On the Hoeffding's combinatorial central limit theorem. *Ann. Inst. Statist. Math.* Vol. 8, 145–54.

Rosén, B. (1964). Limit theorems for sampling from finite populations. *Arkiv for Matematik*, Band 4, nr 28, 383–424.

ON THE THEORY OF
NONPARAMETRIC TESTS

H. WITTING

1 INTRODUCTION AND SUMMARY

It is well known that the practical performance of a rank test and a permutation test is based on the comparison of the same number of points, namely those points which are obtained from the tuple l of ranks respectively from the observed tuple x by appropriate generalized permutations. This stems from the fact, that nonparametric test problems are governed by two groups. The first group is the group G of infinite order which leaves the test problem invariant (i.e. to which the reduction by invariance refers). The second group is the group Q of finite order q, which leaves the distributions belonging to the boundary of the hypotheses invariant (and which permutates the G-orbits in an appropriate way). More precisely, the points to be compared in applying a rank test or a permutation test are exactly those q points which are obtained from the rank tuple l and the observed tuple x, respectively, by the transformations π from the finite group Q, i.e. the q points πl, $\pi \in Q$, or πx, $\pi \in Q$.

Besides these two groups there is a third group which is essential for nonparametric test problems, namely a subgroup Q_1 of Q which leaves each admitted distribution invariant. This implies, by arguments of sufficiency, that—if q_1 denotes the order of Q_1—those q_1 points which can be mapped by transformations from Q_1 lead to the same value of the test statistic. Hence, in applying the test only the $q_2 := q/q_1$ essentially different points are compared.

In the sequel we will emphasize these invariance properties and show that they allow a unified derivation of rank tests as well as (under a suitable additional assumption) of permutation tests at least as far as the general form of the test statistic is concerned. We will not discuss here the problems which arise in connection with the choice of a special class of alternatives, including the problem of proving unbiasedness, cf. [1] and [3]. The invariance properties and the additional assumption are fulfilled for instance for the problems of the comparison of two samples, of testing for symmetry with respect to zero and of testing for independence.

2   GENERALIZED ORDER- AND RANK-STATISTICS; OPTIMAL RANK TESTS

Let $X_1, \ldots, X_n$ be independent k-dimensional random variables with continuous one-dimensional marginal distributions. Hence, the sample space $(\mathscr{X}, \mathscr{B})$ is the kn-dimensional space $\mathscr{R}_{kn}$ with the σ-field $\mathfrak{B}_{kn}$ of Borel subsets, $(\mathscr{X}, \mathscr{B}) = (\mathscr{R}_{kn}, \mathfrak{B}_{kn})$. The distribution of $X := (X_1, \ldots, X_n)$ is denoted by P. The test problem is $P \in \mathscr{P}_H$ against $P \in \mathscr{P}_K$ at level α, where $\mathscr{P}_H$ and $\mathscr{P}_K$ are disjoint subsets of the class $\mathscr{P}$ of the admitted kn-dimensional distributions.

For a unified theory of rank tests we need the following two assumptions:

Assumption 1: There is an infinite group G of measurable transformations g from $(\mathscr{X}, \mathscr{B})$ onto itself with two properties:

(a) The test problem is invariant under G.

(b) The orbits under G are measurable; there is only a finite number of such G-orbits.

In particular, let q denote the number of G-orbits which are not $\mathscr{P}$-null sets, i.e. which are not sets of P-measure zero for all $P \in \mathscr{P}$. Let $\mathscr{X}'$ be the union of those q orbits and L be a maximal invariant statistic with respect to G, defined on $\mathscr{X}'$.

Assumption 2: There is a group Q of finite order q of measurable transformations π from $(\mathscr{X}, \mathscr{B})$ onto itself with two properties:

(a) Q operates transitively on the G-orbits in $\mathscr{X}'$, i.e. the G-orbits in $\mathscr{X}'$ are mapped onto themselves by transformations $\pi \in Q$ and to each pair $(\mathscr{X}_i', \mathscr{X}_j')$ of such orbits there is a transformation $\pi \in Q$ such that $\mathscr{X}_j' = \pi \mathscr{X}_i'$.

(b) The set $\mathscr{P}_J$ of measures P, which are invariant under Q, is a non-void subset $\mathscr{P}_J \subset \mathscr{P}_H$.

It follows from these two assumptions that $\mathscr{X}'$ is an invariant set with respect to G and to Q and that the total probability is concentrated on $\mathscr{X}'$. Therefore, one can replace the test problem on $\mathscr{X}$ by a test problem on $\mathscr{X}'$ and we will do so.

Let V denote a statistic on $\mathscr{X}'$ which is maximal invariant with respect to Q, i.e. a function, which is invariant with respect to the permutations of the G-orbits in $\mathscr{X}'$ and which takes on the same value only for points which can be transformed into each other by such permutations. Hence, in every G-orbit there is exactly one point of every Q-orbit, i.e. one point

y with $V(y) = V(x)$ and the totality of these points is given by $\{\pi x, \pi \in Q\}$.

In choosing the maximal invariant statistics V and L there is some arbitrariness, since only the Q-orbits and the G-orbits are uniquely determined. For example, one can put on one of the G-orbits, say on $\mathscr{X}'_e$, $V(x) = x$. Then, since V is constant for the q points πx, $\pi \in Q$, the maximal invariant statistic is

$$V(x) = \sum_{\pi \in Q} \pi^{-1} x I_{\mathscr{X}'\pi}, \quad \mathscr{X}'_\pi := \pi \mathscr{X}'_e.$$

In the following we denote this special version of $V(x)$ by $x_{(\,)}$. We can interpret $x_{(\,)}$ as a generalized order statistic, since the map $x \to V(x)$ on $\mathscr{X}'_\pi$, i.e. $x \to \pi^{-1}x$, is a generalized ordering of the components of x according to l_π, where l_π denotes the value of L on $\mathscr{X}'_\pi$. Hence, $L(x)$ characterizes the order of the components of x in the tuple x. If $L(x)$ is chosen as a tuple of kn integers, we call L a generalized rank statistic and the components of $L(x)$ the ranks of the components of x. Every point $x \in \mathscr{X}'$ is uniquely determined by the values $L(x)$ and $V(x)$. The point given by the value $x_{(\,)}$ of the order statistic and by l of the rank statistic is denoted by $x_{(l)}$. For the (k-dimensional) components of $x_{(l)}$ we write $x_{(l_1)}, \ldots, x_{(l_n)}$ and for the random variable which originates from X in the same way as $x_{(\,)}$ from x we write $X_{(\,)}$.

Assumptions 1 and 2 already allow the derivation of an optimal rank test. Here a rank test ϕ is defined as a G-invariant test, i.e. because of the maximal invariance of L as a test $\phi(x) = \psi(L(x))$, $x \in \mathscr{X}'$. For this purpose one determines, as usually, a most powerful rank test at level α for $\mathscr{P}_J$ against a simple alternative $\{P_1\} \subset \mathscr{P}_K$. This can be done by means of the Neyman–Pearson lemma, since according to assumption 2 the q non-degenerate G-orbits have the same probability for every $P \in \mathscr{P}_J$, i.e. the distribution of L is the discrete uniform distribution over the q points πl, $\pi \in Q$. Because of $P^L(\{l\}) = P(\mathscr{X}'_{\pi l}) = 1/q$ for all $l \in L(\mathscr{X}')$ the test statistic is $P_1^L(\{l\}) = P_1(\mathscr{X}'_{\pi l})$. Here, $\pi_l \in Q$ is defined by $L(x) = l$ for $x \in \mathscr{X}'_{\pi l}$. If P_1 is dominated by a measure $P_0 \in \mathscr{P}_J$, the distribution of L for $P = P_1 \in \mathscr{P}_K$ can be calculated by a generalized Hoeffding formula

$$P_1^L(\{l\}) = P_1(\mathscr{X}'_{\pi l}) = \frac{1}{q} E_0 \frac{h_1(X_{(l)})}{h_0(X_{(l)})} = \frac{1}{q} E_0 \prod_{j=1}^{n} \frac{f_{1j}(X_{(l_j)})}{f_{0j}(X_{(l_j)})},$$

where h_1 and h_0 are densities of P_1 and P_0 with respect to a dominating measure $\mu^{(n)} = \mu \times \ldots \times \mu$ and f_{1j} and f_{0j} are the corresponding μ-densities of X_j. This formula can easily be proved observing that h_1/h_0 is a P_0-density of P_1 on $\mathscr{X}'\{x: h_0(x) > 0\}$ and qh_0 is a $\mu^{(n)}$-density of $X_{(l)}$ on

$\mathscr{X}'_{\pi_l}$. The latter is true, since the range of the random variable $X_{(l)}$ is $\mathscr{X}'_{\pi_l}$ and, because of the invariance of $P_0 \in \mathscr{P}_J$ under $\pi \in Q$, it holds that
$$P_0(X_{(l)} \in B) = q P_0(X \in B), \quad B \in \mathscr{X}'_{\pi_l} \mathscr{B}.$$

Generally, one does not use this optimal test statistic $P_1^L(\{l\})$, but the test statistic
$$\tilde{\pi}(l) := \frac{dP_\Delta^L(\{l\})}{d\Delta}\bigg|_{\Delta=0} = \sum_{j=1}^{n} E_0 \frac{\frac{\partial f_{\Delta j}}{\partial \Delta}(X_{(lj)})|_{\Delta=0}}{f_{0j}(X_{(lj)})},$$

obtained by differentiation with respect to a parameter Δ, after having replaced P_1 by an element P_Δ of a one-parameter subclass of $\mathscr{P}_K$.

3 TOPOLOGY ON $\mathscr{P}$; OPTIMAL PERMUTATION TESTS

In carrying out a reduction by invariance, the determination of a most powerful test for $\mathscr{P}_H$ against $\mathscr{P}_K$ leads automatically to that of a test similar on $\mathscr{P}_J$, since the distribution of L was seen to be independent of $P \in \mathscr{P}_J$. If a reduction by invariance is not admissible, for an analogous simplification of the test problem it is necessary to assume the unbiasedness of the test. Furthermore the existence of a suitable topology on $\mathscr{P}$ is needed in order that the two notions 'continuity of the power function of a test' and 'boundary of the hypotheses' are defined.

It is well known that unbiased tests for $\mathscr{P}_H$ against $\mathscr{P}_K$ are similar on the boundary $\bar{\mathscr{P}}_H \cap \bar{\mathscr{P}}_K$ of the hypotheses, if the power function is continuous. However, this fact alone does not imply a simplification. In order to be able to determine a most powerful similar test as a test with Neyman structure, it is necessary that there exists a complete sufficient statistic for the boundary. Moreover, the boundary $\bar{\mathscr{P}}_H \cap \bar{\mathscr{P}}_K$ has to be equal to $\mathscr{P}_J$, if one wants to take advantage of the invariance with respect to Q and to get a formal analogy to the rank tests. Therefore we need for the derivation of permutation tests the

Assumption 3. There is a topology on $\mathscr{P}$ with the following properties:

  (a) The power function $\beta_\phi(P) = E_P \phi$ of every test ϕ is continuous in $P \in \mathscr{P}$.
  (b) $\bar{\mathscr{P}}_H \cap \bar{\mathscr{P}}_K = \mathscr{P}_J$, $\bar{\mathscr{P}}_H$ denoting the closed hull of $\mathscr{P}_H$.
  (c) There is a complete sufficient statistic for $P \in \bar{\mathscr{P}}_H \cap \bar{\mathscr{P}}_K$.

(b) justifies to call $\mathscr{P}_J$ the boundary of $\mathscr{P}_H$ and $\mathscr{P}_K$. Because of (b) assumption (c) means that the statistic V, which is sufficient for $\mathscr{P}_J$, is also complete.

In the first lemma of the Appendix we will prove that the topology, defined by the metric of uniform convergence, has the property (a). To prove property (b) for one- and two-sample problems we need a second lemma which establishes the equivalence of the convergence of the distribution functions (in the sense of the uniform convergence of the corresponding k-dimensional distributions) with the convergence of the kn-dimensional distributions on the sample spaces. The proof of property (b) for the symmetry test problem is given in §5, the proof of property (b) for the two other above mentioned test problems is analogous. Furthermore, it is well known, cf. [1] or [3], that in each of these three problems the sufficient statistic V is also complete for the class $\mathscr{P}_J$.

The tests with Neyman structure are conditional tests, given the order statistic $V(x)$. Since V is maximal invariant under Q, i.e.

$$\{y: V(y) = V(x)\} = \{\pi x, \pi \in Q\},$$

each such conditional test is said to be a permutation test.

In order to derive an optimal permutation test for $\mathscr{P}_H$ against $\mathscr{P}_K$ one determines first, as usually, a most powerful permutation test at level α for $\mathscr{P}_J$ against a simple alternative $\{P_1\} \subset \mathscr{P}_K$. This can be done by means of the Neyman–Pearson lemma, since because of the sufficiency the conditional distribution of X, given $V(x)$, can be chosen in such a way that it is independent of $P \in \mathscr{P}_J$. Here the sufficiency is based on the fact that every statistic V which is maximal invariant with respect to a finite group Q is sufficient for any class of measures which are invariant under Q (cf. [2], p. 137).

Also the conditional distribution of X, given $V(x)$, under $P = P_1 \in \mathscr{P}_K$ can be chosen in such a way that it is totally concentrated on these q points πx, $\pi \in Q$. If the dominating measure is so chosen, that it is invariant with respect to Q, for example

$$\nu(B) = \sum_{\pi \in Q} P(\pi B), \quad B \in \mathscr{X}' \mathfrak{B}_{kn},$$

and if h denotes the ν-density of P, it is easy to show that a version of $P^{X|v}$ is given by

$$P^{X|v}(\{x\}) = \frac{h(x)}{\sum\limits_{\pi \in Q} h(\pi^{-1}v)} \quad (x = \pi v, \ \pi \in Q),$$

if

$$h^V(v) := \sum_{\pi \in Q} h(\pi^{-1}v) > 0,$$

and

$$P^{X|v}(\{x\}) = 1/q \quad (x = \pi v, \ \pi \in Q),$$

if

$$h^V(v) := \sum_{\pi \in Q} h(\pi^{-1}v) = 0.$$

In particular, if $P \in \mathscr{P}_J$, h is invariant under Q, i.e.

$$P^{X|v}(\{x\}) = 1/q \quad (x = \pi v, \pi \in Q).$$

Therefore, the test statistic of the conditional test is

$$P_1^{X|v}(\{x\}) = h_1(x)/h_1^V(v)$$

or, equivalently, $h_1(x)$.

4 RANK- AND PERMUTATION-TESTS; REDUCTION BY SUFFICIENCY

In §§ 2 and 3 we have seen that for both types of tests, the q nondegenerate G-orbits are compared, in the case of rank tests as a whole, in the case of permutation tests by means of those q points, which have a fixed value of $V(x)$, and in both cases the null distribution is the discrete uniform distribution. In particular, because of assumption 2, this means that the group G is of importance for the permutation test, though assumption 1 is not explicitly needed for its derivation.

In order that the ranks are defined and the q points, to be compared in applying the permutation test, are distinct, respectively, one has to exclude finitely many hyperplanes, namely the lower dimensional orbits under the above mentioned group G. This is justified, since these orbits are null-sets with respect to all admitted distributions. Furthermore, the degenerate G-orbits are not only those points, for which the ranks are not defined, but also those points, for which the q points πx, $\pi \in Q$, are not distinct.

The question, whether for a concrete problem one should apply a rank- or a permutation-test, is, of course, not touched by these considerations. Besides the fact, that a reduction by invariance is not always justified, the answer to this question depends essentially on how easily one can determine a fractile for the permutation test.

The reduction by sufficiency is based on

Assumption 4. There is a subgroup Q_1 of Q of order q_1, such that each distribution $P \in \mathscr{P}$ is invariant under Q_1.

Hence, as indicated in § 3, every statistic T, which is maximal invariant with respect to Q_1 (and which can be chosen analogously to $x_{()}$), is also sufficient for $\mathscr{P}$. Therefore, before deriving a permutation test a reduction by sufficiency is possible. Since the transitivity of Q implies in particular, that Q—and therefore Q_1, too—is compatible with the maximal invariant statistic L, cf. [4], one can carry out a reduction by sufficiency also in the case of rank tests and this on the range of L.

5 EXAMPLE: TEST FOR SYMMETRY WITH RESPECT TO ZERO

Let $X_1, \ldots, X_n$ be independent random variables with the same one-dimensional continuous distribution, i.e. P is an n-dimensional product measure $v^{(n)}$ with the same one-dimensional marginal distribution v, the distribution function of which is denoted by F, i.e. $P = v_F^{(n)}$. If F_- denotes the distribution function of the distribution reflected in the origin, i.e. $F_-(x) := 1 - F(-x)$, the hypothesis is $F \geqslant F_-$ and the alternative $F \leqslant F_-$. Here $F \leqslant F_-$ means, that the inequality $F(x) \leqslant F_-(x)$ holds for all $x \in \mathscr{R}_1$ and this strictly for at least one $x \in \mathscr{R}_1$. Hence

$$\mathscr{P}_H = \{v_F^{(n)}: F \geqslant F_-\}, \quad \mathscr{P}_K = \{v_F^{(n)}: F \leqslant F_-\}.$$

Here G is the group defined by the transformations $\pi x = (\tau x_1, \ldots, \tau x_n)$ from $\mathscr{R}_n$ onto itself, where τ is a continuous, strictly increasing and odd mapping from $\mathscr{R}_1$ onto itself. The G-orbits which are not $\mathscr{P}$-null sets are the $q = 2^n n!$ sets $\{x \in \mathscr{R}_n: 0 < |x_{i_1}| < \ldots < |x_{i_n}|\}$, where $(i_1, \ldots, i_n)$ is a permutation of $(1, \ldots, n)$. As a maximal invariant statistic, which characterizes these $2^n n!$ sectors, we can choose the n-tuple of the signs ϵ_j of the observations x_j combined with the ranks r_j of their absolute values $|x_j|$, i.e. $L(x) = (\epsilon_1 r_1, \ldots, \epsilon_n r_n)$.

Q is the group of the $q = 2^n n!$ transformations, generated by the 2^n permutations of the signs and by the $n!$ permutations of the co-ordinates. This group operates transitively on the G-orbits and keeps those distributions $P = v_F^{(n)}$ invariant, for which $F = F_-$. As a maximal invariant statistic we can choose the n-tuple of $V(x) = (|x|_{[1]}, \ldots, |x|_{[n]})$ of the absolute values. This is of the form $V(x) = x_{\{\}}$ with

$$\mathscr{X}'_e = \{x \in \mathscr{R}_n: 0 < x_1 < \ldots < x_n\}.$$

Obviously every point $x \in \mathscr{X}'$ is uniquely determined by the generalized ranks (i.e. by the signs ϵ_j multiplied with the ranks r_j of the absolute values) and the generalized order statistic (i.e. the order statistic of the absolute values), $x_{\{j\}} = \epsilon_j |x|_{[r_j]}$.

Therefore, if $\tilde{F}_0$ denotes the distribution function of the $|X_i|$, we have for instance for Lehmann-alternatives

$$\tilde{\pi}(l) = \Sigma \epsilon_j E_0 \tilde{F}_0(|X|_{[r_j]}) = \Sigma \epsilon_j r_j / (n+1).$$

This is the well known statistic of the Wilcoxon matched pairs signed rank test.

As indicated in connection with lemma 2, it is still to prove

$$\bar{\mathscr{P}}_H \cap \bar{\mathscr{P}}_K = \mathscr{P}_J,$$

where $\mathscr{P}_J = \{v_F^{(n)}: F = F_-\}$. Because of $\mathscr{P}_J \subset \mathscr{P}_H$ it is sufficient for $\mathscr{P}_J \subset \bar{\mathscr{P}}_H \cap \bar{\mathscr{P}}_K$ that $\mathscr{P}_J \subset \bar{\mathscr{P}}_K$. Let $v_F^{(n)} \in \mathscr{P}_J$, i.e. F a continuous distribution function with $F = F_-$; then

$$F_m := \left(1 - \frac{1}{m}\right)F + \frac{1}{m}F^2$$

is a continuous distribution function with $F_m \leqslant (F_m)_-$ and

$$d(v_{F_m}, v_F) = \frac{1}{m} \sup_{B \in \mathfrak{B}_1} |v_{F^2}(B) - v_F(B)| \to 0 \quad \text{for } m \to \infty.$$

These two properties imply $v_{F_m}^{(n)} \in \mathscr{P}_K$ and $d(v_F^{(n)}, v_{F_m}^{(n)}) \to 0$, i.e. $v_F^{(n)} \in \bar{\mathscr{P}}_K$.

For the proof of $\bar{\mathscr{P}}_H \cap \bar{\mathscr{P}}_K \subset \mathscr{P}_J$ we notice that for all $v_F^{(n)} \in \bar{\mathscr{P}}_H \cap \bar{\mathscr{P}}_K$ there exist sequences $\{F_m\}$ and $\{F'_m\}$ with $v_{F_m}^{(n)} \in \mathscr{P}_H$, $v_{F'_m}^{(n)} \in \mathscr{P}_K$ and $d(v_{F_m}, v_F) \to 0, d(v_{F'_m}, v_F) \to 0$, i.e. in particular $F_m \geqslant (F_m)_-, (F'_m) \leqslant (F'_m)_-$ and $F_m(x) \to F(x), F'_m(x) \to F(x) \; \forall x \in \mathscr{R}_1$. Therefore it holds that $F \geqslant F_-$ and $F \leqslant F_-$, i.e. $F = F_-$ and, consequently, $v_F^{(n)} \in \mathscr{P}_J$.

Finally, the group Q_1 is the group of the $q_1 = n!$ transformations of $\mathscr{R}_n$, generated by the permutations of the n co-ordinates. Here the order statistic $T(x) = x_{[\,]}$ is maximal invariant with respect to Q_1 and therefore sufficient for $P \in \mathscr{P}$. Since the regions of constancy of $L(x) = (\epsilon_1 r_1, \ldots, \epsilon_n r_n)$ are not changed by permutations of the co-ordinates, Q_1 is compatible with L. Therefore, the practical performance of the permutation test and the rank test is based on the comparison of $q_2 := 2^n$ points, which are yielded by the observed tuple x or the rank tuple l by permutation of the signs.

APPENDIX

Lemma 1: *With respect to the topology defined by the metric*

$$d(P, P_0) := \sup_{B \in \mathscr{B}} |P(B) - P_0(B)|$$

the power function of each test is continuous.

Proof: Let ν denote a $\{P_m : m = 0, 1, 2, \ldots\}$-dominating measure and p_m a ν-density of P_m. Then $d(P_m, P_0) \to 0$ is equivalent with $\int |p_m - p_0| \, d\nu \to 0$ according to

$$\int |p_m - p_0| \, d\nu = \int_{\{p_m > p_0\}} (p_m - p_0) \, d\nu + \int_{\{p_m < p_0\}} (p_0 - p_m) \, d\nu$$

$$= 2 \sup_{B \in \mathscr{B}} \left| \int_B (p_m - p_0) \, d\nu \right| = 2 \sup_{B \in \mathscr{B}} |P_m(B) - P_0(B)|.$$

Therefore, if ϕ denotes a test function or more generally a $\mathscr{B}$-measurable function with $|\phi| \leqslant M < \infty$, it follows from $d(P_m, P_0) \to 0$, that

$$\left| \int \phi \, dP_m - \int \phi \, dP_0 \right| = \left| \int \phi (p_m - p_0) \, d\nu \right| \leqslant M \int |p_m - p_0| \, d\nu \to 0.$$

Lemma 2: *Let* P_m *and* P'_m *denote probability measures on measurable spaces* $(\mathscr{X}, \mathscr{B})$ *or* $(\mathscr{X}', \mathscr{B}')$, *respectively,* $m = 0, 1, 2, \ldots,$ *and* n_1 *and* n_2 *fixed naturals. Then the following three statements for* $m \to \infty$ *are equivalent*

(a) $d(P_m, P_0) \to 0, \quad d(P'_m, P'_0) \to 0,$

(b) $d(P_m \times P'_m, P_0 \times P'_0) \to 0,$

(c) $d(P_m^{(n_1)} \times P'^{(n_2)}_m, P_0^{(n_1)} \times P'^{(n_2)}_0) \to 0.$

Proof: Let p_m denote a ν-density of P_m and p'_m a ν'-density of P'_m. Then $p_m p'_m$ is a $\nu \times \nu'$-density of $P_m \times P'_m$.

Therefore (b) follows from (a) and Fubini's theorem according to

$$\int |p_m p'_m - p_0 p'_0| \, d(\nu \times \nu') \leqslant \int p_m \, d\nu \int |p'_m - p'_0| \, d\nu' + \int p'_0 \, d\nu' \int |p_m - p_0| \, d\nu \to 0.$$

Analogously, $d(P_m^{(n)}, P_0^{(n)}) \to 0$ follows from (a) for all natural integers n by induction and therefore (c) from (a), too. On the other hand, (b) follows from (c) for $n_1 = n_2 = 1$ and (a) from (b) for the sets $B \times \mathscr{X}'$, $B \in \mathscr{B}$, and $\mathscr{X} \times B'$, $B' \in \mathscr{B}'$, respectively.

REFERENCES

[1] Lehmann, E. L. *Testing Statistical Hypotheses*. New York, 1959.

[2] Witting, H. *Mathematische Statistik, Eine Einführung in Theorie und Methoden*. Stuttgart, 1966.

[3] Witting, H. und G. Nölle. *Angewandte Mathematische Statistik, Optimale finite und asymptotische Verfahren*. Stuttgart, 1970.

[4] Witting, H. Zur Theorie der Rangtests. To appear in *Metrika*.

DISCUSSION ON WITTING'S PAPER

W. HOEFFDING

Professor Witting's paper brings out the structure of a generalized rank test and the analogous features of a generalized permutation test (test of structure S), in particular of tests of these types that are optimal against specified alternatives. I want to make a few remarks on the relative merits of these two kinds of tests, a topic which Professor Witting deliberately did not elaborate on.

For definiteness, suppose that the observed random vector

$$\mathbf{X} = (X_1, ..., X_n)$$

has a probability density with respect to n-dimensional Lebesgue measure and that we want to test the hypothesis that the components of $\mathbf{X}$ are independent and identically distributed against the alternative that the probability density is of the form

$$p(\mathbf{X}) = \prod_{i=1}^{n} f_i(x_i), \quad f_i(x) = g(x - \eta - \delta c_i), \tag{1}$$

where the c_i are given numbers, $\delta > 0$, η real. In a nonparametric problem we would not like to specify the function g, but let us make the traditional assumption

$$f_i(x) = (2\pi)^{-\frac{1}{2}} \sigma^{-1} \exp\{-(x - \eta - \delta c_i)^2/(2\sigma^2)\}. \tag{2}$$

In some respects the most powerful permutation test (MPP test) against this alternative compares favorably with the most powerful rank test (MPR test) of the same size. In the first place, it does not depend on the three parameters $\delta > 0, \eta, \sigma$, whereas the MPR test depends on $\Delta = \delta/\sigma$. Against the normal alternatives (2) it is more powerful than the latter, for each $\Delta > 0$. The MPP test is uniformly most powerful, among all similar tests of given size, not only against the alternative (2) but against the wider, nonparametric class of alternatives

$$f_i(x) = g(x, \eta + \delta c_i) \quad (\delta > 0),$$

where

$$g(x, \theta) = A(\theta) B(x) e^{\theta x},$$

with $B(x) \geqslant 0$ arbitrary (subject to integrability of $B(x) e^{\theta x}$). The MPR test against (2) is uniformly most powerful, among rank tests, against another nonparametric class of alternatives, namely, that with $f_i(x)$

[50]

replaced by $f_i(h(x))\,h'(x)$, $h'(x) > 0$, arbitrary, $\Delta = \delta/\sigma$ fixed, but these alternatives do not seem to be of much statistical interest.

From the standpoint of applications, the MPP test takes so much time to carry out that it is practically almost useless. The test statistic for the MPR test is also very difficult to compute, with the notable exception of the locally MPR test (against Δ small enough). Against normal alternatives close to the hypothesis (such that the power is bounded away from 1) the latter test is asymptotically as powerful as the appropriate student test (and therefore asymptotically as powerful as the MPP test).

The practical drawback of the MPP test compared with the locally MPR test is accompanied by a theoretical disadvantage. If the true density is of the form (1) with g not normal, the locally MPR test against normal alternatives is known (under mild restrictions) to be asymptotically more powerful than the Student test against 'close' alternatives (Chernoff–Savage, Hájek). On the other hand, the MPP test imitates the Student test so closely that it is asymptotically equivalent to the latter also when g is not normal (*see* W. Hoeffding, *Ann. Math. Statist.* **23** (1952), 169–92).

ON THE DISTRIBUTION OF THE ONE-SAMPLE RANK ORDER STATISTICS†

PRANAB KUMAR SEN

1 SUMMARY AND INTRODUCTION

Let $X_1, ..., X_n$ be independent random variables with continuous cumulative distribution functions (cdf) $F_1(x), ..., F_n(x)$, respectively. Consider the usual one-sample rank order statistic

$$T_n = n^{-1} \sum_{i=1}^{n} E_{n,i} Z_{n,i}, \qquad (1.1)$$

where $E_{n,i} = J_n(i/[n+1])$, $i = 1, ..., n$, are explicitly known rank scores (satisfying the assumptions I, II and III of §2), and $Z_{n,i}$ is one or 0 according as the ith smallest observation among $|X_1|, ..., |X_n|$ corresponds to a positive X or not, $i = 1, ..., n$. For $F_1 = ... = F_n = F$, the distribution theory of T_n has been studied by Govindarajulu [5], Puri and Sen [9] and Pyke and Shorack [10], among others. In the present paper, the results of [9] are extended in various directions. Namely, (i) the asymptotic normality is deduced without imposing the restriction that $F_1 = ... = F_n = F$, (ii) an inequality on the asymptotic variance of T_n is obtained, and (iii) a confidence band for the empirical cdf for nonidentical $F_1, ..., F_n$ is provided. Some robustness properties of T_n are also studied.

2 PRELIMINARY NOTATIONS

Let $c(u)$ be 1 or 0 according as $u \geqslant$ or < 0. Define

$$\bar{F}_n(x) = (1/n) \sum_{i=1}^{n} c(x - X_i), \quad F_n^*(x) = (1/n) \sum_{i=1}^{n} F_i(x) \quad (-\infty < x < \infty); \quad (2.1)$$

$$H_i(x) = P\{|X_i| \leqslant x\} = F_i(x) - F_i(-x) \quad (i = 1, ..., n, \ 0 \leqslant x < \infty); \quad (2.2)$$

$$\bar{H}_n(x) = (1/n) \sum_{i=1}^{n} c(x - |X_i|) \quad \text{and} \quad H_n^*(x) = (1/n) \sum_{i=1}^{n} H_i(x) \quad (0 \leqslant x < \infty). \quad (2.3)$$

Note that $H_i(x)$ $(i = 1, ..., n)$,

$$\bar{H}_n(x) = \bar{F}_n(x) - \bar{F}_n(-x-) \quad \text{and} \quad H_n^*(x) = F_n^*(x) - F_n^*(-x)$$

† Work supported by the National Institute of Health, Grant GM–12868–04.

[53]

have been defined for non-negative x and by definition are zero for $x \leqslant 0$. It may also be noted that the sequence of cdf's $\{F_1, \ldots, F_n\}$ can be replaced by a double sequence $\{F_{n,1}, \ldots, F_{n,n}\}$, $n \geqslant 1$ (see §4 for example). But for the sake of simplicity of notations, the subscript n in $F_{n,i}$'s, $\overline{F}_n$, F_n^*, $\overline{H}_n$ and H_n^* will often be suppressed.

The following result (proved in §5) is of considerable importance and used extensively in the derivation of the results: for any δ: $0 \leqslant \delta < \frac{1}{2}$,

$$\sup_x n^{\frac{1}{2}} |\overline{F}_n(x) - F_n^*(x)| / [F_n^*(x) \, 1 - F_n^*(x)]^{\delta} = O_p(1). \qquad (2.4)$$

Now, the score function $J_n(u)$ is defined only at $u = i/[n+1]$, $i = 1, \ldots, n$. As in Puri and Sen[9], we extend its domain to $(0, 1)$ by letting it have constant values on the half-open intervals $((i-1)/n, i/n]$, $i = 1, \ldots, n$. Then, T_n in (1.1) may be written as

$$T_n = \int_0^{\infty} J_n \left(\frac{n}{n+1} \overline{H}_n(x) \right) d\overline{F}_n(x). \qquad (2.5)$$

Concerning $\{J_n(u)\}$ it is assumed that

(I) $\lim_{n \to \infty} J_n(u) = J(u)$ exists for all $0 < u < 1$ and $J(u)$ is not a constant,

$$\qquad (2.6)$$

(II) $\displaystyle\int_0^{\infty} \left[J_n \left(\frac{n}{n+1} \overline{H}_n \right) - J \left(\frac{n}{n+1} \overline{H}_n \right) \right] d\overline{F}_n = o_p(n^{-\frac{1}{2}})$, $\qquad (2.7)$

and

(III) $J(u)$ is absolutely continuous in u: $0 < u < 1$,

and $\qquad |J^{(r)}(u)| = |(d^r/du^r) J(u)| \leqslant K[u(1-u)]^{-r-\frac{1}{2}+\delta}, \qquad (2.8)$

for $r = 0, 1$ and some $\delta > 0$, where K is a positive constant. These assumptions are adapted from the two-sample case by Chernoff and Savage[1], as further modified in Puri and Sen[9]. Now, let $\mathscr{F}$ be the class of all univariate continuous cdf's, and let

$$\mathscr{F}_n = \{ \mathbf{F}_n : F_i \in \mathscr{F}, \, i = 1, \ldots, n \}. \qquad (2.9)$$

Note that under the conditions of Theorem 2 of Chernoff and Savage[1], (2.7) is $o(n^{-\frac{1}{2}})$, and hence is $o_p(n^{-\frac{1}{2}})$, uniformly in $\{\mathscr{F}_n\}$. Let us now define

$$\mu_n^* = \int_0^{\infty} J[H_n^*(x)] \, dF_n^*(x) \quad \text{and} \quad \mu_{n,i} = \int_0^{\infty} J[H_n^*(x)] \, dF_i(x) \quad (i = 1, \ldots, n).$$
$$\qquad (2.10)$$

Also, by (2.1), (2.2) and (2.3), we obtain that

$$dF_n^*(x) \leqslant dH_n^*(x) \quad \text{and} \quad d\bar{F}_n(x) \leqslant d\bar{H}_n(x) \quad \text{for all } x \geqslant 0. \quad (2.11)$$

Hence, by (2.1) and (2.8)

$$\mu_n^* = (1/n) \sum_{i=1}^{n} \mu_{n,i} \quad \text{and} \quad |\mu_n^*| \leqslant \int_0^1 |J(u)| \, du < \infty, \quad (2.12)$$

uniformly in $\{\mathscr{F}_n\}$. Also, let

$$\gamma_{n,i}^2 = \int_0^\infty J^2[H^*(x)] \, dF_i(x) - \mu_{n,i}^2 + 2\Bigg[\iint_{0 < x < y < \infty} \{H_i(x) [1 - H_i(y)]$$

$$\times J'[H^*(x)] \, J'[H^*(y)] \, dF^*(x) \, dF^*(y) + [1 - H_i(y)]$$

$$\times J[H^*(x)] \, J'[H^*(y)] \, dF_i(x) \, dF^*(y) - H_i(x) \, J'[H^*(x)]$$

$$\times J[H^*(y)] \, dF^*(x) \, dF_i(y)\} \Bigg] \quad (i = 1, ..., n); \quad (2.13)$$

$$(\gamma_n^*)^2 = (1/n) \sum_{i=1}^{n} \gamma_{n,i}^2. \quad (2.14)$$

Then, the main theorem of the paper is the following.

Theorem 2.1. *If* (2.4), (2.6), (2.7) *and* (2.8) *hold and* $\inf_n \gamma_n^* > 0$, *then*

$$\lim_{n \to \infty} P\{n^{\frac{1}{2}}(T_n - \mu_n^*)/\gamma_n^* \leqslant x\} = (2\pi)^{-\frac{1}{2}} \int_{-\infty}^{x} \exp\{-\tfrac{1}{2}t^2\} \, dt$$

$$\text{for all } -\infty < x < \infty. \quad (2.15)$$

3 THE PROOF OF THEOREM 2.1

We express T_n as the sum of a principal term (involving the average over n independent random variables) and four higher order terms. In §6, it is shown that all these higher order terms are negligible for all $F_n \in \mathscr{F}_n$, while the version of the central limit theorem by Esseen[3] as applied on the principal term yields the desired result. We express T_n in its integral form in (2.5), and write $a_n = n/(n+1)$, so that $a_n = 1 - (n+1)^{-1}$. Let us then write $d\bar{F}_n = dF_n^* + d[\bar{F}_n - F_n^*]$, and

$$J_n(a_n\bar{H}) = J(a_n\bar{H}) + [J_n(a_n\bar{H}) - J(a_n\bar{H})]; \quad (3.1)$$

$$J(a_n\bar{H}) = J(H^*) + (a_n\bar{H} - H^*) \, J'(H^*) + [J(a_n\bar{H}) - J(H^*)$$

$$- (a_n\bar{H} - H^*) \, J'(H^*)]. \quad (3.2)$$

Then, from (2.5), (3.1) and (3.2), we obtain after some simplifications that

$$T_n = (1/n) \sum_{i=1}^{n} B_n(X_i) + \sum_{r=1}^{4} C_{r,n}, \quad (3.3)$$

where
$$B_n(X_i) = B_{n,1}(X_i) + B_{n,2}(X_i); \qquad (3.4)$$

$$B_{n,1}(X_i) = J[H^*(|X_i|)]c(X_i), \qquad (3.5)$$

$$B_{n,2}(X_i) = \int_0^\infty [c(x-|X_i|) - H_i(x)]J'[H^*(x)]dF^*(x) \quad (i=1,\dots,n), \qquad (3.6)$$

and

$$C_{1,n} = \int_0^\infty \{J_n[a_n\bar{H}] - J[a_n\bar{H}]\}d\bar{F} = o_p(n^{-\frac{1}{2}}) \quad \text{uniformly in } \{\mathscr{F}_n\}, \text{ by } (2.7); \qquad (3.7)$$

$$C_{2,n} = [-1/(n+1)]\int_0^\infty \bar{H}(x)J'[H^*(x)]d\bar{F}(x); \qquad (3.8)$$

$$C_{3,n} = \int_0^\infty [\bar{H}(x) - H^*(x)]J'[H^*(x)]d[\bar{F}(x) - F^*(x)]; \qquad (3.9)$$

$$C_{4,n} = \int_0^\infty \{J(a_n\bar{H}) - J(H^*) - [a_n\bar{H} - H^*]J'[H^*]\}d\bar{F}. \qquad (3.10)$$

It follows from (2.10), (2.12), (3.4), (3.5) and (3.6) that

$$(1/n)\sum_{i=1}^n E\{B_n(X_i)\} = \mu_n^*, \quad |\mu_n^*| < \infty, \quad \text{uniformly in } \{\mathscr{F}_n\}. \; (3.11)$$

Also, it follows from (2.13), (3.4), (3.5) and (3.6) that

$$n\,\mathrm{var}\left\{(1/n)\sum_{i=1}^n B_n(X_i)\right\} = (1/n)\sum_{i=1}^n \mathrm{var}\{B_n(X_i)\} = (\gamma_n^*)^2, \qquad (3.12)$$

where γ_n^* is defined by (2.13) and (2.14). $C_{r,n}$, $r = 1,2,3,4$ are all higher order terms, and it is shown in §6 that

$$|C_{r,n}| = o_p(n^{-\frac{1}{2}}), \quad \text{for all } F_n \in \mathscr{F}_n, \text{ for } r = 1,\dots,4. \qquad (3.13)$$

Hence, to prove the theorem, it suffices to show that the double sequence of random variables $\{B_n(X_1),\dots,B_n(X_n), n \geq 1\}$ satisfies the conditions of the central limit theorem. For this purpose, we consider the following moment properties of these random variables.

If $F_1 = \dots = F_n = F_n^*$, then, by definition in (2.13),

$$\gamma_{n,1} = \dots = \gamma_{n,n} = \gamma(F_n^*),$$

where

$$\gamma^2(F_n^*) = \int_0^\infty J^2[H^*(x)]dF^*(x) - (\mu_n^*)^2 + 2\Bigg[\iint_{0<x<y<\infty} \{H^*(x)[1-H^*(y)]$$

$$\times J'[H^*(x)]J'[H^*(y)]dF^*(x)dF^*(y) + [1-H^*(y)]$$

$$\times J[H^*(x)]J'[H^*(y)]dF^*(x)dF^*(y) - H^*(x)$$

$$\times J'[H^*(x)]J[H^*(y)]dF^*(x)dF^*(y)\}\Bigg]. \qquad (3.14)$$

Also, let
$$\alpha_{n,i} = \mu_{n,i} - \mu_n^*$$

and
$$\beta_{n,i} = \int_0^\infty [H_i(x) - H^*(x)] J'[H^*(x)] dF^*(x) \quad (i = 1, \ldots, n). \quad (3.15)$$

Lemma 3.1. For all $\mathbf{F}_n \in \mathscr{F}_n$, *and* $n \geqslant 1$,

$$(\gamma_n^*)^2 = \gamma^2(F_n^*) - (1/n) \sum_{i=1}^n (\alpha_{n,i} + \beta_{n,i})^2 \leqslant \gamma^2(F_n^*) < \infty. \quad (3.16)$$

Proof. It follows from (2.8) and some routine computations that

$$\iint_{0<u<v<1} u^r(1-v)^s |J^{(r)}(u)| \, |J^{(s)}(v)| \, du \, dv < \infty \quad \text{for all } r, s = 0, 1. \quad (3.17)$$

Hence, from (2.11), (3.14) and (3.17), it readily follows that $\gamma^2(F_n^*) < \infty$, uniformly in $\{\mathscr{F}_n\}$. From (2.13), (2.14) and (3.14), we obtain that

$$
\begin{aligned}
(\gamma_n^*)^2 - \gamma^2(F_n^*) = {} & (\mu_n^*)^2 - (1/n) \Sigma \mu_{n,i}^2 + 2 \bigg[\iint_{0<x<y<\infty} (\{(1/n) \Sigma H_i(x) \\
& \times [1 - H_i(y)] - H^*(x) [1 - H^*(y)]\} J'[H^*(x)] J'[H^*(y)] \\
& \times dF^*(x) dF^*(y) + J[H^*(x)] J'[H^*(y)] \{(1/n) \Sigma[1 - H_i(y)] \\
& \times dF_i(x) - [1 - H^*(y)] dF^*(x)\} dF^*(y) \\
& - J'[H^*(x)] J[H^*(y)] \{(1/n) \Sigma H_i(x) dF_i(y) \\
& - H^*(x) dF^*(y)\} dF^*(x)) \bigg], \quad (3.18)
\end{aligned}
$$

where the summation Σ extends over all $i = 1, \ldots, n$. Also, we note that

$$(1/n) \Sigma H_i(x) [1 - H_i(y)] = H^*(x) [1 - H^*(y)]$$
$$- (1/n) \Sigma[H_i(x) - H^*(x)] [H_i(y) - H^*(y)], \quad (3.19)$$

$$(1/n) \Sigma(a - H_i(x)) dF_i(y) = (a - H^*(x)) dF^*(y)$$
$$- (1/n) \Sigma[H_i(x) - H^*(x)] d[F_i(y) - F^*(y)] \quad (a = 0, 1). \quad (3.20)$$

Further, from (2.12) we have on using (3.15)

$$(1/n) \Sigma \mu_{n,i}^2 - (\mu_n^*)^2 = (1/n) \Sigma(\mu_{n,i} - \mu_n^*)^2 = (1/n) \Sigma \alpha_{n,i}^2. \quad (3.21)$$

Finally, by direct computations, we obtain the following three identities:

$$
\begin{aligned}
(1/n) \Sigma 2 \bigg\{ & \iint_{0<x<y<\infty} [H_i(x) - H^*(x)] [H_i(y) - H^*(y)] J'[H^*(x)] J'[H^*(y)] \\
& \times dF^*(x) dF^*(y) \bigg\} = (1/n) \Sigma \bigg\{ \int_0^\infty [H_i(x) - H^*(x)] J'[H^*(x)] dF^*(x) \bigg\}^2 \\
& = (1/n) \sum_{i=1}^n \beta_{n,i}^2; \quad (3.22)
\end{aligned}
$$

$$(1/n) \Sigma 2 \left\{ \iint_{0<x<y<\infty} [H_i(x) - H^*(x)] J'[H^*(x)] J[H^*(y)] d[F_i(y) \right.$$

$$\left. - F^*(y)] \times dF^*(x) \right\}$$

$$= (1/n) \Sigma \left\{ \left[\int_0^\infty [H_i(x) - H^*(x)] J'[H^*(x)] dF^*(x) \right] \right.$$

$$\left. \times \left[\int_0^\infty J[H^*(y)] d[F_i(y) - F^*(y)] \right] \right\} = (1/n) \Sigma \alpha_{n,i} \beta_{n,i}; \qquad (3.23)$$

$$(1/n) \Sigma 2 \left\{ \iint_{0<x<y<\infty} [H_i(y) - H^*(y)] J[H^*(x)] J'[H^*(y)] dF^*(y) \right.$$

$$\left. \times d[F_i(x) - F^*(x)] \right\} = (1/n) \Sigma \alpha_{n,i} \beta_{n,i}. \quad (3.24)$$

The rest of the proof of the lemma follows directly from (3.18) through (3.24).

Lemma 3.2. Under assumption (III),

$$(1/n) \sum_{i=1}^n E\{|B_n(X_i)|^{2+\delta}\} < \infty,$$

uniformly in $\{\mathscr{F}_n\}$, where $\delta \ (> 0)$ is defined by (2.8).

Proof. By virtue of (3.4), (3.5), (3.6) and the inequality that

$$|a+b|^{2+\delta} \leqslant 2^{1+\delta}\{|a|^{2+\delta} + |b|^{2+\delta}\},$$

it suffices to show that

$$(1/n) \sum_{i=1}^n E\{|B_{n,j}(X_i)|^{2+\delta}\} < \infty \quad \text{for } j = 1, 2, \text{ uniformly in } \{\mathscr{F}_n\}. \quad (3.25)$$

Upon using assumption (III) and noting that $(2+\delta)(-\frac{1}{2}+\delta) > -1$, it follows from (2.8), (2.11) and (3.5) that

$$(1/n) \sum_{i=1}^n E\{|B_{n,1}(X_i)|^{2+\delta}\} = \int_0^\infty |J[H^*(x)]|^{2+\delta} dF^*(x)$$

$$\leqslant \int_0^\infty |J(u)|^{2+\delta} du < \infty, \quad \text{uniformly in } \{\mathscr{F}_n\}. \quad (3.26)$$

Let now Y_n be a random variable (independent of $X_1, \ldots, X_n$) following the cdf $F_n^*(x)$. Define

$$d_n(X_i, Y_n) = c(Y_n) [c(Y_n - |X_i|) - H_i(Y_n)] J'[H^*(Y_n)]. \quad (3.27)$$

Then it is easy to verify that

$$B_{n,2}(X_i) = E\{d_n(X_i, Y_n) \mid X_i\} \quad (i = 1, \dots, n). \tag{3.28}$$

Consequently, by straightforward computations we obtain that

$$E\{|B_{n,2}(X_i)|^{2+\delta}\} \leqslant E\{[E(|d_n(X_i, Y_n)|^{1+\delta/2} \mid X_i)]^2\}$$

$$= 2E\bigg\{\iint_{0 < x < y < \infty} |\{c(x - |X_i|) - H_i(x)\}\{c(y - |X_i|)$$

$$- H_i(y)\} J'[H^*(x)] J'[H^*(y)]|^{1+\delta/2} dF^*(x) dF^*(y)$$

$$\leqslant 6\iint_{0 < x < y < \infty} H_i(x) [1 - H_i(y)] |J'[H^*(x)] J'[H^*(y)]|^{1+\delta/2}$$

$$\times dF^*(x) dF^*(y)\bigg\} \tag{3.29}$$

as

$$E[|\{c(x - |X_i|) - H_i(x)\}\{c(y - |X_i|) - H_i(y)\}^{1+\delta/2} | x < y]$$

$$= H_i(x) [\{1 - H_i(x)\}\{1 - H_i(y)\}]^{1+\delta/2} + [H_i(y) - H_i(x)]$$

$$\times [H_i(x) \{1 - H_i(y)\}]^{1+\delta/2} + [1 - H_i(y)] [H_i(x) H_i(y)]^{1+\delta/2}$$

$$\leqslant 3H_i(x) [1 - H_i(y)] \quad (\text{as } x < y \text{ and } \delta > 0). \tag{3.30}$$

Hence, on using (2.8), (2.11) and (3.19), we obtain from (3.29) that

$$(1/n) \sum_{i=1}^{n} E\{|B_{n,2}(X_i)|^{2+\delta}\}$$

$$\leqslant 6\bigg[\iint_{0 < x < y < \infty} H^*(x) [1 - H^*(y)] |J'[H^*(x)] J'(H^*(y))]|^{1+\delta/2} dF^*(x)$$

$$\times dF^*(y) - \frac{1}{2n} \sum_{i=1}^{n} \bigg\{\int_0^\infty [H_i(x) - H^*(x)] |J'[H^*(x)]|^{1+\delta/2} dF^*(x)\bigg\}^2\bigg]$$

$$\leqslant 6\bigg[\iint_{0 < x < y < \infty} H^*(x) [1 - H^*(y)] |J'[H^*(x)] J'[H^*(y)]|^{1+\delta/2} dF^*(x)$$

$$\times dF^*(y)\bigg]$$

$$\leqslant 6\bigg[\iint_{0 < u < v < \infty} u(1 - v) |J'(u) J'(v)|^{1+\delta/2} du\, dv\bigg] < \infty,$$

uniformly in $\{\mathscr{F}_n\}$. (3.31)

Hence the lemma follows from (3.25), (3.26) and (3.31). Q.E.D.

Now, to apply the central limit theorem on the double sequence of independent random variables $\{B_n(X_1), \dots, B_n(X_n); n \geqslant 1\}$, we consider the following theorem due to Esseen ([3], Theorem 1, p. 43), slightly modified to suit our purpose.

Let $\{Z_{n1}, ..., Z_{nN}, n \geqslant 1\}$ (where $N \to \infty$ as $n \to \infty$) be a double sequence of (within row) independent random variables, such that $EZ_{ni} = 0$, $EZ_{ni}^2 = \sigma_{ni}^2$ and for some $\delta > 0$, $E|Z_{ni}|^{2+\delta} = \beta_{ni}^{(\delta)}, i = 1, ..., N$, exist for all $n \geqslant 1$. Let then

$$\sigma_{n\cdot}^2 = N^{-1} \sum_{i=1}^{N} \sigma_{ni}^2, \quad \beta_{n\cdot}^{(\delta)} = N^{-1} \sum_{i=1}^{N} \beta_{ni}^{(\delta)} \quad \text{and} \quad \rho_{\delta, n} = \beta_{n\cdot}^{(\delta)}/\sigma_{n\cdot}^{2+\delta}. \tag{3.32}$$

Finally, let $S_n = N^{-\frac{1}{2}} \sum_{i=1}^{N} Z_{ni}$, $G_n(x) = P[S_n/\sigma_n \leqslant x]$, and let $\Phi(x)$ be the standard normal cdf. Then, there exists a finite positive constant C_δ, depending on δ, such that

$$\sup_x |G_n(x) - \Phi(x)| \leqslant C_\delta[n^{-\delta/2}\rho_{\delta, n} + n^{-\frac{1}{2}}\rho_{\delta, n}^{1/\delta}]. \tag{3.33}$$

Writing now $Z_{ni} = B_n(X_i) - \mu_{n, i}$, $i = 1, ..., n$, we have from Lemma 3.2 that

$$\beta_{n\cdot}^{(\delta)} \leqslant 2\left[n^{-1} \sum_{i=1}^{n} E\{|B_n(X_i)|^{2+\delta}\}\right] < \infty, \quad \text{uniformly in } \{\mathscr{F}_n\}. \tag{3.34}$$

Also, by (3.12) and the hypothesis that $\inf_n \gamma_n^* > 0$, we obtain that

$$\sigma_{n\cdot}^2 = (\gamma_n^*)^2, \quad \inf_n \sigma_{n\cdot} > 0. \tag{3.35}$$

Consequently, from (3.33), (3.34) and (3.35), we have

$$\lim_{n \to \infty} P\left[n^{-\frac{1}{2}} \sum_{i=1}^{n} (B_n(X_i) - \mu_{n, i})/\gamma_n^* \leqslant t\right] = (2\pi)^{-\frac{1}{2}} \int_{-\infty}^{t} \exp[(-\tfrac{1}{2})x^2] dx, \tag{3.36}$$

uniformly in t: $-\infty < t < \infty$.

Remark. By virtue of Theorem 2.1, we may term the sequence of normalizing constants $\{(\gamma_n^*)^2\}$ as the sequence of asymptotic variances of the sequence of random variables $\{n^{\frac{1}{2}}(T_n - \mu_n^*)\}$. It follows from Lemma 3.1 that $\gamma_n^* \leqslant \gamma(F_n^*)$, where the equality sign holds if $\alpha_{n, i} + \beta_{n, i} = 0$ for $i = 1, ..., n$. This holds when each sample cdf is symmetric about the origin. Let $\mathscr{F}_0$ be the class of all univariate cdf's symmetric about 0, so that $\mathscr{F}_0 \subset \mathscr{F}$. Now, suppose that

$$\mathbf{F}_n \in \mathscr{F}_n^0, \quad \text{where } \mathscr{F}_n^0 = \{\mathbf{F}_n \in \mathscr{F}_n : F_i \in \mathscr{F}_0, i = 1, ..., n\}. \tag{3.37}$$

(3.37) implies that $F_n^* \in \mathscr{F}_0$, and hence, $H_n^*(x) = 2F_n^*(x) - 1$ for all $x \geqslant 0$. Thus, from (2.10), (2.12) and (3.15) by partial integration, we obtain that

$$\alpha_{n, i} + \beta_{n, i} = 0 \quad \text{for all } i = 1, ..., n, \tag{3.38}$$

and also that

$$\mu_n^* = \frac{1}{2}\int_0^1 J(u)\,du = \tfrac{1}{2}\mu \text{ (say)} \quad \text{and} \quad \gamma^2(F_n^*) = \frac{1}{4}\int_0^1 J^2(u)\,du = \tfrac{1}{4}A^2 \text{ (say)},$$

(3.39)

where by (2.6), $A^2 > 0$. Consequently, when $F_n \in \mathscr{F}_n^0$, then

$$\gamma_n^* = \gamma(F_n^*) = A/2.$$

4   ROBUSTNESS AND EFFICIENCY OF T_n FOR SHIFT ALTERNATIVES

Consider the general hypothesis of symmetry that

$$H_0: F_n \in \mathscr{F}_n^0. \tag{4.1}$$

Let V_n be the vector of signs of $X_1, \dots, X_n$. Under (4.1), all possible 2^n realizations of V_n are equally likely, and this generates the null distribution of T_n. For large n, it follows from Theorem 2.1 and (3.39) that under H_0 in (4.1), $2n^{\frac{1}{2}}(T_n - \mu/2)/A$ converges in law to a standard normal distribution, where μ and A are defined by (3.39). Hence, the exact test for H_0 in (4.1) based on T_n can be approximated for large values of n by a normal theory test based on the asymptotic normality of $2\sqrt{(n)}\,(T_n - \mu/2)$. It may also be noted that for the hypothesis of symmetry, the exact null distributions of the well-known rank statistics (such as the Wilcoxon signed rank statistic) tabulated for the case $F_1 = \dots = F_n = F$ also stand valid for the heterogeneous case considered here.

Consider now the sequence of shift-alternatives $\{K_n\}$, where K_n specifies that $X_1, \dots, X_n$ are independent random variables having absolutely continuous cdf's $F_{n,1}, \dots, F_{n,n}$, respectively, where

$$F_{n,i}(x) = F_i(x - n^{-\frac{1}{2}}\theta), \quad F_i \in \mathscr{F}_0$$

$$(i = 1, \dots, n, \text{ and } \theta \text{ is real and finite}). \tag{4.2}$$

Note that the absolute continuity of $F_1, \dots, F_n$ implies the same of F_n^*.

Concerning the scores $E_{n,i} = J_n(i/[n+1])$, $i = 1, \dots, n$, we assume that $E_{n,i}$ is the expected value of the ith order statistic of a sample of size n from the distribution $\Psi(x)$, where $\Psi(x) = 2\Psi^*(x) - 1$ if x is non-negative and it is equal to 0 for $x < 0$, and $\Psi^*(x) \in \mathscr{F}_0$. Further, let $J(u) = \Psi^{-1}(u)$ and $J^*(u) = \Psi^{*-1}(u)$. Finally, we assume that for some $\delta: 0 < 2\delta < 1$,

$$|(d^r/du^r)J^*(u)| \leqslant K[u(1-u)]^{-r-\frac{1}{2}+\delta} \quad (r = 0, 1), \tag{4.3}$$

where $K < \infty$. Upon noting that

$$J(u) = \Psi^{-1}(u) = \Psi^{*-1}([1+u]/2) = J^*([1+u]/2) \quad \text{and} \quad J(0) = 0, \tag{4.4}$$

it follows as in Theorem 2 of Chernoff and Savage[1] and corollary to Theorem 2 of [6] that under (4.3), the sequence $\{J_n(u)\}$ satisfies all the assumptions I, II and III of § 2. These assumptions are also satisfied when we let $J_n(i/[n+1]) = J(i/[n+1])$, $i = 1, \dots, n$.

Now, under $\{K_n\}$ in (4.2), the expression for $n^{\frac{1}{2}}(\mu_n^* - \mu/2)$ will, in general, depend on the sequence $\{\mathbf{F}_n\}$. It may be simplified considerably if we assume that

$$|(d/dx) J^*[F_n^*(x)]| \quad \text{is bounded as } x \to \pm \infty. \tag{4.5}$$

Under (4.2) and (4.5) it can be shown that

$$\sqrt(n) \, (\mu_n^* - \mu/2) = B(F_n^*) \, \theta/2 + o(1), \tag{4.6}$$

where
$$B(F) = \int_{-\infty}^{\infty} [(d/dx) J^*(F(x))] \, dF(x);$$

$$(\gamma_n^*)^2 = (\tfrac{1}{4}) A^2 + o(1). \tag{4.7}$$

Thus, from Theorem 2.1, (4.6) and (4.7), we obtain that

$$\lim_{n \to \infty} P_{K_n}[2n^{\frac{1}{2}}(T_n - \mu/2)/A - \theta B(F_n^*)/A \leqslant x] = (2\pi)^{-\frac{1}{2}} \int_{-\infty}^{x} \exp[(-\tfrac{1}{2}) t^2] \, dt, \tag{4.8}$$

uniformly in x: $-\infty < x < \infty$, where μ and A^2 are defined by (3.39) and $\{P_{K_n}\}$ stands for the probability computed under the sequence of alternative hypotheses $\{K_n\}$ in (4.2).

Let us now consider the parametric test based on Student's t-statistic

$$t_n = \sqrt(n) \, \bar{X}_n/s_n; \quad \bar{X}_n = (1/n) \sum_{i=1}^{n} X_i \quad \text{and} \quad s_n^2 = \frac{1}{n-1} \sum_{i=1}^{n} (X_i - \bar{X}_n)^2. \tag{4.9}$$

It is assumed that the cdf F_i has finite absolute moment of order $2 + \delta$ ($\delta > 0$), mean zero and a variance σ_i^2, $i = 1, \dots, n$. Let then

$$(\sigma_n^*)^2 = (1/n) \sum_{i=1}^{n} \sigma_i^2, \quad \rho_{\delta, n}^* = (1/n) \sum_{i=1}^{n} E |X_i|^{2+\delta}. \tag{4.10}$$

It is then assumed that

$$\inf_n \sigma_n^* > 0 \quad \text{and} \quad \limsup_n \rho_{\delta, n}^*/n^\delta (\sigma_n^*)^{2+\delta} = 0. \tag{4.11}$$

Then, again using the theorem by Esseen[3] stated in § 3, we obtain that

$$\lim_{n \to \infty} P_{K_n}[n^{\frac{1}{2}} \bar{X}_n/\sigma_n^* - \theta/\sigma_n^* \leqslant x] = (2\pi)^{-\frac{1}{2}} \int_{-\infty}^{x} \exp[(-\tfrac{1}{2}) t^2] \, dt, \tag{4.12}$$

uniformly in x: $-\infty < x < \infty$. Since $X_1, \ldots, X_n$ are all independent, on using (4.11) it is easy to verify that $\xi_{ni} = X_i^2/n$, $i = 1, \ldots, n$, satisfy both the conditions of a theorem on the law of large numbers stated in Gnedenko and Kolmogorov [7], p. 105), and hence,

$$\left| n^{-1} \sum_{i=1}^{n} (X_i - EX_i)^2 - (\sigma_n^*)^2 \right| \overset{p}{\to} 0. \qquad (4.13)$$

Also, upon noting that $E\bar{X}_n = \theta/n^{\frac{1}{2}}$, we obtain from (4.12) that

$$(\bar{X}_n - E\bar{X}_n)^2 \overset{p}{\to} 0,$$

as $n \to \infty$. Consequently, $|s_n^2 - (\sigma_n^*)^2| \overset{p}{\to} 0$, as $n \to \infty$. Thus, from (4.9) and (4.12) we obtain that

$$\lim_{n \to \infty} P_{K_n}[t_n - \theta/\sigma_n^* \leqslant x] = (2\pi)^{-\frac{1}{2}} . \int_{-\infty}^{x} \exp\left[(-\tfrac{1}{2}) t^2\right] dt, \qquad (4.14)$$

uniformly in x: $-\infty < x < \infty$.

Now, for the study of the asymptotic relative efficiency (ARE) of the rank order and the Student t statistics we require the assumption that the cdf F_n^* converges to a cdf F^* as $n \to \infty$. In an earlier paper [12], the author has considered some models for which this assumption holds. It can also hold even when these models do not apply. Under this assumption, the functions $B(F_n^*)$ and $(\sigma_n^*)^2$ (which is the variance of the cdf F_n^*) converge respectively to $B(F^*)$ and $\sigma^2(F^*)$. Hence, from (4.8) and (4.14), we obtain that the ARE of the test based on T_n with respect to the one based on t_n is equal to

$$e_{T,t}(\{F_n\}) = e_{T,t}(F^*) = \sigma^2(F^*) B^2(F^*)/A^2, \qquad (4.15)$$

which agrees with the ARE of the two-sample Chernoff–Savage [1] statistic with respect to the Student t-statistic when the underlying cdf is F^*. Thus, we can again use the results by Chernoff and Savage [1], with the only change that the sequence of cdf's $\{F_n\}$ is replaced by a homogeneous sequence with the common cdf F^*. Incidentally, if we consider the normal scores statistic [Fraser [4], where $J(u)$, defined by (4.4), is the inverse of the chi-distribution with one degree of freedom], it follows from the results of Chernoff and Savage [1] that (4.15) is always greater than or equal to 1, where the equality sign holds only when F^* is a normal cdf. In particular, when $F_1, \ldots, F_n$ are all normal cdf's, differing only in their variances, F_n^* cannot be normal unless all these variances are equal. Thus, for a normal family of cdf's, if $\{F_i\}$ can be decomposed into two or more subsequences with varying variances, F_n^* cannot be normal, and hence, (4.15) cannot be equal to one. For the Wilcoxon [15] signed rank statistic some allied results are already studied in Sen [11].

5 ASYMPTOTIC BEHAVIOUR OF THE EMPIRICAL CDF

We shall now present a formal proof of (2.4). Note that when

$$F_1 = \ldots = F_n = F,$$

the result follows from Lemma 7 of [6] though a more elementary proof can be given by using the results of §§ 4 and 5 of Doob [2]. We define

$$Y_{ni} = F_n^*(X_i) \quad \text{and} \quad G_{ni}(t) = F_i[F_n^{*-1}(t)] \quad (i = 1, \ldots, n). \quad (5.1)$$

Note that $n^{-1} \sum_{i=1}^{n} G_{ni}(t) = t$, $0 \leqslant t \leqslant 1$, and the G_{ni} are defined over the unit interval $(0, 1)$.

Theorem 5.1. (2.4) *holds for the entire class of* $\{F_n\}$ *of the continuous type.*

Proof. Let $\bar{G}_n(t) = n^{-1} \sum_{i=1}^{n} c(t - Y_{ni})$ be the empirical cdf of the Y_{ni}. Then, $\bar{G}_n(t) = \bar{F}_n[F_n^{*-1}(t)]$, $0 \leqslant t \leqslant 1$. Thus, it suffices to prove (2.4) for the empirical cdf $\bar{G}_n$. We define a stochastic process $[V_n(t): 0 \leqslant t \leqslant 1]$ by $V_n(t) = n^{\frac{1}{2}}[\bar{G}_n(t) - t]$, and for notational simplicity write $G_{ni}(t) = t_i$, $i = 1, \ldots, n$. Then, by direct computations, we have

$$E[V_n(t)] = 0$$

and $\qquad E[V_n(t) V_n(s) \,|\, s \leqslant t] = n^{-1} \Sigma s_i (1 - t_i) \quad \text{for } 0 \leqslant s \leqslant t \leqslant 1. \quad (5.2)$

This leads to

$$E[V_n(t) - V_n(s)]^2 = n^{-1} \Sigma (t_i - s_i)(1 - t_i + s_i) \leqslant (t - s) \quad (0 \leqslant s \leqslant t \leqslant 1),$$

and, in general, it can be shown that for any positive integer k,

$$E[V_n(t) - V_n(s)]^{2k} \leqslant M_k (t - s)^k$$

and $\qquad E[V_n(t)]^{2k} \leqslant M_k [t(1 - t)]^k \quad (0 \leqslant s \leqslant t \leqslant 1), \quad (5.3)$

where M_k is finite for any finite k and is independent of n. We also write $g(t) = [t(1 - t)]^{\frac{1}{2} - \delta'}$, $0 \leqslant t \leqslant 1$. Then, it is easy to show that for $0 \leqslant s \leqslant t \leqslant \frac{1}{2}$ or $\frac{1}{2} \leqslant t \leqslant s \leqslant 1$,

$$[1 - g(s)/g(t)] \leqslant (\tfrac{1}{2} - g')(t - s)/[t(1 - t)]. \quad (5.4)$$

Define then the stochastic process $V_n^*(t) = V_n(t)/g(t)$, $0 \leqslant t \leqslant 1$. From (5.3) and (5.4) it follows that for $0 \leqslant s \leqslant t \leqslant \frac{1}{2}$ or $\frac{1}{2} \leqslant t \leqslant s \leqslant 1$,

$$E[V_n^*(t) - V_n^*(s)]^{2k}$$

$$\leqslant 2^{2k-1} \{ g^{-2k}(t) E[V_n(t) - V_n(s)]^{2k} + [1 - g(s)/g(t)]^{2k} E[V_n(s)/g(s)]^{2k} \}$$

$$\leqslant M_k^* (t - s)^{2k\delta'}, \quad \text{where } M_k^* \, (< \infty) \text{ is independent of } n. \quad (5.5)$$

Now, for any fixed $\delta'(0 < \delta' < \frac{1}{2})$, we can always select a value of k, such that $k\delta' \geqslant 1$. Then the right-hand side of (5.5) is bounded above by $M_\delta^*(t-s)^2$. Also, we define now a stochastic process $V_{n,N}^*(t)$, $0 \leqslant t \leqslant 1$, by

$$V_{n,N}^*(t) = V_n^*\left(\frac{i-1}{N}\right) + [Nt - (i-1)]\left[V_n^*\left(\frac{i}{N}\right) - V_n^*\left(\frac{i-1}{N}\right)\right]$$

$$((i-1)/N \leqslant t \leqslant i/N), \quad (5.6)$$

for $i = 1, ..., N$, where N is a positive integer, to be selected later on. Thus, $V_{n,N}^*(t)$ is a stochastic process with continuous sample paths, and at the $N+1$ points $i/N, i = 0, 1, ..., N, V_{n,N}^*(t) = V_n^*(t)$. Proceeding then precisely on the same line as in Theorem b (on pp. 177–9) of Hájek and Šidák [8] it follows that for every $\epsilon > 0$,

$$\lim_{\eta \to 0} \liminf_n P[\max_{|t-s| < \eta} |V_n^*(t) - V_n^*(s)| < \epsilon] = 1. \quad (5.7)$$

Consider now a set of $m (\geqslant 1)$ points $0 \leqslant t_1 < ... < t_m \leqslant 1$ and real coefficients $\lambda_1, ..., \lambda_m$. Let $\Phi_n(x; \mathbf{t}, \boldsymbol{\lambda})$ be the true cdf of $\Sigma\lambda_j V_n^*(t_j)$. Also, consider a sequence of Gaussian processes $\{[Z_n(t): 0 \leqslant t \leqslant 1], n \geqslant 1\}$, where $EZ_n(t) = 0$ and

$$E[Z_n(s) Z_n(t) | s \leqslant t] = E[V_n^*(s) V_n^*(t) | s \leqslant t] \quad (0 \leqslant s \leqslant t \leqslant 1),$$

and let $\Psi_n(x; \mathbf{t}, \boldsymbol{\lambda})$ be the cdf of $\Sigma\lambda_j Z_n(t_j)$. Then, by the central limit theorem, we have

$$\limsup_n |\Phi_n(x; \mathbf{t}, \boldsymbol{\lambda}) - \Psi_n(x; \mathbf{t}, \boldsymbol{\lambda})| = 0 \quad \text{for all} \quad -\infty < x < \infty, \quad (5.8)$$

and for all $\boldsymbol{\lambda}$ and $\mathbf{t}$. From (5.7) and (5.8), we obtain on using a theorem in Hájek and Šidák [8], p. 180, that

$$\sup_{0 \leqslant t \leqslant 1} |V_n^*(t) - Z_n(t)| \quad \text{converges in law to 0 as } n \to \infty. \quad (5.9)$$

Thus, it suffices to show that for every $\epsilon > 0$ and $0 < \delta' < \frac{1}{2}$,

$$\limsup_n P[\sup_{0 \leqslant t \leqslant 1} |Z_n(t)| > c(\epsilon, \delta')] \leqslant \epsilon. \quad (5.10)$$

Consider now the Gaussian Process $[Z(t): 0 \leqslant t \leqslant 1]$ with $EZ(t) = 0$, and

$$E[Z(s) Z(t) | s \leqslant t] = s(1-t)/g(s) g(t) \quad (0 \leqslant s \leqslant t \leqslant 1). \quad (5.11)$$

It then follows from the results of Lemma 7 of [6] that

$$P\{\sup_{0 \leqslant t \leqslant 1} |Z(t)| > c(\epsilon, \delta)\} < \epsilon, \quad \text{for all } \epsilon > 0. \quad (5.12)$$

Also, let $t^{(j)} = j/(m+1), j = 1, \ldots, m$, be a set of m points on $(0, 1)$. Then, it can also be shown that

$$\lim_{m \to \infty} P\{ \max_{1 \leqslant j \leqslant m} |Z(t^{(j)})| > K \} = P\{ \sup_{0 \leqslant t \leqslant 1} |Z(t)| > K \}$$

for every finite K. (5.13)

Similarly, it can be shown that

$$\lim_{m \to \infty} P\{ \max_{1 \leqslant j \leqslant m} |Z_n(t^{(j)})| > K \} = P\{ \sup_{0 \leqslant t \leqslant 1} |Z_n(t)| > K \}$$

for every finite K. (5.14)

Hence, our desired result would follow if we can show that for every finite m and $t^{(1)}, \ldots, t^{(m)}$,

$$P\{ \max_{1 \leqslant j \leqslant m} |Z_n(t^{(j)})| \leqslant K \} \geqslant P\{ \max_{1 \leqslant j \leqslant m} |Z(t^{(j)})| \leqslant K \}$$

for every finite K. (5.15)

Let, $\mathbf{D}_n^{(m)}$ and $\mathbf{D}^{(m)}$ be the covariance matrices of $[Z_n(t^{(1)}), \ldots, Z_n(t^{(m)})]$ and $[Z(t^{(1)}), \ldots, Z(t^{(m)})]$, respectively. Then, it is easy to verify that $\mathbf{D}^{(m)}$ is positive definite for every m, while $\mathbf{D}_n^{(m)}$ is either positive definite or semidefinite and

$$\mathbf{D}^{(m)} - \mathbf{D}_n^{(m)} = \mathbf{D}_{n*}^{(m)} \text{ (say) is positive semidefinite or positive definite.}$$

(5.16)

Thus, if $\nu_{n,m}^{(1)}, \ldots, \nu_{n,m}^{(m)}$ be the characteristic roots of $\mathbf{D}_n^{(m)}[\mathbf{D}^{(m)}]^{-1}$, it follows that

$$0 \leqslant \nu_{n,m}^{(j)} \leqslant 1 \quad \text{for all } j = 1, \ldots, m, \text{ and every } m \geqslant 1. \quad (5.17)$$

Consequently, the proof of (5.15) follows directly from Lemma 4.4 of Sen et al.[13], after noting that both the vectors have multinormal distributions.

We also note that H_i, $\bar{H}_n$ or H_n^* are also cdf of the absolute values of the sample observations. Hence, as in Theorem 5.1, we obtain the following.

Theorem 5.2. For any fixed $\delta': 0 < 2\delta' < 1$, and for every $\epsilon > 0$, there exists a finite $c(\epsilon, \delta') (> 0)$, such that

$$P\left\{ \sup_x \frac{n^{\frac{1}{2}} |\bar{H}_n(x) - H_n^*(x)|}{\{H_n^*(x)[1 - H_n^*(x)]\}^{\frac{1}{2} - \delta'}} > c(\epsilon, \delta') \right\} < \epsilon \quad \text{for all } \mathbf{F}_n \text{ in } \mathscr{F}_n. \quad (5.18)$$

Corollary 5.2.1. For any $\delta: 0 < 2\delta < 1$, let δ' and δ'' be so chosen that $0 < 2\delta' < \delta'' < \delta$. Then, for every $\epsilon > 0$, there exists a pair (β_1, β_2): $0 < \beta_1 < 1 < \beta_2$, such that

$$P\{\beta_1 < \bar{H}_n(x)/H_n^*(x) < \beta_2 \quad \text{for all } x \text{ for which } H_n^*(x) > cn^{-(1-\delta'')}\} \, 1 - \epsilon,$$

(5.19)

$$P\{\beta_1 < [1 - \bar{H}_n(x)]/[1 - H_n^*(x)] < \beta_2$$

$$\text{for all } x \text{ for which } H_n^*(x) < 1 - cn^{-(1-\delta'')}\} > 1 - \epsilon, \quad (5.20)$$

for all $\mathbf{F}_n$ in $\mathscr{F}_n$, where $c > 0$.

Proof. It suffices to prove (5.19) as (5.20) will follow similarly. From (5.18), we obtain that

$$P\left\{\sup_{H_n^*(x)<0}\left|\frac{\bar{H}_n(x)}{H_n^*(x)}-1\right| < \frac{c(\epsilon,\delta')\,[1-H_n^*(x)]^{\frac{1}{2}-\delta'}}{n^{\frac{1}{2}}[H_n^*(x)]^{\frac{1}{2}+\delta'}}\right\} > 1-\epsilon \quad \text{for all } \mathbf{F}_n \text{ in } \mathscr{F}_n^*.$$

(5.21)

Now for all $H_n^*(x) > cn^{-(1-\delta'')}$,

$$n^{\frac{1}{2}}[H_n^*(x)]^{\frac{1}{2}+\delta'} \geqslant (c^{\frac{1}{2}+\delta'})\,n^\gamma; \quad \gamma = \tfrac{1}{2}\delta'' - \delta' + \delta'\delta'' > 0. \qquad (5.22)$$

Therefore, the left-hand side of (5.22) can be made arbitrarily large when n is taken large. Hence, (5.19) follows from (5.21) and (5.22).

Let now $Z_{n,1} < \ldots < Z_{n,n}$ be the ordered variables corresponding to the random variables $|X_1|,\ldots,|X_n|$, and let $k_n = [n^\delta]$, where $[s]$ is the integral part of s and δ'' is defined in the corollary 5.2.1. Then, we have the following.

Corollary 5.2.2. For every $\epsilon > 0$, there exists a $d(\epsilon) > 0$, such that

$$P\{H_n^*(Z_{n,k_n}) > d(\epsilon)\,n^{-(1-\delta'')}\} > 1-\epsilon \quad \text{for all } \mathbf{F}_n \text{ in } \mathscr{F}_n, \qquad (5.23)$$

$$P\{H_n^*(Z_{n,n-k_n+1}) < 1-d(\epsilon)\,n^{-(1-\delta'')}\} > 1-\epsilon \quad \text{for all } \mathbf{F}_n \text{ in } \mathscr{F}_n. \qquad (5.24)$$

Proof. Here also, we only prove (5.23) as (5.24) will follow similarly. For proving (5.23) it suffices to consider $d(\epsilon) < 1$. Now, by definition,

$$\bar{H}_n(Z_{n,k_n}) - n^{-1}k_n \sim n^{-(1-\delta'')}.$$

Also, from any x, $H_n^*(x) \leqslant d(\epsilon)\,n^{-(1-\delta'')}$ entails that

$$n^{\frac{1}{2}}[H_n^*(x)\{1-H_n^*(x)\}]^{\frac{1}{2}-\delta'} \leqslant [d(\epsilon)]^{\frac{1}{2}-\delta'}\,n^{(\frac{1}{2}\delta''+\delta'-\delta'\delta'')}.$$

Hence, for $H_n^*(Z_{n,k_n}) < d(\epsilon)\,n^{-(1-\delta'')}$,

$$n^{\frac{1}{2}}[\bar{H}_n(Z_{n,k_n})-H_n^*(Z_{n,k_n})]/\{H_n^*(Z_{n,k_n})\,[1-H_n^*(Z_{n,k_n})]\}^{\frac{1}{2}-\delta'}$$

can be made greater than $n^\gamma[1-d(\epsilon)]/\{[d(\epsilon)]^{\frac{1}{2}-\delta'}\}$, where

$$\gamma = \tfrac{1}{2}\delta'' - \delta' + \delta'\delta'' > 0.$$

Since, the last quantity can be made greater than $c(\epsilon,\delta')$, by suitable choice of $d(\epsilon)$, (5.23) follows readily from (5.18). Q.E.D.

6 APPENDIX: TREATMENT OF HIGHER ORDER TERMS

Let (a_n^*, b_n^*) be the interval $S_{n,\eta}$ such that

$$S_{n,\eta} = \{x : H_n^*(x)\,[1-H_n^*(x)] > n^{-1}\eta_\epsilon\}, \qquad (6.1)$$

where $\epsilon(>0)$ is arbitrary and $\eta_\epsilon(>0)$ depends on ϵ. Since $H_n^*(x)$ has been defined for $x \geqslant 0$ and by definition is zero for $x < 0$, (6.1) implies that $0 < a_n^* < b_n^* < \infty$. Also, from (6.1), we have

$$n^{-1}\eta_\epsilon < [H_n^*(a_n^*), 1 - H_n^*(b_n^*)] < n^{-1}\eta_\epsilon[1 + 2n^{-1}\eta_\epsilon]. \tag{6.2}$$

Hence, we can always select η_ϵ in such a way that

$$n[1 - H_n^*(b_n^*) + H_n^*(a_n^*)] \leqslant \epsilon. \tag{6.3}$$

Upon noting that $|X_i|$ has the cdf H_i, $i = 1, \ldots, n$, we have

$$P\{|X_i| \in S_{n,\eta} \quad (i=1,\ldots,n)\} = P\{a_n^* \leqslant |X_i| \leqslant b_n^* \quad (i=1,\ldots,n)\}$$

$$= \prod_{i=1}^{n} \{1 - [1 - H_i(b_n^*) + H_i(a_n^*)]\} \geqslant 1 - \sum_{i=1}^{n} [1 - H_i(b_n^*) + H_i(a_n^*)]$$

$$= 1 - n[1 - H_n^*(b_n^*) + H_n^*(a_n^*)]. \tag{6.4}$$

Hence, by (6.3) and (6.4),

$$P\{|X_i| \in S_{n,\eta} \quad (i=1,\ldots,n)\} \geqslant 1 - \epsilon, \quad \text{uniformly in } \{\mathscr{F}_n\}. \tag{6.5}$$

Thus, by (3.8) and (6.5), we have with probability greater than or equal to $1 - \epsilon$,

$$|C_{2,n}| \leqslant \frac{1}{n+1} \int_{S_{n,\eta}} |J'[H_n^*(x)]| \, d\bar{F}_n(x) < \frac{1}{n} \int_{S_{n,\eta}} |J'[H_n^*(x)]| \, d\bar{H}_n(x). \tag{6.6}$$

Now, from (2.8) and (6.1), for all $x \in S_{n,\eta}$

$$n^{-\frac{1}{2}(1-\delta)} |J'[H_n^*(x)]| \leqslant K^*\{H_n^*(x)[1 - H_n^*(x)]\}^{-1+\delta/2} \quad (K^* < \infty).$$

Thus, the right-hand side of (6.6) is bounded above by

$$n^{-(\frac{1}{2}+\delta/2)} K^* Q_n, \quad \text{where } Q_n = \int_{S_{n,\eta}} \{H_n^*(x)[1 - H_n^*(x)]\}^{-1+\delta/2} d\bar{H}_n(x). \tag{6.7}$$

Now, by (2.3),

$$E(Q_n) = \int_{S_{n,\eta}} \{H_n^*(x)[1 - H_n^*(x)]\}^{-1+\delta/2} dH_n^*(x) < \int_0^1 [u(1-u)]^{-1+\delta/2} < \infty, \tag{6.8}$$

and hence, from (6.6), (6.7) and the Markov inequality, we obtain that

$$n^{\frac{1}{2}} |C_{2,n}| = o_p(1) \quad \text{uniformly in } \{\mathscr{F}_n\}. \tag{6.9}$$

Let us now consider the higher order terms $C_{3,n}$ and $C_{4,n}$. By virtue of (2.8) and Theorem 5.2, we have some $0 < 2\delta' < 1$,

$$P\{n^{\frac{1}{2}} | [\bar{H}_n(x) - H_n^*(x)] J'[H_n^*(x)] |$$

$$\leqslant K^*[H_n^*(x)\{1 - H_n^*(x)\}]^{-1+\delta-\delta'}, \forall x \geqslant 0\} > 1 - \epsilon, \tag{6.10}$$

for all $\mathbf{F}_n$ in $\mathscr{F}_n$, where $K^* = KC(\epsilon, \delta') < \infty$, and we can always take $\delta - \delta' = \delta^* > 0$. Thus, from (3.9) and (6.10) it follows that a sufficient condition for $|C_{3,n}| = o_p(n^{-\frac{1}{2}})$ (for all $\mathbf{F}_n$ in $\mathscr{F}_n$) is that

$$\int_0^1 \{H_n^*(x)\,[1-H_n^*(x)]\}^{-1+\delta^*}\, d[\bar{H}_n(x) - H_n^*(x)] \xrightarrow{p} 0 \quad \text{for all } \mathbf{F}_n \text{ in } \mathscr{F}_n.$$
(6.11)

Now,

$$\int_0^1 \{H_n^*(x)\,[1-H_n^*(x)]\}^{-1+\delta^*}\, d\bar{H}_n(x) = n^{-1}\sum_{i=1}^n \{H_n^*(|X_i|)\,[1-H_n^*(|X_i|)]\}^{-1+\delta^*}.$$

Hence, using the law of large numbers in Gnedenko and Kolmogorov ([7], p. 105), (6.11) can be proved. Hence,

$$n^{\frac{1}{2}}\,|C_{3,n}| = o_p(1) \quad \text{for all } \mathbf{F}_n \text{ in } \mathscr{F}_n.$$
(6.12)

Now, from (2.11) and (3.10), we have

$$|C_{4,n}| \leqslant n^{-1}\sum_{i=1}^n \left| J\left(\frac{i}{n+1}\right) - J(H_n^*(Z_{n,i})) - \left[\frac{i}{n+1} - H_n^*(Z_{n,i})\right] \right.$$
$$\left. \times J'[H_n^*(Z_{n,i})] \right|, \quad (6.13)$$

where $Z_{n,1}, ..., Z_{n,n}$ are defined just before the Corollary 5.2.2. As in there, let $k_n = [n^{\delta''}]$, where δ'' is defined in the Corollary 5.2.1. Then

$$n^{-1}\sum_{i=1}^{k_n} \left| J\left(\frac{i}{n+1}\right) - J(H_n^*(Z_{n,i})) - \left[\frac{i}{n+1} - H_n^*(Z_{n,i})\right] J'[H_n^*(Z_{n,i})] \right|$$
$$\leqslant n^{-1}\sum_{i=1}^{k_n} |J(i/[n+1])| + n^{-1}\sum_{i=1}^{k_n} |J[H_n^*(Z_{n,i})]|$$
$$+ n^{-1}\sum_{i=1}^{k_n} |[i/[n+1] - H_n^*(Z_{n,i})]\, J'[H_n^*(Z_{n,i})]|.$$
(6.14)

Further, by (2.8),

$$n^{-1}\sum_{i=1}^{k_n} |J(i/[n+1])| \leqslant (n^{-1}k_n)\, K[n/(n+1)^2]^{\frac{1}{2}-\delta}$$
$$= o(n^{-\frac{1}{2}-\delta+\delta''}) = o(n^{-\frac{1}{2}}) \quad \text{as } \delta - \delta'' > \delta' > 0.$$

Also, from (6.1) and (6.5), we have

$$P\{H_n^*(Z_{n,i}) > n^{-1}\eta_\epsilon \quad \text{for all } i = 1, ..., n\} > 1 - \epsilon,$$

uniformly in $\{\mathscr{F}_n\}$. Consequently, with probability $> 1 - \epsilon$, the second term on the right-hand side of (6.14) is less than

$$(n^{-1}k_n)\, K[\eta_\epsilon/n]^{-\frac{1}{2}+\delta} = (K\eta_\epsilon^{-\frac{1}{2}+\delta})\,(n^{-\frac{1}{2}-\delta+\delta''}) = o(n^{-\frac{1}{2}}),$$

uniformly in $\{\mathscr{F}_n\}$. Finally, using (6.1), (6.5) and (6.10), it is easily seen that the third term on the right-hand side is also $o(n^{-\frac{1}{2}})$, with a probability $> 1 - \epsilon$, for all $\mathbf{F}_n$ in $\mathscr{F}_n$. Consequently, (6.14) is $o_p(n^{-\frac{1}{2}})$, for all $\mathbf{F}_n$ in $\mathscr{F}_n$. In a similar manner, the upper tail of (6.13) [i.e. the contribution of the terms $(n - k_n + 1, \ldots, n)$ in (6.13)] can also be made $o_p(n^{-\frac{1}{2}})$, for all $\mathbf{F}_n$ in $\mathscr{F}_n$. Hence, it suffices to show that

$$n^{-\frac{1}{2}} \sum_{i=k_n+1}^{n-k_n} \left| J\left(\frac{i}{n+1}\right) - J(H_n^*(Z_{n,i})) - \left[\frac{i}{n+1}\right. \right.$$
$$\left. \left. - H_n^*(Z_{n,i})\right] J'[H_n^*(Z_{n,i})] \right| = o_p(1), \quad (6.15)$$

for all $\mathbf{F}_n$ in $\mathscr{F}_n$. With this end, we define for any $\tau: 0 < \tau < 1$,

$$S_{n,\eta}^{(1)}(\tau) = \{x: \tau \leqslant H_n^*(x) \leqslant 1 - \tau\},$$
$$S_{n,\eta}^{(2)}(\tau) = \{x: Z_{n,k_n} < x < H_n^{*-1}(\tau)\},$$
$$S_{n,\eta}^{(3)}(\tau) = \{x: H_n^{*-1}(1-\tau) < x < Z_{n,n-k_n+1}\},$$

and rewrite (6.15) as

$$\sum_{j=1}^{3} \int_{S_{n,\eta}^{(j)}(\tau)} n^{\frac{1}{2}} \left| J(a_n \bar{H}_n) - J(H_n^*) - [a_n \bar{H}_n - H_n^*] J'[H_n^*] \right| d\bar{H}_n. \quad (6.16)$$

Now, by the definition of the derivative of $J(u)$ and by (2.8)

$$\sup_{|v| \leqslant c} \sup_{\tau \leqslant u \leqslant 1-\tau} \sqrt{(n)} \left| J(u + v/n^{\frac{1}{2}}) - J(u) - (v/n^{\frac{1}{2}}) J'(u) \right| \to 0 \quad \text{as } n \to \infty. \quad (6.17)$$

Hence, from Theorem 5.2 (which limits $n^{\frac{1}{2}} |a_n \bar{H}_n - H_n^*| < c(\epsilon, \delta')$) and (6.17), we have with probability $> 1 - \epsilon$

$$\int_{S_{n,\eta}^{(1)}(\tau)} n^{\frac{1}{2}} \left| J(a_n \bar{H}_n) - J(H_n^*) - [a_n \bar{H}_n - H_n^*] J'[H_n^*] \right| d\bar{H}_n < \xi/2, \quad (6.18)$$

for all $\mathbf{F}_n$ in $\mathscr{F}_n$, where $\xi\ (> 0)$ is a preassigned small quantity. Also, we may write

$$\int_{S_{n,\eta}^{(j)}(\tau)} \sqrt{(n)} \left| J(a_n \bar{H}_n) - J(H_n^*) - [a_n \bar{H}_n - H_n^*] J'[H_n^*] \right| d\bar{H}_n$$
$$\leqslant \int_{S_{n,\eta}^{(j)}(\tau)} \sqrt{(n)} |a_n \bar{H}_n - H_n^*| \left| J[\phi H_n^* + (1-\phi) a_n \bar{H}_n] - J[H_n^*] \right| d\bar{H}_n \quad (j = 2, 3), \quad (6.19)$$

where $0 < \phi < 1$. Since

$$[\phi H_n^* + (1-\phi) a_n \bar{H}_n]/H_n^* = \phi + (1-\phi) a_n \bar{H}_n/H_n^*$$

and

$$\{1 - \{\phi H_n^* + (1-\phi)a_n\bar{H}_n\}\}/[1-H_n^*] = \phi + (1-\phi)(1-a_n\bar{H}_n)/(1-H_n^*),$$

we obtain from the Corollaries 5.2.1 and 5.2.2 that with probability $> 1 - 2\epsilon$,

$$\inf_{x \in S_{n,\eta}^{(j)}(\tau)} \frac{[\phi H_n^* + (1-\phi)a_n\bar{H}_n][1 - \{\phi H_n^* + (1-\phi)a_n\bar{H}_n\}]}{H_n^*(1-H_n^*)} > \beta_1^2(n/[n+1])^2$$

$$(j = 2, 3), \quad (6.20)$$

for all $\mathbf{F}_n$ in $\mathscr{F}_n^*$. From (6.20) Theorem 5.2 and (2.8), we obtain that with probability $> 1 - 2\epsilon$, (6.19) is less than

$$c(\epsilon, \delta') K [1 + \beta_{1n}^{-3+2\delta}] \int_{S_{n,\eta}^{(j)}(\tau)} \{H_n^*(x)[1-H_n^*(x)]\}^{-1+\delta^*} d\bar{H}_n(x) \quad (j = 2, 3),$$

$$(6.21)$$

where $\beta_{1n} = n\beta_1/(n+1)$ and $\delta^* = \delta - \delta'' > 0$ (by the Corollary 5.21). Now,

$$\sum_{j=2}^{3} \int_{S_{n,\eta}^{(j)}(\tau)} \{H_n^*(x)[1-H_n^*(x)]\}^{-1+\delta^*} d\bar{H}_n$$

$$\leqslant \int_0^\tau + \int_{1-\tau}^1 \{H_n^*(x)[1-H_n^*(x)]\}^{-1+\delta^*} d\bar{H}_n(x), \quad (6.22)$$

and it follows easily that for any $\tau < \frac{1}{2}$, the expected value of the right-hand side of (6.22) is

$$\int_0^\tau + \int_{1-\tau}^1 \{H_n^*(x)[1-H_n^*(x)]\}^{-1+\delta^*} dH_n^*(x) \leqslant (2^{1+\delta^*}/\delta^*)\tau^{\delta^*}$$

$$\text{uniformly in } \{\mathscr{F}_n\}. \quad (6.23)$$

Thus, applying the Markov inequality, we obtain that the left-hand side of (6.22) is bounded, in probability, by $(1^{1+\delta^*}/\delta^*)\tau^{\delta^*}$, for all $\mathbf{F}_n$ in $\mathscr{F}_n$. Now, for any fixed ϵ (> 0) and δ', τ may be so chosen that

$$[2^{1+\delta^*}/\delta^*]c(\epsilon, \delta') K[1 + \beta_{1n}^{-3+2\delta}]\tau^{\delta^*} < \xi/2.$$

Combining this with (6.21), (6.22) and (6.23), we conclude that the sum of the two terms (for $j = 2, 3$) in (6.19) is bounded above by $\xi/2$, in probability, uniformly in $\mathscr{F}_n$. Thus, from (6.18), we conclude that (6.15) is bounded by ξ, in probability, for all $\mathbf{F}_n \in \mathscr{F}_n$. Hence, $P_{4,n} = o_p(n^{-\frac{1}{2}})$, for all $\mathbf{F}_n \in \mathscr{F}_n$.

7 ACKNOWLEDGEMENTS

It is a pleasure to acknowledge the fruitful discussions the author had with Professors V. Ďupác and J. Hájek concerning the contents of §5.

REFERENCES

[1] Chernoff, H. and Savage, I. R. (1958). Asymptotic normality and efficiency of certain nonparametric test statistics. *Ann. Math. Statist.* **29**, 972–94.

[2] Doob, J. L. (1949). Heuristic approach to the Kolmogorov–Smirnov theorems. *Ann. Math. Statist.* **20**, 393–403.

[3] Esseen, C. G. (1945). Fourier analysis of distribution functions. *Acta Math.* **77**, 1–125.

[4] Fraser, D. A. S. (1957). Most powerful rank type tests. *Ann. Math. Statist.* **28**, 1040–3.

[5] Govindarajulu, Z. (1960). Central limit theorem and asymptotic efficiency of one sample nonparametric procedures. *Tech. Report* 11, *Dept. Statist. Univ. Minnesota.*

[6] Govindarajulu, Z., LeCam, L. and Raghavachari, M. (1966). Generalizations of theorems of Chernoff and Savage on the asymptotic normality of test statistics. *Proc. fifth Berkeley Symp. Math. Statist. Prob.* **1**, 608–38.

[7] Gnedenko, B. V. and Kolmogorov, A. N. (1954). *Limit theorems for sums of independent random variables.* (Translated by K. L. Chung.) Reading, Mass: Addison–Wesley.

[8] Hájek, J and Šidák, Z. (1967). *Theory of rank tests.* New York: Academic Press.

[9] Puri, M. L. and Sen, P. K. (1969). On the asymptotic normality of one sample rank order statistics. *Teoria veroyatnostey i ee primenyia.* **14**, 167–72.

[10] Pyke, R. and Shorack, G. R. (1968). Weak convergence and a Chernoff–Savage theorem for random sample sizes. *Ann. Math. Statist.* **39**, 1675–85.

[11] Sen, P. K. (1968a). On a further robustness property of the test and estimator based on Wilcoxon's signed rank statistic. *Ann. Math. Statist.* **39**, 282–5.

[12] Sen, P. K. (1968b). Robustness of some nonparametric procedures in linear models *Ann. Math. Statist.* **39**, 1913–22.

[13] Sen, P. K., Bhattacharyya, B. B. and Suh, M. W. (1969). Limiting behaviour of the extremum of some sample functions. *Inst. Statist. Univ. North Carolina Mimeo Report. No.* 628.

[14] Sen, P. K and Puri, M. L. (1967). On the theory of rank order tests for location in the multivariate one sample problem. *Ann. Math. Statist.* **38**, 1216–28.

[15] Wilcoxon, F. (1949). *Some rapid approximate statistical procedures.* American Cyanamid Co. Stamford. Conn.

DISCUSSION ON SEN'S PAPER

VÁCLAV DUPAČ

I think it is always interesting and instructive to compare similar results obtained independently and by completely different proving methods. I would like to sketch here such a comparison. Essentially the same topic, i.e. the asymptotic normality of the one-sample rank order statistics, was studied by a postgraduate student of Professor Hájek, named M. Hušková, in her thesis, defended in December 1968 and submitted then to the *Zeitschrift für Wahrscheinlichkeitsrechnung und ihre Grenzgebiete*.

Whereas Professor Sen's proving method (of the main theorem) follows to certain extent the original Chernoff–Savage paper and further, when dealing with the higher order terms, it utilizes some deep properties of empirical distribution functions (considered as stochastic processes), M. Hušková follows entirely the elementary (though rather involved) proving methods developed by Professor Hájek in his AMS 1968 paper. In that paper, a simple linear rank statistic is considered which includes the two sample and c-sample case as special cases, but which does not cover the one-sample case. M. Hušková succeeded in adapting all the basic tools of Hájek's method to the one-sample case and thus obtained results parallel to those of Hájek.

I'll mention only her main theorem here. (I formulate it with use of symbols introduced by Professor Sen, though M. Hušková used a different notation; I have also transformed her theorems stated in the ϵ-form into limit theorems.) As to the distribution functions, the theorem (as well as the theorem of Professor Sen) does not impose any condition on them, excepting the condition of continuity, i.e. it holds true for all $\{\mathbf{F}_n\} \in \mathscr{F}$.

The condition connecting the scores-generating functions J_n with the limiting J is more restrictive than that of Professor Sen, it is required that the J_n's are all generated by J in one of the following ways:

$$\text{either} \quad J_n\left(\frac{i}{n+1}\right) = J\left(\frac{i}{n+1}\right) \quad \text{or} \quad J_n\left(\frac{i}{n+1}\right) = EJ(U_n^{(i)})$$

(where $U_n^{(i)}$ is the ordered sample of size n from the uniform distribution on $(0, 1)$). On the other hand the condition on J itself is milder: it is required that

$$J(u) = {}_1J(u) - {}_2J(u),$$

[73]

where both $_rJ(u)$, $r = 1, 2$, are non-decreasing, square integrable over $(0, 1)$ and absolutely continuous on $(\epsilon, 1-\epsilon)$, for each $0 < \epsilon < \frac{1}{2}$.

Now, if the sequence $\{\gamma_n^*\}$ is bounded away from 0 (or, equivalently, if $\{\text{var}(n^{\frac{1}{2}}T_n)\}$ is b.a.f. 0) then

$$\mathscr{L}\left(\frac{T_n - ET_n}{(\text{var }T_n)^{\frac{1}{2}}}\right) \to \mathscr{N}(0, 1)$$

as well as
$$\mathscr{L}\left(\frac{T_n - ET_n}{\gamma_n^* n^{-\frac{1}{2}}}\right) \to \mathscr{N}(0, 1).$$

On the other hand, nothing is stated about the possibility of replacing ET_n by a simpler expression μ_n^*.

Finally, it should be remarked that Hušková considers a little bit more general statistics: Professor Sen's T_n may be written as

$$T_n = n^{-1} \sum_{i=1}^{n} J_n\left(\frac{R_i^+}{n+1}\right) e(X_i),$$

where R_i^+ is the rank of $|X_i|$ in the order sample of the absolute values of the X_i's and $e(u) = \begin{cases} 0 \dots u < 0 \\ 1 \dots u \geqq 0 \end{cases}$. On the other hand, Hušková uses the statistic (say)

$$T_n' = n^{-1} \sum_{i=1}^{n} c_i J_n\left(\frac{R_i^+}{n+1}\right) \text{sgn } X_i,$$

where c_i, $1 \leqq i \leqq n$, are arbitrary constants.

If $c_i = 1$ for all $1 \leqq i \leqq n$, then her statistic T_n' differs from T_n by a nonrandom constant only.

A CONTRIBUTION TO THE
ASYMPTOTIC NORMALITY OF SIMPLE
LINEAR RANK STATISTICS

VÁCLAV DUPAČ

1.

This study is closely related to the fundamental paper by Hájek (1968) and to its continuation by Dupač and Hájek (1969). Formulas and theorems of these two papers will be quoted by the abbreviations H and DH respectively, followed by their corresponding numbers. Both papers concern the asymptotic normality of the statistic

$$S = \sum_{i=1}^{N} c_i a(R_i), \qquad (1.1)$$

where $c_1, \ldots, c_N$ are real numbers, $R_1, \ldots, R_N$ are the ranks of independent random variables $X_1, \ldots, X_N$ with continuous distribution functions $F_1, \ldots, F_N$, and the scores $a(i)$ are generated, in a way specified in individual theorems, by a function $\phi(t)$, $0 < t < 1$, supposed to be (in the most general case) the difference of two nondecreasing and square-integrable functions.

Hájek's approach to the problem, applied in both papers, is based on the method of projection (Lemma H, 4.1), and utilizes further as an effective tool an inequality (Theorem H, 3.1) which implies, in particular, that

$$E(S - ES)^2 \leq \text{const.} \, N \max_{1 \leq i \leq N} (c_i - \bar{c})^2 \int_0^1 \phi^2(t) \, dt \qquad (1.2)$$

for $a(i) = \phi\left(\dfrac{i}{N+1}\right)$, ϕ monotone. In the DH paper, a separate study of the case of the unit step function ϕ, is another important tool. In both papers, the asymptotic normality with natural parameters $(ES, \text{var} \, S)$ as well as with a simpler asymptotic variance σ^2 is studied, but the question of a simpler centering constant μ^* has been left open (except for a remark in H, p. 330).

The permissibility of replacing ES by a certain simpler μ^* in Theorem H, 2.3 was shown by Hoeffding (1968) under slightly strengthened assumptions concerning ϕ. Hoeffding's Proposition 2 implies that

$$(ES - \mu^*)^2 \leq \text{const.} \, N \max_{1 \leq i \leq N} (c_i - \bar{c})^2 \left(\int_0^1 |\phi(t)| \, t^{-\frac{1}{2}} (1-t)^{-\frac{1}{2}} dt \right)^2 \qquad (1.3)$$

[75]

for $a(i) = \phi(i/(N+1))$, ϕ monotone; comparing this with (1.2), the strengthening of the assumption

'$\phi = \phi_1 - \phi_2$ with ϕ_α nondecreasing and square-integrable, $\alpha = 1, 2$' in H and DH to Hoeffding's,

'$\phi = \phi_1 - \phi_2$ with ϕ_α nondecreasing and $\phi_\alpha(t) t^{-\frac{1}{2}} (1-t)^{-\frac{1}{2}}$ integrable, $\alpha = 1, 2$'

becomes easily understandable.

However, the strengthened assumption is applied, in the course of the proof, only to the 'irregular' component of the function ϕ, that can be made arbitrarily small. Thus, it is actually needed only in Theorems H, 2.3 and DH, 2, but not in Theorems H, 2.1, DH, 1, and their combination DH, 3, where we can extend the conclusions in order to allow for simpler centering constants, without additional assumptions. On the other hand, in all theorems we shall make use of Hoeffding's idea to center S by a constant different from the usual

$$\mu = \sum_{i=1}^{N} c_i \int_0^1 \phi(t)\, dL_i(t).$$

(See (2.1) for the meaning of L_i.) Hoeffding's version of Theorem H, 2.3 combined with complemented Theorem DH, 1 will then yield modifications of Theorems DH, 2, DH, 4 and DH, 5.

There is one more problem that has been left open in DH paper: Whether some degree of degeneracy of var S (compared with the variance of S under the null hypothesis) may be allowed in the case of unit step function ϕ. The analogous question for ϕ possessing a bounded second derivative was answered affirmatively in Theorem H, 2.1, where the ratio of variances was allowed to tend to zero, provided this convergence was slower than that of $\max_{1 \le i \le N} (c_i - \bar{c})^2 \Big/ \sum_{i=1}^{N} (c_i - \bar{c})^2$.

We shall show in §3 that the answer is positive also in the case of Theorem DH, 1, if the conditions concerning the distribution functions are slightly strengthened. The ratio of variances is now allowed to tend to zero at most at the rate of $N^{-\frac{1}{2}+\epsilon}$ for some $\epsilon > 0$, whereas

$$\max_{1 \le i \le N} (c_i - \bar{c})^2 \Big/ \sum_{i=1}^{N} (c_i - \bar{c})^2$$

is supposed to be $O(N^{-\frac{1}{2}})$. As an example, the median test statistic is investigated under divergent location alternatives.

2.

We shall first continue the list of symbols introduced in § 1. We put

$$
\left.\begin{aligned}
H(x) &= N^{-1} \sum_{i=1}^{N} F_i(x) \quad (-\infty < x < +\infty); \\
H^{-1}(t) &= \inf\{x : H(x) > t\} \quad (0 < t < 1); \\
L_i(t) &= F_i(H^{-1}(t)) \quad (0 < t < 1,\; 1 \le i \le N);
\end{aligned}\right\} \tag{2.1}
$$

$$
\sigma^2 = \sum_{i=1}^{N} \operatorname{var} \int_{\frac{1}{2}}^{H(X_i)} (\tilde{c}(v) - c_i)\, d\phi(v), \tag{2.2}
$$

where we put
$$
\tilde{c}(v) = N^{-1} \sum_{j=1}^{N} c_j L_j'(v), \tag{2.3}
$$

if ϕ is absolutely continuous; if ϕ possesses a singular component, we change the definition (2.3) to

$$
\tilde{c}(v) = N^{-1} \sum_{j=1}^{N} c_j l_j(v) \tag{2.4}
$$

on a measurable set B' containing the singular set of ϕ, but arbitrary otherwise, provided that, for all $v \in B'$, the additional conditions DH, (2.12)–(2.15) are satisfied. (The complement of B' coincides with the set B in DH.)

Further we put

$$
\bar{c} = N^{-1} \sum_{i=1}^{N} c_i, \quad \bar{a} = N^{-1} \sum_{i=1}^{N} a(i), \quad \bar{\phi} = \int_0^1 \phi(t)\, dt
$$

and
$$
\mu = \sum_{i=1}^{N} c_i E\phi(H(X_i)) = \sum_{i=1}^{N} c_i \int_0^1 \phi(t)\, dL_i(t), \tag{2.5}
$$

$$
\mu' = \mu + N\bar{c}(\bar{a} - \bar{\phi}). \tag{2.6}
$$

Now, we shall present the modified versions of the mentioned theorems. To unify the style, we formulate them as limit theorems for sequences. We emphasize that by asymptotic normality of S with parameters (μ^*, σ^2) we shall understand, throughout the paper, the convergence in distribution *together* with convergence of the first two moments, i.e.

$$
\mathscr{L}((S - \mu^*)/\sigma) \to \mathscr{N}(0, 1) \quad and \quad E((S - \mu^*)/\sigma) \to 0, \quad E((S - \mu^*)/\sigma)^2 \to 1.
$$

Note that Theorem H, 2.1 will be slightly generalized with regard to the choice of the scores $a(i)$.

Theorem H, 2.1. Assume that ϕ has a bounded second derivative and that the scores satisfy the relation

$$
\sum_{i=1}^{N} |a(i) - \phi(i/(N+1))| = O(1). \tag{2.7}
$$

Then either of the conditions

$$\lim_{N} \operatorname{var} S / \max_{1 \leq i \leq N} (c_i - \bar{c})^2 = +\infty, \tag{2.8}$$

$$\lim_{N} \sigma^2 / \max_{1 \leq i \leq N} (c_i - \bar{c})^2 = +\infty, \tag{2.9}$$

implies asymptotic normality of S with natural parameters as well as with parameters (μ', σ^2); and under the additional assumption

$$\bar{c} = O(\max |c_i - \bar{c}|) \tag{2.10}$$

also with parameters (μ, σ^2).

Remark 1. As shown in H, p. 341, line 5 from below, the scores $a(i) = E(\phi(U^{(i)}))$, where the $U^{(i)}$, $1 \leq i \leq N$, represent the ordered sample from the uniform distribution on $[0, 1]$, satisfy the assumption (2.7).

In the following theorem, the set B' in the definition of $\tilde{c}$ can be reduced to the one-point set $\{v\}$.

Theorem DH, 1. Suppose that

$$\begin{aligned} \phi(t) &= 0 \quad (0 < t < v), \\ &= 1 \quad (v \leq t < 1), \end{aligned} \right\} \tag{2.11}$$

that the scores satisfy (2.7) and that

$$\max_{1 \leq i \leq N} (c_i - \bar{c})^2 = o\left(\sum_{i=1}^{N} (c_i - \bar{c})^2 \right). \tag{2.12}$$

Then either of the conditions

$$\liminf_{N \to \infty} \operatorname{var} S \Big/ \sum_{i=1}^{N} (c_i - \bar{c})^2 > 0, \tag{2.13}$$

$$\liminf_{N \to \infty} \sigma^2 \Big/ \sum_{i=1}^{N} (c_i - \bar{c})^2 > 0 \tag{2.14}$$

implies asymptotic normality of S with natural parameters as well as with parameters (μ', σ^2); and under the additional assumption (2.10) also with parameters (μ, σ^2).

In the following theorem, the interval $(0, 1)$ can be chosen for the set B', as the assumptions DH, (2.17) (or DH, (2.18) or DH, (2.19)) imply the satisfaction of DH, (2.12) through DH, (2.15) for all $0 < v < 1$.

Theorem DH, 3. Suppose that $\phi = \phi_1 + \phi_2$, where ϕ_1 is constant but for a finite number of jumps and ϕ_2 has a bounded second derivative. Further assume that (2.7), (2.12) and DH, (2.17) or DH, (2.18) or DH, (2.19) hold.

Then either of the conditions (2.13), (2.14) *implies asymptotic normality of* S *with natural parameters as well as with parameters* (μ', σ^2); *and under the additional assumption* (2.10) *also with* (μ, σ^2).

The following theorem in comparison to the previous ones is more general with respect to scores and scores-generating function; on the other hand, it is more restrictive with respect to the asymptotic order of variance (and, implicitly, with respect to the c_i's).

Theorem DH, 2. *Suppose that* $\phi = \phi_1 - \phi_2$, *where*

$$\int_0^1 |\phi_\alpha(t)|\, t^{-\frac{1}{2}}(1-t)^{-\frac{1}{2}}\, dt < +\infty \quad (\alpha = 1, 2), \tag{2.15}$$

and that the scores satisfy the relation

$$\sum_{i=1}^N |a(i) - \phi(i/(N+1))| = o(N^{\frac{1}{2}}). \tag{2.16}$$

Then either of the conditions

$$\liminf_{N\to\infty} \operatorname{var} S/N \max_{1\le i\le N} (c_i - \bar{c})^2 > 0, \tag{2.17}$$

$$\liminf_{N\to\infty} \sigma^2/N \max_{1\le i\le N} (c_i - \bar{c})^2 > 0 \tag{2.18}$$

implies asymptotic normality of S *with natural parameters as well as with parameters* (μ', σ^2); *under the additional assumption* (2.10) *also with* (μ, σ^2).

Remark 2. In Theorems DH, 4 and DH, 5, concerning the two-sample case, condition (2.10) is evidently satisfied, so that the asymptotic normality with centering parameters μ follows, provided (2.15) holds only.

3.

In this section, we shall retain the notation and the overall assumptions introduced in §§ 1 and 2, except for the definition of $\tilde{c}(v)$ and the text connected with it. First we present a version of Theorem DH, 1, which admits some degeneracy of $\operatorname{var} S$.

Modified Theorem DH, 1. *Suppose that* ϕ *is given by* (2.11), *that the scores satisfy* (2.7) *and that*

$$\max_{1\le i\le N} (c_i - \bar{c})^2 = O\left(N^{-\frac{1}{2}} \sum_{i=1}^N (c_i - \bar{c})^2\right). \tag{3.1}$$

Define (for v fixed by (2.11))

$$\tilde{c}(v) = N^{-1} \sum_{i=1}^N c_i L_i'(v),$$

and assume that

(i) *for each $K > 0$, the derivatives $L_i'(t)$ exist in the interval*

$$|t-v| \leq KN^{-\frac{1}{2}}\lg^{\frac{1}{2}} N \tag{3.2}$$

and the relations

$$L_i'(v) = O(1), \quad L_i'(t) = L_i'(v) + O(N^{-\frac{1}{2}}\lg^{\frac{1}{2}} N)$$

hold uniformly in $1 \leq i \leq N$, $N \geq 1$, and t satisfying (3.2);

(ii) $$\liminf_{N\to\infty} N^{-1} \sum_{i=1}^{N} L_i(v)\,(1 - L_i(v)) > 0.$$

Then either of the conditions

$$\liminf_{N\to\infty} \operatorname{var} S/N^{-\frac{1}{2}+\epsilon} \sum_{i=1}^{N} (c_i - \bar{c})^2 > 0 \quad \text{for some } \epsilon > 0, \tag{3.3}$$

$$\liminf_{N\to\infty} \sum_{i=1}^{N} (c_i - \tilde{c}(v))^2 L_i(v)\,(1 - L_i(v))/N^{-\frac{1}{2}+\epsilon} \sum_{i=1}^{N} (c_i - \bar{c})^2 > 0 \text{ for some } \epsilon > 0, \tag{3.4}$$

implies the conclusions of Theorem DH, 1, as stated in § 2.

Remark 3. The result can be extended in an obvious way to a function ϕ which is constant but for a finite number of jumps and, in combination with Theorem H, 2.1, to the sum of such a function and a function with bounded second derivative.

Remark 4. The following condition (similar to DH, (2.17)) implies (i), (ii), and is more easily verifiable: Denote $I_\epsilon = (H^{-1}(v) - \epsilon, H^{-1}(v) + \epsilon)$ and suppose there exist constants $\epsilon_j > 0$, $j = 1, 2, 3$, such that $F_i''(x)$ are uniformly bounded (in x, i, N) on I_{ϵ_1} and

$$N^{-1} \operatorname{card}\{i\colon \inf_{x \in I_{\epsilon_1}} F_i'(x) > \epsilon_2\} > \epsilon_3 \quad \text{for all } N \geq 1.$$

Example. Let us consider the median test statistic in the two-sample location case with normal underlying density. In a formal way, let

$$\phi(t) = \begin{cases} 0 & (0 < t < \frac{1}{2}), \\ 1 & (\frac{1}{2} \leq t < 1), \end{cases}$$

let the scores be given by $a(i) = \phi(i/(N+1))$, $1 \leq i \leq N$, the constants c_i by

$$c_i = \begin{cases} 1 & (1 \leq i \leq m), \\ 0 & (m < i \leq N), \end{cases}$$

and the distributions by

$$F_i = \begin{cases} \Phi & (1 \leq i \leq m), \\ \Phi_\Delta & (m < i \leq N), \end{cases}$$

where

$$\Phi(x) = (2\pi)^{-\frac{1}{2}} \int_{-\infty}^{x} e^{-t^2/2}\,dt, \quad \Phi_\Delta(x) = \Phi(x-\Delta)\ (\Delta = \Delta_N > 0).$$

Then the corresponding statistic S is asymptotically normal, provided Δ remains bounded and both m, $N-m$ tend to infinity. (See DH, § 2, Example.)

The modified Theorem DH, 1 can be used to investigate whether the asymptotic normality is preserved in case $\Delta \to +\infty$ (sufficiently slowly). We shall denote $\lambda = m/N$ and assume that $\lambda \to \lambda_0$, $0 < \lambda_0 < \frac{1}{2}$. (We would obtain a symmetrical result for $\frac{1}{2} < \lambda_0 < 1$, whereas the method fails for $\lambda_0 = \frac{1}{2}$.) The theorem then implies asymptotic normality of S, if

$$\Delta \to +\infty \text{ in such a way that } \limsup_{N\to\infty} \Delta/\lg^{\frac{1}{2}} N < 1, \tag{3.5}$$

whereas a different argument shows that S is not asymptotically normal for any choice of centering and (positive) norming constants if

$$\limsup_{N\to\infty} \Delta/2^{\frac{1}{2}}\lg^{\frac{1}{2}} N > 1. \tag{3.6}$$

(If might be of some interest to compare this result with results obtained for the Wilcoxon statistic in Dupač–Hájek 1969.)

4.

Now we shall give the proofs. We introduce the symbols S_ϕ and μ'_ϕ for the statistic (1.1) and for the constant (2.6) if the scores are given by $a(i) = \phi(i/(N+1))$, leaving S and μ' for the general case. We start with a lemma.

Lemma 1. We have

$$(ES - \mu')^2 \le 2(ES_\phi - \mu'_\phi)^2 + 2 \max_{1\le i\le N} (c_i - \bar{c})^2 \left(\sum_{i=1}^{N} |a(i) - \phi(i/(N+1))| \right)^2 \tag{4.1}$$

for arbitrary $c_i, a(i), 1 \le i \le N$, and integrable $\phi(t), o < t < 1$.

Proof: From the definitions it follows

$$\begin{aligned}
S - \mu' &= \sum_{i=1}^{N} c_i \left\{ a(R_i) - \int_0^1 \phi\,dL_i - \bar{a} + \bar{\phi} \right\} \\
&= \sum_{i=1}^{N} (c_i - \bar{c}) \left[a(R_i) - \int_0^1 \phi\,dL_i \right],
\end{aligned} \tag{4.2}$$

since $\sum\limits_{i=1}^{N} \{...\} = 0$. Setting here $a(i) = \phi(i/(N+1))$, $1 \le i \le N$, and subtracting the equality thus obtained from (4.2), we get

$$S - \mu' = S_\phi - \mu'_\phi + \sum_{i=1}^{N} (c_i - \bar{c})[a(R_i) - \phi(R_i/(N+1))]. \qquad (4.3)$$

Taking expectations and then bounds for the squares, we get

$$(ES - \mu')^2 \le 2(ES_\phi - \mu'_\phi)^2 + 2\left(\sum_{i=1}^{N} (c_i - \bar{c})\, E[a(R_i) - \phi(R_i/(N+1))]\right)^2$$

which already implies (4.1), as

$$\left|\sum_{i=1}^{N} (c_i - \bar{c})\, E[...]\right| \le \max_{1 \le i \le N} |c_i - \bar{c}| \sum_{i=1}^{N} E(|a(R_i) - \phi(R_i/(N+1))|)$$

$$= \max_{1 \le i \le N} |c_i - \bar{c}| \cdot \sum_{i=1}^{N} |a(i) - \phi(i/(N+1))|.$$

Now we are prepared to prove the added conclusions in the theorems of § 2.

To Theorem H, 2.1. The replacement of

$$a(i) = \phi(i/(N+1)) \quad \text{or} \quad a(i) = E\phi(U^{(i)})$$

in the original version by a more general $a(i)$ satisfying (2.7) is justified by the relation

$$E(S - ES - [S_\phi - ES_\phi])^2 = O(\max_{1 \le i \le N} (c_i - \bar{c})^2)$$

which follows from

$$|S - ES - [S_\phi - ES_\phi]| \le 2 \max_{1 \le i \le N} |c_i - \bar{c}| \sum_{i=1}^{N} |a(i) - \phi(i/(N+1))|$$

and entails $\qquad E(S - ES - [S_\phi - ES_\phi])^2/\mathrm{var}\, S \to 0.$

As to the assertion concerning the constants μ', it is sufficient to prove that

$$(ES_\phi - \mu'_\phi)^2 = O(\max (c_i - \bar{c})^2); \qquad (4.4)$$

the rest then follows from Lemma 1. But

$$ES_\phi - \mu'_\phi = \Sigma(c_i - \bar{c})\left[E\phi(R_i/(N+1)) - \int_0^1 \phi\, dL_i\right] \qquad (4.5)$$

(put $a(i) = \phi(i/(N+1))$ in (4.2)); and (4.4) follows from the inequality

$$\left|E\phi(R_i/(N+1)) - \int_0^1 \phi\, dL_i\right| \le KN^{-1},$$

which is proved in H, p. 340, line 11 from above. As to the constants μ, observe first, that

$$|\mu' - \mu| \leq |\bar{c}| \left[|\Sigma a(i) - \Sigma \phi(i/(N+1))| + |\Sigma \phi(i/(N+1)) - N \int_0^1 \phi(t) \, dt| \right],$$
(4.6)

where both differences in the brackets are bounded: the first by (2.7) whereas the boundedness of the second one is an easy consequence of the boundedness of the derivative ϕ'.

Thus, in view of (2.10), we have

$$(\mu' - \mu)^2 = O(\max (c_i - \bar{c})^2),$$

which concludes the proof.

To Theorem DH, 1. As to the assertion on μ', it suffices to prove that

$$(ES_\phi - \mu'_\phi)^2 = o \left(\sum_{i=1}^N (c_i - \bar{c})^2 \right);$$
(4.7)

the rest then follows from (4.1), (2.12) and (2.7). Setting $a(i) = \phi(i/(N+1))$ in (4.2), taking expectations and squaring, we get

$$(ES_\phi - \mu'_\phi)^2 \leq \sum_{i=1}^N (c_i - \bar{c})^2 \sum_{i=1}^N \left(E\phi(R_i/(N+1)) - \int_0^1 \phi \, dL_i \right)^2.$$
(4.8)

Thus, it is sufficient to prove that

$$E\phi(R_i/(N+1)) - \int_0^1 \phi \, dL_i = o(N^{-\frac{1}{2}})$$
(4.9)

uniformly in $1 \leq i \leq N$.

Now, by similar considerations as used in the derivation of DH, (4.11), and DH, (4.12), we obtain

$$E(\phi(R_i/(N+1)) \mid X_i = x) \begin{cases} < N^{-K_1} & \text{for} \quad v - H(x) > K_2 N^{-\frac{1}{2}} \lg^{\frac{1}{2}} N, \\ = \Phi\left(\dfrac{H(x)-v}{DN^{-\frac{1}{2}}}\right) + \vartheta N^{-\frac{1}{2}} \lg^{\frac{1}{2}} N & \text{for} \\ & |v - H(x)| \leq K_2 N^{-\frac{1}{2}} \lg^{\frac{1}{2}} N, \\ > 1 - N^{-K_1} & \text{for} \quad v - H(x) \\ & < - K_2 N^{-\frac{1}{2}} \lg^{\frac{1}{2}} N, \end{cases}$$
(4.10)

where
$$K_1 > 1 \quad \text{for} \quad K_2 > 2,$$

where
$$D^2 = N^{-1} \sum_{i=1}^N L_i(v) \, (1 - L_i(v)),$$

$$\Phi(t) = (2\pi)^{-\frac{1}{2}} \int_{-\infty}^t e^{-s^2/2} ds \quad \text{and} \quad |\vartheta| \leq K_3.$$

(We number the constants K_i in order of their appearance in the present paper.) Also the further course of the proof is similar to the proof of Lemma DH, 7. We first observe that the *equality* in (4.10) remains true even in the larger interval

$$|v - H(x)| \leq K_4 DN^{-\frac{1}{2}} \lg^{\frac{1}{2}} N,$$

where K_4 is such that $D \geq K_2/K_4$ for all $N \geq 1$. Thus we get (writing $u(t)$ for the unit step function)

$$E\phi(R_i/(N+1)) - E\phi(H(X_i))$$

$$= \int_{|v-H(x)| \leq K_4 DN^{-\frac{1}{2}} \lg^{\frac{1}{2}} N} \{\Phi((H(x) - v)/DN^{-\frac{1}{2}}) - u(H(x) - v)$$

$$+ \vartheta N^{-\frac{1}{2}} \lg^{\frac{1}{2}} N\} \, dF_i(x) + O(N^{-K_1})$$

$$= \int_{|p| \leq K_4 \lg^{\frac{1}{2}} N} \{\Phi(p) - u(p)\} \, d_p L_i(v + DN^{-\frac{1}{2}}p) + N^{-\frac{1}{2}} \lg^{\frac{1}{2}} N$$

$$\times \int_{|p| \leq K_4 \lg^{\frac{1}{2}} N} \vartheta \, d_p L_i(v + DN^{-\frac{1}{2}}p) + O(N^{-K_1}) \quad (4.11)$$

where we substituted $(H(x) - v)/DN^{-\frac{1}{2}} = p$. The middle term in the last line of (4.11) is of the order of magnitude $O(N^{-1} \lg N)$ as follows from the assumption DH, (2.12); it remains to estimate the first term. Denote it by I and split it into

$$I = \int_{|p| \leq K_5} \{\Phi - u\} \, d_p L_i + \int_{K_5 \leq |p| \leq K_4 \lg^{\frac{1}{2}} N} \{\Phi - u\} \, d_p L_i = I' + I''.$$

Evaluating I', we use the expansion

$$L_i(v + DN^{-\frac{1}{2}}p) = L_i(v) + l_i(v) DN^{-\frac{1}{2}}p + \Lambda_i(p),$$

with Λ_i of order $o(N^{-\frac{1}{2}})$ uniformly in $[-K_5, K_5]$ and in $1 \leq i \leq N$. We get

$$I' = N^{-\frac{1}{2}} Dl_i(v) \int_{-K_5}^{K_5} \{\Phi(p) - u(p)\} \, dp + \int_{-K_5}^{K_5} \{\Phi(p) - u(p)\} \, d\Lambda_i(p),$$

where the first integral vanishes and the second is of order $o(N^{-\frac{1}{2}})$ as follows by integration by parts.

I'' is a sum of two integrals, the first of which satisfies the inequality

$$\int_{-K_4 \lg^{\frac{1}{2}} N}^{-K_5} \Phi(p) \, d_p L_i(v + DN^{-\frac{1}{2}}p) < K_6 DN^{-\frac{1}{2}} \int_{-\infty}^{-K_5} |p| \, \Phi'(p) \, dp$$

as follows by integrating by parts; the same for the second one. Thus $I'' < N^{-\frac{1}{2}}$ for arbitrarily small $\epsilon > 0$ if K_5 is chosen sufficiently large. This completes the proof of the first assertion (on μ').

The proof of the second assertion (on μ) makes again use of (4.6), where the boundedness of the last difference follows now from the special form of $\phi(t) = u(t-v)$. This together with (2.15) gives

$$(\mu' - \mu)^2 = o\left(\sum_{i=1}^{N} (c_i - \bar{c})^2\right),$$

which concludes the proof.

To Theorem DH, 3. The theorem follows by combination of the two previous ones (H, 2.1 and DH, 1).

To Theorem DH, 2. The inequality

$$\left(\int_0^1 \phi^2(t)\,dt\right)^{\frac{1}{2}} \leq \left(\int_0^1 |\phi(t)|\, t^{-\frac{1}{2}}(1-t)^{-\frac{1}{2}}\,dt\right),$$

holding true for each monotone ϕ (Hoeffding), shows that (2.16) actually implies the satisfaction of DH, (2.4).

As to the centering constants μ', it is sufficient, according to Lemma 1, to prove that

$$(ES_\phi - \mu'_\phi)^2 = o(N \max_{1 \leq i \leq N} (c_i - \bar{c})^2). \qquad (4.12)$$

For this purpose, to each $\alpha > 0$ we may decompose ϕ similarly as in DH, (6.8):

$$\phi = \psi - \lambda + \gamma - \eta + g - h, \qquad (4.13)$$

where all functions on the right are nondecreasing, satisfy (2.16) (at the place of ϕ_α), and

ψ, λ are a.c. such that $\int_{B'} (d\psi + d\lambda) = 0$;

γ, η are bounded such that $\int_B (d\gamma + d\eta) = 0$;

g, h are such that

$$\int_B (dg + dh) = 0 \quad \text{and} \quad \int_0^1 (|g(t)| + |h(t)|)\, t^{-\frac{1}{2}}(1-t)^{-\frac{1}{2}}\,dt < \alpha.$$

Decomposing S_ϕ, μ'_ϕ correspondingly, we have

$$(ES_\phi - \mu'_\phi)^2 \leq 6[(ES_\phi - \mu'_\phi)^2 + \ldots + (ES_h - \mu'_h)^2].$$

Now, the terms on the right, corresponding to ψ, λ, are of order (4.12) according to Hoeffding (1968), his formula (5.10). The same holds true for g, h, according to Hoeffding (1968), Propos. 2, and to our (4.5). Finally, for γ, η we have

$$ES_\gamma - \mu'_\gamma = \int_0^1 (ES_v - \mu'_v)\, d\gamma(v) \qquad (4.14)$$

(and analogously for η), where we denoted by S_v, μ'_v, the S_ϕ, μ'_ϕ from Theorem DH, 1, i.e. with $\phi(t) = u(t-v)$, thus

$$(ES_\gamma - \mu'_\gamma)^2 \leq [\gamma(1-) - \gamma(0+)] \int_0^1 (ES_v - \mu'_v)^2 \, d\lambda(v)$$

which is again of order (4.12) according to (4.7).

The assertion concerning μ follows again from (4.6), where now both differences are of the order $o(N^{\frac{1}{2}})$. For the first of them, it is assumed by (2.17), where eas for the second one it follows from Hoeffding (1968), his formula (4.12).

5.

Now we shall prove the theorem of § 3. We shall follow the original proof of Theorem DH, 1 (where it is at first assumed that $a(i) = \phi(i/(N+1))$), and making use of the new assumption (i) instead of DH, (2.12) and DH, (2.13) we shall strengthen the results of lemmas the proof is derived from, with respect to the rate of remainder terms.

First we observe that in Lemma DH, 5, it may be assumed that $K_{10} \geq \frac{3}{2}$ for $K_9 > 6\frac{1}{2}$ and for $N > N_0(K_9)$. (Here we have retained the original numbering of the K's.) Next in Lemma DH, 7, the order of the remainder may be sharpened to $o(N^{-\frac{3}{2}+\epsilon})$ for each $\epsilon > 0$, under (i). The proof (of the modified Lemma DH, 7) remains unaltered up to the formula DH, (4.13), with $K_{10} \geq \frac{3}{2}$ now, which can now be continued as

$$D^2 N^{-1} \iint_{-K_{11} \lg^{\frac{1}{2}} N \leq p < q \leq K_4 \lg^{\frac{1}{2}} N} \Phi(p)(1-\Phi(q)) L'_i(v+DN^{-\frac{1}{2}}p)$$

$$\times L'_j(v+DN^{-\frac{1}{2}}q) \, dp \, dq + D^2 N^{-1} \iint_{-K_{11} \lg^{\frac{1}{2}} N \leq q < p \leq K_{11} \lg^{\frac{1}{2}} N} \Phi(q)$$

$$\times (1-\Phi(p)) L'_i(v+DN^{-\frac{1}{2}}p) L'_j(v+DN^{-\frac{1}{2}}q) \, dp \, dq + D^2 N^{-\frac{3}{2}} \lg^{\frac{1}{2}} N$$

$$\times \iint_{I'} \vartheta_1 L'_i(v+DN^{-\frac{1}{2}}p) L'_j(v+DN^{-\frac{1}{2}}q) \, dp \, dq + \vartheta_2 N^{-K_{10}} \tag{5.1}$$

with ϑ_i uniformly bounded. As the products $\Phi(p)(1-\Phi(q))$ and $\Phi(q)(1-\Phi(p))$ are also of the order $O(N^{-K_{10}})$ outside the domain of integration, and owing to (i), (5.1) is further equal to

$$D^2 N^{-1} L'_i(v) L'_j(v) \left(\iint_{\{p<q\}} \Phi(p)(1-\Phi(q)) \, dp \, dq \right.$$

$$\left. + \iint_{\{p>q\}} \Phi(q)(1-\Phi(p)) \, dp \, dq \right) + O(N^{-2} \lg^2 N) + O(N^{-\frac{3}{2}} \lg^{\frac{3}{2}} N) + O(N^{-K_{10}}).$$

Taking into account that the sum of the two integrals in parentheses is equal to one, we obtain the desired sharpening.

Similarly, the remainder can be diminished to the order $o(N^{-\frac{3}{2}+\epsilon})$ in Lemma DH, 8 and DH, 9 and to the order $o(N^{-\frac{1}{2}+\epsilon})$ in Lemma DH, 10. Consequently, by the same argument as in Lemma DH, 11 and DH, 12, the upper order estimates

$$E(S - \hat{S})^2 = o(N^{-\frac{1}{2}+\epsilon} \sum_{i=1}^{N} (c_i - \bar{c})^2)$$

and

$$\operatorname{var}\left(S - \sum_{i=1}^{N} Z_i\right) = o\left(N^{-\frac{1}{2}+\epsilon} \sum_{i=1}^{N} (c_i - \bar{c})^2\right)$$

hold true. Thus, it remains to prove the asymptotic normality of $\sum_{i=1}^{N} Z_i$; but this follows from the boundedness of Z_i by $2 \max_{1 \leq j \leq N} |c_j - \bar{c}|$, from (3.1), (3.4) and from the Lindeberg Theorem. The additional statement on centering constants follows essentially from (3.10) which can be handled in a similar way as (5.1). The extension of all results to scores given by (2.7) is immediate. This concludes the proof of the theorem.

The proof of Remark 4 is quite similar to the proof of the implication $C_3 \Rightarrow C_1$ in DH, § 5.

As to the Example, we first observe that $H^{-1}(\frac{1}{2})$ is asymptotically equal to

$$\Delta + \Phi^{-1}\left(\frac{1 - 2\lambda_0}{2(1 - \lambda_0)}\right), \quad \text{where} \quad 0 < \frac{1 - 2\lambda_0}{2(1 - \lambda_0)} < \frac{1}{2}.$$

Hence, the satisfaction of the condition stated in Remark 4 easily follows, while the condition (3.1) is satisfied owing to the assumption on λ. Now, suppose (3.5) holds; then

$$\tilde{c}(\tfrac{1}{2}) = \frac{\lambda \Phi'(H^{-1}(\tfrac{1}{2}))}{\lambda \Phi'(H^{-1}(\tfrac{1}{2})) + (1 - \lambda) \Phi'_\Delta(H^{-1}(\tfrac{1}{2}))} \to 0$$

and

$$\sum_{i=1}^{N} (c_i - \tilde{c})^2 L_i(\tfrac{1}{2}) (1 - L_i(\tfrac{1}{2})) > m(1 - \tilde{c})^2 \Phi(H^{-1}(\tfrac{1}{2})) (1 - \Phi(H^{-1}(\tfrac{1}{2})))$$

$$> \text{const.} \, N e^{-\Delta^2/2} > N^{\frac{1}{2}+\epsilon} \quad \text{for some } \epsilon > 0.$$

This together with $\sum_{i=1}^{N} (c_i - \bar{c})^2 \simeq N\lambda_0(1 - \lambda_0)$

entails the satisfaction of (3.4).

Suppose, finally, that (3.6) holds; passing eventually to a subsequence, we may replace lim sup in (3.6) by lim inf. To prove the non-normality of S it suffices to prove that $P(S = 0)$ is bounded away from 0. This in turn will be entailed by proving that the probability of the simultaneous satisfaction of the inequalities $X_i < \Delta + \Phi^{-1}((1-2\lambda)/2(1-\lambda)), 1 \leqq i \leqq m$, as well as the probability that at most $N(\frac{1}{2}-\lambda)$ among the X_i's, $m \leqq i \leqq N$, are less than $\Delta + \Phi^{-1}((1-2\lambda)/2(1-\lambda))$, are both bounded away from 0.

But the former probability is equal to (for sufficiently small $\epsilon > 0$)

$$[\Phi(\Delta + \Phi^{-1}((1-2\lambda)/2(1-\lambda)))]^m > (1 - e^{-\Delta^2(1-\epsilon)/2})^m$$

$$> (1 - (1/N))^m \cong e^{-\lambda_0},$$

whereas the latter is given by the binomial distribution function with parameters $N(1-\lambda)$ and $(1-2\lambda)/2(1-\lambda)$ at the point $N(\frac{1}{2}-\lambda)$, and it is thus asymptotically equal to $\frac{1}{2}$.

REFERENCES

Dupač, V. and Hájek, J. (1969). Asymptotic normality of simple linear rank statistics under alternatives, II. *Ann. Math. Statist.* **40**, 1992–2017.

Dupač, V. and Hájek, J. (1969a). Asymptotic normality of the Wilcoxon statistic under divergent alternatives. *Applicationes Mathem.* **10**, 171–8.

Hájek, J. (1968). Asymptotic normality of simple linear rank statistics under alternatives. *Ann. Math. Statist.* **39**, 325–46.

Hoeffding, W. (1968). On the centering of a simple linear rank statistic. *Inst. of Statist. Mimeo Series No.* 585, University of North Carolina.

ON ROBUST RANK TESTS†

JOSEPH L. GASTWIRTH

1 INTRODUCTION

This paper continues our approach[6] to the problem of obtaining efficiency-robust rank tests when the experimenter is willing to assume that the density function of the population sampled is a member of a family $\mathscr{F} = \{f_i\}$ of densities. We compare the performance of any rank test D to the asymptotically most powerful rank test (a.m.p.r.t.) T_i for observations from the density f_i. Under fairly general conditions we showed that a rank test R exists which maximizes the minimum value or the ARE (asymptotic relative efficiency[16, 17]) of R to T_i on data from the density f_i over all the members f_i in $\mathscr{F}$ and over all rank tests. Our paper is complemented by the results of Birnbaum and Laska[2].

In general, the actual calculation of the maximin efficiency-robust rank test (m.e.r.r.t.), R, for a specific family $\mathscr{F}$ of densities is not easy. Sometimes there will be two densities in $\mathscr{F}$ which are most different and the m.e.r.r.t. for samples from either of these densities will be the m.e.r.r.t. for the entire family $\mathscr{F}$. When this occurs we call the two densities the extreme pair and $\mathscr{F}$ an extreme dominated family. In §2 we present some simple theorems which enable us to decide, under fairly general conditions, when a family is extreme-dominated. The last part of § 2 is devoted to the discussion of several examples of families of densities which illustrate the use of these results. In particular, we present an infinite family of densities for which the extreme pair can be analytically shown to satisfy the conditions of our results. Numerical studies indicate that the t-family of densities is extreme-dominated. Moreover, if one excludes the Cauchy density so that all the t densities have finite first moment, the normal and t_2 densities become the extreme pair and the minimum efficiency of the m.e.r.r.t. is over 90 % for all members of this family.

Several notions of ordering densities according to the mass in their tails have been proposed and studied[21], [11], [8]. In § 3 we discuss the applications of these notions to the development of efficiency robust rank

---

† Research supported by the National Science Foundation Research Grant No. GP 7118 awarded to the Department of Statistics, The Johns Hopkins University. This paper in whole or in part may be reproduced for any purpose of the United States Government.

[89]

tests for families which are totally ordered with respect to the strongest ordering notion considered. Examples are given illustrating that a family $\mathscr{F}$ can be extreme-dominated but not totally ordered and vice versa.

Section 4 is devoted to extending our results to the problem of censored and grouped data. While the theory extends readily one cannot convert the m.e.r.r.t. for data from complete samples to the m.e.r.r.t. for censored samples in any simple manner because censoring affects each member of $\mathscr{F}$ differently. Examples are given to illustrate the difficulties involved. Finally, we apply the results concerning grouped rank tests to obtain 'quick' or simple tests which are nearly as good as the m.e.r.r.t. One of the quick tests discussed is nearly 80 % as efficient as each a.m.p.r.t. for a family $\mathscr{F}$ containing the normal, Cauchy, double-exponential, logistic and t_2 densities, while the corresponding m.e.r.r.t. has minimum ARE 82·8 % for this family.

2   THE MAXIMIN-EFFICIENCY-ROBUST RANK TEST FOR THE TWO-SAMPLE PROBLEM

In this section we shall review and extend the theory presented earlier [6]. We begin by recalling some background material.

Let $x_1, \dots, x_m$ and $y_1, \dots, y_n$ be two independent random samples from absolutely continuous cdf's $F(x)$ and $G(x)$ respectively. We desire to test the null hypothesis $H_0: F(x) = G(x)$ against the alternative $H_1: F(x) \neq G(x)$ with a rank test. Let $Z_1, \dots, Z_N$ denote the combined sample, ordered from smallest to largest. The indicator r.v.'s δ_i, $i = 1, \dots, N$ are defined by

$$\delta_i = \begin{cases} 1 & \text{if } Z_i \text{ is a } y \text{ observation,} \\ 0 & \text{otherwise.} \end{cases} \tag{2.1}$$

Following Chernoff and Savage [4], a rank test can be specified by a function $J(u)$ defined on the interval $0 < u < 1$, by defining the test statistic as

$$T_N = \sum_{i=1}^{N} a_{iN} \delta_i, \tag{2.2}$$

where $a_{iN} = J(i/(N+1))$. In order that the statistic T_N has an asymptotic variance, when suitably normalized, we assume that $J(u)$ is in $L^2[0, 1]$, i.e. $\int_0^1 J^2(u)\, du < \infty$. Without loss of generality, we may assume that

$$\int_0^1 J(u)\, du = 0 \quad \text{and} \quad \int_0^1 J^2(u)\, du = 1.$$

We denote this set of normalized weight functions by Γ. Hájek[9] showed that the limiting Pitman efficiency of two tests based on the (normalized) weighting functions $J(u)$ and $L(u)$ is given by ρ^2 where

$$\rho = \langle J, L \rangle = \int_0^1 L(u) J(u) \, du, \qquad (2.3)$$

provided that $\rho > 0$ and that one of the two tests is the a.m.p.r.t. for the problem considered. For the two-sample location problem with

$$G(x) = F(x - \theta),$$

the a.m.p.r.t. is specified by the function

$$J(u) = -I_l^{-\frac{1}{2}} \frac{f'[F^{-1}(u)]}{f[F^{-1}(u)]}, \qquad (2.4)$$

where $\quad f(x) = F'(x) \quad$ and $\quad I_l = \displaystyle\int_{-\infty}^{\infty} [f'(x)/f(x)]^2 f(x) \, dx$

is Fisher's information about the location parameter.

The criterion of robustness we adopt combines the minimax principle and Pitman efficiency as follows:

Definition 2.1. Let $\mathscr{F} = \{f_i, i \in I\}$, where I is an arbitrary index set, be a family of density functions with a.m.p.r.t.'s specified by weight functions $\{J_i\}$ respectively. The rank test specified by the function $R(u)$ is called the maximin efficiency robust rank test (m.e.r.r.t.) if

$$\sup_{L \in \Gamma} \inf_{i \in I} \langle L, J_i \rangle^2 = r^2 \qquad (2.5)$$

is achieved when $L(u) = R(u)$.

The basic existence and uniqueness theorem is as follows:

Theorem 2.1. *If there is a vector S in Γ satisfying* $\liminf \langle S, J_i \rangle = \zeta > 0$, *then a unique function $R(u)$ in Γ exists achieving the*

$$\sup_{L \in \Gamma} \inf_{i \in I} \langle L, J_i \rangle = r. \qquad (2.6)$$

The test based on $R(u)$ is the m.e.r.r.t. Moreover, R is a renormalized member of the convex hull generated by the functions $\{J_i\}$.

The existence part of Theorem 2.1 was proved earlier[6]. A simple proof of uniqueness is the following. Suppose R and Q are two distinct functions in Γ whose rank tests have the same maximin efficiency (or correlation r). A test based on the function $P = (1 - \epsilon) R + \epsilon Q$, where $0 < \epsilon < 1$ has inner product $\geqslant r$ with every J_i. Moreover,

$$\| (1 - \epsilon) R + \epsilon Q \|^2 = (1 - \epsilon)^2 + \epsilon^2 + 2\epsilon(1 - \epsilon) \langle R, Q \rangle. \qquad (2.7)$$

As R and Q are distinct vectors with norm 1, the Schwarz inequality implies that $\langle R, Q \rangle < 1$, i.e. $\|P\| < 1$. Upon renormalizing P to have norm 1 we obtain a test with a higher minimum efficiency than R and Q have.

In the examples discussed in [6] and [2], it often happened that the m.e.r.r.t. for the two 'most different' members of $\mathscr{F}$ remained the m.e.r.r.t. over all $\mathscr{F}$. In order to formalize the notion of 'most different' members of $\mathscr{F}$, we make the following definition.

Definition 2.2. Let $\mathscr{F} = \{f_i\}$ be a family of densities whose corresponding a.m.p.r.t.'s are generated by the functions $\{J_i\}$. If a pair of densities f_1 and f_2 say, satisfy
$$\rho_{12} = \min_{i,j} \rho_{ij} > 0, \tag{2.8}$$
where $\rho_{ij} = \langle J_i, J_j \rangle$, they will be called an *extreme pair* of densities in $\mathscr{F}$.

Although a family $\mathscr{F}$ of densities can have several (or no) *extreme pairs*, the most interesting situation occurs when a unique extreme pair exists. We are now ready to prove Theorem 2.2.

Theorem 2.2. *Let $\mathscr{F} = \{f_i\}$ be a family of possible densities and let f_1 and f_2 be an extreme pair. In order that the m.e.r.r.t. for samples from either f_1 or f_2, which is based on the function*
$$R_{12}(u) = [2(1+\rho_{12})]^{-\frac{1}{2}} (J_1(u) + J_2(u)), \tag{2.9}$$
be the m.e.r.r.t. for the whole family $\mathscr{F}$ with the same minimum efficiency, $(1+\rho_{12})/2$, it has on data from f_1 or f_2 it is necessary and sufficient that for each $i \in I$
$$\rho_{1i} + \rho_{2i} \geqslant 1 + \rho_{12}. \tag{2.10}$$

Proof. Suppose (2.10) holds. Since R_{12} is the m.e.r.r.t. for samples from either f_1 or f_2 no test can have greater minimum efficiency than $(1+\rho_{12})/2$ over $\mathscr{F}$. From (2.10), it follows that for any J_i
$$\langle R_{12}, J_i \rangle = \frac{\rho_{1i} + \rho_{2i}}{\sqrt{(2(1+\rho_{12}))}} \geqslant \frac{1+\rho_{12}}{\sqrt{(2(1+\rho_{12}))}} = [(1+\rho_{12})/2]^{\frac{1}{2}}, \tag{2.11}$$
i.e. $\langle R_{12}, J_i \rangle \geqslant \langle R_{12}, J_1 \rangle = \langle R_{12}, J_2 \rangle$. Thus, R_{12} is the m.e.r.r.t. for $\mathscr{F}$ and attains its minimum efficiency on data from either f_1 of f_2. Conversely, if R_{12} is the m.e.r.r.t. for $\mathscr{F}$ with minimum efficiency $((1+\rho_{12})/2)$, then for any J_i,
$$\langle R_{12}, J_i \rangle = (\rho_{1i} + \rho_{2i})/[2(1+\rho_{12})]^{\frac{1}{2}} \geqslant [(1+\rho_{12})/2]^{\frac{1}{2}} \tag{2.12}$$
which implies that (2.10) holds.

A natural question to ask is whether the m.e.r.r.t. for an extreme pair can be the m.e.r.r.t. for the entire family $\mathscr{F}$ with a minimum efficiency less than $(1+\rho_{12})/2$. The following result indicates that this can happen only under very special circumstances. Precisely, we have Theorem 2.3.

Theorem 2.3. If $\mathscr{F}$ is a finite *family of densities satisfying the conditions of Theorem 2.1, whose corresponding J functions are linearly independent, then when the test based on the function R_{12} is maximin over $\mathscr{F}$, its minimum efficiency is $(1+\rho_{12})/2$. Furthermore, if R_{12} is maximin over $\mathscr{F}$ with minimum efficiency less than $(1+\rho_{12})/2$, then the J functions are linearly dependent.*

Proof. Consider the family $\mathscr{F}'$ consisting of the densities of $\mathscr{F}$ with f_1 and f_2 deleted. By Theorem 2.1, there is a unique test R' with minimum efficiency $(r')^2$ over $\mathscr{F}'$. Suppose $r' \geqslant [(1+\rho_{12})/2]^{\frac{1}{2}}$. If R_{12} were maximin over $\mathscr{F}$ with minimum correlation $r < [(1+\rho_{12})/2]^{\frac{1}{2}}$, this value, r, would be achieved at a member of $\mathscr{F}'$. For a suitably chosen $\epsilon > 0$, the test based on the function $(1-\epsilon)R_{12}+\epsilon R'$ (renormalized) would have minimum efficiency over $\mathscr{F}$ larger then r^2 contradicting the assumption that R_{12} is maximin. Next, suppose that $r' < [(1+\rho_{12})/2]^{\frac{1}{2}}$. The assumption of independence implies that $R_{12} \neq R'$ so $r < r'$ and again a suitable test based on $(1-\epsilon)R_{12}+\epsilon R'$ has higher minimum efficiency over $\mathscr{F}$ than R_{12} has. Finally, suppose that R_{12} is maximin over $\mathscr{F}$ with minimum efficiency $r^2 < [(1+\rho_{12})/2]$. This implies that $r = r'$ (otherwise a test based on $(1-\epsilon)R_{12}+\epsilon R'$ dominates R_{12}). By uniqueness of the maximin test R' over $\mathscr{F}'$, $R_{12} = R'$ which means that the J functions are dependent.

While examples (with at least five members in $\mathscr{F}$) can be constructed whose m.e.r.r.t. is the m.e.r.r.t. for the extreme pair with maximin efficiency less than $(1+\rho_{12})/2$, the condition of linear independence is not restrictive from a practical point of view. Also, if there is a dependence relation $J = \Sigma c_i J_i$, with $c_i \geqslant 0$, then any test R which has minimum correlation r with every J_i has correlation at least r with J as

$$\langle R, J \rangle \geqslant r(\Sigma c_i) \quad \text{and} \quad 1 = \|J\| = \Sigma\Sigma c_i \rho_{ij} c_j \leqslant (\Sigma c_i)^2.$$

Moreover, the restriction to finite families is not essential. Thus, we can state Corollary 2.1.

Corollary 2.1. If the conditions of Theorem 2.2 fail to hold and if R_{12} is not in the closed linear subspace of $L^2[0, 1]$ generated by the remaining J functions, then R_{12} is not the m.e.r.r.t. for $\mathscr{F}$.

The reader may ask why we consider only the extreme pair of densities. This is answered by Proposition 2.1.

Proposition 2.1. Under the assumptions of Theorem 2.3, the m.e.r.r.t. R_{34} for a non-extreme pair of densities f_3 and f_4, say, cannot be the m.e.r.r.t. over $\mathscr{F}$.

Proof. Clearly the maximin efficiency of R_{34} over $\mathscr{F}$ is

$$\leqslant (1+\rho_{12})/2 < (1+\rho_{34})/2.$$

Letting $\mathscr{F}''$ be $\mathscr{F}$ with f_3 and f_4 deleted the result follows by the same reasoning used in proving Theorem 2.3.

The theory developed above is general and it holds for rank tests for location change or scale change. For the remainder of this section we shall consider the two-sample shift (location) problem and assume that the family $\mathscr{F}$ of possible densities is a subset of the class of symmetric and unimodal densities. Usually we guess which pair of densities is the extreme pair in $\mathscr{F}$ and then see if Theorem 2.2 is applicable. Our first example is an infinite family of densities for which condition (2.10) can be verified analytically.

Example 2.1. Consider the sequence $\{f_n\}$ of densities defined by

$$f_n(x) = \frac{1}{2}\frac{n-1}{n}\frac{1}{\left(1+\dfrac{|x|}{n}\right)^n} \quad (n = 2, 3, \ldots); \quad f_\infty(x) = \tfrac{1}{2}e^{-|x|}. \quad (2.13)$$

As $\lim\limits_{n\to\infty} f_n(x) = f_\infty(x)\,\forall x$, the double-exponential distribution is a natural member of $\mathscr{F}$. Routine calculation shows that the J functions generating the a.m.p.r.t.'s are given by

$$J_n(u) = [(n+1)/(n-1)]^{\frac{1}{2}}[2(1-u)]^{1/(n-1)},$$

for $u > \frac{1}{2}$ and $-[(n+1)/(n-1)]^{\frac{1}{2}}[2u]^{1/(n-1)}$ for $u < \frac{1}{2}$; $n = 1, 2, 3, \ldots$ and $J_\infty(u) = +1$ if $u > \frac{1}{2}$ and -1 if $u < \frac{1}{2}$. (Note that $J_n(u) \to J_\infty(u)$ as $n \to \infty$.) Finally, we need the fact that

$$\rho_{nm} = \langle J_n, J_m \rangle = [(n^2-1)(m^2-1)]^{\frac{1}{2}}/[nm-1]. \quad (2.14)$$

In particular,

$$\rho_{2m} = \sqrt{3}\sqrt{(m^2-1)}/\sqrt{(2m-1)}, \quad \rho_{\infty m} = (m^2-1)^{\frac{1}{2}}/m \quad \text{and} \quad \rho_{2\infty} = \sqrt{3}/2.$$

Since $\rho_{\infty m}$ is an increasing function of m, it attains its minimum when $m = 2$. To show that f_2 and f_∞ form the extreme pair it suffices to show that $\rho_{nm} \geqslant \rho_{\infty m}$ whenever $n > m$. Since

$$(m^2-1)^{\frac{1}{2}} \geqslant (m-1)^{\frac{1}{2}} \quad \text{and} \quad (n-1/m)^2 \geqslant n^2-1$$

for $m \geqslant 2$,

$$\rho_{nm} = [m^2-1]^{\frac{1}{2}}[n^2-1]^{\frac{1}{2}}/[m(n-1/m)] \geqslant (m^2-1)^{\frac{1}{2}}/m = \rho_{\infty m}. \quad (2.15)$$

The m.e.r.r.t. for observations from either f_2 or f_∞ is generated by the weight function $R(u) = [J_2(u) + J_\infty(u)]/[2 + \sqrt{3}]^{-\frac{1}{2}}$. The verification that $R(u)$ is the m.e.r.r.t. for $\mathscr{F}$ reduces to showing that

$$1 + \tfrac{1}{2}\sqrt{3} \leqslant (m^2 - 1)^{\frac{1}{2}} [(\sqrt{3}/2m - 1) + m^{-1}]. \tag{2.16}$$

Squaring both sides of (2.16) and rearranging terms shows that (2.16) holds if and only if

$$g(m) = m^3 \sqrt{3}(2 + \sqrt{3}) - [31 + 20\sqrt{3}] m^2/4 - 2(2 + \sqrt{3}) m + 1 \geqslant 0 \quad \forall m. \tag{2.17}$$

When $m = 2$, $g(m) = 0$. Differentiation shows that $g''(m) > 0$ for $m > [31 + 20\sqrt{3}]/[12(3 + \sqrt{3})]$ so that g is convex in the region $[2, \infty)$. As $g'(2) > 0$, convexity implies that g assumes its minimum in the region $[2, \infty)$ at $m = 2$.

Example 2.2. A statistically natural class of symmetric densities to study is the t-family. They vary from the heavy-tailed Cauchy density to the normal density and defined by

$$f_k(x) = \frac{1}{\sqrt{k\pi}} \frac{\Gamma(\frac{1}{2}k + \frac{1}{2})}{\Gamma(\frac{1}{2}k)} (1 + x^2/k)^{-(k+1)/2} \quad (k = 1, 2, \ldots); \quad f_\infty(x) = (2\pi)^{-\frac{1}{2}} e^{-x^2/2}. \tag{2.18}$$

In view of the fact that $f_k(x) \to f_\infty(x)$ as $k \to \infty$, one might hope that the m.e.r.r.t. for the extreme pair (Cauchy and normal) remains the m.e.r.r.t. for the entire family $\mathscr{F}$. We have not been able to prove this analytically, but have computed numerically the correlations of the Cauchy scores test (C), normal scores test (N), and the m.e.r.r.t. for samples from either the Cauchy or normal data $(R^* = (1\cdot819)^{-1}(C+N))$ for various values for the degrees of freedom (k). These computations, given in Table 2.1,

Table 2.1. *Correlations and ARE's of the tests,*
C, N, R^* on the t-family

k (df)	$\langle N, T_k \rangle$	$\langle N, T_k \rangle^2$	$\langle C, T_k \rangle$	$\langle C, T_k \rangle^2$	$\langle R^*, T_k \rangle^2$	$\langle R^*, T_k \rangle$	$\frac{\langle N, T_k \rangle}{+ \langle C, T_k \rangle}$
1	0·6559	0·4302	1·000	1·000	0·9103	0·8286	1·6559
2	0·8370	0·700	0·940	0·8846	0·9769	0·9543	1·7770
3	0·9053	0·8196	0·8801	0·7746	0·9815	0·9633	1·7854
4	0·9382	0·8802	0·8389	0·7038	0·9770	0·9545	1·7771
6	0·9677	0·9364	0·7888	0·6222	0·9656	0·9324	1·7565
8	0·9802	0·9608	0·7600	0·5776	0·9567	0·9153	1·7402
10	0·9867	0·9736	0·7414	0·5497	0·9500	0·9025	1·7281
20	0·9963	0·9926	0·7010	0·4914	0·9331	0·8707	1·6973
30	0·9983	0·9966	0·6865	0·4713	0·9262	0·8578	1·6848
100	0·9999	0·9998	0·6654	0·4428	0·9155	0·8381	1·6653
∞	1·000	1·000	0·6559	0·4302	0·9103	0·8286	1·6559

strongly suggest that the test R^* is the m.e.r.r.t. for the entire t-family. When $k \neq 1$, we shall denote the a.m.p.r.t. for data from the t_k density by T_k and the function specifying it by T_k or $T_k(u)$. Since the last column in Table 2.1 is always $\geqslant 1 \cdot 6559$, Theorem 2.2 applies to the finite set of t densities we used in the computation. Furthermore, the numbers in the next to last column which are the ARE's of R^* versus T_k when the observations are from the t_k density decrease (when $k \geqslant 6$) to their value at $k = \infty$. This pattern indicates that the test R^* is the m.e.r.r.t. for the entire family. Since explicit expressions for the weight functions specifying the a.m.p.r.t.'s for the members of the t family are difficult to obtain, the correlations were computed from the integrals for the efficacy of the various tests. However, we were able to obtain the weight function $(T_2(u))$ generating the a.m.p.r.t. for data from the t_2 density. Indeed

$$T_2(u) = \sqrt{30}\,(u - \tfrac{1}{2}\sqrt{(1 - 4(u - \tfrac{1}{2})^2)}). \qquad (2.19)$$

The form of $T_2(u)$ is quite interesting as for u near $\tfrac{1}{2}$, $T_2(u)$ has the same shape as the Wilcoxon test (which is generated by $W(u) = 2\sqrt{3}\,(u - \tfrac{1}{2})$) but as $u \to 1$, $T_2(u)$ approaches 0. For future reference we list the correlations of T_2 with some common rank tests: Median ($\sqrt{5/6} = 0 \cdot 9129$), Wilcoxon ($\pi\sqrt{90}/32 = 0 \cdot 9314$), Normal scores ($0 \cdot 8118$), Cauchy scores ($\sqrt{15}\,J_2(\pi)/2 = 0 \cdot 9400$), Psi test ($\pi^{-2}\sqrt{6}\,\mathrm{cin}\,(2\pi) = 0 \cdot 60499$) and shall abbreviate the tests mentioned by M, W, N, C, and ψ, respectively. The fact that the ARE of the T_2 test versus the normal scores test on normal data is substantially higher ($0 \cdot 66$ compared to $0 \cdot 43$) than the corresponding ARE of Cauchy scores test suggests that we consider Example 2.3.

Example 2.3. Consider the t densities which have finite mean, i.e. $\mathscr{F}$ consists of the t_k densities for $k \geqslant 2$, i.e. only the Cauchy density has been deleted from the previous example. The m.e.r.r.t. R' for samples from either the t_2 or normal densities has efficiency $0 \cdot 9185$. Moreover, numerical computation again indicates that the test $R' = c(N + T_2)$, where c is a normalizing constant, is the m.e.r.r.t. for the entire family. Table 2.2 gives the correlations $\langle T_2, T_k \rangle$ and $\langle N, T_k \rangle$.

Table 2.2. *The correlations* $\langle T_2, T_k \rangle$ *and* $\langle N, T_k \rangle$

k	2	3	4	5	6	8	10	20	30	100	∞
$\langle T_2, T_k \rangle$	1	0·987	0·969	0·953	0·940	0·921	0·907	0·875	0·863	0·845	0·837
$\langle N, T_k \rangle$	0·837	0·9053	0·938	0·956	0·967	0·980	0·986	0·996	0·998	0·998	1

Again it appears that the t_2 and normal densities will be the extreme pair and Theorem 2.2 is satisfied for the densities in Table 2.2. This example is of interest because the minimum relative efficiency is quite high (91·8 %).

For future brevity we shall call any family $\mathscr{F}$ of densities whose m.e.r.r.t. is the m.e.r.r.t. for its extreme pair, extreme-dominated. The results in [6] imply that the following finite families of densities are extreme-dominated.

Example 2.4. Let $\mathscr{F}$ consist of the double-exponential, logistic and normal densities. The normal and double-exponential densities form the extreme pair and the minimum efficiency is 0·8989.

Example 2.5. Let $\mathscr{F}$ consist of the Cauchy, t_2, double-exponential, logistic and normal densities. The extreme pair is the normal and Cauchy and its minimum efficiency over $\mathscr{F}$ is 82·8 %. Thus, we can add the double-exponential and logistic densities to the t-family (Example 2.2) and the simple m.e.r.r.t., R^*, for Cauchy or normal data remains the m.e.r.r.t. for the entire family.

So far we have emphasized examples of extreme-dominated families. As a general rule most families will not be extreme-dominated. Three simple examples are the following:

Example 2.6. Let $\mathscr{F}$ contain the double-exponential, Cauchy and logistic densities. The Cauchy and logistic densities form the extreme pair. (The correlation $\langle W, C \rangle = (6/\pi^2)^{\frac{1}{2}} \simeq 0.7797$ which corrects a misprint in Table 1 of [6].) Since $\langle W, M \rangle + \langle M, C \rangle = 1.766 < 1.78$, Theorem 2.2 fails to hold.

Example 2.7. Let $\mathscr{F}$ consist of the double-exponential, t_2 and logistic densities. The logistic and double-exponential densities form the extreme pair but Theorem 2.2 fails to hold.

Example 2.8. Let $\mathscr{F}$ consist of the normal, t_2 and double-exponential densities. The extreme pair is the double-exponential and normal but condition (2.10) is not satisfied.

Our next example shows that even if we restrict our attention to families of densities which are strongly unimodal, i.e. $-\log f(x)$ is convex, Theorem 2.2 can fail to hold.

Example 2.9. Let $f(x)$ be any symmetric strongly unimodal density whose a.m.p.r.t. is specified by $J(u)$ and assume that $J(u)$ is strictly increasing on $(\frac{1}{2}, 1)$, i.e. $-\log f(x)$ is strictly convex. For observations from $f(x)$

4

which are censored at the αth and $(1-\alpha)$th fractiles of the combined sample, the a.m.p.r.t. is determined by [6]

$$J_\alpha(u) \begin{cases} c, \quad 1-\alpha < u < 1; \quad -c, 0 < u < \alpha; \\ -\alpha f'[F^{-1}(u)]/f[F^{-1}(u)], \quad \alpha < u < 1-\alpha; \end{cases} \qquad (2.20)$$

where $\gamma = I(\alpha)^{-\frac{1}{2}}$, $c = \gamma\alpha^{-1}f[F^{-1}(1-\alpha)]$ and $I(\alpha)$ is the information concerning the location parameter contained in the order observations between αth and $(1-\alpha)$th sample fractiles. As α varies from 0 to $\frac{1}{2}$, $J_\alpha(u)$ varies from $J(u)$, the a.m.p.r.t. for $f(x)$, to the median test, the a.m.p.r.t. for double-exponential data. Moreover, $J_\alpha(u)$ specifies the a.m.p.r.t. for a density $f_\alpha(x)$ which is a rescaled version of $f(x)$ in the middle with exponential tails [7]. The exact form of $f_\alpha(x)$ is obtained by applying Lemma 1 in Chapter I of Hájek and Šidák's treatise [10] or its equivalent [6, p. 938] and is given by

$$f_\alpha(x) = \begin{cases} \frac{1}{2}ce^{-c|x|}, \quad |x| > F_\alpha^{-1}(1-\alpha) = -c^{-1}\ln(2\alpha); \\ \gamma f(\gamma x), \quad |x| < F_\alpha^{-1}(1-\alpha). \end{cases} \qquad (2.21)$$

As α varies we obtain a family of symmetric strongly unimodal densities which lie 'in between' $f(x)$ and the double-exponential. For convenience $I(\frac{1}{2})$ will denote the information contained by the median and M will denote the function specifying the median test.

In general, if $\alpha < \beta$, $\langle J_\alpha, J_\beta \rangle = [I(\beta)/I(\alpha)]^{\frac{1}{2}}$. Since $I(\beta)$ is monotone the smallest value of $\langle J_\alpha, J_\beta \rangle$ occurs when $\alpha = 0$ and $\beta = \frac{1}{2}$ so that the double-exponential and $f(x)$ are the extreme pair.

In order for the family $\mathscr{F}$ to be extreme dominated the following inequality would have to be satisfied for all α:

$$\langle M, J_\alpha \rangle + \langle J, J_\alpha \rangle \geqslant 1 + \langle M, J \rangle. \qquad (2.22)$$

Algebraic manipulation shows that (2.22) is equivalent to

$$I^{\frac{1}{2}}(\alpha) \leqslant I^{\frac{1}{2}}(\tfrac{1}{2}). \qquad (2.23)$$

Since $I(\frac{1}{2}) \leqslant I(\alpha)$ and equality for any α holds only when the median contains as much information as all the ordered observations between the αth and $(1-\alpha)$th fractiles, (2.23) cannot hold for any α provided that $J(u)$ is strictly increasing on $(\frac{1}{2}, 1)$. Thus, the inclusion of any f_α destroys the possibility of having an extreme-dominated family $\mathscr{F}$.

In an interesting numerical study, Birnbaum et al. [3] obtain the maximin efficient linear unbiased estimators (m.l.u.e.'s) over families of densities for small sample sizes. Often the m.l.u.e. for a family of densities was the m.l.u.e. for some pair of shapes. Sometimes the extreme pair of densities for the rank test problem is the extreme pair for their problem but this does not hold generally.

3 THE RELATIONSHIP OF VARIOUS NOTIONS OF ORDERED FAMILIES OF SYMMETRIC DENSITIES TO EFFICIENCY ROBUSTNESS PROBLEMS

Many of the 'quick' statistical procedures are based on the 'folklore' that the lighter the tail of a density, the greater is the proportion of information contained in the extreme observations. In order to formalize this we require a reasonable way of comparing the tail probabilities of various densities. Both van Zwet[21] and Hájek[11] have proposed different partial orderings of symmetric densities according to their tail probabilities. Hájek[11], and independently the author[8], have considered a related partial ordering of rank tests according to the emphasis they place on the extreme ranks relative to the middle ones. In this section we give several examples illustrating the various concepts and suggest that while Hájek's notion of order of rank tests coupled with a dual notion of order of *symmetric* densities may be useful for efficiency robustness problems concerning the *location* parameter, no notion of order yet proposed appears to be entirely satisfactory (see Theorem 3.2).

Hájek orders rank tests T_1 and T_2 according to the relative growth of the weight function J_1 and J_2 which specify them. Formally, he makes the following definition.

Definition 3.1 (Hájek). $J_1(u)$ *increases more rapidly than* $J_2(u)$ if

$$J_1(u) = b(u) J_2(u) \quad (\tfrac{1}{2} < u < 1), \tag{3.1}$$

where $b(u)$ is nondecreasing and nonconstant.

We now list three notions of ordering symmetric densities according to the mass in their tails:

Definition 3.2 (Hájek). 'F has shorter tails than $H(F <' H)$' if

$$F^{-1}(u) = a(u) H^{-1}(u) \quad (u > \tfrac{1}{2}), \tag{3.2}$$

where $a(u)$ is a nonincreasing function which is not constant.

Definition 3.3 (van Zwet). F is s-smaller than $H(F \underset{s}{<} H)$ if

$$c(u) = \frac{H^{-1}(u)'}{F^{-1}(u)'} = \frac{f[F^{-1}(u)]}{h[H^{-1}(u)]} \tag{3.3}$$

is nondecreasing on $\tfrac{1}{2} < u < 1$.

Definition 3.4 [8]. F has more information (about the location parameter) in the tails than H ($F < H$) if J_f increases more rapidly than J_h, where J_f and J_h are the functions specifying the a.m.p.r.t.'s for f and h, respectively.

In his text Hájek essentially shows that $F < H$ implies $F \underset{s}{<} H$ which implies $F <' H$. The converses fail to hold. Indeed the family of densities $f_\alpha(x)$ discussed in Example 2.9 are ordered in van Zwet's sense but not according to Definition 3.4.

In [8], the author proved the following:

Theorem 3.1. *If G and H are symmetric unimodal densities with finite Fisher information such that $G < H$ and if the rank tests T_1 and T_2 are based on the weight functions J_1 and J_2 where J_1 increases more rapidly (in Hájek's sense) than J_2 and both J_1 and J_2 specify a.m.p.r.t.'s for some symmetric unimodal density, then*

$$A[T_2, T_1 | G] \leqslant A[T_2, T_1 | H], \tag{3.4}$$

where $A[T_2, T_1 | G]$ is the asymptotic relative efficiency of the test T_2 compared to the test T_1 when the observations are from $G(x)$.

Intuitively, Theorem 3.1 says that if T_2 places more emphasis on the middle ranks that T_1, T_2 performs (relatively) better on data from heavy tailed distributions. Applying Theorem 3.1 with $h = h$, $g = g$, $T_1 =$ the test based on J_f and $T_2 =$ the test based on J_g yields the following

Corollary 3.1. *If $f < g < h$ are symmetric unimodal densities whose a.m.p.r.t.'s are generated by J_f, J_g and J_h, respectively, then*

$$\langle J_f, J_h \rangle \leqslant \langle J_g, J_h \rangle \langle J_f, J_g \rangle \leqslant \min \{ \langle J_g, J_h \rangle, \langle J_f, J_g \rangle \}. \tag{3.5}$$

We are now ready to establish some connections between the notion of order given in Definition 3.4 and the theory presented earlier. In particular, we have

Theorem 3.2. *If $\mathscr{F} = \{f_i\}$ is a family of densities which is totally ordered y the relation $<$ and if J_f and J_h specify the a.m.p.r.t.'s for a pair of densities which are extreme with respect to the $<$ ordering, i.e. $\forall i \in \mathscr{F} f < f_i < h$ or equivalently J_f/J_i and J_i/J_h are non-decreasing, then f and h are the extreme pair of densities according to Definition 2.2, i.e.*

$$\langle J_f, J_h \rangle = \min \langle J_i, J_j \rangle.$$

Proof. Corollary 3.1 implies that both $\langle J_i, J_f \rangle$ and $\langle J_i, J_f \rangle$ are $\geqslant \langle J_f, J_h \rangle$. For any pair of densities f_i and $f_j \in \mathscr{F}$ either

$$f < f_i < f_j < h \quad \text{or} \quad f < f_j < f_i < h$$

as $\mathscr{F}$ is totally ordered. It suffices to consider the first case. Applying Corollary 3.1 to $f < f_i < f_j$ yields $\langle J_i, J_f \rangle \geqslant \langle J_j, J_f \rangle$. Similarly

$$\langle J_j, J_f \rangle \geqslant \langle J_f, J_h \rangle \quad \text{so that} \quad \langle J_i, J_f \rangle \geqslant \langle J_f, J_h \rangle.$$

For families which are totally ordered by the $<$ relation we have the following sufficient condition for the maximin test for the extreme pair to be maximin for the entire family with the same minimum efficiency:

Theorem 3.3. Under the conditions of Theorem 3.2, let the test R based on

$$R(u) = k[J_f + J_h], \tag{3.6}$$

where $k = [2(1 + \langle J_f, J_h \rangle)]^{-\frac{1}{2}}$ be the a.m.p.r.t. for a density r. If r can be added to $\mathscr{F}$ and the enlarged family remains totally ordered by the $<$ relation, then R remains the maximin test for the enlarged family.

Proof. The result follows from Theorem 2.2 once it is verified that

$$\langle J_i, J_f \rangle + \langle J_i, J_f \rangle \geqslant 1 + \langle J_f, J_h \rangle. \tag{3.7}$$

For any $f_i \in \mathscr{F}$ either $f < f_i < r$ or $r < r_i < h$. It suffices to treat the first case. Corollary 3.1 implies that $\langle J_i, J_f \rangle \langle J_j, R \rangle \geqslant \langle R, J_f \rangle$. Computing the correlations yield

$$\langle J_i, J_f \rangle [\langle J_i, J_f \rangle + \langle J_i, J_h \rangle] \geqslant 1 + \langle J_h, J_f \rangle. \tag{3.8}$$

As $\langle J_i, J_f \rangle \leqslant 1$, (3.7) holds.

Originally, I had hoped that there would be several relationships between various notions of order and efficiency robustness problems. Even when a family is *extreme-dominated*, however, it need not be totally ordered by the $<$ relation. The family discussed in Example 2.5 illustrates this as the t_2 density is not comparable to the double-exponential in the sense of $<$ or van Zwet's ordering.

We now give an example of a family of densities which is totally ordered by the $<$ relation and is not extreme dominated. We shall use the percentile-modified tests [5] which are based on the functions $J_p(u)$, where $J_p(u) = [3/2p^3]^{\frac{1}{2}}(u - (1-p))$ for $1-p < u < 1$, $p \leqslant \frac{1}{2}$; 0 if $p < u < 1-p$; and $[3/2p^3]^{\frac{1}{2}}(u-p)$ for $0 < u < p$. Let f_p denote the density function for which J_p generates the a.m.p.r.t. We consider the family consisting of f_p, the logistic, and the double-exponential density. Clearly, double-exponential $<$ logistic $< f_p$ and the relevant correlations are: $\langle T_p, M \rangle = [3p/2]^{\frac{1}{2}}$, $\langle T_p, W \rangle = (3 - 2p)[p/2]^{\frac{1}{2}}$ and $\langle W, M \rangle = [3/2]^{\frac{1}{2}}$. Thus, for any triple consisting of the logistic, the double-exponential and an f_p, $p \neq \frac{1}{2}$, the double-exponential and the f_p density form the extreme pair. However, the choice of p determines whether or not the extreme pair dominates the family, since the extreme pair dominates if and only if

$$1 + (3p/2)^{\frac{1}{2}} \leqslant [3/2]^{\frac{1}{2}} + (3 - 2p)\,(p/2)^{\frac{1}{2}}. \qquad (3.9)$$

If $p = \frac{1}{3}$, (3.9) holds but when p is sufficiently small, e.g. $p \leqslant 0\cdot01$, (3.9) fails to hold. Thus, neither total ordering of $\mathscr{F}$ or extreme domination imply the other. Finally, we remark that when $p = \frac{1}{3}$, the maximin test does not satisfy the condition of Theorem 3.3 as it is flat in $(\frac{1}{2}, \frac{2}{3})$ so that the Wilcoxon test rises faster than it does in that interval but not when u is $> \frac{2}{3}$. Thus, the condition given in Theorem 3.3 is not necessary and probably not particularly useful.

4   ROBUST RANK TESTS FOR CENSORED AND GROUPED DATA

In some area of applied statistics, such as clinical trials (life testing), an experimenter often obtains censored data. In an earlier paper [7] we showed how to obtain the a.m.p.r.t. for censored data from a particular density $f(x)$ from the a.m.p.r.t. for a complete sample from $f(x)$. Indeed, the a.m.p.r.t. for censored data uses the same scores as the a.m.p.r.t. for complete data for the uncensored portion of the data and scores each missing observation by the average of the scores the original a.m.p.r.t. places on the censored part. In [6], we noted that the a.m.p.r.t. for grouped data, where all the observations between the λ_ith and λ_{i+1}th fractiles of the combined sample receive the same score, is obtained similarly. The purpose of this section is to show that while the general theory developed earlier extends to censored or grouped data, the m.e.r.r.t. for these cases cannot be obtained from the m.e.r.r.t. for complete data in the same manner as the a.m.p.r.t. for censored (or grouped) data is obtained from the a.m.p.r.t. for complete data. This is due to the fact that censoring (or grouping) affects each of the members of a family $\mathscr{F}$ of densities differently.

The tests which are a.m.p.r.t.'s for grouped data are very simple and can be regarded as generalizations of the median test. We discuss this briefly and show that some simple grouped rank tests may be quite useful in practice. Moreover, our results show why the modifications of the median test suggested by Massey[13] and Mathisen[14] do not result in a substantial increase in power on normal data[1].

a. Extension of the theory

In order to sketch the necessary modifications of our approach which are required to cover the cases of censored and grouped data, we make the following definition.

Definition 4.1. Let $\mathcal{F} = \{f_i, i \in I\}$, I an index set, be a family of density functions whose a.m.p.r.t.'s (for complete samples) are specified by functions $\{J_i\}$. Let K be any subspace of $L^2(0, 1)$ and let K^* be the set of tests specified by the elements of K after normalization. The rank test R^* is the m.e.r.r.t. in K^* if

$$\sup_{L \in K^*} \inf_{i \in I} \langle L, J_i \rangle = \inf_{i \in I} \langle R^*, J_i \rangle > 0. \tag{4.1}$$

(We have identified a test with the function specifying it.)

For the problem of censored data the subspace K of $L^2[0, 1]$ consists of those members of $L^2[0, 1]$ which are constant on the censored intervals. For grouped observations where the grouping is accomplished by the λ_ith fractiles, $0 = \lambda_0 < \lambda_1 < ... < \lambda_k < \lambda_{k+1} = 1$, K is the subspace of functions which are constant on each of the intervals $(\lambda_i, \lambda_{i+1})$, $i = 0$, $1, ..., k$, i.e. each member of K assumes at most $k + 1$ different values.

The only difference between the problem of finding the function R^* satisfying (4.1) and the problem discussed in § 2 is that R^* is restricted to lie in a subspace of $L^2(0, 1)$. Paralleling Theorem 2.1, we have

Theorem 4.1. *If there exists a member S of K^* satisfying*

$$\liminf_i \langle K_i, S \rangle = \zeta > 0, \tag{4.2}$$

then there is a unique m.e.r.r.t. *within the class K^* of tests.*

Rather than paraphrase the results of § 2, we shall discuss some examples which we hope will illuminate the difference between the present problem and the original one. In particular, we shall see that it is possible for an a.m.p.r.t. for censored data from one member of $\mathcal{F}$ to be the m.e.r.r.t. for censored data from any member of $\mathcal{F}$. A simple example of this phenomenon occurs when $\mathcal{F}$ consists of $f(x)$ and the density g which is the density for which the a.m.p.r.t. for censored data from $f(x)$ is the a.m.p.r.t. for complete samples.

b. Examples

Our first example illustrates the effect of censoring.

Example 4.1. Consider data censored at the median of the combined sample where the observations come from either a logistic or double-exponential density. As the median test is the a.m.p.r.t. for complete samples from a double-exponential density, it remains the a.m.p.r.t. for such data censored at the median. The a.m.p.r.t. for logistic data censored at the median is specified by the weight function [7], $W_{\frac{1}{2}}(u)$ which is $(6/7)^{\frac{1}{2}}$ on $(\frac{1}{2}, 1)$ and is defined by $4(6/7)^{\frac{1}{2}} (u - \frac{1}{2})$ on $u < \frac{1}{2}$. The test based on $W_{\frac{1}{2}}(u)$ has ARE $6/7$ on double exponential data (compared to the median test, $M(u)$) while the ordinary Wilcoxon has ARE 0.75. Also the efficiency of $W_{\frac{1}{2}}$ on logistic data (relative to the Wilcoxon test on complete data) is $7/8$. To find the m.e.r.r.t. for censored data we consider all convex combinations $pM(u) + qW_{\frac{1}{2}}(u)$, where $p + q = 1$, p and $q \geqslant 0$. The ARE of the test based on $pM(u) + qW_{\frac{1}{2}}(u)$ compared to the median test on double exponential data is $(p + q(6/7)^{\frac{1}{2}})^2/[p^2 + q^2 + 2pq(6/7)^{\frac{1}{2}}]$ and is $[p(3/4)^{\frac{1}{2}} + q(7/8)^{\frac{1}{2}}]^2/[p^2 + q^2 + 2pq(6/7)^{\frac{1}{2}}]$ compared to the Wilcoxon test on logistic data. These ARE's are equal when

$$p = [(7/8)^{\frac{1}{2}} - (6/7)^{\frac{1}{2}}]/[1 - (6/7)^{\frac{1}{2}} + (7/8)^{\frac{1}{2}} - (3/4)^{\frac{1}{2}}] \cong 0.0668$$

and the test has maximin efficiency 0.8745.

The m.e.r.r.t., for data censored at the median, weights the censored Wilcoxon much more heavily than it does the median test. This was to be expected as the censored version of the original Wilcoxon test is closer to the median test than the original Wilcoxon test. In general, the m.e.r.r.t. for censored data will place more weight on the a.m.p.r.t.'s (for censored data) on those densities of a family $\mathscr{F}$ for which the censored portion of the data contains the largest amount of information.

It is interesting to compare the behavior of the m.e.r.r.t. for censored data with the test which places equal weight on each of the best tests for censored data (M and $W_{\frac{1}{2}}$) and also to the projection of the m.e.r.r.t. for complete samples on the space K of functions in $L^2(0, 1)$ which are constant on $(\frac{1}{2}, 1)$. The test based on $\frac{1}{2}(M(u) + W_{\frac{1}{2}}(u))$ has ARE 0.963 relative to the median test on double-exponential data and ARE 0.844 relative to the Wilcoxon test on logistic data. The renormalized projection of the m.e.r.r.t. for complete samples on the space K, has ARE $\simeq 0.965$ relative to the median test on double-exponential data and ARE $\simeq 0.8405$ relative to the Wilcoxon test on logistic data.

Another feature of the problem of efficiency robustness in the case of censored data is due to the difference in the effect of losing various portions of data. This has been treated in great detail by Sarndal [18] so we restrict ourselves to one illustration. We again consider samples from either the logistic or double-exponential densities but now assume that we have only the middle pth to $(1 - p)$th fraction of the observations available. As usual, we know how many observations of each group are in each of the censored portions of the data. We still have complete information if the observations are from the double-exponential density. It turns out that for large p $(p \geqslant 0\cdot41)$ the m.e.r.r.t. is the a.m.p.r.t. for censored data from the logistic density. This results from the fact that as p approaches $\frac{1}{2}$, the function specifying the optimum test for censored logistic data approaches $M(u)$ (in the L^2 sense). This example also illustrates the *influence* that the *class of tests* one is allowed to use has on the problem. In the case of complete samples we used rank tests based on any function in $L_2(0, 1)$. The a.m.p.r.t.'s for censored data form a subspace of $L_2(0, 1)$ and the a.m.p.r.t. for censored logistic data is the m.e.r.r.t. within the family of rank tests specified by square integrable functions on $(0, 1)$ which are constant on $(0, p)$ and $(1-p, 1)$, when $0\cdot41 < p < \frac{1}{2}$. This phenomena is also exhibited by our next example.

Example 4.2. We consider a simple family of grouped rank tests which can be regarded as a modification of the median test in the same way as the percentile-modified rank tests [5] changed the Wilcoxon and Ansari–Bradley–Freund tests. Define the S_p tests to be based on the functions $Sp(u)$ which equal $-(2p)^{-\frac{1}{2}}$ in $(0, p)$; 0 on $(p, 1-p)$; and $(2p)^{-\frac{1}{2}}$ on $(1-p, 1)$. The S_p test statistic is the number of y's in the upper pth fraction of the combined sample *minus* the number of y's in the lower pth fraction. $S_{\frac{1}{2}}(u)$ is the ordinary median test. It follows from Theorem 4.1 of [6] that for any particular density, the value of p which yields the most powerful test of this type also gives the minimum variance estimator in the class of averages of two symmetrically placed sample fractiles. Thus, for normal data the best choice of p is $0\cdot27$ [14]. The ARE of the S_p test relative to the appropriate a.m.p.r.t. for several densities is given in Table 4.1.

This family of tests is interesting because of their simplicity. Moreover, for normal data, the ARE is a very smooth function of p near the optimum value of p $(0\cdot27)$ so that any choice of p between $0\cdot2$ and $0\cdot33$ yields a reasonable test which is a significant improvement over the median test. Suppose we desire to find the m.e.r.r.t. in this simple family of tests for either *normal or logistic data*. The best choice of p is $0\cdot27$ for normal data [15] and $1/3$ for logistic data. However, the ARE of $S_{0\cdot27}$ relative to the Wilcoxon test on logistic data is higher than its ARE compared to the normal scores test on normal data. Thus the best test in our class of tests for a specific density again becomes the m.e.r.r.t. for the entire family. Notice that when the double-exponential density is

Table 4.1. *The ARE of S_p test compared to the a.m.p.r.t.*

p/density	Normal	Cauchy	Logistic	t_2	Double exponential
0·05	0·426	0·005	0·271	0·057	0·10
0·10	0·616	0·037	0·486	0·195	0·20
0·15	0·725	0·115	0·650	0·369	0·30
0·20	0·784	0·242	0·768	0·546	0·40
0·25	0·809	0·405	0·844	0·703	0·50
0·27	0·8098	0·475	0·863	0·754	0·54
0·30	0·806	0·579	0·882	0·823	0·60
0·333	0·794	0·684	0·888	0·878	0·667
0·35	0·784	0·730	0·887	0·897	0·70
0·38	0·763	0·797	0·876	0·918	0·76
0·40	0·746	0·829	0·864	0·922	0·80
0·45	0·696	0·857	0·817	0·898	0·90
0·50	0·637	0·811	0·750	0·833	1·00

added to our family the m.e.r.r.t. occurs when $p = 0.38$ and is no longer the best test in our family for any one member. When we add the Cauchy density to our family the test $S_{0.38}$ remains the maximin test with minimum efficiency 0·76 attained (approximately) on normal or double-exponential data. This example shows that the concept of an extreme pair of densities *depends on the class of statistical procedures we are allowed to use* as well as on the family of possible densities. Hence, the notion of 'order' in the class of symmetric densities which is most relevant for a particular efficiency robustness problem will also depend on the class of allowed statistical procedures. Indeed, this example reflects the fact that the Cauchy density is not $<$ the double-exponential density.

c. Some remarks on quick tests

The S_p tests discussed above are related to tests based on the number of exceedances [10, Ch. III, 1, 2], i.e. the number of y's greater (or less) than all the x's and the number of x's greater (or less) than all of the y's. For example, Tukey [20] proposed the statistic the number of y's greater than all the x's plus the number of x's less than all the y's as a quick test for the two-sample shift problem. Sen [19] showed that the tests based on exceedances are consistent only if the density of the underlying population decays sufficiently rapidly at $\pm\infty$. In particular Tukey's test is not consistent for data from densities which decay at a rate $|x|^{-r}$ for any fixed r. The S_p tests do not suffer from this defect because they are based on a fixed fraction ($2p$) of the observations rather than on the extreme members of both samples. In this section we shall discuss other properties of grouped rank tests [6], stressing how they can be used to approximate a more complicated rank test with relatively little loss in efficiency.

In 1951, Massey suggested an extension of Mood's test which is analysed in detail by Chakravati, Leone and Alanan [1]. These authors discuss the statistic which is similar to a grouped rank test. Letting V_1 denote the number of y's in the first quarter of combined sample and V_2 the number on the second quarter and V_3 ($= n - V_1 - V_2$) the number in the upper half, Massey's test is essentially based on

$$V = \frac{N^2}{mn} \left[\frac{V_1^2}{N/4} + \frac{V_2^2}{N/4} + \frac{V_3^2}{N/2} - \frac{n^2}{N} \right]. \tag{4.3}$$

The results reported in [1] indicate that the ARE of the test based on V to the median test on normal data is approximately one, i.e. by using three groups instead of two (the median test can be regarded as decomposing the sample into its two halves) no gain in ARE is achieved. The reason for the poor power of this version of Massey's test on normal data is not solely due to the fact that the grouping is not optimal but that V_1 and V_2 are weighted equally. Since the a.m.p.r.t. for observations from any strongly unimodal density is a monotonic non-decreasing function on $(0, 1)$, one should always weight V_1 more than V_2, i.e. test statistic of the form $aV_1 + bV_2 + cV_3$, where $V_3 = n - (V_1 + V_2)$ should have a and b of the same sign with $|a| \geq |b|$. For the sake of comparison the best Wilcoxon test for this grouping is based on the function, $W_{\frac{1}{4}, \frac{1}{2}}(u)$ which is $-\sqrt{2}$ on $(0, \frac{1}{4})$, $-\sqrt{2/3}$ on $(\frac{1}{4}, \frac{1}{2})$ and equals $2\sqrt{2/3}$ on $(\frac{1}{2}, 1)$, or, equivalently, $a = -3$, $b = -1$ and $c = 2$. This test has ARE 0.889 relative to the median test on double-exponential data, $\frac{27}{32} \sim 0.844$ relative to the Wilcoxon test on logistic data and ARE 0.746 relative to the normal scores test on normal data. In particular its ARE compared to the median test on normal data is 1.16. We chose the best Wilcoxon test because we thought it unfair to compare Massey's test to the best grouped rank test for normal data.

As our final example, we discuss the grouped rank test which generates the 'quick estimator' studied earlier [6]. This test is based on the (unnormalized) function $J(u)$ which is $+1$ on $(\frac{2}{3}, 1)$, $+0.3$ on $(\frac{1}{2}, \frac{2}{3})$, -0.3 on $(\frac{1}{3}, \frac{1}{2})$ and -1 on $(0, \frac{1}{3})$. Values of its ARE, compared to the appropriate a.m.p.r.t. for some densities of interest are the following: normal (0.804), logistic (0.916), double-exponential (0.844), Cauchy (0.792), t_2 (0.807), t_4 (0.956), t_6 (0.919), t_8 (0.901) and t_{10} (0.885).

This test has a minimum ARE of 79.2% over the above family of densities and *probably* over the entire t-family as well. Since the maximin efficiency of the m.e.r.r.t. for this family is $82 \cdot 8 \%$, the simple test is nearly as efficient as the m.e.r.r.t. and is much easier to use.

Of course one may ask whether the 'quick test' based on four groups is a substantial improvement over the $S_{0 \cdot 38}$ test. The answer should probably depend on small sample studies as the asymptotic efficiency of the S_p test on Cauchy or normal data is rather sensitive to the value of p.

The development of simple tests with high maximin relative efficiency is one of the main objectives of the approach we have adopted. In general for any problem (with complete or censored data) we first try to obtain the m.e.r.r.t. and its maximin ARE, then we try to approximate it by a simple grouped rank test which is easy to use. Usually, a good approximation can be achieved with relatively few groupings. We hope our approach will yield other simple procedures in the near future.

ACKNOWLEDGEMENTS

I would like to take this opportunity to thank Mr J. T. Smith for performing the numerical integrations involved in the tables. I am grateful to him and also to Professors L. J. Gleser and M. V. Johns, Jr. for reading over various parts of the manuscript and suggesting several improvements. It is a pleasure to thank Miss P. Grindle for typing the various versions of the manuscript. Finally, I wish to thank Dr M. L. Puri and the University of Indiana for making the conference both enjoyable and scientifically stimulating.

REFERENCES

[1] Alanen, J. D., Chakravarti, I. M. and Leone, F. C. Asymptotic relative efficiency of Mood's and Massey's two sample tests against some parametric alternatives, *Ann. Math. Statist.* **33** (1962) 1375–83.

[2] Birnbaum, A. and Laska, E. Efficiency robust two-sample rank tests, *J. Am. Statist. Ass.* **62** (1967) 1241–57.

[3] Birnbaum, A., Laska, E. and Meisner, M. Optimally robust linear estimators of location. Submitted for publication (1968).

[4] Chernoff, H. and Savage, I. R. Asymptotic normality and efficiency of certain non-parametric test statistics, *Ann. Math Statist.* **29** (1958) 972–44.

[5] Gastwirth, J. L. Percentile modified rank tests, *J. Am. Statist. Ass.* **60** (1965) 1127–41.

[6] Gastwirth, J. L. On robust procedures. *J. Am. Statist. Ass.* **61** (1966) 929–48.

[7] Gastwirth, J. L. Asymptotically most powerful rank tests for censored data, *Ann. Math. Statist.* **36** (1965) 1243–7.

[8] Gastwirth, J. L. On asymptotic relative efficiencies of a class of rank tests, *Tech. Rep.* 108, *Dept. of Stat.* The Johns Hopkins University (1969).

[9] Hájek, J. Asymptotically most powerful rank-order tests. *Ann. Math. Statist.* **33** (1962) 1124–47.

[10] Hájek, J. and Šidák, Z. *Theory of rank tests*, New York: Academic Press (1967).

[11] Hájek, J. *A course in non-parametric statistics*, San Francisco: Holden–Day (1969).

[12] Lehmann, E. L. *Testing statistical hypothesis*, New York: Wiley (1959).

[13] Massey, E. J. A note on a two sample test, *Ann. Math. Statist.* **22** (1951) 304–6.

[14] Mathisen, H. C. A method of testing the hypothesis that two samples are from the same population, *Ann. Math. Statist.* **14** (1943) 188–94.

[15] Mosteller, F. On some useful 'inefficient' statistics, *Ann. Math. Statist.* **17** (1946) 377–408.

[16] Noether, G. E. On a theorem of Pitman, *Ann. Math. Statist.* **26** (1955) 64–8.

[17] Pitman, E. J. G. Non-parametric inference (unpublished notes of lectures given at Columbia University 1949).

[18] Sarndal, C. E. *Information from censored samples Stockholm*, Almquist and Wiksell.

[19] Sen, P. K. On some asymptotic properties of a class of non-parametric tests based on the number of rate exceedances, *Ann. Inst. Statist. Math.* **17** (1965) 233–55.

[20] Tukey, J. W. A quick, compact, two sample test to Duckworth's specifications, *Technometrics* **1** (1959) 31–48.

[21] Van Zwet, W. R. *Convex transformations of random variables*, Amsterdam: Math. Centre (1964).

MULTIVARIATE PAIRED COMPARISONS: SOME LARGE-SAMPLE RESULTS ON ESTIMATION AND TESTS OF EQUALITY OF PREFERENCE†

ROGER R. DAVIDSON AND RALPH A. BRADLEY

SUMMARY

This is a second paper on multivariate paired comparisons. The first paper contained the methodology and this paper contains some large-sample results on estimation and test of preference equality. The limiting normal distribution is derived for the joint distribution of estimators for attribute preferences and correlations; the dispersion matrix derived indicates desired asymptotic variances and covariances. The likelihood ratio test of equality of preference parameters has the noncentral chi-square limiting distribution and its noncentrality parameter is given under local alternatives. Asymptotic efficiencies of the method of paired comparisons relative to a multiple multivariate sign test are computed.

1 INTRODUCTION

A multivariate model for paired comparisons has been given by Davidson and Bradley (1969). This model is an extension of the univariate model developed by Bradley and Terry (1952) to situations in which there is interest in p attributes or characteristics. One considers a set of t treatments presented in pairs to obtain n_{ij} responses in the comparison of treatments i and j, $i < j, i, j = 1, ..., t$. Each response to the treatment pair (i,j) is a vector of preferences, $\mathbf{s} = (s_1, ..., s_p)$ with $s_\alpha = i$ or j as treatment i or j is chosen on attribute α, $\alpha = 1, ..., p$. The multivariate model for paired comparisons specifies the probability $p(\mathbf{s}|i,j)$ for preference vector $\mathbf{s}$ and treatment pair (i,j);

$$p(\mathbf{s}|i,j) = p^{(1)}(\mathbf{s}|i,j)\, h(\mathbf{s}|i,j), \tag{1}$$

where

$$\left. \begin{aligned} p^{(1)}(\mathbf{s}|i,j) &= \prod_{\alpha=1}^{p} \pi_{\alpha s_\alpha}/(\pi_{\alpha i}+\pi_{\alpha j}), \\ h(\mathbf{s}|i,j) &= 1 + \sum_{\alpha<\beta}\sum \delta(s_\alpha, s_\beta)\rho_{\alpha\beta}(\pi_{\alpha i}/\pi_{\alpha j})^{-\frac{1}{2}\delta(i,\,s_\alpha)}(\pi_{\beta i}/\pi_{\beta j})^{-\frac{1}{2}\delta(i,\,s_\beta)}, \end{aligned} \right\} \tag{2}$$

† Research supported by the Army, Navy and Air Force under an Office of Naval Research Contract. Reproduction in whole or in part is permitted for any purpose of the United States Government.

where $s_\alpha = i, j, \alpha = 1, \dots, p$, and where $\delta(\cdot, \cdot) = \pm 1$, the sign being positive if the two arguments are equal and negative otherwise. The preference parameters, $\pi = \{\pi_{\alpha i}; i = 1, \dots, t, \alpha = 1, \dots, p\}$, are restrained by

$$\sum_{i=1}^{t} \pi_{\alpha i} = 1 \quad (\alpha = 1, \dots, p)$$

and $\rho = \{\rho_{\alpha\beta}; \alpha < \beta, \alpha, \beta = 1, \dots, p\}$ is such that $h(s|i,j) \geqslant 0$ for each of the 2^p cells associated with each of the $\binom{t}{2}$ treatment comparisons. Note that $\rho = 0$ implies independence.

The large-sample properties of the univariate model have been discussed by Bradley (1955). The asymptotic distributions of the maximum likelihood estimators and of the likelihood ratio statistic for testing the hypothesis of equal treatment preferences against local alternatives were derived. In this study a similar development is given for the multivariate model (1, 2).

Bradley also compared the power of the test of equal preferences based on the univariate model to one based on a multi-binomial model. An extension is given to multivariate models. Bradley's discussion of asymptotic relative efficiency is clarified and extended.

2 ESTIMATION

The limiting distribution of the maximum likelihood (m.l.) estimators† p of π and $\hat{\rho}$ of ρ may be obtained by appealing to the results of Bradley and Gart (1962). In addition, we indicate the procedure for obtaining the matrix of variances and covariances for the limiting distribution.

Some new notation is necessary. For each pair (i,j) let

$$f_{ij}(\mathbf{x}, \pi, \rho) = \prod_{s} p(s|i,j)^{x(s|i,j)}, \tag{3}$$

where $p(s|i,j)$ is defined in (1, 2) in terms of π and ρ, $\mathbf{x}$ is the 2^p-dimensional array $\{x(s|i,j); s_\alpha = i, j, \alpha = 1, \dots, p\}$, and $x(s|i,j)$ is unity if $s = (s_1, \dots, s_p)$ occurs as the vector of treatment preferences in a single comparison of treatments i and j and is zero otherwise. $\sum_{s} x(s|i,j) = 1$.

Thus, if one lets $\mathbf{x}^{(k)}$ denote the array $\mathbf{x}$ on the kth of n_{ij} repetitions of the comparison of treatments i and j, it is clear that the likelihood function becomes

$$L = \prod_{i<j} \prod_{k=1}^{n_{ij}} f_{ij}(\mathbf{x}^{(k)}, \pi, \rho).$$

---

† Details on these estimators are given by Davidson and Bradley (1969).

Observe that $f_{ij}(\mathbf{x}, \boldsymbol{\pi}, \boldsymbol{\rho})$ is the probability function for the treatment pair (i,j) so that, for the purpose of applying the results of Bradley and Gart, the set $\{f_{ij}(\mathbf{x}, \boldsymbol{\pi}, \boldsymbol{\rho}); \ i < j, i,j = 1, \ldots, t\}$ gives the probability distributions for a set of $\binom{t}{2}$ associated populations.

The preference parameters $\pi_{\alpha 1}, \ldots, \pi_{\alpha t}$ are such that

$$\sum_{i=1}^{t} \pi_{\alpha i} = 1 \quad (\alpha = 1, \ldots, p)$$

and thus are not functionally independent. Accordingly, write

$$\boldsymbol{\pi}^* = \{\pi_{\alpha 1}, \ldots, \pi_{\alpha, t-1}; \alpha = 1, \ldots, p\}$$

and

$$\mathbf{p}^* = \{p_{\alpha 1}, \ldots, p_{\alpha, t-1}; \alpha = 1, \ldots, p\}$$

and take $\quad \pi_{\alpha t} = 1 - \sum_{i=1}^{t-1} \pi_{\alpha i}, \quad p_{\alpha t} = 1 - \sum_{i=1}^{t-1} p_{\alpha i} \quad (\alpha = 1, \ldots, p).$

It follows from Theorem 3, Bradley and Gart (1962), subject to weak regularity conditions, that $\sqrt{(N)}\,(\mathbf{p}^* - \boldsymbol{\pi}^*)$ and $\sqrt{(N)}\,(\hat{\boldsymbol{\rho}} - \boldsymbol{\rho})$ have a joint limiting multivariate normal distribution in $p(t-1) + \binom{p}{2}$ variates with zero mean vector and dispersion matrix $\boldsymbol{\Sigma}^*$ developed below. Here

$$N = \sum_{i<j}\sum n_{ij}, \quad n_{ij} = \mu_{ij}^{(N)} N, \quad \lim_{N \to \infty} \mu_{ij}^{(N)} = \mu_{ij}$$

and the limiting normal distribution results as $N \to \infty$.

We develop $\boldsymbol{\Sigma}^*$. To simplify the development, we take $n_{ij} = \mu_{ij} N$, a minor loss of refinement. We also take $n_{ji} = n_{ij}$ and remove the requirement that $i < j$ in the definition of n_{ij}. We require various matrices of partial derivatives for the Bradley–Gart application. Let

$$\boldsymbol{\Lambda}_{\alpha\beta} = [\lambda_{\alpha\beta, kl}; k, l = 1, \ldots, t] \quad (\alpha, \beta = 1, \ldots, p), \tag{4}$$

where

$$\lambda_{\alpha\beta, kk} = -\sum_{i}' \mu_{ik} E\left[\frac{\partial^2 \ln f_{ik}}{\partial \pi_{\alpha k}\, \partial \pi_{\beta k}}\right], \tag{5}$$

$$\lambda_{\alpha\beta, kl} = -\mu_{kl} E\left[\frac{\partial^2 \ln f_{kl}}{\partial \pi_{\alpha k}\, \partial \pi_{\beta l}}\right] \quad (k \neq l), \tag{6}$$

and let

$$\boldsymbol{\Delta}_{\alpha} = [\Delta_{\alpha, (\gamma\delta), k}; k = 1, \ldots, t, \gamma < \delta, \gamma, \delta = 1, \ldots, p] \quad (\alpha = 1, \ldots, p), \tag{7}$$

where

$$\Delta_{\alpha, (\gamma\delta), k} = -\sum_{i}' \mu_{ik} E\left[\frac{\partial^2 \ln f_{ik}}{\partial \pi_{\alpha k}\, \partial \rho_{\gamma\delta}}\right]. \tag{8}$$

In (5) and (8) and in subsequent expressions $\sum_i'$ indicates a sum, $i = 1, ..., t$, but excluding $i = k$ or other second similar subscript. Let

$$\mathbf{\Lambda}^*_{\alpha\beta} = [\lambda^*_{\alpha\beta,\, kl}, k, l = 1, ..., (t-1)], \tag{9}$$

where
$$\lambda^*_{\alpha\beta,\, kl} = \lambda_{\alpha\beta,\, kl} - \lambda_{\alpha\beta,\, kt} - \lambda_{\alpha\beta,\, tl} + \lambda_{\alpha\beta,\, tt}$$

and let
$$\mathbf{\Lambda}^* = [\mathbf{\Lambda}^*_{\alpha\beta}, \alpha, \beta = 1, ..., p].$$

Let
$$\mathbf{\Delta}^*_\alpha = [\Delta^*_{\alpha,\, (\gamma\delta),\, k}; k = 1, ..., (t-1), \gamma < \delta, \gamma, \delta = 1, ..., p], \tag{10}$$

where
$$\Delta^*_{\alpha,\, (\gamma\delta),\, k} = \Delta_{\alpha,\, (\gamma\delta),\, k} - \Delta_{\alpha,\, (\gamma\delta),\, t}$$

and let
$$\mathbf{\Delta}^{*\prime} = [\mathbf{\Delta}^{*\prime}_1, ..., \mathbf{\Delta}^{*\prime}_p].$$

Finally, let

$$\mathbf{\Gamma} = -\left[\sum_{i<j}\sum \mu_{ij} E\left\{ \frac{\partial^2 \ln f_{ij}}{\partial \rho_{\alpha\beta} \partial \rho_{\gamma\delta}} \right\}; \alpha < \beta, \gamma < \delta, \alpha, \beta, \gamma, \delta = 1, ..., p \right], \tag{11}$$

with the final relevant matrix being

$$\mathbf{C}^* = \begin{bmatrix} \mathbf{\Lambda}^* & \mathbf{\Delta}^* \\ \mathbf{\Delta}^{*\prime} & \mathbf{\Gamma} \end{bmatrix}. \tag{12}$$

The dispersion matrix $\mathbf{\Sigma}^* = \mathbf{C}^{*-1}$. The expectations needed to evaluate (5), (8) and (11) are given in an Appendix.

$\mathbf{\Sigma}^*$ is the asymptotic dispersion matrix for $\sqrt{(N)}(\mathbf{p}^* - \mathbf{\pi}^*)$ and $\sqrt{(N)}(\hat{\mathbf{\rho}} - \mathbf{\rho})$. Remaining variances and covariances associated with $\sqrt{(N)}(p_{\alpha t} - \pi_{\alpha t})$, $\alpha = 1, ..., p$, may also be determined. Suppose that $\mathbf{\Sigma}^*$ is partitioned in the same way as $\mathbf{C}^*$ in (12) so that

$$\mathbf{\Sigma}^* = \begin{bmatrix} \mathbf{\Sigma}^*_{11} & \mathbf{\Sigma}^*_{12} \\ \mathbf{\Sigma}^*_{21} & \mathbf{\Sigma}^*_{22} \end{bmatrix}.$$

Now,

$$\mathbf{\Sigma}^*_{11} = [\sigma_{\alpha\beta,\, ij}; i, j = 1, ..., (t-1), \alpha, \beta = 1, ..., p], \tag{13}$$

$$\left. \begin{aligned} \mathbf{\Sigma}^{*\prime}_{12} &= \mathbf{\Sigma}^*_{21} = [\sigma_{\alpha,\, (\gamma\delta),\, i}; i = 1, ..., (t-1), \gamma < \delta, \alpha, \gamma, \delta = 1, ..., p], \\ \mathbf{\Sigma}_{22} &= [\sigma_{(\alpha\beta),\, (\gamma\delta)}; \alpha < \beta, \gamma < \delta, \alpha, \beta, \gamma, \delta = 1, ..., p]. \end{aligned} \right\} \tag{14}$$

Note that $\sigma_{\alpha\beta,\, ij}$ is the covariance of $\sqrt{(N)}(p_{\alpha i} - \pi_{\alpha i})$ and $\sqrt{(N)}(p_{\beta j} - \pi_{\beta j})$, $\sigma_{\alpha,\, (\gamma\delta),\, i}$ is the covariance of $\sqrt{(N)}(p_{\alpha i} - \pi_{\alpha i})$ and $\sqrt{(N)}(\hat{\rho}_{\gamma\delta} - \rho_{\gamma\delta})$, and $\sigma_{(\alpha\beta),\, (\gamma\delta)}$ is the covariance of $\sqrt{(N)}(\hat{\rho}_{\alpha\beta} - \rho_{\alpha\beta})$ and $\sqrt{(N)}(\hat{\rho}_{\gamma\delta} - \rho_{\gamma\delta})$. Since

$$\sqrt{(N)}(p_{\alpha t} - \pi_{\alpha t}) = -\sum_{i=1}^{t-1} \sqrt{(N)}(p_{\alpha i} - \pi_{\alpha i}) \quad (\alpha = 1, ..., p),$$

we have
$$\sigma_{\alpha\beta,\, tt} = \sum_{i=1}^{t-1}\sum_{j=1}^{t-1} \sigma_{\alpha\beta,\, ij} \quad \text{and} \quad \sigma_{\alpha\beta,\, it} = -\sum_{j=1}^{t-1} \sigma_{\alpha\beta,\, ij},$$

where $i = 1, ..., (t-1), \alpha, \beta = 1, ..., p$, and

$$\sigma_{\alpha, (\gamma\delta), t} = -\sum_{i=1}^{t-1} \sigma_{\alpha, (\gamma\delta), i} \quad (\gamma < \delta, \alpha, \gamma, \delta = 1, ..., p).$$

Let Σ_{11} and $\Sigma_{12}' = \Sigma_{21}$ correspond respectively to Σ_{11}^* and $\Sigma_{12}^{*'} = \Sigma_{21}^*$ with the ranges of i and j increased to t in (13) and (14) and let

$$\Sigma = \begin{bmatrix} \Sigma_{11} & \Sigma_{12} \\ \Sigma_{21} & \Sigma_{22} \end{bmatrix}.$$

We may summarize by stating: '$\sqrt{(N)}\,(\mathbf{p} - \boldsymbol{\pi})$ and $\sqrt{(N)}\,(\hat{\boldsymbol{\rho}} - \boldsymbol{\rho})$ have, as a limiting distribution, the singular multivariate normal distribution of $p(t-1) + \binom{p}{2}$ dimensions in a space of $pt + \binom{p}{2}$ dimensions with mean vector zero and dispersion matrix Σ.'

A means of obtaining Σ in symmetric fashion may be developed as a generalization of that given by Bradley (1955, pp. 454–5). We give only the result. Let

$$\Lambda = [\Lambda_{\alpha\beta}; \alpha, \beta = 1, ..., p]$$

from (4) and let

$$\Delta' = [\Delta_1', ..., \Delta_p']$$

from (7). With Γ from (11) we take

$$C = \begin{bmatrix} \Lambda & \Delta \\ \Delta' & \Gamma \end{bmatrix}.$$

We define

$$J = \begin{bmatrix} [1] & [0] & \cdots & [0] \\ [0] & [1] & \cdots & [0] \\ \multicolumn{4}{c}{\cdots\cdots\cdots\cdots\cdots} \\ [0] & [0] & \cdots & [1] \end{bmatrix},$$

a $pt \times p$ matrix where [1] and [0] are column vectors of t elements, ones and zeros respectively, and augment C to obtain

$$C_A = \begin{bmatrix} \Lambda & \Delta & J \\ \Delta' & \Gamma & 0 \\ J' & 0 & 0 \end{bmatrix} = \begin{bmatrix} C & J \\ J' & 0 & 0 \end{bmatrix}, \tag{15}$$

where the $\mathbf{0}$'s are appropriately dimensioned matrices of zeros. The matrix Σ is the principal minor of C_A^{-1} corresponding to C of C_A. This may be verified through the use of elementary row and column operations in obtaining $|C_A|$ and the various cofactors of elements of C in C_A.

Estimates of the variances and covariances of the limiting multivariate normal distribution may be required. Consistent estimators may be obtained through substitution of $\mathbf{p}$ for $\boldsymbol{\pi}$ and $\hat{\boldsymbol{\rho}}$ for $\boldsymbol{\rho}$ in $\mathbf{C}^*$ in (12) or in $\mathbf{C}_A$ in (15) to yield $\hat{\mathbf{C}}^*$ or $\hat{\mathbf{C}}_A$. Then $\hat{\boldsymbol{\Sigma}}$ may be obtained by augmentation of $\hat{\boldsymbol{\Sigma}}^* = \hat{\mathbf{C}}^{*-1}$ or as the appropriate minor of $\hat{\mathbf{C}}_A^{-1}$.

As a concluding comment in this section, note that if each $n_{ij} = n$, $N = n\binom{t}{2}$ and $\mu_{ij} = 1 \Big/ \binom{t}{2}$, $i < j, i,j = 1,\ldots,t$. In this situation, one may be more interested in the limiting distribution of $\sqrt{(n)}\,(\mathbf{p} - \boldsymbol{\pi})$ and $\sqrt{(n)}\,(\hat{\boldsymbol{\rho}} - \boldsymbol{\rho})$ and then $\mathbf{C}^*$ and $\mathbf{C}_A$ should be evaluated with each $\mu_{ij} = 1$.

3 TESTS OF HYPOTHESES

Likelihood ratio tests† have been obtained for a variety of testing situations by Davidson and Bradley (1969). For each case of interest the limiting distribution can be obtained from both the null hypothesis and local alternatives by appealing to the results of Davidson and Lever (1970). The procedures are illustrated for a particular test.

Consider the case in which the null hypothesis specifies equality of treatment parameters on each attribute in the presence of possible correlations. The null hypothesis is

$$H_0: \pi_{\alpha i} = 1/t \quad (i = 1,\ldots,t, \alpha = 1,\ldots,p, \boldsymbol{\rho}\ \text{unspecified}). \quad (16)$$

Local alternatives specify

$$\pi_{\alpha i} = 1/t + \delta_{\alpha i N}/\sqrt{N},$$

where

$$n_{ij} = \mu_{ij} N \quad \text{and} \quad \lim_{N \to \infty} \delta_{\alpha i N} = \delta_{\alpha i} \quad (i = 1,\ldots,t, \alpha = 1,\ldots,p).$$

It follows from Theorem 3′ of Davidson and Lever (1970), subject to weak regularity conditions, that under H_0 of (16) the likelihood ratio statistic given in detail by Davidson and Bradley (1969) has a limiting central chi-square distribution with $p(t-1)$ degrees of freedom. Under the local alternatives, the limiting distribution is again chi-square with $p(t-1)$ degrees of freedom but with noncentrality parameter $\lambda^2 = \boldsymbol{\delta}'\bar{\mathbf{C}}_0^*\boldsymbol{\delta}$,

where $\qquad \boldsymbol{\delta}' = (\boldsymbol{\delta}_1',\ldots,\boldsymbol{\delta}_p')$ with $\boldsymbol{\delta}_\alpha' = (\delta_{\alpha 1},\ldots,\delta_{\alpha,\,t-1})$

and $\qquad\qquad\qquad\qquad \bar{\mathbf{C}}_0^* = \boldsymbol{\Sigma}_{110}^{*-1},$

where $\qquad \boldsymbol{\Sigma}_0^* = \begin{bmatrix} \boldsymbol{\Sigma}_{110}^* & \boldsymbol{\Sigma}_{120}^* \\ \boldsymbol{\Sigma}_{210}^* & \boldsymbol{\Sigma}_{220}^* \end{bmatrix} = \mathbf{C}_0^{*-1}$ and $\mathbf{C}_0^* = \mathbf{C}^*|_{\pi=[1/t]}$,

$\mathbf{C}^*$ defined in (12).

† See the references for details that are not repeated here.

It is shown in the appendix that one may write

$$\mathbf{C}_0^* = \begin{bmatrix} \mathbf{\Lambda}_0^* & \mathbf{0} \\ \mathbf{0} & \mathbf{\Gamma}_0 \end{bmatrix},$$

where $\mathbf{\Gamma}_0 = \mathbf{\Gamma}|_{\pi=[1/t]}$ from (11) and where $\mathbf{\Lambda}_0^* = \mathbf{W} \otimes \mathbf{E}$, the Kronecker product of the matrices

$$\mathbf{W} = [w_{\alpha\beta}; \alpha, \beta = 1, \dots, p] \quad \text{and} \quad \mathbf{E} = [e_{ij}; i, j = 1, \dots, (t-1)],$$

where
$$w_{\alpha\alpha} = 1 + 2^{-p} \sum_s \{1 + \sum_{\alpha < \beta} \sum \rho_{\alpha\beta} \delta(s_\alpha, s_\beta)\}^{-1} \{\sum_\gamma' \delta(i, s_\gamma) \rho_{\alpha\gamma}\}^2 \tag{17}$$

and, for $\alpha \neq \beta$,

$$w_{\alpha\beta} = -\rho_{\alpha\beta} + 2^{-p} \sum_s \{1 + \sum_{\alpha < \beta} \sum \rho_{\alpha\beta} \delta(s_\alpha, s_\beta)\}^{-1} \{\sum_\gamma' \delta(i, s_\gamma) \rho_{\alpha\gamma}\} \{\sum_\gamma' \delta(i, s_\gamma) \rho_{\beta\gamma}\} \tag{18}$$

and where
$$e_{ii} = \frac{t^2}{4} (\sum_k' \mu_{ki} + \sum_k' \mu_{kt} + 2\mu_{it}) \tag{19}$$

and, for $i \neq j$,
$$e_{ij} = \frac{t^2}{4} (\sum_k' \mu_{kt} - \mu_{ij} + \mu_{it} + \mu_{jt}). \tag{20}$$

Since $\mathbf{\Delta}_0^* = \mathbf{\Delta}^*|_{\pi=[1/t]} = \mathbf{0}$, it is clear that $\mathbf{C}_0^* = \mathbf{\Lambda}_0^*$. Thus, the non-centrality parameter may be expressed as

$$
\begin{aligned}
\lambda^2 &= \sum_{\alpha=1}^{p} \sum_{\beta=1}^{p} \sum_{i=1}^{t-1} \sum_{j=1}^{t-1} \delta_{\alpha i} \delta_{\beta j} w_{\alpha\beta} e_{ij} \\
&= \frac{t^2}{4} \sum_{\alpha=1}^{p} \sum_{\beta=1}^{p} w_{\alpha\beta} \left\{ \sum_{i=1}^{t} [\sum_k' \mu_{ki}] \delta_{\alpha i} \delta_{\beta i} - \sum_{i \neq j} \mu_{ij} \delta_{\alpha i} \delta_{\beta j} \right\} \\
&= \frac{t^2}{4} \sum_{\alpha=1}^{p} \sum_{\beta=1}^{p} w_{\alpha\beta} \sum_{i<j} \sum \mu_{ij} (\delta_{\alpha i} - \delta_{\alpha j})(\delta_{\beta i} - \delta_{\beta j}), \tag{21}
\end{aligned}
$$

where the last two forms follow since

$$\sum_{i=1}^{t} \delta_{\alpha i} = 0 \quad (\alpha = 1, \dots, p).$$

In the case where all $n_{ij} = n$, one could consider alternatives

$$\pi_{\alpha i} = 1/t + \gamma_{\alpha i n}/\sqrt{n},$$

where $\lim_{n\to\infty} \gamma_{\alpha i n} = \gamma_{\alpha i}$ so that $\gamma_{\alpha i} = \binom{t}{2}^{-\frac{1}{2}} \delta_{\alpha i}$. Then (21) becomes

$$\lambda^2 = \frac{t^3}{4} \sum_{\alpha=1}^{p} \sum_{\beta=1}^{p} w_{\alpha\beta} \sum_{i=1}^{t} \gamma_{\alpha i} \gamma_{\beta i}. \tag{22}$$

Note that for the univariate model we have from (16)

$$H_0: \pi_i = 1/t \quad (i = 1, \dots, t)$$

and alternatives

$$\pi_i = 1/t + \delta_{iN}/\sqrt{N} \quad \text{and} \quad \mathbf{C}_0^* = \mathbf{E},$$

in that $w_{11} = 1$. Then, under local alternatives, the likelihood ratio statistic has the limiting noncentral chi-square distribution with $(t-1)$ degrees of freedom and

$$\lambda^2 = \frac{t^2}{4} \sum_{i<j} \mu_{ij}(\delta_i - \delta_j)^2.$$

This result has been obtained directly by Bradley (1955, pp. 456–9) for the case in which all $n_{ij} = n$, where from (22),

$$\lambda^2 = \frac{t^3}{4} \sum_{i=1}^{t} \gamma_i^2.$$

4 COMPARISONS OF TESTS OF EQUAL PREFERENCES

It is difficult to find tests for comparison with the multivariate paired comparisons method. We use a multiple multivariate sign test and also look at a method of Sen and David (1968). Bradley (1955) compared the univariate tests for paired comparisons with a multi-binomial test.

We develop a multiple multivariate sign test. Consider a probability model of the form (1, 2) but which differs from it in that it permits a distinct set of relevant parameters π and ρ for each treatment pair. We let these parameters be $\pi_{\alpha ij}, \pi_{\alpha ij}, \alpha = 1, \dots, p$, and $\rho(i, j)$ for the pair (i, j). In this situation we regard experimentation on each treatment pair as a distinct experiment with $t = 2$ and $n_{ij} = \mu_{ij} N$.

The hypothesis for the pair (i, j) is

$$H_0: \pi_{\alpha ij} = \pi_{\alpha ji} = \tfrac{1}{2} \quad (\alpha = 1, \dots, p, \rho(i, j) \text{ unspecified}),$$

against alternatives of the form

$$\pi_{\alpha ij} = \tfrac{1}{2} + \eta_{\alpha ijN}/\sqrt{N}, \quad \pi_{\alpha ji} = \tfrac{1}{2} + \eta_{\alpha jiN}/\sqrt{N} \quad (\alpha = 1, \dots, p),$$

where

$$\eta_{\alpha ijN} \to \eta_{\alpha ij}, \quad \eta_{\alpha jiN} \to \eta_{\alpha ji}, \quad \eta_{\alpha ij} + \eta_{\alpha ji} = 0 \quad \text{as } N \to \infty.$$

If we let the likelihood ratio statistic for this treatment pair be S_{ij}, under local alternatives S_{ij} has the noncentral chi-square limiting distribution with p degrees of freedom and noncentrality parameter

$$\lambda_{ij}^2 = \sum_{\alpha=1}^{p} \sum_{\alpha=1}^{p} w_{\alpha\beta ij} \mu_{ij} (\eta_{\alpha ij} - \eta_{\alpha ji})(\eta_{\beta ij} - \eta_{\beta ji}),$$

where the $w_{\alpha\beta ij}$, $\alpha, \beta = 1, \ldots, p$, are given by (17) and (18) when ρ is replaced by $\rho(i,j)$. An overall test of the equality of the preference parameters $\pi_{\alpha ij}$, $i \neq j$, $i,j = 1, \ldots, t$, $\alpha = 1, \ldots, p$, may be made on the basis of the combined statistic $S = \sum\limits_{i<j} \sum S_{ij}$ and we have a multiple multivariate sign test. S has the noncentral chi-square limiting distribution with $p\binom{t}{2}$ degrees of freedom and noncentrality parameter

$$\bar{\lambda}^2 = \sum_{\alpha=1}^{p} \sum_{\alpha=1}^{p} \sum_{i<j} \sum w_{\alpha\beta ij}\mu_{ij}(\eta_{\alpha ij} - \eta_{\alpha ji})(\eta_{\beta ij} - \eta_{\beta ji})$$

under local alternatives.

The multiple multivariate sign test and multivariate paired comparisons test may be compared under the assumptions for the latter which give $\quad \pi_{\alpha ij} = \pi_{\alpha i}/(\pi_{\alpha i} + \pi_{\alpha j}) \quad$ and $\quad \rho(i,j) = \rho$ common to all treatment pairs (i,j). It follows that

$$\eta_{\alpha ijN} = \frac{t}{4}(\delta_{\alpha iN} - \delta_{\alpha jN}) + o(1),$$

where $o(1)$ denotes a quantity which tends to zero as $N \to \infty$. Then

$$\eta_{\alpha ij} = -\eta_{\alpha ji} = \frac{t}{4}(\delta_{\alpha i} - \delta_{\alpha j}) \quad \text{and} \quad w_{\alpha\beta ij} = w_{\alpha\beta}$$

for all treatment pairs (i,j) so that $\bar{\lambda}^2 = \lambda^2$. Thus the two test statistics have chi-square distributions in the limit for local alternatives with identical noncentrality parameters but with different degrees of freedom.

When it is assumed that $\rho(i,j) = \rho$ for all treatment pairs (i,j), the statistic S for the multivariate sign test involves the use of separate estimates of ρ for each treatment pair. Hence S is not the likelihood ratio statistic. However, it is easily seen that, if ρ is estimated jointly over all treatment pairs, the resulting likelihood ratio statistic has the same limiting distribution as S.

Hoeffding and Rosenblatt (1955) have suggested a general measure of asymptotic efficiency and Noether (1957) has indicated how it may be used in our situation. Suppose that one has a test of the hypothesis $\theta = \theta^0$ against alternatives of the form

$$\theta = \theta^N = \theta^0 + dN^{-r} \tag{23}$$

and suppose that under the null hypothesis the test statistic T has limiting cdf $H_0(t)$ and under the local alternatives (23) T has limiting cdf $H_1(t,d)$. Let t_α be such that $H_0(t_\alpha) = 1 - \alpha$ and let $D(\alpha, \beta)$ be the value of d for which $H_1[t_\alpha, D(\alpha, \beta)] = 1 - \beta$, where α and β are respectively

Type I and Type II error rates. Given two tests, T_1 and T_2, of the form described, the asymptotic efficiency of T_1 to T_2 is defined as

$$E_{T_1 T_2}(\alpha, \beta) = [D_2(\alpha, \beta)/D_1(\alpha, \beta)]^{1/r}. \tag{24}$$

Note that (24) depends on α and β in general.

We apply (24) to our situation. The statistic for multivariate paired comparisons is designated as T and the statistic S for the multiple multivariate sign test is given above. We associate notation by taking

$$\theta^N = \frac{t^2}{4} \sum_{\alpha=1}^{p} \sum_{\beta=1}^{p} w_{\alpha\beta} \sum_{i<j} \sum \mu_{ij}(d_{\alpha iN} - d_{\alpha jN})(d_{\beta iN} - d_{\beta jN}),$$

where $d_{\alpha iN} = \pi_{\alpha i} - 1/t$ and $d_{\alpha iN} \to 0$ as $N \to \infty$

in view of the local alternatives to (16). It is seen that $\theta^N = 0$ is equivalent to H_0 in (16) and hence $\theta^0 = 0$. Further $\theta^N = \lambda^2/N$ and $d = \lambda^2$ ($r = 1$). We specify α and β and find values of λ^2 given the noncentral chi-square distributions with degrees of freedom $p\binom{t}{2}$ and $p(t-1)$ to yield

$$D_1(\alpha, \beta) = \lambda^2 \left[p\binom{t}{2}, \alpha, \beta \right] \quad \text{and} \quad D_2(\alpha, \beta) = \lambda^2[p(t-1), \alpha, \beta].$$

Then $$E_{ST}(\alpha, \beta) = \lambda^2[p(t-1), \alpha, \beta]/\lambda^2 \left[p\binom{t}{2}, \alpha, \beta \right]. \tag{25}$$

A tabulation of asymptotic efficiencies obtained from (25) is given in Table 1 for $\alpha = 0.01$ and 0.05, $\beta = 0.1$, 0.5 and 0.9 for values of p and t for which $p \leqslant 4$ and $p(t-1) \leqslant 12$. The values of $\lambda^2(\nu, \alpha, \beta)$ necessary for these computations have been obtained from the tables of Fix (1949). It is obvious that these efficiencies drop off quite sharply as t increases; the method of multivariate paired comparisons takes advantage of all paired comparisons involving treatment i in the estimation of $\pi_{\alpha i}, i = 1, ..., t$, $\alpha = 1, ..., p$, whereas the multiple multivariate sign test does not.

In a recent study, Sen and David (1968) have given two methods for testing equality of preference in bivariate paired comparisons. They first consider three likelihood ratio tests, two of which correspond to the tests based on S and T when $p = 2$. In a second approach, they introduce a permutationally distribution free test statistic D_N which is a bivariate extension of the univariate statistic (cf. David (1963, p. 30–1, 38, 52–3)). Sen and David (1968, Theorem 4.3) show that, under local alternatives, D_N has a limiting noncentral chi-square distribution with $2(t-1)$ degrees of freedom and noncentrality parameter which in the notation of the present paper becomes

$$\frac{t}{4} \sum_{\alpha=1}^{2} \sum_{\beta=1}^{2} w_{\alpha\beta} \sum_{i=1}^{t} [\sum_j{}' \mu_{ij}^{\frac{1}{2}}(\delta_{\alpha i} - \delta_{\alpha j})][\sum_j{}' \mu_{ij}^{\frac{1}{2}}(\delta_{\beta i} - \delta_{\beta j})]. \tag{26}$$

Table 1. *Asymptotic efficiencies, multiple multivariate sign test to multivariate paired comparisons*

$\alpha = 0.01$

$p=1$ t	$p=2$ t	$p=3$ t	$p=4$ t	ν_S $pt(t-1)/2$	ν_T $p(t-1)$	$E_{ST}(\alpha,\beta)$ $\beta=0.1$	$\beta=0.5$	$\beta=0.9$
2				1	1	1·000	1·000	1·000
3	2			3	2	0·833	0·880	0·905
4		2		6	3	0·728	0·792	0·830
5	3		2	10	4	0·654	0·724	0·769
6				15	5	0·598	0·669	0·717
7	4	3		21	6	0·552	0·625	0·674
8				28	7	0·517	0·587	0·636
9	5		3	36	8	0·487	0·555	0·603
10		4		45	9	0·460	0·527	0·575
11	6			55	10	0·439	0·502	0·550
12				66	11	0·420	0·481	0·527
13	7	5	4	78	12	0·402	0·462	0·507

$\alpha = 0.05$

$p=1$ t	$p=2$ t	$p=3$ t	$p=4$ t	ν_S $pt(t-1)/2$	ν_T $p(t-1)$	$E_{ST}(\alpha,\beta)$ $\beta=0.1$	$\beta=0.5$	$\beta=0.9$
2				1	1	1·000	1·000	1·000
3	2			3	2	0·795	0·861	0·893
4		2		6	3	0·690	0·768	0·813
5	3		2	10	4	0·611	0·699	0·751
6				15	5	0·560	0·644	0·698
7	4	3		21	6	0·515	0·599	0·655
8				28	7	0·480	0·562	0·618
9	5		3	36	8	0·454	0·532	0·586
10		4		45	9	0·428	0·504	0·558
11	6			55	10	0·410	0·481	0·533
12				66	11	0·391	0·461	0·512
13	7	5	4	78	12	0·377	0·442	0·492

In the case where $n_{ij} = n, i < j, i,j = 1, \ldots, t, p = 2$, the noncentrality parameter (26) can be shown to equal λ^2 as given in (22). Thus, in the bivariate case, as in the univariate case, the tests of equal preferences based on T and on D_N in a balanced paired comparisons experiment are asymptotically equivalent. This equivalence breaks down, however, when the paired comparison experiment is not balanced.

5   DISCUSSION

In this paper and in the preceding one by Davidson and Bradley (1969), we have developed the method and properties of a method of multivariate paired comparisons. The multivariate problem is complex and variations of the methods may be required when other hypotheses are desired as specializations of the model (1, 2). Such situations may be considered in the same manner as in these two papers. Davidson (1966) has given some additional likelihood ratio tests.

One unsolved problem is of practical interest. In sensory experimentation, it is often desired that one of the p attributes be 'overall preference'. It would be useful to develop a method analogous to regression to develop a predictive relationship between responses on overall preference and responses on other attributes. A tidy way of doing this has not been found.

APPENDIX

The expectations of the various derivatives of $\ln f_{ij}(\mathbf{x}, \boldsymbol{\pi}, \boldsymbol{\rho})$ of (3) which are needed to evaluate (5), (6), (8) and (11) are developed. These, when used in (4) and (7), will permit evaluation of $\mathbf{C}^*$ in (12).

Let
$$d_{\alpha ij} = (\pi_{\alpha i}/\pi_{\alpha j})^{-\frac{1}{2}\delta(i,\,s_\alpha)} \quad (\alpha = 1, ..., p),$$

for all treatment pairs (i, j). We note that

$$
\left.
\begin{aligned}
&h(\mathbf{s}|i,j) = 1 + \sum_{\alpha < \beta} \sum \delta(s_\alpha, s_\beta) \rho_{\alpha\beta} d_{\alpha ij} d_{\beta ij}, \\[6pt]
&\frac{\partial h(\mathbf{s}|i,j)}{\partial \pi_{\alpha i}} = -d_{\alpha ij} \sum_\beta{}' \delta(i, s_\beta) \rho_{\alpha\beta} d_{\beta ij}/2\pi_{\alpha i}, \\[6pt]
&\frac{\partial h(\mathbf{s}|i,j)}{\partial \rho_{\alpha\beta}} = \delta(s_\alpha, s_\beta) d_{\alpha ij} d_{\beta ij}, \\[6pt]
&\frac{\partial^2 h(\mathbf{s}|i,j)}{\partial \pi_{\alpha i}^2} = d_{\alpha ij} \sum_\beta{}' \{\delta(s_\alpha, s_\beta) + 2\delta(i, s_\beta)\} \rho_{\alpha\beta} d_{\beta ij}/4\pi_{\alpha i}^2, \\[6pt]
&\frac{\partial^2 h(\mathbf{s}|i,j)}{\partial \pi_{\alpha i} \partial \pi_{\alpha j}} = -d_{\alpha ij} \sum_\beta{}' \delta(s_\alpha, s_\beta) \rho_{\alpha\beta} d_{\beta ij}/4\pi_{\alpha i}\pi_{\alpha j}, \\[6pt]
&\frac{\partial^2 h(\mathbf{s}|i,j)}{\partial \pi_{\alpha i} \partial \pi_{\beta i}} = \rho_{\alpha\beta} d_{\alpha ij} d_{\beta ij}/4\pi_{\alpha i}\pi_{\beta i} \quad (\alpha \neq \beta), \\[6pt]
&\frac{\partial^2 h(\mathbf{s}|i,j)}{\partial \pi_{\alpha i} \partial \pi_{\beta j}} = -\rho_{\alpha\beta} d_{\alpha ij} d_{\beta ij}/4\pi_{\alpha i}\pi_{\beta j} \quad (\alpha \neq \beta), \\[6pt]
&\frac{\partial^2 h(\mathbf{s}|i,j)}{\partial \pi_{\alpha i} \partial \rho_{\alpha\beta}} = -\delta(i, s_\beta) d_{\alpha ij} d_{\beta ij}/2\pi_{\alpha i}, \\[6pt]
&\frac{\partial^2 h(\mathbf{s}|i,j)}{\partial \pi_{\alpha i} \partial \rho_{\gamma\delta}} = 0 \quad \text{if} \quad \gamma, \delta \neq \alpha, \\[6pt]
&\frac{\partial^2 h(\mathbf{s}|i,j)}{\partial \rho_{\alpha\beta} \partial \rho_{\gamma\delta}} = 0 \quad (\alpha < \beta, \gamma < \delta, \alpha, \beta, \gamma, \delta = 1, ..., p).
\end{aligned}
\right\} \quad \text{(A 1)}
$$

Note that other second-order partial derivatives of $h(\mathbf{s}|i,j)$ involving differentiation with respect to $\pi_{\alpha k}$ when $k \neq i, j, \alpha = 1, ..., p$, are zero. Many simplifications occur as a consequence of the relationships of the derivatives (A 1) to the quantities $p^{(1)}(\mathbf{s}|i,j)$ in (1, 2) which represent the probability model in the case of independence.

Derivations of the following quantities are given in detail by Davidson (1966). We list required expectations:

(i) for the submatrices $\boldsymbol{\Lambda}_{\alpha\alpha}$ in (4), $\alpha = 1, ..., p$:

$$E[\partial^2 \ln f_{ij}/\partial \pi_{\alpha i}^2] = -(\pi_{\alpha j}/\pi_{\alpha i})(\pi_{\alpha i}+\pi_{\alpha j})^{-2}$$
$$-\sum_{\mathbf{s}} p^{(1)}(\mathbf{s}|i,j) h^{-1}(\mathbf{s}|i,j)[\partial h(\mathbf{s}|i,j)/\partial \pi_{\alpha i}]^2 \quad \text{(A 2)}$$

and

$$E[\partial^2 \ln f_{ij}/\partial \pi_{\alpha i}\partial \pi_{\alpha j}] = (\pi_{\alpha i}+\pi_{\alpha j})^{-2}$$
$$-\sum_{\mathbf{s}} p^{(1)}(\mathbf{s}|i,j) h^{-1}(\mathbf{s}|i,j)[\partial h(\mathbf{s}|i,j)/\partial \pi_{\alpha i}][\partial h(\mathbf{s}|i,j)/\partial \pi_{\alpha j}], \quad \text{(A 3)}$$

(ii) for the submatrices $\boldsymbol{\Lambda}_{\alpha\beta}$ in (4), $\alpha \neq \beta, \alpha, \beta = 1, ..., p$:

$$E[\partial^2 \ln f_{ij}/\partial \pi_{\alpha k}\partial \pi_{\beta l}] = \sum_{\mathbf{s}} p^{(1)}(\mathbf{s}|i,j)[\partial^2 h(\mathbf{s}|i,j)/\partial \pi_{\alpha k}\partial \pi_{\beta l}]$$
$$-\sum_{\mathbf{s}} p^{(1)}(\mathbf{s}|i,j) h^{-1}(\mathbf{s}|i,j)[\partial h(\mathbf{s}|i,j)/\partial \pi_{\alpha k}][\partial h(\mathbf{s}|i,j)/\partial \pi_{\beta l}] \quad (k,l=i,j),$$
$$\text{(A 4)}$$

and is zero otherwise,

(iii) for the submatrices $\boldsymbol{\Delta}_{\alpha}$ in (7), $\alpha = 1, ..., p$:

$$E[\partial^2 \ln f_{ij}/\partial \pi_{\alpha k}\partial \rho_{\gamma\delta}] = -\sum_{\mathbf{s}} p^{(1)}(\mathbf{s}|i,j) h^{-1}(\mathbf{s}|i,j)[\partial h(\mathbf{s}|i,j)/\partial \pi_{\alpha k}]$$
$$\times [\partial h(\mathbf{s}|i,j)/\partial \rho_{\gamma\delta}] \quad (k=i,j, \gamma < \delta, \alpha, \gamma, \delta = 1, ..., p), \quad \text{(A 5)}$$

and is zero otherwise, and

(iv) for the submatrix $\boldsymbol{\Gamma}$ in (11):

$$E[\partial^2 \ln f_{ij}/\partial \rho_{\alpha\beta}\partial \rho_{\gamma\delta}] = -\sum_{\mathbf{s}} p^{(1)}(\mathbf{s}|i,j) h^{-1}(\mathbf{s}|i,j)[\partial h(\mathbf{s}|i,j)/\partial \rho_{\alpha\beta}]$$
$$\times [\partial h(\mathbf{s}|i,j)/\partial \rho_{\gamma\delta}] \quad \text{for} \quad (\alpha < \beta, \gamma < \delta, \alpha, \beta, \gamma, \delta = 1, ..., p). \quad \text{(A 6)}$$

We can now evaluate $\mathbf{C}^*$ in (12).

In order to evaluate $\mathbf{C}_0^* = \mathbf{C}^*|_{\pi=[1/t]}$, one needs the evaluation of $h(\mathbf{s}|i,j)$ and certain of its derivatives when $\boldsymbol{\pi} = [1/t]$. Results are as follows:

$$h_0(\mathbf{s}|i,j) = 1 + \sum_{\alpha<\beta}\sum \delta(s_\alpha, s_\beta)\rho_{\alpha\beta},$$
$$\partial h_0(\mathbf{s}|i,j)/\partial \pi_{\alpha i} = -\frac{t}{2}\sum_\beta{}' \delta(i, s_\beta)\rho_{\alpha\beta} \quad (\alpha=1, ..., p), \quad \text{(A 7)}$$
$$\partial h_0(\mathbf{s}|i,j)/\partial \rho_{\alpha\beta} = \delta(s_\alpha, s_\beta) \quad (\alpha < \beta, \alpha, \beta = 1, ..., p),$$

and

$$\partial^2 h_0(\mathbf{s}|i,j)/\partial\pi_{\alpha i}\partial\pi_{\beta i}$$
$$= -\partial^2 h_0(\mathbf{s}|i,j)/\partial\pi_{\alpha i}\partial\pi_{\beta j} = t^2\rho_{\alpha\beta}/4 \quad (\alpha \neq \beta, \alpha, \beta = 1, ..., p).$$

Substitution from (A 7) in (A 2), (A 3) and (A 4) leads to

$$E_0[\partial^2 \ln f_{ij}/\partial\pi_{\alpha i}\partial\pi_{\beta i}] = -w_{\alpha\beta}t^2/4,$$

and $\qquad E_0[\partial^2 \ln f_{ij}/\partial\pi_{\alpha i}\partial\pi_{\beta j}] = w_{\alpha\beta}t^2/4,$

where $\mathbf{W} = [w_{\alpha\beta}; \alpha, \beta = 1, ..., p]$ as given in (17) and (18). From (A 5) it is seen that

$$E_0[\partial^2 \ln f_{ij}/\partial\pi_{\alpha i}\partial\rho_{\gamma\delta}] = 0 \quad (\gamma < \delta, \alpha, \gamma, \delta = 1, ..., p),$$

and from (A 6) that

$$E_0[\partial^2 \ln f_{ij}/\partial\rho_{\alpha\beta}\partial\rho_{\gamma\delta}] = -2^{-p}\sum_{\mathbf{s}} \delta(s_\alpha, s_\beta)\,\delta(s_\gamma, s_\delta)\,[1 + \sum_{\alpha<\beta}\sum \delta(s_\alpha, s_\beta)\rho_{\alpha\beta}]^{-1}.$$

It follows now that the required special case of (9) reduces to

$$\mathbf{\Lambda}^*_{\alpha\beta 0} = w_{\alpha\beta}\,\mathbf{E} \quad (\alpha, \beta = 1, ..., p),$$

where $\mathbf{E}$ is given in (19) and (20). In addition, from (10),

$$\mathbf{\Delta}^*_{\alpha 0} = \mathbf{0} \quad (\alpha = 1, ..., p),$$

and, from (11),

$$\mathbf{\Gamma}_0 = [2^{-p}\sum_{i<j}\sum \mu_{ij}\delta(s_\alpha, s_\beta)\,\delta(s_\gamma, s_\delta)\,\{1 + \sum_{\alpha<\beta}\sum \delta(s_\alpha, s_\beta)\rho_{\alpha\beta}\}^{-1};$$

$$\alpha < \beta, \gamma < \delta, \alpha, \beta, \gamma, \delta = 1, ..., p].$$

The matrix $\mathbf{C}^*_0$ from (12) is

$$\mathbf{C}^*_0 = \begin{bmatrix} \mathbf{\Lambda}^*_0 & \mathbf{0} \\ \mathbf{0} & \mathbf{\Gamma}_0 \end{bmatrix},$$

where $\mathbf{\Lambda}^*_0 = \mathbf{W} \otimes \mathbf{E}$.

REFERENCES

Bradley, R. A. (1955). Rank analysis of incomplete block designs. III. Some large-sample results on estimation and power for a method of paired comparisons. *Biometrika* **42**, 450–70.

Bradley, R. A. and Gart, J. J. (1962). The asymptotic properties of ML estimators when sampling from associated populations. *Biometrika* **49**, 205–14.

Bradley, R. A. and Terry, M. E. (1952). Rank analysis of incomplete block designs. I. The method of paired comparisons. *Biometrika* **39**, 324–45.

David, H. A. (1963). *The Method of Paired Comparisons.* Hafner: New York.

Davidson, R. R. (1966). *Multivariate Paired Comparisons*. Florida State University Dissertation, Tallahassee.

Davidson, R. R. and Bradley, R. A. (1969). Multivariate paired comparisons: The extension of a univariate model and associated estimation and test procedures. *Biometrika* **56**, 81–95.

Davidson, R. R. and Lever, W. E. (1970). *The Limiting Distribution of the Likelihood Ratio Statistic Under a Class of Local Alternatives*. *Sanhkyā*. (In Press.)

Fix, E. (1949). Tables of noncentral χ^2. *Univ. Calif. Publ. Statistics* **1**, 15–19.

Hoeffding, W. and Rosenblatt, J. (1955). The efficiency of tests. *Ann. Math. Statist.* **26**, 52–63.

Noether, G. E. (1957). *The Efficiency of Nonparametric Methods*. Final Report, Contract NONR 2393(00)–NR 042096, Math. Dept., Boston University, Boston.

Sen, P. K. and David, H. A. (1968). Paired comparisons for paired characteristics. *Ann. Math. Statist.* **39**, 200–8.

DISCUSSION ON
DAVIDSON AND BRADLEY'S PAPER

J. EDWARD JACKSON

This discussion relates not only to the paper just presented by Bradley and Davidson but also to another paper by the same authors which appeared in *Biometrika*[1]. Anyone seriously concerned with multi-variate paired comparisons should read the *Biometrika* paper first as it more or less sets the stage for this present paper. It includes the development of the model, a multivariate extension of the Bradley–Terry model, as well as a discussion of some alternative models which were not used. Significance tests are given for testing the equality of the preference parameters both for the case where the correlations between these parameters are unspecified, the case discussed this afternoon, and the case where the correlations are all assumed to be equal to zero. There is also a significance test for the correlations themselves which may be employed as a preliminary test before testing the means. A goodness-of-fit procedure is given which is a natural extension of Bradley's univariate method and two numerical examples are included.

The *Biometrika* paper also discusses iterative methods of solving the maximum likelihood equations, a problem which created a fair amount of interest in the univariate case and ought to create more now because the problem is just that much larger; one must now estimate

$$pt + p(p-1)/2$$

parameters from a set of that many non-linear equations. As the number of variables increases, the number of these parameters increases quite rapidly. I don't think the size of this system of equations will be as much of a problem as it might appear to be on the surface, however, because of the practical limitation on the observer of the number of attributes for which he can make a pair judgement as well as the number of pair comparisons themselves. Nevertheless, one of our standard investigations involves 13 stimuli[2]. The use of as few as three attributes would require us to estimate 42 parameters; four attributes would require 58.

The authors are to be congratulated for presenting this procedure in the general case (i.e. unequal sample sizes) because this is often the sort of thing one runs into in practice. When the human response is involved, it is fairly easy to end up with missing data; further, in many situations each pair comparison is performed by a different group of observers (Scheffé's paired comparison model makes this one of the assumptions [3]) such as in consumer preference surveys where unequal sample sizes would probably be the rule, not the exception. There are premiums for keeping the sample sizes equal, of course, such as simplicity in operation and presumably maximum power although it might make an interesting conjecture to require larger sample sizes for pair comparisons involving treatments having the larger standard errors (assuming some knowledge of this is available in advance) and see what this and other variations in sample size do to the power.

One of the interesting aspects of this technique is the availability from the maximum likelihood equations of the correlations between the response attributes. These correlations are used as a preliminary test to determine which of two procedures should be employed in testing for the equality of stimuli means since if no correlations exist, one can essentially combine the results of p univariate tests. These correlations can also be used to determine the rank of the system and study the underlying structure of the relationship among these response variables. In one of the examples given in the *Biometrika* paper, a comparison of three different brands of chocolate puddings, responses are given for three attributes: taste or flavor, color, and texture or feel in the mouth. This particular example came up with no significant differences between brands. However, the correlations among the attributes were significantly different from zero (by and large, observers tended to prefer one or the other brand in each pair for all three attributes) and the correlation matrix $\rho_{\alpha\beta}$ for these is:

$$\begin{bmatrix} 1 & 0\cdot675 & 0\cdot654 \\ & 1 & 0\cdot588 \\ & & 1 \end{bmatrix}.$$

Looking at the principal components, we find that the characteristic roots are $2\cdot28$, $0\cdot41$ and $0\cdot31$. The latter two are probably not significantly different but the vector associated with the maximum root explains over 75 % of the trace. This vector has nearly equal positive coefficients. We would conclude (as we could also conclude by looking at the data in this small example) that we could get nearly as much information from one attribute alone. This example is quite small ($t = 3, p = 3$) and the $\pi_{i\alpha}$'s were not significantly different, but on a larger experiment, a rotation of the response parameter space might prove quite useful both as an investigation of the parameter space and as a possible guide toward the elimination of redundant response variables.

Most paired comparison procedures have a goodness-of-fit test associated with them as does the procedure of the authors. These procedures use the distance between each pair of parameter estimates on the response scale to predict the original pair preference scores. Large deviations between predicted and observed scores are a function of *circular triads* (i is preferred over j, j is preferred over k, and k is preferred over i) or *secondary circular triads* (i is preferred strongly over j, j is preferred strongly over k, but i is preferred mildy over k). These triads are a result of:

(1) no significant differences among the stimuli,
(2) inconsistencies on the part of one or more observers, or
(3) interactions between stimuli and their presentation to the observer.

Statistical procedures can do nothing to correct the first two problems (though they may be able to detect them). However, multivariate paired comparisons might be of some use in the third case. If an observer is choosing between pairs on the basis of overall quality, he may unconsciously be considering several attributes simultaneously (such as the three in the pudding example) but he must come up with one answer for each pair. Hence, it is possible that he may weigh one attribute more heavily in one pair comparison than he would in another and this interaction might produce the triads discussed above. Asking for a judgement on each attribute separately would, up to a point, make it easier for the observer and minimize the possibility of these circular triads. One should, of course, guard against too many attributes where human judgements are involved because of the complexities of running the experiment, fatigue, etc.

The high speed computer has been responsible for making available to us many statistical techniques which have been in the literature for some time but little used because of the time and labor involved—power spectrum and principal component analysis are two good examples. The procedures developed by Davidson and Bradley are also going to be very dependent on the computer. Not only is the solution of the aforementioned maximum likelihood equations a formidable challenge because of convergence problems, but the determination of the constants for these equations also involves considerable organization of the data. For the procedures discussed, one will need the covariance matrix of the estimates of the stimuli preference parameters and the correlations among them.

This is obviously going to require a computer also. If one uses the augmented matrix procedure, C_A, the matrix to be inverted is of the order $pt + p(p+1)/2$. The other method requires the inversion of a matrix of order $p(t-1) + p(p-1)/2$; and while this has $2p$ less rows and columns than C_A, it does require some extra manipulation after the inverse has been obtained. The main labor, it appears, is in obtaining the matrices in the first place. As is the case in obtaining maximum likelihood equations, these computations are probably not very time consuming but rather complicated, again consisting of organizing the data vectors. Anyone who is considering multivariate paired comparisons should certainly be prepared to obtain this covariance matrix as well since a measure of the precision of the estimates is almost as important as the estimates themselves.

As a part-time worker in the field of sensory difference experiments, I have been in need of such techniques for some time. This procedure should be a useful addition to our kit of tools.

REFERENCES

[1] Davidson, R. R. and Bradley, R. A. Multivariate paired comparisons: The extenson of a univariate model and associated estimation and test procedures. *Biometrika* **56**, 1969, 81–95.
[2] Jackson, J. E. and Fleckenstein, Mary. An evaluation of some statistical techniques used in the analysis of paired comparison data. *Biometrics* **13**, 1957, 51–64.
[3] Scheffé, H. An analysis of variance for paired comparisons. *J. Am. Statist. Ass.* **47**, 1952, 381–400.

STATISTICAL INFERENCE IN
INCOMPLETE BLOCKS DESIGNS

MADAN L. PURI† AND HAROLD D. SHANE

SUMMARY

In an earlier paper (Shane and Puri, 1969), the authors developed a
class of asymptotically nonparametric tests for a bivariate paired comparison model. This paper unifies and complements the results of the
previous paper by deriving a class of genuinely distribution free tests
for the same problem but under the more general framework of $p(\geq 2)$-
variate situations. This is done by exploiting the theory of permutation
distribution under sign invariant transformations to a class of rank
order statistics. Asymptotic properties of these permutation rank
order tests are studied and certain stochastic equivalence relationship
with a similar class of multisample extensions of the p-variate one sample
rank order tests proposed by Sen and Puri (1967) are derived. The asymptotic power properties of these tests are also studied.

1 INTRODUCTION

Although a great deal of work has appeared in recent years in the area
of univariate paired comparisons, very little attention has been paid to
its development in the multivariate situations, that is, those situations,
in which there is interest in several characteristics or attributes. Recently,
Sen and David (1968) and Bradley and Davidson (1969) have considered
the situations where the responses to the paired comparisons on each of
the characteristics are obtained, and these responses are qualitative in
nature. Both these tests are generalizations of the univariate sign test,
and are asymptotically equivalent. Their ARE (Asymptotic Relative
Efficiency) can be as low as zero relative to the normal theory test when
the parent population is normal. Shane and Puri (1969) considered a
class of asymptotic tests based on the ranks of the observations for the
bivariate situations, and established the asymptotic superiority of some
of the members of the proposed class over the Sen and David procedure
for a wide class of alternatives. The present work extends the previous
paper in two directions. (i) It provides the exact tests for the problem
considered and this fills the gap left in the previous paper, and (ii), it

† Work supported under National Science Foundation Grant No. GP–12462.

extends the results of the previous paper to the p-variate situations. The theory is thus unified and an attempt is made to give it a complete form. We now formulate the problem.

Let us consider t treatments in an experiment which yields paired observations, namely, $(\mathbf{X}_{jl}, \mathbf{Y}_{jl})$, $l = 1, ..., N_{jl}$, obtained by N_{ij} independent paired comparisons for each pair (i, j) of treatments, $1 \leqslant i < j \leqslant t$. We assume that N_{ij} difference scores $\mathbf{Z}_{ij,l} = \mathbf{Y}_{jl} - \mathbf{X}_{il} = (X_{ij,l}^{(1)}, ..., X_{ij,l}^{(p)})$, $l = 1, ..., N_{ij}$, have a common absolutely continuous cdf (cumulative distribution function) $\Pi_{ij}(\mathbf{z})$, $\mathbf{z} = (x_1, ..., x_p)'$. This is the situation, for example, if in the analysis of incomplete blocks experiments with each block of size 2, one makes the assumption of additivity in the usual (multivariate) analysis of variance model. We are interested in testing the hypothesis

$$H : \Pi_{ij}(\mathbf{z}) = \Pi(\mathbf{z}) \quad \text{for all} \quad 1 \leqslant i < j \leqslant t, \tag{1.1}$$

where $\Pi(\mathbf{z})$ is diagonally symmetric about $\mathbf{z} = \mathbf{0}$ (that is, its density $\pi(\mathbf{z})$ is invariant under simultaneous changes of signs of all the coordinate variates), against the alternative that not all Π_{ij} are identical.

2 PERMUTATIONALLY DISTRIBUTION-FREE RANK TESTS

Let $c = \binom{t}{2}$ denote the number of all possible pairs, and label the pair (i, j) by

$$\alpha = (i - 1) t + j - \binom{i+1}{2} \quad (1 \leqslant i < j \leqslant t).$$

Then $\mathbf{Z}_{\alpha 1}, ..., \mathbf{Z}_{\alpha N_\alpha}$ are independent observations corresponding to the αth pair, and they are distributed according to the absolutely continuous cdf

$$\Pi_\alpha(\mathbf{z}) = \Pi_\alpha(x_1, ..., x_p)' \quad (\alpha = 1, ..., c).$$

Let

$$N = \sum_{\alpha=1}^{c} N_\alpha = \frac{1}{2} \sum_{i=1}^{t} \sum_{\substack{j=1 \\ i \neq j}}^{t} N_{ij}.$$

Under the null hypothesis H of no treatment difference,

(i) $\mathbf{Z}_{\alpha r}, r = 1, ..., N_\alpha, \alpha = 1, ..., c$

are N independent and identically distributed random variables each having the cdf $\Pi(\mathbf{z})$, and (ii) $\mathbf{Z}_{\alpha r}$ and $(-1)\mathbf{Z}_{\alpha r}$ have the same cdf $\Pi(\mathbf{z})$ for all $r = 1, ..., N_\alpha$, $\alpha = 1, ..., c$. Let us denote the sample point by

$$\mathbf{Y}_N = (\mathbf{Z}_{11}, ..., \mathbf{Z}_{1N_1}, ..., \mathbf{Z}_{cN_c}, \quad \mathbf{Z}_{\alpha r} = (X_{\alpha r}^{(1)}, ..., X_{\alpha r}^{(p)})', \tag{2.1}$$

and the sample space by y_N. Let $G_N^{(1)}$ be the group of transformations

$g_N^{(1)}$ given by

$$g_N^{(1)} Y_N = ((-1)^{j_1} Z_{11}, \ldots, (-1)^{j_{N_1}} Z_{1N_1}, \ldots, (-1)^{j_{N_c}} Z_{cN_c})$$

$$(j_i = 0, 1; \; i = 1, \ldots, N). \quad (2.2)$$

Where

$$(-1) Z'_{ar} = (-X_{ar}^{(1)}, \ldots, -X_{ar}^{(p)}) \quad (r = 1, \ldots, N_\alpha, \alpha = 1, \ldots, c).$$

Let $G_N^{(2)}$ be another group of transformations $g_N^{(2)}$ given by

$$g_N^{(2)} \mathbf{Y}_N = \text{some permutation of the columns of } \mathbf{Y}_N. \quad (2.3)$$

Finally, let
$$G_N = G_N^{(1)} G_N^{(2)}$$

be the direct product of the groups. Thus G_N is a finite group of $N! 2^N$ transformations which map the sample space onto itself. Under H the joint distribution of $\mathbf{Y}_N$ remains invariant under G_N. Thus for every point $\mathbf{Y}_N$ of $\mathbf{y}_N$, there exists a set $S(\mathbf{Y}_N)$ of $N! 2^N$ points which are obtained by operating G_N on $\mathbf{Y}_N$. Hence, under H, given $\mathbf{Y}_N$, all the $N! 2^N$ sample points generated by G_N are equiprobable, each having the conditional probability $(N! 2^N)^{-1}$. Let us denote this conditional probability measure defined over $\{g_N \mathbf{Y}_N, g_N \in G_N\}$ by $\mathscr{P}_N$. Then, if we consider any test function $\phi(\mathbf{Y}_N)$ depending on the completely specified permutational probability measure P_N, it follows that such a test will be strictly distribution free. Now in actual practice $\phi(\mathbf{Y}_N)$ has to be constructed with special attention to the class of alternatives in mind; and in most of the problems, $\phi(\mathbf{Y}_N)$ depends on $\mathbf{Y}_N$ through a single-valued statistic $S_N = S(\mathbf{Y}_N)$ formulated in a suitable manner. With this end in view, we consider a class of rank order tests for H in (1.1).

Let us rank the N elements in each row of $\mathbf{Y}_N$ in the increasing order of their absolute values, and denote the rank of $|X_{ar}^{(i)}|$ by $R_{ar}^{(i)}$. We obtain a $p \times N$ rank matrix

$$\mathbf{R}_N = \begin{pmatrix} R_{11}^{(1)} & \cdots & R_{1N_1}^{(1)} & \cdots & R_{cN_c}^{(1)} \\ \cdots\cdots\cdots\cdots\cdots\cdots\cdots\cdots\cdots \\ R_{11}^{(i)} & \cdots & R_{1N_1}^{(i)} & \cdots & R_{cN_c}^{(i)} \\ \cdots\cdots\cdots\cdots\cdots\cdots\cdots\cdots\cdots \\ R_{11}^{(p)} & \cdots & R_{1N_1}^{(p)} & \cdots & R_{cN_c}^{(p)} \end{pmatrix}, \quad (2.4)$$

where by virtue of the assumed continuity of Π_{ij}, the possibility of ties is neglected in probability. Let now

$$\{E_{N,\alpha}^{(i)}, \alpha = 1, \ldots, N, i = 1, \ldots, p\}$$

be a sequence of real numbers, and denote the value of $E_{N,\alpha}^{(i)}$ corresponding to $\alpha = R_{ar}^{(i)}$ by $E_{N,R_{ar}^{(i)}}^{(i)}$. We get a $p \times N$ matrix of general scores $\mathbf{E}_N$

corresponding to $\mathbf{R}_N$:

$$
\mathbf{E}_N = \begin{pmatrix}
E^{(1)}_{N, R^{(1)}_{11}} & \cdots & E^{(1)}_{N, R^{(1)}_{1N_1}} & \cdots & E^{(1)}_{N, R^{(1)}_{cN_c}} \\
\cdots\cdots\cdots\cdots\cdots\cdots\cdots\cdots\cdots\cdots\cdots\cdots\cdots \\
E^{(i)}_{N, R^{(i)}_{11}} & \cdots & E^{(i)}_{N, R^{(i)}_{1N_1}} & \cdots & E^{(i)}_{N, R^{(i)}_{cN_c}} \\
\cdots\cdots\cdots\cdots\cdots\cdots\cdots\cdots\cdots\cdots\cdots\cdots\cdots \\
E^{(p)}_{N, R^{(p)}_{11}} & \cdots & E^{(p)}_{N, R^{(p)}_{1N_1}} & \cdots & E^{(p)}_{N, R^{(p)}_{cN_c}}
\end{pmatrix}. \tag{2.5}
$$

Later on, in §3, we shall impose certain restrictions on $\mathbf{E}_N$. Consider now the univariate rank order statistics co-ordinatewise:

$$
T_{Nk(\alpha)} = N_\alpha^{-1} \sum_{r=1}^{N_\alpha} E^{(k)}_{N, R^{(k)}_{\alpha r}} C^{(k)}_{\alpha r} \quad (k = 1, \ldots, p, \alpha = 1, \ldots, c), \tag{2.6}
$$

where

$$
C^{(k)}_{\alpha r} = +1 \quad \text{if} \quad X^{(k)}_{\alpha r} > 0 \quad \text{and} \quad C^{(k)}_{\alpha r} = -1 \quad \text{if} \quad X^{(k)}_{\alpha r} < 0. \tag{2.7}
$$

To consider the test statistic we first find the conditional mean and dispersion matrix of $T_{Nk(\alpha)}$, $k = 1, \ldots, p$, $\alpha = 1, \ldots, c$ under $\mathscr{P}_N$. Omitting the routine computations, they are

$$
E[T_{Nk(\alpha)}|\mathscr{P}_N] = 0 \quad (k = 1, \ldots, p, \alpha = 1, \ldots, c). \tag{2.8}
$$

$$
\mathrm{cov}\,[T_{Nk(\alpha)}, T_{Nl(\beta)}|\mathscr{P}_N] = N_\alpha^{-1} N_\beta^{-1} \sum_{r=1}^{N_\alpha} \sum_{s=1}^{N_\beta} E[E^{(k)}_{N, R^{(k)}_{\alpha r}} E^{(l)}_{N, R^{(l)}_{\beta s}} C^{(k)}_{\alpha r} C^{(l)}_{\beta s}|\mathscr{P}_N]
$$

$$
= N_\alpha^{-1} a^*_{N, kl} \delta_{\alpha\beta} \tag{2.9}
$$

where

$$
a^*_{N, kl} = N^{-1} \sum_{\alpha=1}^{c} \sum_{r=1}^{N_\alpha} E^{(k)}_{N, R^{(k)}_{\alpha r}} E^{(l)}_{N, R^{(l)}_{\alpha l}} C^{(k)}_{\alpha r} C^{(l)}_{\alpha r}. \tag{2.10}
$$

The last equality of (2.9) follows from the fact that

$$
E[E^{(k)}_{N, R^{(k)}_{\alpha r}} E^{(l)}_{N, R^{(l)}_{\beta s}} C^{(k)}_{\alpha r} C^{(l)}_{\beta s}|\mathscr{P}_N] = \delta_{\alpha\beta} \delta_{rs} a^*_{N, kl}, \tag{2.11}
$$

where $\delta_{\alpha\beta}$ and δ_{rs} are the usual Kronecker deltas.

Now let us revert back to the original notation and write the pair (i, j) for α. Denote

$$
U^{(k)}_{N, i} = \sum_{\substack{j=1 \\ j \neq i}}^{p} N^{\frac{1}{2}}_{ij} T_{Nk(ij)} \quad (i \neq j = 1, \ldots, p, k = 1, \ldots, p). \tag{2.12}
$$

Then

$$
E[U^{(k)}_{N, i}|\mathscr{P}_N] = 0, \tag{2.13}
$$

$$
\mathrm{cov}\,[(U^{(k)}_{N, i}, U^{(l)}_{N, i'})] = (t\delta_{ii'} - 1) a^*_{N, kl} \quad (i, i' = 1, \ldots, p; k, l = 1, \ldots, p). \tag{2.14}
$$

Denote

$$
\mathbf{A}^*_N = ((a^*_{N, kl})) \tag{2.15}
$$

and assume that $\mathbf{A}^*_N$ is positive definite. (If $\mathbf{A}^*_N$ is singular, then we can work with the highest order non-singular minor of $\mathbf{A}^*_N$ and work with the corresponding variates). Then we consider the following statistic for testing H:

$$\mathscr{L}_N^* = \frac{1}{t} \sum_{i=1}^{p} \mathbf{U}_{N,i} \mathbf{A}_N^{*-1} \mathbf{U}'_{N,i}, \tag{2.16}$$

where

$$\mathbf{U}_{N,i} = (U_{N,i}^{(1)}, ..., U_{N,i}^{(p)}) \quad \text{and} \quad \mathbf{A}_N^{*-1} \quad \text{is the inverse of } \mathbf{A}_N^*. \tag{2.17}$$

From the remarks made earlier, it follows that given $\mathbf{Y}_N$, the permutation distribution of $\mathscr{L}_N^*$ would be strictly distribution free under H, and hence an exact size α test can be constructed using this permutation distribution of $\mathscr{L}_N^*$. However, to apply the test in practice, we would require to study all the $N! \, 2^N$ possible permuted values of $\mathscr{L}_N^*$ for any given $\mathbf{Y}_N$. Naturally, the labor involved in this procedure increases prohibitively with the increase in the sample sizes. So in large samples we are faced with the problem of approximating the true permutation distribution of $\mathscr{L}_N^*$ by some simple law and to reduce the computational labor by that. This study will be taken up in the next section. In passing we may remark that if $t = 2$, then the statistic $\mathscr{L}_N^*$ is equivalent to a class of multivariate one-sample statistics considered in Sen and Puri (1967) which includes as special cases the multivariate one-sample Wilcoxon and the normal scores statistics among others. Thus the present procedures may be regarded as the multi-sample analogues of the multivariate one sample tests developed by Sen and Puri (1967). [For the corresponding univariate theory, the reader is referred to Puri and Sen (1969).]

3 ASYMPTOTIC DISTRIBUTION THEORY OF PERMUTATION RANK ORDER STATISTICS

We first introduce a few notations.

Let $F_k^{(\alpha)}(x)$, $H_k^{(\alpha)}(x)$ and $F_{kl}^{(\alpha)}(x,y)$ denote the marginal cdf's of $X_{\alpha r}^{(k)}$, $|X_{\alpha r}^{(k)}|$ and $(X_{\alpha r}^{(k)}, X_{\alpha r}^{(l)})$ respectively, $k, l = 1, ..., p; \alpha = 1, ..., c$.

Let $F_{k,N}^{(\alpha)}(x)$ and $H_{k,N}^{(\alpha)}(x)$ be the sample cdf's of $X_{\alpha r}^{(k)}$ and

$$|X_{\alpha r}^{(k)}| \quad (r = 1, ..., N_\alpha)$$

respectively for $\alpha = 1, ..., c$ and $k = 1, ..., p$.

Denote $\rho_N^{(\alpha)} = N_\alpha / N$ and let $\rho_N^{(\alpha)} \to \rho_0^{(\alpha)}$ as $N \to \infty$, and assume that

$$0 < \rho_0^{(\alpha)} < 1 \quad \text{for} \quad \alpha = 1, ..., c.$$

Let $\quad F_{k,N}(x) = \sum_{\alpha=1}^{c} \rho_N^{(\alpha)} F_{k,N}^{(\alpha)}(x), \quad \Pi_k(x) = \sum_{\alpha=1}^{c} \rho_0^{(\alpha)} F_k^{(\alpha)}(x),$

$$H_{k,N}(x) = \sum_{\alpha=1}^{c} \rho_N^{(\alpha)} H_{k,N}^{(\alpha)}(x), \quad H_k(x) = \sum_{\alpha=1}^{c} \rho_0^{(\alpha)} H_k^{(\alpha)}(x).$$

Thus $F_{k,N}(x)$ and $H_{k,N}(x)$ are the combined sample cdf's of $X_{\alpha r}^{(k)}$ and $|X_{\alpha r}^{(k)}|, r = 1, \ldots, N_\alpha, \alpha = 1, \ldots, c$ respectively whose population cdf's are $\Pi_k(x)$ and $H_k(x)$ respectively for each $k = 1, \ldots, p$.

Let $\Pi_{kl,N}^{(\alpha)}(x,y)$ be the sample cdf of $(X_{\alpha r}^{(k)}, X_{\alpha r}^{(l)}), r = 1, \ldots, N_\alpha$ and denote

$$\Pi_{kl,N}(x,y) = \sum_{\alpha=1}^{c} \rho_N^{(\alpha)} \Pi_{kl,N}^{(\alpha)}(x,y), \quad F_{kl}(x,y) = \sum_{\alpha=1}^{c} \rho_N^{(\alpha)} F_{kl}^{(\alpha)}(x,y).$$

Thus $\Pi_{kl,N}(x,y)$ is the combined sample cdf of $(X_{\alpha r}^{(k)}, X_{\alpha r}^{(l)})$ for $r = 1, \ldots, N_\alpha$; $\alpha = 1, \ldots, c$ whose population cdf is $F_{kl}(x,y)$. Finally denote

$$H_{kl}(x,y) = F_{kl}(x,y) + F_{kl}(-x,y) + F_{kl}(x,-y) + F_{kl}(-x,-y)$$

and

$$H_{kl,N}(x,y) = \Pi_{kl,N}(x,y) + \Pi_{kl,N}(-x,y) + \Pi_{kl,N}(x,-y) + \Pi_{kl,N}(-x,-y).$$

Note that under H, $\Pi_\alpha(\mathbf{Z}) = \Pi(\mathbf{Z})$ and the marginal cdf of $X_{\alpha r}^{(k)}$ will be denoted by $F_k(x)$, and that of $(X_{\alpha r l}^{(k)}, X_{\alpha r}^{(l)})$ by $F_{kl}^0(x,y)$. $\Pi_\alpha(\mathbf{z})$ may depend upon N, although this is not stated explicitly for notational convenience.

Using the above notations, we can express $T_{Nk(\alpha)}$ defined by (2.6) equivalently as

$$T_{Nk(\alpha)} = \int_0^\infty J_{Nk}\left[\frac{N}{N+1} H_{k,N}(x)\right] d[F_{k,N}^{(\alpha)}(x) + F_{k,N}^{(\alpha)}(-x)]$$

$$(k = 1, \ldots, p; \alpha = 1, \ldots, c), \quad (3.1)$$

where $\quad J_{Nk}\left[\dfrac{r}{N+1}\right] = E_{Nk,r} \quad (r = 1, \ldots, N; k = 1, \ldots, p).$

Although the functions

$$J_{Nj}\left[\frac{r}{N+1}\right] \quad \text{are defined only at} \quad \frac{1}{N+1}, \ldots, \frac{N}{N+1},$$

we may extend their domain of definition to $(0,1)$ by letting them have constant values over

$$\left(\frac{r}{N+1}, \frac{r+1}{N+1}\right) \quad (r = 0, 1, \ldots, N).$$

Furthermore, we make the following assumptions:

Assumption 1. $J_{Nk}(0) = 0 \quad (k = 1, \ldots, p).$

Assumption 2. $\lim\limits_{N \to \infty} J_{Nk}(u) = J_k(u)$ exists for $0 < u < 1$ and is not constant for all $k = 1, \ldots, p$.

Assumption 3. $J_k(u)$ is absolutely continuous, and

$$|J_k^{(i)}(u)| = |dJ_k(u)|du^i| \leqq K[u(1-u)]^{\delta-i-\frac{1}{2}} \quad (i = 0, 1, 2),$$

for some K and some $\delta > 0$.

Assumption 4. $\displaystyle\int_0^\infty \left\{ J_{Nk}\left[\frac{1}{N+1} H_{k,N}(x)\right] - J_k\left[\frac{N}{N+1} H_{k,N}(x)\right] \right\}$

$$+ d[F_{k,N}^{(\alpha)}(x) + F_{k,N}^{(\alpha)}(-x)] = o_p(N^{-\frac{1}{2}}) \quad (k = 1, ..., p).$$

Assumption 5. $\displaystyle\int_{x=0}^\infty \int_{y=0}^\infty \left\{ J_{Nk}\left[\frac{N}{N+1} H_{k,N}(x)\right] J_{Nl}\left[\frac{N}{N+1} H_{l,N}(y)\right]\right.$

$$\left. - J_k\left[\frac{N}{N+1} H_{k,N}(x)\right] J_l\left[\frac{N}{N+1} H_{l,N}(y)\right]\right\} dH_{kl,N}(x,y) = o_p(1).$$

Let us now define

$$a_{kl}^* = \begin{cases} \displaystyle\int_{x=0}^\infty \int_{y=0}^\infty J_k[H_k(x)] J_l[H_l(y)] dH_{kl}(x,y) & (k \neq l = 1, ..., p), \\ \displaystyle\int_{x=0}^\infty J_k^2[H_k(x)] dH_k(x) & (k = l = 1, ..., p). \end{cases} \tag{3.2}$$

and denote $\qquad \mathbf{A}^* = ((a_{kl}^*)) \quad (k, l = 1, ..., l).$ (3.3)

Assumption 6. $\mathbf{A}^*$ is positive definite.

Remark. Since $H_k(x)$, $H_l(y)$ and $H_{kl}(x, y)$ may depend on N, the matrix $\mathbf{A}^*$ may depend on N. However we shall suppress this fact whenever there is no confusion. Assumptions 2 to 4 are needed to prove the asymptotic normality of the permutation distribution of $N^{\frac{1}{2}}[T_{Nk(\alpha)}, k = 1, ..., p, \alpha = 1, ..., c]$. Assumption 5 is required only for the asymptotic convergence of the permutation covariance matrix $\mathbf{A}_N^*$ defined in (2.15). Assumption 1 is unessential (*see* Puri and Sen (1969*a*)), but is introduced here to simplify certain computations.

Theorem 3.1. Under assumptions 1 to 4, $|\mathbf{A}_N^* - \mathbf{A}^*|$ converges to $\mathbf{0}$ in probability as $N \to \infty$, where $\mathbf{0}$ is a null matrix of order $p \times p$ and $\mathbf{A}_N^*$ and $\mathbf{A}^*$ are defined by (2.15) and, (3.3) respectively.

Proof. By virtue of assumptions 4 and 5, it follows that

$$a_{N,kl}^* = \int_{x=0}^\infty \int_{y=0}^\infty J_{Nk}\left[\frac{N}{N+1} H_{k,N}(x)\right] J_{Nl}\left[\frac{N}{N+1} H_{l,N}(y)\right] dH_{kl,N}(x,y)$$

$$= \int_{x=0}^\infty \int_{y=0}^\infty J_k\left[\frac{N}{N+1} H_{k,N}(x)\right] J_l\left[\frac{N}{N+1} H_{l,N}(y)\right] dH_{kl,N}(x,y) + o_p(1).$$

Now proceeding precisely as in Puri and Sen (1966), it follows after omitting the details of computations, that

$$a^*_{N,kl} = \int_{x=0}^{\infty} \int_{y=0}^{\infty} J_k[H_k(x)]\, J_l[H_l(y)]\, dH_{kl}(x,y) + o_p(1).$$

The proof follows.

Corollary 3.1. If (i) $\Pi_\alpha(z) = \Pi(z + \mu_\alpha N^{-\frac{1}{2}})$ *where* $z = (x_1, \ldots, x_p)'$,

$$\mu_\alpha = (\mu_{1\alpha}, \ldots, \mu_{p\alpha})',$$

and $\Pi(z)$ *is a fixed absolutely continuous cdf diagonally symmetric about* 0, *and* (ii) *the conditions of Theorem 3.1 are satisfied, then* $A^*_N \to A$ *in probability, as* $N \to \infty$, *where* $A = ((a_{kl}))$ *is given by*

$$a_{kl} = \begin{cases} \int_{x=0}^{\infty} \int_{y=0}^{\infty} J_k[2F_k(x)-1]\, J_l[2F_{ll}(y)-1]\, dH^{(0)}_{kl}(x,y) \\ \qquad\qquad\qquad\qquad if \quad (k \neq l = 1, \ldots, p), \\ \int_{x=0}^{1} J_k^2(x)\, dx \quad if \quad (k = l = 1, \ldots, p), \end{cases} \quad (3.4)$$

where $H^{(0)}_{kl}(x,y)$ *is the value of* $H_{kl}(x,y)$ *when* $\mu_\alpha = 0$.

The proof of this corollary is an immediate consequence of the above theorem, and is therefore omitted.

Theorem 3.2. Under the assumptions 1 to 4, the random vectors

$$[N_\alpha^{\frac{1}{2}} T_{N(\alpha)}, \alpha = 1, \ldots, c] \quad where \quad T_{N(\alpha)} = (T_{N1(\alpha)}, \ldots, T_{Np(\alpha)}),$$

are under $\mathscr{P}_N$, *asymptotically independent and identically distributed p-variate normal random variables each having mean* 0, *and covariance matrix* A_N.

Proof. From (2.6) and assumption 4,

$$T_{Nk(\alpha)} = T^*_{Nk(\alpha)} + o_p(N^{-\frac{1}{2}}) \quad (k = 1, \ldots, p, \alpha = 1, \ldots, c), \quad (3.5)$$

where

$$T^*_{Nk(\alpha)} = N_\alpha^{-1} \sum_{r=1}^{N_\alpha} C^{(k)}_{\alpha r} J_k[R^{(k)}_{\alpha r}/(N+1)] \quad (k = 1, \ldots, p, \alpha = 1, \ldots, c). \quad (3.6)$$

Define

$$S_{Nk(\alpha)} = N_\alpha^{-1} \sum_{r=1}^{N_\alpha} C^{(k)}_{\alpha r} J_k[U^{(k)}_{\alpha r}], \quad (3.7)$$

where, for each $k = 1, \ldots, p$, $U^{(k)}_{11}, \ldots, U^{(k)}_{1N_\alpha}, \ldots, U^{(k)}_{cN_c}$ are independent and identically distributed random variables having uniform distribution over $[0, 1]$. Denote

$$U_{\alpha r} = (U^{(1)}_{\alpha r}, \ldots, U^{(p)}_{\alpha r})'. \quad (3.8)$$

Then $\mathbf{U}_{\alpha r}, r = 1, \ldots, N_\alpha, \alpha = 1, \ldots, c$ are independent and identically distributed random vectors distributed according to some p-variate cdf, say $G(\mathbf{x}), \mathbf{x} \in R^p$. (Note that if $\mathbf{Z}_{\alpha r} = (Z_{\alpha r}^{(1)}, \ldots, Z_{\alpha r}^{(p)})'$ has the cumulative distribution function $\Pi(\mathbf{x})$, $\mathbf{x} \in R^p$, and if we let $\mathbf{Y} = (Y_1, \ldots, Y_p)'$, where

$$Y_k = F_k(Z_{\alpha r}^{(k)}), \tag{3.9}$$

F_k being the marginal cumulative distribution function of $Z_{\alpha r}^{(k)}$, then (i) the marginal cdf of each Y_k is uniform over $[0, 1]$, and (ii) the cdf $G(\mathbf{x})$ of $\mathbf{U}_{\alpha r}$ is the same as that of the $\mathbf{Y} = (Y_1, \ldots, Y_p)$.)

Using Theorem 4.3.1 of Ghosh (1969), it follows that

$$N_\alpha^{\frac{1}{2}} |T_{Nk(\alpha)}^* - S_{Nk(\alpha)}| = o_p(1) \quad (k = 1, \ldots, p, \alpha = 1, \ldots, c). \tag{3.10}$$

Thus from (3.5) and (3.10), the limit distribution of $N_\alpha^{\frac{1}{2}} \mathbf{T}_{N(\alpha)}, \alpha = 1, \ldots, c$, is the same as that of $N_\alpha^{\frac{1}{2}} \mathbf{S}_{N(\alpha)}$, where $\mathbf{S}_{N(\alpha)} = (S_{N1(\alpha)}, \ldots, S_{Np(\alpha)})'$, if the latter exists. Thus to prove the asymptotic normality of $N_\alpha^{\frac{1}{2}} \mathbf{T}_{N(\alpha)}$, $\alpha = 1, \ldots, c$, it suffices to show that, for any real

$$\delta_{k\alpha}, k = 1, \ldots, p, \alpha = 1, \ldots, c,$$

not all zero, the linear combination

$$W_N = \sum_{\alpha=1}^{c} N_\alpha^{\frac{1}{2}} \sum_{k=1}^{p} \delta_{k\alpha} S_{Nk(\alpha)} \tag{3.11}$$

is asymptotically normal.

Using (3.7), we can rewrite W_N as

$$W_N = \sum_{\alpha=1}^{c} N_\alpha^{\frac{1}{2}} \sum_{k=1}^{p} \delta_{k\alpha} N_\alpha^{-1} \sum_{r=1}^{N_\alpha} C_{\alpha r}^{(k)} J_k[U_{\alpha r}^{(k)}], \tag{3.12}$$

$$= \sum_{\alpha=1}^{c} \left[N_\alpha^{-\frac{1}{2}} \sum_{r=1}^{N_\alpha} V_{\alpha r} \right], \tag{3.13}$$

where

$$V_{\alpha r} = \sum_{k=1}^{p} \delta_{k\alpha} C_{\alpha r}^{(k)} J_k[U_{\alpha r}^{(k)}]. \tag{3.14}$$

(3.13) represents c-summations which involved independent random variables $V_{\alpha r}$ each of which (by repeated use of C_r-inequality) can be shown to have finite absolute moment of order $2 + \delta'$, $0 < \delta' \le 1$. Hence by the central limit theorem (Liapounoff), each sum properly normalised has normal distribution in the limit, with the result that the sum of c-summations will have normal distribution in the limit. This proves the joint asymptotic normality of $\mathbf{T}_{N(\alpha)}, \alpha = 1, \ldots, c$. Now using (2.9), (2.10), (3.5), (3.7) and (3.10), it can easily be verified that the covariances of the limit distribution is zero for $\alpha \ne \beta$, $k, l = 1, \ldots, p$. This coupled with the joint asymptotic normality of the variables $\mathbf{T}_{N(\alpha)}, \alpha = 1, \ldots, c$ establishes their asymptotic independence. Furthermore, since $\mathbf{T}_{N(\alpha)}, \alpha = 1, \ldots, c$ have identical dispersion matrices, the proof follows.

Theorem 3.3. If the assumptions 1 to 4 and 6 hold, then under H_0 in (1.1) the permutation distribution of $\mathscr{L}_N^*$ defined by (2.6) is asymptotically (as $N \to \infty$) the chi-square with $p(t-1)$ degrees of freedom.

The proof of this theorem follows as a consequence of Theorems 3.1 and 3.2, and is therefore omitted.

By virtue of Theorem 3.3, the permutation test procedure based on $\mathscr{L}_N^*$ simplifies in large samples to the following rule.

$$\mathscr{L}_N^* \geqslant \chi^2_{\alpha,\,p(r-1)} \quad \text{reject } H,$$

$$< \chi^2_{\alpha,\,p(r-1)} \quad \text{accept } H,$$

where $\chi^2_{\alpha,\,p(t-1)}$ is $100(1-\alpha)\%$ point of the chi-square distribution with $p(t-1)$ degrees of freedom.

In order to study the asymptotic power properties of the test considered above, we require to study the unconditional asymptotic distribution of $\mathscr{L}_N^*$ under appropriate classes of alternatives. This, in turn, requires the study of joint asymptotic distribution of the rank order statistics defined in (2.6). For the case of $p = 2$, the same has been studied in Shane and Puri (1969). Since the methods for the case of any finite p are analogous for those for the case of $p = 2$, the details will be skipped as far as possible. However the efficiency results will be discussed in detail.

4   THE PROPOSED CLASS OF ASYMPTOTICALLY DISTRIBUTION FREE TESTS

To derive a class of asymptotically distribution free tests, we first state the following theorem.

Theorem 4.1. Under the assumptions 1 to 4 of §3, the random variables $\sqrt{(N_\alpha)}\,(T_{N1(\alpha)} - \mu_{N1(\alpha)}, \dots, T_{Np(\alpha)} - \mu_{Np(\alpha)})\ \alpha = 1, \dots, c$ have asymptotically a multivariate normal distribution with zero means and covariance matrix

$$\operatorname{Cov}\left[\sqrt{(N_\alpha)}\,(T_{Nk(\alpha)} - \mu_{Nk(\alpha)}),\ \sqrt{(N_\beta)}\,(T_{Nl(\beta)} - \mu_{Nl(\beta)})\right] = \sigma_{N,\,kl,\,\alpha\beta},$$

where $((\sigma_{N,\,kl,\,\alpha\beta}))$, $k, l = 1, \dots, p$; $\alpha, \beta = 1, \dots, c$ are given by (6.2), (6.3), (6.4) and (6.5) respectively, and

$$\mu_{Nk(\alpha)} = \int_{x=0}^{\infty} J[H_k(x)]\,d[F_k^{(\alpha)}(x) + F_k^{(\alpha)}(-x)]. \tag{4.1}$$

The proof of this theorem is long and involved, and is therefore not given here. For the case of $p = 2$, it is briefly sketched in (Shane and Puri, 1969, Theorem 3.1). For the case of any finite p, the details are given in Shane (1968). In fact it is established in Shane (1968), that subject to certain weak assumptions, the asymptotic normality holds uniformly with respect to $\Pi_\alpha(z)$, and $\rho_N^{(g)}, \alpha = 1, \dots, c$.

We now consider a sequence of admissible alternative hypotheses H_N, which specify that for each $\alpha = 1, ..., c$,

$$\Pi_\alpha(\mathbf{z}) = \Pi\left(\mathbf{z} + \frac{\mu_\alpha}{\sqrt{N}}\right),$$

where $\mathbf{z} = (x_1, ..., x_p)'$, $\mu_\alpha = (\mu_{1\alpha}, ..., \mu_{p\alpha})'$ and $\Pi(\mathbf{z})$ is a fixed continuous p-variate cdf diagonally symmetric about $\mathbf{0}$. We shall also assume that the constant $E_{Ni, r}$, $i = 1, ... p$; $r = 1, ..., N$ is the expected value of the rth order statistics of a sample of size N from a distribution

$$\Psi_i(x) = \begin{cases} \Psi_i^*(x) - \Psi_i^*(-x) & \text{if } x \geqslant 0, \\ 0 & \text{otherwise,} \end{cases} \tag{4.2}$$

where $\Psi_i^*(x)$ is a cdf symmetric about zero. If we denote by $J_i = \Psi_i^{-1}$ and $J_i^* = \Psi_i^{*-1}$ then the above definition of $E_{Ni, r}$ implies that

$$J_i(u) = J_i^*\left(\frac{u+1}{2}\right) \quad (0 < u < 1), \quad J_i^*(u) = J_i(2u - 1) \quad (\tfrac{1}{2} < u < 1) \tag{4.3}$$

and
$$J_i^*(u) = -J_i^*(1 - u).$$

Remark. Two cases of special interest which we shall study in greater detail later are when (i) $\Psi_i^*(x)$ is the rectangular distribution over $(-1, 1)$ and (ii) $\Psi_i^*(x)$ is the standard normal distribution function. For the case (i) the corresponding statistic will be termed the rank sum statistic, and for the case (ii), the corresponding statistic will be termed the absolute normal scores statistic.

The following theorem, the proof of which follows from Theorem 4.1, plays a central role in deriving a class of asymptotically distribution free tests as well as in studying the efficiency properties.

Theorem 4.2. If (i) $\rho_N^{(\alpha)} \to \rho_\alpha$ *as* $N \to \infty$ *and* $0 < \rho_\alpha < 1$, $\alpha = 1, ..., c$.
   (ii) *The conditions of Theorem 4.1 are satisfied.*
   (iii) *For each fixed* N, *the hypothesis* H_N *is true. Then the random variables* $[\sqrt(N_\alpha)\,(T_{Nj(\alpha)} - u_{Nj(\alpha)}), j = 1, ..., p; \alpha = 1, ..., c]$ *have a limiting multivariate normal distribution as* $N \to \infty$ *with means zero and covariance matrix* $\boldsymbol{\tau} = (\tau_{ij, \alpha\beta})$, $i, j = 1, ..., p$; $\alpha, \beta = 1, ..., c$, *where*

$$\tau_{jj, \alpha\beta} = \delta_{\alpha\beta} A_j^2 \quad (j = 1, ..., p; \alpha, \beta = 1, ..., c), \tag{4.4}$$

$$\tau_{ij, \alpha\beta} = \tau_{ji\,\alpha\beta} = \delta_{\alpha\beta} \mathscr{S}_{ij} \quad (i \neq j = 1, ..., p; \alpha, \beta, = 1, ..., c), \tag{4.5}$$

$$A_i^2 = \int_0^1 J_i^2(u)\,du = \int_0^1 [J_i^*(u)]^2\,du \quad (i = 1, ..., p), \tag{4.6}$$

$$\mathscr{S}_{ij} = \int_{x=-\infty}^{+\infty}\int_{x=-\infty}^{+\infty} J_i^*(F_i(x))\,J_j^*(F_j(y))\,dH_{ij}^0(x, y), \tag{4.7}$$

$\delta_{\alpha\beta}$ is the Kronecker Delta and $H_{ij}^0(x,y)$ is the value of $H_{ij}(x,y)$ under H in (1.1).

Corollary 4.2.1. Suppose that the hypothesis H_0 is true. Then under the conditions of Theorem 4.1, the random variables

$$[\sqrt{(N_\alpha)}\,T_{Nj(\alpha)}, j = 1, \ldots, p; \; \alpha = 1, \ldots, c]$$

have a limiting multivariate normal distribution with means zero and covariance matrix $\boldsymbol{\tau} = (\tau_{ij,\alpha\beta}), i, j = 1, \ldots, p; \alpha, \beta = 1, \ldots, c$ given by (4.4) and (4.5).

Reverting back to our original notation, we have:

Corollary 4.2.2. Under the assumptions of Theorem 4.1, the random variables $[\sqrt{(N_{ij})}\,(T_{Nk(i,j)} - \mu_{Nk(i,j)}), k = 1, \ldots, p; \; 1 \leqslant i < j \leqslant t]$, where

$$\mu_{Nk(i,j)} = \int_{x=0}^{\infty} J_k(H_k(x))\,d[F_k^{(ij)}(x) + F_k^{(ij)}(-x)]$$

$$(k = 1, \ldots, p; \; 1 \leqslant i < j \leqslant t) \quad (4.8)$$

have, in the limit as $N \to \infty$, multivariate normal distribution with zero mean and covariance matrix $\tau = (\tau_{kl(ij)(i'j')}), k, l = 1, \ldots, p; \; 1 \leqslant (i,i') < (j,j') \leqslant t$ defined by

$$\tau_{kl(ij)(i'j')} = \delta_{(ij)(i'j')}A_k^2 \quad (k = 1, 2), \quad (4.9)$$

$$\tau_{kl(ij)(i'j')} = \tau_{lk(ij)(i'j')} = \delta_{(ij)(i'j')}\mathscr{S}_{kl} \quad (k \neq l) \quad (4.10)$$

where A_k^2 and $\mathscr{S}_{kl}$ are given by (4.11) and (4.12) respectively and

$$\delta_{(ij)(i'j')} = \begin{cases} +1 & \text{if } i = i', \; j = j', \\ -1 & \text{if } i = j', \; j = i', \\ 0 & \text{otherwise.} \end{cases}$$

Now let

$$U_{Ni}^{(k)} = \sum_{\substack{j=1 \\ j \neq i}}^{t} \sqrt{(N_{ij})}\,T_{Nk(ij)}, \quad \mu_{Nk(i\cdot)} = \sum_{\substack{j=1 \\ j \neq i}}^{t} \sqrt{(N_{ij})}\,\mu_{Nk(ij)}$$

$$(k = 1, \ldots, p; \; i = 1, \ldots, t) \quad (4.11)$$

$$\eta_{k(ij)} = \lim_{N \to \infty} \sqrt{(N_{ij})}\,\mu_{Nk(ij)}, \quad \eta_{k(i\cdot)} = \sum_{\substack{j=1 \\ j \neq i}}^{t} \eta_{k(ij)} \quad (4.12)$$

and assume that $\eta_{k(ij)}$ exists and is finite for $k = 1, \ldots, p$ and $1 \leqslant i < j \leqslant t$. Let us introduce the following partitioned vectors

$$\mathbf{U}_N = (\mathbf{U}_N^{(1)}, \mathbf{U}_N^{(2)}, \ldots, \mathbf{U}_N^{(p)}), \quad (4.13)$$

where $\qquad \mathbf{U}_N^{(k)} = (U_{N1}^{(k)}, ..., U_{Nt}^{(k)}) \quad (k = 1, ..., p);$ $\qquad$ (4.14)

$$\boldsymbol{\mu}_{N(\cdot)} = (\boldsymbol{\mu}_{N1(\cdot)}, ..., \boldsymbol{\mu}_{Np(\cdot)}) \qquad (4.15)$$

$$\boldsymbol{\mu}_{Nk(\cdot)} = (\mu_{Nk(1\cdot)}, ..., \mu_{Nk(t\cdot)}) \quad (k = 1, ..., p); \qquad (4.16)$$

$$\boldsymbol{\eta}_{(\cdot)} = (\boldsymbol{\eta}_{1(\cdot)}, ..., \boldsymbol{\eta}_{p(\cdot)}), \qquad (4.17)$$

where $\qquad \boldsymbol{\eta}_{k(\cdot)} = (\eta_{k(1\cdot)}, ..., \eta_{k(t\cdot)}) \quad (k = 1, ..., p).$ $\qquad$ (4.18)

Then we have the following theorem.

Theorem 4.3. Under the assumptions of Theorem 4.1, the random vector $\mathbf{U}_N = (\mathbf{U}_{N.}^{(1)}, ..., \mathbf{U}_N^{(p)})$ *has asymptotically a multivariate normal distribution with mean vector* $\boldsymbol{\eta}_{(\cdot)} = (\boldsymbol{\eta}_{1(\cdot)}, ..., \boldsymbol{\eta}_{p(\cdot)})$ *and covariance matrix*

$$\mathbf{M} = (\mathbf{M}_{kl}) = \lim_{N \to \infty} E[(\mathbf{U}_N - \boldsymbol{\mu}_{N(\cdot)})' (\mathbf{U}_N - \boldsymbol{\mu}_{N(\cdot)})],$$

where $\qquad\qquad \mathbf{M}_{kl} \quad (k, l = 1, ..., p);$

is a $(t \times t)$ *matrix with entries*

$$\mathbf{M}_{kl.ij} = (t\delta_{ij} - 1) a_{kl} \quad (k, l = 1, ..., p; \; i, j = 1, ..., t), \qquad (4.19)$$

where $\qquad\qquad a_{kl} = \begin{cases} A_k^2 & \text{if} \quad k = l \\ \mathscr{S}_{kl} & \text{if} \quad k \neq l \end{cases}$

and the rank of $\mathbf{M}$ *is* $p(t-1)$ *if and only if the matrix*

$$\mathbf{A} = (a_{kl}). \qquad (4.20)$$

is nonsingular.

Proof. The asymptotic normality follows directly from Theorem 4.1 and Corollary 4.2.2. To compute the covariance matrix, we note that

$$\mathbf{M}_{kl} = \lim_{N \to \infty} E[\mathbf{U}_N^{(k)} - \mu_{Nk(\cdot)})' (\mathbf{U}_N^{(l)} - \mu_{Nl(\cdot)})]$$

so that $\qquad \mathbf{M}_{kl,ij} = \lim_{N \to \infty} E[(U_{Ni}^{(k)} - \mu_{Nk(i\cdot)}) (U_{Nj}^{(l)} - \mu_{Nl(j\cdot)})]$

$$= \lim_{N \to \infty} \text{Cov} [U_{Ni}^{(k)}, U_{Nj}^{(l)}]$$

$$= \lim_{N \to \infty} \text{Cov} \left[\sum_{\substack{r=1 \\ r \neq i}}^{t} \sqrt{(N_{ir})} \, T_{Nk(ir)}, \sum_{\substack{s=1 \\ s \neq j}}^{t} \sqrt{(N_{js})} \, T_{Nl(js)} \right]$$

$$= \sum_{\substack{r=1 \\ r \neq i}}^{t} \sum_{\substack{s=1 \\ s \neq j}}^{t} \lim_{N \to \infty} \sigma_{N, kl(ir)(js)} = \sum_{\substack{r=1 \\ r \neq i}}^{t} \sum_{\substack{s=1 \\ s \neq j}}^{t} \tau_{kl(ir)(js)}$$

$$= (\tau\delta_{ij} - 1) a_{kl}$$

by comparison with (4.9) and (4.10). Let

$$C = (C_{ij}) = (t\delta_{ij} - 1) \quad (i,j = 1, \ldots, t) \tag{4.21}$$

Then
$$M = \begin{pmatrix} a_{11}C & a_{12}C & \cdots & a_{1p}C \\ \vdots & & & \\ a_{p1}C & a_{p2}C & \cdots & a_{pp}C \end{pmatrix}. \tag{4.22}$$

Let $\|M\|$ = Determinant M. In $\|M\|$ subtracting row 1 from each of the rows numbered $2, \ldots, t$, row $(t+1)$ from each of the rows $(t+2), \ldots, 2t$, and so on leaves $\|M\|$ unchanged. Then add columns $2, \ldots, t$ to column 1; $(t+2), \ldots, 2t$ to $(t+1)$ etc. and the resulting determinant has columns 1, $t+1, 2t+1, \ldots, (p-1)t+1$ all zero. Hence the rank of M is at most $p(t-1)$.

Now striking the 1st, $(t+1)$st, $\ldots, [(p-1)t+1]$st rows and columns leaves the minor,

$$\|M^*\| = \begin{vmatrix} ta_{11}I & \cdots & ta_{1p}I \\ \vdots & & \\ ta_{p1}I & \cdots & ta_{pp}I \end{vmatrix},$$

where I is a $(t-1) \times (t-1)$ identity matrix. The fact that $\|M^*\| = 0$ if and only if $\|A\| = 0$ follows from

Lemma 4.1. Let A be any symmetric matrix and let

$$M^* = \begin{pmatrix} a_{11}I & \cdots & a_{1p}I \\ \vdots & & \\ a_{p1}I & \cdots & a_{pp}I \end{pmatrix},$$

where I is a $(k \times k)$ identity matrix. Then $\|M^*\| = \|A\|^k$.
The proof being trivial, is omitted.

Remark: We have proved that the rank of M is $p(t-1)$ if and only if the matrix A, which is the dispersion matrix of $\{J_1^*(F_1(X^{(1)})), \ldots, J_p^*(F_p(X^{(p)}))\}$ is non-singular. In what follows, we make the assumption that the distribution function $\Pi(z)$ and the score function $J_k, k = 1, \ldots, p$ are such that the moment matrix A is non-singular. The moment matrix will be singular if and only if there exists one or more relationships of the form

$$\sum_{j=1}^{p} a_j J_j^*(F_j(X^{(j)})) = b. \text{ a.s. } \Pi.$$

Let us define

$$\mathscr{L}_N^{**} = \frac{1}{t} \sum_{i=1}^{t} (U_{Ni}^{(1)} \cdots, U_{Ni}^{(p)}) A^{-1} (U_{Ni}^{(1)}, \ldots, U_{Ni}^{(p)})'. \tag{4.23}$$

We now have,

Theorem 4.4. Under the assumptions of Theorem 4.1

$$\mathscr{L}_N^{**} = \frac{1}{t} \sum_{i=1}^{t} (U_{Ni}^{(1)}, ..., U_{Ni}^{(p)}) \, \mathbf{A}^{-1}(U_{Ni}^{(1)}, ..., U_{Ni}^{(p)})'$$

has asymptotically a non-central χ^2 distribution with $p(t-1)$ degrees of freedom and non-centrality parameter

$$\Delta_{\mathscr{L}} = \frac{1}{t} \sum_{i=1}^{t} (\eta_{1(i\cdot)}, \eta_{2(i\cdot)}, ..., \eta_{p(i\cdot)}) \, \mathbf{A}^{-1} (\eta_{1(i\cdot)}, ..., \eta_{p(i\cdot)})'. \qquad (4.24)$$

The proof of this theorem follows as an application of Sverdrup's (1952) Theorem and the well known property (cf. Rao (1965), p. 443 (viii)) of the multivariate normal distribution. [For the case of $p = 2$, cf. Shane and Puri (1969).]

Corollary 4.3.1. Suppose that the hypothesis H_0 is true. Then under the assumption of Theorem 4.2, $\mathscr{L}_N^{**}$ has the limiting central chi-square distribution with $p(t-1)$ degrees of freedom.

Now let $\hat{\mathbf{A}}$ be a consistent estimator of $\mathbf{A}$, and denote

$$\mathscr{L}_N = \frac{1}{t} \sum_{i=1}^{t} (U_{Ni}^{(1)}, ..., U_{Ni}^{(p)}) \, \hat{\mathbf{A}}^{-1}(U_{Ni}^{(1)}, ..., U_{Ni}^{(p)}). \qquad (4.25)$$

Then, it follows that $\mathscr{L}_N - \mathscr{L}_N^{**}$ converges to zero in probability as $N \to \infty$. Hence $\mathscr{L}_N$, too, has the limiting central chi-square distribution $p(t-1)$ degrees of freedom and so the critical function

$$\Phi(\mathscr{L}_N) = \begin{cases} 1 & \text{if} \quad \mathscr{L}_N \geq \chi^2_{p(t-1),\,\alpha}, \\ 0 & \text{if} \quad \mathscr{L}_N < \chi^2_{p(t-1),\,\alpha}, \end{cases}$$

where $\chi^2_{r,\,\alpha}$ is the $100\,(1-\alpha)\%$ point of the chi-squared distribution with r degrees of freedom, provides an asymptotically level α test of H.

From Theorem 4.3, it is clear that any consistent estimator of $\mathbf{A}^{-1}$ will preserve the asymptotic distribution of the test statistic. However to establish the existence of the test statistic $\mathscr{L}_N$, at least one consistent estimation of $\mathbf{A}^{-1}$ has to be proposed. From (2.10) and Theorem 3.1, it follows that the permutational covariance matrix $\mathbf{A}_N^{-1*}$ defined in (2.15) can be taken as a consistent estimator of $\mathbf{A}^{-1}$.

Theorem 4.5. The permutation test based on $\mathscr{L}_N$ given by (2.16) and the asymptotically nonparametric test based on $\mathscr{L}_N$ given by (4.25) are asymptotically power equivalent for sequence of alternatives H_N defined above (4.2).

The proof follows from Corollary (3.1), Theorem 4.2 and 4.3. Thus by virtue of the stochastic equivalence of the tests $\mathscr{L}_N^*$ and $\mathscr{L}_N$, we shall consider only the asymptotic properties of the unconditional test based on $\mathscr{L}_N$.

In most cases, the quantities $\eta_{j,\alpha} = \lim\limits_{N\to\infty} N_\alpha^{\frac{1}{2}} \mu_{Nj(\alpha)}$ take on simple forms through the help of the following lemma similar to Lemma 7.2 of Puri (1964).

Lemma 4.2. If (i) $F_j, j = 1, \ldots, p$ *is continuous cdf differentiable in each of the open intervals* $[(0, a_1^{(j)}), \ldots, (a_s^{(j)}, \infty)]$ *and the derivative of F_j is bounded in each of the intervals and* $F_j(x) + F_j(-x) = 1$. (ii) *The function*

$$\frac{d}{dx} J_j [F_j(x) - F_j(-x)] \quad (j = 1, \ldots, p)$$

is bounded as $x \to +\infty$, $x \to 0+$. (iii) $J_j = \Psi_j^{-1}$ *and* Ψ_j^* *defined by* (4.2) *is symmetric and unimodal with density* $\Psi_j^*, j = 1, \ldots, p$. *Then*

$$\left.\begin{aligned}
\eta_{j(\alpha)} &= \lim_{N\to\infty} \sqrt{(N_\alpha)} \int_{x=0}^{\infty} J_j \left[\sum_{\beta=1}^{c} \rho_N^{(\beta)} \left\{ F_j\left(x + \frac{\mu_{j\beta}}{\sqrt{N}}\right) - F_j\left(-x + \frac{\mu_{j\beta}}{\sqrt{N}}\right) \right\} \right], \\
d&\left[F_j\left(x + \frac{\mu_{j\alpha}}{\sqrt{N}}\right) + F_j\left(-x + \frac{\mu_{j\alpha}}{\sqrt{N}}\right) \right] \\
&= -2\sqrt{(\rho_\alpha)}\mu_{j\alpha} \int_{x=0}^{\infty} \frac{d}{dx} J_j [2F_j(x) - 1] \, dF_j(x).
\end{aligned}\right\} \quad (4.26)$$

In case the conditions of Lemma 4.2 are satisfied, then

$$\Delta_{\mathscr{L}} = \frac{1}{t} \sum_{i=1}^{t} \mathcal{O}_i \mathbf{B}^{*-1} \mathcal{O}_i', \tag{4.27}$$

where $\quad \mathcal{O}_i = (\mathcal{O}_{1i}, \ldots, \mathcal{O}_{pi}); \quad \mathcal{O}_{ki} = \sum\limits_{\substack{j=1 \\ j \neq i}}^{t} \rho_{ij}^{\frac{1}{2}} \mu_{kij} \quad (k = 1, \ldots, p), \tag{4.28}$

$$B^* = a_{kl/a_k^* \, a_l^*} \quad (k, l = 1, \ldots, p), \tag{4.29}$$

$$a_k^* = -2 \int_{x=0}^{\infty} \frac{d}{dx} J_k [2F_k(x) - 1] \, dF_k(x) \quad (k = 1, \ldots, p) \tag{4.30}$$

and a_{kl} is defined by (4.20).

Introducing the notations

$$\mathscr{L}_N = \mathscr{L}_N(R) \text{ when } J_k^*(x) = 2x - 1 \quad (k = 1, \ldots, p) \text{ (rank sum test)} \tag{4.31}$$

and

$$\mathscr{L}_N = \mathscr{L}_N(\Phi) \text{ when } J_k^*(x) = \Phi^{-1}(x) \quad (k = 1, \ldots, p) \text{ (absolute normal}$$
$$\text{scores test).} \tag{4.32}$$

We obtain from (4.27),

$$\Delta_{\mathscr{L}(R)} = \frac{1}{t} \sum_{i=1}^{t} \mathcal{O}_i \Lambda_R^{-1} \mathcal{O}_i', \quad \Lambda_R = ((\lambda_{kl}^{(R)})), \tag{4.33}$$

where
$$\lambda_{kl}^{(R)} = \begin{cases} \frac{1}{12} \left[\int_{-\infty}^{+\infty} f_k^2(x)\, dx \right]^{-2} & \text{if } k = l, \\[2ex] \dfrac{\displaystyle\int_{-\infty}^{+\infty} \int_{-\infty}^{+\infty} F_k(x) F_l(y)\, dF_{kl}(x,y) - \frac{1}{4}}{\displaystyle\int_{-\infty}^{+\infty} f_k^2(x)\, dx \int_{-\infty}^{+\infty} f_l^2(x)\, dx} & \text{if } k \neq l, \end{cases} \tag{4.34}$$

$$\Delta_{\mathscr{L}(\Phi)} = \frac{1}{t} \sum_{i=1}^{t} \mathcal{O}_i \Lambda_\Phi^{-1} \mathcal{O}_i', \quad \Lambda_\Phi = ((\lambda_{kl}^{(\Phi)})), \tag{4.35}$$

where
$$\lambda_{kl}^{(\Phi)} = \begin{cases} \left[\displaystyle\int_{-\infty}^{+\infty} \frac{f_k^2(x)\, dx}{\phi[\Phi^{-1}(F_k(x))]} \right]^{-2} & \text{if } k = l, \\[2ex] \dfrac{\displaystyle\int_{-\infty}^{+\infty} \int_{-\infty}^{+\infty} \Phi^{-1}(F_k(x))\, \Phi^{-1}(F_l(y))\, dF_{kl}(x,y)}{\displaystyle\int_{-\infty}^{+\infty} \frac{f_k^2(x)\, dx}{\phi[\Phi^{-1}(F_k(x))]} \int_{-\infty}^{+\infty} \frac{f_l^2(x)\, dx}{\phi([\Phi^{-1}(F_l(x))])}} & \text{if } k \neq l. \end{cases} \tag{4.36}$$

Sen and David (1968) have also considered a permutationally distribution free test statistic D_N which is a p-variate extension of the univariate sign test. They have shown that under the sequence of alternatives H_N, D_N has the limiting noncentral chi-square distribution with $p(t-1)$ degrees of freedom and non-centrality parameter Δ_D which in the notation of the present paper is

$$\Delta_D = \frac{1}{t} \sum_{i=1}^{t} \mathcal{O}_i \mathscr{D}^{-1} \mathcal{O}_i'^{-1}, \tag{4.37}$$

where $\mathscr{D} = (d_{kl})$ is given by

$$d_{kl} = \begin{cases} \dfrac{\Pi_{kl}(0,0) - \frac{1}{4}}{f_k(0) f_l(0)} & \text{if } k \neq l, \\[2ex] \dfrac{1}{4 f_k(0) f_l(0)} & \text{if } k = l. \end{cases} \tag{4.38}$$

In addition, we shall consider the likelihood ratio $\mathscr{F}$-test obtained by assuming that $\mathbf{Z}_{ijl}$ is distributed according to a normal distribution with mean $\mu_i - \mu_j$ and covariance matrix $\mathbf{\Sigma}$. This statistic can be shown to be asymptotically equivalent to

$$\mathscr{F} = \frac{1}{t} \sum_{i=1}^{t} (\hat{\mu}_i^{(1)} \ldots \hat{\mu}_i^{(p)}) \hat{\Sigma}^{-1} (\hat{\mu}_i^{(1)} \ldots \hat{\mu}_i^{(p)})', \tag{4.39}$$

where
$$\hat{\mu}_i^{(k)} = \sum_{\substack{j=1 \\ j \neq i}}^{t} \frac{1}{N_{ij}} \sum_{l=1}^{N_{ij}} X_{ijl}^{(k)} \tag{4.40}$$

and $\hat{\Sigma}$ is the parametric estimator of the covariance matrix

$$\hat{\Sigma} = (\sigma_{kl}) \quad (k, l = 1, \dots, p), \tag{4.41}$$

where
$$\sigma_{kl} = \mathrm{cov}\,(X^{(k)}, X^{(l)}). \tag{4.42}$$

Under H_N, $\mathscr{F}$ has asymptotically a non-central chi-square distribution with $p(t-1)$ degrees of freedom and non-centrality parameter

$$\Delta_{\mathscr{F}} = \frac{1}{t} \sum_{i=1}^{t} \mathcal{O}_i \, \hat{\Sigma}^{-1} \mathcal{O}'_1. \tag{4.43}$$

5 ASYMPTOTIC RELATIVE EFFICIENCY

It is well known (Puri, 1964) that in the situations we are considering the asymptotic efficiency of one statistic relative to another is equal to the ratio of their non-centrality parameters. Thus, denoting e_{T_1, T_2} as the asymptotic efficiency of T_1 relative to T_2, we have

$$\left. \begin{aligned} e_{\mathscr{L}_N(\Phi), \mathscr{F}} &= \Delta_{\mathscr{L}(\Phi)}/\Delta_{\mathscr{F}} : e_{\mathscr{L}_N(R), \mathscr{F}} = \Delta_{\mathscr{L}(R)}/\Delta_{\mathscr{F}}, \\ e_{\mathscr{L}_N(\Phi), D_N} &= \Delta_{\mathscr{L}(\Phi)}/\Delta_D, \end{aligned} \right\} \tag{5.1}$$

and $\quad \Delta_{\mathscr{L}_N(\Phi), \mathscr{L}_N(R)} = \Delta_{\mathscr{L}(\Phi)}/\Delta_{\mathscr{L}(R)}.$

The above expressions do not lend themselves to easy analysis. They depend upon the underlying distribution $\Pi(\mathbf{z})$ and also on the vectors $\mathcal{O}_i$ and t, the number of treatments. Useful information may be obtained, however, for certain special cases.

Case 1. *Multivariate normal case.* Let us assume that the underlying distribution is a non-singular p-variate normal distribution with mean vector zero and covariance matrix $\Sigma = ((\sigma_{kl})), k, l = 1, \dots, p$. Then

$$\lambda_{kl}^{(\Phi)} = \sigma_{kl}, \quad \lambda_{kl}^{(R)} = 2\sigma_k \sigma_l \sin^{-1}\frac{\rho_{kl}}{2} \tag{5.2}$$

where
$$\rho_{kl} = \begin{cases} 1 & \text{if } k = l, \\ \sigma_{kl}\sigma_k^{-1}\sigma_l^{-1} & \text{if } k \neq l. \end{cases} \tag{5.3}$$

In such a case
$$e_{\mathscr{L}_N(\Phi), \mathscr{F}} = 1. \tag{5.4}$$

This means that for underlying non-singular normal distribution the property of the univariate normal scores test relative to the Student's t-test (or the Analysis of Variance $\mathscr{F}$-test) is preserved in the multivariate case.

Next from 6.2 we note that for any i, the expression

$$\frac{\mathcal{O}_i \Lambda_R^{-1} \mathcal{O}_i'}{\mathcal{O}_i \Sigma^{-1} \mathcal{O}_i'} \tag{5.5}$$

is the asymptotic relative efficiency of one sample p-variate rank sum test relative to Hotelling's T^2-test [cf. Sen and Puri (1969)]. Denoting by ϕ^* the class of all non-singular p-variate normal distributions we obtain (Sen and David, 1968; Shane, 1968; Shane and Puri, 1969)

$$\left. \begin{aligned} \inf_{\Pi \in \phi^*} \inf_{\mathcal{O}} e_{\mathscr{L}_N(R), \mathscr{F}} &= 0 \quad \text{for} \quad p \geqslant 3, \\ 3/\pi \leqslant e_{\mathscr{L}_N(R), \mathscr{F}} &\leqslant 0.965 \quad \text{for} \quad p = 2, \end{aligned} \right\} \tag{5.6}$$

and from (5.4) and (5.6) to follow that

$$\sup_{\Pi \in \phi^*} \sup_{\mathcal{O}} e_{\mathscr{L}_N(\Phi), \mathscr{L}_N(R)} = \infty \quad \text{for} \quad p \geqslant 3.$$

Proceeding as in Bhattacharyya (1967),† we obtain

$$\inf_{\Pi \in \phi^*} \inf_{\mathcal{O}} e_{\mathscr{L}_N(\Phi), \mathscr{L}_N(R)} \geqslant 1 \quad \text{for} \quad p \geqslant 2.$$

The above results indicate that when the underlying distribution is normal, the $\mathscr{L}_N(\Phi)$ test is always preferably to the $\mathscr{L}_N(R)$ test. Similarly

$$\sup_{\Pi \in \phi^*} \sup_{\mathcal{O}} e_{\mathscr{L}_N(\Phi), D_N} = \infty \quad \text{for} \quad p \geqslant 3.$$

Case 2. Independent co-ordinates. Let $\Pi(\mathbf{z})$ have independent co-ordinates, then

$$\sigma_{kl} = \Lambda_{kl}^{(\Phi)} = \Lambda_{kl}^{(R)} = 0 \quad (k \neq l).$$

In such a case,
$$e_{\mathscr{L}_N(\Phi), \mathscr{F}} = \frac{\sum\limits_{i=1}^{t} \sum\limits_{k=1}^{p} \mathcal{O}_{k,i}^2 \Lambda_{kk}^{(\Phi)}}{\sum\limits_{i=1}^{t} \sum\limits_{k=1}^{p} \mathcal{O}_{k,i}^2 \sigma_k^{-2}}. \tag{5.7}$$

To obtain bounds for (5.7) we consider

$$e_A = \frac{\sum\limits_{k=1}^{p} \mathcal{O}_{k,i}^2 \Lambda_{k,k}^{(\Phi)}}{\sum\limits_{k=1}^{p} \mathcal{O}_{k,i}^2 \sigma_k^{-2}}. \tag{5.8}$$

Applying Courant's Theorem‡ we see that

$$\inf_{\mathcal{O}} e_A = \min [\sigma_k^2 \Lambda_{k,k}^{(\Phi)}, k = 1, \dots, p]. \tag{5.9}$$

---

† *Ann. Math. Statist.* (1967), **38**, 1753–8.
‡ The maximal and minimal values of $\mathbf{x}'\mathbf{A}\mathbf{x}/\mathbf{x}'\mathbf{B}\mathbf{x}'$ where A and B are non-negative definite and B is nonsingular are given by the maximal and minimal eigenvalues of AB^{-1}.

However, $\sigma_k^2 \Lambda_{kk}^{(\Phi)}$ is simply the asymptotic relative efficiency of the univariate one sample normal scores test relative to Student's t-test for the distribution F_k. As such it is known to be greater than or equal to 1 and is 1 only for F_k normal. Hence

$$\inf_{\mathcal{O}} e_A = 1, \tag{5.10}$$

$$\inf_{\Pi \in \mathcal{F}_0} \inf_{\mathcal{O}} e_{\mathcal{L}_N(\Phi), \mathcal{F}} = 1, \tag{5.11}$$

where $\mathcal{F}_0$ is the class of all absolutely continuous cdf's having independent co-ordinates.

In the same way, using the results of Chatterjee and Sen (1964), and Sen and Puri (1967), it can be shown that

$$\inf_{\Pi \in \mathcal{F}_0} \inf_{\mathcal{O}} e_{\mathcal{L}_N(\Phi) \, \mathcal{L}_N(R)} = \Pi/6, \tag{5.12}$$

$$\inf_{\Pi \in \mathcal{F}_0} \inf_{\mathcal{O}} e_{\mathcal{L}_N(R), \mathcal{F}} = 0{\cdot}864, \tag{5.13}$$

$$\inf_{\Pi \in \mathcal{F}_0^*} \inf_{\mathcal{O}} e_{D_N, \mathcal{F}} = 0{\cdot}33, \tag{5.14}$$

where $\mathcal{F}_0^*$ is the class of all absolutely continuous diagonally symmetric unimodal distributions with independent co-ordinates.

Case 3. *Identical equicorrelated marginals.* Let us assume

$$F_{ij}(x,y) = F(x,y) \quad \text{for} \quad 1 \leqslant i \neq j \leqslant p.$$

Proceeding as earlier, we obtain the

$$\sup_{\mathcal{O}} e_{\mathcal{L}_N(\Phi), \mathcal{F}} = \{\sigma_1^2 a_\Phi^2\} \max \left\{ \frac{1-\rho}{1-\rho_\sigma}, \frac{1+(p-1)\rho}{1+(p-1)\rho_\Phi} \right\}, \tag{5.15}$$

$$\inf_{\mathcal{O}} e_{\mathcal{L}_N(\Phi), \mathcal{F}} = \{\sigma_1^2 a_\Phi^2\} \min \left\{ \frac{1-\rho}{1-\rho_\Phi}, \frac{1+(p-1)\rho}{1+(p-1)\rho_\Phi} \right\}, \tag{5.16}$$

and similarly, one can evaluate $\inf_{\mathcal{O}} e_{\mathcal{L}_N(R), \mathcal{F}}, \sup_{\mathcal{O}} e_{\mathcal{L}_N(R), \mathcal{F}}$ etc. The bounds for these eigenvalues are not in general known. However, for specific distributions, their values may be computed.

6 APPENDIX

The expressions for $((\sigma_{N, kl, \alpha\beta}))$, $k, l = 1, \ldots, p; \alpha, \beta = 1, \ldots, c.$

In this section we give the expressions for $\sigma_{N, k\, l, \alpha\beta}$. For the details of computations, *see* Shane (1968). Denote

$$A_k^{(\alpha)}(x,y) = F_k^{(\alpha)}(x)\,[1 - F_k^{(\alpha)}(y)], \quad B_{kl}(x,y) = dJ_k[H_k(x)]\,dJ_l[H_l(y)],$$

$$C_{kl}^{(\alpha)}(x,y) = d[F_k^{(\alpha)}(x) + F_k^{(\alpha)}(-x)]\,[F_l^{(\alpha)}(y) + F_l^{(\alpha)}(-y)],$$

$$D_{kl}^{(\alpha)}(x,y) = dH_k(x)\,d[F_l^{(\alpha)}(y) + F_l^{(\alpha)}(-y)], \quad M_k^{(\alpha)}(x) = F_k^{(\alpha)}(x) + F_k^{(\alpha)}(-x),$$

$$U_{kl}(x,y) = J_k'[H_k(x)]\,J_l'[H_l(y)],$$

$$E_{kl}^{(\alpha)}(x,y) = F_{kl}^{(\alpha)}(x,y) + F_{kl}^{(\alpha)}(x,-y) - F_{kl}^{(\alpha)}(-x,y) - F_{kl}^{(\alpha)}(-x,-y),$$

$$G_{kl}^{(\alpha)}(x,y) = F_{kl}^{(\alpha)}(x,y) - F_{kl}^{(\alpha)}(x,-y) - F_{kl}^{(\alpha)}(-x,y) + F_{kl}^{(\alpha)}(-x,-y),$$

$$L_{kl}^{(\alpha)}(x,y) = F_{kl}^{(\alpha)}(x,y) + F_{kl}^{(\alpha)}(-x,y) - F_{kl}^{(\alpha)}(x,-y) - F_{kl}^{(\alpha)}(-x,-y),$$

$$N_{kl}^{(\alpha)}(x,y) = F_{kl}^{(\alpha)}(x,y) + F_{kl}^{(\alpha)}(x,-y) - F_{kl}^{(\alpha)}(-x,y) - F_{kl}^{(\alpha)}(-x,-y),$$

$$P_{kl}^{(\alpha)}(x,y) = F_{kl}^{(\alpha)}(x,y) - F_{kl}^{(\alpha)}(x,-y) - F_{kl}^{(\alpha)}(-x,y) + F_{kl}^{(\alpha)}(-x,-y).$$

$$\tag{6.1}$$

Then

$$
\begin{aligned}
\sigma_{N,kk,\alpha\alpha} = 2\Bigg[&\iint_{0<x<y<\infty} [A_k^{(\alpha)}(x,y) + A_k^{(\alpha)}(-y,-x)]\,B_{kk}(x,y) \\
&+ \int_{x=0}^{\infty}\int_{y=0}^{\infty} A_k^{(\alpha)}(-y,x)\,B_{kk}(x,y)\Bigg] + 2\rho_N^{(\alpha)} \sum_{r=1}^{c} \rho_N^{(r)} \\
&\times \Bigg[\iint_{0<x<y<\infty} \{A_k^{(r)}(x,y) + A_k^{(r)}(-y,-x)\}\,U_{kk}(x,y)\,C_{kk}^{(\alpha)}(x,y) \\
&- \int_{x=0}^{\infty}\int_{y=0}^{\infty} A_k^{(r)}(-y,x)\,U_{kk}(x,y)\,C_{kk}^{(\alpha)}(x,y)\Bigg] \\
&- 2\rho_N^{(\alpha)}\Bigg[\iint_{0<x<y<\infty} \{A_k^{(\alpha)}(x,y) + A_k^{(\alpha)}(-x,y) - A_k^{(\alpha)}(-y,x) \\
&- A_k^{(\alpha)}(-y,-x)\}\,U_{kk}(x,y)\,D_{kk}^{(\alpha)}(x,y) \\
&+ \iint_{0<x<y<\infty} \{A_k^{(\alpha)}(y,x) + A_k^{(\alpha)}(-x,y) - A_k^{(\alpha)}(-y,x) \\
&- A_k^{(\alpha)}(-x,-y)\}\,U_{kk}(x,y)\,D_{kk}^{(\alpha)}(x,y)\Bigg].
\end{aligned}
$$

$$\tag{6.2}$$

$$
\begin{aligned}
\sigma_{N,kl,\alpha\alpha} = &\int_{x=0}^{\infty}\int_{y=0}^{\infty} \{E_{kl}^{(\alpha)}(x,y) - G_{kl}^{(\alpha)}(x,y)\}\,B_{kl}(x,y) \\
&- \rho_N^{(\alpha)} \int_{x=0}^{\infty}\int_{y=0}^{\infty} [\{L_{kl}^{(\alpha)}(x,y) - M_k^{(\alpha)}(x)\,H_k^{(\alpha)}(y)\}\,B_{kl}(x,y) \\
&- \{N_{kl}^{(\alpha)}(x,y) - M_l^{(\alpha)}(y)\,H_k^{(\alpha)}(x)\}\,U_{kl}(x,y)\,D_{kl}^{(\alpha)}(x,y)] - \rho_N^{(\alpha)} \sum_{r=1}^{c} \rho_N^{(r)} \\
&+ \int_{x=0}^{\infty}\int_{y=0}^{\infty} \{P_{kl}^{(r)}(x,y) - H_k^{(r)}(x)\,H_l^{(r)}(y)\}\,U_{kl}(x,y)\,C_{kl}^{(\alpha)}(x,y).
\end{aligned}
$$

$$\tag{6.3}$$

$$\sigma_{N,kl,\alpha\beta} = -\sqrt{(\rho_N^{(\alpha)}\rho_N^{(\beta)})}\left[\int_{x=0}^{\infty}\int_{y=0}^{\infty}\{N_{kl}^{(\alpha)}(x,y) - M_k^{(\alpha)}(x)\,H_l^{(\alpha)}(y)\}\,U_{kl}(x,y)\right.$$

$$\times D_{lk}^{(\beta)}(x,y) + \{N_{kl}^{(\beta)}(x,y) - M_l^{(\beta)}(y)\,H_k^{(\beta)}(x)\}\,U_{kl}(x,y)\,D_{kl}^{(\alpha)}(x,y)$$

$$+ \sum_{r=1}^{c}\rho_N^{(r)}\int_{x=0}^{\infty}\int_{y=0}^{\infty}\{P_{kl}^{(r)}(x,y) - H_{kl}^{(r)}(x)\,H_l^{(r)}(y)\}$$

$$\left. \times U_{kl}(x,y)\,dM_k^{(\alpha)}(x)\,dM_l^{(\beta)}(y)\right]. \quad (6.4)$$

$$\sigma_{N,kk,\alpha\beta} = -\sqrt{(\rho_N^{(\alpha)}\rho_N^{(\beta)})}\iint_{0<x<y<\infty}\{A_k^{(\alpha)}(x,y) - A_k^{(\alpha)}(-y,-x)$$

$$+ A_k^{(\alpha)}(-y,x) - A_k^{(\alpha)}(-x,y)\}\,U_{kk}(x,y)\,D_{kk}^{(\beta)}(x,y)$$

$$- \iint_{0<y<x<\infty}\{A_k^{(\alpha)}(y,x) - A_k^{(\alpha)}(-x,-y) - A_k^{(\alpha)}(-y,x)$$

$$+ A_k^{(\alpha)}(-x,y)\}\,U_{kk}(x,y)\,D_{kk}^{(\beta)}(x,y)$$

$$- \iint_{0<x<y<\infty}\{A_k^{(\beta)}(x,y) - A_k^{(\beta)}(-y,-x) - A_k^{(\beta)}(-x,y)$$

$$+ A_k^{(\beta)}(-y,x)\}\,U_{kk}(x,y)\,D_{kk}^{(\alpha)}(x,y)$$

$$- \iint_{0<y<x<\infty}\{A_k^{(\beta)}(y,x) - A_k^{(\beta)}(-x,-y) - A_k^{(\beta)}(-x,y)$$

$$+ A_k^{(\beta)}(-y,x)\}\,U_{kk}(x,y)\,D_{kk}^{(\alpha)}(x,y) - \sum_{r=1}^{c}\rho_N^{(r)}$$

$$\times\left[\iint_{0<x<y<\infty}H_k^{(r)}(x)\,[1 - H_k^{(r)}(y)]\,U_{kk}(x,y)\,dM_k^{(\alpha)}(x)\,dM_k^{(\beta)}(y)\right.$$

$$\left. + \iint_{0<y<x<\infty}H_k^{(r)}(y)\,[1 - H_k^{(r)}(x)]\,U_{kk}(x,y)\,dM_k^{(\alpha)}(x)\,dM_k^{(\beta)}(y)\right].$$

$$(6.5)$$

ACKNOWLEDGEMENT

The authors would like to express their sincere thanks to Professor P. K. Sen for the careful examination of the paper, and for supplying a new proof of Theorem 3.2 which simplified considerably the author's original proof. They are also grateful to Professor H. T. David for pointing out an error in an original version of the paper.

REFERENCES

Anderson, T. W. (1958). *An Introduction to Multivariate Statistical Analysis*. New York: John Wiley and Sons.

Chatterjee, S. K. and Sen, P. K. (1964). Nonparametric tests for the bivariate two sample location problem. *Calcutta Statist. Ass. Bull.* **13**, 18–58.

David, H. A. (1963). *The Method of Paired Comparisons*. Charles Griffin and Co., Ltd.

Davidson, R. R. and Bradley, R. A. (1970). Multivariate paired comparisons: some large-sample results on estimation and tests of equality of preference. *Nonparametric. Techniques in Statistical Inference*. Cambridge University Press, pp. 111–25.

Durbin, J. (1951). Incomplete blocks in ranking experiments. *Brit. J. Psych.* 4, 85–90.

Essen, C. G. (1945). Fourier analysis of distribution functions. *Acta. Math.* 77, 1–125.

Ghosh, M. (1969). Asymptotically optimal nonparametric tests for miscellaneous problems of linear regression. Ph.D. Thesis, University of North Carolina.

Hájek, Jaroslav (1961). Some extensions of the Wald–Wolfowitz–Noether Theorem. *Ann. Math. Statist.* 32, 506–25.

Noether, G. E. (1960). Remarks about a paired comparison model. *Psychometrika* 25, 357–67.

Puri, M. L. (1964). Asymptotic efficiency of a class of c-sample tests. *Ann. Math. Statist.* 35, 102–21.

Puri, M. L. and Sen, P. K. (1966). On a class of multivariate-multisample rank order tests. *Sankhyā* 28, 353–76.

Puri, M. L. and Sen, P. K. (1960). On the asymptotic normality of one sample rank order test statistics. *Teoria veroyatnostey iee Primenenyia* 14, 167–72.

Puri, M. L. and Sen, P. K. (1967). On some optimum nonparametric procedures in two-way layouts. *J. Am. statist. Ass.* 63, 1214–29.

Puri, M. L. and Sen, P. K. (1969). The asymptotic theory of rank order tests for experiments involving paired comparisons. *Ann. Inst. Statist. Math.* 21, 163–73.

Rao, C. R. (1965). *Linear Statistical Inference and its Applications*. New York: John Wiley and Sons.

Scheffe, H. (1952). An analysis of variance for paired comparisons. *J. Am. Statist. Ass.* 47, 381–400.

Sen, P. K. and David, H. A. (1968). Paired comparisons for paired characteristics. *Ann. Math. Statist.* 39, 200–8.

Sen, P. K. and Puri, M. L. (1967). On the theory of rank order tests for location in the multivariate one sample problem. *Ann. Math. Statist.* 38, 1216–28.

Shane, H. D. (1968). Multivariate multisample analogues of some one sample tests. Ph.D. Thesis, New York University.

Shane, H. D. and Puri, M. L. (1969). Rank Order Tests for Multivariate Paired Comparisons. *Ann. Math. Statist.* 40, 2101–17.

Sverdrup, E. (1952). The limit distribution of a continuous function of several random variables. *Skand. Aktuariedskrift* 35, 1–10.

Wald, A. and Wolfowitz, J. (1944). Statistical tests based on permutations of the observations. *Ann. Math. Statist.* 15, 358–72.

DISCUSSION ON
PURI AND SHANE'S PAPER

H. T. DAVID

This paper, as much as any of the Symposium, shows the importance of nonparametric method for non-subjective statistics. Nonparametric methods require a minimum of probabilistic modelling—only the discrete uniform distribution. In this they seem to meet 'data analyzers' halfway and others, outside the profession, wary of detailed probabilistic assumptions. Yet, as is so well demonstrated in this paper, nonparametric methods manage to compete successfully in the world of parametric hypotheses; there is in addition the attractive deductive link of permutation tests to experimental randomization, pointed out by Fisher and since stressed by Kempthorne. In this connection note that Puri and Shane's group G_N arises from the following experimental randomization $\mathscr{R}$: Randomly assign the N treatment pairs to the N blocks (randomization $\mathscr{R}_1$), then, independently, the individual treatments of a particular pair to the two plots within a block (randomization $\mathscr{R}_2$). Let H_N indicate permutation of paired vectors. Then the implied joint null permutation distribution of the $2N$ vector observations (joint distribution of $2pN$ scalars in all) is uniform over the relevant orbit of $H_N G_N^{(2)} \equiv G_N^*$. The induced joint null permutation distribution of the N vector differences $\mathbf{Z}\alpha, l$ (joint distributions of pN scalars in all) is then uniform over the orbit of G_N containing the observed Z. Thus $\mathscr{R}$ leads naturally to an intra-block permutation analysis, say based on the permutation distribution of (4.39) under G_N. This suggests several comments. To begin with, suppose that the individual vector observations $\mathbf{X}\alpha, l$ and $\mathbf{Y}\alpha, l$ are indeed available, in addition to the differences $\mathbf{Z}\alpha, l = \mathbf{Y}\alpha, l - \mathbf{X}\alpha, l$. Let now T_i be the average of *all* vector observations from all blocks featuring treatment i. Then G_N^* leads equally to an inter-block permutation analysis, say based (at least when the N_{ij} all are equal) on the permutation distribution under G_N^* (equivalently, $G_N^{(2)}$) of a suitable one-way MANOVA statistic involving the t vectors T_i. It is not clear how best to combine the intra- and inter-block analyses; however, replacing an intra-block statistic such as (4.39) by its conditional expectation given $g_N^{(1)}$ at least insures the conditional independence of the two analyses. A further point is that, for blocks of size $k > 2$, G_N^*, and in particular H_N, remains of course appropriate under $\mathscr{R}$. The group applicable to paired treatment differences within blocks is then pinpointed as the cyclic group of order $\binom{k}{2}$ multiplied by a modification of $G_N^{(1)}$ whose elements reverse the order of the $\binom{k}{2}$ difference vectors, in addition to the change of sign. A final point is that some experimenters, faced with the prospect of constant N_{ij} (say $N_{ij} = m > 1$) might not opt for $\mathscr{R}$ at all, but rather for separate randomization in each of m replicates of c blocks. I imagine that this or other alternatives to $\mathscr{R}$ (and hence to G_N^*) would be gauged by parametric models with random rep effects; these in addition to the random block effects that would be introduced in any power computations for the inter-block analysis.

PART 2

TESTING AND ESTIMATION 2

ASYMPTOTIC PROPERTIES OF ISOTONIC ESTIMATORS FOR THE GENERALIZED FAILURE RATE FUNCTION

Part I: Strong Consistency

RICHARD E. BARLOW AND WILLEM R. VAN ZWET

1 ISOTONIC ESTIMATION IN THE CASE OF CONVEX ORDERING

Let $\mathscr{F}$ be the class of absolutely continuous distribution functions F on R^1 with positive and right- (or left-) continuous density f on the *interval* where $0 < F < 1$. It follows that the inverse function F^{-1} is uniquely defined on $(0, 1)$. We take $F^{-1}(0)$ and $F^{-1}(1)$ to be equal to the left- and right-hand endpoints of the support of F (possibly $-\infty$ or $+\infty$). For $F, G \in \mathscr{F}$ we say that F is *c-ordered* (convex ordered) with respect to $G (F \underset{c}{<} G)$ if and only if $G^{-1}F$ is convex on the interval where $0 < F < 1$ [van Zwet (1964)]. Denoting the densities of F and G by f and g, we find that $F \underset{c}{<} G$ implies that

$$r(x) = \frac{d}{dx} G^{-1} F(x) = \frac{f(x)}{gG^{-1}F(x)} \tag{1.1}$$

is nondecreasing in x on the interval where $0 < F < 1$ (if $G^{-1}F$ is not differentiable at x, we take the right- or left-hand derivative and assume f and g to be continuous from the same side). We assume G known, $F \underset{c}{<} G$ and consider the problem of estimating r. Note that if G is the exponential or the uniform distribution, the problem reduces to estimation of a nondecreasing failure rate $f/(1 - F)$ or a nondecreasing density f. Maximum likelihood estimators (MLE) for r in these cases have been investigated by Grenander (1956), Marshall and Proschan (1965), Prakasa Rao (1966) and Robertson (1967).

Let $X_1, X_2, \ldots$ be independent with distribution function F, let $X_{1:n} < X_{2:n} < \ldots < X_{n:n}$ and F_n denote the order statistics and the empirical distribution function corresponding to $X_1, X_2, \ldots, X_n$ and let ρ_n denote an initial or *basic estimator* for r based on $X_1, X_2, \ldots, X_n$. For each n, we define a *grid* on R^1, i.e. a finite or infinite sequence

$$\ldots < w_{-2, n} < w_{-1, n} < w_{0, n} < w_{1, n} < \ldots.$$

We call an interval $[w_{j, n}, w_{j+1, n})$ a *window*. In each window, we choose a point $w_{j,n} \leqslant \xi_{j,n} < w_{j+1,n}$ and to each point $\xi_{j,n}$ we assign a nonnegative weight $\mu_{j,n}$. In applications $\xi_{j,n}$ will usually be $\frac{1}{2}(w_{j,n}+w_{j+1,n})$. Following Brunk, we call

$$r_n(x) = \inf_{t \geqslant i+1} \sup_{s \leqslant i} \left[\frac{\sum\limits_{j=s}^{t-1} \rho_n(\xi_{j,n})\mu_{j,n}}{\sum\limits_{j=s}^{t-1} \mu_{j,n}} \right] \quad (w_{i,n} \leqslant x < w_{i+1,n}), \quad (1.2)$$

the *monotonic regression* or more generally the *isotonic regression of* ρ_n with respect to the discrete measure μ_n. Note that r_n is a nondecreasing step function and that $r_n = \rho_n$ whenever $\rho_n(\xi_{j,n})$ happens to be non-decreasing in j.

If $X_{1:n} \leqslant w_{i,n} < w_{i+1,n} \leqslant X_{n:n}$, then $gG^{-1}F_n(\xi_{i,n}) > 0$ and we may estimate $r(x)$ for $w_{i,n} \leqslant x < w_{i+1,n}$ by

$$\hat{\rho}_n(x) = \frac{f_n(\xi_{i,n})}{gG^{-1}F_n(\xi_{i,n})} = \frac{F_n(w_{i+1,n})-F_n(w_{i,n})}{gG^{-1}F_n(\xi_{i,n})(w_{i+1,n}-w_{i,n})}, \quad (1.3)$$

the *naive* estimator for r. It has been extensively studied by Parzen (1962) and others in the case when G is the uniform distribution and by Watson and Leadbetter (1964) and others in the case where G is the exponential distribution. With $\hat{\rho}_n$ for the basic estimator and weights

$$\mu_{j,n} = gG^{-1}F_n(\xi_{j,n})(w_{j+1,n}-w_{j,n}),$$

the isotonic estimator $\hat{r}_n$ is defined for $X_{1:n} \leqslant w_{i,n} \leqslant x < w_{i+1,n} \leqslant X_{n:n}$ by

$$\hat{r}_n(x) = \inf_{\substack{t \geqslant i+1 \\ w_{i,n} \leqslant X_{n:n}}} \sup_{\substack{s \leqslant i \\ w_{i,n} \geqslant X_{1:n}}} \frac{F_n(w_{t,n})-F_n(w_{s,n})}{\sum\limits_{j=s}^{t-1} gG^{-1}F_n(\xi_{j,n})(w_{j+1,n}-w_{j,n})}. \quad (1.4)$$

Note that the conditions on $w_{i,n}$, $w_{i+1,n}$, $w_{s,n}$ and $w_{t,n}$ originate from the fact that $\hat{\rho}_n(\xi_{j,n})$ is defined only if $X_{1:n} \leqslant w_{j,n} < w_{j+1,n} \leqslant X_{n:n}$. An interesting special case arises when we consider the random grid

$$w_{j,n} = X_{j:n}, \quad j = 1, 2, ..., n,$$

determined by the order statistics. For this grid, (1.4) becomes

$$\hat{r}_n(x) = \inf_{i+1 \leqslant t \leqslant n} \sup_{1 \leqslant s \leqslant i} \frac{t-s}{n \sum\limits_{j=s}^{t-1} gG^{-1}(j/n)(X_{j+1:n}-X_{j:n})}, \quad (1.5)$$

for $X_{i:n} \leqslant x < X_{i+1:n}, i = 1, 2, ..., n-1$. This is the MLE in the case when G is the exponential or the uniform distribution. Note that if G is the exponential distribution we have

$$ngG^{-1}(j/n)(X_{j+1:n}-X_{j:n}) = (n-j)(X_{j+1:n}-X_{j:n})$$

and hence the weights $\mu_{j,n}$ in (1.5) are proportional to the total time on test between $X_{j:n}$ and $X_{j+1:n}$. Since the total time on test for an interval is a measure of our information over that interval, this choice of weights is intuitively appealing in this case. We shall call $gG^{-1}(j/n)(X_{j+1:n} - X_{j:n})$ the *total time on test weights* for general G also.

An alternative estimator r_n^* is obtained using the same basic estimator $\rho_n^* = \hat{\rho}_n$ with weights $\mu_{j,n} = w_{j+1,n} - w_{j,n}$, thus

$$r_n^*(x) = \inf_{\substack{t \geqslant i+1 \\ w_{t,n} \leqslant X_{n:n}}} \sup_{\substack{s \leqslant i \\ w_{s,n} \geqslant X_{1:n}}} \frac{1}{w_{t,n} - w_{s,n}} \sum_{j=s}^{t-1} \frac{F_n(w_{j+1,n}) - F_n(w_{j,n})}{gG^{-1}F_n(\xi_{j,n})}, \quad (1.6)$$

for $X_{1:n} \leqslant w_{i,n} \leqslant x < w_{i+1,n} \leqslant X_{n:n}$.

If we consider the second member of (1.1) rather than the third one, a logical choice for the basic estimator is the *graphical* estimator

$$\tilde{\rho}_n(x) = \frac{G^{-1}F_n(w_{i+1,n}) - G^{-1}F_n(w_{i,n})}{w_{i+1,n} - w_{i,n}}, \quad (1.7)$$

for $w_{i,n} \leqslant x < w_{i+1,n}$ with $X_{1:n} \leqslant w_{i,n} < w_{i+1,n} \leqslant X_{n:n}$. With weights $\mu_{j,n} = w_{j+1,n} - w_{j,n}$, we obtain the isotonic estimator

$$\tilde{r}_n(x) = \inf_{\substack{t \geqslant i+1 \\ w_{t,n} \leqslant X_{n:n}}} \sup_{\substack{s \leqslant i \\ w_{s,n} \geqslant X_{1:n}}} \frac{G^{-1}F_n(w_{t,n}) - G^{-1}F_n(w_{s,n})}{w_{t,n} - w_{s,n}} \quad (1.8)$$

for $X_{1:n} \leqslant w_{i,n} \leqslant x < w_{i+1,n} \leqslant X_{n:n}$. Notice that if $G^{-1}(1) = \infty$ and $w_{i+1,n} = X_{n:n}$, then $\tilde{\rho}_n(x) = \tilde{r}_n(x) = \infty$ for $w_{i,n} \leqslant x < w_{i+1,n}$. When G is the uniform distribution, $\hat{r}_n = r_n^* = \tilde{r}_n$.

In this paper, we shall be concerned with the strong consistency of the isotonic estimator $\hat{r}_n$. In a companion paper (Barlow and van Zwet), we obtain the asymptotic distributions of $\hat{r}_n, r_n^*$ and $\tilde{r}_n$.

2 CONVERGENCE OF TOTAL TIME ON TEST DISTRIBUTIONS

Let $F^{-1}(0) < \xi < F^{-1}(1)$ and define

$$\Phi_{F_n, \xi}(x) = \int_{\xi}^{x} gG^{-1}F_n(u) \, du \quad (X_{1:n} \leqslant x \leqslant X_{n:n}), \quad (2.1)$$

with the usual convention that $\int_{\xi}^{x} = -\int_{x}^{\xi}$ for $x < \xi$. Since

$$\Phi_{F_n, \xi}(X_{t:n}) - \Phi_{F_n, \xi}(X_{s:n}) = \sum_{j=s}^{t-1} gG^{-1}(j/n)(X_{j+1:n} - X_{j:n}) \quad (2.2)$$

is a partial sum of the total time on test weights employed in (1.5), $\Phi_{F_n, \xi}$ will be called the *total time on test distribution*. Notice that the measure induced by $\Phi_{F_n, \xi}$ is in fact only a continuous analogue of the discrete measure used in (1.5) that puts mass $\mu_{j,n} = gG^{-1}(j/n)(X_{j+1:n} - X_{j:n})$ at points $\xi_{j,n} \in [X_{j:n}, X_{j+1:n})$.

For general grids $\{w_{j,n}\}$, we define

$$F_n^*(x) = F_n(\xi_{j,n}) \quad \text{for} \quad w_{j,n} \leqslant x < w_{j+1,n}, \tag{2.3}$$

$$\Phi_{F_n^*, \xi}(x) = \int_\xi^x gG^{-1}F_n^*(u)\, du \quad (U_n \leqslant x \leqslant V_n), \tag{2.4}$$

where U_n and V_n denote the infimum and the supremum of the gridpoints $w_{j,n}$ in the interval $[X_{1:n}, X_{n:n}]$. We have

$$\Phi_{F_n^*, \xi}(w_{t,n}) - \Phi_{F_n^*, \xi}(w_{s,n}) = \sum_{j=s}^{t-1} gG^{-1}F_n(\xi_{j,n})(w_{j+1,n} - w_{j,n}) \tag{2.5}$$

for $X_{1:n} \leqslant w_{s,n} < w_{t,n} \leqslant X_{n:n}$, which is a partial sum of the weights employed in (1.4).

A crucial step in our proof of strong consistency of $\hat{r}_n$ consists of showing that with probability 1, $\Phi_{F_n, \xi}$ and $\Phi_{F_n^*, \xi}$ converge uniformly on $[\xi, X_{n:n}]$ and $[\xi, V_n]$ to $\Phi_{F, \xi}$, where

$$\Phi_{F, \xi}(x) = \int_\xi^x gG^{-1}F(u)\, du \quad (F^{-1}(0) \leqslant x \leqslant F^{-1}(1)). \tag{2.6}$$

If G is the exponential distribution then

$$\Phi_{F_n, \xi}(X_{i:n}) - \Phi_{F_n, \xi}(X_{1:n}) = \int_{X_{1:n}}^{X_{i:n}} (1 - F_n(x))\, dx$$

$$= \frac{1}{n} \sum_{j=1}^{i-1} (n-j)(X_{j+1:n} - X_{j:n}).$$

This transform was introduced in Marshall and Proschan (1965) and exploited by Prakasa Rao (1966) in his study of the MLE for the failure rate function of monotone failure rate distributions. Let $\mu = \int x\, dF(x)$ be finite and assume for simplicity that $F(0) = 0$. Strong uniform convergence of $\Phi_{F_n, \xi}$ to $\Phi_{F, \xi}$ in this case is an easy consequence of the Glivenko–Cantelli theorem and the strong law since

$$\Phi_{F_n, \xi}(X_{n:n}) - \Phi_{F_n, \xi}(X_{1:n}) = \frac{1}{n} \sum_{j=1}^{n} X_{j:n} - X_{1:n} \overset{\text{a.s.}}{\to} \mu - F^{-1}(0)$$

$$= \Phi_{F, \xi}(F^{-1}(1)) - \Phi_{F, \xi}(F^{-1}(0)).$$

For general G, our first convergence proof for $\Phi_{F_n, \xi}$ given in Theorem 2.1 is based on this idea too.

To show convergence of $\Phi_{F_n^*,\xi}$, we shall obviously have to impose conditions on the asymptotic behavior of the grid. We shall say that $\{w_{j,n}\}$ *becomes dense* in an interval I, if for any pair $x_1 < x_2$ in I there exists an integer N such that for $n \geqslant N$ a gridpoint $w_{j,n}$ exists with $x_1 < w_{j,n} < x_2$. We say that $\{w_{j,n}\}$ *becomes subexponential* in the right tail of F, if numbers $N, z > 0$ and $c > 1$ exist such that for $n \geqslant N$, $z \leqslant w_{j,n} < w_{j+1,n} < F^{-1}(1)$ implies $w_{j+1,n} \leqslant cw_{j,n}$; similarly $\{w_{j,n}\}$ becomes subexponential in the left tail of F if $n \geqslant N$ and $F^{-1}(0) < w_{j-1,n} < w_{j,n} \leqslant -z$ implies $w_{j-1,n} \geqslant cw_{j,n}$. Note that if the support of F is bounded on the right (left), then any grid trivially becomes subexponential in the right (left) tail of F since z may be chosen arbitrarily large. We say that $\{w_{j,n}\}$ becomes subexponential near $+\infty$ ($-\infty$) if it becomes subexponential in the right (left) tail of a distribution F with $F^{-1}(1) = \infty$ ($F^{-1}(0) = -\infty$). In Lemma 2.1, we prove a property of grids that we shall need in the sequel.

Lemma 2.1. Let $\{w_{j,n}\}$ become dense in the support of F and subexponential in the right tail of F. Then numbers $N, z_0 > 0$ and $c > 1$ exist such that for $n \geqslant N$ and $z_0 \leqslant x < V_n$, $F_n^*(x) \geqslant F_n(x/c)$.

Proof. If $F^{-1}(1) < \infty$, the lemma is trivially true for $z_0 \geqslant F^{-1}(1) \geqslant V_n$. Assume $F^{-1}(1) = \infty$. By definition, $N', z > 0$ and $c > 1$ exist such that $w_{j+1,n} > w_{j,n} \geqslant z$ and $n \geqslant N'$ implies $w_{j+1,n} \leqslant cw_{j,n}$. Choose $z_0 > z$ with $z_0 > F^{-1}(0)$. Since $\{w_{j,n}\}$ becomes dense in the support of F, an integer $N \geqslant N'$ exists such that for $n \geqslant N$ a gridpoint between z and z_0 exists. Together with the definition of V_n, this implies that for $n \geqslant N$ and for every $x \in [z_0, V_n)$ gridpoints $w_{j,n}$ and $w_{j+1,n}$ exist with

$$0 < z \leqslant w_{j,n} \leqslant x < w_{j+1,n},$$
and hence

$$F_n^*(x) = F_n(\xi_{j,n}) \geqslant F_n(w_{j,n}) \geqslant F_n(w_{j+1,n}/c) \geqslant F_n(x/c).$$

Theorem 2.1. Let $F, G \in \mathscr{F}$, let $EX^+ = \int_0^\infty x\,dF(x) < \infty$ and let gG^{-1} be uniformly continuous on $(0,1)$. Assume that either $F^{-1}(1) < \infty$ or $gG^{-1}(y)/(1-y)$ is bounded on $(0,1)$. Then, for any fixed $\xi \in (F^{-1}(0), F^{-1}(1))$, $\Phi_{F,\xi}(F^{-1}(1)) < \infty$ and for $n \to \infty$,

$$\sup_{\xi \leqslant x \leqslant X_{n:n}} |\Phi_{F_n,\xi}(x) - \Phi_{F,\xi}(x)| \to 0 \quad \text{almost surely.}$$

If, moreover, the grid $\{w_{j,n}\}$ becomes dense in the support of F and subexponential in the right tail of F, then for $n \to \infty$

$$\sup_{\xi \leqslant x \leqslant V_n} |\Phi_{F_n^*,\xi}(x) - \Phi_{F,\xi}(x)| \to 0 \quad \text{almost surely.}$$

Proof. Throughout the proofs in this section, we omit the subscript ξ. If $F^{-1}(1) < \infty$, then $\Phi_F(F^{-1}(1))$ is an integral of a bounded function over a finite interval and therefore finite. If $gG^{-1}(y) \leq a(1-y)$ on $(0,1)$, then

$$
\begin{aligned}
\Phi_F(F^{-1}(1)) &= \int_{\xi}^{F^{-1}(1)} gG^{-1}F(x)\,dx \\
&\leq a \int_{\xi}^{F^{-1}(1)} (1 - F(x))\,dx = aE(X - \xi)^+ < \infty.
\end{aligned}
$$

Since $X_{1:n}$ and $X_{n:n}$ tend a.s. to $F^{-1}(0)$ and $F^{-1}(1)$ by the strong law, $\Phi_{F_n}(x)$ is a.s. defined for any fixed $x \in [\xi, F^{-1}(1))$ for sufficiently large n. Since both Φ_F and Φ_{F_n} are nondecreasing,

$$
\Phi_F(\xi) = 0 \quad \text{and} \quad \Phi_F(F^{-1}(1)) < \infty,
$$

it is obviously sufficient to show that Φ_{F_n} converges pointwise to Φ_F on $[\xi, F^{-1}(1))$ and that for any $\epsilon > 0$ we can choose $x_0 \in (\xi, F^{-1}(1))$ in such a way that

$$
\limsup_n \int_{\xi}^{X_{n:n}} gG^{-1}F_n(x)\,dx < \epsilon \quad \text{almost surely.} \tag{2.7}
$$

Convergence for fixed $x \in [\xi, F^{-1}(1))$ is immediate as the interval $[\xi, x]$ is finite and a.s. contained in the set where $0 < F < 1$ and $0 < F_n < 1$ for sufficiently large n, and hence

$$
\int_{\xi}^{x} (gG^{-1}F_n(u) - gG^{-1}F(u))\,du \to 0
$$

almost surely by the Glivenko–Cantelli theorem and the uniform continuity of gG^{-1} on $(0,1)$.

It remains to prove (2.7). If $F^{-1}(1) < \infty$, then, because gG^{-1} is bounded on $(0,1)$, (2.7) may be satisfied by choosing x_0 close to $F^{-1}(1)$. If

$$
F^{-1}(1) = \infty \quad \text{and} \quad gG^{-1}(y) \leq a(1-y)
$$

on $(0,1)$ the assumption $EX^+ < \infty$ enables us to choose x_0 in such a way that

$$
\int_{x_0}^{\infty} (1 - F(x))\,dx < \frac{\epsilon}{a},
$$

and hence

$$
\begin{aligned}
\limsup_n \int_{x_0}^{X_{n:n}} gG^{-1}F_n(x)\,dx &\leq \limsup_n a \int_{x_0}^{X_{n:n}} (1 - F_n(x))\,dx \\
&= a \limsup_n \frac{1}{n} \sum_{i=1}^{n} (X_i - x_0)^+ = aE(X - x_0)^+ = a \int_{x_0}^{\infty} (1 - F(x))\,dx < \epsilon
\end{aligned}
$$

almost surely by the strong law. This proves (2.7).

The proof of the strong uniform convergence of $\Phi_{F_n^*}$ follows very much the same pattern with F_n replaced by F_n^*, $X_{1:n}$ by U_n, and $X_{n:n}$ by V_n. We list the modifications. Like $X_{1:n}$ and $X_{n:n}$, U_n and V_n tend a.s. to $F^{-1}(0)$ and $F^{-1}(1)$ because we assume that the grid becomes dense in the support of F. The same assumption guarantees pointwise a.s. convergence of F_n^* to F on the interval where $0 < F < 1$, even though F_n^* may be defective for all n (e.g. if no gridpoint $w_{j,n} > X_{n:n}$ exists for any n). As F is a continuous probability distribution function, F_n^* is nondecreasing and $0 \leqslant F_n^* \leqslant 1$, a standard argument shows that the pointwise a.s. convergence implies that $\sup|F_n^* - F| \xrightarrow{\text{a.s.}} 0$. This replaces the Glivenko–Cantelli theorem for F_n.

It remains to prove the analogue of (2.7), viz. for any

$$\epsilon > 0, \quad x_0 \in (\xi, F^{-1}(1))$$

exists such that

$$\limsup_n \int_{x_0}^{V_n} gG^{-1}F_n^*(x)\,dx < \epsilon \quad \text{almost surely.} \tag{2.8}$$

We need only consider the case when $F^{-1}(1) = \infty$ and $g(G^{-1}(y)) \leqslant a(1-y)$ on $(0,1)$. By Lemma 2.1, $N, z_0 > 0$ and $c > 1$ exist such that

$$F_n^*(x) \geqslant F_n(x/c) \quad \text{for} \quad n \geqslant N \quad \text{and} \quad x \in [z_0, V_n).$$

Choose $x_0 \geqslant z_0$ in such a way that

$$\int_{x_0/c}^{\infty} (1-F(x))\,dx < \frac{\epsilon}{ac},$$

then

$$\limsup_n \int_{x_0}^{V_n} gG^{-1}F_n^*(x)\,dx \leqslant \limsup_n a \int_{x_0}^{V_n} (1-F_n^*(x))\,dx$$

$$\leqslant a \limsup_n \int_{x_0}^{V_n} (1-F_n(x/c))\,dx \leqslant ac \limsup_n \int_{x_0/c}^{\infty} (1-F_n(x))\,dx$$

$$= ac \int_{x_0/c}^{\infty} (1-F(x))\,dx < \epsilon,$$

by the strong law. This completes the proof of Theorem 2.1.

We note that in Theorem 2.1 it is not assumed that $F \underset{c}{<} G$. The theorem was first proved by Brunk with a different proof for the case when $F(0) = 0$ and $\xi = F^{-1}(0)$ and under the slightly stronger condition that gG^{-1} is continuously differentiable on $[0, 1]$ or $F^{-1}(1) < \infty$.

To prove a second convergence theorem under different conditions on F and G, we shall need the following lemma.

Lemma 2.2. Let $X_1, X_2, \ldots$ be independent and identically distributed with continuous distribution function F and let $X_{1:n} < X_{2:n} < \ldots < X_{n:n}$ and F_n denote the order statistics and the empirical distribution function corresponding to $X_1, X_2, \ldots, X_n$. Then

$$\lim_{n \to \infty} \frac{n^{\frac{1}{2}}}{\log n} \sup_x |F_n(x) - F(x)| = 0 \quad \text{almost surely,}$$

$$\limsup_n \frac{-\log(1 - F(X_{n:n}))}{\log n} = 1 \quad \text{almost surely.}$$

Proof. It is shown in Dvoretzky, Kiefer and Wolfowitz (1956) that

$$P\left(\sup_x |F_n(x) - F(x)| > z\right) < ce^{-2nz^2}$$

for a positive constant c independent of F, n and z, and all $z \geqslant 0$. The first part of the lemma now follows from the Borel–Cantelli lemma.

Since $X_1, X_2, \ldots$ are independent and

$$\sum_{n=1}^{\infty} P\left(\frac{-\log(1 - F(X_n))}{\log n} \geqslant 1 + \delta\right) = \sum_{n=1}^{\infty} P(F(X_n) \geqslant 1 - n^{-(1+\delta)}) = \sum_{n=1}^{\infty} n^{-(1+\delta)}$$

converges for $\delta > 0$ and diverges for $\delta = 0$, we have

$$\limsup_n \frac{-\log(1 - F(X_n))}{\log n} = 1 \quad \text{almost surely,}$$

according to the Borel zero-one criterion. For $\delta \geqslant 0$ infinitely many of the events $\{F(X_{n:n}) \geqslant 1 - n^{-(1+\delta)}\}$ occur if and only if infinitely many of the events $\{F(X_n) \geqslant 1 - n^{-(1+\delta)}\}$ occur, which proves the second assertion of the lemma.

Theorem 2.2. For $F, G \in \mathscr{F}$, let $F \underset{c}{<} G$ and let gG^{-1} be uniformly continuous on $(0, 1)$. Assume that there exists a number $0 < \eta < 1$ such that for

$$\eta \leqslant y < 1, \quad gG^{-1}(y)$$

is nonincreasing and $gG^{-1}(y)/(1 - y)$ is nondecreasing in y. Then for any fixed $\xi \in (F^{-1}(0), F^{-1}(1))$, $\Phi_{F, \xi}(F^{-1}(1)) < \infty$ and for $n \to \infty$

$$\sup_{\xi \leqslant x \leqslant X_{n:n}} |\Phi_{F_n, \xi}(x) - \Phi_{F, \xi}(x)| \to 0 \quad \text{almost surely.}$$

If, moreover, the grid $\{w_{j,n}\}$ becomes dense in the support of F and subexponential in the right tail of F, then for $n \to \infty$

$$\sup_{\xi \leqslant x \leqslant V_n} |\Phi_{F_n^*, \xi}(x) - \Phi_{F, \xi}(x)| \to 0 \quad \text{almost surely.}$$

Proof. Since $F \underset{c}{<} G$,

$$\Phi_F(F^{-1}(1)) = \int_\xi^{F^{-1}(1)} gG^{-1}F(x)\,dx = \int_\xi^{F^{-1}(1)} \frac{f(x)}{r(x)}\,dx \leqslant \frac{1-F(\xi)}{r(\xi)} < \infty,$$

as $r > 0$ on $(F^{-1}(0), F^{-1}(1))$. To prove the theorem, it is sufficient to show that (2.7) and (2.8) hold for the case when $F^{-1}(1) = \infty$. The remainder of the proof may be copied from the proof of Theorem 2.1.

The relation $F \underset{c}{<} G$ and $F^{-1}(1) = \infty$ imply that $G^{-1}(1) = \infty$ and since gG^{-1} is nonincreasing on $[\eta, 1)$ we have $\lim\limits_{y \to 1} gG^{-1}(y) = 0$. Combined with the assumption that $gG^{-1}(y)$ is nonincreasing and $gG^{-1}(y)/(1-y)$ is nondecreasing for $\eta \leqslant y < 1$, this in turn implies that for any two points y_1 and y_2 in $[\eta, 1)$

$$|gG^{-1}(y_1) - gG^{-1}(y_2)| \leqslant |y_1 - y_2| \frac{gG^{-1}(y_2)}{1 - y_2}. \tag{2.9}$$

Choose $x_0 \in (\xi, \infty)$ in such a way that $F(x_0) > \eta$ and

$$\int_{x_0}^\infty gG^{-1}F(x)\,dx < \epsilon. \tag{2.10}$$

Then with probability 1 there exists an integer N such that $F_n(x_0) > \eta$ and $X_{n:n} > x_0$ for $n \geqslant N$. Using (2.9) and the relation $F \underset{c}{<} G$, we find for $n \geqslant N$

$$\left| \int_{x_0}^{X_{n:n}} (gG^{-1}F_n(x) - gG^{-1}F(x))\,dx \right| \leqslant \sup_x |F_n(x) - F(x)| \cdot \int_{x_0}^{X_{n:n}} \frac{gG^{-1}F(x)}{1 - F(x)}\,dx$$

$$\leqslant \sup_x |F_n(x) - F(x)| \cdot \frac{1}{r(x_0)} \int_{x_0}^{X_{n:n}} \frac{f(x)}{1 - F(x)}\,dx$$

$$\leqslant \frac{1}{r(x_0)} \sup_x |F_n(x) - F(x)| \cdot (-\log(1 - F(X_{n:n}))),$$

where the right-hand member tends to zero a.s. by Lemma 2.2. Together with (2.10) this proves (2.7).

To prove (2.8), we note that Lemma 2.1 and (2.7) guarantee the existence of $x_0 \in (\xi, \infty)$, $c > 1$ and N such that $F_n^*(x) \geqslant F_n(x/c)$ for $n \geqslant N$ and $x \in [x_0, V_n)$, $F(x_0/c) > \eta$ and

$$\limsup_n \int_{x_0/c}^{X_{n:n}} gG^{-1}F_n(x)\,dx < \frac{\epsilon}{c}.$$

Then a.s. $F_n(x_0/c) > \eta$ and $V_n > x_0$ for sufficiently large n and since gG^{-1} is nonincreasing on $[\eta, 1)$,

$$\limsup_n \int_{x_0}^{V_n} gG^{-1}F_n^*(x)\,dx \leqslant \limsup_n \int_{x_0}^{V_n} gG^{-1}F_n(x/c)\,dx$$

$$\leqslant c \limsup_n \int_{x_0/c}^{X_{n:n}} gG^{-1}F_n(x)\,dx < \epsilon.$$

This completes the proof of Theorem 2.2.

If in Theorems 2.1 and 2.2, we replace $F(x)$ and $G(x)$ by $1 - F(-x)$ and $1 - G(-x)$, then left and right tails are interchanged and we immediately obtain dual versions of these theorems. Although we shall not make use of these dual theorems in this paper, we give these results here because of their relevance for problems concerning estimation of r under conditions other than $F \underset{c}{<} G$.

Theorem 2.3. Let $F, G \in \mathscr{F}$, let $EX^- = -\int_{-\infty}^{0} x\,dF(x) < \infty$ and let gG^{-1} be uniformly continuous on $(0, 1)$. Assume that either $F^{-1}(0) > -\infty$ or $gG^{-1}(y)/y$ is bounded on $(0, 1)$. Then for any fixed

$$\xi \in (F^{-1}(0), F^{-1}(1)), \quad \Phi_{F, \xi}(F^{-1}(0)) > -\infty$$

and for $n \to \infty$

$$\sup_{X_{1:n} \leqslant x \leqslant \xi} |\Phi_{F_n, \xi}(x) - \Phi_{F, \xi}(x)| \to 0 \quad \text{almost surely.}$$

If, moreover, the grid $\{w_{j,n}\}$ becomes dense in the support of F and subexponential in the left tail of F, then for $n \to \infty$

$$\sup_{U_n \leqslant x \leqslant \xi} |\Phi_{F_n^*, \xi}(x) - \Phi_{F, \xi}(x)| \to 0 \quad \text{almost surely.}$$

Theorem 2.4. For $F, G \in \mathscr{F}$, let $G \underset{c}{<} F$ and let gG^{-1} be uniformly continuous on $(0, 1)$. Assume that there exists a number $0 < \eta < 1$ such that for $0 < y \leqslant \eta$, $gG^{-1}(y)$ is nondecreasing and $gG^{-1}(y)/y$ is nonincreasing in y. Then for any fixed

$$\xi \in (F^{-1}(0), F^{-1}(1)), \quad \Phi_{F, \xi}(F^{-1}(0)) > -\infty$$

and for $n \to \infty$

$$\sup_{X_{1:n} \leqslant x \leqslant \xi} |\Phi_{F_n, \xi}(x) - \Phi_{F, \xi}(x)| \to 0 \quad \text{almost surely.}$$

If, moreover, the grid $\{w_{j,n}\}$ becomes dense in the support of F and subexponential in the left tail of F, then for $n \to \infty$

$$\sup_{U_n \leqslant x \leqslant \xi} |\Phi_{F_n^*, \xi}(x) - \Phi_{F, \xi}(x)| \to 0 \quad \text{almost surely.}$$

The result for the left tail that we need in this paper is very much weaker. We consider the special case $F = G$. Note that for $x < \xi$, $\Phi_{G,\xi}$, $\Phi_{G_n,\xi}$ and $\Phi_{G_n^*,\xi}$ are negative and that $\Phi_{G,\xi}(x) = -(G(\xi) - G(x))$.

Theorem 2.5. Let $G \in \mathscr{F}$ and let gG^{-1} be uniformly continuous on $(0,1)$. Then for any fixed $\xi \in (G^{-1}(0), G^{-1}(1))$

$$\liminf_n [\inf_{X_{1:n} \leqslant x \leqslant \xi} (-\Phi_{G_n,\xi}(x) + \Phi_{G,\xi}(x))] \geqslant 0 \quad \text{almost surely.}$$

If, moreover, the grid $\{w_{j,n}\}$ becomes dense in the support of G, then

$$\liminf_n [\inf_{U_n \leqslant x \leqslant \xi} (-\Phi_{G_n^*,\xi}(x) + \Phi_{G,\xi}(x))] \geqslant 0 \quad \text{almost surely.}$$

Proof. Let $x_0 < \xi$ with $0 < G(x_0) < \epsilon$. Then for all $X_{1:n} \leqslant x \leqslant \xi$,

$$-\Phi_{G_n}(x) + \Phi_G(x) = \int_x^\xi (gG^{-1}G_n(u) - gG^{-1}G(u))\, du$$

$$\geqslant -\int_{x_0}^\xi |gG^{-1}G_n(x) - gG^{-1}G(x)|\, dx - G(x_0) \overset{\text{a.s.}}{\to} -G(x_0) \geqslant -\epsilon$$

by the Glivenko–Cantelli theorem, the uniform continuity of gG^{-1} on $(0,1)$ and the fact that $[x, \xi]$ is a.s. contained in the interval where

$$0 < F < 1 \quad \text{and} \quad 0 < F_n < 1$$

for sufficiently large n. Since $\epsilon > 0$ may be taken arbitrarily small, the result for Φ_{G_n} follows. The modifications needed to prove the result for $\Phi_{G_n^*}$ are indicated in the proof of Theorem 2.1.

3 STRONG CONSISTENCY OF $\hat{r}_n$

We assume $F \underset{c}{<} G$ and show that, under regularity conditions, $\hat{r}_n(x)$ defined by (1.4) and (1.5) tends to $r(x)$ with probability 1 at continuity points x of r. We use the strong uniform convergence of $\Phi_{F_n,\xi}$ and $\Phi_{F_n^*,\xi}$ to $\Phi_{F,\xi}$ on the interval bounded below by ξ. This generalizes consistency proofs for order statistics grids of Marshall and Proschan (1965) for the case $G(x) = 1 - e^{-x}$ and Robertson (1967) for the case $G(x) = x$. A large part of our proof is similar to that of Marshall and Proschan while the use of the strong convergence of Φ_{F_n} was suggested by the proof of Robertson. For the grid based on order statistics, the estimator $\hat{\rho}_n$ is *not* consistent. However, as Prakasa Rao (1966) has shown, isotonizing effectively widens the windows so that $\hat{r}_n$ is strongly consistent.

We shall impose the following condition on the right tail of G.

Condition R. Either $G^{-1}(1) < \infty$; or $gG^{-1}(y)/(1-y)$ is bounded on $(0,1)$; or there exists a number $0 < \eta < 1$ such that for $\eta \leqslant y < 1$, $gG^{-1}(y)$ is nonincreasing and $gG^{-1}(y)/(1-y)$ is nondecreasing in y.

Note that this condition will usually be satisfied in practice, the only exceptions being distributions for which g or $g/(1-G)$ oscillate near $+\infty$.

Theorem 3.1. For $F, G \in \mathscr{F}$, let $F < G$, let $\displaystyle\int_0^\infty x\,dG(x) < \infty$ and let gG^{-1} be uniformly continuous on $(0,1)$. Assume that G satisfies the right tail condition R and that either $\{w_{j,n}\}$ is a nonrandom grid that becomes dense in R^1 and subexponential in the right tail of G, or $\{w_{j,n}\} = \{X_{j:n}\}$ is the grid determined by the order statistics. Then for $\hat{r}_n$ defined by (1.4) (or (1.5)) we have for every fixed x_0 with $0 < F(x_0) < 1$,

$$r(x_0 - 0) \leqslant \liminf_n \hat{r}_n(x_0) \leqslant \limsup_n \hat{r}_n(x_0) \leqslant r(x_0 + 0) \quad \text{almost surely.}$$

Proof. Let ξ be an arbitrary point satisfying $F^{-1}(0) < \xi < x_0$ and let $w_{a_n,n} \geqslant \xi$ be a sequence of gridpoints converging to ξ for $n \to \infty$. With probability 1, there exists an integer N such that for $n \geqslant N$,

$$\xi \leqslant w_{a_n,n} \leqslant x_0 < V_n,$$

and hence

$$\hat{r}_n(x_0) = \inf_{x_0 < w_{t,n} \leqslant X_{n:n}} \sup_{X_{1:n} \leqslant w_{s,n} \leqslant x_0} \frac{F_n(w_{t,n}) - F_n(w_{s,n})}{\sum_{j=s}^{t-1} gG^{-1}F_n(\xi_{j,n})\,(w_{j+1,n} - w_{j,n})}$$

$$\geqslant \inf_{x_0 \leqslant x \leqslant V_n} \frac{F_n(x) - F_n(w_{a_n,n})}{\Phi_{F_n^*,\xi}(x) - \Phi_{F_n^*,\xi}(w_{a_n,n})}.$$

Since $F < G$, one easily verifies that $\displaystyle\int_c^\infty x\,dG(x) < \infty$ implies that

$$\int_0^\infty x\,dF(x) < \infty; \quad \text{also} \quad F^{-1}(1) < \infty \quad \text{if} \quad G^{-1}(1) < \infty,$$

and it follows that if the grid becomes subexponential in the right tail of G, then it will also become subexponential in the right tail of F. Hence, the conditions of either Theorem 2.1 or Theorem 2.2 are satisfied and as a result $\Phi_{F_n^*,\xi}$ converges a.s. uniformly on $[\xi, V_n]$ to $\Phi_{F,\xi}$. Since also F_n converges a.s. uniformly to F by the Glivenko–Cantelli theorem,

$$w_{a_n,n} \to \xi \quad \text{and} \quad x_0 - \xi > 0,$$

we have

$$\liminf_n \hat{r}_n(x_0) \geqslant \inf_{x_0 \leqslant x < F^{-1}(1)} \frac{F(x) - F(\xi)}{\Phi_{F,\xi}(x)} \geqslant r(\xi) \quad \text{almost surely,}$$

where the second inequality follows from

$$\Phi_{F,\xi}(x) = \int_\xi^x gG^{-1}F(u)\,du = \int_\xi^x \frac{f(u)}{r(u)}\,du \leqslant \frac{1}{r(\xi)}(F(x) - F(\xi))$$

for $x > \xi$ since r is nondecreasing. Since $\xi \in (F^{-1}(0), x_0)$ was arbitrary, this proves the left-hand inequality of the theorem. Note that the proof applies also in the case $w_{j,n} = X_{j:n}$, where $F_n^* = F_n$ and $V_n = X_{n:n}$.

To prove the right-hand inequality, we let ξ be an arbitrary point satisfying $x_0 < \xi < F^{-1}(1)$; let $w_{b_n,n} \leqslant \xi$ be a sequence of gridpoints converging to ξ for $n \to \infty$. With probability 1, there exists an integer N such that for $n \geqslant N$, $U_n < x_0 < w_{b_n,n} \leqslant \xi$, and hence

$$\hat{r}_n(x_0) \leqslant \sup_{U_n \leqslant w_{s,n} \leqslant x_0} \frac{F_n(w_{b_n,n}) - F_n(w_{s,n})}{\sum\limits_{j=s}^{b_n-1} gG^{-1}F_n(\xi_{j,n})(w_{j+1,n} - w_{j,n})}.$$

Consider the random variables $Y_1, Y_2, \ldots$ defined by $Y_i = G^{-1}F(X_i)$. These random variables are independent with distribution function G. The order statistics and the empirical distribution function corresponding to $Y_1, Y_2, \ldots, Y_n$ are $Y_{i:n} = G^{-1}F(X_{i:n})$ and $G_n = F_n F^{-1}G$. Define $\tilde{x}_0 = G^{-1}F(x_0)$, $\tilde{\xi} = G^{-1}F(\xi)$, $\tilde{w}_{j,n} = G^{-1}F(w_{j,n})$, $\tilde{\xi}_{j,n} = G^{-1}F(\xi_{j,n})$ and $\tilde{U}_n = G^{-1}F(U_n)$. Note that $0 < G(\tilde{x}_0) < G(\tilde{\xi}) < 1$ and that $\tilde{U}_n$ is the infimum of the transformed gridpoints $\tilde{w}_{j,n}$ in the interval $[Y_{1:n}, Y_{n:n}]$. Also $F_n(w_{j,n}) = G_n(\tilde{w}_{j,n})$, $F_n(\xi_{j,n}) = G_n(\tilde{\xi}_{j,n})$ and for $w_{j+1,n} \leqslant \xi$ we have

$$\tilde{w}_{j+1,n} - \tilde{w}_{j,n} = \int_{w_{j,n}}^{w_{j+1,n}} r(x)\,dx \leqslant r(\xi)(w_{j+1,n} - w_{j,n}).$$

Hence for $n \geqslant N$

$$\hat{r}_n(x_0) \leqslant r(\xi) \sup_{\tilde{U}_n \leqslant \tilde{w}_{s,n} \leqslant \tilde{x}_0} \frac{G_n(\tilde{w}_{b_n,n}) - G_n(\tilde{w}_{s,n})}{\sum\limits_{j=s}^{b_n-1} gG^{-1}G_n(\tilde{\xi}_{j,n})(\tilde{w}_{j+1,n} - \tilde{w}_{j,n})}$$

$$\leqslant r(\xi) \sup_{\tilde{U}_n \leqslant x \leqslant \tilde{x}_0} \frac{G_n(\tilde{w}_{b_n,n}) - G_n(x)}{\Phi_{G_n^*,\tilde{\xi}}(\tilde{w}_{b_n,n}) - \Phi_{G_n^*,\tilde{\xi}}(x)},$$

where $G_n^*(x) = G_n(\tilde{\xi}_{j,n})$ for $\tilde{w}_{j,n} \leqslant x < \tilde{w}_{j+1,n}$. Thus, we have transformed the problem to the case when $F = G$. If $\{w_{j,n}\}$ becomes dense in R^1, then $\{\tilde{w}_{j,n}\}$ becomes dense in the support of G; if $w_{j,n} = X_{j:n}$, then $\tilde{w}_{j,n} = Y_{j:n}$. Hence Theorem 2.5, the Glivenko–Cantelli theorem, the convergence of $\tilde{w}_{b_n,n}$ to $\tilde{\xi}$ and the fact that $\xi - x_0 > 0$, yield

$$\limsup_n \hat{r}_n(x_0) \leqslant r(\xi) \sup_{G^{-1}(0) < x \leqslant \tilde{x}_0} \frac{G(\tilde{\xi}) - G(x)}{\Phi_{G,\tilde{\xi}}(x)} = r(\xi) \quad \text{almost surely,}$$

since $\Phi_{G,\xi}(x) = G(\tilde{\xi}) - G(x)$. Since $\xi \in (x_0, F^{-1}(1))$ was arbitrary, the proof of the theorem is completed.

Remark 1. The reason why the two parts of the proof are dealt with in a different manner may not be immediately obvious. In the first part, we can give a direct proof without transforming F into G because we are concerned with the right tail and Theorems 2.1 and 2.2 yield convergence for arbitrary $F \underset{c}{<} G$. In the second part, we have to transform F into G because Theorem 2.5 was proved for this case only. However, there are also reasons of a different nature. The first part simply cannot be proved by the methods employed here after first transforming F into G, because in the right tail the transformed grid $\{\tilde{w}_{j,n}\}$ does not necessarily become subexponential even if $\{w_{j,n}\}$ is subexponential. The second part could, of course, be proved directly with the aid of a theorem analogous to Theorem 2.5 for $F \underset{c}{<} G$. However, without employing the transformation $G^{-1}F$, it would seem hard to avoid a condition in such a theorem that ensures that $\int_{-\infty}^{0} gG^{-1}F(x)\,dx < \infty$. This is unattractive since F is unknown.

Remark 2. A similar consistency proof can be given for the case where the c-ordering is reversed and $G \underset{c}{<} F$. In this case, the estimator becomes

$$\hat{r}_n(x) = \sup_{\substack{t \geqslant i+1 \\ w_{t,n} \leqslant X_{n:n}}} \quad \inf_{\substack{s \leqslant i \\ w_{s,n} \geqslant X_{1:n}}} \frac{F_n(w_{t,n}) - F_n(w_{s,n})}{\sum_{j=s}^{t-1} gG^{-1}F_n(\xi_{j,n})(w_{j+1,n} - w_{j,n})}$$

for $w_{i,n} \leqslant w_{i+1,n}$. Consistency of $\hat{r}_n$ may now be proved by applying Theorems 2.3 and 2.4 and a dual version of Theorem 2.5 obtained by reflection (replace $F(x)$ by $1 - F(-x)$ and $G(x)$ by $1 - G(-x)$). We may also immediately obtain the desired result by reflection from Theorem 3.1. The conditions for strong consistency of $\hat{r}_n$ at continuity points of r are thus seen to be: $F, G \in \mathcal{F}$; $G \underset{c}{<} F$; $\int_{-\infty}^{0} x\,dG(x) > -\infty$; gG^{-1} uniformly continuous on $(0,1)$; either $G^{-1}(0) > -\infty$, or $gG^{-1}(y)/y$ bounded on $(0,1)$ or existence of $0 < \eta < 1$ such that $gG^{-1}(y)$ nondecreasing and $gG^{-1}(y)/y$ nonincreasing for $0 < y \leqslant \eta$; either $\{w_{j,n}\}$ a nonrandom grid that becomes dense in R^1 and subexponential in the left tail of G, or $w_{j,n} = X_{j:n}$.

ACKNOWLEDGEMENT

We would like to gratefully acknowledge the help and advice of Professor H. D. Brunk. Professor Ron Pyke kindly supplied the reference to Dvoretzky, Kiefer and Wolfowitz.

REFERENCES

Barlow, R. E. and W. R. van Zwet (1970). Asymptotic Properties of Isotonic Estimators for the Generalized Failure Rate Function. Part II: Asymptotic Distributions. Unpublished manuscript.

Brunk, H. D. (1965). Conditional Expectation Given a σ-Lattice and Applications, *Ann. Math. Statist.* **36**, 1339–50.

Dvoretzky, A., J. Kiefer and J. Wolfowitz, (1956). Asymptotic Minimax Character of the Sample Distribution Function and of the Classical Multinomial Estimator, *Ann. Math. Statist.* **27**, 642–69.

Grenander, Ulf, (1956). On the Theory of Mortality Measurement, Part II, *Skand. Aktuarietidskr.*, **39**, 125–53.

Marshall, A. W. and F. Proschan, (1965). Maximum Likelihood Estimation for Distributions with Monotone Failure Rate, *Ann. Math. Statist.* **36**, 69–77.

Nadaraya, E. A. (1965). On Non-Parametric Estimates of Density Functions and Regression Curves, *Theory of Probability and its Applications*, **10**, 186–90.

Parzen, E. (1962). On Estimation of a Probability Density and Mode. *Ann. Math. Statist.* **33**, 1065–76.

Prakasa Rao, B. L. S., (1966). Asymptotic Distributions in Some Non-regular Statistical Problems, Technical Report No. 9, Department of Statistics and Probability, Michigan State University.

Robertson, Tim, (1967). On Estimating a Density Which is Measurable with Respect to a σ-Lattice, *Ann. Math. Statist.* **38**, no. 2, 482–93.

van Zwet, W. R. (1964). Convex Transformations of Random Variables, Mathematical Centre, Amsterdam.

Watson, G. S. and M. R. Leadbetter, (1964). Hazard Analysis, Part II. *Sankhyā*, **26**, 101–16.

DISCUSSION ON
BARLOW AND VAN ZWET'S PAPER

A. W. MARSHALL

I would like to thank Professors Barlow and van Zwet for a most interesting paper. They have very significantly advanced earlier related work, and I am confident that researchers who follow them will find this paper an invaluable source of ideas.

The following comments are relatively minor, but they seem to apply equally well to a number of isotonic estimation problems.

In choosing an estimation procedure for a restricted family of distributions, one might want to begin by considering as candidates for selection a class of estimators *not* depending upon which of several equivalent descriptions is given for the restricted family. Of course, the assumption that ϕ is increasing is equivalent to the assumption that the composition $u\phi v$ is increasing, where u and v are strictly increasing continuous functions with appropriate domains and ranges. The class of estimators one considers under the assumption that ϕ is increasing perhaps ought to be the same as under the assumption that $u\phi v$ is increasing.

To estimate ϕ, Barlow and van Zwet begin with a sequence $\{\phi_n\}$ of piecewise constant 'basic' estimators:

(a) $\phi_n(x) = \phi_n(w_{j,n}), \quad w_{j,n} \leqslant x < w_{j+1,n};$

then they compute the isotonic regression

(b) $\overline{\phi}_n(x) = \inf_{t \geqslant i} \sup_{s \leqslant i} \sum_{j=s}^{t} \phi_n(w_{j,n}) (w_{j+1,n} - w_{j,n})/(w_{t+1,n} - w_{s,n})$
$$(w_{i,n} \leqslant x < w_{i+1,n}).$$

If instead, they had started with the corresponding basic estimator $u\phi_n v$ for $u\phi v$, and then used the same kind of isotonic regression, they would have obtained the estimator $\overline{u\phi_n v}$ for $u\phi v$. Transforming this estimate in order to estimate ϕ, one obtains

(c) $u^{-1}\overline{u\phi_n v}v^{-1}(x) = \inf_{t \geqslant i} \sup_{s \leqslant i} u^{-1}\left(\dfrac{\sum\limits_{j=s}^{t} u\phi_n(w_{j,n}) (w_{j+1,n}^* - w_{j,n}^*)}{w_{t+1,n}^* - w_{s,n}^*} \right),$

where $w_{j,n}^* = v^{-1}(w_{j,n})$. Choice of the function v is a choice of arbitrary averaging weights, as considered by Barlow and van Zwet. But choice

[174]

of the function u is a choice of a quite general averaging, whereas Barlow and van Zwet consider only arithmetic averaging.

Other work on isotonic regression has also been limited to arithmetic averaging. However, in the case of an increasing failure rate, the maximum likelihood principle chooses harmonic averaging. Although Barlow and van Zwet rewrite this as an arithmetic mean, in so doing they have to accept the rather unusual feature that weights used in averaging depend upon the quantities to be averaged.

It may be useful to observe that if one regards u and v as initially chosen and incorporated in the definition of ϕ, then one can restrict consideration to estimators of the relatively simple form (b).

To estimate an increasing function, there is an advantage in initially estimating its (convex) integral. E.g. Barlow and van Zwet could have estimated $G^{-1}F(x) = \int_{-\infty}^{x} r(z)\, dz$ in place of the generalized failure rate r. This eliminates problems which arise when natural basic estimators for $G^{-1}F$ (like $G^{-1}F_n$) do not have derivatives, and do not correspond to basic estimators for r. After the appropriate operation is performed to obtain a convex estimator for $G^{-1}F$ from such a basic estimator, there are no serious problems in differentiating to obtain an increasing estimator for r, and in fact consistent convex estimators for $G^{-1}F$ yield consistent monotone estimators for r.

Lemma A. If h and $h_n, n = 1, 2, \ldots,$ are defined and convex on an open interval I, and if $\lim_{n \to \infty} h_n(x) = h(x)$ for all I, then

$$h^-(x) \leqslant \varliminf h_n^-(x) \leqslant \varlimsup h_n^+(x) \leqslant h^+(x) \quad (x \in I),$$

where h^- (h^+) is the left- (right-) hand derivative of h. This easily proven lemma is surely not new.

If one chooses to make use of this lemma by first estimating $G^{-1}F$ rather than its derivative r, the choice of a scaling function u becomes a choice of which convex function to estimate. Barlow and van Zwet are concerned primarily with

$$-\Phi_{F,\xi}(F^{-1}(y)) = -\int_{F(\xi)}^{y} \frac{gG^{-1}(z)}{fF^{-1}(z)}\, dz,$$

which is convex if and only if $G^{-1}F$ is convex. Other convex functions are obtained by integrating increasing functions u of the generalized failure rate. Possibly the following lemma can sometimes serve as a guide for choosing an appropriate convex function.

Lemma B. Let Ψ be convex on $[0, 1]$, and let Φ be a continuous real valued function on $[0, 1]$. Let $\bar\Phi(x) = \sup\{h(x): h\ \text{is convex and}\ h(z) \leqslant \Phi(z)\ \text{for all}$ $z \in [0, 1]\}$. Then

$$\sup_{0 \leqslant x \leqslant 1} |\Phi(x) - \Psi(x)| \geqslant \sup_{0 \leqslant x \leqslant 1} |\bar\Phi(x) - \Psi(x)|.$$

Of course, the right derivative of $\bar\Phi$ is given by (b) whenever the right derivative of Φ is given by (a). This lemma exhibits a metric which is decreased via isotonic regression.

To prove Lemma B, note first that for all $y \in [0, 1]$, either $\bar\Phi(y) = \Phi(y)$, or y is an interior point of a closed interval I over which $\bar\Phi$ is linear. For such an interval, either $\sup_{x \in I} |\bar\Phi(z) - \Psi(x)|$ is attained at an endpoint of I (where $\bar\Phi = \Phi$), or it is attained at an interior point, where $\Psi < \bar\Phi$. Since $\bar\Phi \leqslant \Phi$ on $[0, 1]$ it follows that

$$\sup_{x \in I} |\bar\Phi(x) - \Psi(x)| \leqslant \sup_{x \in I} |\Phi(x) - \Psi(x)|.$$

Note that Lemma B coupled with Barlow and van Zwet's Theorem 2.1 provides an immediate proof of the uniform strong consistency of the convex minorant of $\Phi_{F_n, \xi}$ as an estimator of $\Phi_{F, \xi}$. This together with Lemma A yields an alternate proof of Theorem 3.1.

Lemma B can also be adapted for proving consistency when $G^{-1}F$ is not convex, but only starshaped (i.e. $G^{-1}F(x)/x$ is increasing in $x > 0$). This is because Lemma B remains valid if everywhere the word 'convex' is replaced by 'starshaped'. An important class of life distributions is formed by those F for which $G^{-1}F$ is starshaped when G is the exponential distribution.

ESTIMATION OF ISOTONIC REGRESSION

H. D. BRUNK

1 INTRODUCTION

Chernoff (1964) remarks: 'The problems of estimating the density and the mode of a distribution are rather delicate and this delicacy is related to the fact that the density may be changed considerably over a short range without affecting the probability distribution substantially.' On the other hand, of course, the empirical distribution function of a random sample estimates the distribution function well. Similar remarks apply to the estimating of a regression function. Unless an unknown regression function is assumed to belong to a severely restricted class of functions (such as the class of linear combinations of a few given functions), observations at a finite number of points in an interval yield little information about the regression function. But in the presence of mild regularity conditions such as continuity of the regression function, its indefinite integral can be estimated; such estimation is the subject of §§2 and 3. The analogy with estimation of a probability distribution function is striking, and it is pointed out in §2 that the two situations can be regarded as extreme instances of an 'independent observations regression model'. At one extreme, the observation points are degenerate random variables, i.e. they are given, and independent observations are made on distributions associated with them. At the other extreme, the observations points are random variables, but the 'observations' made at them are degenerate random variables, each identically equal to 1. In the former case of given observations points, a theorem giving conditions for uniformly consistent estimation of the integral regression function with probability 1 is essentially only a restatement of a known variant of the strong law of large numbers. The theorem may be interpreted as an analogue of the Glivenko–Cantelli Theorem. It is applied to yield uniformly strongly consistent estimation of the integral regression function in the more general situation of an independent observations regression model.

In §3, Prohorov's (1956) generalization of Donsker's Theorem yields the asymptotic distribution of the estimator, suitably normalized, of the integral regression function. This can be used to obtain asymptotic distributions of various statistics which might be used to test the simple hypothesis that the regression function is a specified function. (Galen

[177]

Shorack and Ronald Pyke have kindly called the author's attention to a related paper by Hájek (1956). Hájek deals with the linear regression model, and proposes a statistic based on ranks for testing the composite hypothesis that the slope is zero. He shows that the asymptotic distribution is that of the Kolmogorov–Smirnov statistic.)

Bearing in mind Chernoff's remark quoted above, it is worthy of note that estimators of nonincreasing (or unimodal) densities and of densities required to satisfy order restrictions of other kinds (e.g. Grenander, 1956; Marshall and Proschan, 1965; Barlow, 1968), have been proposed and studied. In further illustration of a correspondence between estimation of probability distributions and estimation of regression functions are studies of analogous estimators of regression functions required to be monotonic (e.g. Eeden, 1957a; Ayer, Brunk, Ewing, Reid and Silverman, 1955; Bartholomew, 1959, 1961; for other references see Brunk, 1965). Section 4 gives a uniform strong consistency theorem for such a situation which is only trivially different from (a corrected version of) Theorem 6.2 of Brunk (1958), which strengthened and generalized a consistency theorem in Ayer, Brunk, Ewing, Reid and Silverman (1955). It is included here to clarify the relationship between it and the consistency explored in §§2 of the estimator of the integral regression function.

Chernoff (1964) determined the asymptotic distribution of an estimator of a mode, and Prakasa Rao (1968) that of the maximum likelihood estimator of a unimodal density. Their methods are applied in §5 to obtain the asymptotic distribution of the estimator of a nonincreasing regression function; it is essentially the same as that found by Rao in the density case.

The theorem of §6 represents a slight extension of known results on maximum likelihood estimation of an isotonic regression function in sampling from members of an exponential family.

2 UNIFORMLY CONSISTENT ESTIMATION OF THE INTEGRAL REGRESSION FUNCTION

Suppose that associated with each $t \in [0, 1]$ is a univariate distribution $D(t)$ with mean $\mu(t)$; $\mu(\cdot)$ is the *regression function*. Let $\{t_n\}$ be a sequence of numbers in $[0, 1]$, not necessarily distinct, to be called *observation points*. For each n, let $Y_n(t_n)$ denote a random variable having the distribution associated with t_n, so that $EY(t_n) = \mu(t_n)$; and let the random variables $\{Y_n(t_n), n = 1, 2, \ldots\}$ be independent.

Let $\psi_j(\cdot)$ denote the indicator of the interval $[t_j, \infty)$: $\psi_j(t) = 1_{[t_j, \infty)}(t)$. Set

$$S_n(t) = \sum_{j=1}^{n} Y_j(t_j)\,\psi_j(t) \quad (t \in [0, 1]). \tag{2.1}$$

Let $F_n(\cdot)$ denote the 'empirical distribution function' of the set

$$\{t_1, \ldots, t_n\} \colon F_n(t) = \sum_{j=1}^{n} \psi_j(t)/n. \tag{2.2}$$

For a given probability distribution function F with support in $[0, 1]$, set

$$M(t) = \int_{[0, t]} \mu(v)\, dF(v) \quad (t \in [0, 1]), \tag{2.3}$$

the *integral regression function*. Let

$$\left.\begin{aligned}
M_n(t) &= ES_n(t)/n = \sum_{j=1}^{n} \mu(t_j)\,\psi_j(t)/n, \\
&= \int_{[0, t]} \mu(v)\, dF_n(v).
\end{aligned}\right\} \tag{2.4}$$

If F_n converges weakly to F, then $M_n(t) \to M(t)$ as $n \to \infty$ at each continuity point t of F, if $\mu(\cdot)$ is continuous. In this section conditions are given sufficient in order that $S_n(\cdot)/n$ converge uniformly on $[0, 1]$ to $M(\cdot)$ with probability 1. The essence of the proof is contained in Theorem 6.1 of Brunk (1958). Proposition 2.2 below is little more than a restatement of that theorem in different terminology.

2.1. *Definition.* Let r be a fixed positive number. A sequence $\{Y_n\}$ of independent random variables will be said to satisfy the *r-order Kolmogorov condition* if $\sum_n E|Y_n - EY_n|^{2r}/n^{r+1} < \infty$.

2.2. *Proposition.* Let $\{t_n\}$ be a sequence of real numbers in $[0, 1]$. Let $\mu(\cdot)$ be continuous on $[0, 1]$. Let there exist $r > 0$ such that the sequence $\{Y_n(t_n)\}$ of independent random variables satisfies the r-order Kolmogorov condition. Then

$$P\{\lim_n \sup_{0 \leqslant t \leqslant 1} |S_n(t)/t - M_n(t)| = 0\} = 1.$$

It will be clear from the proof of Proposition 2.2 that $S_n(\cdot)$ may be replaced by the function which coincides at points $t_1, \ldots, t_n$ with $S_n(\cdot)$ defined by (2.1), and interpolates linearly between adjacent observation points.

Proof. The random function $S_n(\cdot)/n - M_n(\cdot)$ achieves its maximum in $[0, 1]$ on the set $\{t_1, \ldots, t_n\}$. Thus the statement

$$P\{\lim_n \sup_{0 \leqslant t \leqslant 1} |S_n(t)/n - M_n(t)| = 0\} = 1$$

is equivalent to the statement

$$P\{\lim_{n}\ \max_{1\leqslant j\leqslant n}\ (1/n)|S_n(t_j)-ES_n(t_j)| = 0\} = 1.$$

But this latter is an immediate consequence of Theorem 6.1 of Brunk (1958).

2.3. *Theorem. Let $\{t_n\}$ be a sequence of real numbers in $[0,1]$. Let $\mu(\cdot)$ be continuous on $[0,1]$. Let the random variables $\{Y_n(t_n)\}$ be independent with $EY_n(t_n) = \mu(t_n)$ and with bounded variances. Then*

$$P\{\lim_{n}\ \sup_{0<t<1}\ |S_n(t)/n - M_n(t)| = 0\} = 1.$$

If further the empirical distribution function $F_n(\cdot)$ of $\{t_1, ..., t_n\}$ converges uniformly to a probability distribution function $F(\cdot)$, then

$$P\{\lim_{n}\ \sup_{0\leqslant t\leqslant 1}\ |S_n(t)/n - M(t)| = 0\} = 1.$$

Proof. The first statement is immediate from Proposition 2.2, on setting $r = 1$. Also, it is readily verified by approximating the continuous function $\mu(\cdot)$ by a step function that the hypothesis on F_n implies

$$\sup_{0<t<1} |M_n(t) - M(t)| \to 0.$$

The second conclusion follows.

A general kind of regression situation will now be considered, in which $\{T_n\}$ as well as $\{Y_n\}$ are random variables. It is convenient to utilize a structure formalizing the notion that $\{T_n\}$ is a discrete parameter stochastic process, and that conditional on a realization $\{t_n\}$ of this process the $\{Y_n\}$ are independent random variables such that Y_n has a given distribution $D(t_n)$, $n = 1, 2, ...$. (Such a distribution $D(t)$ may be specified by a distribution function $G(y; t) = P[Y_n \leqslant y | T_n = t]$.) Let Ω_1 denote the space of all sequences of reals in $[0, 1]$, $\mathscr{A}_1$ a σ-field of subsets of Ω_1, and P_1 a probability measure on $\mathscr{A}_1$. The probability space $(\Omega_1, \mathscr{A}_1, P_1)$ is interpreted as the probability distribution of the stochastic process $T = (T_1, T_2, ...)$. Let Ω_2 denote the space of all sequences of reals, and $\mathscr{A}_2$ the σ-field generated by the finite-dimensional (cylindrical) sets. The space Ω_2 is interpreted as the range space of the process

$$Y = (Y_1, Y_2, ...).$$

Let P_2^1 be a transition probability (Neveu, 1965, p. 73) with the following properties. For $\omega_1 = (t_1, t_2, ...) \in \Omega_1$, let $P_2^1(\omega_1, \cdot)$ denote the product measure of the distributions $D(t_1), D(t_2), ...$. That is, $P_2^1(\cdot, \cdot)$ is defined so that if $\omega_1 = (t_1, t_2, \ldots)$ and if A is a finite-dimensional set in Ω_2,

$$A = B_1 \times B_2 \times \ldots \times B_r \times R \times R \ldots,$$

then
$$P_2^1(\omega_1, A) = \prod_{i=1}^{r} \int_{B_i} dG(y; t_i).$$

Suppose also that for every $A_2 \in \mathscr{A}_2, P_2^1(\cdot, A_2)$ is measurable on $(\Omega_1, \mathscr{A}_1)$.

By the *distribution* of the bivariate process $(T, Y) = \{(T_n, Y_n)\}_1^\infty$ is to be understood the probability space $(\Omega_1 \times \Omega_2, \mathscr{A}_1 \otimes \mathscr{A}_2, P)$, where P is the unique probability such that

$$P(A_1 \times A_2) = \int_{A_1} P_2^1(\omega_1, A_2) \, P_1(d\omega_i)$$

(Neveu, 1965, p. 74).

2.4. Definition. Such a bivariate process (T, Y) is an *independent observations regression model*.

If (T, Y) is an independent observations regression model, one may define as before the function $S_n(\cdot)$ on $[0, 1]$ by

$$S_n(t) = \sum_{j=1}^{n} Y_j 1_{[0,t]}(T_j). \tag{2.5}$$

Again, suppose that for $t \in [0, 1]$ the distribution $D(t)$ has mean $\mu(t)$, i.e.

$$E(Y_j | T_j = t) = \mu(t).$$

Set again
$$M(t) = \int_{[0,t]} \mu(t) \, dF(t).$$

2.5. Corollary. *For each positive n, let $F_n(\cdot)$ denote the empirical distribution function of the random variables $T_1, \ldots, T_n$. Suppose there exists a probability distribution function F such that with probability 1 $\{F_n\}$ converges uniformly on $[0, 1]$ to F. If (T, Y) is an independent observations regression model such that with probability 1 the distributions $D(T_1), D(T_2), \ldots$ have bounded variances, then*

$$P\{\lim_{n} \sup_{0 \leqslant t \leqslant 1} |S_n(t)/t - M(t)| = 0\} = 1. \tag{2.6}$$

Proof. Set $\quad B = \{\omega \in \Omega_1 \times \Omega_2 : \lim_{n} \sup_{0 \leqslant t \leqslant 1} |S_n(t)/t - M(t)| = 0\}$.
Let A_1 denote the subset of Ω_1 consisting of points $\omega_1 = (t_1, t_2, \ldots)$ such that the empirical distribution function of $\{t_1, t_2, \ldots, t_n\}$ converges uniformly to F, and such that the distributions $D(t_1), D(t_2), \ldots$ have bounded variances. By hypothesis, $P_1(A_1) = 1$. For $\omega_1 \in \Omega_1$, let B_{ω_1}

denote the ω_1-section of $B: B_{\omega_1} = \{\omega_2 \in \Omega_2 : (\omega_1, \omega_2) \in B\}$. For $\omega_1 \in A_1$, $P_2^1(\omega_1, B_{\omega_1}) = 1$ by Theorem 2.3. Then

$$P(B) = \int P_2^1(\omega_1, B_{\omega_1}) P_1(d\omega_1) = \int_{A_1} P_2^1(\omega_1, B_{\omega_1}) P_1(d\omega_1) = 1.$$

2.6. *Corollary (Galtonian regression). Let (T, Y) be an independent observations regression model, let $T_1, T_2, \ldots$ be independent identically distributed random variables, and let the distributions $D(T_1), D(T_2), \ldots$ have bounded variances with probability 1. Then (2.6) holds.*

Proof. The Glivenko–Cantelli theorem provides verification of the first hypothesis of Corollary 2.5, and the conclusion follows.

It is interesting to note that while the Glivenko–Cantelli theorem is used in the proof of Corollary 2.6, it is itself an instance: the case in which, for each $t \in [0, 1]$, $D(t)$ is the degenerate distribution concentrating the total probability mass at 1; i.e. $Y(t) \equiv 1$ for each t.

3 ASYMPTOTIC DISTRIBUTION OF THE DIFFERENCE BETWEEN THE INTEGRAL REGRESSION FUNCTION AND ITS ESTIMATOR

Prohorov's generalization ([1956]; cf. also Billingsley, 1968, p. 77) of Donsker's Theorem is applied in this section to yield results concerning the asymptotic distribution of the estimator of the integral regression function as the number of observation points becomes infinite. No sequence of observation points is required for a study of the asymptotic distribution; rather we assume a sequence of finite sets of observation points

$$0 < t_{n,1} < t_{n,2} < \ldots < t_{n,n} \leqslant 1 \quad (n = 1, 2, \ldots).$$

It will be convenient to assume exactly *one* observation is made at each observation point.

In applying Prohorov's Theorem we shall require the following lemma, which is in effect a special case of Billingsley's remarks on random change of time (cf. Billingsley, 1968, p. 145 and Theorem 4.4). Denote by $C_{[a,b]}$ the space of continuous functions on $[a, b]$ with the uniform metric, and set $C = C_{[0,1]}$. We adopt the notation used in [Billingsley, 1968], $X_n \overset{D}{\to} X$, for convergence in distribution of a sequence $\{X_n\}$ of random elements of C to a random element X. We use 'o' to denote composition: $(X \circ \lambda)(t) = X[\lambda(t)]$.

3.1. *Lemma. Let X_n, X be random elements of C such that $X_n \overset{D}{\to} X$. Let*

$\{\lambda_n\}$ *be a sequence of real non-decreasing continuous functions on* $[0,1]$ *such that* $0 \leqslant \lambda_n(t) \leqslant 1$ *for* $t \in [0,1]$ *and such that* $\lambda_n \to \lambda$ *uniformly on* $[0,1]$ *where* λ *is continuous. Then* $X_n \circ \lambda_n \overset{D}{\to} X \circ \lambda$.

As in §2, F_n denotes the empirical distribution function of the set $\{t_{n,i}, i = 1, 2, ..., n\}$:

$$F_n(t) = \sum_{i=1}^{n} \psi_{n,i}(t_{n,i})/n, \qquad (3.1)$$

where $\quad \psi_{n,i}(t) = 1_{[t_{ni}, \infty)}(t) \quad (i = 1, 2, ..., n, \quad n = 1, 2, ...). \qquad (3.2)$

For an application to be made in §5, it is necessary to be somewhat more liberal with the distributions D also: we suppose that associated with each observation point t_{ni} is a distribution $D_n(t_{ni})$ with mean which may be taken to be the value at t_{ni} of a function $\mu_n(\cdot)$ on $[0,1]$: $\mu_n(t_{ni})$ is the mean of $D_n(t_{ni})$, and similarly $\sigma_n^2(t_{ni})$ denotes its variance. The result of an observation on $D_n(t_{ni})$ will be denoted by

$$Y_n(t_{ni}) \quad (i = 1, 2, ..., n, \quad n = 1, 2, ...).$$

Let $1_{n,i}(t)$ denote the indicator of the interval $(t_{n,i-1}, t_{n\,i}]$, and set

$$\left. \begin{aligned} \chi_{n,i}(t) &= \int_0^t 1_{n,i}(v)\, dv, \\ \delta_{n,i} &= t_{n,i} - t_{n,i-1} \quad (i = 1, 2, ..., n). \end{aligned} \right\} \qquad (3.3)$$

Set
$$\Psi_n(t) = \int_{[0,\,t]} \mu_n(v)\, dF(v). \qquad (3.4)$$

We consider as estimator of Ψ_n,

$$\overline{Y}_n(t) = \sum_{i=1}^{n} Y_n(t_{ni})\, \chi_{ni}(t)/n\delta_{ni} \quad (t \in [0,1]). \qquad (3.5)$$

Then
$$E\overline{Y}_n(t) = \sum_{i=1}^{n} \mu_n(t_{ni})\, \chi_{ni}(t)/n\delta_{ni}. \qquad (3.6)$$

We have $\quad E\overline{Y}_n(t_{ni}) = \int_{[0,\,t_{ni}]} \mu_n(v)\, dF_n(v) \quad$ and $\quad E\overline{Y}_n(\cdot)$

linear between adjacent observation points.

For $n = 1, 2, ...,$ let

$$s_n^2 = \sum_{i=1}^{n} \sigma_n^2(t_{ni}) = n \int_{[0,1]} \sigma_n^2(v)\, dF_n(v).$$

We shall impose Lindeberg's condition on the random variables

$$\{Y_n(t_{ni})\}:$$

for each $\epsilon > 0$,

$$\left. (1/s_n^2) \sum_{i=1}^{n} E\{[Y_n(t_{ni}) - \mu_n(t_{ni})]^2 \cdot 1_{\{[Y_n(t_{ni}) - \mu_n(t_{ni})]^2 \geqslant \epsilon^2 s_n^2\}}\} \to 0. \right\} \qquad (3.7)$$

This will hold in particular if there exists a positive number h such that

$$(1/s_n^{2+h}) \sum_{i=1}^{n} E(|Y_n(t_{ni})|^{2+h}) \to 0.$$

3.2. *Proposition. Let the random variables*

$$Y_n(t_{ni}) \quad (i = 1, 2, ..., n)$$

be independent for each n, and satisfy Lindeberg's condition (3.7). *If*

$$\sup_{t \in [0,1]} \left| \sum_{i=1}^{n} \mu_n(t_{ni}) \chi_{ni}(t)/n\delta_{ni} - \int_{[0,t]} \mu_n(v) \, dF(v) \right| = o(n^{-\frac{1}{2}}) \quad as \quad n \to \infty,$$

(3.8)

and if there is a positive function $\sigma^2(\cdot)$ such that

$$\sup_{t \in [0,1]} \left| \sum_{i=1}^{n} \sigma_n^2(t_{ni}) \chi_{ni}(t)/n\delta_{ni} - \int_{[0,t]} \sigma^2(v) \, dF(v) \right| = o(1) \quad as \quad n \to \infty,$$

(3.9)

then
$$(\sqrt{n}/\overline{\sigma})\,[\overline{Y}_n(\cdot) - \overline{\Psi}_n(\cdot)] \xrightarrow{D} W[U(\cdot)],$$
(3.10)

where W denotes the Wiener process on $[0, 1]$ *with $W(0) = 0$, and where*

$$U(t) = (1/\overline{\sigma}^2) \int_{[0,t]} \sigma^2(v) \, dF(v) \quad (t \in [0,1]),$$
(3.11)

$$\overline{\sigma}^2 = \int_{[0,1]} \sigma^2(v) \, dF(v) < \infty.$$
(3.12)

Proof. This theorem is an application of Prohorov's (1956) generalization of Donsker's Theorem. Set

$$s_{n,i}^2 = \sum_{j=1}^{i} \sigma_n^2(t_{nj}),$$

$$u_{n,i} = s_{n,i}^2/s_n^2 = (n/s_n^2) \int_{[0,t_{ni}]} \sigma_n^2(v) \, dF_n(v),$$

$$X_n(u_{ni}) = \sum_{j=1}^{i} [Y_n(t_{nj}) - \mu_n(t_{nj})]/s_n \quad (i = 1, 2, ..., n);$$
(3.13)

and define $X_n(\cdot)$ by linear interpolation between adjacent points $u_{n,i-1}, u_{n,i}$. Prohorov's Theorem states that $X_n \xrightarrow{D} W$. Set

$$U_n(t) = \sum_{i=1}^{n} \sigma_n^2(t_{ni}) \chi_{ni}(t)/s_n^2 \delta_{ni} \quad (t \in [0,1]).$$
(3.14)

Then $U_n(t_{ni}) = u_{ni}$, $U_n(\cdot)$ is linear between adjacent observation points, and is a nondecreasing continuous map from $[0, 1]$ into $[0, 1]$. From (3.9) it follows that $S_n^2/n \to \overline{\sigma}^2$, and also that

$$\sup_{t \in [0, 1]} \left| (s_n^2/n) U_n(t) - \int_{[0, t]} \sigma^2(u) \, dF(u) \right| \to 0.$$

Hence $U_n(\cdot)$ converges uniformly to $U(\cdot)$ on $[0, 1]$. Since $X_n(\cdot) \overset{D}{\to} W(\cdot)$, it now follows from Lemma 3.1 that $X_n[U_n(\cdot)] \overset{D}{\to} W[U(\cdot)]$. But

$$X_n[U_n(t_{ni})] = X_n(u_{ni}) = (n/s_n) [\overline{Y}_n(t_{ni}) - E\overline{Y}_n(t_{ni})]$$

and both $X_n[U_n(\cdot)]$ and $\overline{Y}_n(\cdot) - E\overline{Y}_n(\cdot)$ are linear between adjacent observation points. Since $s_n^2/n \to \overline{\sigma}^2$, we have

$$(\sqrt{n}/\overline{\sigma}) [\overline{Y}_n(\cdot) - E\overline{Y}_n(\cdot)] \overset{D}{\to} W[U(\cdot)].$$

Finally by hypothesis (3.8),

$$\sqrt{n} \sup_{t \in [0, 1]} | E\overline{Y}_n(t) - \Psi_n(t)| \to 0,$$

and the conclusion (3.10) follows (cf. Theorem 4.1, Billingsley, 1968).

In the following lemma the functions $\{\mu_n(\cdot)\}$ will be assumed to satisfy a uniform Libschitz condition: there exists a constant K such that for all n,

$$|\mu_n(u) - \mu_n(v)| \leqslant K|u - v| \quad \text{for} \quad u, v \in [0, 1].$$

This will hold in particular if these functions have uniformly bounded derivatives.

3.3. *Lemma. Suppose*

$$\sum_{i=1}^{n} |(1/n) - [F(t_{ni}) - F(t_{n, i-1})]| = o(n^{-\frac{1}{2}}) \quad as \quad n \to \infty. \quad (3.15)$$

Let the functions $\{\mu_n(\cdot)\}$ be uniformly bounded and satisfy a uniform Libschitz condition, and suppose $\sigma_n^2(\cdot)$ converges uniformly on $[0, 1]$ to a continuous function $\sigma^2(\cdot)$. Then (3.8) and (3.9) are satisfied.

Note that (3.15) is clearly satisfied for uniform spacings, with $F(\cdot)$ the identity, for each summand in (3.15) is zero. Each summand will be zero, of course, if F is strictly increasing and

$$t_{ni} = F^{-1}(i/n) \quad (i = 1, 2, ..., n).$$

The proof is omitted.

3.4. *Theorem. Let the random variables $Y_n(t_{ni})$, $i = 1, 2, ..., n$, be independent for each n, and satisfy Lindeberg's condition (3.7). Let $\sigma_n^2(\cdot)$ converge*

uniformly on $[0, 1]$ *to a continuous function* $\sigma^2(\cdot)$ *and the functions* $\{\mu_n(\cdot)\}$
be uniformly bounded and satisfy a uniform Libschitz condition on $[0, 1]$.
Let the observation points and $F(\cdot)$ *satisfy* (3.15). *Then*

$$(\sqrt{n}/\overline{\sigma})\,[\overline{Y}_n(\cdot) - \Psi_n(\cdot)] \overset{D}{\to} W[U(\cdot)]. \tag{3.10}$$

This is immediate from Proposition 3.2 and Lemma 3.3.

3.5. *Corollary. Let the random variables* $Y_n(t_{ni}), i = 1, 2, \dots, n$, *be independent
for each* n, *and satisfy Lindeberg's condition* (3.7). *Let* $\sigma_n^2(\cdot) \equiv \sigma^2(\cdot)$,
$n = 1, 2, \dots$, *which is continuous on* $[0, 1]$. *Let* $\mu_n(\cdot) \equiv \mu(\cdot)$, $n = 1, 2, \dots$,
where $\mu(\cdot)$ *satisfies a Libschitz condition. Let the observation points and*
$F(\cdot)$ *satisfy* (3.15). *Then*

$$(\sqrt{n}/\overline{\sigma})\,[\overline{Y}_n(\cdot) - M(\cdot)] \overset{D}{\to} W[U(\cdot)]. \tag{3.16}$$

For $\mu_n(\cdot) \equiv \mu(\cdot)$ implies $\Psi_n(\cdot) \equiv M(\cdot)$.

Corollary 3.5 yields asymptotic distributions of various statistics for
testing a null hypothesis specifying $\mu(.)$; Corollary 3.6 illustrates this
remark.

3.6. *Corollary. Under the hypotheses of Corollary* 3.5,

$$\sup_t (\sqrt{n}/\overline{\sigma})|\overline{Y}_n(t) - M(t)| \overset{D}{\to} \sup_t |W[U(t)].$$

Proof. The function $\sup_t (\cdot)$ is continuous on C so that the conclusion is
immediate from Corollary 3.5 and Corollary 1 of Chapter 5 in Billingsley
(1968).

Thus we have in principle the asymptotic distribution of a statistic
for testing a hypothesis $H_0 \colon \mu(\cdot) = \mu_0(\cdot)$, i.e. $M(\cdot) = M_0(\cdot)$, where $\mu_0(\cdot)$
is a given function and $M_0(\cdot)$ its indefinite integral. Under H_0, the statistic
$\sup_t (\sqrt{n}/\overline{\sigma})|\overline{Y}_n(t) = M_0(\cdot)|$ has asymptotically the distribution of

$$\sup_t |W[U(t)]|.$$

4   UNIFORMLY CONSISTENT ESTIMATION OF AN
   INCREASING REGRESSION FUNCTION

Let again a probability distribution $D(x)$ with mean $\theta(x)$ be associated
with each $x \in [0, 1]$. Let $\{x_r\}$ be a sequence of 'observation points' in
$[0, 1]$, not necessarily distinct. An estimator of $\theta(\cdot)$ appropriate to a
situation in which $\theta(\cdot)$ is known to be nondecreasing in $[0, 1]$ is described,
and is proved to be uniformly consistent under appropriate hypotheses.
The changes required to estimate a nonincreasing $\theta(\cdot)$ will be obvious.

Let $\{Z_r(x_r)\}$ be a sequence of independent random variables such that $Z_r(x_r)$ has the distribution $D(x_r)$; in particular, $EZ_r(x_r) = \theta(x_r)$. Let $a(\cdot)$ be a given bounded positive function on $[0, 1]$ bounded away from 0; $a(x_r)$ is to be interpreted as a weighting factor to be applied to $Z_r(x_r)$. For a fixed positive integer r, let $0 \leqslant x_{r,1} \leqslant x_{r,2} \leqslant \ldots \leqslant x_{r,k}$ be the $k = k(r)$ distinct numbers among $x_1, \ldots, x_r$, arranged in increasing order. Let $m_i = m_{r,i}$ denote the number of numbers among $\{x_1, \ldots, x_r\}$ which are equal to $x_{r,i}$, and let $w_i = w_{r,i}$ denote the sum of the weights at $x_{r,i}$:

$$m_i = \sum_{j:\, x_j = x_{r,i}} 1, \quad w_i = m_i a(x_{r,i}). \tag{4.1}$$

Set
$$\bar{Z}_r(x_{r,i}) = \sum_{j:\, x_j = x_{r,i}} Z_j(x_{r,i})/m_i, \tag{4.2}$$

$$b_i = b_{ri} = \sum_{j=1}^{i} w_j \quad (i = 1, 2, \ldots, k), \tag{4.3}$$

and
$$\sum_i = \sum_{r,i} = \sum_{j=1}^{i} w_j \bar{Z}_r(x_{r,j}) \quad (i = 1, 2, \ldots, k). \tag{4.4}$$

Consider the greatest convex minorant of the set of points $\{(b_i, \sum_i)\}_{i=1}^{k}$ in the Cartesian plane. The estimator $\hat{\theta}(x_{ri})$ of $\theta(x_{ri})$ is the slope from the left at b_i of this greatest convex minorant. (This graphical interpretation was introduced by W. T. Reid [Brunk, Ewing and Reid, 1954], and independently, for estimating a monotonic density, by U. Grenander (1956). A simple algorithm for the computation when $a(\cdot) = 1$ is given in Ayer, Brunk, Ewing, Reid and Silverman, 1955.) For x between two adjacent observation points, $\theta(x)$ can be estimated by linear interpolation, but it is convenient here to assume the estimator constant between adjacent observation points. One then has the formula:

$$\hat{\theta}_r(x) = \max_{x_{r,q} \leqslant x} \min_{x_{r,s} \geqslant x} \left[\sum_{i=q}^{s} w_i \bar{Z}_r(x_{r,i}) \right] \Big/ \sum_{i=q}^{s} w_i. \tag{4.5}$$

For use in the following theorem, for each r let $N_r(J)$ denote the number of numbers among $x_1, x_2, \ldots, x_r$ which lie in an interval

$$J : N_r(J) = \text{card}\{i : 1 \leqslant i \leqslant r, \ x_i \in J\}.$$

4.1. *Theorem. Let $a(\cdot)$ be a bounded positive function on $[0, 1]$, bounded away from 0. Let $\theta(\cdot)$ be continuous and nondecreasing on $[0, 1]$. Let $\{x_r\}$ be a sequence of observation points, not necessarily distinct, such that for each interval $J \subset (0, 1)$, $\limsup_r r/N_r(J) < \infty$. Let the variances of the independent observed random variables $Z_r(x_r)$ be bounded. If $0 < a < b < 1$, then*
$$P\{\limsup_r \sup_{a \leqslant x \leqslant b} |\hat{\theta}_r(x) - \theta(x)| = 0\} = 1. \tag{4.6}$$

Proof. We prove that for fixed $x \in (0, 1)$, $P[\hat{\theta}_r(x) \to \theta(x)] = 1$. It then follows that $P[\hat{\theta}_r(x_i) \to \theta(x_i)$ for all $i] = 1$. Since $\theta(\cdot)$ is continuous and nondecreasing, and $\hat{\theta}_r(\cdot)$ is nondecreasing, we shall then have the conclusion (4.6).

For fixed $x \in (0, 1)$ and $\in > 0$, choose r sufficiently large that there exists $j_0 \leqslant r$ such that $x_{j_0} < x$ with $|\theta(x_{j_0}) - \theta(x)| < \in$. For fixed r at least this large, define q by $x_{rq} = x_{j_0}$. From (4.5),

$$\hat{\theta}_r(x) - \theta(x) \geqslant \min_{x_{rs} \geqslant x} \sum_{i=q}^{s} w_i [\bar{Z}_r(x_{ri}) - \theta(x_{rq})] \bigg/ \sum_{i=q}^{s} w_i - [\theta(x) - \theta(x_{rq})].$$

Since $\theta(\cdot)$ is nondecreasing,

$$\hat{\theta}_r(x) - \theta(x) \geqslant \min_{x_{rs} \geqslant x} \left\{ \sum_{i=q}^{s} w_i [\bar{Z}_r(x_{ri}) - \theta(x_{ri})] \bigg/ \sum_{i=q}^{s} w_i \right\} - \in.$$

Further, $$\sum_{i=q}^{s} w_i \bigg/ \sum_{i=q}^{s} m_i \geqslant \delta > 0,$$

where δ is a lower bound on $a(\cdot)$. Using '$\alpha \wedge \beta$' to denote the smaller of two numbers α and β, we then have

$$\hat{\theta}_r(x) - \theta(x) \geqslant \min_{x_{rs} \geqslant x} (1/\delta) \left\{ 0 \wedge \sum_{i=q}^{s} w_i [\bar{Z}_r(x_{ri}) - \theta(x_{ri})] \bigg/ \sum_{i=q}^{s} m_i \right\} - \in.$$

(4.7)

In order to apply Theorem 2.3, for $n = 1, 2$, let $r(n)$ be the nth number among $x_1, x_2, \ldots$, which is at least as large as x_{j_0}. Set $t_n = x_{r(n)}$,

$$Y_n(t_n) = a(t_n) [Z_{r(n)}(x_{r(n)}) - \theta(x_{r(n)})],$$

and $$S_n(t) = \sum_{j=1}^{n} Y_j(t_j) \, \psi_j(t).$$

For $$x_{r(n), s-1} < t \leqslant x_{r(n), s},$$

$$S_n(t) = \sum_{i=q}^{s} w_i [\bar{Z}_{r(n)}(x_{r(n), i}) - \theta(x_{r(n), i})].$$

Using the hypothesis that $r/N_r(J)$ is bounded for each nondegenerate interval $J \subset (0, 1)$, we find there is a constant $K = K(x_{j_0}, x)$ such that

$$\left\{ 0 \wedge \sum_{i=q}^{s} w_i [\bar{Z}_{r(n)}(x_{r(n), i}) - \theta(x_{r(n), i})] \bigg/ \sum_{i=q}^{s} m_{r(n), i} \right\}$$

$$\geqslant K \{ 0 \wedge [S_n(t)/n] \} \quad \text{if} \quad x_{r(n), s-1} < t \leqslant x_{r(n), s}.$$

Since $EY_n(t_n) = 0, n = 1, 2, \ldots$, we have from Theorem 2.3 that

$$\lim_{n} \sup_{t \in [0, 1]} |S_n(t)/n| = 0 \quad \text{with probability 1.}$$

The right-hand member of (4.7) as a function of r is constant for r between $r(n)$ and $r(n+1), n = 1, 2, \ldots$, and it follows that

$$\limsup_r \hat{\theta}_r(x) - \theta(x) \geqslant 0 \quad \text{with probability 1.}$$

Similarly $\liminf_r \hat{\theta}_r(x) - \theta(x) \leqslant 0$ with probability 1. Together with the remarks at the beginning of the proof, this completes the proof of the theorem.

(Theorem 4.1 is essentially equivalent to Theorem 6.2 of Brunk (1958), with the hypothesis that for each interval $J \subset (0, 1)$, $r/N_r(J)$ is bounded in place of the weaker hypothesis that $\{x_r\}$ is dense in $[0, 1]$. The proof of Theorem 6.2 in Brunk (1958) does not appear to be valid without some such added restriction. In effect, the ratios

$$\sum_{i=q}^{s} w_i[\bar{Z}(x_{ri}) - \theta(x_{ji})] \bigg/ \sum_{i=q}^{s} m_i$$

are *not* of the form $S_n(t)/n$ as is claimed in that paper, but are rather of the form $[S_n(t)/n]\left(n \bigg/ \sum_{i=q}^{s} m_i\right)$.)

5 ASYMPTOTIC DISTRIBUTION OF THE ESTIMATOR AT A POINT

Methods used by Chernoff (1964) in determining the asymptotic distribution of an estimator of a mode, and by Prakasa Rao (1969) in determining the asymptotic distribution of the maximum likelihood estimator of a unimodal density, apply somewhat more easily in the regression situation, because of independence of increments. To illustrate the method without becoming bogged down in regularity conditions concerned with the spacing of the observation points, let us assume that only one observation is made at each observation point, and that equal weights are used. Here again, in discussing asymptotic distribution, no sequence $\{x_r\}$ is required; rather, there is a sequence of sets $\{x_{r,1}, \ldots, x_{r,r}\}$. $r = 1, 2, \ldots$. For this situation the estimator $\hat{\theta}(x_0)$ is described in §4 as the slope at x_0 of the greatest convex minorant (hereafter abbreviated as 'slogcom') of the set $\{(i, \sum_i)\}_{i=0}^{r}$, where

$$\sum_i = \sum_{r,i} = \sum_{j=1}^{i} Z(x_{r,j}).$$

Since the estimator is described thus, it is convenient to assume that after a preliminary change of variable if necessary, $x_{r,i} = i/r$.

Let $1_i(x) = 1_{r,i}(x)$ denote the indicator of the interval $((i-1)/r, i/r]$. Set

$$\chi_i(x) = \chi_{r,i}(x) = \int_0^x 1_i(v)\, dv, \tag{5.1}$$

and

$$\bar{Z}_r(x) = \sum_{i=1}^n Z(i/r)\chi_i(x) \quad (x \in (0,1]). \tag{5.2}$$

Then $\bar{Z}_r(i/r) = \sum_i /r$, and $\bar{Z}_r(\cdot)$ is linear on $((i-1)/r, i/r]$.

We suppose the distribution $D(x)$ of $Z(x)$ has variance $\phi(x)$ and mean $\theta(x)$, and consider the asymptotic distribution of the estimator $\hat\theta(x_0)$ for a fixed $x_0 \in (0,1)$. We suppose $\theta'(\cdot)$ exists in a neighbourhood of x_0. The first step in following the outline of Prakasa Rao's treatment of a unimodal density is Lemma 5.1 below, of some interest in itself. For fixed $c > 0$, let $\theta_0^* = \theta_{r,i}^*(x_0)$ denote the slope of the greatest convex minorant of $\bar{Z}_r(\cdot)$, restricted to the interval $[x_0 - 2cr^{-\frac{1}{3}}, x_0 + 2cr^{-\frac{1}{3}}]$, evaluated at x_0. Let $J = J(c, r)$ denote the interval $[x_0 - 2cr^{-\frac{1}{3}}, x_0 + 2cr^{-\frac{1}{3}}]$. We write

$$\theta_0^* = \{\operatorname*{slogcom}_J \bar{Z}_r(x)\}_{x=x_0}. \tag{5.3}$$

5.1. *Lemma There is a function* $K(\cdot)$ *such that*

$$\limsup_{r \to \infty} P[\theta_0^* \neq \hat\theta(x_0)] < K(c) \quad \text{and} \quad K(c) \to 0 \quad \text{as} \quad c \to \infty.$$

The proof, which follows the lines of Prakasa Rao's to a certain point but is essentially easier because of the independence of the increments of $\bar{Z}_r(\cdot)$, is omitted.

5.2. *Theorem. For fixed* r, *let one observation be made at each of the observation points* $x_{r,i} = i/r$, $i = 1, 2, \ldots, r$, *on a distribution with mean* $\theta(x_{ri})$ *and variance* $\phi(x_{ri})$. *Let the observations satisfy Lindeberg's condition,* (3.7). *Let* $x_0 \in (0,1)$, *and let there be a neighbourhood of* x_0 *in which* $\phi(\cdot)$ *and the derivative* $\theta'(\cdot)$ *of* $\theta(\cdot)$ *are continuous. Then*

$$[2r/\phi(x_0)\, \theta'(x_0)]^{\frac{1}{3}}\, [\hat\theta(x_0) - \theta(x_0)]$$

converges in distribution to the slope at 0 of the greatest convex minorant of $W(t) + t^2$, *where* $W(\cdot)$ *is the two-sided Wiener–Levy process with mean 0, variance 1 per unit* t, $W(0) = 0$. *Its probability density at* θ *is* $(1/2)\psi(\theta/2)$, *where* $\psi(\theta) = (1/2)u_1(\theta^2, \theta)\, u_1(\theta^2, -\theta)$, *where* $u(\cdot, \cdot)$ *is the solution of the heat equation* $(1/2)u_{11}(xy) = u_2(x, y)$ *for* $x < y^2$, *subject to the boundary conditions* $u(x, y) = 1$ *for* $x \geqslant y, u(x, y) \to 0$ *as* $x \to -\infty$, *and where the subscripts on* u *indicate partial differentiation with respect to corresponding arguments.*

Indication of proof. The change of variable $t = (x - x_0) r^{\frac{1}{3}}/2c$ maps J onto $[-1, 1]$ with x_0 mapping into 0. Consider first the right half of the interval, $[0, 1]$, and let $n = n(r)$ denote the number of (transformed) observation points in $[0, 1]$: $n = 2cr^{\frac{2}{3}} + o(1)$. Let $\{t_{n,i}\}$ denote the transformed observation points, and $Y(t_{ni})$ the corresponding observations. In the notation of §3, we have

$$E\overline{Y}_n(t_{ni}) = (1/n) \sum_{i=1}^{i} \mu_n(t_{ni}),$$

where $$\mu_n(t) = \theta(x_0 + 2cr^{-\frac{1}{3}}t).$$

Set $$\Delta_n(t) = \overline{Y}_n(t) - \int_0^t \mu_n(v)\,dv.$$

One finds that

$$\theta_0^* - \theta(x_0) = \left\{ \operatorname*{slogcom}_{[0,\,1]} \Delta_n(t) + \int_0^t [\mu_n(v) - \mu_n(0)]\,dv \right\}_{t=0}.$$

Applying Theorem 3.4, we have $[n/\phi(x_0)]^{\frac{1}{3}} \Delta_n(\cdot) \xrightarrow{D} W(\cdot)$ on $[0, 1]$. One verifies also that

$$[n/\phi(x_0)]^{\frac{1}{3}} \int_0^t [\mu_n(v) - \mu_n(0)]\,dv = [2c^3/\phi(x_0)]^{\frac{1}{3}} \theta'(x_0) t^2 + o(1) \quad \text{as} \quad n \to \infty.$$

The remainder of the argument coincides with that given by Prakasa Rao (1969).

6 MAXIMUM LIKELIHOOD ESTIMATION OF AN ISOTONIC REGRESSION FUNCTION

Let X be a finite set, and let ' $\lesssim$ ' denote a partial order on X, a binary relation which is reflexive: $x \lesssim x$; transitive: $x \lesssim y$, $y \lesssim z$ implies $x \lesssim z$; and antisymmetric: $x \lesssim y$ and $y \lesssim x$ implies $x = y$. The points of X are to be interpreted as observation points. In the earlier sections they were real numbers, and X was endowed with the natural order. A real function $\mu(\cdot)$ on X is *isotonic* if $x \lesssim y$ implies $\mu(x) \leqslant \mu(y)$.

Let $w(\cdot)$ be a positive function on X.

6.1 *Definition.* If $g(\cdot)$ is a real function on X, the *isotonic regression*, $g^*(\cdot)$, of g with weights $w(\cdot)$ is the isotonic function on X which minimizes

$$\sum_{x \in X} [g(x) - f(x)]^2 w(x)$$

in the class of isotonic functions f on X. It is easy to verify that g^* is uniquely determined and is characterized by:

$$\sum_{x} [g(x) - g^*(x)] f(x) w(x) \leqslant 0$$

for all isotonic functions f on X, with equality when $f = g^*$. The isotonic regression, g^*, of g also satisfies

$$\sum_x g(x)\,w(x) = \sum_x g^*(x)\,w(x).$$

Its calculation is discussed in Eeden (1957 a) and in Brunk (1955), and for special cases in Ayer, Brunk, Ewing, Reid and Silverman (1955), Thompson (1962), and Kruskal (1964).

Suppose now that a distribution $D(x)$ with mean $\mu(x)$ is associated with each $x \in X$, and that $\overline{Y}(x)$ is the mean of a random sample of size $m(x)$ from $D(x)$. When the distributions belong to a common exponential family, it is known (for references, see Brunk, 1965) that the isotonic regression of $\overline{Y}(\cdot)$ using sample sizes as weights furnishes the maximum likelihood estimate of a regression function $\mu(\cdot)$ known to be isotonic on X. (This is the case $h = 1$ of Theorem 6.1 below.) An exponential family involves just one free parameter, while the normal distribution, for example, involves two. Eeden (1957 b) and Bartholomew (1959, 1961) pointed out that if the distributions $D(\cdot)$ are normal with unknown means $\mu(\cdot)$ and variances $\sigma^2(\cdot)$ for which the ratios $\sigma^2(x_1)/\sigma^2(x_2)$, $x_1, x_2 \in X$ are known, then the maximum likelihood estimate of $\mu(\cdot)$ known to be isotonic on X is given by the isotonic regression of the sample means with weights proportional to sample sizes and inversely proportional to variances. The content of Theorem 6.1 is essentially that other infinitely divisible distributions may play the role of the normal here.

We consider a two parameter family of densities with respect to some measure $\nu_h(dy)$ which need not be specified which may depend on the parameter h but not on the parameter θ:

$$f(y; \theta, h) = \exp\{[\Phi(\theta) + (y - \theta)\,\phi(\theta)]\,h\}, \qquad (6.1)$$

where Φ is strictly convex and ϕ is a measurable determination of its derivative, say its derivative from the right, where θ ranges over an interval of real numbers, and where h is positive. Then

$$\int f(y;\,\theta, h)\,\nu_h(dy) = 1.$$

If ϕ has a derivative ϕ' (which is then positive since Φ is strictly convex) and if the formal differentiation is valid, we find that

$$\int (y - \theta)f(y;\,\theta, h)\,\nu_h(dy) = 0, \qquad (6.2)$$

so that θ is the mean of the distribution with density (6.1). A further differentiation of (6.2) yields

$$\int (y - \theta)^2 f(y; \theta, h)\, \nu_h(dy) = 1/h\phi'(\theta); \qquad (6.3)$$

so that providing the formal calculations are justified, $h\phi'(\theta)$ is the reciprocal of the variance, or the *precision* of the distribution with density (6.1).

6.2. *Theorem. Let X be a finite set, partially ordered by '$\precsim$'. For $x \in X$, let the distribution $D(x)$ have density given by (6.1) with $\theta = \mu(x)$ and with $h = a\lambda(x)$, where $\lambda(x)$ is a known positive number for each x, but the positive number a may be unknown. Let independent random samples be taken from these distributions with sizes $m(x) > 0, x \in X$. Then the maximum likelihood estimate of $\mu(\cdot)$ given that $\mu(\cdot)$ is isotonic is furnished uniquely by the isotonic regression of the sample means, with weights $w(x) = \lambda(x)\, m(x), x \in X$.*

Proof. Let $\overline{Y}(x)$ denote the sample mean of the sample from $D(x)$, $x \in X$. Then the log-likelihood function may be written as

$$a \sum_x \{\Phi[\mu(x)] + [\overline{Y}(x) - \mu(x)]\, \phi[\mu(x)]\}\, \lambda(x)\, m(x).$$

But it is known (Brunk, 1965) that this is uniquely maximized in the class of isotonic $\mu(\cdot)$ by the isotonic regression of $\overline{Y}(\cdot)$ with weights $\lambda(\cdot)\, m(\cdot)$.

It can be shown as follows that there are densities of the form (6.1) with arbitrary positive h associated with an infinitely divisible distribution F. Let $M(t) = \exp[\theta(t)]$ be the moment generating function, convergent in a neighbourhood of the origin, of an infinitely divisible distribution F. Then for every positive integer n there exists a moment generating function M_n such that $M(t) = [M_n(t)]^n$. Given $h > 0$, let $\{m(n)\}$ be a sequence of positive integers such that $\lim_{n \to \infty} m(n)/n = h$. For each n, $[M(t)]^{m(n)/n} = [M_n(t)]^{m(n)}$ is the moment generating function of the sum of $m(n)$ independent random variables with moment generating function M_n. It can be shown to follow that the limit as $n \to \infty$, $[M(t)]^h$, is then the moment generating function of a distribution. Thus if F is infinitely divisible and has moment generating function $\exp[\Theta(t)]$ for t in a neighbourhood of the origin, for each $h > 0$ there is a distribution G_h such that

$$e^{h\Theta(t)} = \int e^{yt} G_h(dy)$$

and thus a distribution F_h such that

$$e^{h\Theta(t)} = \int e^{hty} F_h(dy). \qquad (6.4)$$

Then $\exp\{[\Theta(t) - ty]\}$ is a probability density with respect to $F_h(dy)$, and this may be rewritten as (6.1), where Φ is the convex conjugate of Θ, $\Phi(\theta) = \sup_t \{\Theta t - \theta(t)\}$.

6.3. *Example.* The normal distribution. Let F_h have density

$$(h/2\pi)^{\frac{1}{2}} \exp(-hy^2/2)$$

with respect to Lebesgue measure. Then $\Theta(t)$ defined by (6.4) is $\Theta(t) = t^2/2$. We have $\theta(t) = t$, $\phi(\theta) = 0$, $\Phi(\theta) = \theta^2/2$, and

$$f(y;\theta,h) F_h(dy) = (h/2\pi)^{\frac{1}{2}} \exp[-h(y-\theta)^2/2] \, dy,$$

so that a random variable with this distribution is normally distributed with mean θ and variance $1/h$ (precision h).

6.4. *Example.* The gamma distribution. For $h > 0$, set

$$F_h(dy) = h(hy)^{h-1} e^{-hy} \, dy/\Gamma(h) \quad (y > 0).$$

Then $\Theta(t)$ defined by (6.4) is

$$\theta(t) = -\log(1-t) \quad (t < 1).$$

We have

$$\theta(t) = 1/(1-t), \quad \phi(\theta) = 1 - 1/\theta, \quad \Phi(\theta) = \theta - \log\theta - 1 \quad \text{for} \quad \theta > 0,$$

and

$$f(y;\theta,h) F_h(dy) = (hy/\theta)^{h-1} e^{-hy/\theta}(h\,dy/\theta)/\Gamma(h) \quad (y > 0).$$

This is the density of the gamma distribution with mean θ and precision h/θ^2 (variance θ^2/h).

ACKNOWLEDGEMENTS

The author wishes to express his gratitude to R. E. Barlow and D. J. Bartholomew for helpful conversations to which the present paper owes much.

REFERENCES

Ayer, Miriam, Brunk, H. D., Ewing, G. M., Reid, W. T. and Silverman, Edward (1955). An empirical distribution function for sampling with incomplete information. *Ann. Math. Statist.* **26**, 641–7.

Barlow, R. E. (1968). Likelihood ratio tests for restricted families of probability distributions. *Ann. Math. Statist.* **39**, 547–60.

Bartholomew, D. J. (1959). A test of homogeneity for ordered alternatives. *Biometrika*, **46**, 36–48.

Bartholomew, D. J. (1961). A test of homogeneity of means under restricted alternatives. *J. R. statist. Soc. B* 23, 239–81.

Billingsley, Patrick (1968). *Convergence of probability measures.* New York: John Wiley and Sons.

Brunk, H. D. (1955). Maximum likelihood estimation of monotone parameters. *Ann. Math. Statist.* 26, 607–16.

Brunk, H. D. (1958). On the estimation of parameters restricted by inequalities. *Ann. Math. Statist.* 29, 437–54.

Brunk, H. D., Ewing, G. M. and Reid, W. T. (1954). The minimum of a certain definite integral suggested by the maximum likelihood estimate of a distribution function. (Abstract). *Bull. Am. Math. Soc.* 60, 535.

Brunt, H. D. (1965). Conditional expectation given a σ-lattice and applications. *Ann. Math. Statist.* 26, 1339–50.

Chernoff, Herman (1964). Estimation of the mode. *Ann. Inst. Statist. Math.* 16, 31–41.

Eeden, Constance van (1957 a). Maximum likelihood estimation of partially or completely ordered parameters, I and II. *Indag. Math.* 19, 128–36, 201–11.

Eeden, Constance van (1957 a). A least squares inequality for maximum likelihood estimates of ordered parameters. *Indag. Math.* 19, 513–21.

Grenander, Ulf (1956). On the theory of mortality measurement, Part II. *Skand. Akt.* 39, 126–53.

Hájek, J. (1965). Extension of the Kolmogorov–Smirnov test to regression alternatives. *Bernoulli–Bayes–Laplace* Anniversary Volume. New York: Springer-Verlag Inc., 45–60.

Kruskal, J. B. (1964). Nonmetric multidimensional scaling: a numerical method. *Psychometrika* 29, 115–29.

Marshall, A. W. and Proschan, Frank (1965). Maximum likelihood estimation for distributions with monotone failure rate. *Ann. Math. Statist.* 36, 69–77.

Neveu, Jacques (1965). *Mathematical Foundations of the Calculus of Probability.* San Francisco: Holden–Day, Inc.

Prakasa Rao, B. L. S. (1969). Estimation of a unimodal density. *Sankhyā,* Series A 31, 23–36.

Prohorov, Yu. V. (1956). Convergence of random processes and limit theorems in probability theory. *Theory of Prob. and its Applics.* (trsl. of *Teor. Veroyatnost.*) 1, 157–214.

Thompson, W. A., Jr. (1962). The problem of negative estimates of variance components. *Ann. Math. Statist.* 33, 273–89.

DISCUSSION ON BRUNK'S PAPER

RONALD PYKE

In this paper, Professor Brunk describes an interesting model for regression problems which is general enough to include such 'non-regression' processes as the classical empirical distribution function. Although the title refers to estimation of isotonic regression, only the last section of the paper refers explicitly to this situation. In fact, the phrase *isotonic regression* is not defined until this last section. The basic idea of isotonic regression however runs through the entire paper, as it does also throughout the preceding paper by Barlow and van Zwet. The adjective 'isotonic' is a relatively new term which has been introduced by the author of this paper. To several other people in the Northwest corners of the United States, this concept has for a number of years been commonly referred to as *Brunkizing*. Both this paper and the preceding one by Barlow and van Zwet deal in part with problems of estimation of a function which possesses some ordering, such as monotonicity or unimodality. The estimates are obtained by first drawing a rough estimate and then 'Brunkizing' this estimate to make it satisfy the necessary ordering. For example, one fits the greatest convex minorant in many cases of monotonicity. Although the work of Professor Brunk and others in this area goes back fifteen years to 1954, the dissemination of the method has been hampered by the lack of an expository treatise. We are pleased to learn that Professor Brunk together with Professor Bartholomew is now in the process of preparing a monograph of isotonic regression. (It is also hoped that the paper of Prakasa Rao (1966) which presents basic asymptotic methods for certain isotonic estimates will soon be published. It is referred to in the preceding Barlow–van Zwet paper as well as in this paper by Brunk.)

Concerning Professor Brunk's paper, I believe he has presented an interesting unification of problems of estimating distribution functions and integral regression functions. There seems to be one basic difference between the two problems however. When estimating a distribution function it is fairly natural to assume a single sequence of observations $\{t_n\}$. On the other hand, for many regression problems one would more naturally postulate a triangular double array $\{t_{n,i} : 1 \leqslant i \leqslant k_n, n \geqslant 1\}$ of observations on the t-axis. In view of this it might have been natural to include theorems about the uniform convergence *in probability* rather than *almost surely* of the estimator M_n to M under triangular arrays.

In § 2 on uniform consistency, one assumes for (2.3) that μ is integrable. Without loss of generality one could split μ into its positive and negative parts, thereby making M a nondecreasing bounded function. This allows one to make use of the fact that pointwise convergence of distribution functions implies uniform convergence in results such as Theorem 2.3.

The paper is clearly written, and the notation is generally well chosen. This writer would have preferred other notation for indicator functions. (In most instances 1 denotes an indicator function on [0, 1] but in (3.7), 1 denotes an indicator function defined on the underlying probability space.)

Section 3 gives an interesting application of weak convergence to the centered integral regression function. It might be noted that the existence of a positive function $\sigma^2(\cdot)$ could be relaxed to the existence of a monotone function $\Sigma(\cdot)$ satisfying $\sigma^2(\overline{Y}_n(t)) \to \Sigma(t)$ uniformly in t. The author's assumption imposes absolute continuity on Σ.

The remaining sections study both consistency and asymptotic distributions for estimates of the nonintegrated regression function under ordering assumptions like monotoneity. Although the details are parallel to those of existing papers by the author, Rao and others, and thus to a great extent omitted or abbreviated, the theorems are easily understood and interesting.

Professor Brunk has posed a very interesting model, which suggests several directions of generalizations, What can be done for example if some dependence is added to the Y's? It would seem that many examples could arise in which some form of dependence is dictated. Since weak convergence results are available for certain dependent processes (cf. Billingsley (1968)) some of the results in §§ 2 and 3 should extend without too much trouble. The material in §§ 4–6 on the other hand could be very difficult in the nonindependent case. A second question might be to inquire about other functions than the mean. Might it not be possible to have data in which the means of the Y's are constant but their variances are functions of t? There is also the very difficult question of rates of consistency for isotonic estimators of this type.

In summary then, Professor Brunk's paper is both one of considerable substance and one which will generate many new research problems.

DENSITY ESTIMATES AND
MARKOV SEQUENCES†

M. ROSENBLATT

1 INTRODUCTION

Estimates of the density function of a population based on a sample of independent observations have been considered in a number of papers [1, 6–7]. Questions of bias, variance and asymptotic distribution of the estimates have been dealt with at greatest length. Our object is to look at such estimates of the density function when the observations are dependent. The results will not be dealt with in the most general context or under very general conditions. To obtain results in a simple and readily understandable form, the observations are assumed to be sampled from a stationary Markov sequence with a fairly strong condition on the Markov transition operator. However, the extent to which some of the conditions can be obviously relaxed will be indicated.

It should be noted that the asymptotic results we obtain in the case of dependent observations are essentially the same as those in the case of independent observations. This initially is surprising because it is certainly not true when estimating the distribution function by means of the sample distribution function. The more complicated nature of asymptotic results for this problem in the case of dependence can be seen in Billingsley [2]. However, the happy fact that the results we obtain in estimating the density have the same character as in the case of independence is due to the local character of the estimates.

2 REMARKS ON THE CASE OF INDEPENDENT SAMPLING

We just briefly make a few remarks on estimation of the density function of a population with a sample of independent observations. More detailed and precise results can be found [1, 6–7]. Consider a population with continuous positive density function $f(x)$. Let the sample of independent observations be $X_1, X_2, ..., X_n$. Let the estimate of $f(x)$ be

$$\hat{f}_n(x) = n^{-1}b(n)^{-1} \sum_{j=1}^{n} h(b(n)^{-1}(x - X_j)), \tag{1}$$

† This research was supported by the Office of Naval Research.

where $h(x)$ is a given bounded continuous density function and $b(n)$ is a bandwidth parameter such that $b(n) \to 0$ and $nb(n) \to \infty$ as $n \to \infty$. The bias of the estimate is

$$b_n(x) = E\hat{f}_n(x) - f(x) = \int h(\alpha)\{f(x - b(n)\alpha) - f(x)\}\,d\alpha. \qquad (2)$$

If f is continuously differentiable up to second order and the moments $m_i = \int \alpha^i h(\alpha)\,d\alpha, i = 1, 2$, exist then

$$b_n(x) = f'(x)\,m_1 b(n) + f''(x)\,m_2 b(n)^2/2 + o(b(n)^2). \qquad (3)$$

Thus $b_n(x) = 0(b(n)^2)$ if $m_1 = 0$. The variance

$$\sigma^2(\hat{f}_n(x)) \simeq n^{-1}b(n)^{-1}f(x)\int h^2(\alpha)\,d\alpha \quad \text{as} \quad n \to \infty. \qquad (4)$$

It is clear that the estimates are asymptotically normally distributed. Specifically $\sqrt{(nb(n))}[\hat{f}_n(x) - E\hat{f}_n(x)]$ is asymptotically normally distributed with mean zero and variance $f(x)\int h^2(\alpha)\,d\alpha$. It is also interesting to look at the covariance of estimates at different points x, y.

The covariance

$$\text{cov}\,(\hat{f}_n(x), \hat{f}_n(y)) = n^{-1}b(n)^{-2}\,\text{cov}\,(h(b(n)^{-1}(x - X)), h(b(n)^{-1}(y - X))$$

$$= n^{-1}b(n)^{-2}\bigg\{\int h(b(n^{-1}(x - \alpha))\,h(b(n)^{-1}(y - \alpha))f(\alpha)\,d\alpha$$

$$- \int h(b(n)^{-1}(x - \alpha))\,f(\alpha)\,d\alpha \int h(b(n)^{-1}(y - \alpha))\,f(\alpha)\,d\alpha\bigg\}$$

$$= n^{-1}b(n)^{-1}\bigg\{\int h(\beta)\,h(b(n)^{-1}(y - x) + \beta)f(x - \beta b(n))\,d\beta$$

$$- b(n)\int h(\beta)f(x - b(n)\beta)\,d\beta\int h(\beta)f(y - b(n)\beta)\,d\beta\bigg\}.$$

$$(5)$$

This certainly implies that for fixed x, y with $x \neq y$

$$\sqrt{(nb(n))}[\hat{f}_n(x) - E\hat{f}_n(x)] \quad \text{and} \quad \sqrt{(nb(n))}[\hat{f}_n(y) - E\hat{f}_n(y)]$$

are asymptotically independent.

3   THE MARKOV ASSUMPTION

Let us now assume that $X_1, X_2, \ldots, X_n$ is a sequence of observations on a stationary Markov process with stationary continuous density function $f(x) > 0$ and continuous transition probability density function $f(y|x)$. Let $f_k(y|x)$ and $f_k(y, x)$ denote the conditional probability density of

X_{k+1} given X_1 and the joint probability density of X_{k+1} and X_1 respectively, $k = 1, 2, \ldots$. Clearly

$$f_k(y, x) = f_k(y|x) f(x) \tag{6}$$

and
$$f_{k+1}(y|x) = \int f_k(y|z) f(z|x)\, dz \quad (k = 1, 2, \ldots). \tag{7}$$

We use the same sort of estimate $\hat{f}_n(x)$ of the density function $f(x)$ as in the case of independent observations. The bias of the estimate in the case of a Markov sequence is the same as in the case of independent observations. Let us now look at the covariance of the estimate when one has dependence. Clearly

$$\operatorname{cov}[\hat{f}_n(x), \hat{f}_n(y)] = n^{-2} b(n)^{-2} \sum_{j=-n+1}^{n-1} (n - |j|)\operatorname{cov}[h(b(n)^{-1}(x - X_0)), h(b(n)^{-1} \\ \times (y - X_j))]. \tag{8}$$

In §5 we will determine the asymptotic behavior of the covariance and of the distribution of the estimates $\hat{f}_n(x)$ themselves, given certain asymptotic conditions on the character of the dependence of the Markov sequence. Some of these conditions will be introduced and discussed in §4.

4   REMARKS ON THE TRANSITION PROBABILITY OPERATOR

In order to determine the asymptotic behaviour of covariance of density estimates at different points as well as the asymptotic distribution of estimates we shall have to impose certain conditions on the transition probability function of the Markov sequence. The conditions discussed here are undoubtedly too strong. Certainly weaker conditions that ensure the desired results can be determined. However, these conditions have a certain interest even outside the problem of density estimation and the discussion will be given in a broader context. Let T be the transition probability operator of a stationary Markov sequence. In our problem

$$(Th)(y) = \int h(x) f(x|y)\, dx, \tag{9}$$

where h can be thought of as a bounded function. The condition to be imposed on T corresponds roughly to what has been called 'geometric ergodicity' (see Kendall [5]) in some of the probability literature. Let $\|h\|_p$, $1 \leqslant p \leqslant \infty$, be the L^p norm of the function h with respect to

the invariant measure μ $(d\mu = f(x)\,dx$ in our case of density estimation)
so that

$$\|h\|_p = \begin{cases} \left[\int |h(x)|^p \mu(dx)\right]^{1/p} & \text{if } 1 \leqslant p < \infty, \\ \underset{x}{\text{ess sup}} \, |h(x)| & \text{if } p = \infty, \end{cases} \tag{10}$$

where ess sup is understood as referring to the measure μ. The L^p norm
of the operator T^n, $n = 1, 2, \ldots$,

$$\|T^n\|_p = \sup_{h \neq 0} \frac{\|T^n h\|_p}{\|h\|_p} = 1 \tag{11}$$

for $1 \leqslant p \leqslant \infty$ since $T^n 1 = 1$. However, since we are interested in the
rate at which $T^n h \to Eh = \int h(x)\,\mu(dx)$ as $n \to \infty$, the modified norm

$$\begin{aligned} |T^n|_p &= \sup_h \frac{\|T^n h - Eh\|_p}{\|h - Eh\|_p} \\ &= \sup_{h \perp 1} \frac{\|T^n h\|_p}{\|h\|_p} \end{aligned} \tag{12}$$

is introduced where by $h \perp 1$ one means that $\int h(x)\,\mu(dx) = 0$.

The transition probability operator T *is said to satisfy the condition*
G_p $(1 \leqslant p \leqslant \infty)$ *if there is some positive integer* n *such that*

$$|T^n|_p \leqslant \alpha < 1. \tag{13}$$

The reason for 'geometric ergodicity' is apparent since if (13) holds, it
follows that

$$\lim_{m \to \infty} |T^m|_p / \alpha^{m/n} \leqslant 1. \tag{14}$$

Lemma. If T *is a transition probability operator of a stationary Markov
sequence satisfying condition* G_p *for some* $p, 1 < p < \infty$, *then* T *satisfies
the condition* G_q *for all* $q, 1 < q < \infty$.

Let U be such that $Uh = Th - Eh$. Notice that

$$\|h - Eh\|_q \leqslant \|h\|_q + |Eh| \leqslant 2\|h\|_q \tag{15}$$

and that $\qquad U^n h = U^{n-1}(Th - Eh) = T^{n-1}(Th - Eh).$ $\qquad$ (16)

This implies that $\quad \|U^n\|_q \leqslant 2|T^n|_q \leqslant 2, \quad 1 \leqslant q \leqslant \infty.$ $\qquad$ (17)

However $\qquad\qquad |T^n|_q \leqslant \|U^n\|_q.$ $\qquad\qquad\qquad$ (18)

Since $\|U^n\|_1, \|U^n\|_\infty \leqslant 2$ the Riesz convexity theorem [4] can be applied
to this situation. Suppose that for some p, $1 < p < \infty$ (for example
$p = 2$), there is a positive integer n such that

$$|T^n|_p = \alpha < 1. \tag{19}$$

It then follows that $|T^{nk}|_p \leqslant \alpha^k$. Consider any other q with $1 < q < \infty$. For $q > p$ we have $1/q = (p/q)(1/p) + (1 - p/q)$, while if $q < p$ then

$$1/q = \left(\frac{1 - 1/q}{1 - 1/p}\right) 1/p + \left(1 - \frac{1 - 1/q}{1 - 1/p}\right).$$

The Riesz convexity theorem tells us that if $q > p$

$$\log \|U^m\|_q \leqslant \frac{p}{q} \log \|U^m\|_p + \left(1 - \frac{p}{q}\right) \log \|U^m\|_\infty \tag{20}$$

and if $q < p$

$$\log \|U^m\|_q \leqslant \left(\frac{1 - 1/q}{1 - 1/p}\right) \log \|U^m\|_p + \left(1 - \frac{1 - 1/q}{1 - 1/p}\right) \log \|U^m\|_1. \tag{21}$$

Since $\|U^{nk}\|_p \leqslant 2\alpha^k$ and $\|U^m\|_\infty, \|U^m\|_1 \leqslant 2$ it is clear that for large enough k

$$\|U^{nk}\|_q \leqslant \beta < 1 \tag{22}$$

and therefore

$$|T^{nk}|_q \leqslant \beta < 1. \tag{23}$$

It is of some interest to consider the well known Doeblin condition and see how it relates to the conditions G_p. Let $P(x, A)$ be the transition probability function of the stationary Markov sequence defined for each real x and each Borel set A. Then

$$(Th)(x) = \int P(x, dy) h(y) \tag{24}$$

for bounded h and the n step transition probability function $P_n(x, A)$ is defined recursively via

$$P_{n+1}(x, A) = \int P(x, dy) P_n(y, A) \quad (n = 1, 2, \ldots). \tag{25}$$

The Doeblin condition D is said to be satisfied if there is a finite measure ϕ on the Borel sets with positive mass, an integer $n \geqslant 1$, and $\epsilon > 0$ such that

$$P_n(x, A) \leqslant 1 - \epsilon \tag{26}$$

for all x if $\phi(A) \leqslant \epsilon$. Doob (see [3]) considers the more stringent condition D_0 where one assumes that D is satisfied, there is only one ergodic set, and there are no cyclically moving sets. Under condition D_0, in Lemma 7.2 on p. 224 of Doob's book on stochastic processes, it is shown that there are constants γ and ρ, with $0 < \rho < 1$, such that if h is bounded

$$(|h| \leqslant M < \infty)$$

then

$$|(T^k h)(x) - Eh| \leqslant 2\gamma\rho^k. \tag{27}$$

However, it is easily seen that (27) is just the condition G_∞. By the Riesz convexity theorem, it follows that if G_∞ is satisfied then G_p holds for $1 < p \leqslant \infty$. Similarly, if G_1 is satisfied then G_p holds for $1 \leqslant p < \infty$.

Notice that the condition D_0 is satisfied if we have an ergodic aperiodic stationary Markov sequence with continuous stationary probability density $f(x)$ and transition probability density $p(y|x)$ satisfying

$$p(y|x) \leqslant K(1+y^2)^{-\alpha} \tag{28}$$

for some constant K and $\alpha > 1/2$.

However, it is easy to give a simple example which doesn't satisfy D or G_∞ but for which G_p, $1 < p < \infty$, holds. This example is provided by the aperiodic Gaussian stationary Markov sequences. For convenience, assume that the random variables of the sequence have mean zero and variance one. The invariant probability density is then

$$f(x) = \frac{1}{\sqrt{(2\pi)}} \exp(-x^2/2)$$

and the transition probability density is

$$p(y|x) = \frac{1}{\sqrt{(2\pi(1-\rho^2))}} \exp\left(-\frac{(y-\rho x)^2}{2(1-\rho^2)}\right) \tag{29}$$

with ρ a constant such that $|\rho| < 1$. Mehler's formula indicates that (29) can be written

$$p(y|x) = \sum_{\nu=0}^{\infty} \rho^\nu h_\nu(y) h_\nu(x) \frac{e^{-y^2/2}}{\sqrt{(2\pi)}}, \tag{30}$$

where the h_ν are the Hermite polynomials orthonormal with respect to the standard Gaussian density $f(x)$. By using (3) one can see that

$$|T^n|_2 = |\rho|^{2n} \tag{31}$$

so that condition G_2 is certainly satisfied. By the Lemma of this section it follows that G_p is satisfied for $1 < p < \infty$.

5 ASYMPTOTIC BEHAVIOR OF THE COVARIANCE AND DISTRIBUTION OF ESTIMATES

The asymptotic behavior of the covariance of a density estimate at two points x, y will first be determined before the asymptotic distribution is considered. Assume that h is a continuous density function with

$$h(u) = o(|u|^{-1}) \quad \text{as} \quad |u| \to \infty.$$

The density f is taken to be bounded and continuous. Consider the term corresponding to $j = 0$ in the sum (8).

Then $\text{cov}\{h(b(n)^{-1}(x-X)), h(b(n)^{-1}(y-X))\}$

$$= b(n)\int h(u)\,h(b(n)^{-1}(y-x)+u)f(x-b(n)\,u)\,du$$

$$- b(n)^2\int h(u)f(x-b(n)\,u)\,du\int h(u)f(y-b(n)\,u)\,du$$

$$\cong \begin{cases} b(n)f(x)\int h^2(u)\,du & \text{if } x=y \\ 0(b(n)^2) & \text{if } x\neq y \end{cases} \tag{32}$$

when $f(x) > 0$ as $n \to \infty$. The terms with $j \neq 0$ have the following asymptotic behavior

$$\text{cov}\{h(b(n)^{-1}(x-X_0)), h(b(n)^{-1}(y-X_j))\}$$

$$= b(n^2)\int h(u)\,h(v)f_j(x-b(n)\,u, y-b(n)\,v)\,du\,dv$$

$$- b(n)^2\int h(u)f(x-b(n)\,u)\,du\int h(u)f(y-b(n)\,u)\,du$$

$$\cong b(n)^2\{f_j(x,y)-f(x)f(y)\} \tag{33}$$

as $n \to \infty$ if the joint density functions $f_j(\cdot,\cdot)$ are bounded continuous functions. We shall now get a bound on (33) under the assumption that the Markov sequence satisfies condition G_2. Let

$$g_x(\alpha) = h(b(n)^{-1}(x-\alpha)) - \int h(b(n)^{-1}(x-\alpha))f(\alpha)\,d\alpha. \tag{34}$$

Then $|\text{cov}\{h(b(n)^{-1}(x-X_j)), h(b(n)^{-1}(y-X_0))\}|$

$$= |E(g_y(X)\,(T^j g_x)\,(X))|$$

$$\leqslant \{E|g_y(X)|^2\,E|(T^j g_x)\,(X)|^2\}^{\frac{1}{2}}$$

$$\leqslant M\rho^j\{E|g_y(X)|^2\,E|g_x(X)|^2\}^{\frac{1}{2}}$$

$$\leqslant b(n)\,M'\rho^j\int h^2(u)\,du\,\sqrt{(f(x)f(y))} \tag{35}$$

for sufficiently large n with M, M' and ρ constants, $0 < \rho < 1$, if $f(x)$ and $f(y)$ are positive. The estimates (32), (33) and (35) indicate that

$$\lim_{n\to\infty} nb(n)\,\text{cov}\,[\hat{f}_n(x), \hat{f}_n(y)]$$

$$= \begin{cases} f(x)\int h^2(u)\,du & \text{if } x=y \\ 0 & \text{if } x\neq y \end{cases} \tag{36}$$

under the conditions assumed.

We now consider the asymptotic distribution of

$$\sqrt{(nb(n))}[\hat{f}_n(x_j) - E\hat{f}_n(x_j)] \quad (j = 1, \ldots, m),$$

for any finite m-tuple of points x_j. These random variables are asymptotically jointly normal and independent with means zero and variances given by (36) as $n \to \infty$ under the assumptions made. So as to limit the notation the argument on asymptotic normality will be given for a single point, that is, $m = 1$. However, exactly the same argument and estimates can be used to obtain the multivariate result by applying them to any given linear combination

$$\sqrt{(nb(n))} \sum_{j=1}^{m} \alpha_j (\hat{f}_n(x_j) - E\hat{f}_n(x_j))$$

of the random variables. This is the standard reduction of a multivariate limit theorem to a univariate limit theorem. Consider

$$\sqrt{(nb(n))}[\hat{f}_n(x) - E\hat{f}_n(x)].$$

The argument for asymptotic normality will proceed by the classic device of proving such a limit theorem for dependent random variables, that is, by writing the sum as a sum of big blocks separated by small blocks. The contribution due to the small blocks is shown to be negligible and the big blocks approximately independent. A standard central limit theorem for independent random variables (the Liapounov theorem) is then used to get asymptotic normality. Now,

$$\sqrt{(nb(n))}[\hat{f}_n(x) - E\hat{f}_n(x)]$$

$$= \sum_{j=1}^{n} \frac{1}{\sqrt{(nb(n))}} [h(b(n)^{-1}(x - X_j)) - Eh(b(n)^{-1}(x - X_j))]$$

$$= \sum_{l=1}^{k} (A_l + B_l) + H \tag{37}$$

with
$$A_l = \sum_{j=(l-1)(m+r)+1}^{lm+(l-1)r}, \quad B_l = \sum_{j=lm+(l-1)r+1}^{l(m+r)} \tag{38}$$

and
$$H = \sum_{j=k(m+r)+1}^{n}. \tag{39}$$

Here $m = m(n)$ and $r = r(n)$ are the summation ranges of the big and small blocks A_l and B_l respectively. They will be taken so that

$$m(n), r(n) \to \infty \quad \text{as} \quad n \to \infty \quad \text{but} \quad m(n) = o(n) \quad \text{and} \quad r(n) = o(m(n)).$$

Further $k = k(n) = [n/(m+r)]$ is the greatest integer less than or equal to $n/(m+r)$. The term H accounts for the additional few terms at the end not included in the big blocks or small blocks. Notice that $k(n) \to \infty$ as $n \to \infty$ since $m, r = o(n)$. Estimates like those used in (32), (33) and (35) imply that

$$E \left| \sum_{l=1}^{k} B_l + H \right|^2 \leqslant 2 \left(\sum_{l=1}^{k-1} E|B_l|^2 + E|B_k + H|^2 \right)$$

$$\leqslant C \left(\frac{kr(n)}{n} + \frac{m}{n} \right) = o(1), \qquad (40)$$

with C a constant.

Thus $\sum_{l=1}^{k} B_l + H$ can be neglected. Consider the characteristic function of $\sum_{l=1}^{k} A_l$. We use the G_2 property of the Markov sequence to compare this characteristic function with the product of the characteristic functions of the individual A_l's and find that

$$\left| E \left\{ \exp \left(it \sum_{l=1}^{k} A_l \right) \right\} - \prod_{l=1}^{k} E\{\exp (it A_l)\} \right|$$

$$\leqslant \sum_{j=2}^{k} \left| E \left\{ \exp \left(it \sum_{l=1}^{j} A_l \right) \right\} - E\{\exp (it A_j)\} E \left\{ \exp \left(it \sum_{l=1}^{j-1} A_l \right) \right\} \right|$$

$$\leqslant (k-1) M \rho^{r(n)}, \qquad (41)$$

where M is a constant and $0 < \rho < 1$. The sequences $m(n)$, $r(n)$ will be chosen so that $k(n) \rho^{r(n)} \to 0$. However this means that we have asymptotic normality if we can show that $\sum_{l=1}^{k} A_l$ is asymptotically normal when the A_l's are treated as independent random variables with the same marginal distributions. This will be shown to be true by using the Liapounov theorem. For this we require an estimate of the fourth moment of the A_l's. Given that the joint density functions up to fourth order are continuous and bounded, estimates like those obtained in (32) and (33) show that

$$E|A_l|^4 \leqslant \sum_{j=1}^{k} M_j (mb(n))^j, \qquad (42)$$

where the M_j's are constants. But then

$$\frac{kE|A_l|^4}{(E|\sum A_l|^2)^2} \leqslant \frac{M_1}{nb(n)} + M_2 \frac{m}{n} + M_3 \frac{m^2}{n} b(n) + M_4 \frac{m^3}{n} b(n)^2. \qquad (43)$$

If $nb(n) \to \infty$, $b(n) \to 0$, but $m(n) = o(n^{\frac{1}{3}})$ the expression (43) tends to zero as $n \to \infty$ and the Liapounov theorem is applicable. We can still choose $r(n) = o(m(n))$ so that (41) tends to zero as $n \to \infty$.

Theorem 1. *Let* $X_n, n = \ldots, -1, 0, 1, \ldots$, *be a stationary Markov sequence with bounded continuous density function* $f(x)$ *and bounded transition probability density* $p(y|x)$ *continuous in the variables* (y, x). *Consider the estimate* $\hat{f}_n(x)$ *of the density* $f(x)$ *given by*

$$\hat{f}_n(x) = n^{-1} b(n)^{-1} \sum_{j=1}^{n} h(b(n)^{-1}(x - X_j)),$$

where h *is a given uniformly continuous density function. If* f *is continuously differentiable up to second order and the moments* $m_i = \int \alpha^i h(\alpha) \, d\alpha$, $i = 1, 2,$ *of* h *exist then the bias*

$$E\hat{f}_n(x) - f(x) = m_1 f'(x) \, b(n) + m_2 f''(x) \, b(n)^2 + o(b(n))^2,$$

where the bandwidth $b(n) \to 0$ *as* $n \to \infty$. *If the sequence* $\{X_n\}$ *satisfies condition* G_2 *and* $nb(n) \to \infty$ *as* $n \to \infty$ $(b(n) \to 0)$ *any* m-*tuple* (m *finite) of the normalized deviations*

$$\sqrt{(nb(n))}[\hat{f}_n(x_i) - E\hat{f}_n(x_i)] \quad (i = 1, \ldots, m)$$

(x_i *distinct) are asymptotically normal and independent with mean zero and variances*

$$f(x_i) \int h^2(\alpha) \, d\alpha \quad (i = 1, \ldots, m).$$

One could equally well consider estimation of the joint density function of a fixed number of random variables. Under conditions corresponding to those assumed in Theorem 1, one again finds that the asymptotic results are the same as those in the case of independent observations. Actually we shall explicitly note one such result under somewhat broader conditions of dependence. Now $X_n, n = \ldots, -1, 0, 1, \ldots$, is assumed to be a stationary sequence of random variables. *Assume that all joint distributions of a finite number of distinct* X_j's *are absolutely continuous with uniformly bounded continuous density functions.* Let $f(x, x')$ be the joint density function of X_j, X_{j+1}. A natural estimate of $f(x, x')$ is given by

$$f_n(x, x') = n^{-1} b(n)^{-2} \sum_{j=1}^{n-1} h(b(n)^{-1}(x - X_j), b(n)^{-1}(x' - X_{j+1})), \quad (44)$$

where $h(x, x')$ is a bounded continuous density function. Let $\mathscr{B}_n$ be the Borel field generated by X_k, $k \leqslant n$, and $\mathscr{F}_n$ be the Borel generated by $X_k, k \geqslant n$. We shall say that $\{X_m\}$ *satisfies the condition S if for every random variable* g *with* $Eg^2 < \infty$, $Eg = 0$, *that is* $\mathscr{F}_{n+k}$ *measurable it follows that*

$$E|E(g|\mathscr{B}_n)|^2 \leqslant a(k) E|g|^2, \quad (45)$$

where $a(k) = O(k^{-4-\epsilon})$ for some $\epsilon > 0$ as $k \to \infty$. By making estimates quite similar to those in the proof of the Theorem one can prove the following result.

Theorem 2. Let $\{X_m, m = \ldots, -1, 0, 1, \ldots\}$ be stationary with uniformly bounded continuous joint density functions. Consider the estimate $\hat{f}_n(x, x')$ of the joint density $f(x, x')$ of X_j, X_{j+1} given by (44). If f is continuously differentiable up to second order and the moments,

$$m_{i,j} = \int \alpha^i \beta^j h(\alpha, \beta) \, d\alpha \, d\beta \quad (i, j = 0, 1, 2),$$

when $i+j \leqslant 2$ exist then the bias

$$Ef_n(x, x') - f(x, x') = [m_{1,0} f_x(x, x') + m_{0,1} f_{x'}(x, x')] b(n)$$
$$+ \tfrac{1}{2}[m_{2,0} f_{x,x} + 2m_{1,1} f_{x,x'} + m_{0,2} f_{x',x'}] b(n)^2 + o(b(n)^2) \quad (46)$$

as $b(n) \to 0$. If $\{X_m\}$ satisfies the condition S and $nb(n)^2 \to \infty$ as

$$n \to \infty \quad (b(n) \to 0)$$

then any finite m-tuple of the normalized deviations

$$\sqrt{(nb(n)^2)} [\hat{f}_n(x, x') - E\hat{f}_n(x, x')] \quad (47)$$

corresponding to distinct points (x, x') are independent with mean zero and variances

$$f(x, x') \int h^2(\alpha, \beta) \, d\alpha \, d\beta. \quad (48)$$

Regression estimates have been considered in [8] and [10]. Without going into details, one can show that under conditions similar to those specified in [8], the asymptotic behavior of these in the case of dependent observations is the same as the asymptotic behavior in the case of independent observations. In [9] estimation of the transition probability density for Markov sequences is discussed under condition D_0.

6 ACKNOWLEDGEMENT

I should like to thank D. Brillinger and M. Woodroofe for their helpful comments on this paper. A. Garsia suggested that the Riesz convexity theorem might be helpful in comparing the norms considered in §4.

REFERENCES

[1] Bartlett, M. S. (1963). Statistical estimation of density functions. *Sankhyā*, Ser. A **25**, 245–54.

[2] Billingsley, P. (1968). *Convergence of probability measures*. New York: John Wiley.

[3] Doob, J. L. (1953). *Stochastic processes*. New York: John Wiley.

[4] Dunford, N. and Schwartz, J. L. (1958). *Linear Operators*, Part I. Interscience.

[5] Ķendall, D. G. (1959). Unitary dilations of Markov transition operators, and corresponding integral representations for transition probability matrices. In *Probability and statistics* (ed. U. Grenander). New York: John Wiley.

[6] Parzen, E. (1962). On estimation of a probability density and mode, *Ann. Math. Statist.* **33**, 1065–76.

[7] Rosenblatt, M. (1956). Remarks on some nonparametric estimates of a density function, *Ann. Math. Statist.* **27**, 832–5.

[8] Rosenblatt, M. (1969). Conditional probability density and regression estimators. In *Multivariate Analysis*, vol. II. Academic Press.

[9] Roussas, G. (1967). Nonparametric estimation in Markov processes, *Technical Report No.* 110, Department of Statistics, University of Wisconsin, Madison, Wisconsin.

[10] Watson, G. S. (1964). Smooth regression analysis, *Sankhyā*, Ser. A **26**, 359–72.

DISCUSSION ON ROSENBLATT'S PAPER

MICHAEL WOODROOFE

I believe that Professor Rosenblatt's paper constitutes a significant contribution to the nonparametric literature and therefore feel quite honored to be a discussant of it. The paper not only contributes a major result by giving sufficient conditions for the asymptotic normality of density estimates from stationary Markov processes, but should also stimulate more research on the problem of density estimation from such processes. Indeed, it is to be hoped that Professor Rosenblatt's present paper on density estimation from dependent observations will stimulate as much research as did his first paper on density estimation from independent observations [4]. Among other things, I will indicate some directions in which I feel this research might profitably proceed.

In considering the asymptotic normality of the estimates, Professor Rosenblatt has solved a rather difficult problem and has left some less difficult results unstated. Since I feel these results to be important, I will take the liberty of stating some of them together with an indication of their proofs. Let X_n be a stationary process satisfying condition S with $a(k)$ summable. (This includes the case of a stationary Markov Process which satisfies G_2.) The argument of [3] then extends without difficulty to show that

$$M_n = \sup_x |\hat{f}_n(x) - f(x)| \to 0 \tag{1}$$

in probability as $n \to \infty$, provided that f is uniformly continuous, $\sqrt{[nb(n)]} \to \infty$, and the Fourier Transform of h is integrable. Thereafter, it follows, again as in [3], that the sample mode is a consistent estimate of the true mode, provided that the latter is unique. It would also be of interest to know the rate of convergence to zero of M_n and such related random variables as were studied in [6] and to determine conditions for and the exact rate of convergence to zero of the integrated mean square error.

$$I_n = \int_{-\infty}^{\infty} E[(f_n(x) - f(x))^2] \, dx.$$

To the best of my knowledge, the only other work which seriously considers density estimation from Markov processes is the unpublished report [5], in which consistency and asymptotic normality of kernel-type density estimates are obtained under the condition D_0. It is therefore of interest to consider Professor Rosenblatt's condition G_2 with some care. It is quite intuitively appealing in that it requires the dependence between segments of the process to decrease as the distance between them increases. I was therefore intrigued by the G_2 condition and hoped that it might be satisfied by a large enough class of stationary Markov processes to provide a reasonable setting for asymptotic inference from such processes. I still have this hope, but in the course of attempting to verify G_2 for linear processes of the form

$$X_k = \sum_{j=0}^{\infty} \beta^j Y_{k-j} \quad (k = 0, 1, 2, \ldots), \tag{2}$$

where the Y_i are independent and identically distributed and $|\beta| < 1$, I discovered an example which has shaken that hope slightly. If the Y_i have the distribution function

$$F_1(y) = 1 - (\log y)^{-2} : x \geqslant e, \tag{3}$$

then (2) does not satisfy G_2. Indeed, by considering LaPlace transforms and applying the Tauberian theorems of [1], one may show that F, the stationary distribution function, has a slowly varying tail. Letting

$$g_n(x) = (1 - F(n))^{-\frac{1}{2}} \quad \text{for} \quad x > n \quad \text{and} \quad g_n(x) = 0 \quad \text{for} \quad x \leqslant n,$$

it is then easily verified that

$$|T^k g_n|_2^2 = \|T^k g_n\|_2^2 - o(1) \geqslant \left[\frac{1 - F(n\beta^{-k})}{1 - F(n)} \right] - o(1) \to 1 \quad \text{as} \quad n \to \infty,$$

for each fixed k. Since $|g_n|_2 \leqslant \|g_n\|_2^2 = 1$ for all n, G_2 cannot be satisfied.

The distribution function in (3) is, of course, sufficiently pathological to be of little practical interest, but I still find this example disturbing. Since $X_k = \beta^k X_0 + Z$, where Z is independent of X_0, the process defined by (2) has the property that the dependence of X_k on X_0 decreases geometrically as $k \to \infty$, which property led me to believe that any process of the form (2) should be geometrically ergodic, that it should satisfy G_2. I believe the generality of the G_2 condition to be a question on which research could profitably be done. If it should turn out to be reasonably general, it might well simplify subsequent work on asymptotic inference from Markov processes.

Another area in which I feel research could profitably be done is that of density estimation from general linear processes of the form

$$X_k = \sum_{j=0}^{\infty} c_j Y_{k-j} \quad (k = 0, 1, 2, \ldots),$$

where the Y_i are as in (2). What conditions on the coefficients c_j and/or the distribution of Y_i are sufficient to insure the consistency and asymptotic normality of density estimates from such processes? These questions could be attacked either by verifying an appropriate version of condition S or from scratch, perhaps by using the techniques of [2], Part II. The former approach would be more desirable since conditions similar to S appear throughout the probability literature (e.g. in [2]), but the latter approach would probably be the easier.

REFERENCES

[1] Feller, William (1966). *An Introduction to probability theory and its applications*, 2. New York: John Wiley.

[2] Ibragimov, I. A. (1962). Some limit theorems for stationary processes, *Theory Probab. Applic.* 7, 349–82.

[3] Parzen, E. (1962). On estimation of a probability density and mode, *Ann. Math. Statist.* 33, 1065–76.

[4] Rosenblatt, M. (1956). Remarks on some nonparametric estimates of a density, *Ann. Math. Statist.* 27, 832–5.

[5] Roussas, G. (1967). Non-parametric estimation in Markov Processes, *Technical Report No.* 110, Department of Statistics, University of Wisconsin, Madison, Wisconsin.

[6] Woodroofe, M. (1967). On the maximum deviation of the sample density, *Ann. Math. Statist.* 38, 475–81.

SOME NONPARAMETRIC TESTS FOR STOCHASTIC PROCESSES

C. B. BELL, M. WOODROOFE AND T. V. AVADHANI

1 INTRODUCTION AND SUMMARY

The object here is to extend known nonparametric test theory in the univariate and multivariate cases to stochastic processes. The major obstacle to such a development is the fact that very little is known of dependence. The authors would have liked to have treated martingales, submartingales, nth-order Markov chains, etc. However, the only processes in which the authors could handle the dependence were those processes from which one could generate iid random variables. For the Poisson processes, the stationary independent increment processes, and the processes with monotone marginals, a simple transformation produces iid variables. Exchangeable and spherically exchangeable $(S-E)$ processes are mixtures of iid processes, and one can employ much of the known theory there.

Besides the DF (distribution-free) tests of goodness-of-fit, two-sample tests and tests of the model, one considers invariant tests for the processes named. In each case an attempt is made to characterize all DF tests; 'good' DF tests; all invariant DF tests, etc.

2 PRELIMINARIES AND KNOWN RESULTS

The notation and known DF test theory of this paper are essentially found in Bell and Smith (1969 a, b) and Bell and Donoghue (1969).

2.1 Types of distribution freeness

The two major types of DF statistics are (a) those based on measure isomorphisms with the parameter explicitly given, and (b) those based on similar sets and not involving the parameter. This distinction can be seen in the normal case for the one-sample and two-sample Student's t-statistic. A third type of DF-ness is related to the invariance of the power function. SDF (strongly distribution-free) statistics are treated in §8.

In all cases DF-ness is relative to a family of distributions. The most important families and groups to be used in the sequel are given below. Ω, Ω', etc. will denote generic families of distribution functions.

[215]

$\Omega_2(A)$ = the family of all nonatomic distributions on A.

$\Omega_2^*(A)$ = the family of all distributions on A with strictly monotonic marginals and conditionals, where A is a subset of some R_k.

$\Omega_s^{(k)}(R_p)$ = the family of continuous distributions on R_{kp} which are invariant under all permutations of the k 'natural' p-tuples.

$\Omega'(N)$ = $\{F^{(N)}: F \text{ is in } \Omega'\}$, that is, the family of random sample distributions from Ω'.

$\Omega(J, \mathscr{G})$ = $\{J_g: g \text{ is in } \mathscr{G}\}$, that is, all '$\mathscr{G}$-translates' of J. For example if $J = \Phi$, and $\mathscr{G}$ is the affine group on R_1, then $\Omega(J, \mathscr{G})$ is the family of all univariate normals.

Other continuous groups of interest are $\mathscr{G}_s$ = the translation group on R_1; $\mathscr{G}_2^*(R_1)$ and $\mathscr{G}_2^*(R_1^+)$, the group of 1-1 strictly monotone transformations of R_1 onto R_1, and R_1^+ onto R_1^+, respectively. [It is seen, for example that $\Omega_2^*(R_1) = \Omega(J, \mathscr{G}_2^*(R_1))$ for all J in $\Omega_2^*(R_1)$; and similarly for R_1^+.]

From the point of view of constructing and characterizing statistics, the classes of DF statistics are (A) those statistics based on Rosenblatt's (1952) transformation; (B) those based on finite permutation groups; and (C) those based on non-countable permutation groups, e.g. Q_k, the orthogonal group on R_k. Type (A) concerns each of the processes treated, type (B) concerns all but the S–E processes, which are related to type (C). The major known results for finite permutation groups are given below.

2.2 Orbits and finite permutation groups

The theory of DF hypothesis testing as given orginally by Pitman (1937 a, b), Scheffe (1943), Lehmann and Stein (1949) and others is based on finding the largest permutation group w.r.t. which the null hypotheses class is invariant. The permutation groups of interest in the sequel are:

$\mathscr{S}_n$ = the symmetric group of $n!$ permutations of the integers $(1, \ldots, n)$; and $\mathscr{S}_k \mathrm{Wr} \mathscr{S}_n$, the *wreath product* (see § 6) of $\mathscr{S}_k$ with $\mathscr{S}_n$, to be employed in analyzing symmetry;

$\overset{k}{\underset{1}{\mathrm{X}}}\mathscr{S}_{n_i}$, the direct product group of symmetric groups; and

$\mathscr{S}^*(n_1, \ldots, n_k)$ = that subgroup of $\mathscr{S}_N$ (with $N = \sum_i n_i$) which leaves the K sets of order statistics $(w_1(1) < \ldots < w_1(n_1); \ldots; w_k(1) < \ldots < w_k)$ unchanged. [This last group is related to the joint distribution of several independent sets of order statistics, which arise in cross-sectional sampling from Poisson processes (§ 4). The hypotheses and related groups are given in Table 2.1 for several hypotheses of interest.]

Table 2.1. *Families and groups for several hypotheses*

H_0	$\Omega'(H_0)$	z	$\mathscr{S}'$	$K(\mathscr{S}')$
$G_1 = G_2 = \ldots = G_k$ (modified k-sample)	$\{G^*\colon G \in \Omega_2(R_1)\}$	$w_1(1), \ldots, w_1(n_1); \ldots;$ $w_k(1), \ldots, w_k(n_k)$, where $w_i(j)$ is in R_1	$\mathscr{S}^*(n_1, \ldots, n_k)$	$(\mathcal{N}!)$ $[\Pi n_i!]$
$F_1 = \ldots = F_n$ (randomness)	$[\Omega_2(R_p)]\,(n)$	$\mathbf{x}_1, \ldots, \mathbf{x}_n$, where $\mathbf{x}_1$ is in R_p	$\mathscr{S}_n$	$n!$
$F(\mathbf{x}_1, \ldots, \mathbf{x}_k)$ $= F(\gamma(\mathbf{x}_1, \ldots, \mathbf{x}_k))$ for all γ in $\mathscr{S}_k$ (symmetry)	$[\Omega_s^{(k)}(R_p)]\,(n)$	$\mathbf{x}_{11}, \ldots, \mathbf{x}_{1k}; \ldots;$ $\mathbf{x}_{n1}, \ldots, \mathbf{x}_{nk}$, where $\mathbf{x}_{ij}$ is in R_p	$\mathscr{S}_k Wr \mathscr{S}_n$	$(n!).$ $(k!)^n$
There exist F_i such that $F(\mathbf{x}_1, \ldots, \mathbf{x}_k)$ $= \pi F_i(\mathbf{x}_i)$ (independence)	$[\overset{k}{\underset{1}{\mathrm{X}}}\,\Omega_2(R_{p_i})]\,(n)$	$\mathbf{x}_{11}, \ldots, \mathbf{x}_{n1}; \ldots;$ $\mathbf{x}_{1k}, \ldots, x_{nk}$, where $\mathbf{x}_{ij}$ is in R_{p_i}	$\overset{k}{\underset{1}{\mathrm{X}}}\,\mathscr{S}_n$	$(n!)^k$

If S' is a permutation group $\{\gamma(z)\colon \gamma \in S'\}$ is called the *orbit* of z. Constructing similar sets consists of choosing a fixed proportion of the points of a.e. orbit. This is accomplished by means of Pitman functions.

Definition 2.1. v is a *B-Pitman function* w.r.t. (finite permutation group) $\mathscr{S}'$ if for all J in $\Omega(H_0)$ and for all non-identity elements γ of $\mathscr{S}'$, $P\{v(\gamma(z)) = v(z)|J\} = 0$.

This means that v assumes $K(\mathscr{S}')$ values on a.e. orbit $\mathscr{S}'(z)$, where $K(\mathscr{S}')$ is the order or cardinality of $\mathscr{S}'$.

Definition 2.2. The B-Pitman statistic induced by B-Pitman function v, is defined by $\tilde{R}(v(z)) = \sum\limits_{\gamma \in \mathscr{S}'} \epsilon\{v(z) - v(\gamma(z))\}$, where $\epsilon(u) = 1$ or 0 according as $u \geqslant 0$ or not.

It is the rank of the v-value of the original data point among the v-values on the orbit. $\tilde{R}(v)$ will assume the values $1, 2, \ldots, K(\mathscr{S}')$ on a.e. orbit.

In the current framework, the subscript on a random variable denotes 'time'. For many applications the chronological order of the data points should not be a factor. For such cases one may wish to 'identify' points which differ only in a permutation of time.

Definition 2.3. $\hat{v}$ is a *BNS-Pitman function* w.r.t. $\mathscr{S}'$ and its subgroup $\mathscr{S}_T$ if
(i) $\hat{v}(z) = v(\gamma(z))$ for all γ in $\mathscr{S}_T$; and
(ii) $P\{\hat{v}(z) = v(\gamma(z))|J\} = 0$ for all J in $\Omega(H_0)$, and for all γ not in $\mathscr{S}_T$.
[Note: NS = 'non-sequential'.]

Example 2.1. Pitman's (1937a) original BNS-Pitman function was $\hat{v}(z) = \bar{x} - \bar{y}$ in the two sample case. Clearly, $\hat{v}$ is invariant under $\mathscr{S}_T = \mathscr{S}_m \times \mathscr{S}_n$ for $z = (x_1, ..., x_m; y_1, ..., y_n)$.

Lemma 2.1. (i) *If $\hat{v}$ is a BNS-Pitman function, then on a.e. orbit it assumes* $K(\mathscr{S}')[K(\mathscr{S}_T)]^{-1}$ *distinct values.*

(ii) *Each B-Pitman function is a BNS-Pitman function (trivially) for* $\mathscr{S}_T = \{l\}$, *where 'l' is the identity element of* $\mathscr{S}_T$.

Definition 2.4. *The BNS-Pitman statistic* induced by *BNS*-Pitman function $\hat{v}$, is defined by $\tilde{R}(\hat{v}(z)) = [K(\mathscr{S}_T)]^{-1} \sum\limits_{\gamma \in \mathscr{S}'} \epsilon\{v(z) - v(\gamma(z))\}$.

This function, then, assumes on a.e. orbit, the values

$$1, 2, ..., [K(\mathscr{S}')] [K(\mathscr{S}_T)]^{-1}.$$

[Equivalent to the definition above, one can define *BNS*-Pitman functions in terms of the quotient group of $\mathscr{S}'$ with the 'time' group. More precisely,

Lemma 2.2. *A BNS-Pitman function is a B-Pitman function over $\mathscr{S}'/\mathscr{S}_T$, where $\mathscr{S}'$ and $\mathscr{S}_T$ the 'time' group are as given below.*

(a) *k-sample:* $\mathscr{S}' = \mathscr{S}_N, \mathscr{S}_T = \overset{k}{\underset{1}{\text{X}}} \mathscr{S}_{n_i}$.

(b) *2-sample:* $\mathscr{S}' = \mathscr{S}_N, \mathscr{S}_T = \mathscr{S}_m X \mathscr{S}_n$.

(c) *Independence:* $\mathscr{S}' = \overset{k}{\underset{1}{\text{X}}} \mathscr{S}_n, \mathscr{S}_T = \mathscr{S}_n$.

(d) *Symmetry:* $\mathscr{S}' = \mathscr{S}_k Wr \mathscr{S}_n, \mathscr{S}_T = \mathscr{S}_n$.

The basic theorem in terms of these functions and groups is given below, where 'MP' = most powerful; 'U' = uniformly, and 'L' = locally.

Theorem 2.3. *The following are valid for the z's, Ω''s and $\mathscr{S}''$s given in Table 1.2.* (i) $\int \phi(z) \, dJ(z) = \alpha$ *for all J in Ω' iff $\Sigma \phi(\gamma(z)) = K(\mathscr{S}') \alpha$ for a.a.z, where the summation is over γ in $\mathscr{S}'$.*

(ii) *A statistic T is* DF *w.r.t. Ω' iff there exists a B-Pitman function (v w.r.t. $\mathscr{S}'$) and a function ψ such that $T \equiv \psi[R(v)]$.*

(iii) *The following three conditions are equivalent.*

(a) *A is similar of size α w.r.t. Ω';*

(b) *A contains $K(\mathscr{S}') \alpha$ elements of a.e. orbit $\{\gamma(z): \gamma \in \mathscr{S}'\}$;*

(c) *There exists a B-Pitman function v(w.r.t. $\mathscr{S}'$) such that*

$$A = \{z: \tilde{R}(v(z)) \leqslant K(\mathscr{S}')\alpha\}.$$

(iv) *Consider a simple alternative likelihood function* $L_1(\cdot)$. *The MPDF test against* L_1 *is of the form*

$$\phi(z) = \begin{cases} 1 & \text{if } \tilde{R}(v_1(z)) > C(\alpha, \mathscr{S}') \\ \delta & = \\ 0 & < \end{cases}$$

where v_1 *is a B-Pitman function w.r.t.* $\mathscr{S}'$, *whose ordering on a.e. orbit is consistent with that of* L_1. *The power of this test is greater than* α *iff* L_1 *is not invariant under* $\mathscr{S}'$.

(v) *Consider the family* $\mathscr{F}' = \{L_1 \colon L_1(z) = \exp[a(\theta)\,v_1(z) + b(z, \theta)]\}$ *where* $a(\theta) > 0$ *for all* θ; *and* v *is a B-Pitman function w.r.t.* $\mathscr{S}'$.

    (a) *The test in* (iv) *above is* UMPDF *w.r.t.* $\mathscr{F}'$ *if* b *is invariant under* $\mathscr{S}'$.

    (b) *Assume that when* $\theta = \theta_0$, L_1 *is in* Ω'; *and that there is sufficient regularity that* $b_2(z, \theta) = o(a(\theta))$ *in some neighborhood of* θ_0, *where* $b = b_1 + b_2$; *and* b_1 *is invariant under* $\mathscr{S}'$. *Then, the test in* (iv) *above is* LMPDF *against* $\mathscr{F}'$.

Proof of Theorem 2.3. [The proofs for the case of randomness, symmetry and independence can be found in one form or the other in the references. The proof here concerns only $S^*(n_1, \ldots, n_k)$.]

Let

$$B_0 = \{(y_1, \ldots, y_N) \colon y_1 < \ldots < y_N\} \quad \text{and} \quad D = \mathop{\mathrm{X}}_{1}^{n} \{w_i(1) < \ldots < w_i(n_i)\}.$$

Then $D = U\gamma(B_0)$ where the union is over all γ in $\mathscr{S}^* = \mathscr{S}^*(n_1, \ldots, n_k)$. Further, the joint distribution of $z = (w_1(1), \ldots, w_k(n_k))$ is given by

$$G^*(z) = \begin{cases} \displaystyle\prod_{1}^{k}(n_i!)\prod_{i=1}^{k}\prod_{j=1}^{n_1} dG(w_i(j)) & \text{if } z \in D, \\ 0, & \text{otherwise.} \end{cases}$$

(i) If $\alpha = \displaystyle\int_D \phi\, dG^*$, then

$$\alpha = \sum_{\gamma \in \mathscr{S}^*} \int_{\gamma^{-1}(B_0)} \phi(z)\, dG^*(z)$$

$$= \sum_{\gamma \in \mathscr{S}^*} \int_{B_0} \phi(\gamma(z))\, dG^*(\gamma(z))$$

$$= \int_{B_0} [\sum_{\gamma \in \mathscr{S}^*} \phi(\gamma(z))]\, dG^*(z).$$

Let

$$\psi(z) = [\sum_{\gamma \in \mathscr{S}^*} \phi(\gamma(z))].$$

Then one has for all continuous G on R_1,

$$\alpha = \int_{B_0} \psi(z)\, dG^*(z) = [N!]^{-1}\prod_{1}^{k}(n_i!)\int_{B_0} \psi(y_1, \ldots, y_N)\left(N!\prod_{1}^{N} dG(y_i)\right).$$

But the order statistics are complete. Hence (i) follows.

(iii) follows if ϕ in (i) is taken to be the indicator function of the set A. The B-Pitman function is used to select the points. (See Bell and Doksum, 1967.)

(ii) follows from (iii) in that each $T^{-1}(B) = A$ must be similar.

(iv) is the result of applying the Neyman–Pearson lemma to each orbit.

(v) (a) follows from noting that the orderings on the orbits of v_1, $a(\theta)v_1(z), a(\theta)v_1(z) + b(z,\theta)$, etc. are identical if $b(z,\theta)$ is invariant under $\mathscr{S}'$.

(v) (b) results from the facts that $v_1(z) + b(z,\theta)[a(\theta)]^{-1}$ is negligible for θ sufficiently close to θ_0.

The above theorem is used in each section of the sequel. However, in some of the development for S–E processes, one uses the orthogonal group. A generalization of the above theorem is, then, necessary and is given in § 7.

3   PROCESSES WITH STATIONARY INDEPENDENT INCREMENT (SII)

If $\{X_t, t \geqslant 0\}$ is a SII process, then for each t' and n

$$Y_1 = X_{t'} - X_0, \quad Y_2 = X_{2t'} - X_{t'}, \quad Y_3 = X_{3t'} - X_{2t'}, \ldots, Y_n = X_{nt'} - X_{(n-1)t'}$$

are independently and identically distributed. The common distribution function F_0 is uniquely determined by the law $\mathscr{L}_0$ of the process and the time interval $(0, t']$, and conversely F_0 and t' determine $\mathscr{L}$ if $X_0 = 0$ (Doob, 1953).

Here, as elsewhere in this paper one should be concerned with how to choose t' and n. The authors feel that this question requires a more thorough study of alternatives than is being treated here.

Since the Y's are iid and univariate, one can make use of known results for the *univariate* goodness-of-fit, two-sample and randomness hypotheses.

3.1. Goodness-of-fit

The hypothesis $H_0: \mathscr{L} = \mathscr{L}_0$ becomes $H_0: F = F_0$ in terms of $Y_1, \ldots, Y_n$. Since this is a simple hypothesis, the usual parametric theory holds.

A large family of DF statistics is obtained by making use of the probability transformation $U = F_0(Y)$.

Theorem 3.1. (i) *Each statistic* $T = \psi[F_0(Y_1), \ldots, F_0(Y_n)]$ *is DF w.r.t.* $\Omega(H_0)$ *in the sense that there exists Q such that*

$$P\{\psi[F_0(Y_1), \ldots, F_0(Y_n)] \leqslant t \mid F_0\} = Q(t)$$

for all F in $\Omega(H_0)$ and for all t.

(ii) $Q(t) = P\{\psi[U_1, ..., U_n] \leqslant t | U_0\}$ for all t where U_0 is the standard uniform distribution.

(iii) There exists a DF statistic which is not of the form in (i) above.

Proof. This follows immediately from Birnbaum and Rubin (1954) and Bell (1964). [These statistics are said to be of structure (d'_n); see §8.]

Example 3.1. (a) $\bar{U} = n^{-1} \sum_{i=1}^{n} F_0(Y_i)$ is DF w.r.t. $\Omega(H_0)$, and $Q(\cdot)$ is the distribution of the mean of a random sample of size n from a uniform distribution. Further, $(\bar{U} - \frac{1}{2}) \sqrt{(12n)}$ is asymptotically normal, with the normal approximation adequate for most purposes with $n \geqslant 11$.

(b) $\bar{Z} = \dfrac{1}{\sqrt{n}} \sum_{i=1}^{n} \Phi^{-1}(F_0(Y_i))$ is DF with Q exactly $\mathcal{N}(0, 1)$. Further if F_0 is $\mathcal{N}(\mu, \sigma^2)$, then $\bar{Z} = \left(\dfrac{\bar{X} - \mu}{\sigma}\right) \sqrt{n}$.

3.2. Two-sample case

$H_0: \mathscr{L}_1 = \mathscr{L}_2$ reduces to $H_0: F_1 = F_2$, where the data is $X'_1, ..., X'_m$ and $Y'_1, ..., Y'_n$ with $X'_1 = X_{it'} - X_{(i-1)t'}$ and Y'_i defined similarly.

Theorem 3.2. (i) $\int \phi \, dJ = \alpha$ for all J in $\Omega(H_0)$ iff

$$\sum_{\gamma \in S_N} (x_1, ..., x_m; y_1, ..., y_n) = (N!)\alpha$$

for a.a. $(x_1, ..., y_n)$ where $N = m + n$.

(ii) A statistic T is DF w.r.t. $\Omega(H_0)$ iff there exists a B-Pitman function v w.r.t. $\mathscr{S}_N$ and a function W such that $T = W[\tilde{R}(v)]$.

Proof. This follows immediately from Bell and Smith (1969a).

Since the families of statistics and test functions is quite large, one seeks methods of selecting tests and/or statistics which are in some sense 'good'. One such method is related to invariance and the SDF property. This is treated in §8.

For a simple alternative and a Koopman–Pitman-type class related to it one can obtain a uniformly most powerful DF test.

Theorem 3.3. Let $Z = (X'_1, ..., Y'_n)$ and let v_1 a B-Pitman function; then the UMPDF test function against the alternative family

$$\{(\mathscr{L}_1, \mathscr{L}_2): L(z) = \exp\{av_1(z) + h(z)\} \quad \text{with } a > 0$$

and $h(z)$ invariant under $\mathscr{S}_N\}$ is of the form

$$\psi(z) = \begin{cases} 1 & \text{if } \tilde{R}(v_1(z)) > k(\alpha, m, n) \\ \delta & \quad\quad = \\ 0 & \quad\quad < \end{cases}$$

Proof. This follows from Bell and Smith (1969*a*).

By a slight modification one can obtain a LMPDF test for such a family of alternatives. This is done for the next hypothesis.

At this time one should consider examples illustrating the two theorems above.

Example 3.2. (*a*) Let $W_1, \ldots, W_N, N = n+m$, be standard normal random sample independent of the two independent stochastic processes. Let $W(R(X'_i))$ be that W whose rank in the W-sample is equal to that of X'_i in the combined sample

$$[Z_1, \ldots, Z_N] = \{X'_1, \ldots, X'_m, Y'_1, \ldots, Y'_n\};$$

and let $W(R(Y'_j))$ be analogously defined. Form

$$T = \frac{1}{m} \sum_1^m W(R(X'_i)) - \frac{1}{n} \sum_1^n W(R(Y'_j)).$$

For each fixed set of W's, T is a DF statistic in the usual sense; that is, T is conditionally a DF statistic. When the W's constitute a random normal sample, T is a randomized DF statistic (Bell and Doksum, 1965). The null distribution of T is $\mathcal{N}(0, m^{-1}+n^{-1})$.

$$(b) \quad T_1 = \int [F_m(z) - G_n(z)]^2 \, d\left(\frac{mF_m + nG_n}{N}\right)$$

$$= N^{-1}m^{-2}n^{-2} \sum_{r=1}^N [NnG_n(Z(r)) - rn]^2$$

is a (non-randomized) statistic which is DF.

3.3 Tests for the model

One wishes to determine whether or not the process being observed is a SII process. On restricting oneself to the continuous case, as usual, one finds for the increments

$$X'_i = X_{it} - X_{(i-1)t} \quad (1 \leqslant i \leqslant n),$$

the null hypothesis class is that of randomness. Hence, one has from Bell and Donoghue (1969):

Theorem 3.4. (i) *If $\mathcal{L}_1$ is a law such that $L_1(x_1, \ldots, x_n) = \prod_1^n f(x_i)$, or is otherwise invariant under $\mathcal{S}_n$, then the MPDF test of size α has power α.*

Further, consider the family of processes with laws L yielding

$$\{L_1: L_1(z) = \exp[a(\theta)\, v_1(z) + b(z, \theta)]$$

with $a(\theta) > 0$, and v a fixed B-Pitman function}. For this family

(ii) $\tilde{R}(v)$ *is* UMPDF *if* $b(\,\cdot\,; \theta)$ *is invariant under* $\mathscr{S}_n$ *for all* θ; *and*

(iii) $\tilde{R}(v)$ *is* LMPDF *if* $b = b_1 + b_2$, *where* b_1 *is invariant under* $\mathscr{S}_n$ *and* $b_2(z, \theta) = o(a(\theta))$, *where* $a(\theta)$ *is continuous and* $a(\theta_0) = 0$.

Example 3.3. Let the process be such that its increments $X'_1, \ldots, X'_n$ form a first order Markov process with the following properties:

(a) $X'_1 = aX'_0 + Y_1, X'_2 = aX'_1 + Y_2, \ldots, X'_n = aX'_{n-1} + Y_n$;

(b) $\{Y_1, \ldots, Y_n\}$ are iid $\mathcal{N}(0, \sigma^2)$, and

(c) X_0 is arbitrarily distributed.

Then the Pitman function which generates the MPDF test is

$$v = \sum_{i=1}^{n} x_i x_{i-1}.$$

An exponential family w.r.t. which $\tilde{R}(v)$ is UMPDF is

$$\left\{ L_1 \colon L_1 = \exp\left[a(\theta) \sum_{i=1}^{n} x_i x_{i-1} + b(\theta, z) + \sum_{i=1}^{n} g(x_i) \right], \right.$$

where $b(\,\cdot\,, \theta)$ is invariant under $\mathscr{S}_n$ for all $\left. \theta \right\}$.

Further $\tilde{R}(v)$ generates the LMPDF test against any family of processes whose likelihood functions are of the form

$$\left\{ \exp\left[a(\theta) \sum_{i=1}^{n} x_i x_{i-1} + b(\theta, z) + \sum_{i=1}^{n} g(x_i) + Q(z, \theta, i) \right] \right\},$$

where $Q(z, \theta, i) = o(a(\theta))$ for all i and for a.a. z.

The preceding theorem and examples simply indicate that if the alternative has a tractable analytic structure, 'good' DF tests can be obtained.

4 HOMOGENEOUS AND NONHOMOGENEOUS POISSON PROCESSES

A homogeneous Poisson Processes has stationary independent increments. Further, each nonhomogeneous Poisson process can be transformed into a homogeneous Poisson process. However, in these cases the increments have discrete distributions. Hence, the DF techniques of the preceding section do not apply.

On the other hand the interarrival times for homogeneous processes are iid. The methodology of the preceding section would apply to these interarrival times, but one obtains more powerful results on making use of the fact that they have a common exponential distribution.

For mathematical convenience one restricts attention to Poisson processes $\{N(t), t \geq 0\}$ whose mean functions $\mu(\cdot)$ are strictly monotone, continuous and satisfy $\mu(0) = 0$ and $\mu(\infty) = \infty$. (However, most of the results of this section hold for Poisson processes with nondecreasing mean functions. This latter statement is not true for the results of §9.)

4.1. Structure results

One can now give two useful structure and distribution theorems, which can for the most part be found in Parzen (1962) and are well known.

Theorem 4.1. (i) *There is a 1–1 correspondence between the Poisson processes with strictly monotone mean functions (satisfying $\mu(0) = 0$ and $\mu(\infty) = \infty$) and the group $\mathcal{G}_2^*(R_1^+)$.*

(ii) *Let $\{N(t),\ t \geq 0\}$ have mean function $\mu_0(\cdot)$ and dual waiting time process $\{W_n: n = 1, 2, \ldots\}$. Let $M(t) = N(\mu_1(t))$. Then $\{M(t): t \geq 0\}$ is a Poisson process with mean function $\mu_0(\mu_1(\cdot))$ and waiting time process $\{V_n: n = 1, 2, \ldots\}$; where $V_n = \mu_1^{-1}(W_n)$. In particular, if $\mu_0(\cdot) = b\mu_2(\cdot)$ and $\mu_1(\cdot) = \mu_2^{-1}(\cdot)$ for some constant $b > 0$, then $\{M(\mathcal{U}), \mathcal{U} \geq 0\}$ is a homogeneous Poisson process with $\mu(t) = bt$.*

(iii) *If $\mu(t) = bt$ for some $b > 0$, then*

(a) *the interarrival times $\tau_1, \ldots, \tau_n, \ldots$ are iid with common density function $f(t) = be^{-bt}, t \geq 0$;*

(b) *the waiting time $W_n = \sum_1^n \tau_j$ has density*

$$f_n(t) = e^{-bw}(bw)^{n-1}[(n-1)!]^{-1}b \quad (w \geq 0).$$

Further from Parzen (1962) and J. Mycielski (1956) one has

Theorem 4.2. (i) *If $\mu(\cdot)$ is not linear with $\mu(0) = 0$, then the τ's are neither independently nor identically distributed.*

(ii) *For arbitrary $\mu(\cdot)$ (in $\mathcal{G}_2^*(R_1^+)$):*

(a) *for fixed t', the conditional distribution of the waiting times $W_1, \ldots, W_k$ (given $k = N(t')$) is that of the order statistics from a random sample of size k with common distribution function*

$$G(t) = \frac{\mu(t)}{\mu(t')} \quad (0 \leq t \leq t'),$$

(b) *for fixed k, the conditional distribution of $W_1, \ldots, W_{k-1}$, given $W_k = w'$, is that of the order statistics from a random sample of size $(k-1)$ with common distribution*

$$G(t) = \frac{\mu(t)}{\mu(w')} \quad (0 \leq t \leq w');$$

Theorem 4.3. *Let* $W_1, ..., W_n$ *denote times of occurrence of the first n events in a non-homogeneous Poisson process with mean function* μ. *If* ϕ *is a measurable function on* R^n *for which*

$$E[\phi(W_1, ..., W_n)|\mu] = 0$$

for all μ; *then* $\phi(W_1, ..., W_n) = 0$ *w.p. one for all* μ.

Proof. Fix μ_0 and let $\mu = a\mu_0$, $a > 0$; then

$$0 = E[\phi(W_1, ..., W_n)|\mu]$$

$$= E[E[\phi(W_1, ..., W_{n-1}, W_n)|W_n, \mu_0]|a\mu_0]$$

for all $a > 0$ implies (since μ_0 is restricted to $\mathscr{G}_2^*(R_1^+)$)

$$E[\phi(W_1, ..., W_{n-1}, W_n)|W_n, \mu_0] = 0$$

for a.a. W_n and all μ_0. But the conditional distribution of $W_1, ..., W_{n-1}$ given W_n and μ_0 is that of order statistics of a random sample from

$$G_0 = \mu_0(t)/\mu_0(W_n) \quad (0 \leqslant t \leqslant W_n);$$

the latter distributions being complete we get

$$\phi(W_1, ..., W_{n-1}, W_n) = 0$$

w.p. one, for a.a. W_n for all μ_0.

One can now treat the hypotheses of interest.

4.2 Goodness of fit

In some sense it is more desirable from a practical point of view to fix t', that is, the time interval $[0, t']$ and test $H_0: \mu(\cdot) = \mu_0(\cdot)$ on the basis of the arrivals in that time interval.

Theorem 4.4. *The* MP *size* α *test of* $H_0: \mu(\cdot) = \mu_0$, $H_1: \mu(\cdot) = \mu_1(\cdot)$ *has a critical region of the form*

$$\left\{ [\mu_0(t')]^k [\mu_1(t')]^{-k} \prod_1^k \mu_1'(w_i) [\mu_0'(w_i)]^{-1} > C(\alpha, t', k) \right\},$$

where $k = N(t')$ *and the* μ's *are assumed to be differentiable.*

One danger of this approach is that $k = N(t')$, may be quite 'small'. Probabilistically, this can be avoided by choosing t' such that $\mu_0(t')$ is 'large'.

Another way to avoid this situation is to terminate observation after the nth arrival or waiting time. In such a case, the original data $W_1, ..., W_n$ is transformed first to $W_1', ..., W_n'$, where $W_i' = \mu_0(W_i)$; and then to $\tau_1', ..., \tau_n'$, where $\tau_i' = W_i' - W_{i-1}'$ and $W_0' = 0$. The original hypothesis then becomes $H_0: F(t) = 1 - e^{-t}$, $t \geqslant 0$ for the τ's.

In these terms, analogous to the preceding theorem one obtains *Corollary* 4.5. *A critical region of the form*

$$\left\{[L_1(t_1, ..., t_n)]\exp\sum_1^n t_i > k(\alpha, n)\right\}$$

is MP *against an alternative process* $\mathscr{L}_1$ *with* $\mathscr{L}_1(t_1, ..., t_n)$ *as the likelihood function of* $\tau_1', ..., \tau_n'$.

To use either the theorem or equivalently, the corollary above, one needs to have a specific simple alternative in mind. A large class of statistics is obtained from the following theorem for which no alternative be explicitly stated.

Theorem 4.6. (i) *For the goodness-of-fit hypothesis* H_0: $\mu = \mu_0$, *each statistic of the form*

$$T(\mu_0; W_1, ..., W_n) = \tilde\psi[\mu_0(W_1), ..., \mu_0(W_n)] = \psi[\tau_1', ..., \tau_n'],$$

where $\tau_i' = \mu_0(W_i) - \mu_0(W_{i-1})$, *is* DF.

(ii) $$P\{T(\mu_0; W_1, ..., W_n) \leqslant s | \mathscr{L}_{\mu_0}\} = P\{\psi[v_1, ..., v_n] \leqslant s | J_1^{(n)}\}$$

for all s, *where* $\mathscr{L}_{\mu_0}$ *is the law of the Poisson process with mean function* $\mu_0(\cdot)$; *and* $J_1^{(n)}$ *is the n-fold power distribution of a standard exponential distribution,* J_1.

Proof. These are the statistics of structure (d_n') treated in § 8.

Example 4.1. The MP test of H_0: $\mu(t) = t^3$ vs H_1: $\mu(t) = t^4$ for the time interval $[0, 20]$ has by Theorem 4.4 a critical region of the form

$$\left\{(8000)^k (160{,}000)^{-k}\prod_1^k [4w_i^3][3w_i^2]^{-1} > C(\alpha, 20, k)\right\}$$

$$= \left\{20^{-k}\prod_1^k w_i > C(\alpha, 20, k)\right\}.$$

Example 4.2. Let G_n be the empirical distribution function and let

$$T_1 = n\int [G_n(t) - G_0(t)]^2\, dG_0(t) = \frac{1}{12n} + \sum_{i=1}^n \left[\frac{2i-1}{2n} - \left(\frac{\mu_0(w_i)}{\mu_0(t)}\right)\right]^2,$$

where $(0, t']$ is the interval of time over which the process is observed. T_i is DF and under H_0 has the distribution of a Cramér–von Mises statistic.

Example 4.3. Let $T = 2\mu_0(W_n)$. Then T is DF and has a chi-square distribution with $2n$ degrees of freedom.

For the two-sample case, the mean function is unknown. Hence, the methodology is different from that above.

4.3. Two-sample case

One very practical problem is the collection of data. If one decides on a time interval $(0, t']$, there is always the possibility of too few occurrences. If one fixes the number of arrivals at m and n, say, then, in general, for the waiting times $W_m \neq V_n$; and the joint distribution of

$$\{W_1, ..., W_m; V_1, ..., V_n\}$$

is not tractable.

Theorem 4.7. *Under* H_0: $\mathscr{L}_1 = \mathscr{L}_2$, *(in fact if* $\mu_1 = a\mu_2$, $a > 0$*) the conditional distributions of* $R(W_1), ..., R(W_m), R(V_1), ..., R(V_n)$ *given*

$$N_1(t') = m \quad and \quad N_2(t') = n$$

as given by $\quad P[R(W_1) = r_{11}, ..., R(V_n) = r_{2n}] = \binom{m+n}{n}^{-1}$

for all permutations $(r_{11}, ..., r_{2n})$ *of* $(1, ..., m+n)$ *for which* $r_{11} < ... < r_{1m}$ *and* $r_{21} < ... < r_{2n}$ *and* $= 0$ *for other permutations. Moreover, let* $\phi(Z)$, *where* $Z = (W_1, ..., W_m, V_1, ..., V_n)$ *be a test function similar of size* α *or equivalently,* $\quad E[\phi(W_1, ..., V_n) | N_1(t') = m, \quad N_2(t') = n] = \alpha$

for all μ, *that is, for all* G^* *where* G *is monotone on* $(0, t']$. *Then,*

$$\Sigma\phi(\gamma(z)) = \alpha \binom{m+n}{n}$$

w.p. *one where the summation extends over all* $\gamma \in \mathscr{S}^*(m, n)$.

Proof. This follows immediately from Theorems 2.2 and 4.2.

Corollary 4.8. (i) *A statistic* T *is* DF *w.r.t.* $\Omega(H_0)$ *iff there exists a B-Pitman function* v *w.r.t.* $\mathscr{S}^*(m, n)$ *and a function* ψ *such that* $T = \psi[\tilde{R}(v)]$.

(ii) *A test function* ϕ *is similar of size* α *w.r.t.* $\Omega(H_0)$ *iff* $\Sigma\phi(\gamma(z)) = \binom{N}{m}\alpha$ *for a.a.* z *where the summation is over* γ *in* $\mathscr{S}^*(m, n)$, *and* $N = m+n$.

(ii) *Theorem* 3.3 *is valid* $z = (w_1, ..., w_m; v_1, ..., v_n)$ *and '$\mathscr{S}_N$' is replaced by '$\mathscr{S}^*(m, n)$'.*

Example 4.4. Observe the process until time t'. Let $m = N_1(t')$, $n = N_2(t')$ and reject for large values of $\sum_1^m h(R(W_i))$, where $h(r)$ is the expectation of the rth order statistic of a random sample of size m from a standard exponential population. This statistic was studied by Savage (1956) and Hájek and Šidák (1967, p. 97).

Example 4.5. Against the simple alternative characterized by the mean functions $\mu_1(t) = at$ and $\mu_2(t) = bt^2$, the MPDF test is based on $R^*(v^*)$, where $v^*(z) = \prod_1^n t_j$, a BNS-Pitman function. This is because the conditional density is

$$L_1(z) = L_1(w_1, \ldots, w_m; v_1, \ldots, v_n) = m!\,n!\,a^m(2b)^n\,(t')^{-m-2n}\left(\prod_1^n w_j\right).$$

Another approach is available through use of the following result which is easily proved.

Lemma 4.9. (i) $P\{N_1(t') = r\,|\,N_1(t') + N_2(t') = k\} = \dbinom{k}{r} p^r q^{k-r}$,

where $p = \mu_1(t')\,[\mu(t') + \mu_2(t')]^{-1}$ *and* $q = 1 - p$.

(ii) *For cross-sectional data one has*

$$N_{11}(t'), \ldots, N_{1m}(t')   \text{and}   N_{21}(t'), \ldots, N_{2n}(t').$$

The $P\left\{\sum_1^m N_{1i}(t') = r \,\middle|\, \sum_1^m N_{1i}(t') + \sum_1^n N_{2j}(t) = k\right\} = \dbinom{k}{r} p^r q^{k-r}$,

where $q = 1 - p$ and $p = m\mu_1(t')\,[m\mu_1(t') + n\mu_2(t')]^{-1}$.

The tests based on this theorem are then condition tests of the hypothesis $p = m/(m+n)$. However, there are power difficulties since $\mu_1(t') = \mu_2(t')$ does not mean that the two mean functions are equal for all t.

Theorem 4.10. *Let $T = \sum_{i=1}^m N_{1i}(t')$. (a) Then under H_0, T has conditionally a binomial distribution, with parameters k and $\dfrac{m}{m+n}$. (b) Further, any size α test based only on T, will have power α for all pairs $\mathscr{L}_1$ and $\mathscr{L}_2$ of Poisson Processes such that $\mu_1(t') = \mu_2(t')$. (c) The power function of such a test is parameterized solely by*

$$p = \frac{m\mu_1(t')}{[m\mu_1(t') + n\mu_2(t')]}.$$

Proof. This follows immediately from Lemma 4.9.

Example 4.6. Let $m = 10$, $n = 20$, $t' = 15$ and $H_1 \colon \mu_1(\cdot) < \mu_2(\cdot)$. Under H_0, then $p = \frac{1}{3}$; and under $H_1 \colon p < \frac{1}{3}$. Using the Neyman–Pearson Lemma the MP test has a critical region of the form

$$\left\{\sum_1^m N_{1i}(t') < C(\alpha, k, \tfrac{1}{3})\right\}.$$

4.4. Test of the model

The authors would wish to test whether a process is or is not a homogeneous or non-homogeneous Poisson process. It is easily seen that no meaningful test within the current framework is possible for this hypothesis. The authors did, however, find tractable the following fairly 'large' model: H_0: There exists $a > 0$ such that $\mu(\cdot) = a\mu_0(\cdot)$; that is,

$$\Omega(H_0') = \{\mathscr{L}_\mu \colon \mu(\cdot) = a\mu_0(\cdot), a > 0\}.$$

The procedure is to observe a fixed number n of waiting times. In view of Theorem 4.1, the waiting times $W_1, \ldots, W_n$ should be transformed to

$$\tau_1' = \mu_0(W_1), \quad \tau_2' = \mu_0(W_2) - \mu_0(W_1), \quad \ldots, \quad \tau_n' = \mu_0(W_n) - \mu_0(W_{n-1}).$$

For these τ's, the null hypothesis family is then $\Omega(H_0) = \{J_a^{(n)} \colon a > 0\}$, where $J_a(x) = 1 - e^{-ax}, x \geqslant 0$.

In order to characterize the family of similar tests, one needs the following well-known lemma which is related to Theorem 5.1.

Lemma 4.11. (i) For the family $\Omega(H_0) = \{J_a^{(n)} \colon a > 0\}, \mu_0(W_n) = \sum_1^n \tau_j'$ is a complete sufficient statistic;

(ii) Its H_0-distribution has density

$$f_n(w) = e^{-aw}(aw)^{n-1}[(n-1)!]^{-1}a, \quad w \geqslant 0.$$

Then, from the Neyman-Structure Theorem (for example Lehmann, (1959), p. 130 ff.) one has

Theorem 4.12. (A) The following conditions are equivalent:

(i) $E(\phi | \mu_0(W_n) = t') = \alpha$ for a.a. t'.

(ii) $\int \phi\left(\tau_1, \ldots, \tau_{n-1}, w - \sum_1^{n-1} \tau_i\right) \prod_1^{n-1} d\tau_i$

$$= \alpha \frac{w^{n-1}}{(n-1)!} \quad \text{for a.a. } w > 0 \left\{\sum_1^{n-1} \tau_i < w, \tau_i > 0\right\}.$$

(iii) $\int \phi(w_1, w_2 - w_1, \ldots, w_{n-1} - w_{n-2}, w - w_{n-1})$

$$+ \prod_1^n dw_i\{0 < w_1 < \ldots < w_{n-1} < w\} = \alpha w^{n-1}[(n-1)!]^{-1}.$$

(iv) $\int \phi(\tau_1, \ldots, \tau_n) a^n e^{-a} \sum_1^n \tau_i \prod_1^n d\tau_i = \alpha \quad \text{for } a > 0.$

(B) T is a statistic DF w.r.t. $\Omega(H_0)$ iff $P\{T \leqslant t\} = P\{T \leqslant t | \mu_0(w_n) = s\}$ for a.a. s, for all t, and for all $\mathscr{L}$ in $\Omega(H_0)$.

The hyperplane sections $\left\{ \sum_1^n \tau_j' = w, \tau_j' > 0 \right\}$ correspond to the orbits in the two-sample case. The MPDF tests are then obtained by applying the Neyman–Pearson Lemma to a.e. hyperplane section.

Theorem 4.13. (i) *The* UMPDF *test function* w.r.t. *the family*

$$\{ \mathscr{L} : L(z) = \exp[ac_1(z) + b(z)] \quad \text{for } a > 0,$$

$b(\cdot)$ *a.e. constant on a.e. hyperplane section and* $z = (\tau_1', \dots, \tau_n')\}$ *is of the form*

$$\psi = \begin{cases} 1 & \text{if } L_1(z) > k(\alpha, n), \\ 0 & \text{otherwise,} \end{cases}$$

where $L_1(\cdot)$ *is a monotone function of* $c_1(\cdot)$.

(ii) *Further, if the μ's are differentiable, then, the* UMPDF *test against the family* $\{ \mathscr{L} : \mu(\cdot) = b\mu_1(\cdot), b > 0 \}$ *is of the form*

$$\psi = \begin{cases} 1 & \text{if } \prod_1^{n-1} \dfrac{\mu_1'(w_i)}{\mu_0'(w_i)} > k(\alpha, n, W_n) \\ 0 & \text{otherwise} \end{cases}$$

in terms of the original data and conditional on W_n.

Example 4.7. If $\mu_0(t) = t$ and $\mu_1(t) = e^t - 1$, then

$$\prod_{i=1}^{n-1} \frac{\mu_1'(w_i)}{\mu_0'(w_1)} = \exp\left\{ \sum_1^{n-1} w_i \right\}.$$

For $n \geqslant 1d$, the cut off point $k(\alpha, n, w)$ may be approximated by the central limit theorem to be

$$\log k(\alpha, n, w) = \frac{(n-1)w}{2} + w \sqrt{\left(\frac{n-1}{12} \Phi^{-1}(1 - \alpha) \right)},$$

where Φ denotes the standard normal distribution function.

One should note that the character of the problem changes significantly if one fixes the cut off time for observation at, say, t'. In this case $n = N(t')$ is a Poisson random variable with mean $a\mu_0(t')$. Further, conditional on $n = N(t')$, $W_1, \dots, W_n$ are distributed as order statistics from a population with distribution function $G_0(t) = \mu_0(t)[\mu_0(t')]^{-1}$, $0 \leqslant t \leqslant t'$. (See Theorem 4.2.)

Because of completeness difficulties the family of all DF tests for this goodness-of-fit hypothesis is not known. However, one can state the following theorem.

Theorem 4.14. *Let ψ_n be a measurable real-valued function on the unit hypercube in n-dimensions for $n = 1, 2, \ldots$; and*

$$T(\mu_0; t'; w_1, \ldots, w_n) = \psi_n[G_0(w_1), \ldots, G_0(w_n)]$$

when $n = N(t')$. Then, (i) *T is DF w.r.t.* $\Omega(H_0)$.

(ii) *Let $P\{\psi_n(U_1, \ldots, U_n) \in C_n\} = \alpha$ when $U_1, \ldots, U_n$ constitute a random sample from a standard uniform distribution; and*

$$\phi(\mu_0; t'; w_1, \ldots, w_n) = \begin{cases} 1 & \text{if } T(\mu_0; t'; w_1, \ldots, w_n) \in C_n, \\ 0 & \text{otherwise}. \end{cases}$$

Then, the size of the test is

$$E(\phi) = \sum_{n=0}^{\infty} \alpha_n e^{-\mu_0(t')}[\mu_0(t')]^n (n!)^{-1},$$

where α_0 is the probability of rejecting H_0 if no observations occur in the allotted time.

Proof. This follows from the goodness-of-fit theorems of §3, 4 and 8.

Example 4.8. (*a*) Observe the process for t' time units and let $n = N(t')$, $W_{n+1} = t'$, $W_0 = 0$ and

$$S = \sum_{i=1}^{n+1} \left| \frac{W_i}{t'} - \frac{W_{i-1}}{t'} - \frac{1}{n+1} \right| = (t')^{-1} \sum_{i=1}^{n+1} \left| W_i - W_{i-1} - \frac{t'}{n+1} \right|.$$

Then S is DF and has the distribution of Sherman's (1950) spacing statistic.

(*b*) $\mu_0(W_r)[\mu_0(W_n) - \mu_0(W_r)]^{-1}$ is DF and has an F-distribution with $2r$ and $(2n - 2r)$ degrees of freedom.

5 PROCESSES WITH MONOTONE MARGINALS

For these processes only the one-sample and two-sample hypotheses are treated. Testing the model, that is, that the process being observed is a process with monotone marginals, has no meaning in the context of this paper.

For the two hypotheses to be tested the procedure will be to choose a finite set of times $t_1, \ldots, t_N$. Then the variables $X_{t_1}, \ldots, X_{t_N}$ have a (joint) multivariate distribution, and will be treated by known multivariate techniques. The methods to be derived apply for any choice of times. However, in practice one must be concerned with the maximum time, that is, the length of time intervals, periodicities, that is, whether or not the times should be equally spaced, etc.

5.1. Goodness-of-fit

To test the hypothesis $H_0 \colon \mathcal{L} = \mathcal{L}_0$, one observes $X_{t_1}, \ldots, X_{t_n}$, whose joint distribution is J_0. Using the Rosenblatt (1952) transformation one forms

$$U_1 = J_0(X_{t_1}), \quad U_2 = J_0(X_{t_2} | X_{t_1}), \quad \ldots, \quad U_n = J_0(X_{t_n} | X_{t_1}, \ldots, X_{t_{n-1}})$$

or $\qquad [U_1, \ldots, U_n] = \delta_{J_0}(X_{t_1}, \ldots, X_{t_n}).$

For $U_1, \ldots, U_n$ the hypothesis becomes $H_0' \colon F = U_0$, the standard uniform distribution. $\Omega(H_0)$, then consists of the single element $U_0^{(n)}$.

Since this is a simple hypothesis one can get a variety of tests, say, using the Neyman–Pearson Lemma with different alternatives. Since the major properties of $U_1, \ldots, U_n$ are uniformity, randomness, symmetry, and independence, one seeks tests specifically designed for these hypotheses. Some such tests will be given in the examples. However, the structure theorem giving a large class of test statistics is simply,

Theorem 5.1. (i) *Each statistic*

$$T(\mathcal{L}_0; X_{t_1}, \ldots, X_{t_n}) = \psi[\delta_{J_0}(X_{r_1}, \ldots, X_{t_n})]$$

is DF w.r.t. *the goodness-of-fit hypothesis.*

(ii) $\qquad P\{T(\mathcal{L}_0, X_{t_1}, \ldots, X_{t_n}) \leqslant s | \mathcal{L}_0\} = Q_\psi(s),$

where $\qquad Q_\psi(s) = P\{\psi[U_1, \ldots, U_n] \leqslant s | U_0^{(n)}\} \quad$ *for all* s.

(iii) *If a process has law $\mathcal{L}_1$ yielding the same joint distribution J_0 as $\mathcal{L}_0$ for $X_{t_1}, \ldots, X_{t_n}$, then the power of any test based on T is equal to its size.*

Proof. This follows from Bell and Smith (1969a).

Example 5.1. (a) $\pi = -2 \sum_{i=1}^{n} \ln U_i$ and $\pi' = -2 \sum_{i=1}^{n} \ln(1 - U_i)$, statistics due to Fisher and Pearson, are DF statistics directed toward deviations from uniformity

(b) $T' = \sum_{i=1}^{n} iR(U_i)$ is a DF statistic directed at trend alternatives to randomness.

To test independence and/or symmetry in the current context, it seems most feasible to employ cross-sectional data:

$$X_{t_{11}}, \ldots, X_{t_{1k}}; \quad X_{t_{21}}, \ldots, X_{t_{2k}}, \ldots, X_{t_{n1}}, \ldots, X_{t_{nk}}.$$

For applying the Rosenblatt transformation to each k-tuple it is not at all necessary that the k sets of t's be identical. One obtains $U_0^{(nk)}$ as the joint distribution of

$$[U_{11}, ..., U_{kn}] = [\delta_{J_1}(X_{t_{11}}, ..., X_{t_{1k}}), ..., \delta_{J_k}(X_{t_{n1}}, ..., X_{t_{nk}})],$$

where J_i is the joint distribution of $(X_{t_{i1}}, ..., X_{t_{ik}})$ when $\mathscr{L} = \mathscr{L}_0$. Also there may be practical reasons for not choosing the n t-sets identical. Such a case might occur if the process has periodicities.

Example 5.2. Let ϕ_1 be a test function satisfying

$$\Sigma\phi(\gamma(u_{11}u_{21}, ..., u_{n1}; ...; u_{1k}, ..., u_{nk})) = (k!)^n \alpha$$

for a.a. u's where the sum is over all γ in $\underset{1}{\overset{n}{\text{X}}} \mathscr{S}_k$. Let ϕ_2 be a test function satisfying

$$\Sigma\phi_2(\gamma(u_{11}, u_{12}, ..., u_{1k}; ...; u_{n1}, ..., u_{nk})) = (n!)\,(k!)^n \alpha$$

for a.a. u's where the sum is over all γ in $\mathscr{S}_k \text{Wr} \mathscr{S}_n$, the wreath product. Then by the results of Bell and Smith (1969a,b) concerning tests of symmetry and tests of independence, ϕ_1 and ϕ_2 are similar test functions.

In closing this subsection, one should consider the following interesting example which indicates one of the possible difficulties that can arise on employing the tests in the preceding example.

Example 5.3. Let $H_0: \mathscr{L} = \mathscr{L}_0$, where $\mathscr{L}_0$ is the law of a non-degenerate Gaussian process with $\mu(t) \equiv 0$, and covariance function $\Gamma_0(s, t)$. Let $H_1: \mathscr{L} = \mathscr{L}_1$, where $\mathscr{L}_1$ is the law of a non-degenerate Gaussian process with the same covariance function but a different mean function, say, $\mu(t) \neq 0$ for all t. Now fix the times $t_1, ..., t_n$, and apply the Rosenblatt transformation $\delta_{J_0}(X_{t_1}, ..., X_{t_n}) = [U_1, ..., U_n]$. One finds

$$U_1 = \Phi(X_t \sigma_1^{-1})\,U_2 = \Phi[X_{t_2} - \rho\sigma_2\sigma_1^{-1}X_{t_1})\,\sigma_2^{-1}(1-\rho^2)^{-\frac{1}{2}}], \quad \text{etc.},$$

where Φ is the standard normal cumulative. It turns out that the U's are mutually independent both under H_0 and H_1. Hence, any test of independence based on δ_{J_0} will have power α.

5.2. Two-sample case

One employs cross-sectional data here and it is imperative that the sets of times be identical. The data, then, for $H_0: \mathscr{L}_1 = \mathscr{L}_2$, is

$$X_{t_1}^{(1)}, \quad ..., \quad X_{t_k}^{(1)}; \quad ...; \quad X_{t_1}^{(m)}, \quad ..., \quad X_{t_k}^{(m)};$$

and

$$Y_{t_1}^{(1)}, \quad ..., \quad Y_{t_k}^{(1)}; \quad ...; \quad Y_{t_1}^{(n)}, \quad ..., \quad Y_{t_k}^{(n)}.$$

It is somewhat more convenient to write the data in vector form:

$$\tilde{Z} = [\mathbf{X}^{(1)}, \quad \ldots, \quad \mathbf{X}^{(m)}; \quad \mathbf{Y}^{(1)}, \quad \ldots, \quad \mathbf{Y}^{(n)}],$$

where $\qquad \mathbf{X}^{(i)} = (X_{t_1}^{(i)}, \quad \ldots, \quad X_{t_k}^{(i)}), \quad$ etc.

The multivariate two-sample theory now applies and one has (as in § 2),

Theorem 5.2. (*For the given choices of t's.*)

(i) T *is DF w.r.t.* $\Omega(H_0)$ *iff there exists a B-Pitman function v and a function ψ such that $T \equiv \psi[\tilde{R}(v)]$.*

(ii) *A test function ϕ is similar of size α w.r.t.* $\Omega(H_0)$ *iff for a.a.* $\tilde{z}$, $\Sigma\phi(\gamma(\tilde{z})) = N!\,\alpha$ *where the sum is over γ in $\mathscr{S}_N$, and $N = m+n$.*

(iii) *For a simple alternative $\mathscr{L}_1$, with likelihood function $L_1(\tilde{z})$, the MPDF test is based on $\tilde{R}(v_1(\tilde{z}))$, where v_1 is a B-Pitman function whose ordering on a.e. orbit is consistent with that of L_1.*

(iv) *Against the family*

$$\{\mathscr{L}_1 \colon L_1(\tilde{z}) = \exp\,[a(\theta)\,v_1(\tilde{z}) + b(\tilde{z},\theta)],\, a(\theta) > 0\}$$

the test of (iii) is UMPDF if $b(\cdot,\cdot)$ is invariant under $\mathscr{S}_N$ for all θ; and is LMPDF if $b = b_1 + b_2$, where b, is invariant under $\mathscr{S}_N$ and $b_2(\tilde{z},\theta) = o(a(\theta))$, $a(\cdot)$ is continuous and $a(\theta_0) = 0$.

Example 5.4. Let

$$\|\mathbf{X}^{(i)}\|^2 = \sum_{j=1}^{k} [X_{tj}^{(i)}]^2 \quad \text{and} \quad T = \sum_{i=1}^{m} R(\|\mathbf{X}^{(i)}\|).$$

T is DF w.r.t. $\Omega(H_0)$, but one might have some doubts as to the power of a test based on it since each k-tuple is replaced by a single number.

In this vein one might note that in some application, for example, radioastronomy, the individual observations are integrated over time intervals and the resulting data is treated by the usual techniques.

Theorem 5.3. *Let the function $a(\cdot)$ and the process be sufficiently regular that for fixed (predetermined) t'*

(i) $W_i = \int_0^{t'} X_0^{(i)}\,da(t) \;$ *and* $\; V_i = \int_0^{t'} Y_t^{(i)}\,da(t)$ *exist for a.a. sample function (under H_0); and*

(ii) *the common distribution of W_i and V_i (under H_0) is continuous. Then $\phi(w_1, \ldots, w_m, v_1, \ldots, v_n)$ is similar of size α iff $\Sigma\Phi(\gamma(z)) = N!\,\alpha$ for a.e. $z = (w_1, \ldots, v_n)$ where the sum is over all γ in $\mathscr{S}_N$.*

Example 5.5. Let

$$a(t) = k^{-1}\sum_{1}^{k} \in (t - t_j), \quad \text{that is,} \quad W_i = \frac{1}{k_j}\Sigma X_{tj}^{(i)}, \quad \text{etc.}$$

Then $$T = \sup_t |F_m(t) - G_n(t)|,$$

where F_m is the empirical distribution of the W's, etc., is DF w.r.t. $\Omega(H_0)$, and have a two-sample Kolmogorov–Smirnov distribution.

In neither of the two examples is the essentially multivariate character of the data explicitly used. This is an unfortunate characteristic of current tests in multivariate analysis and in the analysis of stochastic processes, although one can concoct tests which do have a more multivariate flavor.

Example 5.6. If the alternatives of interest differ in location, one might consider a multivariate analogue of the original Pitman (1937) statistic. Let $\hat{v}(\tilde{z}) = \|\bar{x} - \bar{y}\|$. Then $\hat{v}$ is a *BNS*-Pitman function and $\hat{R}(\hat{v})$ is DF w.r.t. $\Omega(H_0)$.

6 EXCHANGEABLE PROCESSES

We will now consider inference from processes which are exchangeable in the sense of the following definition.

Definition 6.1. A discrete parameter process $\{X_n, n \geq 1\}$ is called exchangeable iff for each fixed $k \geq 1$, the joint distribution of $X_{i_1}, \dots, X_{i_k}$ is the same for all choices of distinct positive integers $i_1, \dots, i_k$.

Example 6.1. (*a*) Any process of iid random variables is exchangeable. Hence exchangeable processes may be thought of as generalizations of iid processes.

(*b*) Let Z be an exponentially distributed random variable; and conditionally, given $Z = z > 0$, let $X_1, X_2, \dots$ be iid $\mathcal{N}(0, z)$. Then $[X_n : n \geq 1]$ is an exchangeable process; indeed, if $i_1, \dots, i_k$ are distinct positive integers

$$P[X_{i_1} \leq x_1, \dots, X_{i_k} \leq x_k] = \int_0^\infty a \prod_{i=1}^k \Phi(x_i z^{-\frac{1}{2}}) e^{-az} dz.$$

6.1. Structure theorems

The property of conditional independence exhibited by Example 6.1 (*b*) is typical of exchangeable processes. In fact, we have the following theorem of DeFinetii (Loeve (1960), p. 365).

Theorem 6.1. *Let* $\{X_n; n \geq 1\}$ *be an exchangeable process on a probability space* $(\mathcal{X}, \mathcal{A}, P)$; *then there is a sub-$\sigma$-algebra* $\mathcal{A}_0 \subset \mathcal{A}$ *such that* $X_1, X_2, \dots$ *are conditionally iid given* $\mathcal{A}_0$.

A related theorem due to Hewitt and Savage (1955) will also be useful. Let R^∞ denote the countably infinite product of lines with elements denoted $\mathbf{X} = (X_1, X_2, \ldots)$ and let $\mathscr{B}_\infty$ be the Borel sets of R^∞, then any discrete parameter stochastic process $[X_n; n \geqslant 1]$ induces a measure P^* on $\mathscr{B}_\infty$ by

$$P^*(B) = P[(X_1, X_2, \ldots) \in B], \ B \in \mathscr{B}_\infty.$$

Obviously the process is exchangeable iff P^* is *symmetric*—that is, invariant under permutations of any finite number of co-ordinates of R^∞. The following is a special case of a theorem of Hewitt and Savage (1955).

Theorem 6.2. Let $\mathscr{F}$ denote the class of all univariate distribution functions and let $\mathscr{F}$ be endowed with its weak topology. For each $F \in \mathscr{F}$ let π_F denote the product measure induced on $\mathscr{B}_\infty$ by F. If P^* is any symmetric probability measure on $\mathscr{B}_\infty$, then there is a uniquely determined probability measure λ on the Borel sets of $\mathscr{F}$ for which

$$P^*(B) = \int_{\mathscr{F}} \pi_F(B)\,\lambda(dF). \tag{*}$$

In particular, * holds if P^* is the law of an exchangeable process. In this case we will write $P^* = P_\lambda$ and call λ the *mixing measure* of the process.

One consequence of Theorems 6.1 and 6.2 is that a historical sample $X_1, \ldots, X_k$ from an exchangeable process may be thought as follows: a random distribution function $F \in \mathscr{F}$ is generated according to λ; then k independent random variables with common distribution function F are observed. For cross-sectional samples, of course, the above procedure is simply repeated. Another consequence is that, subject to certain continuity conditions, exchangeable processes and iid processes have the same similar sets. More precisely, we have the following corollary.

Corollary 6.3. Let Λ_0 be set of $\lambda \in \Lambda$ for which $P_\lambda([\mathbf{x}: x_1 = x_2]) = 0$. If $B \in \mathscr{B}_\infty$ then $P_\lambda(B) = \alpha$ for all $\lambda \in \Lambda_0$ iff $\pi_F(B) = \alpha$ for all continuous $F \in \mathscr{F}$.

Proof. If F is continuous then π_F is a P_λ with $\lambda \in \Lambda_0$, so the 'only if' assertion is trivial. To prove the 'if' assertion, it will suffice to say that for $\lambda \in \Lambda_0$ a.e. F is continuous (λ). This follows from

$$0 = P_\lambda([\mathbf{x}: x_1 = x_2])$$
$$= \int_{\mathscr{F}} \left[\int_{-\infty}^{\infty} (F(x+0) - F(x-0))\,dF(x) \right] \lambda(dF)$$

which implies that for a.e. $F(\lambda)$, $F(x+0) = F(x-0)$ for a.e. $x(F)$.

Remark. Corollary 6.3 is false if Λ_0 is replaced by the class Λ_0^* of all $\lambda \in \Lambda$ for which $P_\lambda([x: x_1 = C]) = 0$ for all $C \in R^1$. Indeed, Λ_1 includes P_λ for which $P_\lambda([\mathbf{x}: x_1 = x_2 = \ldots]) = 1$.

For the remainder of this section we will consider only processes whose mixing measure λ belongs to Λ_0.

Corollary 6.3 then has the following application. Let $X_1, \ldots, X_k$ be a historical sample from such an exchangeable process; then the distribution of any rank or permutation statistic will be the same as if $X_1, \ldots, X_k$ were a random sample from a continuous distribution function.

Example 6.2. To test the null hypothesis that $X_1, \ldots, X_k$ is a historical sample from an exchangeable process with unspecified mixing measure $\lambda \in \Lambda_0$ against alternatives of an increasing trend, an appropriate test would be to reject for large value of the Spearman Rank Correlation Coefficient

$$\rho = \frac{R}{k^3 - k} \sum_{i=1}^{k} i \left(R_i - \frac{k+1}{2} \right),$$

where R_i denotes the rank of $X_i, i = 1, \ldots, k$. The cut-off points may be read from existing tables.

6.2. H_0: λ is degenerate

This is equivalent to the randomness hypothesis, that is, that the X's are iid. However, any exchangeable sample function contains conditionally iid variables. Hence, one needs cross-section data $X_{11}, \ldots, X_{1k}, \ldots;$ $X_{n1}, \ldots, X_{nk}$. Then following the treatment of randomness in §§ 2 and 3 one has

Theorem 6.4. (i) *A statistic T is DF w.r.t.* $\Omega(H_0)$ *iff there exists a B-Pitman function v (w.r.t. $\mathscr{S}_{nk}$) and a function ψ such that* $T \equiv \psi[\tilde{R}(v)]$.

(ii) *A test function ϕ is similar of size α iff*

$$\Sigma \phi(\gamma(z)) = (nk)! \, \alpha \text{ for a.a. } z = (x_1, \ldots, x_n),$$

where the sum is over all γ in $\mathscr{S}_{nk}$.

(iii) *The* MPDF *test against a simple alternative with likelihood function L_1 is of the form*

$$\psi(z) = \begin{cases} 1 & \text{if } \tilde{R}(v_1(z)) > C(\alpha, nk) \\ \delta & = \\ 0 & < \end{cases}$$

where v_1 is a B-Pitman function which on a.e. orbit gives an ordering consistent with that of L_1.

Proof. This follows from the randomness results of § 2.

As in §§ 2 and 3 the usual results concerning UMPDF tests, LMPDF tests, and Koopman–Pitman-type exponential families hold.

Example 6.3. Consider a normal regression alternative, that is that $X_{i1}, \ldots, X_{ik}$ are independently normal with constant σ^2 and

$$\mu_i = \mu + ib \quad \text{for } 1 \leqslant i \leqslant n.$$

(a) Then the MPDF test is based on the likelihood function

$$L_1(z) = (2\pi\sigma^2)^{-\frac{1}{2}(nk)}\exp\left(-\tfrac{1}{2}\sigma^{-2}\right)\sum_{i,j}(x_{ij}-\mu_i)^2.$$

Two BNS-Pitman functions giving the same ordering on the orbits are

$$V_1^* = \sum_1^n \mu_i\bar{x}_i. \quad\text{and}\quad v_1^* = \sum_1^n i\bar{x}_i.$$

(b) A DF statistic simpler to compute and directed against the same type of alternatives is $T = \sum_i iR(\bar{x}_i.)$.

6.3. Tests of the model; $\mathscr{L}$ is exchangeable

Let $X_{i1},...,X_{ik}, i = 1,...,n$ denote n independent observations on an initial segment of a stochastic process $X_1, X_2, ...$. We wish to test the hypothesis that the process is exchangeable with an unspecified mixing measure $\lambda \in \Lambda_0$ the null hypothesis class is therefore

$$\Omega(H_0) = [(P_{\lambda,k}^*)^{(n)}: \lambda\in\Lambda_0],$$

where $P_{\lambda,k}^*$ denotes the projection of P_λ^* onto the Borel sets of R_k under the null hypothesis, the vectors $(X_{i1},...,X_{ik})$ will behave like random samples from different (random) distribution functions $F_i, i = 1,...,n$.

From the symmetry development of § 2 and Corollary 6.3 one has

Theorem 6.5. (i) T *is DF w.r.t.* $\Omega(H_0)$ iff *there exists a B-Pitman function* v(w.r.t. $\mathscr{S}_k\,\mathrm{Wr}\,\mathscr{S}_n$) *and a function* ψ *such that* $T \equiv \psi[R(v)]$.

(ii) ϕ *is a test function similar of size* α *w.r.t.* $\Omega(H_0)$ iff

$$\Sigma\phi(\gamma(x_{11},...,x_{1k};...,x_{n1},...,x_{nk})) = (k!)^n n!\,\alpha$$

for a.a. $(x_{11},...,x_{nk})$, *where the sum is over* γ *in* $\mathscr{S}_k\,\mathrm{Wr}\,\mathscr{S}_n$.

Of course, the usual results concerning UMPDF and LMPDF tests hold here.

Example 6.4. Consider a Gaussian alternative with mean function $\mu(\cdot)$ and covariance function $\Gamma(\cdot,\cdot)$. Let $\Sigma = (\Gamma(s,t))$, where $1 \leqslant s, t \leqslant k$ and $\mu = [\mu(1),...,\mu(k)]$; and the cross-section data be as usual. The likelihood function is

$$L_1(z) = K(n,k)\exp\left\{(-\tfrac{1}{2})\sum_{i=1}^n(\mathbf{x}^{(i)}-\mu)'\Sigma^{-1}(\mathbf{x}^{(i)}-\mu)\right\},$$

where $x^{(i)} = [x_{i1},...,x_{in}]$. A BNS-Pitman function giving the same ordering on the orbits is

$$v_1(z) = \sum_{i=1}^n(\mathbf{x}^{(i)})'\Sigma^{-1}\mu.$$

The MPDF test is based on $R^*(v_1(z))$. By the usual Koopman–Pitman-type derivation one arrives at an exponential family against which the same test is UMPDF. Then by adding an asymptotically negligible term to certain members of the family one finds a new family w.r.t. which the test is LMPDF.

One notes that if $\mu(t) \equiv 0$, then the resulting Gaussian process is exchangeable; L_1 is constant over a.e. orbit; $v_1 \equiv 0$; and, of course, tests based on these have power equal to size.

6.4. Goodness-of-fit tests

We now consider drawing inferences about the mixing measure λ of an exchangeable process and, as in the previous sub-section, we will consider cross-sectional data, $X_{i1}, ..., X_{ik}, i = 1, ..., n$. We will be particularly interested in hypotheses of the form

$$H_0: \lambda = \lambda_0, \quad H_1: \lambda \in \Lambda_1,$$

where λ_0 is specified and Λ_1 consists of all λ having a specified support. It turns out that tests of H_1 may be used as an initial step in making tests of H_0. To fix ideas let us consider simple mixtures of the form

$$\lambda(\{F\}) = p \quad \text{and} \quad \lambda(\{G\}) = q = 1 - p,$$

where F and G are specified, continuous distribution functions but p is not specified. We wish to test H_1 where Λ_1 consists of all λ of the above form where $0 \leqslant p \leqslant 1$. One method of testing H_1 is the following. For each $i = 1, ..., n$, let $\hat{F}_i$ denote the sample distribution function of $X_{i1}, ..., X_{ik}$, and let $\quad \phi(H, J) = \sup_x |H(x) - J(x)|$

for distribution functions H and J. For suitable choices of γ, $0 < \gamma < 1$, it is then possible to find positive numbers ϵ, depending only on k and γ, for which

$$P[\rho(F, \hat{F}_i) < \epsilon | F] = \gamma = P[\rho(G, \hat{F}_i) < \epsilon | G] \quad \text{for } i = 1, ..., k.$$

We will assume that k has been chosen so large that $\epsilon \leqslant \frac{1}{2}[\rho(F, G)]$. For each $i = 1, ..., k$ let A_i be the event that $\rho(F, \hat{F}_i) < \epsilon$, let B_i be the event that $\rho(G, \hat{F}_i) < \epsilon$, and let C_i be the compliment of $A_i \cup B_i$. We propose the following test: reject H_1 if and only if too many of the C_i occur. More precisely, for each $r = 1, ..., n$ we may define a test ϕ_r by agreeing to reject H_1 if and only if at least r of the C_i occur. The exact levels of the ϕ_r tests are difficult to obtain, but it is easy to find upper bounds. Indeed, we have

$$P(C_i) = pP_F(C_i) + qP_G(C_i)$$
$$\leqslant pP_F(A'_i) + qP_G(B'_i) = \gamma$$

so that for $r > n\gamma$, the size of ϕ_r is at most

$$\sum_{j=r}^{n} \binom{n}{j} \gamma^i (1-\gamma)^{n-j}.$$

The choices of γ and r in a particular situation would presumably be based on power considerations and subject to the constraint $\rho(F, G) \geqslant 2\epsilon$. This question awaits further investigation.

Letting λ_0 be defined by (*) where $p = p_0$ as specified, we will now consider the hypothesis H_0. Our approach is basically the following: we will first test H_1 and then, if H_1 is accepted, we will test that $p = p_0$. More precisely, we will reject H_0 if either (1) at least one of the C_i occurs, or (2) none of the C_i occur but $|\hat{p}-p_0| \geqslant Z > 0$ where $\hat{p}$ denotes the proportion of the samples on which A_i occurs and Z is a design constant. To deduce an upper bound for the size of our test, the following lemma will be useful.

Lemma 6.6. *Let Z denote the number of correct classifications and W the number of samples from F_1; if H_1 is true, then W and Z are independent.*

Proof. We have $W = W_1 + \ldots + W_n$ and $Z = Z_1 + \ldots + Z_n$, where $W_i = 1$ or 0 accordingly as the ith sample came from F or G and $Z_i = 1$ or 0 accordingly as the ith sample was correctly classified. Moreover, if H_1 is true and $\lambda \in \Lambda_1$, then

$$P_\lambda[Z_i = 1 | W_i = 1] = P_F(A_i)$$

$$= \gamma$$

$$= P_G(B_i) = P_\lambda[Z_i = 1 | W_i = 0]$$

so that W_i and Z_i are independent. Since W_i and Z_i are obviously independent for $i \neq j$, the lemma follows.

To obtain an upper bound on the size of our test, we may now proceed as follows. First observe that the rejection of H_1 implies that $Z < n$; therefore

$$P_{\lambda_0}[\text{reject } H_0] \leqslant P_{\lambda_0}[Z < n] + P_{\lambda_0}[|\hat{p}-p_0| \geqslant z, Z = n]$$

$$= 1 - \gamma^n + \gamma^n P_{\lambda_0}\left[\left|\frac{W}{n} - p_0\right| \geqslant z\right].$$

Since W has the Binomial distribution, the latter term may be read from Binomial Tables if n is small or approximated by the Central Limit Theorem if n is large. In either case the parameters γ and z may be chosen to make the right side of the above inequality small.

The techniques of the previous paragraph extend without difficulty to yield tests of the hypotheses

$$\lambda(\{F_i\}) = p_i \quad (i = 1, \ldots, m),$$

where $F_1, \ldots, F_m$ are specified continuous distribution functions and $p_1, \ldots, p_m$ are either specified or unspecified non-negative real numbers for which $p_1 + \ldots + p_m = 1$.

The *general approach* may also be applied to continuous λ's as we will now illustrate.

Example 6.5. Suppose, for example, that we are interested in testing the hypothesis that μ_λ of the form

$$\mu_\lambda(B) = \int_0^\infty \pi_\theta(B) \, \lambda^*(d\theta) \quad (B \in \mathcal{B}_\infty), \qquad (**)$$

where for each $\theta > 0, \pi_\theta$ is the measure on $\mathcal{B}_\infty$ making the co-ordinate functions independent *exponential random variables* with mean θ, and λ is a specified or unspecified measure on the Borel sets of $(0, \infty)$. To test the hypothesis H_1 that λ is of the form $(**)$ with λ^* unspecified on the basis of cross-sectional data $X_{i1}, \ldots, X_{ik}, i = 1, \ldots, n$ we simply test that each of the samples $X_{i1}, \ldots, X_{ik}$ are random samples from an exponential with unspecified mean and reject H_1 if we find that any (or too many) of them are not. If k is large this may be done by using the results of Darling (1955), and if k is small by using some special property of the exponential, such as the fact that

$$\left(\frac{X_{i1} + \ldots + X_{im}}{m} \right) \left(\frac{X_{im+1} + \ldots + X_{ik}}{k - m} \right)^{-1}$$

has an F distribution if H_1 is true. If we accept H_1, we may then test H_0 that $\lambda^* = \lambda_0^*$ (specified) by testing that the common distribution function of

$$\theta_i = \frac{X_{i1} + \ldots + X_{ik}}{k}$$

is

$$G(x; \lambda_0^*) = \int_0^\infty F_k(kx/\theta) \, \lambda_0^*(d\theta),$$

where F_k is the Gamma distribution function

$$F_k(x) = \int_0^\infty \frac{y^{k-1} e^{-y}}{\Gamma(k)} \, dy.$$

As in the previous section, the latter hypothesis may be tested with a variety of standard techniques.

7 SPHERICALLY EXCHANGEABLE (S-E) PROCESSES

These processes represent generalization of iid Gaussian processes, that is, discrete parameter Gaussian processes with mean function $\mu(k) \equiv 0$ and covariance function $\Gamma(r, k) = 0$ if $r \neq k$; $= \sigma^2$ if $r = k$. The formal definition is

Definition 7.1. A stochastic process $\{X_n: n = 1, 2, ...\}$ is called S–E (spherically exchangeable) if there exists a function ψ on R_1^* such that for each finite set $(i_1, ..., i_k)$ of natural numbers, the joint characteristic function of $X_{i_1}, ..., X_{i_k}$ satisfies

$$\tilde{\psi}(t_1, ..., t_k) = \psi\left(\sum_1^k t_i^2\right).$$

Example 7.1. (*a*) If $\{X_n: n = 1, 2, ...\}$ are iid $\mathcal{N}(0, \sigma^2)$, then the process is S–E.

(*b*) Let η be a random variable with distribution λ in $\Omega_2^*(R_1^+)$; and $\{Y_n: n = 1, 2, ...\}$ be a process such that conditionally given $\eta = \sigma_0$, the Y's are iid $\mathcal{N}(0, \sigma_0^2)$. Then, the (unconditional) process is S–E. In fact,

$$\tilde{\psi}(t_1, ..., t_s) = \int_0^\infty \exp\left(-\tfrac{1}{2}\right) \sigma^{-2} \sum_1^k t_i^2 \lambda(d\sigma).$$

Immediately from the definition one has the following theorem. (See Kelker (1968); Smith (1969).)

Theorem 7.1. (i) *Each* S–E *process is exchangeable.*

(ii) *Let* $\{X_n: n = 1, 2, ...\}$ *be an exchangeable process. Then, the process is spherically exchangeable iff for each* k *the polar co-ordinates* $\left(\sum_1^k X_i^2, \theta^{(k)}\right)$ *of the initial* k-*segment are independent with* $\theta^{(k)}$ *being uniformly distributed over the surface of the unit hypersphere in* k-*dimensions.*

(iii) *Further,* $\{X_n: n = 1, 2, ...\}$ *is spherically exchangeable iff for each* k *the joint distribution of* $(X_1, ..., X_k)$ *is identical with that of* $C_k(X_1, ..., X_k)$ *for each* k-*dimensional orthogonal transformation* C_k.

Kelker (1968) and Smith (1969) have treated finite families of spherically exchangeable random variables. Many of the properties developed by Kelker and Smith carry over to stochastic processes; but the basic structure theorem below is not valid for finite collections.

7.1. Structure theorems

In this sub-section we show that S–E processes are mixtures of zero mean iid Gaussian processes in the same sense that exchangeable processes are mixtures of iid processes. Lemma 7.2 follows directly from a theorem of Schönberg (1938) and the well known fact that a completely monotone function is the La Place transform of a measure on $(0, \infty)$.

Lemma 7.2. If $\{X_n : n \geqslant 1\}$ is a S–E process, then the function ψ of Definition 7.1 is completely monotone and may therefore be written in the form

$$\psi(t) = \int_0^\infty e^{-ts} \lambda_1(ds),$$

where λ_1 is a probability measure on $[0, \infty)$.

Theorem 7.3. Let $\{X_n : n \geqslant 1\}$ be a S–E process and for each $\sigma \geqslant 0$ let $\pi'_\sigma = \pi_{F_\sigma}$, where F_σ is the $\mathcal{N}(0, \sigma^2)$ distribution function. There is a uniquely determined measure λ on the Borel sets of $[0, \infty)$ for which

$$(*) \quad P(B) = \int_0^\infty \pi'_\sigma(B) \, \lambda(d\sigma)$$

for all $B \in \mathscr{B}_\infty$. λ will be called the mixing measure of the process and the right side of (*) denoted by $P_\lambda(B)$.

Proof. The change of variable $\mathcal{S} = \tfrac{1}{2}\sigma^2$ shows that ψ may be written

$$\psi(t) = \int_0^\infty e^{-\frac{1}{2}t\sigma^2} \lambda(dt),$$

where λ is a measure on the Borel sets of $[0, \infty)$. Since mixtures of characteristic functions correspond uniquely to mixtures of distributions, it now follows that (*) holds wherever B is a cylinder set. The extension from cylinder sets to all Borel sets is a straightforward application of the monotone class theorem. Finally, the uniqueness follows from the uniqueness of λ in Theorem 6.2.

λ can be viewed as a parameter for the family of spherically interchangeable distributions. A sufficient statistic for this parameter is given by

Lemma 7.4. For an initial segment $(X_1, \ldots, X_k)$ of a spherically interchangeable process,

(i) $S_k^2 = k^{-1} \sum_1^k X_i^2$ *is a sufficient statistic; and*

(ii) *the conditional distribution of $(X_1, \ldots, X_k)$ given $S_k^2 = s^2$ is a uniform distribution over the hypersphere of radius ks^2 centered at the origin.*

Proof. This follows from the symmetry development in § 2. It will be more convenient to work with the distribution of S_k^2 rather than with λ in some of the problems below.

Theorem 7.5. *Consider a spherically exchangeable process with mixing measure* λ; S_k^2 *for some fixed* $k \geqslant 1$; *and* $G_k(\cdot; \lambda)$, *the distribution function of* S_k^2.

(i) $G_k(y; \lambda) = \int_0^\infty P\{S_k^2 \leqslant y | \mathcal{N}(0, \sigma^2)\} \lambda(d\sigma)$ for all y; and

(ii) $G_k(\cdot; \lambda') = G_k(\cdot; \lambda)$ iff $\lambda = \lambda'$.

Proof. (i) is clear. (ii) may be established by differentiating the identity $G_k(\cdot; \lambda) = G_k(\cdot; \lambda')$ and applying unicity theorems for LaPlace transforms.

Corollary 7.6. (i) *Let* $\{X_n : n = 1, 2, ...\}$ *and* $\{X'_n : n = 1, 2, ...\}$ *be* spherically exchangeable *processes*. X_1 *and* X'_1 *have identical distributions, iff the two processes have identical laws.*

(ii) *There exist two* exchangeable *processes which do not have the property in* (i).

Proof. (i) Let λ and λ' denote the mixing measure. If X_1 and X'_1 have the same distribution, then $G_1(\cdot; \lambda) = G_1(\cdot; \lambda')$, so that $\lambda = \lambda'$ and therefore $P_\lambda = P'_\lambda$. (ii) If $X_1, X_2, \ldots$ is any S–E process with non-degenerate λ, then we may construct an iid process $X'_1, X'_2, \ldots$ having the common unconditional distribution of X_1. The two processes have different laws, but X_1 and X'_1 have the same distribution.

One is now in a position to treat a variety of related hypotheses concerning λ's, $G_k(\cdot; \lambda)$'s and $\mathcal{L}$'s.

7.2. $H_0: \mathcal{L} = \mathcal{L}_0$ Goodness-of-fit

For this case one needs cross-sectional data, in order to make meaningful inference. The data for n first segments each of length k is

$$X_{11}, ..., X_{1k}; \quad ...; \quad X_{n1}, ..., X_{nk},$$

which is reduced to $S_1^2, ..., S_n^2$, where $S_i^2 = k^{-1} \sum_1^k X_{ij}^2$.

The new goodness-of-fit hypothesis and a family of tests for it are now given by

Theorem 7.7. (i) The $S_1^2, ..., S_n^2$ are distributed as a random sample of size n from $G(\cdot; \lambda_0)$.

(ii) Each statistic of the form

$$T = \Psi[G(S_1^2; \lambda_0), ..., G(S_n^2; \lambda_0)] i$$

is DF w.r.t. the goodness-of-fit hypothesis.

Proof. (i) follows from the definitions and Theorem 7.5. (ii) This is the result of theorems in §§ 2 and 8 on statistics of structure (d'_n).

There are, of course, other DF statistics, see Birnbaum and Rubin (1954); but these are the only ones known by the authors to have been used.

Example 7.2. Let
$$T = n^{-1} \sum_{i=1}^{n} J^{-1}(G(S_i^2; \lambda_0)),$$

where J is a strictly monotone distribution on R_1. Then T is DF, and the null distribution is that of the mean of a random sample of size n from a population with distribution J.

7.3. Two-sample case, $H_0: \mathscr{L}_1 = \mathscr{L}_2$

Again one needs cross-sectional data $X_{11}, ..., X_{mk}; Y_{11}, ..., Y_{nk}$ which is reduced to $S_1^2, ..., S_m^2$, where $S_i^2 = k^{-1} \sum_1^k X_{ij}^2$, and $V_1^2, ..., V_n^2$, where $V_1^2 = k^{-1} \sum_1^k Y_{ij}^2$. The hypothesis is, then, $H_0: G_k(\cdot; \lambda_1) = G_k(\cdot; \lambda_2)$. The only difference between this and the general two-sample (univariate) problem of § 2, is that the S^2's and V^2's only assume values on R_1^+.

Theorem 7.8. Theorems 3.2 and 3.3 hold verbatim here.

Example 7.3. Let F_m and J_n be the empirical distribution functions of the S^2's and V^2's, respectively
$$[Z_1, ..., Z_N] = [S_1^2, ..., S_m^2; V_1^2, ..., V_n^2]; \quad 0 = q_0 < q_1 < \cdots < q_{s+1} = 1,$$
be such that each Nq_i is a positive integer;
$$n_{2i} = n[J_n(Z(Nq_i)) - J_n(Z(Nq_{i-1}))];$$
and
$$n_{1i} = m[F_m(X(Nq_i)) - F_m(Z(Nq_{i-1}))];$$
$$\hat{T} = 2 - 2(nm)^{-\frac{1}{2}} \sum_1^{s+1} (n_{1i} n_{2i})^{\frac{1}{2}}$$
$$= \sum_{i=1}^{s+1} [(n_{1i} m^{-1})^{\frac{1}{2}} - (n_{2i} n^{-1})^{\frac{1}{2}}]^2.$$

Then, $\hat{T}$ (developed by Matusita, 1955) is DF w.r.t. $\Omega(H_0)$. Further, $\hat{T}$ is asymptotically chi-square with s degrees of freedom.

7.4 $H_0: \mathscr{L}$ is degenerate

This means there exists some unknown $\sigma > 0$ such that the variables are iid $\mathscr{N}(0, \sigma^2)$. Again one needs cross-sectional data
$$X_{11}, ..., X_{1k}; \quad ...; \quad X_{n1}, ..., X_{nk},$$
which are reduced to $S_1^2, ..., S_n^2$. Under H_0, the S_i^2 are iid estimates of σ^2. Three major families of tests are given in Theorem 7.9.

Theorem 7.9. (i) *In terms of the* $S_1^2, \ldots, S_n^2$, *if* v *is a B-Pitman function w.r.t. the quotient group* $S_{nk}/(S_k \mathrm{Wr} S_n)$, *then each* $\psi[\tilde{R}(v)]$ *is a statistic DF w.r.t.* $\Omega(H_0)$;

(ii) $\{\mathcal{N}(0, \sigma^2): \sigma > 0\} = \{Fg: g$ *is in the scale group,* $G_s\}$ *for each zero mean normal distribution* F. *Consequently each statistic* T *invariant under* $G_s^{(n)}$, *is DF w.r.t.* $\Omega(H_0)$; *and*

(iii) $\sum\limits_{i,j} X_{ij}^2 = k \sum\limits_{1}^{n} S_i^2$ *is a complete, sufficient statistic. Hence, each similar test function of size* α *must satisfy* $E\left(\phi \mid k\sum\limits_{1}^{n} S_n^2 = t\right) = \alpha$ *for a.a.t.* > 0.

Proof. (i) follows from the theory of BNS-Pitman functions in § 2,

(ii) follows from Bell (1964 a),

(iii) is the Neyman Structure Theorem (Lehmann 1959, p. 130).

Example 7.4. Let $Z = (X_{11}, \ldots, X_{nk})$; T_1 be the Kruskal Wallis statistic

$$\frac{12}{N(N+1)} \sum_{i=1}^{k} n_i^{-1} \left[\sum_{j=1}^{n_i} R(X_{ij}) \right]^2 - 3(N+1), \quad \text{where } N = nk;$$

$$T_2 = \left[\sum_{i=1}^{r} S_i^2 \right] \left[\sum_{j=r+1}^{n} S_j^2 \right]^{-1};$$

and
$$\phi_1 = \begin{cases} 1 & \text{if } \max\,(S_i^2) > b\sum\limits_{1}^{n} S_i^2, \\ 0 & \text{otherwise.} \end{cases}$$

Then T_1, T_2, and ϕ_1 satisfy the respective conditions of Theorem 7.8.

7.5. Test of the model H_0: $\mathcal{L}$ is spherically exchangeable

Here one transforms the cross-sectional data $X_{11}, \ldots, X_{nk}$ to polar co-ordinates $(kS_1^2, \theta_1), \ldots, (kS_n^2, \theta_n) = (r_1^2, \theta_1, \ldots, r_n^2, \theta_n)$,

where θ_i is a point on the surface of the unit hypersphere in k-dimensions. Since the segments are finite, the development follows closely that of Smith (1969) and Bell and Smith (1969 a).

The first tests are those suggested by Theorem 7.1, that is, the independence of S^2 and θ and the uniformity of θ over the surface of the k-dimensional unit hypersphere.

Example 7.5. Divide the unit hypersphere into r congruent regions $A_1, \ldots, A_r$. Let n_i = the number of θ's which fall in A_i.

$R^*(\theta) = r$ if θ in A_r;

$$T_1 = \sum_{=1}^{n} R(S_i^2)\, R^*(\theta_i); \quad T_2 = rn^{-1} \sum_{1}^{r} (n_i - r^{-1}n)^2 = rn^{-1}\left(\sum_{1}^{r} n_i^2 \right) - n.$$

Then under H_0: (a) $(n_1, \ldots, n_r)$ has a multinomial distribution with parameters $(n; r^{-1}, \ldots, r^{-1})$, (b) T_1 and T_2 have constant distributions, that is, are DF w.r.t. $\Omega(H_0)$, (c) T_2 is asymptotically chi-square with $r - 1$ degrees of freedom.

In some sense, T_1 is a statistic for a tests of independence and T_2, for uniformity. Further, one notes that if R^* were the statistically equivalent block rank (for example, Anderson, 1966; or Ahmad, 1969), then T_1 and T_2 would be independent.

To obtain the family of all tests, and the 'good' tests one needs the approach of Smith (1969) and Bell and Smith (1969b). This means forming orbits and generalized B-Pitman functions; using these functions to select points from the orbits and form statistics. The 'good' tests are then based on generalized B-Pitman functions related to the likelihood function of an alternative.

Definition 7.2. (i) A real-valued function $v^*(z)$ is called a *generalized B-Pitman function* if $P\{v^*(z) = v^*(\gamma(z))|J\} = 0$ for all $J \in \Omega(H_0)$ and for all non-identity elements γ of $\mathscr{S}'' = Q_p \operatorname{Wr} \mathscr{S}_n$, and v^* is continuous w.r.t. γ.

(ii) A real-valued function $\hat{v}^*(z)$ is a generalized BNS-Pitman function if $P\{\hat{v}^*(z) = \hat{v}^*(\gamma(z))|J\} = 1$ if non-identity γ is in $\mathscr{S}_n$; $= 0$, otherwise for J in $\Omega(H_0)$; and $\hat{v}^*$ is continuous w.r.t. γ.

These functions must be continuous since the orbits are continuous. Similarly, the basic statistics induced by these functions will have a continuous distribution.

Definition 7.3. (i) The generalized B-Pitman statistic induced by B-Pitman function v^* is defined by

$$R^*(v^*(z)) = P\{v^*(\gamma(z)) \leqslant v^*(z)|\mathscr{S}''(z)\}.$$

(ii) The generalized BNS-Pitman statistic $\hat{R}^*(\hat{v}^*(z))$ is defined by

$$\hat{R}^*(\hat{v}^*(z)) = P\{\hat{v}^*(\gamma(z)) \leqslant \hat{v}^*(z)|\mathscr{S}''(z)\}.$$

Bell and Smith (1969b), then prove

Theorem 7.10. (i) *Under H_0, $R^*(v^*(z))$ has a standard uniform distribution.*

(ii) *A statistic T is DF w.r.t. $\Omega(H_0)$ iff there exists a function ψ and a generalized B-Pitman function v^* such that $T = \psi[R^*(x^*)]$.*

(iii) *Given any univariate distribution G_0, there exists a DF statistic with H_0 distribution, G_0.*

(iv) *The MPDF size α test against a simple alternative with likelihood L_1 is given by $\phi(z) = 1$ if $R^*(v_1^*) \geqslant C(\alpha, n)$; $= 0$ if $R^*(v^*) < C(\alpha, n)$, where v_1^* is a generalized B-Pitman function satisfying $L_1(z) < L_1(\gamma(z))$ iff $v_1^*(z) < v_1^*(\gamma(z))$ for all γ in $\mathscr{S}''$ and for a.a. z.*

Theorem 7.11. *Let* $(X_{11}, \ldots, X_{1k}; \ldots, X_{n1}, \ldots, X_{nk})$ *be* n *initial segments from cross-sectional data; and*

$$[kS_1^2, \boldsymbol{\theta}_1; \ldots; kS_n^2, \boldsymbol{\theta}_k] = [r_1^2, \boldsymbol{\theta}_1, \ldots, r_n^2, \boldsymbol{\theta}_1]$$

be their polar co-ordinate representation.

(i) $[r(1), \ldots, r(n)]$ is a complete sufficient statistic for $\Omega(H_0)$;

(ii) *a test function* ϕ *is similar of size* α *iff* $E\{\phi \mid \mathscr{G}^*(\mathbf{z})\} = \alpha$ *for a.a.* $\mathbf{z} = (r_1^2, \boldsymbol{\theta}_1, \ldots, r_n^2, \boldsymbol{\theta}_n)$, *where* $\mathscr{G}^*$ *is the wreath product* $Q_k \operatorname{Wr} S_n$ *of the orthogonal group in* k-*dimensions with* S_n;

(iii) T *is DF w.r.t.* $\Omega(H_0)$ *iff there exists a function* ψ *and a generalised B-Pitman function* v^* *such that* $T \equiv \psi[R^*(v^*)]$.

(iv) *The most powerful size* α *similar test function against the simple alternative* L_1 (*or* λ_1) *is of the form*

$$\phi(\mathbf{z}) = \begin{cases} 1 & \text{if } R^*(v^*(\mathbf{z})) > C(\alpha, n), \\ 0 & \text{otherwise,} \end{cases}$$

where v^* *is a generalized B-Pitman function with an ordering on a.e. orbit consistent with that of* L_1, *the alternative likelihood function.*

Proof. Smith (1969) and Bell and Smith (1969 b).

One notes that LMPDF and UMPDF tests can be obtained by the usual methods.

Example 7.6. Consider the case of initial segments of length $k = 2$; and a non S–E Gaussian process with mean function $\mu(t) \equiv 0$, and covariance function

$$\Gamma(r, k) = \begin{cases} 1 & \text{if } r = k \quad \text{odd}, \\ (1+a)^{-1} & \text{if } r = k \quad \text{even}, \\ 0 & \text{otherwise.} \end{cases}$$

The joint density of $(X_{11}, X_{12}, \ldots, X_{n1}, X_{n2})$ in terms of the polar co-ordinates $(r_1^2, \boldsymbol{\theta}_1, \ldots, r_n^2, \boldsymbol{\theta}_n)$ is

$$L_1(z) = (1+a)^{\frac{1}{2}n} (2\pi)^{-n} \left(\prod_1^n r_i \right) \exp\left[-\frac{1}{2} \sum_{i=1}^n (r_i^2 + r_i^2 a \sin^2 \boldsymbol{\theta}_i) \right].$$

(One notes that each $\boldsymbol{\theta}$ here is a point on the unit circle and can be viewed as an angle in 2-space.)

$\hat{v}_1(\mathbf{z}) = \operatorname{sgn}(-a) \sum_1^n r_i^2 \sin^2 \theta_i$ is a generalized *BNS*-Pitman function whose ordering on each orbit is equivalent to that of L_1. The MPDF size α test is then $\psi(\mathbf{z}) = 1$ or 0 according as $\hat{R}(\hat{v}(z)) > C(\alpha, n)$ or not. An approximation to this test is obtained as follows. Let

$$A_n = \left\{ z \colon \left[\sum_1^n r_i^2 (\sin^2 \theta_i - \tfrac{1}{2}) \right] \left[\sum_1^m r_i^4 \right]^{-\frac{1}{2}} 8^{\frac{1}{2}} > \Phi^{-1}(1-\alpha) \right\}.$$

Then $\lim_{n \to \infty} P\{A_n \triangle [\psi = 1]\} = 0$ for all laws in $\Omega(H_0)$.

8 INVARIANCE AND THE STRONGLY DISTRIBUTION-FREE (SDF) PROPERTY

Birnbaum and Rubin (1954) extended a power property of the normals to goodness-of-fit tests. Several authors have extended these ideas to other hypotheses, e.g. Berk and Bickel (1968) and Bell (1964a). An adaptation to stochastic processes is given below.

8.1. One-sample or goodness-of-fit

Definition 8.1. A one-sample statistic T is SDF w.r.t. Ω'.

(*a*) for SII processes if there exists a family of univariate distribution function $\{Q(\cdot; \cdot)\}$ such that

$$P\{T(\mathscr{L}_0; X_t, X_{2t}, ..., X_{nt}) \leqslant s | \mathscr{L}_1\} = Q(F_1 F_0^{-1}; s)$$

for all s and for all F_0 and F_1 in $\Omega' = \Omega_2^*(R_1)$, the family of distributions of $X_{2t} - X_t$, where F_i corresponds to $\mathscr{L}_i$ for $i = 0, 1$;

(*b*) for *Poisson Processes* if there exists a family of univariate distribution functions $\{Q(\cdot, \cdot)\}$ such that

$$P\{T(\mathscr{L}_0; W_1, ..., W_n) \leqslant s | \mathscr{L}_1\} = Q(\mu_1 \mu_0^{-1}; s)$$

for all s and for all mean functions μ_0 and μ_1 in $\Omega' = \Omega_2^*(R_1^+)$ which μ_i corresponds to $\mathscr{L}_i$ for $i = 0, 1$;

(*c*) for S–E *Processes* if there exists a family $\{Q(\cdot; \cdot)\}$ of univariate distribution functions such that

$$P\{T(\mathscr{L}_0; S_1^2, ..., S_n^2) \leqslant s | \mathscr{L}\} = Q(G_1 G_0^{-1}; s)$$

for all s and for all G_0 and G_1 in $\Omega' = \Omega_2^*(R_1^+)$ the family of distributions of $S_i^2 = \sum_1^k X_{ij}$, where G_i corresponds to $\mathscr{L}_i$ for $i = 0, 1$; and

(*d*) *for Processes with Monotone Marginals* if there exists a family $\{Q(\cdot; \cdot)\}$ of univariate distribution functions such that

$$P\{T(\mathscr{L}_0; X_{t_1}, ..., X_{t_n}) \leqslant s | \mathscr{L}_1\} = Q(\delta_{J_1} \delta_{J_0}^{-1}; s)$$

for all s and for all J_0 and J_1 in $\Omega' = \Omega_2^*(R_n)$, the family of joint distributions of $(X_{t_1}, ..., X_{t_n})$, where J_0 and J_1 corresponds to $\mathscr{L}_0$ and $\mathscr{L}_1$, respectively.

250   C. B. BELL, M. WOODROOFE & T. V. AVADHANI

For the structure theorem utilizing this power property, one needs the definition below.

Definition 8.2. A one-sample statistic T is of structure (d'_n) as it can be written in the form

(a) $\psi[F_0(Y_1), \ldots, F_0(Y_n)]$, where $Y_i = X_{it} - X_{(i-1)t}$ and F_0 is the H_0-distribution of Y_i for SII processes;

(b) $\psi[\mu_0(W_1), \ldots, \mu_0(W_n)]$, where μ_0 is the mean function under H_0, for Poisson processes;

(c) $\psi[G_0(S_1^2), \ldots, G_0(S_n^2)]$, where G_0 is the H_0-distribution of S_i^2, for S–E processes, and

(d) $\psi[\delta_{J_0}(X_{t_1}, \ldots, X_{t_n})]$, where J_0 is the H_0-distribution of $(X_{t_1}, \ldots, X_{t_n})$ for processes with monotone marginals.

Paralleling the proofs in Bell (1960), and Bell and Smith (1969), and making use of the completeness result in Theorem 4.3, one can demonstrate

Theorem 8.1. (i) *For SII, S–E, and Monotone Marginal Processes*

(a) *if T has structure (d'_n), then T is SDF;*

(b) *if T is SDF, there exists a statistic $\hat{T}$ of structure (d'_n) such that*

$$P\{T(\mathscr{L}_0; Z) \leqslant s | \mathscr{L}_1\} = P\{\hat{T}(\mathscr{L}_0; Z) \leqslant s | \mathscr{L}_1\}$$

for all $\mathscr{L}_0$ and $\mathscr{L}_1$ in Ω' (of Definition 9.1);

(c) *if T is a function of the data only through the order statistics (of $\{Y_i\}$ or $\{S_i^2\}$ or $\{U_1, \ldots, U_n\} = \delta_{J_0}(X_1, \ldots, X_n)$, respectively), then T is SDF iff T is equivalent to a statistic of structure (d'_n).*

(ii) *For Poisson Processes* (i) (a) *and* (i) (b) *above hold with no modification, and* (c) *T is SDF iff T is equivalent to a statistic of structure (d'_n).*

One notes that (c) follows from (b) if the family of distributions involved is symmetrically complete, that is, if the order statistic is a complete statistic. For Poisson Processes, (c) follows from (b) and Theorem 4.3. Further this latter result is valid for the SII, S–E and monotone marginal processes if the class of alternatives is sufficiently 'rich'.

In applying these results one should note that the

$$\{F_0(X_t - X_0), F_0(X_{2t} - X_t), \ldots, F_0(X_{nt} - X_{(n-1)t})\},$$

the $\{G_0(S_1^2), \ldots, G_0(S_n^2)\}$ and $\{U_1, \ldots, U_n\} = \{\delta_{J_0}(X_{t_1}, \ldots, X_{t_n})\}$

all behave as random samples from a uniform distribution. On the other hand one would have to modify the

$$\{\mu_0(W_1), \ldots, \mu_0(W_n)\} \quad \text{to} \quad \{J_1^*(W_1'), J_1^*(W_2' - W_1'), \ldots, J_1^*(W_n' - W_{n-1}')\},$$

where $W_i' = \mu_0(W_i)$ and $J_1^*(x) = 1 - e^{-x} \, (x \geqslant 0)$, in order to attain uniformity.

Example 8.1. The following statistics have structure (d_n'); are SDF; and have the indicated null distributions.

(a) $\dfrac{1}{\sqrt{(n)}} \sum\limits_{i=1}^{n} \Phi^{-1}(F_0(Y_i))$ is DF w.r.t. SII Processes and has null distribution $\mathcal{N}(0, 1)$.

(b) $2\mu_0(w_n)$ is DF w.r.t. *Poisson processes* and has as null distribution a chi-square distribution with $2n$ degrees of freedom.

(c) $\sup\limits_{t} |G_n(t) - G_0(t)|$ is DF w.r.t. *S–E Processes* and a 1-sample Kolmogorov distribution as its H_0-distribution.

(d) $\sum\limits_{i=1}^{n+1} \left| U(i) - U(i-1) - \dfrac{1}{n+1} \right|$ is DF w.r.t. *Processes with Monotone Marginals* has under H_0 has the distribution of Sherman's (1950) spacing statistic.

One notes in closing this subsection that for S–E processes, the data in the above treatment must be cross-section data, while this is not true for the other processes. However, if in the other cases the data is cross-section data, only a slight modification lends it to the treatment above.

8.2. Two-sample case: $H_0\colon L_1 = L_2$

When the Birnbaum–Rubin (1954) idea is extended to similar sets, e.g. Bell and Doksum (1967), it is more convenient to express this idea in terms of group invariance

Definition 8.3. A statistic T is SDF w.r.t. $\Omega(H_0 \cup H_1)$ and group $\mathscr{G}'$ if

(i) $\Omega(H_0) = \{Jg \colon g \in \mathscr{G}'\}$ for all J in $\Omega(H_0)$; and

(ii) $P\{T \leqslant t | J\} = P\{T \leqslant t | Jg\}$ for all J in $\Omega(H_0 \cup H_1)$ and for all g in $\mathscr{G}'$ and for all t.

The first condition says that $\Omega(H_0)$ must be one of the equivalence classes of $\Omega(H_0 \cup H_1)$ under $\mathscr{G}'$. The second condition says that the power function is constant over each equivalence class. The groups and families to be considered are given in Table 8.1.

The object here is to show that the SDF statistics and the invariant statistics are closely related.

The invariant statistics are, of course, the rank statistics:

$$\psi[R(X_1'), \ldots, R(Y_n')] \quad \text{or} \quad \psi[R(W_1), \ldots, R(W_n')] \quad \text{or} \quad \psi[R(S_1^2), \ldots, R(V_n^2)].$$

Following Bell and Doksum (1967), one can show

Table 8.1. *Groups and families for the two-sample hypothesis*
$$(N = m+n)$$

	SII	Poisson	S–E
Data	$X_t, X_{2t}, ..., X_{mt}$	$W_1, ..., W_m$	$X_{11}, ..., X_{1k}; ...;$ $X_{m1}, ..., X_{mk}$
	$Y_t, T_{2t}, ..., Y_{nt}$	$W'_1, ..., W'_n$	$Y_{11}, ..., Y_{1k}; ...;$ $Y_{n1}, ..., Y_{nk}$
Transformed data	$X'_1, ..., X'_m$ $Y'_1, ..., Y'_n$	Same as above	$S^2_1, ..., S^2_m$ $V^2_1, ..., V^2_n$
	where		where $S^2_i = \sum_{j=1}^{k} X^2_{ij}$, etc.
	$X_i = X_{it} - X_{(i-1)t}$, etc.		
$\mathscr{G}'$	$[\mathscr{G}^*_2(R_1)]\,(N)$	$[\mathscr{G}^*_2(R^+_1)]\,(N)$	$[\mathscr{G}^*_2(R^+_1)]\,(N)$
$\Omega(H_0)$	$[\Omega^*_2(R_1)]\,(N)$	$\{G^*: G \in \Omega^*_2(R_1)\}$ (see Sections 2 and 4)	$[\Omega^*_2(R^+_1)]\,(N)$
$\Omega(H_0 \cup H_1)$	$[\Omega^*_2(R_1)]\,(m)$ $\times [\Omega^*_2(R_1)]\,(n)$	$\{F^*_1 \cdot F^*_2 \cdot F_i \in \Omega^*_2(R_1)\}$	$[\Omega^*_2(R^+_1)]\,(m)$ $\times [\Omega^*(R^1_+)]\,(n)$

Theorem 8.2. *For the* SII, *Poisson, and* S–E *processes* (*with the transformed data*)

(i) *if* T *is a rank statistic, then* T *is* SDF;

(ii) *if* $\tilde{T}$ *is* SDF, *there exists a rank statistic* T *satisfying*

$$P\{\tilde{T} \leqslant t|J\} = P\{T \leqslant t|J\} \quad \text{for all } t \text{ and for all } J \text{ in } \Omega(H_0 \cup H_1);$$

(iii) *If* $\tilde{T}$ *is a function of the transformed data only through the order statistics of each sample, then* $\tilde{T}$ *is* SDF *iff* $\tilde{T}$ *is equivalent to a rank statistic.* (See, for example, Bell (1964b).)

[One notes that the waiting times for a Poisson process satisfy the desired order statistic property automatically.]

Example 8.2. (a) For SII processes, $\sum_{i=1}^{m} R(X'_i)$ is SDF, DF, invariant, and has a Mann–Whitney–Wilcoxon distribution under H_0.

(b) For Poisson processes, $\sup_x |F_m(x) - G_n(x)|$ is SDF, DF, invariant, and has a two-sample Komogorov–Smirnov distribution under H_0, where

$$F_m(x) = \frac{1}{m}\sum_{j=1}^{m} \epsilon(x - W_j) \quad \text{and} \quad G_n(x) = \frac{1}{n}\sum_{j=1}^{n} \epsilon(x - V_j).$$

(c) For S–E processes let $T = $ the number of runs in the combined sample $\{S^2_1, ..., S^2_m; V^2_1, ..., V^2_n\}$. T' is SDF, DF and invariant and has a two-sample runs distribution as its H_0-distribution.

In closing this discussion of the two-sample problem one should mention two results communicated to the authors by P. Smith in reference to Bell and Smith (1969 a, b) and Smith (1969).

Let $\overline{\mathscr{G}} = \{\delta_F^{-1} \delta_G : F, G \in \Omega_2^*(R_n)\}$, where δ_F is as usual the Rosenblatt transformation.

Theorem 8.3. (i) $\overline{\mathscr{G}}$ *is a group of 1-1 transformations of* R_n *onto* R_n;

(ii) $\Omega_2^*(R_n) = \Omega(J, \overline{\mathscr{G}})$ *for all* $J \in \Omega_2^*(R_n)$, *that is,* $\overline{\mathscr{G}}$ *generates* $\Omega_2^*(R_n)$; *and*

(iii) *for cross-sectional data*

$$\{X_{1,t_1}, ..., X_{1,t_n}; X_{2,t_1}, ..., X_{2,t_n}; ...; X_{k,t_1}, ..., X_{k,t_n}\}$$

the maximal invariant under $\overline{\mathscr{G}}(k)$ *is*

$$[R(X_{1,t_1}), R(X_{2,t_1}), ..., R(X_{k,t_1})].$$

The result means that Table 8.1 and Theorem 8.2 could have been extended to include Processes with Monotone Marginals with no mathematical difficulties. However, one sees immediately that the maximal invariant involves only the first observation of each sample function. It seems, then, impractical to use statistics based on the maximal invariant. Further, one concludes that this SDF property, which allows one to parametrize the power function, is not necessarily desirable in all cases.

8.3. Tests of fit for the model

For the three classes of processes below one will be concerned with the groups and families in Table 8.2 one should recall the following special notation

θ = a generic point on the unit hypersphere in R_k,

$S_k(1)$ = the unit hypersphere in R_k,

$\tilde{F}(\cdot | W_n; \mu_0)$ = the joint distribution of the order statistic of a sample of size $n-1$ from a population with distribution,

$$F(t) = \frac{m_0(t)}{m_0(W_n)}, \quad 0 \leqslant t \leqslant W_n. \quad \mathscr{U}(S_k(1)) \text{ the uniform distribution over}$$

the surface of $S_k(1)$.

Definition 8.4. The definition of SDF here is verbatim that in Definition 8.3. [However, the families $\Omega(H_0)$ and $\Omega(H_0 \cup H_1)$ are as in Table 8.2.]

The main theorem here is proved in a manner completely analogous to the proof of Theorem 9.2.

Table 8.2. SDF *statistics for tests of the models*

	SII	Poisson	S–E	For §9.4
Data	$X_t, X_{2t}, ..., X_{nt}$	$W_1, ..., W_n$	$X_{11}, ..., X_{1k}; ...;$ $X_{n1}, ..., X_{nk}$	$X_{11}, ..., X_{1k}; ...$ $X_{n1}, ..., X_{nk}$
Transformed data	$X_1', ..., X_n'$ where $X_i' = X_{it}$ $- X_{(i-1)t}'$	$\tau_1, ..., \tau_n$ where $\tau_i = \mu_0(W_i) - \mu_0(W_{i-1})$	$S_1^2, ..., S_n^2;$ $\theta_1, ..., \theta_n$	Same as above
$\Omega(H_0)$	$[\Omega_2^*(R_1)](n)$	$\Omega''(n)$, where Ω'' is the family of the exponentials	$[\Omega_2^*(R_1^+)](n)$ $\times \overset{n}{\underset{1}{X}} U(S_k(1))$	$\hat\Omega(nk)$, where $\hat\Omega = \{\mathcal{N}(0, \sigma^2):$ $\sigma > 0\}$
$\mathcal{G}'$	$[\mathcal{G}_2^*(R_1)](n)$	$\mathcal{G}_s(n)$	$[\mathcal{G}_2^*(R_1^+)](n)$ $\times \overset{n}{\underset{1}{X}} \{l\}$, where l is the identity of Q_k	$\mathcal{G}_s(nk)$
Maximal invariant	$[R(X_1'), ..., R(X_n')]$	$[\tau_1 \tau_n^{-1}, ..., \tau_{n-1}\tau_n^{-1}]$	$[R(S_1^2), ..., R(S_n^2);$ $\theta_1, ..., \theta_n]$	$[X_1 X_{nk}^{-1}, ...,$ $X_{nk-1}X_{nk}^{-1}]$
$\Omega(H_0 \cup H_1)$	$\Omega_2^*(R_n)$	$\Omega_2^*(R_n)$	$[\Omega_2(R_k)](n)$	$\Omega_k(n)$ where Ω_k is the family of normal variance mixtures on R_k

Theorem 8.4. (i) *For* SII *Processes and the transformed data* T *is*

$$\text{SDF iff } T \equiv \psi[R(X_1'), ..., R(X_n')]$$

for some ψ.

(ii) *The statements in* (i) *above are true if one replaces* 'SII' *by* 'Poisson' ['S–E'] *and* '$\psi[R(X_1'), ..., R(X_n')]$' *by*

$$'\psi[\tau_1 \tau_n^{-1}, ..., \tau_{n-1}\tau_n^{-1}]', \quad ['\psi[R(S_1^2), ..., R(S_n^2); \theta_1, ..., \theta_n]'].$$

Proof. The proof follows from Berk and Bickel (1968).

Example 8.3. (a) $T_1 = \sum_{i=1}^{n} iR(X_i')$ is a statistic SDF for SII processes. Intuitively, its test should have high power against those processes for which successive increments are stochastically increasing or decreasing.

(b) Divide the surface of the unit hypersphere $S_k(1)$ into C_1 congruent regions $A_1, ..., A_{k_1}$. Then, for each i, let $S_{i1}^2, ..., S_{in_i}^2$ be those S^2's whose angles fall in A_i. Let $T_2 =$ the Kruskal–Wallis statistic, that is

$$T_2 = \frac{12}{n(n+1)} \sum_{i=1}^{c_1} n_i^{-1} \left(\sum_{j=1}^{n_i} R(S_{ij}^2) \right)^2 - 3(n+1).$$

This statistic is SDF w.r.t. S–E processes. (One should note that in practice the test function should be adjusted to cope with n_i which are zero.)

(c) Let

$$T_3 = \sum_1^r x_j \left(\sum_{r+1}^n x_j \right)^{-1}.$$

Then T_3 is SDF w.r.t.

$$\{\mathscr{L}_\mu : \mu(\cdot) = a\mu_0(\cdot), a > 0\}.$$

8.4. H_0: λ is degenerate

This hypothesis states that all sample functions correspond to iid zero mean normals with the same variance. This was treated in § 7.4. For the cross-section data $X_{11}, ..., X_{1k}; ...; X_{nk}$, one wishes to test the hypothesis $\sigma_1^2 = ... = \sigma_n^2$. $\mathscr{G}'$, $\Omega(H_0)$, $\Omega(H_0 \cup H_1)$ and the maximal invariant are as given in Table 8.2.

The SDF theorem is then

Theorem 8.5. *T is SDF iff for some* ψ, $T \equiv \psi[X_1 X_{nk}^{-1}, ..., X_{nk-1} X_{nk}^{-1}]$.

Proof. The proof is analogous to that of Theorem 8.4.

Example 8.4. Example 7.4 gives an SDF statistic for this hypothesis.

9 CONCLUDING REMARKS AND OPEN PROBLEMS

The authors wish to acknowledge K. A. Doksum's unpublished class notes in which some of the early papers on DF techniques for stochastic processes are informally discussed. His references are indicated by * in the list of references.

We also wish to thank I. R. Savage and H. Rubin for their comments. The problems they posed as well as others are mentioned below.

(I) *Useful approximations to the exact tests.* Since the permutation tests are cumbersome, one needs some tractable approximations.

(II) *Techniques not based on* iid *data.* The paper contains almost no idea not from known DF test theory. There is a need for methodology which is 'stochastic-process'-oriented.

(III) *Characterization of* DF *statistics for exchangeable processes.* The parameter space here is 'larger' than in the other cases. Perhaps, one should work with a dense subset.

(IV) SDF *statistics for exchangeable processes.* Is there a group which generates the null-hypothesis class?

(V) '*Reasonable' subfamilies of Poisson processes and processes with monotone marginals.* Due to completeness, tests of the model are trivial. For which subfamilies can these tests be made?

(VI) *Characterization of 'goodness-of-fit' tests.* The structure (d'_n) tests are not the only ones. Which others are tractable?

(VII) *Dependence.* Which other families of stochastic processes can be adapted to the iid data methods of this paper?

(VIII) *SEB-rank procedures.* Rank procedures are quite practical in the univariate cases. Is there some practical use of the ideas of Anderson (1966) and Ahmad (1969) in stochastic processes?

REFERENCES

Ahmad, R. (1969). *Some nonparametric texts for multivariate linear hypotheses.* Ph.D. thesis, Case Western Reserve University.

Anderson, T. W. (1966). Some nonparametric multivariate procedures based on statistically equivalent blocks, in *Multivariate Analysis*, ed. by Krishnaiah. New York: Academic Press, pp. 5–28.

Bell, C. B. (1960). On the structure of distribution-free statistics. *Ann. Math. Statist.* **31**, 703–9.

Bell, C. B. (1964a). Some basic theorems of distribution-free statistics. *Ann. Math. Statist.* **35**, 150–6.

Bell, C. B. (1964b). A characterization of multisample distribution-free statistics. *Ann. Math. Statist.* **35**, 735–8.

Bell, C. B., Doksum, K. A. (1965). Some new distribution-free statistics. *Ann. Math. Statist.* **36**, 203–14.

Bell, C. B., Doksum, K. A. (1967). Distribution-free tests of independence. *Ann. Math. Statist.* **37**, 619–28.

Bell, C. B., Donoghue, J. F. (1969). Distribution-free tests of randomness (to appear in *Sankhyā*).

Bell, C. B., Haller, H. S. (1969). Bivariate Symmetry texts: parametric and nonparametric. *Ann. Math. Statist.* **40**, 259–69.

Bell, C. B., Smith, P. J. (1969a). Some nonparametric tests for the multivariate goodness-of-fit, multisample, independence and symmetry problems, in *Multivariate Analysis*, ed. by P. R. Krishnaiah. New. York: Academic Press.

Bell, C. B., Smith, P. J. (1969b). Models of Symmetry: some nonparametric tests (submitted for publication).

Berk, R. H., Bickel, P. J. (1968). On invariance and almost invariance. *Ann. Math. Statist.* **39**, 1573–6.

Birnbaum, A. (1954). Statistical methods for Poisson processes and exponential populations. *J. Am. Statist. Ass.* **49**, 245–66.

Birnbaum, Z. W., Rubin, H. (1954). On distribution free statistics. *Ann. Math Statist.* **25**, 593–8.

*Darling, D. A. (1955). The Cramér–Von Mises statistic in the parametric case. *Ann. Math. Statist.* **26**, 1–20.

DeFinetti, B. (1929). Sulle funzioni a incremento aleatorio. *Rend. Accad. Naz. Lincei*, **10**.

Doksum, K. A. (1963): 'Distribution-free statistics for stochastic processes', unpublished class notes, San Diego State College.

Doob, J. L. (1953). *Stochastic Processes*. New York: Wiley.

*Edington, E. S. (1961). Probability tables for number of runs of signs of first differences in ordered series. *J. Am. Statist. Ass.* **56**, 156–9.

*Goodman, L. A., Grunfeld, Y. (1961). Some nonparametric tests for common events between time series. *J. Am. Statist. Ass.* **56**, 11–27.

Hájek, J. and Šidak, Z. (1967). *Theory of Rank Tests*. New York: Academic Press.

Hewitt, E., Savage, L. J. (1955). Symmetric measures on Cartesian products. *Trans. Amer. Math. Soc.* **80**, 470–501.

Kelker, D. (1968). *Distribution Theory of Spherical Distributions and Some Characterization Theorems*. Michigan State University Technical Report RM–210DK–1.

Kendall, M. G., Stewart, A. (1961). *The Advanced Theory of Statistics*. New York: Hafner Publishing Co.

Lange, O. (1955). Statistical estimation of parameters in Markov processes. *Colloquium Math.* III, 147–60.

Lehmann, E. L. (1959). *Testing Statistical Hypotheses*. New York: John Wiley.

Lehmann, E. L., Stein, C. (1949). On the theory of some nonparametric hypotheses. *Ann. Math. Statist.* **20**, 28–45.

Loeve, M. (1960). *Probability Theory*. New York: Van Nostrand Co.

Matusita, K. (1955). Decision rules, based on the distance, For problems of fit, two-samples, and estimation. *Ann. Math. Statist.* **26**, 631–40.

Mycielski, J. (1956). On the distance between signals in the non-homogeneous Poisson stochastic processes. *Studia Math.* **15**, 300–13.

Parzen, E. (1962). *Stochastic Processes*. San Francisco: Holden–Day, Inc.

Pitman, E. J. G. (1937a). Significance tests which may be applied to samples from any population. *Suppl. J. R. Statist. Soc.* **4**, 119–30.

Pitman, E. J. G. (1937b). Significance tests which may be applied to samples from any population, II. *Suppl. J. R. Statist. Soc.* **4**, 225–32.

Pitman, E. J. G. (1938). Significance tests which may be applied to samples from any population, III. *Biometrika* **29**, 322–35.

Rosenblatt, M. (1952). Remarks on the multivariate transformation. *Ann. Math. Statist.* **23**, 470–72.

Savage, I. R. (1956). Contributions to the theory of rank order statistics—The two-sample case. *Ann. Math. Statist.* **27**, 590–615.

Scheffe, H. (1943). On a measure problem arising in the theory of nonparametric tests. *Ann. Math. Statist.* **14**, 227–33.

Schönberg, I. J. (1938). Metric spaces and completely monotone functions. *Ann. Math. Statist.* **39**, 811–41.

Sherman, B. (1950). A random variable related to the spacing of a sample values. *Ann. Math. Statist.* **21**, 339–61.

Smith, P. J. (1969). *Structure of Nonparametric Tests of Some Multivariate Hypotheses*. Ph.D. thesis, Case Western Reserve University.

*Stewart, A. (1952). The power of two difference sign tests. *J. Am. Statist. Ass.* **47**, 416–24.

*Wallis, W. A., Moore, G. H. (1941). A significance test for time series analysis. *J. Am. Statist. Ass.* **36**, 401–9.

*Wallis, W. A., Moore, G. H. (1943). Time series significance tests based on signs of differences. *J. Am. Statist. Ass.* **38**, 153–64.

ON ASYMPTOTICALLY
OPTIMAL NONPARAMETRIC CRITERIA†

A. A. BOROVKOV AND N. M. SYCHEVA

The present paper is closely connected with article [1] and to some extent it is its continuation.

We shall repeat briefly the statement of the problem.

Let X be a simple sample of size n from a population with a continuous distribution function $F(x)$; let $F_n(x)$ denote the corresponding empirical distribution function. It is required to test the hypothesis $H_0 = \{F(x) = F^0(x)\}$ against the alternative H_1. We shall consider under H_1 the totality of continuous functions $\mathscr{F}$ which are at some fixed distance from $F^0(x)$ in the sense of a metric defined on the space C of continuous functions. The elements $F(x)$ of $\mathscr{F}$ can be regarded as pure strategies of nature. Along with this we shall consider mixed strategies defined by assigning a measure μ on the space $(C, \mathscr{M})$, where $\mathscr{M}$ is a σ-algebra, generated, let us say, by cylindrical sets. In this case the choice of the competing distribution is generated randomly with probability μ, and F can be considered as a realization of a random process. In accordance with the preceding, the measure μ is all concentrated on the set $\mathscr{F} \in \mathscr{M}$ of distribution functions and for some $\epsilon > 0$ satisfies the condition

$$\mu\{F: \sup |F(t) - F^0(t)| \geqslant \epsilon\} = 1. \tag{1}$$

In real problems the distribution μ is usually unknown. In view of this for the class of nonparametric criteria based on generalized Kolmogorov–Smirnov statistics

$$G_g(X) = \sqrt{n} \sup_{M(\theta_1, \theta_2)} \frac{F^0(t) - F_n(t)}{g(F^0(t))}; \quad \bar{G}_g(X) = \sqrt{n} \sup_{M(\theta_1, \theta_2)} \frac{|F^0(t) - F_n(t)|}{g(F^0(t))} \tag{2}$$

in the article [1] we studied the problem of the existence of representatives in these classes which are independent of μ, but for each distribution μ is close to the most powerful criteria (in its class). In the formulas (2) θ_1, θ_2 are fixed; $M(\theta_1, \theta_2) = \{t: \theta_1 \leqslant F^0(t) \leqslant \theta_2\}$, $g(t) \in C_+$, the set of continuous functions strictly positive in the interval $[0, 1]$.

It was shown that if the critical region $G_g(X) > x_g$, for example, for the first statistic of (2) is such that the probability of error

$$P(G_g(x) > x_g | H_0) = \epsilon_n \to 0$$

† Translated by Professor Esther Seiden.

[259]

9-2

as $n \to \infty$ slower than the exponential, then the required 'almost optimal' strategies exist and they are defined by the functions

$$g(t) = \Psi'(t) = \sqrt{(t(t-1))}.$$

In the present paper we consider the case when $\epsilon_n \to 0$ at the rate of exponential. Here the results are somewhat different and their proofs more transparent than in [1]. Without loss of generality we shall again assume that $F^0(t) = t$; $0 \leqslant t \leqslant 1$ and $F(0) = 0$, $F(1) = 1$ for all $F \in \mathscr{F}$. The same arguments as in [1] allow us to restrict our consideration to the first statistic in (2) only, assuming that

$$\mu\{F: \sup(t - F(t)) > \epsilon\} = 1. \tag{3}$$

We shall choose the critical region

$$G_g(X_n) > x_g \tag{4}$$

(i.e. the parameter x_g) so that for a fixed sequence ϵ_n such that

$$\lim\left(-\frac{\ln \epsilon_n}{n}\right) = \alpha > 0 \tag{5}$$

the $P(G_g(X > x_g|H_0) \sim \epsilon_n$ holds.† This amounts to $x_g \sim h_g\sqrt{n}$, where h_g is independent of n, $0 \leqslant h_g \leqslant 1$. Clearly we may assume, without loss of generality, that $h_g = 1$ (it is enough to consider $g_0 = gh_g$).

The problem we are concerned with is, whether one can for a fixed sequence ϵ_n, choose in advance $g \in C_+$ in such a way that the power of the criterion (4) be close to the most powerful test. More precisely, considering the choice of g as the strategy of the statistician we shall assume that the loss function has the form

$$\mathscr{L}(g, F) = \frac{1}{n}\ln P(G_g(X) \leqslant x_g|F)$$

($P(G_g(X) < x_g|F)$ represents the probability of type II error).

The question is whether the statistician has at his disposal an asymtotically minimax pure strategy Ψ such that $\mathscr{L}(\Psi, F)$ for each F is nearly $\inf_g \mathscr{L}(g, F)$. If such a strategy does exist then it will clearly preserve the above mentioned property, also, in the case when nature will use a randomized strategy μ with the loss function

$$\mathscr{L}(g, \mu) = \frac{1}{n}\ln P(G_g(X) \leqslant x_g|\mu).$$

† $a_n \sim b_n$, if $a_n/b_n \to 1$ as $n \to \infty$.

In this case Ψ will also be asymptotically minimax; $\mathscr{L}(\Psi, \mu)$ will be nearly $\inf_{g} \mathscr{L}(g, \mu)$ for all μ. This means that Ψ will be also an asymptotically Bayes strategy.

We shall say that $\Psi \in C_+$ is asymptotically uniformly optimal (a.u.o.) if for arbitrary functions

$$g \in C_+ \quad \text{and} \quad F \in \mathscr{F},$$

$$\mathscr{L}(g, F) \geqslant \mathscr{L}(\Psi, F)(1 + \delta_n),$$

where $\delta_n = \delta(n, g, F) \to 0$ for $n \to \infty$. (Here Ψ, g and F are always independent of n.)†

The problem of the existence of an a.u.o. function will be made clear in the following theorem.

Theorem 1. *Let* $\quad \lim \left(-\dfrac{\ln \epsilon_n}{n} \right) = \alpha > 0, \quad q \in [0, 1],$

$$H(p, q) = p \ln \frac{p}{q} + (1 - p) \ln \frac{1 - p}{1 - q} \quad for \quad p \geqslant 0$$

and $\qquad\qquad H(p, q) = \infty \quad for \quad p < 0.$

Then for any fixed θ_1, θ_2 $(0 \leqslant \theta_1 < \theta_2 \leqslant 1)$, *there exists a unique (up to a constant factor) a.u.o. function* $\Psi_\alpha(t)$ $(h_{\Psi_\alpha} = 1)$ *which is a unique positive solution of the equation*

$$H(t - \Psi_\alpha(t), t) = \alpha \quad (0 \leqslant t \leqslant 1). \tag{6}$$

The statement about the uniqueness of the solution has to be made more precise. It follows from [2] (cf. Lemma 2) that the function

$$\Lambda(t, \Psi) = H(t - \Psi, t)$$

for all $t \in [0, 1]$ is monotonically and indefinitely increasing with the increase of Ψ, $\Lambda(t, 0) = 0$. In addition $\Lambda(t, \Psi)$ is an analytic function of Ψ except at the point $\Psi = t$, where $\Lambda(t, \Psi)$ has a point of discontinuity, $\Lambda(t, t) = \Lambda(t, t - 0) = \ln \dfrac{1}{1 - t}$. Therefore for every t such that

$$\alpha < \ln \frac{1}{1 - t} \quad (t > 1 - e^{-\alpha})$$

the equation (6) has a unique and moreover an analytical solution in the usual sense. For $t \leqslant 1 - e^{-\alpha}$ the level α will consist of the jump of the function $\Lambda(t, \Psi)$ for $\Psi = t$. In this case we shall call the value

† It is possible to state the problem in another way when the functions Ψ could depend on n. In this case there will exist an a.u.o. in a more refined sense when $\delta_n \to 0$ at a more rapid rate.

$\Psi'_\alpha(t) = t$ the solution of (6). At the point $t = 1 - e^{-\alpha}$ the two analytical parts of the function $\Psi'_\alpha(t)$ are glued together with the continuous first derivative. The function $\Psi'_\alpha(t)$ is convex in $[0, 1]$, $\Psi'_\alpha(1) = 0$, $\Psi''_\alpha(1) = \infty$.

We remark that the assertion of Theorem 1 and the very concept of a.u.o. function become meaningful when α and F are such that

$$\sup_{F} \inf_{u} (\Psi'_\alpha(F^{-1}(u)) + u - F^{-1}(u)) < 0.$$

(This condition guaranteeing the exponential rate of decrease of

$$P(G_{\Psi_\alpha}(X) \leqslant x_{\Psi_\alpha}|F)$$

is always satisfied provided that α is sufficiently small since $\Psi'_\alpha(t) \to 0$ as $\alpha \to 0$.)

The function $\Psi'_\alpha(t)$ has also other remarkable properties. If we choose the function g in such a way that $\rho(g, \Psi_\alpha) > 0$ (we assume that $h_g = h_{\Psi_\alpha} = 1$), where

$$\rho(g, f) = \sup_{t \in [\theta_1, \theta_2]} |g(t) - f(t)|$$

then under some conditions on μ of general type, there exist a δ such that the average of the ratio $\mathscr{L}(g, F)/\mathscr{L}(\Psi_\alpha, F)$ according to the measure μ does not exceed $1 - \delta$. The exact formulation of this assertion is contained in the next theorem. Let $l_{\Psi_\alpha, F}, l_{g, F}$ be the lower limits of the convex hull of the sets of points (here again $h_g = h_{\Psi_\alpha} = 1$) of

$$\{(0, 0); (u, \Psi'_\alpha(F^{-1}(u)) + u - F^{-1}(u)), F(\theta_1) \leqslant u \leqslant F(\theta_2); (1, 0)\}$$

and

$$\{(0, 0); (u, g(F^{-1}(u)) + u - F^{-1}(u)), F(\theta_1) \leqslant u \leqslant F(\theta_2); (1, 0)\}$$

respectively. These are clearly convex absolutely continuous functions. We assume

$$\Omega_g = \{F : \rho(l_{\Psi_\alpha, F}, l_{g, F}) > 0\} \in \mathscr{M}.$$

Theorem 2. If (5) is satisfied and $\mu(\Omega_g) > 0$ then there exists $\delta > 0$ such that

$$\int \frac{\mathscr{L}(g, F)}{\mathscr{L}(\Psi_\alpha, F)} d\mu < 1 - \delta$$

for sufficiently large n.

We shall call the $\lim_{n \to \infty} \mathscr{L}(g, F)$, when (5) holds, the α level of the test at the point F. Then, roughly speaking, one can summarize the above formulated assertions in the following manner. If in accordance with the criterion $G_g(X) > x_g$ the true hypothesis F is not close to F^0 then the a.u.o. test based on the function $g = \Psi_\alpha$ will guarantee the uniformly most powerful α level test for all values of F.

Analogous statement holds for the region $\bar{G}_g(x) > \bar{x}_g$. In this case assumption (3) will be replaced by (1) and the a.u.o. function will be replaced by $\phi_\alpha(t) = \max(\Psi'_\alpha(t), \Psi'_\alpha(-t)) + \Psi_\alpha(t)$. But this means that augmenting the class of statistics $\bar{G}_g(X)$, for example, to the class of statistics

$$G_{g_1, g_2}(X) = \sqrt{n}\max\left[\sup_{M(\theta_1,\,\theta_2)} \frac{F^0(t) - F_n(t)}{g_1(F^0(t))}, \sup_{M(\theta_1,\,\theta_2)} \frac{-F^0(t) + F_n(t)}{g_2(F^0(t))}\right],$$

where $g_1(t) = \Psi'_\alpha(t)$, $g_2(t) = \Psi'_\alpha(-t)$, we can obtain better values for the α levels.

Let us return now to the statistic $G_g(X)$. In order to estimate the parameters of the criteria (4) it is necessary to be able to evaluate

$$P(G_{\Psi_\alpha}(X) > x|F^0) \quad \text{and} \quad P(G_{\Psi_\alpha}(X) \leqslant x|F)$$

for all different $F \in \mathscr{F}$. The form of the asymptotic expression

$$P(G_g(X) > x|F^0)$$

for an arbitrary $g \in C_+$ can be obtained from the next theorem. Let

$$\zeta(t) = \sqrt{n}\left(t - \frac{\xi(t)}{n}\right), \quad \text{where } \xi(t), \xi(0) = 0, \quad t \geqslant 0$$

is a 'conditional' continuous from the right Poisson process with the parameter n obtained by fixing the value $\xi(1) = n$. Further let

$$\tilde{g}(t) = \begin{cases} g(t) & t \in [\theta_1, \theta_2], \\ \infty & t \bar{\in} [\theta_1, \theta_2]. \end{cases}$$

Then as it is known that

$$P(G_g(X) > x|F^0) = P\left(\sup_{0 \leqslant t \leqslant 1} \frac{\zeta(t)}{\tilde{g}(t)} > x\right).$$

Theorem 3. If $x \sim \sqrt{n}$, then

$$\frac{1}{n}\ln P\left(\sup_{0 \leqslant t \leqslant 1} \frac{\zeta(t)}{\tilde{g}(t)} > x\right) = -\inf_t \Lambda(t, \tilde{g}(t))(1 + o(1)).$$

In order to estimate the type I error, when $g = \Psi_\alpha$ this assertion could be refined (for the application it is enough to consider the case $x_{\Psi_\alpha} = \sqrt{n}$).

Theorem 4. When $\theta_2 < 1$, then

$$P\left(\sup_{0 \leqslant t \leqslant 1} \frac{\zeta(t)}{\Psi'_\alpha(t)} > \sqrt{n}\right) = A\sqrt{n}\exp(-n\alpha)(1 + o(1)),$$

where $\quad A = \int_{\theta_1}^{\theta_2} \frac{\Psi'_\alpha(u) - u\Psi'_\alpha{}'(u)}{\sqrt{[2\pi(u - \Psi_\alpha(u))(1 - u + \Psi_\alpha(u))]}}\,du.$

The proof of this theorem is analogous to the proofs of [1] and it will be published later. In the present paper we do not make use of this assertion.

The type II error can be estimated with the help of the following theorem:

Theorem 5. For arbitrary $F \in \mathscr{F}$ and $g \in C_+$

$$\ln P(G_g(X) \leqslant x|F) = -n \int_0^1 (1 - l'_{g,F}(t)) \ln (1 - l'_{g,F}(t)) \, dt (1 + o(1)),$$

where $l_{g,F}$ is defined as above.

Theorems 3, 5 are consequences of Theorem 3, 11 of the article [2]. Assertions analogous to Theorem 3 are also contained in [4].

Proof of Theorems 1 and 2. As a preliminary we shall prove the following lemma:

Lemma 1. Let $B_\alpha \subset C_+$ be the set of all functions g for which

$$\lim \frac{x_g}{\sqrt{n}} = h_g = 1; \quad \lim \left(-\frac{\ln \epsilon_n}{n}\right) = \alpha > 0.$$

Then the lower envelope of the set $\phi(t) = \inf_{B_\alpha} g(t)$ on $[\theta_1, \theta_2]$ coincides with $\Psi_\alpha(t)$.

Proof. Clearly $\phi(t) \leqslant \Psi_\alpha(t)$ since by Theorem 3 $\Psi_\alpha \in B_\alpha$. We shall show now that $\phi(x) \geqslant \Psi'(t)$. Suppose the contrary. Then there exist $\delta > 0$ and $t_0 \in [\theta_1, \theta_2]$ such that $\phi(t_0) < \Psi_\alpha(t_0) - 2\delta$. Consequently we can find $g_0 \in B_\alpha$ such that $g(t_0) < \Psi_\alpha(t_0) - \delta$. Moreover as we already remarked for each t, $\Lambda(t, u)$ is monotonically increasing with u and equation (6) has a unique positive solution. Therefore by Theorem 3

$$\lim_{n \to \infty} \left(-\frac{1}{n} \ln P(G_{g_0}(X) > x_{g_0}|F^0)\right) = \inf_t \Lambda(t_0, \tilde{g}_0(t))$$

$$\leqslant \Lambda(t_0, g_0(t_0)) \leqslant \Lambda(t_0, \Psi_\alpha(t_0) - \delta) < \Lambda(t_0, \Psi_\alpha(t_0)) = \alpha.$$

These relations lead to contradiction since it follows from them that $g_0 \bar{\in} B_\alpha$.

We shall turn now to the proof of Theorems 1 and 2. By Lemma 1 the inequality

$$g(t) \geqslant \Psi_\alpha(t)$$

holds for an arbitrary function $g \in B_\alpha$ on $[\theta_1, \theta_2]$.

Hence $l_{g,F} \geqslant l_{\Psi_\alpha, F}$ for every F and consequently†

$$\int_0^1 (1 - l'_{g,F}(t)) \ln (1 - l'_{g,F}(t))\, dt \leqslant \int_0^1 (1 - l'_{\Psi_\alpha, F}(t)) \ln (1 - l'_{\Psi_\alpha, F}(t))\, dt.$$

The assertion of Theorem 1 follows now from Theorem 5. The uniqueness of the function $\Psi_\alpha(t)$ will follow from Theorem 2.

Proof of Theorem 2. Suppose that the condition $\mu(\Omega_g) > 0$ holds. In other words

$$\mu\{F: \sup_{t \in [0,\, 1]} (l_{g,F}(t) - l_{\Psi_\alpha, F}(t)) > 0\} > 0.$$

Then since $l_{g,F}$ and $l_{\Psi_\alpha, F}$ are absolutely continuous and convex functions there exist numbers t_1 and δ such that

$$\mu\{F: l_{g,F}(t_1) - l_{\Psi_\alpha, F}(t_1) > \delta\} > 0. \tag{7}$$

Denote the set of function in (7) by $\mathscr{F}_\delta$. For $F \in \mathscr{F}_\delta$ there exists

$$\epsilon = \epsilon(\delta) > 0$$

the same for all $F \in \mathscr{F}_\delta$, such that

$$\int_0^1 (1 - l'_{g,F}(t)) \ln (1 - l'_{g,F}(t))\, dt \leqslant \int_0^1 (1 - l'_{\Psi_\alpha, F}(t)) \ln (1 - l'_{\Psi_\alpha, F}(t))\, dt(1 - 2\epsilon).$$

Then it follows from Theorem 5 that for any $F \in \mathscr{F}_\delta$ and sufficiently large n

$$\frac{\mathscr{L}(g,F)}{\mathscr{L}(\Psi_\alpha, F)} \leqslant 1 - \epsilon.$$

Since $l_{g,F} \geqslant l_{\Psi_\alpha, F}$ for an arbitrary $F \in \mathscr{F}$ then

$$\int_{\mathscr{F}} \frac{\mathscr{L}(g,F)}{\mathscr{L}(\Psi_\alpha, F)}\, d\mu \leqslant (1 + o(1))\mu(\mathscr{F} - \mathscr{F}_\delta) + \int_{\mathscr{F}_\delta} \frac{\mathscr{L}(g,F)}{\mathscr{L}(\Psi_\alpha, F)}\, d\mu$$

$$\leqslant 1 - \epsilon\mu(\mathscr{F}_\delta) + o(1) \leqslant 1 - \frac{\epsilon}{2}\mu(\mathscr{F}_\delta)$$

for sufficiently large n. The proof is completed.

† One can use here, for example, the following inequality:

$$\int_0^1 p(x) \ln p(x)\, dx < \int_0^1 g(x) \ln g(x)\, dx$$

provided that p and g are monotonically increasing densities in $[0, 1]$ such that $\int_0^x p(t)\, dt \geqslant \int_0^x g(t)\, dt$ with strict inequality for at least one $x \in [0, 1]$.

REFERENCES

[1] A. A. Borovkov and N. M. Sycheva. On some asymptotically optimal non-parametric criteria. *Teor. Veroy at. Primen.* XIII, 3 (1968), 385–418.

[2] A. A. Borovkov. The analysis of large deviation in boundary-value problems with arbitrary boundaries, I, II. *Sib. Math. T.* 5, 2 (1964) 253–89; 5, 4 (1964) 750–76 (in Russian).

[3] A. A. Borovkov. Boundary-value problems for random walks and large deviations in function spaces. *Theory Probab. Applic.* 12 (1967) 575–95 (English translation).

[4] J. G. Abrahamson. Exact Bahadur efficiencies for the Kolmogorov–Smirnov and Kuiper one- and two-sample statistics. *Ann. Math. Statist.* 38, 5 (1967) 1475–90.

EFFICIENT LINEARIZED ESTIMATES
BASED ON RANKS

CHARLES H. KRAFT AND CONSTANCE VAN EEDEN

1 INTRODUCTION

Efficient estimates, of location parameters, which are based on rank tests
and which can be computed from a single ranking of $n+m$ numbers in
a two-sample problem (or, of m numbers in a one-sample problem) are
proposed here. In some cases, these estimates can be considered as
linearized versions of a class of robust 'non-parametric' estimates proposed by Hodges and Lehmann[3]. Their estimates are defined as roots
(defined as the average of inf $\{\theta\colon h(\theta) \leqslant 0\}$ and sup $\{\theta\colon h(\theta) \geqslant 0\}$) of certain
non-linear equations $h(\theta) = 0$ where the function $h(\theta)$ depends upon the
observations and F. That the functions $h(\theta)$ can be approximated by
linear functions was shown by Jureckova[4]. However, to our knowledge, the theorem of Jureckova has not been applied to obtain simply
computable estimates.

The general definition of these linearized estimates and the proofs of
their principal properties are given in § 2. Their efficiencies for distributions other than that assumed for their construction are discussed in § 3.

2 DEFINITION AND PROPERTIES OF LINEARIZED RANK ESTIMATES

Let $F(x/b)$ be the common distribution of $X_1 - \theta, X_2 - \theta, ..., X_m - \theta,$
$Y_1, ..., Y_n$ (or of $X_1 - \theta, ..., X_m - \theta$ for one-sample problems). We assume
here the regularity conditions of Hájek[1] (or see Hájek and Šidak[2]).
It is assumed that $F(x)$ has an absolutely continuous density f which has
a derivative f'. Let

$$\phi(u) \quad (0 \leqslant u \leqslant 1), \quad \text{be defined by} \quad \phi(u) = -\frac{f'}{f}(F^{-1}(u)).$$

It is assumed that $\phi = \psi_1 + \psi_2$, where ψ_1 is nondecreasing and ψ_2 is nonincreasing. Further it is assumed that

$$\int_0^1 \psi_1^2(u)\,du < \infty \quad \text{and that} \quad \int_0^1 \psi_2^2(u)\,du < \infty.$$

[267]

Consider estimates of the form

$$\theta = \theta_1 + b_N \frac{h(\theta_1)}{NK\lambda},$$

where $b_N \to b$ in probability, and where,

(a) for a two-sample problem, $N = m+n$, $\lambda = \dfrac{m \times n}{N^2}$; for a one-sample problem, $N = m$, $\lambda = 1$.

(b) θ_1 is an estimate so that $(\theta_1 - \theta)$ has a limiting distribution (i.e. $P(\sqrt{(N\lambda)}\,(\theta_1 - \theta) \leqslant t) \to L(t)$ for some distribution L), and

$$\theta_1\left(\frac{X_1-a}{b}, \ldots, \frac{X_m-a}{b}, \frac{Y_1}{b}, \ldots, \frac{Y_n}{b}\right) = \frac{\theta_1(X_1, \ldots, X_m, Y_1, \ldots, Y_n) - a}{b}.$$

(c) h is a test statistic for the null-hypothesis θ and h is of the form

(i) $h(\theta) = \sum\limits_{i=1}^{m} \phi\left(\dfrac{R_{X_i-\theta}}{N+1}\right)$ for two samples;

(ii) $h(\theta) = \sum\limits_{X_i-\theta<0} \phi\left(\dfrac{1}{2} \times \dfrac{R_{|X_i-\theta|}}{N+1} + \dfrac{1}{2}\right) - \sum\limits_{X_i-\theta<0} \phi\left(\dfrac{1}{2} \times \dfrac{R_{|X_i-\theta|}}{N+1} + \dfrac{1}{2}\right)$

(d) $K = \displaystyle\int_0^1 \phi^2(u)\,du.$ for one sample.

It will be shown now that the root to the line $b_N h(\theta_1) - NK\lambda(\xi - \theta_1)$ has the same limiting distribution as $(bh(0)/NK\lambda) + \theta$. Hodges and Lehmann [3] showed, for somewhat different regularity conditions and somewhat different monotone functions h, that the root to $h(\xi) = 0$ has the limiting distribution of $(bh(0)/NK\lambda) + \theta$. Van Eeden [7] pointed out that the asymptotic distribution of the root to $h(\xi) = 0$, for h as defined here, has the same asymptotic variance as the maximum likelihood estimate, namely, that given by the Cramer–Rao bound. Under general conditions the equation $h(\xi) = 0$ is very nearly the likelihood equation so that the estimates proposed here can be regarded as robust and readily computable forms of the maximum likelihood estimates. Since h has the property that $h(\xi - a, X - a, Y) = h(\xi, X, Y)$ and θ_1 satisfies

$$\theta_1(X-a, Y) = \theta_1(X, Y) - a,$$

it will hold that $\theta(X - \theta, Y) = \theta(X, Y) - \theta$. Hence if $\theta(X - \theta, Y)$ is, for each θ, an efficient estimate of zero, it follows that $\theta(X, Y)$ is an efficient estimate of θ. That is, it can be supposed that $\theta = 0$. Then, since θ_1 has a limiting distribution it will be, with high probability, in a $1/\sqrt{(N\lambda)}$-neighborhood of the origin and it will be sufficient to have

$$b_N h(\theta_1) - NK\lambda(\xi - \theta_1)$$

sufficiently close to $bh(\xi)$ near $\xi = 0$.

The remainder of this section will be written for the two-sample problem and assuming that b is known and equal to 1. The modifications for the one-sample problem and when b is estimated from the observations by b_N are straight forward.

The following special case of a theorem of Jureckova will be needed.

Theorem 1 (Jureckova [4]). *If $\theta = 0$ and c is a fixed number, and*

$$K' = \int_0^1 \phi(u)\, \phi_G(u)\, du,$$

then $$P_G\left(\sup_{|\xi| \leqslant c/\sqrt{(N\lambda)}} \frac{1}{\sqrt{(N\lambda)}} |h(\xi) - h(0) + N\xi K'\lambda| > \epsilon\right)$$

converges to zero for each $\epsilon > 0$.

That the root to $h(\theta_1) - NK\lambda(\xi - \theta_1) = 0$ has the same asymptotic distribution as $\dfrac{h(0)}{NK\lambda}$, and hence the same as that root of $h(\xi) = 0$ which is closest to θ_1, is shown in Theorem 2.

Theorem 2. $\sqrt{(N\lambda)}\left[\theta - \dfrac{h(0)}{NK\lambda}\right]$ *converges in probability to zero.*

Proof. Let L be the limit distribution function of $\sqrt{(N\lambda)}\,\theta_1$. Choose $c > 0$ so that $L(c) - L(-c) \geqslant 1 - \frac{1}{2}\epsilon$ for a given $\epsilon > 0$. Then, since

$$P(\sqrt{N\lambda}\,|\theta_1| > c)$$

converges to $1 - [L(c) - L(-c)]$, there exists $N_{\epsilon,c}$ such that

$$P(\sqrt{(N\lambda)}\,|\theta_1| \leqslant c) \geqslant 1 - \epsilon, \quad \text{for all } N > N_{\epsilon,c}.$$

For given $\Delta > 0$ and $\delta > 0$, there exists, by Theorem 1, an $N_{\Delta,\delta,c}$ such that, for $N > N_{\Delta,\delta,c}$

$$P\{\sup_{|\xi| \leqslant c/\sqrt{(N\lambda)}} |h(\xi) - h(0) + N\xi K\lambda| \leqslant \Delta\sqrt{(N\lambda)}\} \geqslant 1 - \delta.$$

Hence, for $N > \max[N_{\epsilon,c}, N_{\Delta,\delta,c}]$ it follows that

$$P\{|h(\theta_1) - h(0) + N\theta_1 K\lambda| \leqslant \Delta\sqrt{(N\lambda)}\} \geqslant 1 - \delta - \epsilon.$$

However, $$\theta - \frac{h(0)}{NK\lambda} = \theta_1 + \frac{h(\theta_1)}{NK\lambda} - \frac{h(0)}{NK\lambda}$$

so that

$$P\left\{\sqrt{(N\lambda)}\left|\theta - \frac{h(0)}{NK\lambda}\right| \leqslant \frac{\Delta}{K}\right\} = P\{|NK\lambda\theta_1 + h(\theta_1) - h(0)|$$
$$\leqslant \Delta\sqrt{(N\lambda)}\} \geqslant 1 - \delta - \epsilon. \quad \text{Q.E.D.}$$

It is a consequence of Theorem 2 that θ and $h(0)/NK\lambda$ have the same asymptotic distribution. Hájek[1] has shown that $h(0)/KN\lambda$ is asymptotically normal with mean 0 and variance $\left[\int_0^1 \phi^2(u)\,du\right]^{-1}$. When b is not known and is estimated, consistently, from the observations by b_N, then θ and $b_N h(0)/KN\lambda$ have the same asymptotic distribution, namely, normal with mean 0 and variance $b^2\left[\int_0^1 \phi^2(u)\,du\right]^{-1}$.

3   RELATIVE EFFICIENCIES OF LINEARIZED RANK ESTIMATES

Relative efficiencies are discussed here for the two-sample problem. The numerical values are the same for the one-sample problems although their calculation differs slightly.

The estimates above have their stated asymptotic variance if the observations are from populations $F((x-\theta)/b)$ and $F(x/b)$ for some b. If they are from distributions $G(x-\theta)$ and $G(x)$ the asymptotic variance and, hence, the relative efficiency of the linearized estimates can depend upon the initial estimate θ_1 or on b_N.

It will be helpful to indicate the dependence of h, ϕ, and K on $F(x)$ by subscripts. Hence, for a distribution $F(x)$,

$$\phi_F(u) = -\frac{f_F'}{f_F}(F^{-1}(u)), \quad h_F(\xi) = \sum_{i=1}^{m} \phi_F\left(\frac{R_{X_i-\xi}}{N+1}\right)$$

and
$$K_{FF} = \int_0^1 \phi_F^2(u)\,du.$$

Also,
$$K_{FG} = \int_0^1 \phi_F(u)\,\phi_G(u)\,du.$$

It is assumed that the distributions F, G, and S of this section satisfy the regularity conditions of § 2.

The dependence on θ_1 of the relative efficiency of

$$\theta = \theta_1 + \frac{h_F(\theta_1)\,b_N}{NK_{FF}\lambda}$$

can be most readily seen when θ_1 is asymptotically equivalent to the root of an equation $h_S(\xi) = 0$. For this case, the result is given by Theorem 3.

Theorem 3. *The asymptotic variance of*

$$\theta = \theta_1 + \frac{b_N h_F(\theta_1)}{NK_{FF}\lambda} \quad when \quad X_1-\theta, ..., X_m-\theta, Y_1, ..., Y_n$$

are independent observations from $G(x)$ is

$$\sigma^2 = \frac{K_{SS}}{K_{SG}^2}\left[1 - \frac{bK_{FG}}{K_{FF}}\right]^2 + \frac{2K_{SF}b}{K_{SG}K_{FF}}\left[1 - \frac{bK_{FG}}{K_{FF}}\right] + \frac{b^2}{K_{FF}},$$

where $b = P_G - \lim b_N$.

Proof. Assuming $\theta = 0$ and b is known and equal to 1, it follows from the theorem of Jureckova that $\hat{\theta}_1$ can be written as

$$\frac{h_S(0) \pm \epsilon_1 \sqrt{(N\lambda)}}{N\lambda K_{SG}}$$

and $h_F(\hat{\theta}_1)$ can be written as $h_F(0) - \hat{\theta}_1 \lambda N K_{FG} \pm \epsilon_2\sqrt{(N\lambda)}$. It follows that θ can be written

$$\theta = \frac{h_S(0)}{\lambda N K_{SG}}\left[1 - \frac{K_{FG}}{K_{FF}}\right] + \frac{h_F(0)}{N\lambda K_{FF}} \pm \frac{\epsilon_2}{\sqrt{(N\lambda)}\,K_{FF}} \pm \frac{\epsilon_1}{\sqrt{(N\lambda)}\,K_{SG}}\left[1 - \frac{K_{FG}}{K_{FF}}\right].$$

From this last expression for θ it is easily seen that its asymptotic variance is given by σ^2 above. It is immediate from Slutsky's theorem that the Theorem holds if $b_N \xrightarrow{P_G} b$.   Q.E.D.

Several relative efficiencies for the linearized estimates θ are given in Table 1 for two choices of the initial estimate.

Table 1. *Efficiencies of* $\hat{\theta}_1 + \dfrac{h_F(\hat{\theta}_1)b_N}{NK_{FF}\lambda}$ *relative to* $\hat{\theta}_1 + \dfrac{h_G(\hat{\theta}_1)b_N}{NK_{GG}\lambda}$ *when the observations are from* $G\left(\dfrac{x-\theta}{b}\right)$ *and* $G\left(\dfrac{x}{b}\right)$ *and b_N is estimated from the sample standard deviations*

$$\hat{\theta}_1 = \text{med}(X) - \text{med}(Y)$$

$G \backslash F$	Normal	Logistic
Normal	1·00	0·95
Logistic	0·95	1·00

$$\hat{\theta}_1 = \bar{X} - \bar{Y}$$

$G \backslash F$	Normal	Logistic
Normal	1·00	0·96
Logistic	0·96	1·00

For comparison, in Table 2 are efficiencies of the actual root to $h_F(\xi) = 0$ relative to the actual root of $h_G(\xi) = 0$.

Table 2. *Efficiencies of the root of* $h_F(\xi) = 0$ *relative to the root of* $h_G(\xi) = 0$
when the observations are from $G(x - \theta)$ *and* $G(x)$

$G \backslash F$	Normal	Logistic
Normal	1·00	$\dfrac{3}{\pi} = 0\cdot95$
Logistic	$\dfrac{3}{\pi} = 0\cdot95$	1·00

Linearized estimates which have the efficiencies of Table 2 can be constructed. They require two rankings of translated observations. Let $A_N = A/\sqrt{N}$ for a fixed $A > 0$. Then if the slope of the line approximating $h_F(\xi)$ is estimated by $\dfrac{h_F(\theta_1 + A_N) - h_F(\theta_1 - A_N)}{2A_N}$ the root to the line through the point $(\theta_1 + A_N, h_F(\theta_1 + A_N))$ with this estimated slope gives

$$\theta' = \theta_1 - A_N \frac{h_F(\theta_1 + A_N) + h_F(\theta_1 - A_N)}{h_F(\theta_1 + A_N) - h_F(\theta_1 - A_N)}.$$

That $\sqrt{(N\lambda)}\,(\theta' - \theta_0)$, where θ_0 is the root to $h_F(\xi) = 0$, converges to zero in probability when the observations are from $G(x - \theta)$ and $G(x)$ follows readily from Jureckova's theorem.

ACKNOWLEDGEMENTS

The authors wish to thank Professor H. L. Koul for pointing out to us in [5] the work of Jureckova. We wish to thank Mr Jaafar Al-Abdulla of the Mathematics Research Center for computing numerically one of the integrals for Table 1. We also wish to thank the referee for calling to our attention the paper of Sen [6] in which the estimate based on the Wilcoxon test is studied.

This work was done while the authors were on leave from the University of Montreal and visiting the Mathematics Research Center, United States Army, Madison, Wisconsin, under Contract No.: DA–31–124–ARO–D–462.

REFERENCES

[1] Hájek, J. Asymptotically most powerful rank-order tests, *Ann. Math. Statist.* **33** (1962), 1124–47.

[2] Hájek, J. and Šidak, Z. *Theory of rank tests.* New York: Academic Press, 1967.

EFFICIENT LINEARIZED ESTIMATES BASED ON RANKS **273**

[3] Hodges, J. L. and Lehmann, E. L. Estimates of location based on rank tests, *Ann. Math. Statist.* **34** (1963), 598–611.
[4] Jureckova, J. Asymptotic linearity of a rank statistic in regression parameter, *Ann. Math. Statist.* **40** (1969), 1889–1900.
[5] Koul, H. L. Asymptotic behaviour of a class of regions based on ranks. Michigan State University, January 1969, Technical Report.
[6] Sen, P. K. On the estimation of relative potency in dilution (-direct) assays by distribution-free methods, *Biometrics* **19** (1963), 532–52.
[7] van Eeden, C. Nonparametric estimation, Lecture notes, Seventh Session of the 'Séminaire de Mathématiques Supérieures,' Les Presses de l'Université de Montréal, 1968.

ON ESTIMATING POINTS OF LOCAL MAXIMA AND MINIMA OF DENSITY FUNCTIONS

B. P. LIENTZ

1 INTRODUCTION AND SUMMARY

In this paper we consider the problem of estimating points at which a density function attains a local maximum or minimum. We examine two separate cases. In the first case there is assumed to be a known disjoint cover of E^1 such that in each set of the cover the density assumes a unique maximum or minimum. In the second case this cover is not known. Estimation procedures and properties are given for both cases. Finally, we investigate some specific applications to reliability theory and traffic flow analysis.

2 POINTS OF LOCAL MAXIMUM

Let X be a random variable with continuous density function f and distribution function F. Suppose that $X_1, ..., X_n$ are a random sample of size n from X. We are also given a closed, bounded interval $D = [a, b]$. We make the following definition.

Definition 1. θ is a local mode with respect to f if and only if there exists a $\delta > 0$ such that $f(\theta) \geqslant f(x)$ for all x in the interval $(\theta - \delta, \theta + \delta)$.

We will assume that f has k local modes in the set D. Let these be denoted by $\theta_1, ..., \theta_k$ with $a < \theta_1 < ... < \theta_k < b$.

Let $\{a_i\}_{i=1}^k$ be a collection of points with $a_0 = a, a_k = b$ and $a_{i-1} < \theta_i < a_i$ for $1 \leqslant i \leqslant k$. Two cases will be considered. In the first the $\{a_i\}$ will be assumed known. In the second the $\{a_i\}$ will not be assumed to be known.

2.1. $\{a_i\}$ known

Suppose that the order statistics (increasingly ordered) are given by $Y, ..., Y_n$. We now partition the order statistics over the intervals $\{(a_{i-1}, a_i)\}$ so that

$$\left.\begin{aligned}
a_0 = a < Y_{m_1} < ... < Y_{m_2 - 1} < a_1, \\
a_1 < Y_{m_2} \quad < ... < Y_{m_3 - 1} < a_2, \\
\vdots \qquad\qquad \vdots \\
a_{k-1} < Y_{m_k} < ... < Y_{m_{k+1} - 1} < b = a_k,
\end{aligned}\right\} \quad (2.1.1)$$

where $Y_{m_1 - 1}$ is the largest order statistic smaller than a and $Y_{m_{k+1}}$ is the

[275]

smallest order statistic greater than b. The inequalities in (2.1.1) define the values $m_1, ..., m_{k+1}$. We will now examine the general interval $[a_{i-1}, a_i]$ and the order statistics lying in the interval (i.e. $Y_{m_i}, ..., Y_{m_{i+1}-1}$). Recall that θ_i is the point at which f attains its maximum in (a_{i-1}, a_i).

Let K be a kernel function in the sense of Parzen[3]. That is, $K(\cdot)$ is a nonnegative valued function on E^1 which is symmetric about the origin, is nonincreasing for positive values of the argument, and integrates to 1 on the real line. (Parzen does not use this last property but assumes K is absolutely integrable and that $\lim_{y \to \infty} |yK(y)| = 0$. However, for most practical situations K is usually a density.)

Let $\{h_n\}$ be a sequence of real numbers satisfying

$$\lim_{n \to \infty} h_n = \infty \quad \text{and} \quad \lim_{n \to \infty} h_n/n = 0.$$

Define $f_n(\cdot)$ by $\qquad f_n(x) = h_n \int K(h_n(y-x))\, dF_n(y),$ \hfill (2.1.2)

where F_n is the empirical distribution function of the sample. It was shown in [3] that f_n is consistent and is asymptotically unbiased for estimating f. Furthermore, he showed that if f is uniformly continuous, then $f_n \to f$ uniformly—a.e. as $n \to \infty$.

Let $\theta_{i,n}$ be the local mode with respect to f_n in the interval (a_{i-1}, a_i). Then we have the following result for consistency. In cases where $\theta_{i,n}$ may not be the unique mode in (a_{i-1}, a_i), $\theta_{i,n}$ can be selected as one of the local modes of f_n in (a_{i-1}, a_i).

Theorem 1. *Suppose* $m_{i+1} - m_i \to \infty$ *as* $n \to \infty$. *If* θ_i *is the unique local mode of* f *in the interval* (a_{i-1}, a_i) *and if* f *is continuous on* $[a_{i-1}, a_i]$ *then for all* $\epsilon > 0$
$$\lim_{n \to \infty} P[|\theta_{i,n} - \theta_i| \geq \epsilon] = 0.$$

Proof. Since f is continuous on the closed interval $[a_{i-1}, a_i]$, it is uniformly continuous there. Hence, $f_n \to f$ uniformly—a.e. on $[a_{i-1}, a_i]$. Since, $\theta_{i,n}$ and θ_i are the local modes of f_n and f respectively, then we can show that $f_n(\theta_{i,n}) \overset{P}{\to} f(\theta_i)$ and thereby show from the triangle inequality that $f(\theta_{i,n}) \overset{P}{\to} f(\theta_i)$. Also, $\{\theta_{i,n}\}$ is a sequence in the compact set $[a_{i-1}, a_i]$ and so it has an accumulation point θ'. But if $\{\theta_{i,n_k}\}$ is a subsequence of $\{\theta_{i,n}\}$ with $\theta_{i,n_k} \overset{P}{\to} \theta'$ then from the continuity of f, $f(\theta_{i,n_k}) \overset{P}{\to} f(\theta')$. But $f(\theta_{i,n}) \overset{P}{\to} f(\theta_i)$ and θ_i is the unique mode of f in $[a_{i-1}, a_i]$ so that $\theta' = \theta_i$.

Therefore, the only accumulation point of $\{\theta_{i,n}\}$ is θ_i. The last step is to argue that $f(\theta_{i,n}) \xrightarrow{P} f(\theta_i)$ implies that $\theta_{i,n} \xrightarrow{P} \theta_i$. Suppose there exists $\epsilon > 0$ such that for all k there exists θ_{i,n_k} with

$$|\theta_{i,n_k} - \theta_i| \geqslant \epsilon \quad \text{and} \quad |f(\theta_{i,n_k}) - f(\theta_i)| < 1/n_k.$$

But $\{\theta_{i,n_k}\}$ is a sequence with an accumulation point. Let the convergent subsequence be $\{\theta_{i,n_{k_m}}\}$. Then we have from the above that $\theta_{i,n_{k_m}} \xrightarrow{P} \theta_i$. This produces the desired contradiction.

The above result can be extended to half infinite interval by constructing the Stone–Čech compactification of E^1 and noting that $f(x) \to 0$ as $x \to \pm\infty$.

The asymptotic normality of $\theta_{i,n} - \theta_i$ can be shown in the usual fashion by assuming the kernel and density functions obey certain regularity conditions (i.e., f has continuous second derivative, K is a density, and the random variable associated with K has a finite $2 + \delta$ moment for $0 < \delta < 1$).

For the kernel function which is a uniform density on $(-a, a)$ Chernoff [1] derived the consistency and other asymptotic properties of the estimate of f, f_n, formed by this kernel function. Using arguments similar to the above, results for this kernel can be obtained for local modes.

We might remark on the choice of kernel functions. Heretofore, the kernel functions themselves have been independent of the sample size. (The function $K(h_n(y))$ depends on n only through h_n.) It would seem in some applications that the size of the window would depend on n. This is especially true if the cost of sampling was high. As a function of n, the kernel should decrease in range as $n \to \infty$. A logical choice for the limiting function would be a kernel of the type considered by Parzen and Chernoff. The following result is in this vein.

Theorem 2. Let $\{K_n : n \geqslant 1\}$ and K be kernel functions and let $\hat{f}_n$ be as in (2.1.2) with K replaced by K_n. Suppose $K_n \to K$ uniformly, then at every point of continuity x of f

$$\lim_{n \to \infty} \hat{f}_n(x) = \lim_{n \to \infty} h_n \int_{-\infty}^{+\infty} K_n(h_n(y-x))\, dF_n(y) = f(x). \qquad (2.1.3)$$

Proof. Let x be a point of continuity of f. We first observe that if f_n is the density given in (2.1.2) with kernel K then

$$|\hat{f}_n(x) - f(x)| \leqslant |\hat{f}_n(x) - f_n(x)| + |f_n(x) - f(x)|. \qquad (2.1.4)$$

We know $f_n(x) \to f(x)$ as $n \to \infty$. For the first term of the right-hand side of (2.1.4) we have

$$|\hat{f}_n(x) - f_n(x)| \leqslant h_n \int_{-\infty}^{+\infty} |K_n(h_n(y-x)) - K(h_n(y-x))| \, dF_n(y).$$

Taking the limit as $n \to \infty$ and applying the Lebesgue dominated convergence theorem, then since $K_n \to K$ uniformly, $|\hat{f}_n(x) - f_n(x)| \to 0$. This completes the proof.

Corollary 1. *With $\hat{f}_n$ as above, $\hat{f}_n$ is an asymptotically unbiased estimate of f at points of continuity of f.*

Proof. This follows from Theorem 2 and the dominated convergence theorem.

Theorem 3. *For f_n and K_n as above*

$$\lim_{n \to \infty} (n/h_n) \operatorname{var}(f_n(x)) = f(x) \int_{-\infty}^{+\infty} K^2(y) \, dy. \qquad (2.1.5)$$

Proof. The variance of $\hat{f}_n$ is given by

$$\operatorname{var}(\hat{f}_n(x)) = (1/n) \operatorname{var}(h_n K_n(h_n(x-X))). \qquad (2.1.6)$$

By corollary 1 it suffices to consider the second moment of $\hat{f}_n$. We have

$$h_n^{-1} E(h_n(K_n(h_n(x-X))))^2 = h_n \int_{-\infty}^{+\infty} K_n^2(h_n(x-y)) f(y) \, dy. \qquad (2.1.7)$$

Applying Theorem 1 to (2.1.7) we obtain, in the limit as $n \to \infty$,

$$f(x) \int_{-\infty}^{+\infty} K^2(y) \, dy.$$

Using the above results and Theorem 3A in [3] we have

Theorem 4. *If f is uniformly continuous, then $\hat{f}_n \to f$ uniformly*—a.s. *as $n \to \infty$.*

The results pertaining to estimation of the mode and local modes follow directly from Theorems 1–4.

Venter [5] considered a different method of estimating modes. Venter's method of estimation is to first find the interval of minimum length which contains $2r_n + 1$ order statistics for a sequence $\{r_n\}$. The two estimates of the mode are the middle order statistic in the smallest such interval and the mean of the first and the last statistic in this interval. For particular choices of $\{r_n\}$ Venter obtained interesting results for consistency, rates of convergence, and asymptotic distributions of the estimates of the mode. As a note we remark that these results can be localized to an interval (a_{i-1}, a_i).

At this time there appears to be no definite preference for any of these methods or those of Prakasa Rao[4] and others. There are several reasons for this. First, the results are asymptotic. Second, some of the restrictions imposed by the authors are excessive in terms of what information is available in experiments. For example, bounds, continuity, and existence of high-order derivatives of the density are assumed known. The author is presently investigating the finite sample case.

2.2 $\{a_i\}$ unknown

With the $\{a_i\}$ unknown the assumptions are that the density is continuous and that somewhere on the real line are exactly k local modes, $\theta_i, \dots, \theta_k$ that are separated from one another.

If the cost of sampling were small and the accuracy of the asymptotic results known, then a procedure that could be used would be for given ϵ and α to sample until the probability that the local modes of f_n (or $\hat{f}_n$) and those of f differed by greater than ϵ with probability not exceeding α. However, since this is not the case, we might turn to a discussion of concentration of area under the density function.

Define for $\beta\epsilon(0, 1)$ and x the function $S(x, \beta)$ as the smallest non-negative quantity such that

$$\beta \leqslant \int_{x-S(x,\,\beta)}^{x+S(x,\,\beta)} f(y)\,dy = F(x+S(x,\beta)) - F(x-S(x,\beta)). \quad (2.2.1)$$

If equality in (2.2.1) is impossible for some f, then we will select the smallest value of S which gives a value in (2.2.1) as close to β as possible. The interval $[x - S(x, \beta), x + S(x, \beta)]$ is referred to as a β-modal interval of x with respect to f. Let $S_n(x, \beta)$ be defined as in (2.2.1) with f_n replacing f. S and S_n are indicators of the minimum width of the interval centered about a given point so that the rise in height of F and $\tilde{F}_n$ (corresponding to f_n) over the interval is β. In [2] the convergence of S_n to S was established along with other properties of S and S_n such as continuity and bounds in probability.

If it can be established that for certain values of β $S(\cdot, \beta)$ attains its infimum at one of the local modes, then we can investigate $S_n(\cdot, \beta)$ and from it estimate the local modes. The following theorem fulfills this purpose.

Theorem 5. Let S and $\{\theta_i\}$ be as above. Then there exist a $\beta_0 > 0$ such that for all $\beta < \beta_0$

$$\inf_x S(x, \beta) = \min_{1 \leqslant i \leqslant k} S(\theta_i, \beta). \quad (2.2.2)$$

Proof. Let $\theta_i, ..., \theta_k$ be the local modes with respect to f and let $\tau_i, ..., \tau_k$ be the points at which f achieves a local minimum with $\theta_i < \tau_i < \theta_{i+1}$. The set $\{\tau_i\}$ will be examined in more detail later. We will show that there exists a β_{i+1} such that for all $\beta < \beta_{i+1}$

$$\inf_{\tau_1 \leqslant x \leqslant \tau_{i+1}} S(x, \beta) = S(\theta_i, \beta) \qquad (2.2.3)$$

for $i = 0, ..., k$. Here $\tau_0 = -\infty$ and $\tau_k = \infty$. We will consider the general case of the interval $[\tau_i, \tau_{i+1}]$.

From the continuity of f and since θ_{i+1} is the mode of f in the interval $[\tau_i, \tau_{i+1}]$, we can find an ϵ-neighborhood about θ_{i+1}, say $Ne(\theta_{i+1}, \epsilon)$ such that there exists a β, say β'_0, with the property that the infimum of $S(\cdot, \beta)$ over $Ne(\theta_{i+1}, \epsilon)$ is attained at θ_{i+1} for $\beta < \beta'_0$.

There exists β'_1 and β'_2 such that for $\beta < \beta'_1$ we have $S(\tau_i, \beta) > S(\theta_{i+1}, \beta)$ and for $\beta < \beta'_2$ we have $S(\tau_{i+1}, \beta) > S(\theta_{i+1}, \beta)$. This is because

$$f(\theta_{i+1}) > f(\tau_i), \quad f(\theta_{i+1}) > f(\tau_{i+1}),$$

and f is continuous. Formally, we can locate δ_1- and δ_2-neighborhoods of τ_i and τ_{i+1} respectively, and a δ-neighborhood of θ_{i+1} such that for every point z in $Ne(\theta_{i+1}, \delta)$, q in $Ne(\tau_i, \delta_1)$, and y in $Ne(\tau_{i+1}, \delta_2)$ we have $f(z) > f(q)$ and $f(z) > f(y)$. β'_1 and β'_2 can now be selected so that $S(\tau_i, \beta) > S(\theta_{i+1}, \beta)$ and $S(\tau_{i+1}, \beta) > S(\theta_{i+1}, \beta)$ for $\beta \leqslant \min\{\beta'_1, \beta'_2\}$ with $S(\tau_i, \beta_1) < \delta_1$ and $S(\tau_{i+1}, \beta'_2) < \delta_2$.

Finally, we select β_{i+1} to be a positive number not exceeding the minimum of β'_0, β'_1, and β'_2 such that

$$x - S(x, \beta_{i+1}) \geqslant \tau_i - \delta_1 \quad \text{and} \quad x + S(x, \beta_{i+1}) \leqslant \tau_{i+1} + \delta_2$$

for $x \in [\tau_i, \tau_{i+1}]$. By this procedure we have established that for $\beta < \beta_{i+1}$

$$\inf_{\tau_i \leqslant x \leqslant \tau_{i+1}} S(x, \beta) = S(\theta_{i+1}, \beta).$$

To complete the proof let $\beta_0 = \min_{i \leqslant i \leqslant k} \{\beta_i\}$ and note that the intervals $\{[\tau_i, \tau_{i+1}]\}_{i=0}^{k}$ form a cover of E^1.

The procedure based on the above is to first estimate how small the first choice of β, β_1 must be. This can often be done from prior information based on previous experience. If f is quite flat, then β_1 must be chosen quite small; and if f has decided peaks and troughs, then β_1 need not be close to zero. After the selection of an initial value, $S_{n_1}(\cdot, \beta_1)$ can be computed for an initial sample size n_1. By the above theorem we select the k points, $q_{1,1}, q_{1,2}, ..., q_{1,k}$ for which $S_{n_1}(\cdot, \beta_1)$ attains its minimum so that the intervals $(q_{1,i} - S_{n_1}(q_{1,i}, \beta_1), q_{1,i} + S_{n_1}(q_{1,i}, \beta_1))$ are disjoint. If they are not disjoint, then a smaller β can be selected and S_{n_1} recalculated. We can now continue sampling and reapplying this technique. The decision to stop will be made on the basis of costs of observation, and the behavior of S_n.

3 POINTS OF LOCAL MINIMUM

Let X be a random variable with continuous density f and distribution function F. Let D be a closed, bounded interval in E^1.

Definition 2. τ is a local minimum with respect to f if and only if there exists a $\delta > 0$ with $f(\tau) \leqslant f(x)$ for all x in the interval $(x - \delta, x + \delta)$.

Assume that there are k local minima in the interval $D, \tau_1, ..., \tau_k$ with $\tau_1 < \tau_2 < ... < \tau_k$. As with the case of local modes, we can consider the two cases. We will merely established some of the initial results and remark that the outline in the previous section can be followed.

From a random sample of size n and a kernel function K let f_n be as constructed previously. Suppose $\tau_1, ..., \tau_k$ are the local minima with respect to f and that a set of constants $\{b_i\}_{i=0}^k$ are known with

$$-\infty < b_0 < \tau_1 < b_1 < ... < b_{k-1} < \tau_k < b_k < +\infty.$$

Furthermore, assume that τ_i is the unique point at which f attains a minimum in (b_{i-1}, b_i). Let $\{\tau_{i,n}\}$ be the local minima of f_n in the intervals $\{(b_{i-1}, b_i)\}$ respectively. With these assumptions we obtain the following theorem by methods similar to those employed in Theorem 1.

Theorem 6. If the number of order statistics in (b_{i-1}, b_i) for $i = 1, ..., k$ tends to ∞ as $n \to \infty$, then $\tau_{i,n} \to \tau_i$ in probability as $n \to \infty$.

The other asymptotic results follow analogously.

Let us now consider the instance where the values $\{b_i\}$ are unknown. For $\beta \in (0, 1)$, f and x let $S(x, \beta)$ be as before. To estimate local modes we looked at points where X attained its minimum. Thus, for local minima one seeks to find the points where S attains its maximum. The procedure and results are similar to those in Part 2 of the previous section and so are omitted.

4 APPLICATIONS AND FURTHER REMARKS

Let us suppose we are testing a component whose life-length distribution satisfies the properties of the previous sections. Based on knowledge acquired from earlier experiments on similar components, we are led to expect that failure will likely occur in certain time intervals which are known. To estimate the maxima of the failure density we can apply the results in the first part of § 1.

Now consider a statistical analysis of traffic in a certain region. Although there are some crude estimates of periods of peak traffic, these estimates are unreliable due to the hours that employees work at surrounding plants and due to new construction. In this problem we are led to the methods of the second part of § 1.

In a later paper we will consider extensions of the estimation problem to higher dimensional cases. Also, we will examine rates of convergence when f belongs to particular classes of functions.

5 ACKNOWLEDGEMENT

I am grateful to the referee for his valuable suggestions.

REFERENCES

[1] Chernoff, H. (1964). Estimation of the mode, *Ann. Inst. Statist. Math.* **16**, 31–41.

[2] Lientz, B. P. (1969). On nonparametric modal intervals. System Devel. Corporation document.

[3] Parzen, E. (1962). On estimation of a probability density function and mode, *Ann. Math. Statist.* **33**, 1065–76.

[4] Prakasa Rao, B. L. S. (1968). Estimation of the location of the cusp of a continuous density, *Ann. Math. Statist.* **39**, 76–87.

[5] Venter, J. H. (1967). On estimation of the mode, *Ann. Math. Statist.* **38**, 1446–55.

ASYMPTOTICALLY EFFICIENT TESTS FOR LOCATION: NONPARAMETRIC AND ASYMPTOTICALLY NONPARAMETRIC†

KEI TAKEUCHI

INTRODUCTION

Recently [4] the author obtained a robust estimator of a location parameter which is shown to be asymptotically fully efficient for a wide class of continuous distributions and, as is indicated by Monte Carlo experiments, has high relative efficiency for various types of distributions even for such small sample sizes as $n = 10$ or 20. The basic idea of that method is to estimate the coefficients of the best linear unbiased estimator (BLUE) from the sample, by considering a fictitious sub-sample of size k, and estimating the coefficients of BLUE for the sample of size k from the sample of size n. In this paper the same idea is applied to a testing situation for the two-sample and one-sample location parameter cases, and it is shown that asymptotically it gives a uniformly efficient (most powerful) procedure. In order to show this, it is necessary to develop some asymptotic theory of tests based on linear statistics in the two-sample situation. This is done in the first section, and then our own method is discussed.

The theory developed in § 1 is applicable to both of the cases where the coefficients c_α are fixed and where they are determined from the sample. When the sample size is not too large, tests based on such 'robust' estimators as the trimmed means would show fairly 'robust' property in terms of the power. This is closely related to the problem of 'studentization' discussed by Huber [8] in this volume. It should also be noted that the relation between the efficient tests and estimators is parallel to that of Gastwirth [9].

Uniformly efficient tests are also discussed by Stein [6] and Hajék [7], and the regularity conditions in Hajék [7] seem to be a little weaker. However, our method has the advantage in that it works fairly well even if the sample size is as small as 10 or 20, at least for fairly regular distributions. This is confirmed by extensive Monte Carlo experiments for the case of estimation [4]. For testing cases the efficiency in small samples

† This investigation was supported by PHS Grant No. GM–16202–01 (National Institute of General Medical Sciences).

seems to be generally smaller than the corresponding estimators, because the test statistics are subject to fluctuations both in the estimator in the numerator and in the estimator of the variance in the denominator. But the exact comparison is not possible since very little is known about the efficiencies of parametric tests for non-normal cases especially when the scale is unknown.

1 ASYMPTOTIC CONDITIONAL NORMALITY OF TEST STATISTICS

First we shall discuss the test based on a linear statistic for the hypothesis about a location parameter in two sample problems.

Suppose that we have independent samples $X_1 \ldots X_n$ and $Y_1 \ldots Y_m$ from continuous populations with densities $f(x)$ and $g(y)$ respectively. We would like to test the hypothesis that the two distributions are the same, i.e.
$$H: f(x) = g(x)$$
against a location shift,
$$K: g(x) = f(x - \theta) \quad (\theta \neq 0).$$

Let $X_{(1)} < \ldots < X_{(n)}$ be the order statistics obtained from X values and consider the statistics of type
$$L = \sum_{\alpha=1}^{n} c_\alpha X_{(\alpha)} \tag{1}$$
as a test statistic.

We shall consider the 'permutation' test based on L. Let
$$O_{m+n} = \{Z_1 < \ldots < Z_{m+n}\}$$
be the set of order statistics obtained from X's and Y's put together. Then under the null hypothesis, the conditional distribution of $X_{(i)}$'s, hence that of L is independent of the shape of the distribution, and
$$P_n\{X_{(1)} = Z_{i_1} \ldots X_{(n)} = Z_{i_n} | O_{m+n}\} = \frac{1}{{}_{m+n}C_n} \quad \text{for any } i_1 < \ldots < i_n. \tag{2}$$
Hence if we denote by $l_{(k)}$ the kth largest values of $\sum_{\alpha=1}^{n} c_\alpha Z_{(i_\alpha)}$ among all possible ${}_{m+n}C_n$ choices of $C_{(i_\alpha)}$'s, we can reject the hypothesis if
$$L \geqslant l_{(k_0)}$$
$$\leqslant l_{(1-k_0+1)}, \quad \text{where} \quad k_0 = \left[\frac{\alpha}{2} {}_{m+n}C_n\right]. \tag{3}$$

This procedure is strictly nonparametric, in the sense that the size of the test is independent of the shape of distribution. Usually it is troublesome to compute the conditional distribution of L given O_{m+n}, but we can easily compute the conditional mean and variance of L by the following formulae:

$$E(L|O_{m+n}) = \sum_{\alpha=1}^{p} c_\alpha T_\alpha, \tag{4}$$

where
$$T_\alpha = E(X_{(\alpha)}|O_{m+n})$$

$$= \sum_{i=1}^{m+n} P_\alpha^i Z_i,$$

and
$$P_\alpha^i = \frac{{}_{i-1}C_{\alpha-1} \cdot {}_{m+n-i}C_{n-\alpha}}{{}_{m+n}C_n},$$

$$V(L|O_{m+n}) = \sum_\alpha \sum_\beta c_\alpha c_\beta S_{\alpha\beta}, \tag{5}$$

where
$$S_{\alpha\beta} = E(X_{(\alpha)}X_{(\beta)}|O_{m+n}) - T_\alpha T_\beta$$

$$= \sum_i P_\alpha^i Z_i^2 - T_\alpha^2 \qquad \text{when } \alpha = \beta$$

$$= \sum_{i<j}\sum P_{\alpha\beta}^{ij} Z_i Z_j - T_\alpha T_\beta \qquad \text{when } \alpha < \beta$$

$$= S_{\beta\alpha} \qquad \text{when } \alpha > \beta$$

and
$$P_{\alpha\beta}^{ij} = \frac{{}_{i-1}C_{\alpha-1} \cdot {}_{j-i-1}C_{\beta-\alpha-1} \cdot {}_{m-n-j}C_{n-\beta}}{{}_{m+n}C_n} \quad \text{for } \alpha < \beta,\, i < j.$$

Hence if we can assume that the conditional distribution is approximately normal, we can obtain an approximate test by rejecting the hypothesis if

$$|\Sigma c_\alpha(X_{(\alpha)} - T_\alpha)| \geqslant k_\alpha(\Sigma\Sigma c_\alpha c_\beta S_{\alpha\beta})^{\frac{1}{2}}, \tag{6}$$

where k_α is the percentile of the unit normal distribution.

Approximation of the conditional distribution by the normal is justified by proving the asymptotic normality in the following way. Let $a_{Ni}, i = 1, ..., N, N = 1, 2, ...$, be a double sequence of real numbers, and $X_j^N, j = 1, ..., n_N$ be a random sample of size n_N chosen from the set $\{a_{Ni} : i\}$ with uniform probabilities and without replacement. Let c_j^N, $j = 1, ..., n_N$ be a sequence of constants which satisfies the condition that

$$c_j^N = \frac{1}{n_N} J\left(\frac{j}{n_N}\right), \tag{7}$$

where $J(u)$ is a function defined for $0 \leqslant u \leqslant 1$.

Now denote the order statistics obtained from X_j^N by

$$X_{(1)}^N < ... < X_{(n_N)}^N$$

and consider the statistic of the type

$$L_N = \sum_{j=1}^{n_N} c_j^N X_{(j)}^N.$$

We shall assume the following.

(i) Define $F_N(t) = 1/N$ {number of the values among a_{Ni} not greater than t}; then there exists a continuous distribution function $F(t)$ and a constant K such that

$$|F_N(t) - F(t)| < \frac{K}{\sqrt{N}} \quad \text{for all } t.$$

(ii) $\int |t|\, dF(t) < \infty.$

(iii) $\lim_{N\to\infty} \sqrt{N} \left(\frac{1}{N} \sum_{i=1}^{N} b_{Ni} \right) \to 0$ and $\lim_{N\to\infty} \max_{1\leqslant i\leqslant N} \dfrac{b_{Ni}^2}{\sum b_{Nj}^2} = 0,$

where $\quad b_{Ni} = \displaystyle\int_{\alpha_{Ni}}^{\infty} J(F(x))\,(1-F(x))\,dx - \int_{-\infty}^{\alpha_{Ni}} J(F(x))\,F(x)\,dx.$

(iv) $J(u)$ is continuous and continuously differentiable, and $J'(u)$ is of bounded variation on the closed interval $[0, 1]$. Then we can prove the following

Theorem 1. *Under the assumptions* (i)–(iv), *the conditional distribution of* $\sqrt{(n_N)}\,(L_N - \mu)$ *is asymptotically normal with mean zero and variance* $(1 - n_N/N)\,\sigma^2$, *where*

$$\mu_N = \int t J(F(t))\, dF(t)$$

and $\qquad \sigma^2 = 2 \displaystyle\iint_{s<t} J(F(s))\,J(F(t))\,F(s)\,[1 - F(t)]\, ds\, dt \qquad (8)$

provided that $\sigma^2 < \infty$, *and* $n_N \to \infty$.

Proof. The proof is exactly analogous to that of Moore[3] for the asymptotic normality of linear functions of order statistics. Let

$$S_N(t) = \frac{1}{n_N} \{\text{number of values of } X_{(j)}^N\text{'s not greater than } t\}.$$

Then L_N can be expressed

$$L_N = \int x J(S_N(x))\, dS_N(x). \qquad (9)$$

By the mean value theorem, we have

$$J(S_N(x)) - J(F(x)) = J'(V_n(x))\,(S_N(x) - F(x)),$$

where $\quad V_n(x) = \theta S_N(x) + (1-\theta)\,F(x) \quad \text{for some } 0 < \theta < 1.$

Therefore we can write

$$\sqrt{n_N}(L_N - \mu) = A_N + B_N + C_N,$$

where

$$A_N = \sqrt{n_N} \int x J'(F(x))\,(S_N(x) - F(x))\,dF(x)$$

$$+ \sqrt{n_N} \int x J(F(x))\,d(S_N(x) - F(x)),$$

$$B_N = \sqrt{n_N} \int x[J'(V_N(x)) - J'(F(x))]\,(S_N(x) - F(x))\,dS_N(x),$$

$$C_N = \sqrt{n_N} \int x J'(F(x))\,(S_N(x) - F(x))\,d(S_N(x) - F(x)).$$

Then, since

$$|S_N(x) - F(x)| \leqslant |S_N(x) - F_N(x)| + |F_n(x) - F(x)| = O_p(n_N^{-\frac{1}{2}}),$$

uniformly in x, it can be shown in exactly the same way as in Moore that

$$B_N \to 0, \quad C_N \to 0 \quad \text{in probability}.$$

Then integrating A_N by parts, we have

$$A_N = -\sqrt{n_N} \int J(F(x))\,(S_N(x) - F(x))\,dx$$

$$= -\frac{1}{\sqrt{n_N}} \sum_{i=1}^{n_N} R(X_i),$$

where $\quad R(X) = \int_X^\infty J(F(x))\,(1 - F(x))\,dx + \int_{-\infty}^X J(F(x))\,F(x)\,dx.$ (10)

Hence if the condition (3) holds true, we can apply the central limit theorem (see Hájek[2]) for the mean of the sample from a finite population and obtain the desired result.

Next we shall consider the problem of estimating μ and σ^2. First we shall transform σ^2,

$$\sigma^2 = \int_{-\infty}^\infty J(F(t))\,(1 - F(t))\,dt \left[\int_{-\infty}^t J(F(s))\,F(s)\,ds \right]$$

$$+ \int_{-\infty}^\infty J(F(s))\,F(s)\,ds \left[\int_s^\infty J(F(t))\,(1 - F(t))\,dt \right].$$

By repeated integration by parts,

$$\sigma^2 = \int_{-\infty}^{\infty} tJ(F(t))\,(1-F(t))\,J(F(t))\,F(t)\,dt$$

$$-\iint_{s<t} sJ(F(t))\,(1-F(t))\,\{J'(F(s))\,F(s)+J(F(s))\}\,dF(s)\,dt$$

$$-\int_{-\infty}^{\infty} sJ(F(s))\,F(s)\,J(F(s))\,(1-F(s))\,ds$$

$$-\iint_{s<t} tJ(F(s))\,F(s)\,\{J'(F(t))\,(1-F(t))-J(F(t))\}\,dF(t)\,ds$$

$$= \int s^2 J(F(s))\,(1-F(s))\,\{J'(F(s))\,F(s)+J(F(s))\}\,dF(s)$$

$$-\int t^2 J(F(t))\,(F(t))\,\{J'(F(t))\,(1-F(t))-J(F(t))\}\,dF(t)$$

$$+2\iint_{s<t} st\{J'(F(s))\,F(s)+J(F(s))\}$$
$$\times\{J'(F(t))\,(1-F(t))-J(F(t))\}\,dF(s)\,dF(t)$$

$$= \int t^2 J^2(F(t))\,dF(t) + 2\iint_{s<t} st\{J'(F(s))\,F(s)+J(F(s))\}$$
$$\times\{J'(F(t))\,(1-F(t))-J(F(t))\}\,dF(s)\,dF(t). \quad (11)$$

Integration by parts in (11) is valid, when

$$\lim_{x\to=\pm\infty} x^2 J(F(x))^2\,F(x)\,(1-F(x))\,dx \to 0$$

which should be true if the last expression in (11) is finite. We denote by $\hat{\sigma}^2$ the quantity obtained from (11) by substituting the empirical cdf for $F_N(t)$ in (11), and also denote

$$\hat{\mu} = \int xJ(F_N(x))\,dF_N(x)).$$

Then we can prove the following theorem about the unconditional asymptotic distribution of the test statistic.

Theorem 2. Suppose that the hypothesis is true, and the distribution has the finite second order moment. If J satisfies the assumption (iv) above, and σ^2 in (8) is finite then

$$\left(\frac{Nn}{N-n}\right)^{\frac{1}{2}}\frac{(L-\hat{\mu})}{\hat{\sigma}} \quad (12)$$

is asymptotically normal with mean 1, as $N \to \infty$ and $n \to \infty$.

Proof. Since $L-\hat{\mu}$ is also a linear function of order statistics

$$\left(\frac{Nn}{N-n}\right)^{\frac{1}{2}}\frac{(L-\hat{\mu})}{\sigma}$$

is shown to be asymptotically normal with mean 0 and variance 1. And since J and J' are uniformly continuous,

$$J(F_N(t)) \to J(F(t)), \quad J'(F_N(t)) \to J'(F(t))$$

uniformly in probability, and $\hat{\sigma}^2 \to \sigma^2$ in probability, which establishes the theorem.

The assumption of the existence of the second order moment may seem to be irrelevant and unnecessary, but it is actually necessary in order to simplify the statement of the theorem. Otherwise,

$$\int t^2 J^2(F_n(t))\,dF_N(t) \to \int t^2 J^2(F(t))\,dF(t)$$

may not necessarily follow. It is possible, however, to weaken the condition if we modify the estimator slightly.

Also it would be interesting to see whether it is possible to replace μ and σ^2 in Theorem 1 or $\hat{\mu}$ and $\hat{\sigma}^2$ in Theorem 2 by $E(L|O_{m+n})$ and $V(L|O_{m+n})$. It can easily be shown that this is not always the case for Theorem 1, for the asymptotic mean or variance are not necessarily equal to the asymptotic values of the mean or variance. We can find a counter example by taking a suitably irregular sequence $\{F_N(t)\}$.

Now note that

$$E(L|O_{m+n}) = \sum_\alpha C_\alpha T_\alpha$$

$$= \frac{1}{N}\sum_{i=1}^N K_N\left(\frac{i}{N}\right)Z_{(i)},$$

where

$$K_N\left(\frac{i}{N}\right) = \frac{N}{n}\Sigma P_\alpha^i J\left(\frac{\alpha}{n}\right).$$

It is easily shown that

$$\sum_\alpha P_\alpha^i = \frac{n}{N},\quad \Sigma P_\alpha^i a = \frac{n}{N}i,\quad \Sigma P_\alpha^i\left(\alpha-\frac{n}{N}i\right)^2 = O\left(\frac{1}{N}\right).$$

Hence if we assume that

(iv) $J(u)'$ is twice differentiable and $J''(u)$ is bounded, then

$$K\left(\frac{i}{N}\right)-J\left(\frac{i}{N}\right) = \frac{1}{2}\sum_\alpha J''(\xi_\alpha)P_\alpha^i\left(\alpha-\frac{n}{N}i\right)^2 = O\left(\frac{1}{N}\right),$$

where

$$\left|\xi_\alpha-\frac{n_i}{N}\right| \leqslant \left|\alpha-\frac{n}{N}i\right|.$$

Therefore
$$\sum_\alpha c_\alpha T_\alpha - \hat{\mu} = \Sigma c_\alpha T_\alpha - \frac{1}{N} \Sigma J\left(\frac{i}{N}\right) Z_{(i)}$$

$$= O\left(\frac{1}{N^2} \Sigma |Z_{(i)}|\right) = O_p\left(\frac{1}{N}\right)$$

assuming $E(|Z|) < \infty$. Hence

$$\sqrt{n^{\frac{1}{2}}} |\Sigma c_\alpha T_\alpha - \hat{\mu}| \to 0 \quad \text{in probability as } N \to \infty.$$

Also,
$$\sum_{\alpha<\beta}\sum P_{\alpha\beta}^{ij} = \frac{n(n-1)}{N(N-1)} \quad \text{for } i < j,$$

$$\sum_{\alpha<\beta}\sum P_{\alpha\beta}^{ij}\alpha = \frac{n(n-1)}{N(N-1)} i \sum_{\alpha<p}\sum P_{\alpha p}^{ij}\beta = \frac{n(n-1)}{N(N-1)} j,$$

$$\sum_{\alpha<\beta}\sum P_{\beta\alpha}^{ij}\left(\alpha - \frac{n^2}{N^2}i\right)\left(\beta - \frac{n^2}{N^2}j\right) = \frac{1}{N}.$$

From these relations it can also be proved that if we assume that $J''(u)$ is continuous and $E(Z^2) < \infty$,

$$nV(L|O_{m+n}) \to \left(1 - \frac{n}{N}\right)\sigma^2$$

in probability. Thus if we add some regularity conditions then $\hat{\sigma}^2$ and $\hat{\mu}$ in Theorem 2 can be replaced by $E(L|O_{m+n})$ and $V(L|O_{m+n})$.

Similarly, if we define
$$L^* = \sum_{\beta=1}^{m} c_\beta^* Y_{(\beta)}, \tag{13}$$

where $Y_{(\beta)}$ denotes the order statistic of Y values, we can consider the conditional distribution of L^* and also of $L - L^*$ in an exactly similar way as above.

And if c_β^* is also expressed by
$$c_\beta^* = J^*\left(\frac{\beta}{m_N}\right), \tag{14}$$

where J^* also satisfies the condition (4), then we have

Theorem 3. *Under the assumptions* (i)–(iv), $\sqrt{N}\{(L - L^*) - (\mu_1 - \mu_2)\}$ *tends conditionally to the normal distribution with mean a, variance σ^2, where*

$$\mu_1 = \int tJ(F(t))\,dF(t), \quad \mu_2 = \int tJ^*(F(t))\,dF(t)$$

and
$$\sigma^2 = \frac{2}{\lambda(1-\lambda)} \iint_{s<t} \{(1-\lambda)\,J(F(s)) + \lambda J^*(F(s))\}$$

$$\times \{(1-\lambda)\,J(F(t)) + \lambda J^*(F(t))\}\,F(s)\,(1-F(t))\,ds\,dt \tag{15}$$

provided that $n_N/N \to \lambda$ *and* $\sigma^2 < \infty$.

Proof. $\sqrt{N}(L-L^*)$ is asymptotically equivalent to

$$-\frac{\sqrt{N}}{n_N}\sum_i R(X_i)+\frac{\sqrt{N}}{m_N}\sum_i R^*(Y_j)$$

$$=-\frac{\sqrt{N}}{n_N}\Sigma\left\{\left(R(X_i)+\frac{n_N}{m_N}R^*(X_i)\right)\right\}+\frac{\sqrt{N}}{m_N}\sum_{i=1}^{N}R^*(Z_i),\quad(16)$$

where $R^*(Y_i)$ is defined in the same way as $R(X)$ in (10) replacing T by T^*. Since the last term of (13) is constant given the order statistic, the theorem follows.

In Theorem 3 also, σ^2 can be replaced by its estimator. Especially if we put $J \equiv J^*$, it will be given by

$$\hat{\sigma}^2 = \frac{1}{\lambda}\hat{\sigma}_1^2+\frac{1}{1-\lambda}\hat{\sigma}_2^2,$$

where $\hat{\sigma}_1^2$ and $\hat{\sigma}_2^2$ are obtained from the expression (11) replacing $F(t)$ by empirical cdf's of X and Y values respectively. And we have the asymptotic test by rejecting the hypothesis if

$$\frac{\sqrt{N}(L-L^*)}{\hat{\sigma}} > k_\alpha.\quad(17)$$

The asymptotic power of the test can also be obtained. Consider the sequence of alternatives,

$$G_N(x) = F\left(x-\frac{\theta_0}{\sqrt{N}}\right)\quad(N=1,2,\dots).\quad(18)$$

Then under the sequence of alternatives (18), we have,

Theorem 4. $\sqrt{N}(L-L^*)$ *is asymptotically normal with mean θ_0 and variance $\sigma^2/\lambda(1-\lambda)$ where σ^2 is given in (8).*

Proof. Since L and L^* are linear combinations of X's and Y's it is easily proved in the usual way that asymptotically

$$\sqrt{n_N}(L-\mu)\quad\text{and}\quad\sqrt{m_N}(L^*-\mu_N^*)$$

are independently and normally distributed with mean 0 and variance σ^2, where

$$\mu = \int tJ(F(t))\,dF(t)\quad\text{and}\quad\mu_N^* = \int tJ(G_N(x))\,dG_N(x) = \mu+\frac{\theta_0}{\sqrt{N}}.$$

Then the theorem immediately follows.

From this theorem it is seen that the asymptotic power of the test (17) under the sequence of alternatives (18) is given by

$$\Phi[k_\alpha + \sqrt{(\lambda(1-\lambda))}\,(\theta_0/\sigma)] + \Phi[-k_\sigma + \sqrt{(\lambda(1-\lambda))}(\theta_0\sigma)],$$

where Φ is the cdf of the unit normal distribution.

It is well known [1] that if the density $f(x) = d/dx\, F(x)$ is twice differentiable and $xf'(x) \to 0$ as $x \to \pm\infty$, by putting

$$J(u) = -\frac{d^2}{dx^2} \log f(x),$$

we have
$$\sigma^2 = \frac{1}{\lambda(1-\lambda)}\, I,$$

where I is the Fisher Information.

Consequently we have as in [9],

Theorem 5. The test given by (17) *is asymptotically equivalent to the (locally) most powerful unbiased test when*

$$J(u) = -\frac{d^2}{dx^2} \log f(x)\big|_{x=F^{-1}(u)} \quad and \quad c_\alpha = J\left(\frac{\alpha}{n}\right).$$

2   UNIFORMLY ASYMPTOTICALLY EFFICIENT TESTS

It is to be noted that the conditional asymptotic normality of the test statistic is not affected if the coefficients c_α themselves are functions of order statistics O_{m+n}. Hence they can be 'estimated' from the sample themselves. And if it is also proved that under the sequence of alternatives

$$\sqrt{n}(\Sigma(\hat{c}_\alpha - c_\alpha^*)\, X_\alpha) \to 0 \quad \text{and} \quad \sqrt{n}(\Sigma(\hat{c}_\alpha - c_\alpha^*)) \to 0$$

in probability, where $\hat{c}_\alpha$ are estimated from the sample and c_α^* are constants, then the test based on $\Sigma c_\alpha X_\alpha$ is asymptotically equivalent to the test based on $\Sigma c_\alpha^* X_\alpha$.

A method which provides a uniformly asymptotically efficient estimator in the sense that it is asymptotically equivalent to the best linear estimator for the location parameter for a class of distributions with some regularity conditions, is given by the author [4], and is also given in the following way.

Let k be some integer smaller than n and m. Suppose that a random subsample is of size k from the order statistics of Z's and denote the order statistics obtained from the subsample by

$$W_{(1)} < \dots < W_{(k)}.$$

Define a linear combination of $W_{(i)}$ by

$$\sum_{i=1}^{k} a_t W_{(t)}, \tag{19}$$

with $\Sigma a_t = 1$.

The conditional distribution of $\Sigma a_t W_{(t)}$ can be computed in exactly the same way as was discussed in § 1, and the conditional variance can be computed.

We minimize the conditional variance of (19) under the condition $\Sigma a_t = 1$, then the coefficients are given

$$a_t^* = \sum_\beta s^{t\beta} / \sum_\alpha \sum_\beta s^{\alpha\beta},$$

$$\{s^{\alpha\beta}\} = \{s_{\alpha\beta}\}^{-1}, \quad \text{where } s_{\alpha\beta} \text{ is given in § 1}.$$

Now denote by $V_{(t)}$ the conditional expectation of $W_{(t)}$ given that the subsample be drawn from X values, and put

$$L = \sum_{t=1}^{k} a_t^* V_{(t)} = \Sigma c_\alpha X_{(\alpha)}. \tag{20}$$

Then it was shown [4] that L is asymptotically equivalent to the BLUE of the location parameter based on X values, hence is asymptotically efficient.

Thus if we apply the conditional procedure given in § 1 to L given by (20) we can obtain an asymptotically uniformly efficient test under a set of regularity conditions. (The regularity conditions under which L converges in probability to the BLUE are discussed by the author in another paper [5].)

A second procedure is to compute the coefficients separately for X and Y in the following way.

Suppose that we take subsamples of size k separately from the sets of X values and Y values. We determine it so as to minimize,

$$V \left(\sum_{i=1}^{k} b_t W_{(t)} \Big| X_{(i)} < \ldots < X_{(n)} \right)$$

under the condition that $\sum_{t=1}^{k} b_t = 1$, and also determine c_s so as to minimize

$$V \left(\sum_{s=1}^{k} c_s W_{(s)} \Big| Y_{(1)} < \ldots < Y_{(m)} \right)$$

under the same condition. Then we put

$$\Sigma d_\alpha X_{(\alpha)} = E(\Sigma b_t W_{(t)} | X_{(1)} < \ldots < X_{(n)}),$$

$$\Sigma d_\beta' Y_{(\beta)} = E(\Sigma c_s W_{(s)} | Y_{(1)} < \ldots < Y_{(m)})$$

and also $\quad V_X = \dfrac{n-k}{n} \Sigma \Sigma b_t b_{t'} \, V(W_t, W_{t'} | X_{(1)} < \ldots < X_{(n)}),$

$$V_Y = \dfrac{m-k}{m} \Sigma \Sigma c_s c_{s'} \, V(W_s, W_{s'} | Y_{(1)} < \ldots < Y_{(m)})$$

and we can make use of the following as a test statistic:

$$\frac{\Sigma d_\alpha X_{(\alpha)} - \Sigma d_{\beta'} Y_{(\beta)}}{\sqrt{(V_X + V_Y)}}. \tag{21}$$

If we fix k, and let $n, m \to \infty$, it is shown from the results in § 1 that the asymptotic distribution of the above statistic is (under some set of regularity conditions) normal, but the small sample distribution or appropriate approximation to it is yet to be found.

3 ONE-SAMPLE TEST

The same idea as above can be applied to the one-sample problem. Suppose that $X_1 \ldots X_n$ are distributed independently and identically according to a continuous distribution with density $f(x - \theta)$. We further assume that the distribution is symmetric, i.e.

$$f(-x) \equiv f(x).$$

We shall consider tests of the hypothesis

$$H\colon \theta = 0.$$

Again, we can make use of a linear combination of the order statistics,

$$L = \Sigma c_\alpha X_{(\alpha)}$$

as a test statistic.

Under the hypothesis, conditional distribution of the order statistics given $Q_n = \{|X_{(1)}| \ldots |X_{(n)}|\}$ is symmetric, and if we denote by

$$0 < Z_1 < Z_2 < \ldots < Z_n$$

the absolute values of the X_i's arranged according to their magnitudes, then

$$\Pr\{(X_{(1)}, \ldots, X_{(n)}) = (\epsilon_1 Z_1, \ldots, \epsilon_n Z_n)|Q_n\} = \tfrac{1}{2}^n$$

for any choice of $\epsilon_i = \pm 1$, $i = 1, \ldots n$. Hence

$$\left.\begin{aligned}
\Pr\{X_{(\alpha)} = Z_i|Q_n\} &= \tfrac{n-i C_{n-\alpha}}{2^{n-i+1}} \quad \text{for } \alpha \geqslant i, \\[2mm]
&= 0 \qquad\qquad \text{otherwise,} \\[3mm]
\Pr\{X_{(\alpha)} = -Z_i|Q_n\} &= \tfrac{n-i C_{\alpha-1}}{2^{n-i+1}} \quad \text{for } \alpha \geqslant n-i+1, \\[2mm]
&= 0 \qquad\qquad \text{otherwise.}
\end{aligned}\right\} \tag{22}$$

Also,

$$
\left.\begin{aligned}
\Pr\{X_{(\alpha)} = Z_i, X_{(\beta)} = Z_j\} &= 2^{-n+i-1}\,{}_{n-j}C_{n-\beta}\cdot{}_{j-i-1}C_{\beta-\alpha-1} \\
&\qquad\text{for } \alpha < \beta,\, i < j, \\
\Pr\{X_{(\alpha)} = -Z_i, X_{(\beta)} = Z_j\} &= 2^{-n+i-1}\,{}_{n-j}C_{n-\beta}\cdot{}_{j-i-1}C_{\beta-\alpha-i} \\
&\qquad\text{for } \alpha < \beta,\, i < j, \\
\Pr\{X_{(\alpha)} = -Z_i, X_{(\beta)} = Z_j\} &= 2^{-n+j-1}\,{}_{n-i}C_{\alpha-1}\cdot{}_{i-j-1}C_{\beta-\alpha-j} \\
&\qquad\text{for } \alpha < \beta,\, i > j, \\
\Pr\{X_{(\alpha)} = -Z_i, X_{(\beta)} = -Z_j\} &= 2^{-n+j-1}\,{}_{n-i}C_{\alpha-1}\cdot{}_{i-j-1}C_{\beta-\alpha-1} \\
&\qquad\text{for } \alpha < \beta,\, i > j.
\end{aligned}\right\}
\tag{23}
$$

From these formulae we can obtain $E(L|Q_n) = 0$ and the value of $V(L|Q_n)$. Asymptotic normality of L can be proved similarly as in the two-sample case and we obtain an approximate (or asymptotic) test criterion by rejecting the hypothesis if

$$
|L| \geqslant k_\alpha \sqrt{(V(L|Q_n))}.
\tag{24}
$$

A more accurate approximation to the conditional distribution may be obtained by computing the fourth order conditional moment $M_4(L|Q_n)$ and approximating $L/\sqrt{(V(L)Q_n) - L^2}$ by a t-distribution with f degrees of freedom where f is given by

$$
\frac{2f}{f+3} = \frac{M_4(L|Q_n)}{(V(L|Q_n))^2} - 1.
$$

The asymptotic (Pitman) efficiency of such tests under the sequence of alternatives $\theta = \theta_0/\sqrt{n}$ (θ_0, fixed) is equal to the asymptotic efficiency of L as an estimator of θ measured by the asymptotic variance of $\sqrt{n}\,(L-\theta)$, under some regularity conditions. Thus, if L is the best linear unbiased estimator, the test based on L is asymptotically equivalent to the most powerful (unbiased) test.

For the case when the shape of the distribution is unknown, an asymptotically (uniformly) efficient estimator is obtained by a similar but a little simpler way than the two-sample case. That is, we first fix some $k < n$, and determine a_t so as to minimize

$$
V\left(\sum_{i=1}^{k} a_t Y_{(t)} \middle| X_{(1)} < \ldots < X_{(n)}\right),
$$

where $Y_{(t)}$ denotes the order statistic of the randomly chosen subsample of size k. Then the conditional expectation

$$
L^* = E(\Sigma a_t Y_{(t)} | X_{(1)} < \ldots < X_{(n)})
$$

is shown [4] to be asymptotically equivalent to the best linear unbiased estimator.

The whole discussion above can also be applied to L^* and we obtain a uniformly asymptotically efficient (most powerful unbiased) test criterion.

A second, quicker, method of testing would be to make use of

$$L^*/\sqrt{V}, \tag{25}$$

where
$$V = \frac{n-k}{n}\Sigma\Sigma\, a_t a_{t'}(Y_t, Y_{t'} | X_{(1)} < \ldots < X_{(n)})$$

and approximate the distribution under the hypothesis by an appropriate t-distribution. An approximation discussed in [4] is to equate

$$\frac{L^*}{\sqrt{\dfrac{(n-k+1)}{n-1}V}}$$

to the t-distribution with $n-k+1$ degrees of freedom.

ACKNOWLEDGEMENTS

The author is grateful to Professor M. L. Puri, for the chance to attend the meeting and to present this paper. He is also indebted to Professor A. Birnbaum, Professor P. Huber, Professor J. L. Gastwirth for helpful discussions.

REFERENCES

[1] Chernoff, H., Gastwirth, J. L. and Johns, M. V. Asymptotic distribution of linear combination of functions of order statistics with applications to estimation. *Ann. Math. Statist.* 38, 1967.

[2] Hájek, J. Some extensions of the Wald–Wolfowitz–Noether Theorem. *Ann. Math. Statist.* 32, 1961.

[3] Moore, D. S. An elementary proof of asymptotic normality of linear functions of order statistics. *Ann. Math. Statist.* 39, 1968.

[4] Takeuchi, K. Uniformly asymptotically efficient robust estimators for a location parameter. *Tech. Rep. of Courant Inst. math. Sci.* 1969.

[5] Takeuchi, K. A polynomial approximation of the coefficient function of BLUE'S (to appear).

[6] Stein, C. Efficient non-parametric test and estimation. *Proc. Third Berkeley Symp. on Math. Statist. and Probab.* 1, 1956.

[7] Hájek, J. Asymptotically most powerful rank order tests. *Ann. Math. Statist.* 37, 1962.

[8] Huber, P. Studentization of robust estimators. This volume.

[9] Gastwirth, J. L. On robust procedures. *J. Am. Statist. Ass.* 61 (1966).

PART 3

ORDER STATISTICS AND
ALLIED PROBLEMS

DEVIATIONS BETWEEN
THE SAMPLE QUANTILE PROCESS
AND THE SAMPLE DF

J. KIEFER†

1 INTRODUCTION AND SUMMARY

Let $X_1, X_2, \ldots$ be iid with common twice differentiable univariate df F on the unit interval I. We assume $\inf_{x \in I} F'(x) > 0$ and $\sup_{x \in I} F''(x) < \infty$, and write $\xi_p = F^{-1}(p)$ for the p-tile of F. Let S_n be the sample df and let $Y_{p,n}$ be the sample p-tile, both based on $(X_1, \ldots, X_n)$; i.e.

$$nS_n(x) = [\text{number of } X_i \leqslant x, 1 \leqslant i \leqslant n],$$

and

$$Y_{p,n} = \inf\{x : S_n(x) = p\}.$$

(The choice we have made in this last definition when np is an integer, is immaterial.) Write $\sigma_p = [p(1-p)]^{\frac{1}{2}}$. Let

$$R_n(p) = Y_{p,n} - \xi_p + [S_n(\xi_p) - p]/F'(\xi_p). \tag{1}$$

Bahadur (1966) initiated the study of $R_n(p)$, and it was shown in [8], among other things, that, for $u > 0$,

$$\lim_{n \to \infty} P\{n^{\frac{3}{4}} F'(\xi_p) R_n(p) \leqslant u\} = 2 \int_0^\infty \Phi(k^{-\frac{1}{2}}u) \, d_k \Phi(k/\sigma_p)$$

(where Φ is the standard normal df), and that, for either choice of sign, $\# = +$ or $-$,

$$\limsup_{n \to \infty} \# \ F'(\xi_p) R_n(p)/[2^5 3^{-3} \sigma_p^2 n^{-3} (\log \log n)^3]^{\frac{1}{4}} = 1 \text{ wp } 1.$$

The behavior of $\sup_{|t| < T} R_n(p + n^{-\frac{1}{2}}t)$ was also studied, but results on the behavior of

$$
\left.
\begin{aligned}
R_n^{\#} &= \sup_{p \in I} \# \ F'(\xi_p) R_n(p), \quad \text{where } \# = + \text{ or } -, \\
R_n^* &= \max(R_n^+, R_n^-),
\end{aligned}
\right\} \tag{2}
$$

and the problem of studying these rv's, were only mentioned, without inclusion of proofs (§ 6.2 of [8]). In the present paper details will be given of the proofs of the following two results:

† Prepared under Contract No. NONR 401(50) for presentation at First International Conference on Nonparametric Inference, which the author was then unfortunately unable to attend. Reproduction is permitted for any purpose of the U.S. Government.

[299]

Theorem 1. *For* # = +, − *or* *, *and for* $t > 0$,

$$\lim_{n\to\infty} P\{n^{\frac{3}{4}}(\log n)^{-\frac{1}{2}}R_n^{\#} > t\} = 2\sum_{m=1}^{\infty}(-1)^{m+1}e^{-2m^2t^4}. \tag{3}$$

Theorem 2. *For* # = +, −, *or* *,

$$\limsup_{n\to\infty} n^{\frac{3}{4}}(\log n)^{-\frac{1}{2}}(\log\log n)^{-\frac{1}{4}}R_n^{\#} = 2^{-\frac{1}{4}}\text{wp } 1. \tag{4}$$

Readers familiar with the Kolmogorov–Smirnov statistic

$$D_n = n^{\frac{1}{2}}\sup_x |S_n(x) - F(x)|$$

will have noticed that (3) states that $n^{\frac{3}{4}}(\log n)^{-\frac{1}{2}}R_n^{\#}$ has the same limiting df as $D_n^{\frac{1}{2}}$. Indeed, we shall prove the more fundamental

Theorem 1*A*. *For* # = +, −, *or* *, *as* $n \to \infty$,

$$n^{\frac{3}{4}}R_n^{\#}/(D_n\log_n)^{\frac{1}{2}} \to 1 \quad \text{in probability}, \tag{5}$$

from which Theorem 1 follows at once.

There is also a strong law corresponding to Theorem 1 A, and Theorem 2 follows at once from it and the law of the iterated logarithm (LIL) for D_n [5]. However, my proof of this strong law is at present so long and tedious that I will omit it here and prove Theorem 2 directly. The reader familiar with LIL proofs will see that (4) is easy to *guess* from (3), in that the right side of (3) for $t = [2^{-1}(1+\epsilon)\log\log n_r]^{\frac{1}{4}}$, with $n_r \sim \gamma^r$ and $\gamma > 1$, is summable in r if and only if $\epsilon > 0$; thus, roughly, one hopes as usual to prove (4) by first treating such a result with $\{n\}$ replaced by $\{n_r\}$.

Equation (5) explains why, perhaps unexpectedly, $n^{\frac{3}{4}}(\log n)^{-\frac{1}{2}}R_n^{\#}$ has the same limiting law in Theorem 1 as each of the two rv's of which it is the maximum.

Some of the consequences of Theorem 2 are treated in [9] and [10]. They include

Theorem 3. (*A consequence of Theorem* 2 *and* [5].) *If* $\lambda_n \uparrow \infty$, *then*

$$P\{\sup_{0\leqslant p\leqslant 1} F'(\xi_p)\,|Y_{p,n} - \xi_p| > \lambda_n n^{-\frac{1}{2}}\text{i.o.}\}$$

$$= \begin{cases} 1 \Leftrightarrow \infty = \\ 0 \Leftrightarrow \infty > \end{cases} \sum_n n^{-1}\lambda_n^2 e^{-2\lambda_n^2}.$$

(The analogue for a single p is Theorem 4 of [8].)

Theorem 4. (*A consequence of Theorem* 2 *and* [6].) *If* $p_n \downarrow 0$ *and*

$$p_n > (\log\log n)/o(n),$$

then

$$P\{\limsup_{n\to\infty} [n^{-1}2p_n\log\log(np_n)]^{-\frac{1}{2}}F'(\xi_{p_n})[p_n - Y_{p_n,n}] = 1\text{ wp }1\}.$$

Theorem 5. In the Breiman–Brillinger Brownian bridge representation of the sample df [3], [4], [10] *with error* $\epsilon_n(p)$ *(say), there is a finite positive constant* C^* *such that*

$$\limsup_n \sup_p |\epsilon_n(p)| / [n(\log n)^2 (\log \log n)]^{\frac{1}{4}} = C^* \text{ wp } 1.$$

(Brillinger [4] showed that the order of $\sup_p |\epsilon_n(p)|$ was *no greater* than that just stated. The proof of this in [10] is much shorter and more elementary, and Theorem 2 makes it possible. Shortcomings of this representation are discussed in [9] and [10].)

The next paragraph justifies the fact that THE STATEMENT OF (4) AND THE STATEMENT OF (5) (AND HENCE, OF (3)) DO NOT DEPEND ON F OR THE 'SIGN' #.

We now mention a space-saving reduction. As in Lemma 1 of [8], we use Taylor's Theorem with remainder to obtain for the function R_n', defined to be the R_n of (1) when the X_i are replaced by the uniformly distributed rv's $F(X_i)$,

$$R_n'(p) - F'(\xi_p) R_n(p) = F''(\xi_p') (Y_{p,n} - \xi_p)^2 / 2F'(\xi_p), \tag{6}$$

where ξ_p' is a chance value between $Y_{p,n}$ and ξ_p.

The LIL for the sample df [5] easily shows that the right side of (6), uniformly in p, is $O(n^{-1}\log\log n)$ wp 1 as $n \to \infty$. Hence, IN THE PROOFS OF THEOREMS 1A AND 2 WE CAN AND DO HEREAFTER ASSUME F TO BE THE UNIFORM DF ON [0, 1]. MOREOVER, WE ONLY PROVE THE STATED THEOREMS FOR THE CASE # = +. The last is justified by noting that the statement of the theorems for R_n^+ computed from $\{X_i\}$ is equivalent to their statement for R_n^- computed from $\{1 - X_i\}$, and the statements for R_n^+ and R_n^- imply that for R_n^*.

We recall some definitions and preliminary results from [8]. We define

$$K_n(p) = n^{\frac{1}{2}}[p - S_n(p)],$$

$$T_n(x, p) = \begin{cases} S_n(x+p) - S_n(p) - x = S_n(x+p) - (x+p) + n^{-\frac{1}{2}}K_n(p) \\ \qquad\qquad\qquad\qquad\qquad\qquad\qquad\qquad \text{if } (x+p) \in I, \\ 0 \quad \text{if } (x+p) \notin I. \end{cases} \tag{7}$$

(The second line in the definition of T_n permits brevity in writing results like (10).) Then for F uniform it is clear, as in (1.12) of [8], that for $d > 0$, wp 1,

$$\begin{aligned} R_n(p) > d &\Leftrightarrow T_n(n^{-\frac{1}{2}}K_n(p) + d, p) < -d, \\ R_n(p) < -d &\Leftrightarrow T_n(n^{-\frac{1}{2}}K_n(p) - d, p) \geqslant d. \end{aligned} \tag{8}$$

If Z_n is binomially distributed with parameters N and π_N, the central limit theorem asserts, as in Lemma 2 of [8] (obtained from pp. 168–79 of [7]), as $N \to \infty$,

$$\{N\pi_N \to \infty; \limsup_N \pi_N < 1; \Delta_N > 0; \Delta_N^2/N\pi_N \to \infty; \Delta_N^3/N^2\pi_N^2 \to 0\}$$
$$\Rightarrow \log P\{Z_n - N\pi_N > \Delta_N\} \sim -\Delta_N^2/2N\pi_N(1-\pi_N). \quad (9)$$

Finally, Lemma 5 of [8] contains the result

$$\{0 < \sigma < 1; a > 0; w > 0; [n/a(1+w)]^{\frac{1}{2}} w(\sigma-a) \geqslant 2\}$$
$$\Rightarrow P\{\sup_{0\leqslant v\leqslant a} |T_n(v,p)| > w\} \leqslant \tfrac{4}{3}P\{|T_n(a^{(p)},p)| > w(1-\sigma)\}, \quad (10)$$

where $a^{(p)} = \min(a, 1-p)$. The last probabilistic expression is in fact a binomial probability. Using also the corresponding result for the rv's $1 - X_i$, we can obtain from (10) and (9), for δ fixed ($0 < \delta < 1$) and $n \to \infty$,

$$\{0 \leqslant p \leqslant 1; 0 < a_n \to 0; 0 < w_n \to 0; na_n \to \infty; w_n^2 n/a_n \to \infty; w_n^3 n/a_n^2 \to 0\}$$
$$\Rightarrow \log P\{\sup_{0\leqslant|v|\leqslant a_n} |T_n(v,p)| > w_n\} < -(1-\delta)^3 w_n^2 n/2a_n \quad (11)$$

for all $n > N_0$ (say), where N_0 depends on $\{a_n\}$ and $\{w_n\}$ and δ, but not on p. In obtaining this result, we use the fifth condition of (11) to verify the fourth condition of both (9) and (10); and we use the fact that, if $a_n + p > 1$, then the conditions of (11) imply that, for $n >$ an N_0 of the above form, the last expression of (10) is no greater than twice that obtained by replacing $a_n^{(p)}$ by $-a_n$.

We abbreviate 'A_n infinitely often' (= 'infinitely many of the events $A_1, A_2, \ldots$ occur') by 'A_n i.o. (n)', or simply 'A_n i.o.' if no ambiguity is possible, and 'A_n for almost all n' (= 'all but a finite number of the events $A_1, A_2, \ldots$ occur') by 'A_n for a.a.n.' The complement of an event A is denoted $\bar{A}$. By $\mathrm{int}\{L\}$ we mean the greatest integer $\leqslant L$. Orders such as $O(n)$ or $o(r)$ hold as $n \to +\infty$ or $r \to +\infty$ (the latter when, in §§ 4 and 5, we deal with a subsequence $\{n_r\}$ of the natural numbers).

In §§ 2 and 3 we shall prove, respectively, the statement obtained by considering only the right and left inequality of

$$\lim_{n\to\infty} P\{(1-\bar{\epsilon})D_n^{\frac{1}{2}} < n^{\frac{3}{4}}(\log n)^{-\frac{1}{2}} R_n^+ < (1+\bar{\epsilon})D_n^{\frac{1}{2}}\} = 1 \quad (12)$$

for $\bar{\epsilon} > 0$. This and the paragraph containing (6) yield Theorem 1A. Similarly, we obtain Theorem 2 by proving in §§ 4 and 5, respectively, the right and left halves of

$$P\{(1-\bar{\epsilon})2^{-\frac{1}{4}} < \limsup_{n\to\infty} n^{\frac{3}{4}}(\log n)^{-\frac{1}{2}}(\log\log n)^{-\frac{1}{4}} R_n^+ < (1+\bar{\epsilon})2^{-\frac{1}{4}}\} = 1. \quad (13)$$

Further comments (including remarks on the methods of proof) are contained in § 6.

2 WEAK UPPER BOUND

Fix $A > 0$. For $0 \leqslant p \leqslant 1$ and $d > 0$ and $L > 0$, we define the events

$$
\left.
\begin{aligned}
B_n(p, d) &= \{\sup_{|t| \leqslant A} R_n(p + n^{-\frac{1}{2}}t) > n^{-\frac{3}{4}}d\}, \\
C_n(L) &= \{\sup_{p \in I} |K_n(p)| \leqslant L\} = \{D_n \leqslant L\},
\end{aligned}
\right\}
\tag{14}
$$

and the set

$$
M_n = \{p \colon p = 1 \text{ or } p = 2jn^{-\frac{1}{2}}A, j \text{ integral}, 0 \leqslant p < 1\}.
$$

For $\# = +$, the statement obtained by considering only the right hand inequality in (12) will be shown in the next paragraph to follow from the fact (which we shall verify in this section) that, for each $L > 0$ and $c > L^{\frac{1}{2}}$,

$$
\lim_{A \to 0} \lim_{n \to \infty} P\{ \bigcup_{p \in M_n} [B_n(p, c[\log n]^{\frac{1}{2}}) \text{ and } |K_n(p)| \leqslant L]\} = 0. \tag{15}
$$

We define

$$
\psi(L) = \lim_{n \to \infty} P\{\overline{C_n(L)}\} = 2 \sum_{m=1}^{\infty} (-1)^{m+1} e^{-2m^2 L^2}.
$$

This is of course the tail probability of Kolmogorov's limiting df. To see that (15) implies the right half of (12), let $\epsilon' > 0$ be given and let $h > 0$ be such that

$$
\psi(h) - \psi(h^{-1}) > 1 - \epsilon'. \tag{16}
$$

The subset $\quad Q = \{(x_1, x_2) \colon h \leqslant x_1 \leqslant h^{-1}, x_2 \geqslant (1 + \bar{\epsilon}) x_1\} \tag{17}$

of the plane can clearly be covered by a finite number of sets Q_i of the form

$$
Q_i = \{(x_1, x_2) \colon x_1 \leqslant c_i, x_2 > (1 + \bar{\epsilon}/2) c_i\} \tag{18}
$$

with $c_i > 0$. Since $\{D_n \leqslant L\} \subset \{|K_n(p)| \leqslant L\}$ for each p, we see that (15) with $L^{\frac{1}{2}} = c_i$ and $c = (1 + \bar{\epsilon}/2) L^{\frac{1}{2}}$ implies that

$$
\lim_{n \to \infty} P\{(D_n^{\frac{1}{2}}, n^{\frac{3}{4}}(\log n)^{-\frac{1}{2}} R_n^+) \in Q_i\} = 0. \tag{19}
$$

Since finitely many Q_i cover Q, (19) holds with Q_i replaced by Q. Since ϵ' is arbitrarily small in (16), this establishes the right half of (12).

It remains to prove (15). Given $c > L^{\frac{1}{2}}$, choose $A > 0$ and δ $(0 < \delta < 1)$ so that

$$
c^2(1 - \delta)^4 \delta^2 > A \quad \text{and} \quad c^2(1 - \delta)^6 > A + L. \tag{20}
$$

Using first the first line of (8), and then the relations

$$T_n(x, p+t) = T_n(x+t, p) - T_n(t, p)$$

and

$$K_n(p+t) = K_n(p) - n^{\frac{1}{2}} T_n(t, p),$$

and lastly a trivial inclusion, we have

$$B_n(p, d) = \{\exists t, |t| \leqslant A, \text{with } T_n(n^{-\frac{1}{2}} K_n(p + n^{-\frac{1}{2}}t) + n^{-\frac{3}{4}}d, p + n^{-\frac{1}{2}}t) < -n^{-\frac{3}{4}}d\}$$

$$= \{\exists t, |t| \leqslant A, \text{with } T_n(n^{-\frac{1}{2}}t + n^{-\frac{1}{2}} K_n(p) - T_n(n^{-\frac{1}{2}}t, p)$$

$$+ n^{-\frac{3}{4}}d, p) - T_n(n^{-\frac{1}{2}}t, p) < -n^{-\frac{3}{4}}d\}$$

$$\subset \{\sup_{|t| \leqslant A} |T_n(n^{-\frac{1}{2}}t, p)| \geqslant n^{-\frac{3}{4}} \delta d\} \cup$$

$$\{\sup_{|t| \leqslant A + |K_n(p)| + n^{-\frac{1}{2}}(1+\delta)d} |T_n(n^{-\frac{1}{2}}t, p)| > n^{-\frac{3}{4}}(1-\delta)d\}$$

$$= G_n(p, d) \cup H_n(p, d) \quad \text{(say)}. \tag{21}$$

Applying (11) with $a_n = An^{-\frac{1}{2}}$ and $w_n = n^{-\frac{3}{4}} \delta c(\log n)^{\frac{1}{2}}$, we have

$$\log G_n(p, c[\log n]^{\frac{1}{2}}) < -[(1-\delta)^4 \delta^2 c^2 \log n]/2A \tag{22}$$

for all sufficiently large n, uniformly in p. On the other hand, applying (11) with

$$a_n = n^{-\frac{1}{2}}[A + L + n^{-\frac{1}{4}}(1+\delta) c(\log n)^{\frac{1}{2}}] \quad \text{and} \quad w_n = n^{-\frac{3}{4}}(1-\delta) c(\log n)^{\frac{1}{2}},$$

we have

$$\log P\{H_n(p, c[\log n]^{\frac{1}{2}}) \text{ and } |K_n(p)| \leqslant L\} < -[(1-\delta)^6 c^2 \log n]/2[A + L] \tag{23}$$

for all sufficiently large n, uniformly in p.

Since there are at most $(n^{\frac{1}{2}}/2A) + 2$ elements in M_n, equation (15) is a consequence of (22), (23), (20), and the inclusion between the first and last lines of (21).

3   WEAK LOWER BOUND

We now consider the expression consisting of only the lower bound inequality on R_n^+ in (12).

Let $\bar{\epsilon}$ be specified in (12) $(0 < \bar{\epsilon} < 1)$, and let g, H, b, ϵ be positive values, and let α be a positive integer, satisfying

$$H > 2; \quad b \geqslant g^2/(1 - 2H^{-1}) \geqslant g^2(1 - \bar{\epsilon}/2)^{-2}; \quad \epsilon < 1/(\alpha + 1)bH. \tag{24}$$

We shall devote this section to proving that

$$\lim_{n \to \infty} P\{D_n^{\frac{1}{2}} \geqslant b^{\frac{1}{2}}; n^{\frac{1}{4}}(\log n)^{-\frac{1}{2}} R_n^+ \leqslant g\} = 0. \tag{25}$$

Since $g \leqslant (1 - \bar{\epsilon}/2) b^{\frac{1}{2}}$ by (24), the lower half of (12) follows from (25) in exactly the manner that the upper half followed from (19).

Let

$$
\left.
\begin{aligned}
M_n^* &= \{j: 0 \leqslant j \leqslant \epsilon n^{\frac{1}{2}}, j \text{ integral}\}, \\
M_n^{*\prime} &= \{j: 0 \leqslant j \leqslant \epsilon n^{\frac{1}{2}} - 1, j \text{ integral}\}, \\
p_{n,j}^i &= i/(\alpha+1) + n^{-\frac{1}{2}} jbH \quad \text{for} \quad j \in M_n^* \quad \text{and} \quad 1 \leqslant i \leqslant \alpha, \\
p_{n,j}^{*i} &= p_{n,j}^i + n^{-\frac{1}{2}} K_n(p_{n,j}^i) + n^{-\frac{3}{4}} g(\log n)^{\frac{1}{2}}.
\end{aligned}
\right\} \quad (26)
$$

We write k_n^i for a vector with real components $k_{n,j}^i$, $j \in M_n^*$, abbreviate $K_n(p_{n,j}^i)$ as $K_{n,j}^i$, and write K_n^i for the random vector with components $K_{n,j}^i$, $j \in M_n^*$. We define the set

$$
\Lambda_n^i = \{k_n^i: 0 < b \leqslant k_{n,j}^i < bH/2 \quad \text{for} \quad j \in M_n^*\}, \quad (27)
$$

and the events

$$
\left.
\begin{aligned}
\Lambda_n^{i*} &= \{K_n^i \in \Lambda_n^i\}, \quad \Lambda_n^{i**} = \{K_n^i \text{ or } -K_n^i \in \Lambda_n^i\}, \\
\Lambda_n^{**} &= \bigcup_{1 \leqslant i \leqslant \alpha} \Lambda_n^{i**}, \\
A_{n,j}^i &= \{R_n(p_{n,j}^i) > n^{-\frac{3}{4}} g(\log n)^{\frac{1}{2}}\} \quad \text{for} \quad j \in M_n^* \quad \text{and} \quad 1 \leqslant i \leqslant \alpha; \\
A_n^i &= \bigcup_{j \in M_n^{*\prime}} A_{n,j}^i.
\end{aligned}
\right\} \quad (28)
$$

If Λ_n^{i*} occurs, then by (24), (26), and (27), for all sufficiently large n we have $p_{n,j}^i < p_{n,j}^{*i} < p_{n,j+1}^i$ whenever $j \in M_n^{*\prime}$. Hence, for such j, from (8) and (7),

$$
\begin{aligned}
P\{A_{n,j}^i | \Lambda_n^{i*}; K_n^i = k_n^i\} \\
= P\{S_n(p_{n,j}^{*i}) - p_{n,j}^{*i} + n^{-\frac{1}{2}} k_{n,j}^i \\
< -n^{-\frac{3}{4}} g(\log n)^{\frac{1}{2}} | K_{n,j}^i = k_{n,j}^i; K_{n,j+1}^i = k_{n,j+1}^i\}. \quad (29)
\end{aligned}
$$

This last is a binomial probability; in the notation of (9), under the conditioning of (29), the number of X_t ($1 \leqslant t \leqslant n$) falling between $p_{n,j}^i$ and $p_{n,j+1}^i$ is

$$
N = n^{\frac{1}{2}}(bH + u_{n,j}^i), \quad (30)
$$

where we have written (and will write)

$$
u_{n,j}^i = k_{n,j}^i - k_{n,j+1}^i; \quad U_{n,j}^i = K_{n,j}^i - K_{n,j+1}^i; \quad (31)
$$

the (conditional) probability that any of these N conditionally independent rv's X_t falls between $p_{n,j}^{*i}$ and $p_{n,j+1}^i$ is

$$
\pi_N = (p_{n,j+1}^i - p_{n,j}^{*i})/(p_{n,j+1}^i - p_{n,j}^i) = 1 - (bH)^{-1}[k_{n,j}^i + n^{-\frac{1}{4}} g(\log n)^{\frac{1}{2}}]; \quad (32)
$$

and the associated binomial rv is

$$Z_N = n[S_n(p_{n,\,j+1}^i) - S_n(p_{n,\,j}^{*i})].$$

If we define

$$\Delta_N = (bH)^{-1}[n^{\frac{1}{2}}k_{n,\,j}^i u_{n,\,j}^i + (bH + u_{n,\,j}^i)\,n^{\frac{1}{4}}g(\log n)^{\frac{1}{2}}], \qquad (33)$$

we can rewrite (29) in the notation of (9) as

$$P\{S_n(p_{n,\,j+1}^i) - S_n(p_{n,\,j}^{*i})$$

$$> n^{-\frac{1}{2}}(bH - k_{n,\,j+1}^i)|K_{n,\,j}^i = k_{n,\,j}^i;\; K_{n,\,j+1}^i = k_{n,\,j+1}^i\} = P\{Z_N - N\pi_N > \Delta_N\}. \qquad (34)$$

Define the subset of M_n^*,

$$J_n^i(k_n^i) = \{j: j \in M_n^{*\prime};\; u_{n,\,j}^i \leqslant 0\},$$

and let $|J_n^i(k_n^i)|$ denote the cardinality of this set. Also define the set and the events

$$\left.\begin{array}{l} C_n^i = \{k_n^i: |J_n^i(k_n^i)| \geqslant \epsilon n^{\frac{1}{2}}/3\}, \\[2mm] C_n^{i*} = \{K_n^i \in C_n^i\}, \quad C_n^* = \bigcap_{1 \leqslant i \leqslant \alpha} C_n^{i*}. \end{array}\right\} \qquad (35)$$

At the end of this section we shall prove that, if $n^{-\frac{1}{2}}bH < \frac{1}{4}$, then

$$P\{\overline{C_n^{i*}}\} \leqslant c_0[\epsilon^{-1}n^{-\frac{1}{2}} + (bH)^{-\frac{1}{2}}n^{-\frac{1}{4}}], \qquad (36)$$

where c_0 is a universal constant.

In the next two sentences we consider N, π_N, Δ_N as functions of k and u without regard to the relation (31) between the latter. For fixed $k_{n,\,j}^i$, let $c_{n,\,j}^i$ be the value of $u_{n,\,j}^{i\prime}$ which makes the expression (33) equal zero. It is easy to verify that, for $k_n^i \in \Lambda_n^i$, the expression $\Delta_N^2/N\pi_N(1-\pi_N)$ obtained from (30)–(33) is strictly increasing in $u_{n,\,j}^i$ for $u_{n,\,j}^i \geqslant c_{n,\,j}^i$. Consequently, for $k_n^i \in \Lambda_n^{i\prime}$ and $j \in J_n^i(k_n^i)$, either $u_{n,\,j}^i < c_{n,\,j}^i$, in which case the probability of (34) is $< \frac{3}{4}$ for all sufficiently large n, or else $c_{n,\,j}^i \leqslant u_{n,\,j}^i \leqslant 0$, in which case (substituting this upper bound on $u_{n,\,j}^i$ into $\Delta_N^2/N\pi_N(1-\pi_N)$),

$$\Delta_N^2/N\pi_N(1-\pi_N) < g^2(\log n)\,[1+o(1)]/k_{n,\,j}^i(1 - k_{n,\,j}^i/bH)$$

$$< g^2(\log n)\,[1+o(1)]/b(1 - H^{-1}), \qquad (37)$$

the last inequality following from the fact that the concave function $k(1 - k/bH)$ on the interval $b \leqslant k \leqslant bH/2$ attains its minimum at an end point, and that $0 < b(1 - H^{-1}) < bH/4$ by (24). Moreover, it is easily seen that in the bounds of (37) the $o(1)$ term as $n \to \infty$ is uniform for $k_n^i \in \Lambda_n^i$ and j satisfying $c_{n,\,j}^i \leqslant u_{n,\,j}^i \leqslant 0$. This uniformity persists in the application of (9). Since the events $\{A_{n,\,j}^i, j \in M_n^{*\prime}\}$ for fixed n are conditionally independent, conditioned on the event $\{K_n^i = k_n^i\}$, we obtain from (35) and (37) (and the fact that half the last expression there exceeds $-\log(\tfrac{3}{4})$ for large n) for n sufficiently large, uniformly for $k_n^i \in \Lambda_n^i \cap C_n^i$,

$$P\{\overline{A_n^i}|\Lambda_n^{i*}; C_n^{i*}; K_n^i = k_n^i\} = P\{\bigcap_{j \in M_n^{\bullet'}} \overline{A_{n,j}^i}|\Lambda_n^{i*}; C_n^{i*}; K_n^i = k_n^i\}$$

$$\leqslant \prod_{j \in J_n^i(k_n^i)} P\{\overline{A_{n,j}^i}|\Lambda_n^{i*}; C_n^{i*}; K_n^i = k_n^i\}$$

$$\leqslant [1 - n^{-g^2[1+o(1)]/2b(1-H^{-1})}]^{\epsilon n^{\frac{1}{2}}/3}. \tag{38}$$

The last expression approaches 0 as $n \to \infty$, provided that

$$g^2/2b(1 - H^{-1}) < \tfrac{1}{2}, \tag{39}$$

and (39) follows at once from (24).

We note, using the second line of (28) and the elementary relation $(\bigcup_i \Lambda_i) \cap (\bigcup_i A_i) \supset \bigcup_i \Lambda_i - \bigcup_i (\Lambda_i \cap \overline{A_i})$, the simple probabilistic relations

$$P\{\Lambda_n^{**} \cap \bigcup_{1 \leqslant i \leqslant \alpha} A_n^i\} \geqslant P\{\Lambda_n^{**}\} - P\{\bigcup_{1 \leqslant i \leqslant \alpha} [\overline{\Lambda_n^{i**}} \cap A_n^i]\}$$

$$\geqslant P\{\Lambda_n^{**}\} - \left[P\{\overline{C_n^*}\} + \sum_{i=1}^{\alpha} P\{\overline{A_n^i}|\Lambda_n^{i**}; C_n^{i*}\} \right]. \tag{40}$$

All of the above development was for fixed $g, H, b, \epsilon, \alpha$ satisfying (24), and of course the events appearing in (38) and (40) depend on those values. Still fixing those values, let $n \to \infty$ in (40). The previous paragraph, its analogue when Λ_n^{i*} is replaced by $\Lambda_n^{i**} - \Lambda_n^{i*}$, and (36), imply that the expression in square brackets, in the last line of (40), approaches zero. Thus, (40) yields

$$\lim_n P\{\Lambda_n^{**} \cap \overline{\bigcup_{1 \leqslant i \leqslant \alpha} A_n^i}\} = 0, \tag{41}$$

and hence, by inclusion,

$$\lim_n P\{\Lambda_n^{**} \cap \{R_n^+ \leqslant n^{-\frac{3}{4}}g(\log n)^{\frac{1}{2}}\}\} = 0. \tag{42}$$

It is well known that

$$\lim_{\epsilon \downarrow 0} \lim_n [P\{\bigcup_{1 \leqslant i \leqslant \alpha} \{b \leqslant |K_n(i/(\alpha+1))| < bH/2\}\} - P\{\Lambda_n^{**}\}] = 0. \tag{43}$$

Since Λ_n^{**} is a subset of the union in (43), we obtain from (43) on letting $\epsilon \downarrow 0$ in (42),

$$\lim_n P\{\{R_n^+ \leqslant n^{-\frac{3}{4}}g(\log n)^{\frac{1}{2}}\} \cap \bigcup_{1 \leqslant i \leqslant \alpha} \{b \leqslant |K_n(i/(\alpha+1))| < bH/2\}\} = 0. \tag{44}$$

Letting $H \to \infty$ in (44), and using the fact that the event expressed as a union there has a probability whose limit in H is attained uniformly in n, we obtain

$$\lim_n P\{R_n^+ \leqslant n^{-\frac{3}{4}} g(\log n)^{\frac{1}{2}}; \ \max_{1 \leqslant i \leqslant \alpha} |K_n(i/(\alpha+1))| \geqslant b\} = 0. \tag{45}$$

The event concerning K_n in the second half of (45) is a subset of the event $\{D_n \geqslant b\}$ and (invariance principle) their difference has a limiting probability, as $n \to \infty$, which can be made arbitrarily close to 0 by taking α sufficiently large. This fact and (45) yield (25).

This completes the proof of the lower inequality of (12), except for the proof of (36), to which we now turn.

Throughout this paragraph, symbols θ_i denote values satisfying $|\theta_i| \leqslant 1$ and symbols c_i denote universal constants. We write

$$\pi = n^{-\frac{1}{2}} bH = p_{n,j+1}^i - p_{n,j}^i$$

and assume $\qquad \pi < \frac{1}{4}, \ c_1(n\pi)^{-\frac{1}{2}} < \frac{1}{12}, \ n^{\frac{1}{2}}\epsilon > 10, \tag{46}$

where c_1 is the absolute constant of (48) below; the first restriction of (46) is that assumed just above (36), and the other two may be assumed because (perhaps with an enlarged c_0) (36) is trivial if either is violated. $U_{n,j}^i$ is of the form $n^{-\frac{1}{2}}(V - EV)$, where V is binomial, the sum of n iid Bernoulli rv's, each with expectation π. For $\pi < \frac{1}{4}$ (as specified in (46)), such a Bernoulli rv W_j satisfies

$$E|W_j - \pi|^3/[\mathrm{var}\,(W_j)]^{\frac{3}{2}} = [\pi^3(1-\pi) + \pi(1-\pi)^3]/[\pi(1-\pi)]^{\frac{3}{2}} \leqslant \pi^{-\frac{1}{2}}, \tag{47}$$

since $[\pi^2 + (1-\pi)^2](1-\pi)^{-\frac{1}{2}} \leqslant 1$ for $0 \leqslant \pi \leqslant \frac{1}{4}$. The Berry–Esseen bound and (47) yield, for $\pi < \frac{1}{4}$,

$$P\{U_{n,j}^i \leqslant 0\} = \frac{1}{2} + c_1\theta_1(n\pi)^{-\frac{1}{2}}. \tag{48}$$

For j_1, j_2 distinct integers in $M_n^{*\prime}$, if W_j is the indicator of the event $p_{n,j}^i < X_1 < p_{n,j+1}^i$, we have $E\{W_{j_1} W_{j_2}\} = 0$ and hence $\mathrm{cov}\,(W_{j_1}, W_{j_2}) = -\pi^2$. The inverse of the covariance matrix of W_{j_1}, W_{j_2} is hence

$$\begin{pmatrix} \pi(1-\pi) & -\pi^2 \\ -\pi^2 & \pi(1-\pi) \end{pmatrix}^{-1} = (1-2\pi)^{-1} \begin{pmatrix} \pi^{-1}-1 & 1 \\ 1 & \pi^{-1}-1 \end{pmatrix}, \tag{49}$$

which has maximum element $\Delta = (1-2\pi)^{-1}(\pi^{-1}-1)$. The bivariate normal law with means zero, variances $\pi(1-\pi)$, and covariance $-\pi^2$, assigns probability $P_\pi(\text{say}) < \frac{1}{4}$ to the first quadrant. (Elementary trigonometry in fact yields $P_\pi = \frac{1}{4} - c_5|\theta_5|\pi$, but this is neither helpful nor needed.) Bergstrom's[2] multivariate generalization of the Berry–Esseen bound asserts that

$$P\{U^i_{n,j_1} \leqslant 0,\ U^i_{n,j_2} \leqslant 0\} = P_\pi + c_2\theta_2\Delta^{\frac{3}{2}} \sum_{i=1}^{2} E\,|W_{j_i} - \pi|^3 n^{-\frac{1}{2}}$$

$$< \tfrac{1}{4} + c_3\theta_3(n\pi)^{-\frac{1}{2}}, \quad (50)$$

since $\Delta^{\frac{3}{2}} < 4\pi^{-\frac{3}{2}}$ for $0 < \pi < \tfrac{1}{4}$, and $E\,|W_{j_t} - \pi|^3$ is as in (47). If $Z^i_{n,j}$ is the indicator of the event $\{U^i_{n,j} \leqslant 0\}$, then (see (35)) the rv $|J^i_n(K^i_n)|$, hereafter denoted $|J|$, is simply $\sum_{j\in M^{*\prime}_n} Z^i_{n,j}$. For fixed n and i, the $Z^i_{n,j}$ are not independent; however, by (48) and (50), we can assert, for $j, j_1, j_2 \in M^{*\prime}_n$ and $j_1 \neq j_2$,

$$\left.\begin{aligned}
E\{Z^i_{n,j}\} &= \tfrac{1}{2} + c_1\theta_1(n\pi)^{-\frac{1}{2}}, \\
\operatorname{var}(Z^i_{n,j}) &= \tfrac{1}{4} - c_1^2\theta_1^2(n\pi)^{-1} \leqslant \tfrac{1}{4}, \\
\operatorname{cov}(Z^i_{n,j_1}, Z^i_{n,j_2}) &< \tfrac{1}{4} + c_3\theta_3(n\pi)^{-\frac{1}{2}} - [\tfrac{1}{2} + c_1\theta_1(n\pi)^{-\frac{1}{2}}]^2 \\
&= c_4\theta_4(n\pi)^{-\frac{1}{2}} - c_1^2\theta_1^2(n\pi)^{-1} \leqslant c_4(n\pi)^{-\frac{1}{2}}.
\end{aligned}\right\} \quad (51)$$

Since the number of elements m in $M^{*\prime}_n$ satisfies $n^{\frac{1}{2}}\epsilon - 1 \leqslant m \leqslant n^{\frac{1}{2}}\epsilon$, we have (using the last two inequalities of (46) in the next line)

$$\left.\begin{aligned}
E\{|J|\} &= mE\{Z^i_{n,0}\} \geqslant (n^{\frac{1}{2}}\epsilon - 1)(\tfrac{1}{2} - c_1(n\pi)^{-\frac{1}{2}}) \geqslant 3n^{\frac{1}{2}}\epsilon/8; \\
\operatorname{var}(|J|) &< n^{\frac{1}{2}}\epsilon/4 + (n^{\frac{1}{2}}\epsilon)^2 c_4(n\pi)^{-\frac{1}{2}}.
\end{aligned}\right\} \quad (52)$$

Hence,

$$\operatorname{var}(|J|)/[E\{|J| - n^{\frac{1}{2}}\epsilon/3\}]^2 < 576[n^{-\frac{1}{2}}\epsilon^{-1}/4 + c_4 n^{-\frac{1}{4}}(bH)^{-\frac{1}{2}}], \quad (53)$$

which, with Chebyshev's inequality, establishes (36).

4 STRONG UPPER BOUND

We now verify the statement obtained by considering only the right hand inequality of (13). Given a value $\bar{\epsilon} > 0$ there, fix $A > 0$ and choose $\delta\,(0 < \delta < 1)$ and $\gamma > 1$ to satisfy

$$(1-\delta)^8(1+\bar{\epsilon})^2/(1+2\delta) > 1 \quad \text{and} \quad \delta^2(1-\delta)^7(1+\bar{\epsilon})^2/4(\gamma-1)(1+2\delta) > 1. \quad (54)$$

We let $\{n_r,\ r = 1, 2, \ldots\}$ be any increasing sequence of positive integers for which, as $r \to \infty$,

$$n_r \sim \gamma^r. \quad (55)$$

We define $S_{n', n''}$ for $n' < n''$ as the sample df based on

$$(X_{n'+1}, X_{n'+2}, \ldots, X_{n''}),$$

and $T_{n', n''}$ is then obtained by replacing S_n by $S_{n', n''}$ in the first form on the right in the second line of (7). We define M_n as in § 2, and

$$
\left.
\begin{aligned}
I_r &= \{n: n_r < n \leqslant n_{r+1}, \; n \text{ integral}\}, \\
d_n &= [2^{-1}(\log n)^2 \log \log n]^{\frac{1}{2}} (1 + \bar{e}), \\
\Gamma_n(p) &= n_{r_L}^{-\frac{1}{2}} A + n^{-\frac{1}{2}} |K_n(p)| + n^{-\frac{3}{4}}(1 + \delta) \, d_n \quad \text{for } n \in I_r, \\
\Gamma_r &= (1 + 2\delta) \, [(2n_r)^{-1} \log \log n_r]^{\frac{1}{2}}.
\end{aligned}
\right\}
\tag{56}
$$

Recalling (8), (14), and the first equation of (21), we see that our desired upper class result will follow from

$$
\bigcup_{n \in I_r} \bigcup_{p \in M_{n_r}} \{\exists t, |t| \leqslant A, \text{ with } T_n(n^{-\frac{1}{2}} K_n(p + n_r^{-\frac{1}{2}}t) + n^{-\frac{3}{4}} d_n, p + n_r^{-\frac{1}{2}}t)
$$
$$
< -n^{-\frac{3}{4}} d_n\} \text{ occurs only finitely often } (r), \text{ wp } 1. \tag{57}
$$

(Note in (57) and in the sequel that the same sets

$$
\{t: |n_r^{-\frac{1}{2}}t - p| \leqslant A\}, \quad p \in M_{n_r},
$$

are used in studying what occurs for all $n \in I_r$.) From the remaining relations of (21), modified to take into account the departure of (57) from the second expression of (21), we see that (57) is a consequence of

$$
P\{\bigcup_{n \in I_r} \bigcup_{p \in M_{n_r}} \{\sup_{|t| \leqslant A} |T_n(n_r^{-\frac{1}{2}}t, p)| \geqslant n^{-\frac{3}{4}} \delta d_n\} \text{ i.o. } (r)\} = 0 \tag{58}
$$

and of

$$
P\{\bigcup_{n \in I_r} \bigcup_{p \in M_{n_r}} \{\sup_{|t| \leqslant \Gamma_n(p)} |T_n(t, p)| > n^{-\frac{3}{4}}(1 - \delta) \, d_n\} \text{ i.o. } (r)\} = 0. \tag{59}
$$

Moreover, since $\sup_{p \in I} |K_n(p)| < (1 + \delta)(2^{-1} \log \log n)^{\frac{1}{2}}$ for a.a. n wp 1 by the LIL for the sample df [5], elementary estimation yields

$$
\sup_{n \in I_r, \, p \in M_{n_r}} \Gamma_n(p) < \Gamma_r \quad \text{for a.a. } r, \text{ wp } 1. \tag{60}
$$

Since $n^{-\frac{3}{4}} d_n > n^{-1} n_r^{\frac{1}{4}} d_{n_r}$ for $n \in I_r$ and large r, we see that (59) will follow from

$$
P\{\bigcup_{p \in M_{n_r}} \{\sup_{|t| \leqslant \Gamma_r} |T_{n_r}(t, p)| > n_r^{-\frac{3}{4}}(1 - \delta)^2 d_{n_r}\} \text{ i.o. } (r)\} = 0, \tag{61}
$$

$$
P\{\bigcup_{p \in M_{n_r}} \{\max_{n \in I_r} \sup_{|t| \leqslant \Gamma_r} (n - n_r) |T_{n_r, n}(t, p)| \geqslant \delta(1 - \delta) n_r^{\frac{1}{4}} d_{n_r}\} \text{ i.o. } (r)\} = 0, \tag{62}
$$

and the obvious relation

$$
(n - n_r) \, T_{n_r, n} + n_r T_{n_r} = n T_n \quad \text{for } n \in I_r. \tag{63}
$$

It remains to prove (58), (61), and (62). In each of these (11) will be applied for a single sequence of pairs (a_n, w_n), so there is no difficulty from the increasing size of M_{n_r}. For fixed p and n, the probability of the event in the inside braces of (58) is estimated, as in (22) but now using $a_n = An_r^{-\frac{1}{2}}$ and $w_n = n^{-\frac{3}{4}} \delta d_n$ in (11), to be no greater than

$$\exp\left\{ -(1-\delta)^4 \delta^2 (1+\bar{\varepsilon})^2 \left[2^{-1} n^{-1} \log\log n \right]^{\frac{1}{2}} (\log n)/2An_r^{-\frac{1}{2}} \right\} \qquad (64)$$

for all sufficiently large n (where, of course, $n \in I_r$ in (64)). The product of (64) with the upper bound $(n_r^{\frac{1}{2}}/2A) + 2$ on the number of elements in M_{n_r} is clearly summable in n. This and the Borel–Cantelli lemma yield (58).

As for (61), the expression in braces there for fixed p and r is similarly seen to have probability no greater, for large r, than

$$\exp\left\{ -(1-\delta)^8 n_r^{-\frac{1}{2}} d_{n_r}^2 / 2\Gamma_r \right\} = \exp\left\{ -(1-\delta)^8 (1+\bar{\varepsilon})^2 (\log n_r)/2(1+2\delta) \right\}. \qquad (65)$$

The product of (65) with $(n_r^{\frac{1}{2}}/2A) + 2$ is summable in r by the first condition of (54), yielding (61).

Finally, we turn to (62). We shall carry through the details for one quarter of the proof, restricting attention to positive t and T in (62) and proving that

$$\bigcup_{p \in M_{n_r}} \left\{ \max_{n \in I_r} \sup_{0 \leqslant t \leqslant \Gamma_r} (n - n_r) T_{n_r, n}(t, p) \geqslant \delta(1-\delta) n_r^{\frac{1}{4}} d_{n_r} \right\} \qquad (66)$$

occurs only finitely often (r) wp 1; the other three choices of sign are handled similarly, and (62) can be violated only if, for one of the four cases (of which (66) is one), the counterpart of (66) occurs i.o. (r) wp > 0.

Fix r and $p \in M_{n_r}$, and suppose the event in braces in (66) occurs. Call this event $D_r(p)$. Let ν be the smallest n in I_r for which the supremum in (66) satisfies the inequality there, and then let the rv Z be the smallest positive value of z ($0 < z \leqslant 1$) for which

$$(\nu - n_r) T_{n_r, \nu}(z\Gamma_r, p) \geqslant \delta(1-\delta) n_r^{\frac{1}{4}} d_{n_r}. \qquad (67)$$

The conditional probability

$$P\{(n_{r+1} - n) T_{n, n_{r+1}}(z\Gamma_r, p) < -\delta(1-\delta) n_r^{\frac{1}{4}} d_{n_r}/2 \,|\, D_r(p); \nu = n; Z = z\} \qquad (68)$$

is a binomial probability $P\{Z_N < N\pi_N + \Delta_N\}$, where the binomial variate Z_N is the sum of N iid Bernoulli variates, each of expectation π_N, and

$$N = n_{r+1} - n, \quad \pi_N = z\Gamma_r, \quad \Delta_N = -\delta(1-\delta) n_r^{\frac{1}{4}} d_{n_r}/2,$$

so that $1 - \pi_N \to 1$ uniformly in $z \in (0, 1]$ as $r \to \infty$, and

$$\Delta_N^2/N\pi_N = \delta^2(1-\delta)^2 (1+\bar{\varepsilon})^2 (\log n_r) n_r/4(n_{r+1} - n) z(1+2\delta)$$

approaches infinity uniformly in $n \in I_r$ and $z \in (0, 1]$ (and also in p) as $r \to \infty$. Hence, by Chebyshev's inequality, the probability (68) is $\leqslant \frac{1}{4}$

for all large r, uniformly in $n \in I_r$ and z and p. (The probability is zero if $n = n_{r+1}$.) From (67) and the event complementary to that of (68) we obtain, for large r,

$$P\{(n_{r+1}-n_r)\sup_{0\leqslant t\leqslant\Gamma_r}T_{n_r,\,n_{r+1}}(t,p) > \delta(1-\delta)\,n_r^{\frac{1}{2}}\,d_{n_r}/2\,|\,D_r(p)\} \geqslant \tfrac{3}{4}. \quad (69)$$

Multiplying both sides of (69) by $P\{D_r(p)\}$, we obtain by inclusion,

$$P\{D_r(p)\} \leqslant (\tfrac{4}{3})\,P\{(n_{r+1}-n_r)\sup_{0\leqslant t\leqslant\Gamma_r}T_{n_r,\,n_{r+1}}(t,p) > \delta(1-\delta)\,n_r^{\frac{1}{2}}\,d_{n_r}/2\}. \quad (70)$$

The probability on the right side of (70) is, by (11), no greater than

$$\exp\{-\delta^2(1-\delta)^6\,n_r^{\frac{1}{2}}\,d_{n_r}^2/8(n_{r+1}-n_r)\,\Gamma_r\}$$

$$< \exp\{-\delta^2(1-\delta)^7\,(1+\bar{\epsilon})^2\,(\log n_r)/8(\gamma-1)\,(1+2\delta)\} \quad (71)$$

for large r, uniformly in p. Multiplying this by $(n_r^{\frac{1}{2}}/2A)+2$, the result is summable in r by the second condition of (54). Thus, (70) and (71) imply that (66) occurs i.o. (r) wp 0, and the proof of this section is complete.

5 STRONG LOWER BOUND

For typographical simplification we write 4ϵ for the $\bar{\epsilon}$ of (13) and suppose ϵ is specified,
$$0 < \epsilon = \bar{\epsilon}/4 < \tfrac{1}{72}.$$

Let δ and γ be positive values satisfying the following conditions (written somewhat redundantly for the sake of easy reference):

$$\left.\begin{array}{l} \delta < 1;\ \gamma > \epsilon^{-1};\ \gamma > 2^{\frac{5}{2}}\epsilon^{-2}(1-\delta)^{-7};\ (1-\gamma^{-1})^{\frac{3}{2}} > 2\epsilon; \\[4pt] \gamma > 36\epsilon^{-6};\ \gamma^{-1} < \epsilon^3/4;\ |\epsilon^4(1-\gamma^{-1})-\gamma^{-1}| < \epsilon^3/4. \end{array}\right\} \quad (72)$$

Let $\{n_r, r = 1, 2, ...\}$, as in §4, be any increasing sequence of positive integers satisfying (as $r \to \infty$)

$$n_r \sim \gamma_r. \quad (73)$$

Abbreviating the notation introduced below (55) (since n', n'' will always be n_r, n_{r+1} here), we write $m_r = n_{r+1}-n_r$, and let S_r' be the sample df based on the $n_{r+1}-n_r$ observations $X_{n_r+1}, X_{n_r+2}, ..., X_{n_{r+1}}$. Similarly, let $K_r'(p)$ be defined by (7) with S_n replaced by S_r' and $n^{\frac{1}{2}}$ replaced by $m_r^{\frac{1}{2}}$, and let T_r' be obtained by altering S_n, $n^{-\frac{1}{2}}$, and K_n in this way in (7). We shall repeatedly use asymptotic ($r \to \infty$) relations such as

$$m_r \sim (1-\gamma^{-1})\,n_{r+1}, \quad \log n_r \sim \log n_{r+1}, \quad \log\log m_r \sim \log r, \quad \text{etc.},$$

without further comment.

Let
$$d'_n = 2^{-\frac{1}{4}}n^{-\frac{3}{4}}(\log n)^{\frac{1}{2}}(\log\log n)^{\frac{1}{4}}.\tag{74}$$
We shall show
$$P\{\inf_{p\in I} T_{n_{r+1}}(n_{r+1}^{-\frac{1}{2}}K_{n_{r+1}}(p)+(1-4\epsilon)\,d'_{n_{r+1}},p)<-(1-4\epsilon)\,d'_{n_{r+1}}\text{ i.o.}\}=1,\tag{75}$$
which by (8) implies the left inequality of (13).

To obtain (75), we shall show
$$P\{\inf_{p\in I} T'_r(n_{r+1}^{-\frac{1}{2}}K_{n_{r+1}}(p)+(1-4\epsilon)\,d'_{n_{r+1}},p)<-(1-2\epsilon)\,d'_{n_{r+1}}\text{ i.o.}\}=1\tag{76}$$
and
$$P\{\sup_{p\in I} T_{n_r}(n_{r+1}^{-\frac{1}{2}}K_{n_{r+1}}(p)+(1-4\epsilon)\,d'_{n_{r+1}},p)\leqslant\epsilon\gamma d'_{n_{r+1}}\text{ for a.a. }r\}=1;\tag{77}$$

(75) is an obvious consequence of (76), (77), and the facts that (as in (63)
$$n_r T_{n_r}(x,p)+m_r T'_r(x,p)=n_{r+1} T_{n_{r+1}}(x,p)\tag{78}$$
and (a consequence of the second condition of (72)), for large r,
$$n_r\epsilon\gamma+m_r(-1+2\epsilon)<-n_{r+1}(1-4\epsilon).\tag{79}$$

We first prove (77) (by a calculation similar to that used to obtain (61)). Consider the event in the first line of (80) below. Writing
$$L=\gamma^{-\frac{1}{2}}n_r^{-\frac{1}{2}}(\log\log n_r)^{\frac{1}{2}}+(1-4\epsilon)\,d'_{n_{r+1}}\quad\text{and}\quad\rho=\gamma\epsilon d'_{n_{r+1}},$$
we see that this event is contained in the event
$$\{\sup_{p\in I}\sup_{|\theta|\leqslant 1}|T_{n_r}(\theta L,p)|>\rho\},$$
which in turn entails the occurrence, for at least one p in the set
$$M''=\{p\colon p=1\text{ or }p=jL,\ 0\leqslant p<1,j\text{ integral}\},$$
of the event $\{\sup_{|\theta|\leqslant 1}|T_{n_r}(\theta L,p)|>\rho/2\}$. Computing an upper bound on this last event by applying (11) with (n,a_n,w_n) of (11) $=(n_r,L,\rho/2)$ here, and noting that M'' contains fewer than $n_r^{\frac{1}{2}}$ elements when r is large and that $d'_{n_{r+1}}=o(L)$; we obtain, for r sufficiently large,
$$P\{\sup_{p\in I}\sup_{|\theta|\leqslant 1}|T_{n_r}(\theta\gamma^{-\frac{1}{2}}n_r^{-\frac{1}{2}}(\log\log n_r)^{\frac{1}{2}}+(1-4\epsilon)\,d'_{n_{r+1}},p)|>\gamma\epsilon d'_{n_{r+1}}\}$$
$$\leqslant n_r^{\frac{1}{2}}\exp\{-(1-\delta)^6\,n_r(\gamma\epsilon d'_{n_{r+1}}/2)^2/2\gamma^{-\frac{1}{2}}(\log\log n_r)^{\frac{1}{2}}n_r^{-\frac{1}{2}}\}$$
$$\leqslant n_r^{\frac{1}{2}}\exp\{-\gamma\epsilon^2(\log n_r)(1-\delta)^7/2^{\frac{7}{2}}\}.\tag{80}$$

By the third condition of (72), the last expression is summable in r, which by the Borel–Cantelli lemma proves that the event of the first line of (80) occurs i.o. (r) wp 0. But for each large r the event of the first line of (80) contains the *complement* of the event of (77), because

$$\sup_{p} \left| n_{r+1}^{-\frac{1}{2}} K_{n_{r+1}}(p) \right| < [(\log\log n_r)(1+\delta)/2\gamma n_r]^{\frac{1}{2}}$$

for a.a. r wp 1 (by the LIL for the sample df) and $[(1+\delta)/2]^{\frac{1}{2}} < 1$ (by (72)). Hence, the complement of the event of (77) occurs i.o. (r) wp 0. Thus, (77) is verified.

It remains to prove (76). To this end we modify the definitions of (26)–(28) as follows (in order that we may consider values of K'_r of order $(\log\log m_r)^{\frac{1}{2}}$):

$$\left.\begin{aligned}
M'_r &= \{j : 0 \leqslant j < \epsilon^8 m_r^{\frac{1}{2}}/(\log\log m_r)^{\frac{1}{2}}, j \text{ integral}\}, \\
p'_{r,j} &= 2^{-1} + m_r^{-\frac{1}{2}} j b \epsilon^{-1}(\log\log m_r)^{\frac{1}{2}} \quad \text{for } j \in M'_r, \\
p^{*\prime}_{r,j} &= p'_{r,j} + b m_r^{-\frac{1}{2}}(\log\log m_r)^{\frac{1}{2}}, \\
\Lambda^{*\prime}_r &= \{b < K'_r(\tfrac{1}{2})(\log\log m_r)^{-\frac{1}{2}} < b(1+\epsilon^4)\}, \\
A'_{r,j} &= \{T'_r(b[m_r^{-1}\log\log m_r]^{\frac{1}{2}}, p'_{r,j}) < -(1-\epsilon)d'_{n_{r+1}}\}.
\end{aligned}\right\} \quad (81)$$

Here the value ϵ is as specified above (72), and b is chosen so that

$$2^{-\frac{1}{2}} > b > 2^{-\frac{1}{2}}(1-\gamma^{-1})^{\frac{3}{2}}, \quad b > \tfrac{1}{2}, \quad (82)$$

where γ was chosen to satisfy (72). We shall prove the following three results, from which (76) follows at once:

$$\{P\Lambda^{*\prime}_r \cap \bigcup_{j \in M'_r} A'_{r,j} \text{ occurs i.o. } (r)\} = 1; \quad (83)$$

$$P\{ \sup_{|\theta| \leqslant 1, j \in M'_r} [T'_r(b[1+\theta\epsilon^3][m_r^{-1}\log\log m_r]^{\frac{1}{2}}, p'_{r,j})$$
$$- T'_r(b[m_r^{-1}\log\log m_r]^{\frac{1}{2}}, p'_{r,j})] \leqslant \epsilon d'_{n_{r+1}} \text{ for a.a. } r\} = 1; \quad (84)$$

$$P\{\Lambda^{*\prime}_r \cap \{\max_{j \in M'_r} |[n_{r+1}^{-\frac{1}{2}} K_{n_{r+1}}(p'_{r,j}) + (1-4\epsilon)d'_{n_{r+1}}]/b[m_r^{-1}\log\log m_r]^{\frac{1}{2}}$$
$$- 1| > \epsilon^3\} \text{ i.o. } (r)\} = 0. \quad (85)$$

To prove (83), we follow a development like (but simpler than) that leading from (29) to (45). Given that $\Lambda^{*\prime}_r$ occurs and that $K'_r(p'_{r,j}) = k'_{r,j}$ for $j \in M'_r$, we have $p'_{r,j} < p^{*\prime}_{r,j} < p'_{r,j+1}$, and $A'_{r,j}$ is conditionally a binomial event with

$$\left.\begin{aligned}
N &= m_r^{\frac{1}{2}}(u'_{r,j} + b\epsilon^{-1}(\log\log m_r)^{\frac{1}{2}}), \\
u'_{r,j} &= k'_{r,j} - k'_{r,j+1}, \quad U'_{r,j} = K'_{r,j} - K'_{r,j+1}, \\
\pi_N &= (p'_{r,j+1} - p^{*\prime}_{r,j})/(p'_{r,j+1} - p'_{r,j}) = 1 - \epsilon, \\
Z_N &= m_r[S'_r(p'_{r,j+1}) - S'_r(p^{*\prime}_{r,j})], \\
A'_{r,j} &\Leftrightarrow m_r^{-1} Z_N > (1-\epsilon)d'_{n_{r+1}} + m_r^{-\frac{1}{2}} u'_{r,j} + b(\epsilon^{-1}-1)[m_r^{-1}\log\log m_r]^{\frac{1}{2}} \\
&\Leftrightarrow Z_N - N\pi_N > (1-\epsilon)m_r d'_{n_{r+1}} + \epsilon m_r^{\frac{1}{2}} u'_{r,j} \equiv \Delta_N.
\end{aligned}\right\} \quad (86)$$

Define

$$
\left.
\begin{aligned}
&\pi'_r = m_r^{-\frac{1}{2}} b \epsilon^{-1} (\log \log m_r)^{\frac{1}{2}}, \\
&\mu_r = \epsilon^8 m_r^{\frac{1}{2}} (\log \log m_r)^{-\frac{1}{2}}, \\
&|J_r| = \text{number of nonpositive } U'_{r,j} \text{ for } j \text{ satisfying } j, j+1 \in M'_r, \\
&C_r^{*\prime} = \{|J_r| \geqslant \mu_r/3\}.
\end{aligned}
\right\}
\tag{87}
$$

The argument of the last paragraph of § 3 applies, essentially intact, with $p^i_{n,j}$, $U^i_{n,j}$, π, n, $n^{\frac{1}{2}}\epsilon$ of that section replaced by $p'_{r,j}$, $U'_{r,j}$, π'_r, m_r, μ_r, here. (The analogues of (46) cause no difficulty, since $\pi'_r \to 0$, $(m_r \pi'_r)^{-\frac{1}{2}} \to 0$, and $\mu_r \to \infty$, as $r \to 0$.) One obtains, in analogy with (52), that for all sufficiently large r,

$$
\left.
\begin{aligned}
&E|J_r| \geqslant 3\mu_r/8, \\
&\operatorname{var}(|J_r|) \leqslant \mu_r/4 + \mu_r^2 c_4 (m_r \pi'_r)^{-\frac{1}{2}}.
\end{aligned}
\right\}
\tag{88}
$$

Hence, the analogue of (53)–(36) is that

$$
P\{\overline{C_r^{*\prime}}\} \leqslant c'_0 \{\mu_r^{-1} + (m_r \pi'_r)^{-\frac{1}{2}}\}
\tag{89}
$$

for some constant c'_0, and therefore

$$
\Sigma_r P\{\overline{C_r^{*\prime}}\} < \infty.
\tag{90}
$$

Just as in § 3, it is again easily established that $\Delta_N^2/N\pi_N(1-\pi_N)$ is a strictly increasing function of $u'_{r,j}$, for $u'_{r,j} \geqq$ the value $c'_{r,j}$ of $u'_{r,j}$ at which this function vanishes. If $u'_{r,j} \leqslant 0$, then by using (82) we have for large r and for $c'_{r,j} \leqslant u'_{r,j}$, upon substituting into $\Delta_N^2/N\pi_N(1-\pi_N)$ the upper bound $0 \geqslant u'_{r,j}$ obtained above (in parallel with (37)),

$$
\begin{aligned}
\Delta_N^2/N\pi_N(1-\pi_N) &\leqq 2^{-\frac{1}{2}}(1-\epsilon)(1-\gamma^{-1})^{\frac{3}{2}} b^{-1}[1+o(1)]\log n_{r+1} \\
&\leqq (1-\epsilon)\log n_{r+1}.
\end{aligned}
\tag{91}
$$

From the relation between the first expression of (40) and the last line of (40) we thus have for r large, by using a calculation like that of (38),

$$
\begin{aligned}
P\{\Lambda_r^{*\prime} \cap \bigcap_{j \in M'_r} A'_{r,j}\} &\geqq P\{\Lambda_r^{*\prime}\} - P\{\bigcap_{j \in M'_r} \overline{A'_{r,j}} | \Lambda_r^{*\prime}; C_r^{*\prime}\} - P\{\overline{C_r^{*\prime}}\} \\
&\geqq P\{\Lambda_r^{*\prime}\} - [1 - e^{-2^{-1}(1-\epsilon/2)\log n_{r+1}}]^{\epsilon^8 m_r^{\frac{1}{2}}/3(\log\log m_r)^{\frac{1}{2}}} \\
&\qquad\qquad\qquad\qquad\qquad\qquad\qquad - P\{\overline{C_r^{*\prime}}\}. \tag{92}
\end{aligned}
$$

By the first inequality of (82) and (9) (with

$$
N = m_r, \pi_N = \tfrac{1}{2}, \quad Z_N = m_r[1 - S'_r(\tfrac{1}{2})],
$$

$$
\Delta_N[m_r \log\log m_r]^{-\frac{1}{2}} = b \quad \text{or} \quad b(1+\epsilon^4)),
$$

we see that $\sum_r P\{\Lambda_r^{*\prime}\} = +\infty$; on the other hand (recall (90)), the remaining terms of the last line of (92) are summable. Since the events of (83) for different values of r depend on different X_i's, they are independent, and hence (83) follows from (92) and the Borel–Cantelli lemma.

As for (84), the event considered there for fixed r and $\underset{j}{j}$ (the sup operation being temporarily deleted) can be written, by using the T_n relation just above (21), as

$$\sup_{|\theta|\leqslant 1} T_r'(\theta b\epsilon^3[m_r^{-1}\log\log m_r]^{\frac{1}{2}}, p_{r,j}' + b[m_r^{-1}\log\log m_r]^{\frac{1}{2}}) \leqslant \epsilon d_{n_{r+1}}'. \quad (93)$$

By (11) (with n replaced by m_r and $(1-\delta)^3$ of (11) equal to $\frac{1}{2}$) the probability of the *complement* of the event (93), for all sufficiently large r and for each $j \in M_r'$, is at most

$$\exp\{-m_r(\epsilon d_{n_{r+1}}')^2/4b\epsilon^3[m_r^{-1}\log\log m_r]^{\frac{1}{2}}\}$$
$$= \exp\{-b^{-1}\epsilon^{-1}(32)^{-\frac{1}{2}}(1-\gamma^{-1})^{\frac{3}{2}}[1+o(1)]\log n_{r+1}\}. \quad (94)$$

There are fewer than $n_{r+1}^{\frac{1}{2}}$ elements in M_r' for large r, and we see by the fourth condition of (72) and the first condition of (82) that $n_{r+1}^{\frac{1}{2}}$ times the last expression of (94) is summable. Hence, by the Borel–Cantelli lemma, the *complement* of the event considered in (84) for fixed r (but including now the sup operation) occurs for only finitely many r, wp 1. This proves (84).

It remains to prove (85). This will clearly follow from adding the inequalities in the following three results which we shall prove next:

$$\sup_{0\leqslant t\leqslant \epsilon^7 b} n_{r+1}^{-\frac{1}{2}}|K_{n_{r+1}}(\tfrac{1}{2}+t) - K_{n_{r+1}}(\tfrac{1}{2})|$$
$$\leqslant 3^{-1}b[m_r^{-1}\log\log m_r]^{\frac{1}{2}}\epsilon^3 \quad \text{for a.a.} \, r, \text{wp 1}; \quad (95)$$

$$|n_{r+1}^{-\frac{1}{2}}K_{n_{r+1}}(\tfrac{1}{2}) - m_r^{\frac{1}{2}}n_{r+1}^{-1}K_r'(\tfrac{1}{2})|$$
$$< 3^{-1}b[m_r^{-1}\log\log m_r]^{\frac{1}{2}}\epsilon^3 \quad \text{for a.a.} \, r, \text{wp 1}; \quad (96)$$

$$\Lambda_r^{*\prime} \Rightarrow |m_r^{\frac{1}{2}}n_{r+1}^{-1}K_r'(\tfrac{1}{2}) + (1-4\epsilon)d_{n_{r+1}}' - b[m_r^{-1}\log\log m_r]^{\frac{1}{2}}|$$
$$< 3^{-1}b[m_r^{-1}\log\log m_r]^{\frac{1}{2}}\epsilon^3 \quad \text{for a.a.} \, r. \quad (97)$$

For fixed r, the *complement* of the event of (95) can be written

$$\sup_{0\leqslant t\leqslant \epsilon^7 b} |T_{n_{r+1}}(t,\tfrac{1}{2})| > 3^{-1}\epsilon^3 b[m_r^{-1}\log\log m_r]^{\frac{1}{2}}, \quad (98)$$

and by (9) and (10) (with the δ of (10), not that of (72), now chosen to satisfy $(1-\epsilon^7 b)^{-1}(1-\delta)^2 > \frac{1}{2}$) the probability of the event (98) for large r is at most

$$\exp\{-[n_{r+1}/4\epsilon^7 b]\,3^{-2}\epsilon^6 b^2[m_r^{-1}\log\log m_r]\}, \quad (99)$$

which is summable since $\epsilon < \frac{1}{72}$ and $b > \frac{1}{2}$. This and the Borel–Cantelli lemma yield (95).

As for (96), since $n_{r+1}^{\frac{1}{2}} K_{n_{r+1}} = n_r^{\frac{1}{2}} K_{n_r} + m_r^{\frac{1}{2}} K_r'$, the rv considered on the left side of (96) is

$$\left| n_{r+1}^{-\frac{1}{2}} K_{n_{r+1}}(\tfrac{1}{2}) - m_r^{\frac{1}{2}} n_{r+1}^{-1} K_r'(\tfrac{1}{2}) \right| = \left| n_r^{\frac{1}{2}} n_{r+1}^{-1} K_{n_r}(\tfrac{1}{2}) \right|. \tag{100}$$

By the LIL, $|K_{n_r}(\tfrac{1}{2})| < [\log\log n_r]^{\frac{1}{2}}$ for a.a. r, wp 1. This fact, together with (100), the fifth condition of (72), and the last condition of (82), yields (96).

Finally, the event $\Lambda_r^{*\prime}$ can be written

$$m_r n_{r+1}^{-1} - 1 < \frac{m_r^{\frac{1}{2}} n_{r+1}^{-1} K_r'(\tfrac{1}{2}) - b[m_r^{-1}\log\log m_r]^{\frac{1}{2}}}{b[m_r^{-1}\log\log m_r]^{\frac{1}{2}}} < (1+\epsilon^4)\, m_r n_{r+1}^{-1} - 1. \tag{101}$$

Thus, the last two conditions of (72), and the fact that $d'_{n_{r+1}} = o(m_r^{-\frac{1}{2}})$, yield (97), completing the proof of this section, and hence of Theorem 2.

6 COMMENTS

A. On the proofs

1. While the weak law of this paper is more difficult to obtain than the straightforward limit law (displayed below (1)) for a single p, the strong law here requires less delicate technique than that of [8] (displayed above (2)). The reason for the latter is that the critical values of K_n here can be obtained crudely from the LIL, while in [8] they are only a fraction as large, and the value of the fraction is crucial.

2. The crude device of (16)–(19), used to prove (12) from (15) and (25), is used to avoid the more difficult computation of relevant conditional probabilities given D_n. The latter computation is not helped by passing to the limiting Brownian bridge, since the limiting R_n deviations are not then present; this reflects the failure, mentioned in paragraph 6.2 of [8], of usual invariance arguments in the case of the R_n process. We remark that the complications attending the use of the device of (16)–(19) when the number of sets Q_i one must consider is unbounded in n, accounts for part of the difficulty, mentioned in § 1, of finding a simple proof of a strong counterpart of (5).

3. We note that, as in the case of the LIL for D_n, large deviations of $R_n^{\#}$ in the strong law (4) are accounted for essentially by behavior near $p = \frac{1}{2}$; this is reflected in the definitions of (81).

4. The right *order of magnitude* in the strong upper class conclusion of (4) can be obtained in a few lines, in contrast with the development of

§ 4; one considers (21), (22), (23) as before and notes that, with d_n multiplied by $3^{\frac{1}{2}}$, the expression (65) with n for n_r, when multiplied by $n^{\frac{1}{2}}$, is summable in n (as is, even more obviously, (64)). This is of interest for establishing a brief treatment [10] of the subject of [3], [4] mentioned in Theorem 5.

5. Since the crude bound of (36)–(89) suffices for the proofs of §§ 3 and 5, we have not taken the space to give sharper (exponential) bounds, but it is not hard to do so.

6. The reduction which uses (6) breaks down without the assumptions of the first paragraph, but it is not clear to what extent the *conclusions* of [1], [8], and the present paper fail.

B. Open problems

1. The previous paragraph A 6 suggests some obvious problems, of studying $R_n^{\#}$ for F not satisfying the assumptions used herein. Other variants to be studied include the results of [8] quoted in the first paragraph of Section 1 but with fixed p replaced by $\{p_n\}$. (For example, with F uniform, $R_n(\frac{1}{2}n)$ is of order n^{-1} rather than $n^{-\frac{3}{4}}$, in probability.) The consideration of (2) with I replaced by $\{0 \leqslant p \leqslant p_n\}$ is relevant to the extension of Theorem 4 to cases where $p_n \downarrow 0$ at a faster rate.

2. Similarly, one can study (2) without multiplying by F' there.

3. Computations like that of paragraph B 2 just above, and of other variants of $R_n^{\#}$ analogous to test statistics like ω^2 or other functionals of sample spacings, are of interest for purposes of inference. It is striking that, at least in the sense of (5), $R_n^{\#}$ does not exhibit the drastic loss of information evident for $R_n(p)$ for a single p, as discussed in § 6.6 of [8].

4. The finer characterization of functions of upper and lower class, for $R_n(p)$ as well as $R_n^{\#}$, still seems beyond the present methods.

5. A theoretical development which allows functionals of the R_n process to be treated in terms of a limiting process (see paragraph A 2 of this section, and paragraph 6.2 of [8]) would be very interesting.

C. Corrections of [8]

It seems in place to record these here:

In (1.4), lower '$\frac{1}{2}$' to yield subscript $p + n^{-\frac{1}{2}}t$ on ξ.

In proof of Lemma 1, write U_1, U_2 for H_1, H_2.

Just above (1.20), write $\lambda > 0$ for $\lambda > 1$.

In (1.22), write $\cap$ for $\cup$.

In (4.2), definition of D_r, the $+$ sign should be a superscript.

Seven lines below (2.2), replace '(2.1)' by '(2.2)'.

REFERENCES

[1] R. R. Bahadur. A note on quantiles in large samples. *Ann. Math. Statist.* **37**, 577–80 (1966).

[2] H. Bergstrom. On the central limit theorem in the space R^k, $k > 1$. *Sk. and Akt.* **28**, 106–27 (1945).

[3] L. Breiman. *Probability.* Addison-Wesley (1968).

[4] D. R. Brillinger. An asymptotic representation of the sample df. *Bull. AMS* **75**, 545–7 (1969).

[5] K. L. Chung. An estimate concerning the Kolmogoroff limit distribution. *Trans. Amer. Math. Soc.* **67**, 36–50 (1949).

[6] F. Eicker. A log log law for double sequences of random variables. (To appear.)

[7] W. Feller. *An introduction to probability theory and its applications*, vol. 1, second edn. (1957); vol. 2, first edn. (1966). New York: John Wiley and Sons.

[8] J. Kiefer. On Bahadur's representation of sample quantiles. *Ann. Math. Statist.* **38**, 1323–42 (1967).

[9] J. Kiefer. Old and new methods for studying order statistics and sample quantiles. This volume, pp. 349–357. (Discussion of Weiss and Eicker papers.)

[10] J. Kiefer. On a Brownian bridge approximation to the sample df. (To appear.)

A NEW PROOF OF THE
BAHADUR–KIEFER REPRESENTATION
OF SAMPLE QUANTILES

F. EICKER

1 INTRODUCTION AND SUMMARY

Throughout the paper let $X_1, X_2, \ldots$ be independent identically distributed random variables (iidrv's) with uniform distribution $U(0, 1)$ over $[0, 1]$. The latter assumption can easily be generalized (see e.g. [1] or [6]). Let $p \equiv 1 - q \in (0, 1)$ be a given constant, $Y_n \equiv Y_n(p)$ the sample p-quantile based on $(X_1, \ldots, X_n)$ [i.e. the $([np] + 1)$st order statistic $X_{n, [np]+1}$ in that sample], and

$$F_n(x) := n^{-1} \sum_{j=1}^{n} I(X_j < x) \quad (0 \leqslant x \leqslant 1), \tag{1.1}$$

the sample distribution function (df) [hence

$$F_n(Y_n) = n^{-1}[np] \in (p - n^{-1}, p] \text{ a.s.}; \tag{1.2}$$

$I(A)$ is the indicator function of the set A]. Let

$$H_n \equiv H_n(p) := 2p - F_n(p) - Y_n \tag{1.3}$$

and $\log_2 \equiv \log \log$. In [6] Kiefer proved

$$\limsup \pm H_n/\eta_n = 1 \text{ a.s.}, \tag{1.4}$$

$$\eta_n := \sqrt{2} \, (pq)^{\frac{1}{4}} \left(\frac{2}{3} \frac{\log_2 n}{n} \right)^{\frac{3}{4}}, \tag{1.5}$$

sharpening an earlier result of Bahadur [1]. Kiefer's proof is considerably more sophisticated than Bahadur's. Although basically the same line of thought is used as in proving the classical law of the iterated logarithm (LIL) for Bernoulli rv's (see e.g. [4]), many of the details are changed and new ideas are developed.

The present paper gives another proof of the upper class half of (1.4) [with the plus sign], utilizing a new, general principle. The lower class case is not proved in this paper, but it appears that the same method can again be applied. The characteristic new tool is an exact expression for the probability that a sample df from the uniform df over $[0, 1]$ lies below a line segment (see Appendix B for the relevant and some related formulae. See also [2]). With these formulae the proof of (1.4) is fairly straightforward, using standard techniques. The usefulness of the

II [321] PNT

principle in other contexts is illustrated by a proof of a log log-law for triangular arrays of Bernoulli rv's [3]. Considerable generalizations of this result should be feasible.

The results presented in Appendix B are of interest in themselves independent of the context of the present paper for the following two reasons. First, in relation to probabilities of corresponding events defined in terms of functionals of the Brownian bridge (conditioned Wiener) process, the question of the rate of the weak convergence involved is of imminent interest, but it is not discussed here. Secondly, Appendix B generalizes in a certain way the Kolmogorov–Smirnov and the Renyi–Csörgö statistic (more about this in [2]).

Characterizing in summary the present approach as compared to Kiefer's it may be said that it emphasizes combinatorial (finite) considerations more and that consequently asymptotic evaluations enter at a later stage. This fact may well be responsible for the flexibility of the method.

Appendices A and B collect approximation formulae for binomial probabilities and probabilities on functionals of sample probabilities, respectively.

Further notations.

$S_n(x) := nF_n(x) =$ number of $X_j < x$ $(0 \leqslant x \leqslant 1)$,

$G_n(x) := (S_n(x) > np)$ $(0 \leqslant x \leqslant 1)$;

$A_n \equiv A_n(p) := (H_n > \delta\eta_n)$ $(\delta > 0$ a constant$)$;

$F_n(B) := n^{-1} \sum_{j=1}^{n} I(X_j \in B)$ $(B$ any Borel set of $[0, 1])$;

$A^c =$ complement of the set A;

$R \equiv (-\infty, \infty)$ the real line; $\Omega =$ basic sample space;

$I(A) =$ indicator function defined on Ω of a set $A \subset \Omega$;

$U(a, b) =$ uniform distribution over $(a, b) \subset R$ given by the density equal to $(b - a)^{-1}$ for $x \in (a, b)$ and zero elsewhere;

$b(j, n, p) = \binom{n}{j} p^j q^{n-j}$;

$B(j, n, p) = \sum_{k=0}^{[j]} b(k, n, p)$ $(j \in R_1)$;

if $j \notin [0, n]$ the B-function is defined by 0 or 1 resp.;

Σ is used as the union ($\cup$) symbol for disjoint events;

$x \vee y := \max(x, y)$, $x \wedge y := \min(x, y)$ $(x, y \in R_1)$;

$x := y$ or $y =: x$ means x is defined by y;

$[x]$ = greatest integer $\leqslant x \in R_1$;

$[x, y]$, $[x, y)$ = closed resp. halfclosed interval, $-\infty < x \leqslant y < +\infty$;

wrt = with respect to;

$\uparrow$, $\downarrow$ monotone in- or decreasing (wide sense);

iid = independent identically distributed;

rv = random variable;

df = distribution function;

$\log_2$ = $\log \log$;

$\sim$ = asymptotically equal or distributed as;

$\forall$ = for all.

2 PROOF OF THE UPPER CLASS CASE

(i) For $\delta > 1$ it has to be shown

$$P(\limsup_{n \to \infty} A_n) = 0 \quad \text{or} \quad \lim_{m \to \infty} P(\bigcup_{n > m} A_n) = 0. \tag{2.1}$$

With the identity (in the following equalities between events or rv's often hold only upto sets of probability zero)

$$(x > Y_n) = (S_n(x) > np) =: G_n(x) \quad (0 \leqslant x \leqslant 1) \tag{2.2}$$

it follows, putting

$$x_{n,k} := 2p - \delta\eta_n - \frac{k}{n}, \; (k = 0, 1, \ldots, n \quad n = 1, 2, \ldots);$$

$$A_n = (2p - F_n(p) - \delta\eta_n > Y_n) = \sum_{k=0}^{n} (S_n(p) = k) G_n(x_{n,k}) \tag{2.3a}$$

$$= \bigcup_{k=0}^{n} (S_n(p) \leqslant k) G_n(x_{n,k}) \tag{2.3b}$$

$$\left.\begin{array}{l} = \sum_{\kappa < -\delta\eta_n n} (S_n(p) = [np] + \kappa, \, S_n[p, x_{n,k}) > |\kappa|) \\[2mm] + \sum_{\kappa \geqslant 0} (S_n(p) = [np] + \kappa, \, S_n[x_{n,k}, p) < \kappa); \end{array}\right\} \tag{2.3c}$$

here $$S_n[x, p] := \sum_{j=1}^{n} I(x \leqslant X_j < p) \quad (0 \leqslant x < p \leqslant 1), \tag{2.4}$$

and $$x_{n,k} = p - \delta\eta_n - \frac{\kappa}{n} \quad (\kappa := k - [np]) \tag{2.5}$$

(up to an error of at most n^{-1} which may be taken care of by replacing δ by $\delta + o(1)$ for $n \to \infty$. However, the $o(1)$ can and will be ignored subsequently).

In proving (2.3b) note $S_n(x_{n,k}) \leqslant S_n(x_{n,k-1})$, consequently

$$G_n(x_{n,k}) \subset G_n(x_{n,k-1}) \quad \text{and} \quad G_n(x_{nk}) = \bigcup_{j=k}^{n} G_n(x_{nj}).$$

As is common in proofs of $\log_2$-laws, the set of naturals will be subdivided according to a geometric progression d^r, $r = 1, 2, \ldots, d > 1$ a constant to be determined in the course of the proof by a set of conditions involving only constants like δ and reals. For $r = r_0, r_0 + 1, \ldots$ with r_0 a suitable integer, put

$$n_r := [d^r] + 1, \quad N_r := \{n \text{ integer}: d^{r-1} \leqslant n < d^r\},$$

$$B_{n_{r-1}} := A_{n_{r-1}}, \quad B_n := A_n \bigcap_{m=n_{r-1}}^{n-1} A_m^c, \quad n > n_{r-1}. \tag{2.6}$$

Then

$$\bigcup_{n \in N_r} A_n = \sum_{n \in N_r} B_n. \tag{2.7}$$

Repeatedly in the proof the ranges of occurring rv's will be restricted to those that are assumed with high probability; this considerably simplifies later computations. Thus it suffices in proving (2.1) to consider only κ-values satisfying

$$\underline{\kappa}_r := c \frac{n_r}{r} < |\kappa| < (2\delta n_{r+1} pq \log r)^{\frac{1}{2}} =: \overline{\kappa}_r \quad \text{for } n \in N_r, \tag{2.8}$$

with any constant $c > 0$.

To justify first the right inequality, note that by (2.7) and (2.3a)

$$P(\bigcup_{n \in N_r} A_n) = \sum_{n \in N_r} \sum_{k=0}^{n} P(B_n(S_n(p) = k) G_n(x_{n,k})). \tag{2.9}$$

The (n, k)th summand, for $k < np$, is multiplied by $P\Delta_n \approx \frac{1}{2}$, where

$$\Delta_n := (S_{n_{r+1}}(p) - S_n(p) < \Delta np) \quad (\Delta n := n_{r+1} - n) \tag{2.10}$$

and by $P\Delta_n^c$ for $k \geqslant np$. By independence of Δ_n and B_n, the product of the probabilities equals the probability of the intersection. Now for $\kappa < -\overline{\kappa}_r$

$$(S_n(p) = k) \Delta_n = (S_n(p) = k) (S_{n_{r+1}}(p) < n_{q+1} p + \kappa)$$

$$\subset (S_n(p) = k) (S_{n_{r+1}}(p) < n_{r+1} p - (2\delta n_{r+1} pq \log r)^{\frac{1}{2}})$$

$$\subset (S_n(p) = k) L_r. \tag{2.11}$$

Here the event $L_r := (|S_{n_{r+1}}(p) - n_{r+1} p| > \overline{\kappa}_r)$ is independent of k and of $n \in N_r$ and by Lemma A1 $PL_r = O(r^{-\delta})$. A similar computation can be made for $\kappa > \overline{\kappa}_r$.

Now in

$$P(\bigcup_{n\in N_r} A_n) = \sum_n P[B_n(|S_n(p)-np| < \bar\kappa_r)] + \sum_n P[B_n(|S_n(p)-np| \geqslant \bar\kappa_r)]$$

(2.12)

the second sum on the right is

$$< 3\sum_n \sum_{|\kappa|\geqslant\bar\kappa_r} P[B_n(S_n(p) = k)L_r] \leqslant 3PL_r,$$

and this is summable wrt r. Hence only the first sum in (2.12) must be bounded above, and there the restriction to $|\kappa| < \bar\kappa_r$ is permitted in any one of the representations (2.3) of A_n.

To justify the left inequality of (2.8) one uses (2.3c). For, say,

$$0 > \kappa > -\underline{\kappa}_n$$

$$P(S_n[p, x_{n,k}) > |\kappa|) = 1 - B\left(|\kappa|, n, -\delta\eta_n + \frac{|\kappa|}{n}\right)$$

$$< n^{-c' \log_2 n}, \quad \text{some constant } c' > 0; \quad (2.13)$$

this may be inferred from (A 12) (see Appendix): one notes that

$$-\delta\eta_n + \frac{|\kappa|}{n} \to 0$$

uniformly in κ. Hence the variance of the B-df is $\sim |\kappa|$.

Further $$|\kappa| - n\left(-\delta\eta_n + \frac{|\kappa|}{n}\right) = \delta n\eta_n,$$

hence $$\delta n\eta_n |\kappa|^{-\frac{1}{2}} > c''(\log n \log_2 n)^{\frac{1}{2}} \quad \text{if } |\kappa| < \frac{\sqrt\kappa}{\log n}, \quad (2.14)$$

with some constant $c'' > 0$. The assumptions of Lemma A 2 are easily checked, hence (2.13). Now there are $O((n \log n)^{\frac{1}{2}})$ terms in the sum (2.3c) the probability of each being bounded by (2.13) if $|\kappa| < \underline{\kappa}_n$. Hence their total contribution is $< n^{-2}$ for large n, which is summable. If $\kappa > 0$, a similar argument holds.

(ii) After these preliminaries we now set out to show first that the sum over all n of the probabilities of the events $B_n \cap (-\bar\kappa_r < S_n(p) < -\underline{\kappa}_n)$, which are responsible for 'half' of the summands of the first sum of (2.12), converges. The method of proof is similar in part to that of the upper class case of the classical law of the iterated logarithm (comp. e.g. the lemma on pp. 192–3 of [4]) in that each summand still of interest (i.e. with κ satisfying (2.8)) in the right most member of

$$P(\bigcup_{n\in N_r} A_n) = \sum_{n\in N_r} PB_n = \sum_{n\in N_r}\sum_\kappa P(B_n(S_n(p) = k)) \quad (2.15)$$

is multiplied by the probability of an event $C_{\kappa,\,n}$ (see (2.20) below) which is independent of $B_n \cap (S_n(p) = k)$ and whose probabilities are bounded below by a slowly decreasing function of r (see (2.24)) uniformly in κ and n on their respective ranges.

Next it is shown (always for $\kappa < 0$ in this subsection) that

$$B_n(S_n(p) = k)\,C_{\kappa,\,n} \subset A_r^* \qquad (2.16)$$

with a suitable event A_r^* depending only on n_r [but *not* equal to A_{n_r} as in the classical case or as in Kiefer's proof]. Consequently, since the

$$B_n(S_n(p) = k) \qquad (2.17)$$

for different pairs (n, k) are disjoint,

$$\sum_{n \in N_r}\ \sum_{-\bar\kappa_r < \kappa < -\underline\kappa_r} B_n(S_n(p) = k)\,C_{\kappa,\,n} \subset A_r^*. \qquad (2.18)$$

Hence $P(\bigcup_{u \in N_r} A_n)$ is bounded above by $(PA_r^*)\,r^{-O(d^2-1)}$ plus

$$P(\sum_{n \in N_r} B_n(S_n(p) \geqslant np)),$$

the latter to be dealt with in (v). It will be shown

$$PA_r^* = O(r^{-\delta'}),\ \ \delta' > 1 \qquad (2.19)$$

[see (2.39)ff.] which is summable wrt r even after multiplication by (2.24) if $d^2 - 1$ is sufficiently small.

In detail let for $n \in N_{r-1}$ and $\kappa < 0$ subject to (2.8)

$$C_{\kappa,\,n} := \left(\Delta S(p) < \Delta n \left(p - \frac{|\kappa|}{n}\right),\ \Delta S[p, x_{n,\,k}] > \Delta n\left(\frac{|\kappa|}{n} - \delta\eta_n\right)\right) \qquad (2.20)$$

[where $\Delta S := S_{n_r} - S_n,\ n \in N_{r-1};\ \Delta n := n_r - n$]. Then

$$C_{\kappa,\,n}(S_n(p) = k,\ S_n[p, x_{n,\,k}] > |\kappa|)$$

$$\subset \left(S_{n_r}(p) < n_r\left(p - \frac{|\kappa|}{n}\right),\ S_{n_r}[p, x_{n,\,k}] > n_r\frac{|\kappa|}{n} - \Delta n\delta\eta_n\right) \qquad (2.21)$$

$$\subset (F_{n_r}(p) < p - x,\ F_{n_r}[p, p - \delta\eta_{n_r} + x] > x - (d^2 - 1)\delta\eta_{n_r}) =: U_x V_x, \qquad (2.22)$$

where $\dfrac{|\kappa|}{n} =: x$. By (2.8)

$$cn^{-\frac{1}{2}}\log^{-1}n < x < d\left(2\delta\frac{pq}{n}\log r\right)^{\frac{1}{2}} \quad (n \in N_r,\ c > 0). \qquad (2.23)$$

Note first

$$PC_{\kappa,\,n} > \sum_{-2\Delta nx < j < -n\Delta x} b(\Delta np + j;\ \Delta n, p)$$

$$\times \left(1 - B\left(\Delta n(x - \delta\eta_n);\ \Delta nq - j,\ \frac{x_{nk} - p}{q}\right)\right) > \tfrac{1}{4}B(\Delta n(p - x);\ \Delta n, p)$$

$$> cr^{-\delta(d^2-1)d^2},\tag{2.24}$$

$c > 0$ a constant, uniformly on the x-range. This may be inferred from Berry–Esseen's theorem (e.g. [5], p. 515) in case of the conditional probability $1 - B(\ldots)$ since

$$x_{n,\,k} - p = x - \delta\eta_n \sim x \to 0,\quad \max|j| = o(n)$$

and $\quad \Delta n(x - \delta_n) - \Delta n(x_{n,\,k} - p) + j\dfrac{x_{n,\,k} - p}{q} = j\dfrac{x - \delta\eta_n}{q}\begin{cases} < 0 \\ = o(nx)\end{cases}$ (2.25)

and hence $1 - B(\ldots) \to \tfrac{1}{2}$ uniformly in x and j as $n \to \infty$. To prove the last inequality of (2.24), (A 9 a) of Lemma A 1 may be applied. The assumptions (A 1/2/8) are satisfied

[note $0 < p < 1$ and $\min|j| \to \infty$],

and $[\Delta n(p - x) - \Delta np]\,(\Delta npq)^{-\frac{1}{2}} > -d(2\delta(d^2 - 1)\log r)^{\frac{1}{2}} \to -\infty$

(note still that $B(\Delta n(p - 2x)\,\Delta n, p)$ is of smaller order than this, and that the factor $(\log r)^{-\frac{1}{2}}$ causes merely in the exponent of r a factor $1 + o(1)$).

The lower bound (2.24), although not bounded away from zero uniformly as is the case with the analogue of the classical law of the iterated logarithm (see (5.5) on p. 192 of [4]), tends to zero so slowly that the sequence $r^{-\delta}$, say, remains summable wrt r even after multiplication with $r^{\delta d^2(d^2-1)}$ if only $d^2 - 1$ is sufficiently small, which however can always be assumed from the beginning.

Now, the events $U_x V_x$ [defined by (2.22)] are contained in their union which is independent of κ and n, and this union can be used as the A^*_{r-1} introduced earlier (x, κ, n on their respective ranges).

The probabilities of these unions in dependence on r will be shown to be $O(r^{-\delta'})$, some $\delta' > 1$, and are thus summable wrt r even after multiplication with (2.24). Each union will be split in two according to (2.27) below, and the second one will be decomposed according to (2.30). The probability of the union will then be bounded above by the sum of the probabilities of (2.29) and of (2.30).

Put $x_0 := n^{-\frac{1}{2}}r^{-1}$ (due to (2.8); same for x_2);

$$x_1 := \left(\frac{pq}{n}\log r\right)^{\frac{1}{2}}(\tfrac{2}{3})^{\frac{3}{2}};\quad x_2 := d\left(2\delta\frac{pq}{n}\log r\right)^{\frac{1}{2}},\tag{2.26}$$

where from now on n stands for n_r.

Then the union over the x-values of the events (2.22) is contained in

$$(\bigcup_{x_0 < x < x_1} V_x) \cup \bigcup_{x_1 < x < x_2} [U_x \cap (\bigcup_{x_1 < \xi < x} V_\xi)], \qquad (2.27)$$

and $\bigcup_{x_1 < x < x_2} [\ldots]$ is contained in

$$U_{x_2} + \sum_{x_1 < j/n < x_2,\, j \text{ integer}} \left(F_n(p) = p - \frac{j}{n} \right) \cap (\bigcup_{x_1 < \xi < j/n} V_\xi). \qquad (2.28)$$

The first union of (2.27) equals

$$\bigcup_{x_0 < x < x_1} V_x = \bigcup_{p - \delta\eta_n + x_0 < x < p - \delta\eta_n + x_1} (F_n[p, x] > x - p + (2 - d^2)\,\delta\eta_n). \qquad (2.29)$$

In (2.28)

$$PU_{x_2} = B(np - d(2\delta npq \log r)^{\frac{1}{2}}, n, p) = O(r^{-d^2\delta})$$

by (A 11). These terms are summable wrt r. Hence in (2.28) only the sum over the j is still to be considered. It may be rewritten as (upon substituting $j = -\rho$)

$$\sum_{-nx_1 > \rho > -nx_2} \left(F_n(p) = p + \frac{\rho}{n} \right) \bigcup_{\alpha < x < \beta_\rho} (F_n[p, x] > x - p + (2 - d^2)\,\delta\eta_n), \qquad (2.30)$$

where $\qquad \alpha := p - \delta\eta_n + x_1, \quad \beta_\rho := p - \delta\eta_n - \dfrac{\rho}{n} \quad (\rho \text{ integer}). \qquad (2.31)$

The probabilities of (2.29) and (2.30) are determined from (B 12) and (B 8a), respectively, of Appendix B.

(iii) First the probability of (2.29) will be computed. The quantities in (B 12) are identified from (2.29) as follows:

$$p := p, \ \ \alpha := p - \delta\eta_n + x_0 = p - \delta\eta_n + n^{-\frac{1}{2}}r^{-1}, \qquad (2.32)$$

$$\beta := p - \delta\eta_n + x_1 = p - \delta\eta_n + \left(\frac{pq}{n} \log r \right)^{\frac{1}{2}} (\tfrac{2}{3})^{\frac{1}{2}}, \qquad (2.33)$$

$$\delta := -p + (2 - d^2)\,\delta\eta_n \ (\delta \text{ in two different meanings!}), \qquad (2.34)$$

$$\gamma := 1, \qquad (2.35)$$

$$\kappa' \equiv n(\delta + \gamma\alpha) := -(d^2 - 1)\,n\delta\eta_n + n^{\frac{1}{2}}r^{-1}, \qquad (2.36)$$

$$\lambda' \equiv n(\delta + \gamma\beta) := -(d^2 - 1)\,n\delta\eta_n + (npq \log r)^{\frac{1}{2}} (\tfrac{2}{3})^{\frac{1}{2}} = O(\sqrt{(n \log r)}), \qquad (2.37)$$

$$\lambda_k^* = \begin{cases} n - k & \text{if } \lambda' > n - k, \\ [\lambda'] & \text{if } \lambda' < n - k \text{ and } \lambda' \neq \text{integer}, \\ \lambda' - 1 & \text{if } \lambda' < n - k \text{ and } \lambda' \text{ integer}. \end{cases} \qquad (2.38)$$

Due to (2.8), $\lambda' = o(n - k)$ uniformly in k. Consequently, $|\lambda^* - \lambda'| \leqslant 1$, so that λ_k^* may be replaced by λ' everywhere. Now, first in $\{1 - [\ldots]\}$ of (B 12),

$$1 - B\left(\lambda_k^*; n - k, \frac{\beta - p}{q}\right) = O(r^{-\delta'}), \tag{2.39}$$

some $\delta' > 1$, uniformly in k, by (A 12), Lemma A 2, because

$$\lambda_k^* - (n - k)\frac{\beta - p}{q} \sim (2 - d^2)\, \delta n \eta_n \begin{cases} > 0, \\ o \text{ (variance)}, \end{cases} \tag{2.40}$$

and the variance of the B-function being

$$\sim (npq \log r)^{\frac{1}{2}} (\tfrac{2}{3})^{\frac{3}{2}}. \tag{2.41}$$

The t^2-value becomes, with $\eta_{n_r} \sim \sqrt{2}\,(pq)^{\frac{1}{4}}\left(\frac{2}{3}\frac{\log r}{n_r}\right)^{\frac{1}{4}}$, $(2 - d^2)^2\,\delta^2 2\log r$, and this yields (2.39) if d is close enough to 1. Hence the contributions of the terms (2.39) in (B 12) are summable wrt r; we bound $\sum_k b(k; n, p) < 1$.

Similarly, the two sums over s arising in $[\ldots]$ of (B 11) are at most of order (2.39): in the first sum, putting $k := np + \rho$, $|\rho| < (2\delta npq \log r)^{\frac{1}{2}}$,

$$n(\delta + \gamma - 1) + k = (2 - d^2)\,\delta n \eta_n + \rho = O((n\log r)^{\frac{1}{2}}), \tag{2.42}$$

$$n(\delta + \gamma) - s \sim nq \quad (\text{since } s = O((n\log r)^{\frac{1}{2}})). \tag{2.43}$$

The ratio of these two expressions is

$$O(d^{-r/2}\log^{\frac{1}{2}} r). \tag{2.44}$$

The number of terms in $\underset{s}{\sum}$ is $O(\lambda') = O(d^{r/2}\log^{\frac{1}{2}} r)$. The variance of the first b-term is $\sim s \gtrsim d^{+r/3}$. The inverse root of this multiplied by the two preceding O-terms gives

$$O(s^{-r/6}) \tag{2.45}$$

which is smaller than (2.39) [the b-term has been bounded by $O(\text{var}^{-\frac{1}{2}})$]. Next the probabilities

$$b(\lambda^* - s; n - k - s - 1, p(s)) \tag{2.46}$$

in the curly bracket in (B 8a) are to be evaluated. Since

$$p(s) \sim (\lambda' - s)/nq = O(n^{-\frac{1}{2}}(\log r)^{\frac{1}{2}}) \to 0,$$

(B 8a) now shows an advantage over (B 8) because the range of

$$(s - n(\delta + \gamma p))/n\gamma(\beta - p)$$

would be roughly $(0, 1)$. The variance of (2.46) is [remembering (2.43) and $n - k - s - 1 \sim nq$]

$$\sim \lambda' - s. \tag{2.47}$$

Only
$$b\left(s; n-k, \frac{s-n(\delta+\gamma p)}{n\gamma q}\right) \tag{2.48}$$

must be evaluated more carefully. First,

$$\frac{s-n(\delta+\gamma p)}{n\gamma q} = \frac{s-(2-d^2)\,\delta n\eta_n}{nq} \sim \frac{s}{nq} = O(n^{-\frac{1}{2}}\log^{\frac{1}{2}}r) \to 0 \tag{2.49}$$

so that the variance is
$$\sim s > \frac{c\sqrt{n}}{r}. \tag{2.50}$$

The numerator of the t-value is

$$s - \frac{(n-k)\,(s-n(\delta+\gamma p))}{n\gamma q} = s - \frac{(nq-\rho)\,(s-(2-d^2)\,\delta n\eta_n)}{nq}$$

$$= (2-d^2)\,\delta n\eta_n + \frac{\rho(s-(2-d^2)\,\delta n\eta_n)}{nq} \sim (2-d^2)\,\delta n\eta_n.$$

Consequently

$$t^2 \sim (2-d^2)^2\,\delta^2 n^2\eta_n^2/\lambda' \sim (2-d^2)^2\,\delta^2 2\log r > 2\delta'\log r,$$

some $\delta' > 1$, and (2.48) is by (A 10)

$$O(s^{-\frac{1}{2}}r^{-\delta'}).$$

Now with (2.47)

$$\sum_{s=\kappa}^{\lambda} [(\lambda'-s)s]^{-\frac{1}{2}} = O\left(\int_{\sqrt{n}}^{\lambda} + \int_{\kappa}^{\sqrt{n}}\right) = O(n^{-\frac{1}{4}}(\lambda-\sqrt{n})^{\frac{1}{2}}$$

$$+ (\lambda'-\sqrt{n})^{-\frac{1}{2}}\,(n^{\frac{1}{4}}-\sqrt{\kappa})) = O(n^{-\frac{1}{4}}\sqrt{\lambda}) = O(\log^{\frac{1}{4}}r),$$

so that the contributions to (B 12) arising from (2.46) are of the same order as (2.39). Thus it is shown that the probabilities of (2.29) are summable wrt r.

(iv) It remains to show the same for the events (2.30), the probability of whose individual term

$$\left(F_n(p) = p - \frac{\rho}{n}\right) \bigcup_{\alpha < x < \beta_\rho} (F_n[p, x] > x - p + (2-d^2)\,\delta\eta_n)$$

may now conveniently be determined with the help of (B 8) rather than (B 8a). In contrast to the evaluation of (B 12), the factors $b(k; n, p)$ are now on the average smaller and their effect has to be taken into account more precisely. By (2.26), the $k\,[\equiv [np]+\rho]$ range now over the smallest values that have to be considered. With (2.26), (2.31), (2.34), (2.35) one has

$$\kappa' := -(d^2-1)\,\delta n\eta_n + (npq\log r)^{\frac{1}{2}}\,(\tfrac{2}{3})^{\frac{3}{2}} \quad (=\text{the former } \lambda' \text{ (2.37)}),$$

$$\lambda' := -(d^2-1)\,\delta n\eta_n - \rho[\epsilon(\kappa', d(2\delta npq\log r)^{\frac{1}{2}})].$$

Since $\rho = O((n \log r)^{\frac{1}{2}})$, λ' is of the same order, and it is allowed again to replace $\lambda*$ by $[\lambda']$. Now, analogous to (2.39) ff.

$$1 - B\left(\lambda'; nq - \rho, \frac{\beta_\rho - p}{q}\right) < \exp\left(\frac{-\frac{1}{2}(2 - d^2)^2 \delta^2 n^2 \eta_n^2}{|\rho|}\right) \{\approx r^{-O(1)}\} \quad (2.51)$$

since $\beta_\rho - p \sim \dfrac{|\rho|}{n}$ and thus

$$\lambda' - \frac{(nq - \rho)(\beta_\rho - p)}{q} = (2 - d^2)\delta n \eta_n - \rho \frac{\left(\dfrac{\rho}{n} + \delta \eta_n\right)}{q} \sim (2 - d^2)\delta n \eta_n;$$

the variance becomes $\sim |\rho|$.

Next, the first sum over s in [...] of (B 8) is the same as that arising in (B 8a). It gives exactly the same bound (2.45) as above and is of smaller order than (2.51).

In the second sum, the factor

$$\frac{n - k - \lambda'}{n - k - s} \sim 1$$

can be ignored. The term $b\left(s; \lambda', \dfrac{s - n(\delta + \gamma p)}{n\gamma(\beta - p)}\right)$ will be bounded above by the inverse root of its variance, the latter being

$$\sim |\rho|\,(s - (2 - d^2)\delta n \eta_n/|\rho|) \sim s > c(n \log r)^{\frac{1}{2}},$$

some $c > 0$. The variance of

$$b\left(\lambda'; n - k, \frac{\beta - p}{q}\right) \quad (2.52)$$

is

$$\sim n(\beta_\rho - p) \sim |\rho| > c(n \log r)^{\frac{1}{2}},$$

some $c > 0$. Hence the product of the inverses of the two standard deviations is

$$O((n \log r)^{-\frac{1}{2}}).$$

Multiplication with the number of terms in

$$\sum_{s=\kappa}^{\lambda}, \quad \text{which is } O(\lambda'), \quad \text{yields } O(1).$$

Hence only the exponential term in the approximation to (2.52) must still be considered which is, however, just the same as in (2.51) (cf. (A 12)).

Hence the probability of (2.30) is

$$O(n^{-\frac{1}{2}} \sum_{-nx_1 > \rho > -nx_2} \exp\left(-\frac{1}{2}[\rho^2/(npq) + \delta' n^2 \eta_n^2/|\rho|]\right))$$

$$= O\left(\int_{(\frac{2}{3})^{\frac{3}{2}} \log^{\frac{1}{2}} r}^{d(2\delta \log r)^{\frac{1}{2}}} dx \exp\left(-\frac{1}{2}[x^2 + 2\delta'((2 \log r)/3)^{\frac{3}{2}}/x]\right)\right). \quad (2.53)$$

The minimum of [...] in the exponent is assumed at

$$x_m = (\delta')^{\frac{1}{3}} ((2 \log r)/3)^{\frac{1}{2}}$$

(a value located in the integration interval). The integrand at x_m becomes just $r^{-(\delta')^{\frac{3}{2}}}$ which when multiplied by the length of the integration interval is still $r^{-\delta''}$, some $\delta'' > 1$, so that (2.53) is of this order and hence summable wrt r.

(v) The computation of the probability of the second sum of (2.3c) $[\kappa > 0]$ turns out to be surprisingly simple, compared to steps (iii) and (iv). The reason for this is easily seen when one graphs typical paths belonging to A_n for $\kappa > 0$ and $\kappa < 0$, respectively. As before, restriction (2.8) of the κ-values for each given n is permitted and will be postulated henceforth.

Instead of (2.20) we now consider the event

$$\tilde{C}_{\kappa, n} := \left(\Delta S(p) > \Delta_n \left(p + \frac{\kappa}{n} \right), \ \Delta S[x_{n, k}, p] < \Delta n \left(\frac{\kappa}{n} + \delta \eta_n \right) \right), \quad (2.54)$$

independent of B_n. Again, with $x := \kappa/n \approx O(n^{-\frac{1}{2}})$ [up to logarithmic terms],

$$P\tilde{C}_{\kappa, n} > \sum_{\Delta n x < j < 2\Delta n x} b(\Delta np + j; \Delta n, p) \, B\left(\Delta n(x + \delta\eta_n); \Delta np + j, \frac{x + \delta\eta_n}{p} \right)$$
$$> \tfrac{1}{4}(1 - B(\Delta n(p + x); \Delta n, p)) > c r^{-\delta(d^2 - 1)d^2},$$

with some constant $c > 0$; this follows because

$$\Delta n(x + \delta\eta_n) - (\Delta np + j)(x + \delta\eta_n)/p \sim -j\frac{x}{p} = o((nx)^{\frac{1}{2}}),$$

where nx is the variance of the B-function in (2.54). Hence the B-terms are greater than $\tfrac{1}{4}$ uniformly in j, x, and n. As to the last inequality of (2.54), one has by $\kappa < \bar{\kappa}_r$

$$\Delta n x (\Delta npq)^{-\frac{1}{2}} < (2\delta(d^2 - 1)\log r)^{\frac{1}{2}}.$$

Instead of (2.21) we now have for $n \in N_{r-1}$

$$\tilde{C}_{\kappa, n}(S_n(p) = k, S_n[x_{n, k}, p] < \kappa)$$

$$\subset \left(S_{n_r}(p) > n_r \left(p + \frac{\kappa}{n} \right), \ S_{n_r}[x_{n, k}, p] < n_r \frac{\kappa}{n} + \Delta n \delta \eta_n \right)$$

$$\subset (F_{n_r}(p) > p + x, \ F_{n_r}[p - \delta\eta_{n_r} - x, p) < x + (d^2 - 1)\delta\eta_{n_r}) =: \tilde{U}_x \tilde{V}_x.$$
$$(2.55)$$

With $x_0 < x_1 < x_2$ as defined in (2.26), one notes that $x_0 \leqslant x \leqslant x_1$ as before (see (2.23)) as κ and n vary. Denoting now n_r by n, the events (2.55) are contained in

$$(\bigcup_{x_0<x<x_1} \bar{V}_x) \cup \bigcup_{x_1<x<x_2} [\bar{U}_x \cap (\bigcup_{x_1<\xi<x} \bar{V}_\xi)]$$

where $\bigcup_{x_0<x<x_1}$ may be written in analogy to (2.28) as

$$\bar{U}_{x_2} \cap (\bigcup_{x_1<\xi<x_2} \bar{V}_\xi) + \sum_{x_1<j/n<x_2} \left(F_n(p) = p+\frac{j}{n}\right) \cap (\bigcup_{x_1<\xi<j/n} \bar{V}_\xi), \qquad (2.56)$$

and

$$\bigcup_{x_0<x<x_1} \bar{V}_x = \bigcup_{p-x_1-\delta\eta_n<x<p-x_0-\delta\eta_n} (F_n[x,p] < p-x-(2-d^2)\,\delta\eta_n).$$

The probability of this can be obtained with the help of (B 25a) if we identify in that formula

$$\left.\begin{aligned}
\delta &:= p-(2-d^2)\,\delta\eta_n; \quad \gamma := 1, \\
\alpha &:= p-x_1-\delta\eta_n; \quad \beta := p-x_0-\delta\eta_n; \quad [\alpha \sim \beta \sim p]
\end{aligned}\right\} \qquad (2.57)$$

(p and n are unchanged);

$$\kappa_k := [k-nx_1-(d^2-1)\,\delta n\eta_n]; \quad \lambda'_k := k-nx_0-(d^2-1)\,\delta n\eta_n;$$

k runs up to $n(x_1+(d^2-1)\,\delta\eta_n) = (npq\log r)^{\frac{1}{2}}(\frac{2}{3})^{\frac{3}{2}}+(d^2-1)\,\delta n\eta_n$.

However, for these k-values $|\kappa| > \bar{\kappa}_r$, and hence they need not be considered. Thus only the probability of the sum in (2.56) must be determined, using (B 24). Again (2.57) holds, but now

$$\alpha := \alpha_j := p-\delta\eta_n-j/n; \quad \beta := p-\delta\eta_n-x_1; \quad (j = k-np);$$

$$\kappa_k := [np-(d^2-1)\,\delta n\eta_n]; \quad \lambda'_k := k-nx_1-(d^2-1)\,\delta n\eta_n.$$

The first term arising from the [...]-bracket of (B 24) is, observing $nx_1 < j < nx_2$,

$$1-B\left(\lambda'_k; k, 1-\frac{x_1+\delta\eta_n}{p}\right) < r^{-\delta^2(2-d^2)^2}, \qquad (2.58)$$

the variance being $\sim nx_1$, and

$$\lambda'_k - k\left(1-\frac{x_1+\delta\eta_n}{p}\right) = -nx_1-(d^2-1)\,\delta n\eta_n+\left(n+\frac{j}{p}\right)(x_1+\delta\eta_n)$$

$$\sim (2-d^2)\,\delta n\eta_n.$$

(2.58) is summable wrt r for $d > 1$ small enough. In the first sum over s one has

$$-n(\delta-\gamma p) := (2-d^2)\,\delta n\eta_n,$$

$$k-n(\delta-\gamma p)-s := k-s+(2-d^2)\,\delta n\eta_n > nx_1$$

so that the ratio of the two up to logarithmic terms is

$$O(n^{-\frac{1}{4}}). \tag{2.59}$$

Next,
$$b\left(s, k, \frac{s-k+n\delta}{n\gamma p}\right) = O(n^{-\frac{1}{2}}), \tag{2.60}$$

the variance being $\sim s - k + n\delta \sim np$. Multiplication of (2.60) by (2.59) and the number of terms, namely $\lambda_k - \kappa_k \approx O(\sqrt{n})$, gives a contribution $O(n^{-\frac{1}{4}}) = O(d^{-\frac{1}{4}r})$ which is summable wrt r, and so are the second sums over s of (B 24). One has $k - \lambda_k^* \sim nx_1$, $k - s \gtrsim nx_1$ with ratio ~ 1. Next

$$b\left(s; \lambda_k^*, \frac{s-k+n\delta}{n\gamma\beta}\right) = O(n^{-\frac{1}{2}})$$

(the variance being asymptotically that of (2.60)), and

$$b\left(\lambda_k^*; k, \frac{\beta}{p}\right) = O(n^{-\frac{1}{2}}) \tag{2.61}$$

with variance $\sim n\beta \sim np$, and

$$\lambda^* - k\beta/p = (n+j/p)(x_1 + \delta\eta_n) - nx_1 - (d^2 - 1)\delta n\eta_n \sim (2 - d^2)\delta n\eta_n.$$

Division by $\sqrt{(np)}$ yields an $o(1)$; (2.61) follows. The number of terms is roughly $O(\sqrt{n})$. Consequently the total contribution of the second sum over s is, say, $O(n^{-\frac{1}{4}}) = O(d^{-\frac{1}{4}r})$ which is summable. Thus the upper class case of the theorem is proved with the plus sign in (1.4).

APPENDIX A

ESTIMATES OF BINOMIAL PROBABILITIES
FOR LARGE DEVIATIONS

The following lemmas are used in the above proofs. They are listed for easy reference. The bound (A 9) is stronger than what seems to be known for the considered parameter ranges. In particular, it cannot be derived from Okamoto's and related inequalities, even in refined form.

The other estimates stated in this appendix are either classical ones or simple consequences of them.

Lemma A 1. *Let N be the set of naturals and let p: $N \times N \to (0, 1)$ with values $p(n, k)$ be any function such that*

$$\frac{k}{np(n, k)} \to 1 \quad and \quad \frac{k - np(n, k)}{n - k} \to 0$$

$$\left[equivalently \ \frac{n-k}{nq(n, k)} \to 1, q = 1 - p\right] \tag{A 1}$$

as $n \to \infty,\ k \to \infty$ *independently or not, but such that* $n - k \to \infty$.   (A 2)

These conditions imply $npq \to \infty$. *Denoting*

$$t := (k - np)/\sqrt{(npq)},\qquad\text{(A 3)}$$

there holds under (A 2)

$$b(k; n, p) = \frac{1 + o(1)}{\sqrt{(2\pi npq)}} \exp\left(-\frac{1 + o(1)}{2} t^2\right),\qquad\text{(A 4)}$$

more precisely

$$b(k; n, p) = \frac{1 + o(1)}{\sqrt{(2\pi npq)}} \exp\left\{-\frac{t^2}{2} + \frac{t^2}{2}\left(1 - \left(\frac{k(n-k)}{npnq}\right)^{-1}\right)\right.$$

$$\left. + \frac{t^3}{3(npq)^{\frac{1}{2}}} (2p - 1 + o(1)\right\}.\qquad\text{(A 5)}$$

In these expressions the dependence of p and q on n and k has been suppressed for simplicity of notation. In terms of $t = t(n, k)$ conditions (1) imply

$$t\sqrt{\frac{q}{np}} \to 0,\quad t\sqrt{\frac{p}{nq}} \to 0,\quad \text{hence } t/\sqrt{(npq)} \to 0\qquad\text{(A 6)}$$

under the limit process (A 2).

Moreover, one has under (A 1), (A 2)

$$(1 + o(1)) \frac{\sqrt{(npq)}}{t} b(k + 1; n, p) < 1 - B(k; n, p)$$

$$< \left(1 + \frac{\sqrt{(npq)}}{t}\right) b(k + 1; n, p),\qquad\text{(A 7)}$$

where the first inequality requires

$$t \to \infty;\qquad\text{(A 8)}$$

consequently under (A 1), (A 2), (A 8)

$$1 - B(k; n, p) = O(1) t^{-1} e^{-t^2}/(2 + o(1)).\qquad\text{(A 9)}$$

In each formula the $o(1)$ and $O(1)$ terms are to be understood as follows: $|o(1)|$ is smaller than any given $\epsilon > 0$ and $O(1)$ is bounded between two positive numbers whenever the limits in (A 1) are approached sufficiently closely and n, k, and $n - k$ are sufficiently large.

If $t^3/\sqrt{(npq)}$ is small and npq large, Uspensky's bound is often useful: for any $k, n, p \equiv 1 - q$ satisfying $npq \geqslant 25$ there holds

$$b(k; n, p) = (2\pi npq)^{-\frac{1}{2}} e^{-t^2/2} \left[1 + \frac{(q - p)(t^3 - 3t)}{6\sqrt{(npq)}}\right] + \Delta,$$

$$|\Delta| < \text{const } (npq)^{-\frac{3}{2}},\qquad\text{(A 10)}$$

where the constant is absolute.

Lemma A 2. *For* $p \in (0,1)$; $n = 1, 2, \ldots, j = 0, 1, \ldots$ *such that*

$$0 < \frac{np-j}{\sigma^2} \leqslant 1,1$$

where $\sigma^2 = npq$, *there holds*

$$B(j; n, p) < \exp\left(-\frac{t^2}{2}\left(1 - \frac{1}{2}\frac{t}{\sigma}\right)\right), t := \frac{np-j}{\sigma}; \qquad (A\,11)$$

if $0 < \sigma^{-2}(j - np) \leqslant 1$, *with* $t = \sigma^{-1}(j - np)$

$$b(j; n, p) \leqslant 1 - B(j-1; n, p) < \exp\left(-\frac{t^2}{2}\left(1 - \frac{t}{2\sigma}\right)\right). \qquad (A\,12)$$

The exponential bound (A 14) is usually stated under the condition $0 \leqslant ec \leqslant 1$ instead of (A 13). However, in our context it is convenient to replace the upper bound on ec by a slightly larger one.

Lemma A 3. *Let* $X_1, \ldots, X_n$ *be i.r.v.'s with* $EX_i = 0$, $0 < \sigma_i^2 := EX_i^2 < \infty$ *for all* i,

$$S := \sum_1^n X_i, \quad s^2 := \sum_1^n \sigma_i^2, \quad \max_i \operatorname{ess\,sup} |X_i|/s =: c < \infty.$$

Then for a constant ϵ *such that*

$$0 \leqslant ec \leqslant 1,1 \qquad (A\,13)$$

there holds $\qquad P\left(\frac{S}{s} > \epsilon\right) < \exp\left(-\frac{\epsilon^2}{2}\left(1 - \frac{ec}{2}\right)\right). \qquad (A\,14)$

(Sharper but more complicated bounds on a wider range may e.g. be obtained from Hoeffding, *JASA* 1963.)

APPENDIX B
ON A PROPERTY OF SAMPLE PROBABILITIES

Definition. Let $X_1, \ldots, X_n$ be iidrv's from the df F defined on $(a, b) \subset R$. Then

$$F_n(B) := n^{-1} \sum_{j=1}^n I(X_j \in B) \quad \text{for } B \in (a, b)\,\mathscr{B}, \qquad (B\,1)$$

with $(a, b)\,\mathscr{B}$ the Borel field on (a, b), is called the *sample probability* of the sample $X_1, \ldots, X_n$. If e.g. $[c, d) \subset (a, b)$ we write $F_n([c, d)) = F_n[c, d)$.

One notes that if the X_j defined on the probability space $(\Omega, \mathscr{A}, P)$ then F_n is a transition probability defined on $\Omega \times [(a, b)\,\mathscr{B}]$ onto $[0, 1]$, i.e., F_n is a probability on $(a, b)\,\mathscr{B}$ for any fixed $X_1(\omega), \ldots, X_n(\omega)$, any $\omega \in \Omega$, and a rv on $(\Omega, \mathscr{A}, P)$ for any fixed $B \in \mathscr{B}$ (comp. e.g. [7], p. 73).

In the following we limit ourselves to the basic case of rv's $X_j \sim U(0,1)$ although the generalization to continuous strictly monotonic or merely continuous df's is easily possible. In Theorem B 1 below, the probability is determined that the values of the sample probability at the intervals $[p, x) \subset [0, 1]$ lie below a line segment $\delta + \gamma x$ simultaneously for all $x \in (\alpha, \beta)$ when

$$0 \leqslant p \leqslant \alpha < \beta \leqslant 1. \tag{B 2}$$

In addition, a condition on the values of $F_n(p)$ may be imposed. The theorem is an easy corollary to Remark 3 of [2] which is needed here in the following form:

Remark 1. Let $X_1, ..., X_n$ be iidrv's from the continuous strictly monotonic df F, and let α, β, γ, δ be constants satisfying

$$0 \leqslant \alpha < \beta \leqslant 1, \quad 0 < \gamma < \infty, \quad 0 \leqslant \delta + \gamma\alpha < 1. \tag{B 3}$$

Then [using expressions $(1.17d')$ and $(1.17d)$ of [2]]

$$P(F_n(x) \leqslant \delta + \gamma F(x) \; \forall x \text{ with } F(x) \in (\alpha, \beta))$$

$$= B(\lambda^*; n, \beta) - \sum_{s=\kappa+1}^{\lambda^*} \frac{n(\delta + \gamma - 1)}{n(\delta + \gamma) - s} b\left(s; n, \frac{s - n\delta}{n\gamma}\right)$$

$$\times B(\lambda^* - s - 1; n - s - 1, p(s)) - b(\lambda^*; n, \beta) \sum_{s=\kappa+1}^{\lambda^*} \frac{n - \lambda^*}{n - s} b\left(s; \lambda^*, \frac{s - n\delta}{n\gamma\beta}\right) \tag{B 4}$$

$$= B(\lambda^*; n, \beta) - \sum_{s=\kappa+1}^{\lambda^*} b\left(s; n, \frac{s - n\delta}{n\gamma}\right)$$

$$\times \left[\frac{n(\delta + \gamma - 1)}{n(\delta + \gamma) - s} B(\lambda^* - s - 1; n - s - 1, p(s))\right.$$

$$\left. + \frac{n\gamma(1 - \beta)}{n(\delta + \gamma) - s} b(\lambda^* - s; n - s - 1, p(s))\right], \tag{B 4a}$$

where

$$\kappa := [n(\delta + \gamma\alpha)], \quad \lambda' := n(\delta + \gamma\beta),$$

$$\lambda^* := \begin{cases} n & \text{if } \lambda' > n, \\ \lambda \equiv [\lambda'] & \text{if } \lambda' < n \text{ and } \lambda' \text{ not an integer,} \\ \lambda' - 1 & \text{if } \lambda' \leqslant n \text{ and } \lambda' \text{ an integer;} \end{cases} \tag{B 5}$$

$$p(s) := (\lambda' - s)/(n(\delta + \gamma) - s).$$

Theorem B 1. Let $X_1, ..., X_n \sim U(0, 1)$, and let (B 2) and (B 3) hold. Put

$$(F_n[p, x) < \delta + \gamma x \; \forall \; x \in (\alpha, \beta)) =: A, \tag{B 6}$$

$$(F_n(p) = k/n) =: H_k \quad (k = 0, 1, ..., n). \tag{B 7}$$

Then for $k = 0, 1, \ldots, n - \kappa - 1$ *[with* $q := 1 - p$ *and notations* (B 5); λ_k^* *given by* (B 23) *below]*

$$P(AH_k) = b(k; n, p) \left[B\left(\lambda_k^*; n - k, \frac{\beta - p}{q}\right) \right.$$

$$- \sum_{s=\kappa+1}^{\lambda_k^{\bullet}} \frac{n(\delta + \gamma - 1) + k}{n(\delta + \gamma) - s} b\left(s; n - k, \frac{s - n(\delta + \gamma p)}{n\gamma q}\right)$$

$$\times B(\lambda_k^* - s - 1; n - k - s - 1, p(s)) - b\left(\lambda_k^*; n - k, \frac{\beta - p}{q}\right)$$

$$\left. \times \sum_{s=\kappa+1}^{\lambda_k^{\bullet}} \frac{n - k - \lambda_k^*}{n - k - s} b\left(s; \lambda_k^*; \frac{s - n(\delta + \gamma p)}{n\gamma(\beta - p)}\right) \right] \quad \text{(B 8)}$$

$$= b(k; n, p) \left[B\left(\lambda_k^*; n - k, \frac{\beta - p}{q}\right) - \sum_{s=\kappa-1}^{\lambda_k^{\bullet}} b\left(s; n - k, \frac{s - n(\delta + \gamma p)}{n\gamma q}\right) \right.$$

$$\times \left\{ \frac{n(\delta + \gamma - 1) + k}{n(\delta + \gamma) - s} B(\lambda_k^* - s - 1; n - k - s - 1, p(s)) \right.$$

$$\left. \left. + \frac{n\gamma(1 - \beta)}{n(\delta + \gamma) - s} b(\lambda_k^* - s; n - k - s - 1, p(s)) \right\} \right],$$

$$\text{(B 8a)}$$

$$p(s) = (\lambda' - s)/(n(\delta + \gamma) - s);$$

$$P(AH_k) = PH_k = b(k; n, p) \quad (k \geqslant n - \kappa). \quad \text{(B 9)}$$

For subsets $\qquad \mathscr{M} \subset \left\{\dfrac{0}{n}, \dfrac{1}{n}, \ldots, \dfrac{n}{n}\right\},$

$$P(A(F_n(p) \in \mathscr{M})) = \text{sum of corresponding terms (B 8/9)}. \quad \text{(B 10)}$$

In particular [with (B 8a)]

$$PA = 1 - B(n - \kappa; n, p) + \sum_{k=0}^{n-\kappa} b(k; n, p) \left[B\left(\lambda_k^*; n - k, \frac{\beta - p}{q}\right) \right.$$

$$- \sum_{s=\kappa+1}^{\lambda_k^{\bullet}} b\left(s; n - k, \frac{s - n(\delta + \gamma p)}{n\gamma q}\right)$$

$$\times \left\{ \frac{n(\delta + \gamma - 1) + k}{n(\delta + \gamma) - s} B(\lambda_k^* - s - 1; n - k - s - 1, p(s)) \right.$$

$$\left. \left. + \frac{n\gamma(1 - \beta)}{n(\delta + \gamma) - s} b(\lambda_k^* - s; n - k - s - 1, p(s)) \right\} \right] \quad \text{(B 11)}$$

and

$$P(A^c) = P\left(\bigcup_{\alpha < x < \beta} (F_n[p, x) > \delta + \gamma x) \right) = \sum_{k=0}^{n-\kappa} b(k; n, p)\{1 - [\ldots]\} \quad \text{(B 12)}$$

with [...] as in (B 11); in both formulae the [...] may also be taken as the square bracket of (B 8a).

The expressions [...] in (B 8/11) may be written in various equivalent forms, as listed in [2].

Proof of (B 9): since

$$F_n[p, x] = F_n(x) - F_n(p), \tag{B 13}$$

$$AH_k = H_k \bigcap_{\alpha < x < \beta} (F_n(x) < \delta + \gamma x + k/n). \tag{B 14}$$

For $k \geqslant n - \kappa$, $\delta + \gamma x + k/n \geqslant n^{-1}(\kappa' + n - \kappa) \geqslant 1$

[where $\kappa' := n(\delta + \gamma \alpha)$], \hfill (B 15)

hence $AH_k = H_k$.

As to (B 8), let $Y_1, ..., Y_n$ be the order statistics of $X_1, ..., X_n$. Since $H_k = (Y_k < p < Y_{k+1})$,

$$AH_k = H_k \bigcap_{\alpha < x < \beta} \left(n^{-1} \sum_{j=k+1}^{n} I(Y_j < x) < \delta + \gamma x \right)$$

$$= H_k \bigcap_{\alpha < x < \beta} \left((n-k)^{-1} \sum_{j=k+1}^{n} I(Y_j < x) < \frac{n}{n-k}(\delta + \gamma x) \right). \tag{B 16}$$

Now the conditional distribution of $Y_{k+1}, ..., Y_n$ under H_k is the same as the unconditional distribution of the order statistics $\tilde{Y}_1, ..., \tilde{Y}_{n-k}$ of a random sample $\tilde{X}_1, ..., \tilde{X}_{n-k}$ from $U(p, 1)$. Hence

$$P(AH_k) = P\left(\bigcap_{\alpha < x < \beta} \left(\tilde{F}_{n-k}(x) < \frac{n}{n-k}(\delta + \gamma x) \right) \right), \tag{B 17}$$

where

$$m\tilde{F}_m(x) = \sum_{j=1}^{m} I(\tilde{Y}_j < x) = \sum_{j=1}^{m} I(\tilde{X}_j < x). \tag{B 18}$$

The probability (B 17) equals (B 4) by the above remark with

$$F(x) = (x - p)/q, \quad p \leqslant x \leqslant 1.$$

However, the event on the right of (B 17) must still be expressed in the form of the event in (B 4) [first line]. Now

$$\alpha < x < \beta \Leftrightarrow \frac{\alpha - p}{q} =: \tilde{\alpha} < F(x) < \frac{\beta - p}{q} =: \tilde{\beta} \tag{B 19}$$

and

$$\tilde{F}_{n-k}(x) < \frac{n}{n-k}(\delta + \gamma x) \Leftrightarrow \tilde{F}_{n-k}(x) < \tilde{\delta} + \tilde{\gamma} F(x) \tag{B 20}$$

if

$$\tilde{\delta} := \frac{n}{n-k}(\delta + \gamma p), \quad \tilde{\gamma} := \frac{n}{n-k}\gamma q. \tag{B 21}$$

Since necessarily $k < n$, the assumptions (B 3) are satisfied by the constants $\tilde{\alpha}, \ldots, \tilde{\delta}$; note here

$$
\left.\begin{array}{l}
\tilde{\kappa}' := (n-k)\,(\tilde{\delta}+\tilde{\gamma}\tilde{\alpha}) = \kappa' < 1, \\
\tilde{\lambda}' := (n-k)\,(\tilde{\delta}+\tilde{\gamma}\tilde{\beta}) = \lambda'.
\end{array}\right\} \tag{B 22}
$$

Moreover,

$$
\lambda_k^* := \left\{\begin{array}{l}
n-k \text{ if } \tilde{\lambda}' > n-k \Leftrightarrow \lambda' > n-k, \\
\lambda := [\lambda'] \text{ if } \lambda' < n-k \text{ and } \lambda' \text{ is not an integer,} \\
\lambda' - 1 \text{ if } \lambda' \leqslant n-k \text{ and } \lambda' \text{ is an integer.}
\end{array}\right\} \tag{B 23}
$$

Hence $\quad P(A|H_k) = P\left(\tilde{F}_{n-k}(x) < \tilde{\delta}+\tilde{\gamma}\dfrac{x-p}{q} \; \forall \; \tilde{\alpha} < \dfrac{x-p}{q} < \tilde{\beta}\right)$

and (B 4) yields (B 8), and (B 4 a) (B 8 a) [after the due substitutions; note $(n-k)\,(\tilde{\delta}+\tilde{\gamma}) = n(\delta+\gamma)$ and $(n-k)\,\tilde{\gamma}(1-\tilde{\beta}) = n\gamma(1-\beta)$]. The other assertions of the theorem are obvious.

Theorem B 2. *Let* $X_1, \ldots, X_n$ *be iidrv's* $\sim U(0,1)$, *let*

$$0 \leqslant \alpha < \beta \leqslant p \leqslant 1; \quad 0 < \gamma < \infty; \quad 0 \leqslant \delta - \gamma\beta < \delta - \gamma\alpha \leqslant 1$$

be given constants, and let

$$B := (F_n[x,p] > \delta - \gamma x \quad \text{for all } \alpha < x < \beta),$$

$$H_k := (F_n(p) = k/n) \quad (k = 0, 1, \ldots, n);$$

then for $k = 1, 2, \ldots, [n(\delta - \gamma\alpha)]$

$$
\begin{aligned}
P(BH_k) &= b(k; n, p)\Bigg[B\left(\lambda_k^*; k, \frac{\beta}{p}\right) - \sum_{s=\kappa_k+1}^{\lambda_k^*} \frac{-n(\delta-\gamma p)}{k-n(\delta-\gamma p)-s} \\
&\quad \times b\left(s; k, \frac{s-k+n\delta}{n\gamma p}\right) B(\lambda_k^*-s-1; k-s-1, p(s)) \\
&\quad -b\left(\lambda_k^*; k, \frac{\beta}{p}\right) \sum_{s=\kappa_k+1}^{\lambda_k^*} \frac{k-\lambda_k}{k-s}\, b\left(s; \lambda_k^*, \frac{s-k+n\delta}{n\gamma\beta}\right)\Bigg] \\[4pt]
&= b(k; n, p)\Bigg[B\left(\lambda_k^*; k, \frac{\beta}{p}\right) - \sum_{s=\kappa_k+1}^{\lambda_k^*} b\left(s; k, \frac{s-k+n\delta}{n\gamma p}\right) \\
&\quad \times \Bigg\{ \frac{-n(\delta-\gamma p)}{k-n(\delta-\gamma p)-s}\, B(\lambda_k^*-s-1; k-s-1, p(s)) \\
&\quad + \frac{n\gamma(p-\beta)}{k-n(\delta-\gamma p)-s}\, b(\lambda_k^*-s; k-s-1, p(s))\Bigg\}\Bigg],
\end{aligned}
$$

(B 24)

(B 25)

where $\quad \kappa_k := [k - n(\delta - \gamma\alpha)], \quad \lambda'_k := k - n(\delta - \gamma\beta)\ [\leqslant k],$

$$\lambda^*_k := \begin{cases} [\lambda'_k] & \textit{if } \lambda'_k \neq \textit{integer} \\ \lambda'_k - 1 & \textit{if } \lambda'_k = \textit{integer}, \end{cases}$$

$$p(s) := (\lambda'_k - s)/\{k - n(\delta - \gamma p) - s\}.$$

$P(BH_k) = 0$ for $k = 0$ [*since* $\delta - \gamma\alpha > 0$ *is assumed*] and for $k > n(\delta - \gamma\alpha)$.
It follows

$$PB = \sum_{k=1}^{[n(\delta - \gamma\alpha)]} P(BH_k). \tag{B 25a}$$

The *proof* parallels that of Theorem B 1. Instead of (B 16) one has for $k = 1, 2, \ldots, [n(\delta - \gamma\alpha)]$

$$BH_k = H_k \bigcap_{\alpha < x < \beta} \left(n^{-1} \sum_{j=1}^{k} I(x < Y_j) > \delta - \gamma x \right)$$

$$= H_k \bigcap_{\alpha < x < \beta} \left(k^{-1} \sum_{j=1}^{k} I(Y_j < x) < 1 - \frac{n}{k}(\delta - \gamma x) \right).$$

To (B 17) corresponds

$$P(BH_k) = P \left(\bigcap_{\alpha < x < \beta} \left(\tilde{F}_k(x) < 1 - \frac{n}{k}(\delta - \gamma x) \right) \right), \tag{B 26}$$

where $\tilde{F}_k$ is the sample df of a k-sample from the df x/p, $0 \leqslant x \leqslant p$. (B 26) can be rewritten like (B 4) as

$$P \left(\bigcap_{\alpha/p < x/p < \beta/p} \left(\tilde{F}_k(x) < 1 - \frac{n}{k}\delta + \frac{n}{k}\gamma p \frac{x}{p} \right) \right). \tag{B 27}$$

(B 4/4 a) applies if we identify therein

$$\alpha := \alpha/p; \quad \beta := \beta/p; \quad n := k; \quad \delta := 1 - n\delta/k;$$

$$\gamma := \frac{n}{k}\gamma p; \quad \kappa' := k - n(\delta - \gamma\alpha); \quad \lambda' := k - n(\delta - \gamma\beta).$$

Note $k(\delta + \gamma) = k - n(\delta - \gamma p)$. The conditions (B 3) are satisfied if $k \leqslant n(\delta - \gamma\alpha)$.

Further symmetry properties. The analogon to Lemma 1.1 of [2] for sample probabilities on intervals is

Lemma B 1. *Let* $X_1, \ldots, X_n$ *be iidrv's* $\sim U(0, 1)$,

$$0 \leqslant \alpha \leqslant \beta < b \leqslant 1 \tag{B 28}$$

constants, and $h: (\alpha, \beta) \to [0, 1]$ *a non-increasing function. Then*

$$P(F_n[x, b] < h(x) \quad \text{for all} \quad x \in (\alpha, \beta))$$
$$= P(F_n[1 - b, x] < h(1 - x) \quad \text{for all} \quad x \in (1 - \beta, 1 - \alpha)). \tag{B 29}$$

The equation remains valid, if ' $<$ ' is replaced by ' $>$ ' both times.

Proof. Let $X_j' := 1 - X_j \sim U(0, 1)$, any j. Then

$$I(x \leqslant X_j < b) = I(1 - b \leqslant X_j' < 1 - x) \quad \text{a.s.,}$$

and for the sample probability F_n' of the X_j'-sample there holds a.s.

$$(F_n[x, b) < h(x)) = (F_n'[1 - b, 1 - x) < h(x)). \qquad \text{(B 30)}$$

Hence a.s.

$$\bigcap_{\alpha < x < \beta} (F_n[x, b) < h(x)) = \bigcap_{1 - \beta < x < 1 - \alpha} (F_n'[1 - b, x) < h(1 - x)).$$

Lemma B 2. *In addition to the assumptions of Lemma* B 1 *let* $p = 1 - q \in (0, 1)$ *and* $k \in \{0, 1, ..., n\}$ *be given. Then*

$$P\left(\left(F_n(p) = \frac{n}{k}\right) \bigcap_{\alpha < x < \beta} (F_n[x, b) > h(x))\right)$$

$$= P\left(\left(F_n(q) = 1 - \frac{k}{n}\right) \bigcap_{1 - \beta < x < 1 - \alpha} (F_n[1 - b, x) < h(1 - x))\right) \quad \text{(B 31)}$$

and

$$P\left(\left(F_n(p) = \frac{k}{n}\right) \bigcap_{\alpha < x < \beta} (F_n[x, b) > h(x))\right)$$

$$= P\left(\left(F_n(q) = 1 - \frac{k}{n}\right) \bigcap_{1 - \beta < x < 1 - \alpha} (F_n[1 - b, x) > h(1 - x))\right). \quad \text{(B 32)}$$

Proof. $F_n(p) = 1 - F_n'(g)$.

REFERENCES

[1]   Bahadur, R. R. A note on quantiles in large samples. *Ann. Math. Statist.* 1966, **37**, 577–80.

[2]   Eicker, F. On the probability that a sample distribution function lies below a line segment. (To appear.)

[3]   Eicker, F. A log log-law for double sequences of random variables. (To appear.)

[4/5] Feller, W. *An introduction to probability theory and its applications.* Vol. 1, 2nd ed. New York: Wiley, 1957. Vol. 2, 1966.

[6]   Kiefer, J. On Bahadur's representation of sample quantiles. *Ann. Math. Statist.* 1967, 1323–42.

[7]   Neveu, J. *Mathematical foundations of the calculus of probability.* San Francisco: Holden–Day, 1965.

ASYMPTOTIC DISTRIBUTIONS OF QUANTILES IN SOME NONSTANDARD CASES

LIONEL WEISS†

1 INTRODUCTION

In deriving the asymptotic distribution of a set of sample quantiles, it is usual to write the joint probability density function of the quantiles for a finite sample size, and study the joint probability density function as the sample size increases. This technique works well enough when the elements of the sample are independent and identically distributed scalar random variables, with their common probability density function satisfying certain regularity conditions. See [1], [2]. However, in less simple cases this technique can get very complicated. For example, in [3] and [4] the technique was used when each element of the sample is itself a k-dimensional random variable, and the problem is to find the joint asymptotic distribution of a quantile of the first co-ordinates in the sample, a quantile of the second co-ordinates in the sample, ..., a quantile of the kth co-ordinates in the sample. The computations were lengthy. In a later paper [5], the results were derived more simply and under less restrictive conditions by studying the joint cumulative distribution function of the quantiles, rather than the joint probability density function. In general, it seems that studying the joint cumulative distribution function of the quantiles is a convenient way to derive asymptotic distributions. The purpose of this paper is to illustrate this with some nonstandard cases.

2 THE CASE OF A DISCONTINUOUS DENSITY FUNCTION

Suppose $X_1, ..., X_n$ are independent and identically distributed random variables, with common probability density function $f(x)$, cumulative distribution function $F(x)$. Let b be a fixed value in the open interval $(0, 1)$, and let $Y(n)$ denote the bth sample quantile. That is, letting $[nb]$ denote the largest integer not greater than nb, $Y(n)$ is equal to that one of the values $X_1, ..., X_n$ such that exactly $[nb]$ of the other values are below it. (We can ignore the possibility of ties among $X_1, ..., X_n$, since with probability one no ties will occur.) Suppose $f(x)$ is discontinuous at

† Research supported by National Science Foundation Grant No. GP–7798.

[343]

$F^{-1}(b)$, but is continuous on the right at $F^{-1}(b)$ and on the left at $F^{-1}(b)$. More specifically, we assume

$$\lim_{\substack{x \to F^{-1}(b) \\ x < F^{-1}(b)}} f(x) = A > 0, \qquad \lim_{\substack{x \to F^{-1}(b) \\ x > F^{-1}(b)}} f(x) = B > 0$$

and $A \neq B$. (If $A = B$, we are in the standard case.)

For any given value z, let $N_n(z)$ denote the number of values $X_1, ..., X_n$ which are greater than z. For any z, the events

$$\{Y(n) \leqslant z\} \quad \text{and} \quad \{N_n(z) < n - [nb]\}$$

are identical, and therefore

$$P\{Y(n) \leqslant z\} = P\{N_n(z) < n - [nb]\}. \tag{1}$$

We note that $N_n(z)$ has a binomial distribution with parameters n, $1 - F(z)$. Thus $N_n(z)$ has mean $u_n(z)$ and standard deviation $s_n(z)$ given as follows:
$$u_n(z) = n(1 - F(z)), \quad s_n(z) = (nF(z)(1 - F(z)))^{\frac{1}{2}}.$$

We define $N_n'(z)$ as the 'standardized' value of $N_n(z)$. That is,

$$N_n'(z) = \frac{N_n(z) - u_n(z)}{s_n(z)}.$$

Next we fix a positive value t. For convenience in writing, let q denote $F^{-1}(b) - A^{-1}n^{-\frac{1}{2}}(b(1-b))^{\frac{1}{2}} t$. Using (1), we have

$$P\{Y(n) \leqslant q\} = P\{N_n(q) < n - [nb]\} = P\left\{ N_n'(q) < \frac{n - [nb] - u_n(q)}{s_n(q)} \right\}. \tag{2}$$

It is easily verified that $F(q)$ can be written as

$$b - n^{-\frac{1}{2}}t(b(1-b))^{\frac{1}{2}} + n^{-\frac{1}{2}}j_n(t),$$

where $j_n(t)$ approaches zero as n increases. Using this fact, and substituting in the third probability in (2), a very simple calculation shows that the third probability in (2) can be written as $P\{N_n'(q) < -t + k_n(t)\}$, where $k_n(t)$ approaches zero as n increases. Since it is easily seen that the cumulative distribution function for $N_n'(q)$ approaches the standard normal cumulative distribution function as n increases, we have shown that

$$\lim_{n \to \infty} P\{An^{\frac{1}{2}}(b(1-b))^{-\frac{1}{2}}(Y(n) - F^{-1}(b)) \leqslant -t\} = (2\pi)^{-\frac{1}{2}} \int_{-\infty}^{-t} e^{-\frac{1}{2}v^2} dy$$

for any positive t.

A completely analogous argument shows that for any positive t,

$$\lim_{n \to \infty} P\{Bn^{\frac{1}{2}}(b(1-b))^{-\frac{1}{2}}(Y(n) - F^{-1}(b)) \leqslant t\} = (2\pi)^{-\frac{1}{2}} \int_{-\infty}^{t} e^{-\frac{1}{2}v^2} dy.$$

Thus the asymptotic cumulative distribution function for

$$n^{\frac{1}{2}}(Y(n) - F^{-1}(b))$$

consists of pieces of two different normal cumulative distribution functions joined together.

3 THE CASE OF A DOUBLE SAMPLE

Suppose $X_1, \ldots, X_m, W_1, \ldots, W_m$ are mutually independent random variables, each of the variables $X_1, \ldots, X_m$ having probability density function $f(x)$ and cumulative distribution function $F(x)$, and each of the variables $W_1, \ldots, W_n$ having probability density function $g(x)$ and cumulative distribution function $G(x)$. b_1, b_2 are fixed values, each in the open interval $(0, 1)$. Let $Y(n)$ denote the b_1th sample quantile of the sample consisting of $X_1, \ldots, X_m$, and let $Z(n)$ denote the b_2th sample quantile of the sample consisting of $X_1, \ldots, X_m, W_1, \ldots, W_n$. m and n will approach infinity, with m/n always equal to a fixed finite positive value c.

We assume that $f(x)$ is positive and continuous at the point $F^{-1}(b_1)$. We also assume that there is a unique solution (v, say) to the equation

$$cF(x) + G(x) = (1 + c)b_2,$$

and that $f(x)$ and $g(x)$ are both continuous at v, and $cf(v) + g(v)$ is positive.

For any z, let $N_m(z)$ denote the number of values $X_1, \ldots, X_m$ which are greater than z, and let $N_n(z)$ denote the number of values $W_1, \ldots, W_n$ which are greater than z. If z, w are any nonrandom values, $N_m(z)$ and $N_n(w)$ are independent binomial random variables, with parameters m, $1 - F(z)$ for $N_m(z)$, and parameters n, $1 - G(w)$ for $N_n(w)$.

Denote $N_m(w) + N_n(w)$ by $T_n(w)$. Denote the mean of $N_m(z)$ by $u_m(z)$, the standard deviation of $N_m(z)$ by $s_m(z)$, the mean of $T_n(w)$ by $u_t(w)$, the standard deviation of $T_n(w)$ by $s_t(w)$, and the covariance between $N_m(z)$ and $T_n(w)$ by $c_n(z, w)$. Thus

$$u_m(z) = m(1 - F(z)),$$
$$s_m(z) = \{mF(z)(1 - F(z))\}^{\frac{1}{2}},$$
$$u_t(w) = m(1 - F(w)) + n(1 - G(w)),$$
$$s_t(w) = \{mF(w)(1 - F(w)) + nG(w)(1 - G(w))\}^{\frac{1}{2}},$$
$$c_n(z, w) = mF(w)(1 - F(z)) \quad \text{if } w \leqslant z,$$
$$c_n(z, w) = mF(z)(1 - F(w)) \quad \text{if } w \geqslant z.$$

Let $N'_m(z)$ denote $\dfrac{N_m(z) - u_m(z)}{s_m(z)}$

and let $T'_n(w)$ denote $\dfrac{T_n(w) - u_t(w)}{s_t(w)}$.

For any nonrandom values y, z, the events

$$\{Y(n) \leqslant y, Z(n) \leqslant z\}$$

and $\qquad \{N_m(y) < m - [mb_1], T_n(z) < m + n - [(m+n)b_2]\}$

are identical, and therefore

$$P\{Y(n) \leqslant y, Z(n) \leqslant z\}$$

$$= P\{N_m(y) < m - [mb_1], T_n(z) < m + n - [(m+n)b_2]\}. \quad (3)$$

Fix any values t, u (positive or negative). For typographical convenience, denote

$$F^{-1}(b_1) + \frac{t(b_1(1-b_1))^{\frac{1}{2}}}{m^{\frac{1}{2}}f(F^{-1}(b_1))}$$

by q_1, and denote

$$v + \frac{u\{cF(v)(1 - F(v)) + G(v)(1 - G(v))\}^{\frac{1}{2}}}{n^{\frac{1}{2}}(cf(v) + g(v))}$$

by q_2. We will compute

$$P\{Y(n) \leqslant q_1, Z(n) \leqslant q_2\} \quad (4)$$

by using the fact that, by (3), (4) is equal to

$$P\left\{N'_m(q_1) < \frac{m - [mb_1] - u_m(q_1)}{s_m(q_1)}, \quad T'_n(q_2) < \frac{m + n - [(m+n)b_2] - u_t(q_2)}{s_t(q_2)}\right\}.$$
$$(5)$$

It is easily verified that we can write $F(q_1)$ as

$$b_1 + m^{-\frac{1}{2}}t(b_1(1-b_1))^{\frac{1}{2}} + m^{-\frac{1}{2}}j_n(t),$$

where $j_n(t)$ approaches zero as n increases. Also, we can write $F(q_2)$ as

$$F(v) + \frac{uf(v)\{cF(v)(1 - F(v)) + G(v)(1 - G(v))\}^{\frac{1}{2}}}{n^{\frac{1}{2}}(cf(v) + g(v))} + n^{-\frac{1}{2}}k_n(u)$$

and $G(q_2)$ as

$$G(v) + \frac{ug(v)\{cF(v)(1 - F(v)) + G(v)(1 - G(v))\}^{\frac{1}{2}}}{n^{\frac{1}{2}}(cf(v) + g(v))} + n^{-\frac{1}{2}}k'_n(u),$$

where $k_n(u)$ and $k'_n(u)$ both approach zero as n increases.

Using these facts, it follows readily that

$$\frac{m - [mb_1] - u_m(q_1)}{s_m(q_1)} = t + A_n,$$

where A_n approaches zero as n increases, and

$$\frac{m + n - [(m+n) b_2] - u_t(q_2)}{s_t(q_2)} = u + B_n,$$

where B_n approaches zero as n increases.

Define $r(b_1, v)$ as follows:

$$r(b_1, v) = \frac{cF(v)(1 - b_1)}{\{cb_1(1 - b_1)\}^{\frac{1}{2}} \{cF(v)(1 - F(v)) + G(v)(1 - G(v))\}^{\frac{1}{2}}} \quad \text{if } v \leqslant F^{-1}(b_1),$$

$$r(b_1, v) = \frac{cb_1(1 - F(v))}{\{cb_1(1 - b_1)\}^{\frac{1}{2}} \{cF(v)(1 - F(v)) + G(v)(1 - G(v))\}^{\frac{1}{2}}} \quad \text{if } v \geqslant F^{-1}(b_1).$$

The joint cumulative distribution function of $N'_m(q_1)$, $T'_n(q_2)$ approaches the bivariate normal cumulative distribution function with zero means, unit variances, and correlation coefficient $r(b_1, v)$, as n increases. Let $h(r, s)$ denote the bivariate normal probability density function with zero means, unit variances, and correlation coefficient $r(b_1, v)$. It follows from (5) and the facts developed immediately following (5), that

$$\lim_{n \to \infty} P \left\{ \begin{array}{l} \dfrac{m^{\frac{1}{2}} f(F^{-1}(b_1)) \{Y(n) - F^{-1}(b_1)\}}{\{b_1(1 - b_1)\}^{\frac{1}{2}}} \leqslant t \\[4mm] \dfrac{n^{\frac{1}{2}} \{cf(v) + g(v)\} \{Z(n) - v\}}{\{cF(v)(1 - F(v)) + G(v)(1 - G(v))\}^{\frac{1}{2}}} \leqslant u \end{array} \right\} = \int_{-\infty}^{u} \int_{-\infty}^{t} h(r, s) \, dr \, ds.$$

4 OTHER CASES

The technique illustrated above can be applied to a great variety of other cases. For example, the two cases above could be combined: we could allow discontinuities in the density functions in a double sample. Or the density function could be zero at a unique population quantile. The resulting asymptotic distributions are more complicated, but the technique for finding them is simple to apply.

REFERENCES

[1] Cramér, H. *Mathematical methods of statistics*. Princeton University Press (1951).

[2] Fisz, M. *Probability theory and mathematical statistics*. New York: John Wiley (1963).

[3] Mood, A. M. On the joint distribution of medians in samples from a multivariate population. *Ann. Math. Statist.* **12**, 268–78 (1941).

[4] Siddiqui, M. M. Distribution of quantiles in samples from a bivariate population. *J. Res. N. B. S.* **64B** (*Math. and Math. Phys.*) No. 3, 145–50 (1960).

[5] Weiss, L. On the asymptotic joint normality of quantiles from a multivariate distribution. *J. Res. N. B. S.* **68B** (*Math. and Math. Phys.*) No. 2, 65–6 (1964).

OLD AND NEW METHODS
FOR STUDYING ORDER STATISTICS
AND SAMPLE QUANTILES

J. KIEFER †

This paper, prepared for presentation with those of Weiss [21] and of Eicker [6] in this volume, contains some comments on those two papers, as well as related remarks on other results and unsolved problems in the area described by the present paper's title.

1 MULTINOMIAL EVENTS EQUIVALENT TO THOSE CONCERNING SAMPLE QUANTILES: DISTRIBUTION THEORY

It is remarkable that the years have seen such great neglect or ignorance of the simple approach used here by Weiss [21], of studying limit probabilities of events concerning sample quantiles by rewriting them as events concerning multinomial rv's. Presented with Weiss's paper, I decided to check once more on the approach used in standard texts in the simplest problem of this type—that of finding the limiting df of a single sample p-tile $Y_{p,n}$ $(0 < p < 1)$ of n iid rv's $X_1, ..., X_n$ when the df F of X_1 has positive continuous density f at the corresponding population p-tile, ξ_p. To my horror, each of the two dozen texts I found on my shelf which treated the subject at all, used a less wieldy proof in terms of the exact sample quantile density rather than one in terms of the limiting suitably normalized binomial df. Perhaps this should not have been too surprising in view of Weiss's comments and the need many instructors must have shared with me for years, of distributing supplementary notes on this topic when it arose in an elementary course; but one would have thought that the approach in Cramér or Wilks might have been abandoned in at least one standard text after all this time, in favor of the simpler approach.

Of course, this simpler approach can be found in earlier research papers of Weiss, as well as in David Moore's recent work on efficient multivariate location-parameter estimation; as Weiss indicates, the

† This paper was prepared as discussion of the papers of Weiss and Eicker at the First International Conference on Nonparametric Inference, under support from ONR Contract No. NONR 401 (50). Reproduction in whole or in part is permitted for any purpose of the United States Government.

computational advantage of using the multinomial technique becomes even greater in these more complex settings. What is the ancient history of the technique, which, while not deep, is so simple and useful? In a 'Classroom Note' to appear in the M.A.A. *Monthly*, in which Moore also gives an exposition of this technique, he mentions the passing reference to the method in Feller [10] (vol. 2, p. 23) and the use in Loève's paper [14]. There are undoubtedly other appearances of the method in research papers. (This morning I noticed a paper in the 1968 *JRSS(B)*, p. 570, by A. M. Walker, in which the multinomial treatment initiates a proof which for some reason then unnecessarily uses the characteristic function.) I do not know who first used the approach, but know that it was taught at Columbia (and, undoubtedly, elsewhere) twenty years ago; the earliest use I know of in print is in Smirnov's 1949 paper [18].

As Weiss remarks in his last section, this technique can also be used when $f(\xi_p) = 0$: indeed, (1) below is still completely trivial, and the arithmetic which follows is solely to illustrate the explicit computation of the approximate inverse df in particular cases. Suppose the df of X_1 is

$$\begin{cases} p - G_1(\xi_p^{(1)} - x) & \text{for} \quad x \leqslant \xi_p^{(1)}, \\ p & \text{for} \quad \xi_p^{(1)} \leqslant x \leqslant \xi_p^{(2)}, \\ p + G_2(x - \xi_p^{(2)}) & \text{for} \quad \xi_p^{(2)} \leqslant x, \end{cases}$$

where $G_i(x) = 0$ only at $x = 0$ (and of course $\xi_p^{(1)} = \xi_p^{(2)}$ in the case of a unique p-tile). If the G_i are suitably regular, the domain of the asymptotic density of $Y_{p,n}$ can be conveniently broken up into two parts, as in Weiss's treatment (in addition to the part between the $\xi_p^{(i)}$ where this density is zero). We need only treat one part, corresponding to positive values of $Y_{p,n} - \xi_p^{(2)}$. Suppose G_2 is continuous and strictly increasing in an interval $[0, \epsilon]$ for some $\epsilon > 0$, and let G_2^{-1} be its inverse in that interval. Write $\sigma_p = [p(1-p)]^{\frac{1}{2}}$. For fixed $t > 0$ one then has, by the binomial equivalence,

$$\lim_{n \to \infty} P\{Y_{p,n} - \xi_p^{(2)} < G_2^{-1}(t\sigma_p n^{-\frac{1}{2}})\} = \Phi(t). \tag{1}$$

If G_2 is sufficiently regular, this can be made more explicit because G_2^{-1} will have a simple approximate form. Suppose $G_2(t)$ is regularly varying of exponent $r > 0$, as $t \downarrow 0$; that is, as $t \downarrow 0$,

$$G_2(t) \sim Ct^r L(t) \tag{2}$$

for some finite positive value C, where $L(t)$ is 'slowly varying', i.e. $\lim_{t \downarrow 0} L(t)/L(ct) = 1$ for each $c > 0$. We suppose also that, for $\alpha > 0$, $c > 0$, and β real,

$$\lim_{t \downarrow 0} L(ct^\alpha [L(t)]^\beta)/L(t) = \gamma_\alpha, \tag{3}$$

where γ_α is a finite positive value not depending on c or β.† Then it is easily verified that, as $t \downarrow 0$,

$$G_2^{-1}(t) \sim [t/C\gamma_r L(t)]^{1/r}. \tag{4}$$

Define the positive value z by setting $zG_2^{-1}(n^{-\frac{1}{2}}\sigma_p \gamma_r \gamma_{1/r}^{-1})$ asymptotically equal to $G_2^{-1}(t\sigma_p n^{-\frac{1}{2}})$ in (1); it will be seen that z can be taken to be constant as $n \to \infty$. Abbreviating $n^{-\frac{1}{2}}\sigma_p \gamma_r \gamma_{1/r}^{-1}$ by u as $u \downarrow 0$, we have

$$G_2(zG_2^{-1}(u)) \sim Cz^r u[C\gamma_r L(u)]^{-1} L(zu^{1/r}[C\gamma_r L(u)]^{-1/r})$$

$$\sim z^r u\gamma_{1/r}/\gamma_r = z^r n^{-\frac{1}{2}}\sigma_p, \tag{5}$$

and from (5) as well as (4) with $t = u$ we obtain, for $z > 0$,

$$\lim_{n\to\infty} P\{[n^{\frac{1}{2}}C\gamma_{1/r}\gamma_{\frac{1}{2}}L(n^{-1})\sigma_p^{-1}]^{1/\alpha}(Y_{p,n} - \xi_p^{(2)}) < z\} = \Phi(z^r). \tag{6}$$

Another way of saying this, if we write a superscript 2 on C, L, r, and γ, and suppose a result corresponding to (6) to hold for $Y_{p,n} - \xi_p^{(1)} < 0$ (with superscripts 1 on C, L, γ, r), is that $\psi_n(Y_{p,n})$ is asymptotically standard normal, where

$$\psi_n(y) = \begin{cases} n^{\frac{1}{2}}C^{(2)}\gamma_{1/r^{(2)}}^{(2)}\gamma_{\frac{1}{2}}^{(2)}L^{(2)}(n^{-1})\sigma_p^{-1}(y - \xi_p^{(2)})^{r^{(2)}} & \text{if } y > \xi_p^{(2)}, \\ 0 & \text{if } \xi_p^{(1)} \leq y \leq \xi_p^{(2)}, \\ -n^{\frac{1}{2}}C^{(1)}\gamma_{1/r^{(1)}}^{(1)}\gamma_{\frac{1}{2}}^{(1)}L^{(1)}(n^{-1})\sigma_p^{-1}(\xi_p^{(1)} - y)^{r^{(1)}} & \text{if } y < \xi_p^{(1)}. \end{cases} \tag{7}$$

An important area for further research is that of establishing bounds on the difference between the actual distribution and the limiting df of a rv involving sample quantiles or order statistics, analogous to estimates of Berry–Esseen type for sums. Jumping ahead to the topic of the next section, we note that an invariance principle with error term, used in studies of certain functionals of the sample df S_n by such authors as Skorokhod [17] and Rosenkrantz [15], [16], can be a powerful tool.

† The condition (3) is not automatically satisfied by a slowly varying function, as can be seen from its failure for the function $L(t) = \exp\{(\log t^{-1})/\log\log t^{-1}\}$. It holds, for example, for $L(t) = \prod_{i=1}^{k}(\log_i t^{-1})^{a_i}$ where as usual $\log_1 = \log$ and $\log_{i+1} = \log\log_i$. For this function L, one has $\gamma_\alpha = \alpha^{a_1}$. Of course, the assumption (3) is only used for special values of α, β, c in the sequel. One can always take $C = 1$ in (2), but might not, for convenience.

The reason for developing this example in detail is to illustrate a simplication which often arises in statistics where inverses are concerned; the case $r = 0$ can be more difficult from this point of view, since in place of the simple expression (4) one must express the inverse of L in that case. The arithmetic of (4) is relevant in such other settings as nonregular ML examples of arbitrary normalization; for example, if the X_i have unknown location parameter θ and $P_\theta\{X_i - \theta \leq x\}$ is $G_2(x)$ for $x \geq 0$ and 0 for $x < 0$, where $G_2(x)$ is concave for $x \geq 0$ and satisfies (2) and (3), then the ML estimator $\hat{T}_n = \min(X_i, ..., X_n)$ satisfies $P\{\hat{T}_n - \theta > G_2^{-1}(t/n)\} \sim e^{-t}$ for $t > 0$ as $n \to \infty$, as is well known, and (4) can be used to simplify this.

2   INVARIANCE PRINCIPLES; STRONG LAWS

Classical uses of the probabilistic invariance principle in limiting distribution theory are by now too familiar to be detailed here; it will suffice to recall a few of the early names such as Erdös, Kac, Donsker, Doob, Prohorov. Recent developments of interest here (in addition to those referred to in the previous paragraph) are (i) the introduction by Strassen [19], [20] of *strong invariance principles* which allow such results as the law-of-the-iterated-logarithm (LIL) to be obtained for sums

$T_n = \sum_1^n U_i$ of fairly general sequences of rv's $\{U_i\}$ ($\{T_n\}$ can be a martingale

satisfying appropriate conditions) from corresponding results for Brownian motion $\{B(t), t \geq 0\}$, as a consequence of a Skorokhod–Strassen representation of the form

$$T_n - ET_n - B(\operatorname{var} T_n) = O(g(\operatorname{var} T_n)) \quad wp\, 1 \quad \text{as} \quad n \to \infty, \tag{8}$$

where $g(m)$ is of suitably small order as $m \to +\infty$; and (ii) the recognition by Bahadur of the close relationship between the *linear* process $\{nS_n(\xi_p)\}$ (a sequence of *sums* of iid rv's) and the *nonlinear* process $\{Y_{p,n}\}$ (no obvious transform of which is a martingale), for fixed p. Analogous to the use just mentioned of the development (i), the development (ii) allows strong laws about $\{nS_n(\xi_p)\}$ to be used to prove corresponding results about $\{Y_{p,n}\}$; this can also be thought of as a strong analogue to the corresponding distribution theory results discussed in § 1, wherein the limiting df of normalized $Y_{p,n}$ is obtained from that of $S_n(\xi_p)$. Since strong convergence of $Y_{p,n}$ (in the case of unique ξ_p) is well known, we shall mention here LIL results.

Let us first backtrack to the direct application of the device of § 1 for each n (as distinct from Bahadur's application of a representation for the whole sequence $\{Y_{p,n}\}$ at once). It is not too hard to see (and is probably fairly well known although not well publicized) that a LIL for $Y_{p,n}$ can be obtained in cases where f is positive and suitably regular† at ξ_p, by making appropriate alterations in the standard proof for Bernoulli rv's (as it appears, for example, in vol. 1 of Feller [10]), using the binomial equivalence of events concerning $Y_{p,n}$. (The main departure is that the study of $S_n(\xi_{p_n})$, with p_n varying slightly, is no longer quite that of events concerning an iid Bernoulli sequence.) Even easier is the ana-

---

† While some LIL-type results can be obtained without difficulty when $f(\xi_p) = 0$ but f is positive (except at ξ_p) and regular in a neighborhood of ξ_p, we omit description of these in this brief discussion.

logous result for an order statistic of fixed order, say $Z_{k,n} = k$th smallest of $X_1, X_2, ..., X_n$; for example, the bottom half is that, if the X_i are independent and uniformly distributed on $[0, 1]$ and $nq_n \downarrow 0$, then

$$P\{Z_{k,n} < q_n \text{ for infinitely many } n\} = 1 \Leftrightarrow \sum_{n=1}^{\infty} n^{k-1} q_n^k = +\infty. \qquad (9)$$

The domain of results between (9) and the LIL for $Y_{p,n}$ (where $|k - np| \leqslant 1$), namely,

where $k \to \infty$ as $n \to \infty$ but $k/n \to 0$,

is not fully explored; we will return to this in § 3. For now, let us mention the much simpler way of proving the LIL for $Y_{p,n}$ which was perceived by Bahadur [1], who obtained it at once from the LIL for the binomial rv's $nS_n(\xi_p)$ upon invoking his result referred to in the previous paragraph, that

$$R_n(p) = Y_{p,n} - \xi_p + [S_n(\xi_p) - p]/f(\xi_p) \qquad (10)$$

is of much smaller order than $n^{-\frac{1}{2}}$, wp 1. In fact, Bahadur's device enables one to obtain easily even the finer characterization of functions of upper and lower class for $\{Y_{p,n}\}$ [11], a result which would be very difficult to obtain (if, in fact, it is so obtainable) by modifying an existing proof [9] in the Bernoulli case in the manner mentioned above for the simple LIL result. The proof is like that at the end of § 3.

It would be very interesting to obtain a representation of the sample quantile process $\{Y_{p,n}\}$ for fixed p in terms of something like Brownian motion, in the manner of Skorokhod and Strassen for sums of rv's as in (8). One can do this *by using Bahadur's result* (see [11]), but that is the wrong direction, since one would like to *prove* Bahadur's result from a start with such a quantile process defined in terms of Brownian motion. A major difficulty, of course, is that $\{Y_{p,n}\}$ is not exactly Markovian, a martingale, or anything else that seems useful. I believe Miss Helen Finkelstein, a student of Strassen at Berkeley, worked on such a representation, but do not know the details. (We will return to this process in (15) below, but there p will also vary.)

One tempting attempt in this direction is to use Breiman's [3] Skorokhod–Strassen type construction of the Brownian bridge representation of $\{n^{\frac{1}{2}}(S_n(t) - t), \; 0 \leqslant t \leqslant 1\}$ in the case where the X_i are independent and uniform on $[0, 1]$; this uses (8) in the case where the U_i are iid and

$$P\{U_i > u\} = e^{-u} \quad \text{for} \quad u > 0.$$

It is well known in that case that $\{T_i/T_{n+1}, 1 \leqslant i \leqslant n\}$ has the same distribution as the order statistics from $\{X_1, X_2, ..., X_n\}$. Let S'_n be the 'sample df' constructed from $\{T_i/T_{n+1}, 1 \leqslant i \leqslant n\}$. Breiman uses this development to show that

$$\rho_n(t) = S'_n(t) - t - n^{-\frac{1}{2}}[B(nt) - tB(n)]$$

is suitably small.† Why, then, should we not use the sequence

$$\{T_{[np]}/T_{n+1}\}$$

(where $[x]$ is the greatest integer $\leqslant x$) and consequently

$$\{n^{-\frac{1}{2}}[B(np) - pB(n)]\} \quad \text{to approximate} \quad \{Y_{p,n}\}?$$

The difficulty is that, although the approximation is satisfactory for each fixed n, the joint distribution of the sequence is not appropriate; one can see this at once upon noting how T_i/T_{n+1} for each fixed i behaves as n varies.

This shortcoming also applies to the possibility of using Breiman's representation to obtain strong laws (such as Chung's LIL [5]) for $\{S_n\}$ from corresponding results for B. The construction of a Skorokhod–Strassen type representation for $\{S_n\}$, which is useful for obtaining strong laws, remains an open problem.

3 RECENT WORK ON BAHADUR'S REPRESENTATION

Eicker is to be commended for the fortitude he has shown in carrying out his delicate calculations. As he states, this delicacy appears in the precise nature of his combinatorial probabilities and in his using approximations only at a later stage of the proof than I did. However, since his proof (of the upper class result, which is all that is provided here) is somewhat longer, we should ask what its benefits are. There are several possibilities. First, perhaps, the finer complete characterization of upper- and lower-class functions (that is, the description of those sequences $\{\phi_n\}$ for which $R_n(p) > \phi_n$ for finitely- or infinitely-many n, wp 1), which is as yet unknown,‡ requires such finer calculations. This is not really clear; on the one hand, both Feller and Chung have made it

---

† Subsequently Brillinger[4], evidently without knowing Breiman's result, used the same development. In addition, he computed in detail an upper bound on the order (wp 1) of $\sup_{0 \leqslant t \leqslant 1} |\rho_n(t)|$. It is shown elsewhere [13] how this last result can be obtained in a few lines from Bahadur's representation and the order of $\sup_p |R_n(p)|$ discussed in the next section, and how $\rho_n(t)$ can thereby be more clearly described, and its exact order computed.

‡ Not to be confused with the corresponding complete characterization for $Y_{p,n}$, which was mentioned in §2.

very clear in analogous complete characterizations in LIL-type theorems they have proved, that asymptotic estimates of tail probabilities, with fairly rough error terms, suffice; on the other hand, the present setting contains notable differences from the earlier LIL problems, in that one is not considering sums of basic rv's but rather a smaller order difference between such a sum and a rv almost equal to that sum.

Secondly, perhaps such exact calculations may help in giving better demonstrations than now exist, of the limiting behavior of

$$R_n^* = \sup_{0 \leqslant p \leqslant 1} |R_n(p)| f(\xi_p).$$

As mentioned in [11], some results are known, but there are difficulties in obtaining such results from convergence results for the process $\{R_n(t), 0 \leqslant t \leqslant 1\}$, which cannot be normalized to approach weakly a nontrivial separable limiting process. The known results, proofs of which appear in [12], are that, if f is bounded away from 0 and f' is bounded on a finite interval to which f assigns probability 1, then

$$\left.\begin{array}{l} \lim_{n \to \infty} P\{R_n^* n^{\frac{3}{4}}(\log n)^{-\frac{1}{2}} > r\} = 2 \sum_{m=1}^{\infty} (-1)^{m+1} e^{-2m^2 r^4} \quad \text{for} \quad r > 0; \\[2mm] \limsup_{n \to \infty} R_n^* n^{\frac{3}{4}}(\log n)^{-\frac{1}{4}} (\log \log n)^{-\frac{1}{4}} = 2^{-\frac{1}{4}} \quad \text{wp 1.} \end{array}\right\} \quad (11)$$

In fact, the first line of (11) is a trivial consequence of the fact that

$$R_n^* n^{\frac{1}{2}}(\log n)^{-\frac{1}{2}} / \sup_x |S_n(x) - F(x)|^{\frac{1}{2}} \to 1 \quad \text{in probability,}$$

which more fundamental result is proved in [12]. The second line of (11) follows from a corresponding strong law and the LIL [5] for the sample df deviations.

In connection with Eicker's Appendix B and his paper [7] quoted here, one should mentioned the related work on Takács and others on ballot problems.

Finally, let us describe the domain of usefulness of Bahadur's representation in studying the LIL for $\{Y_{p_n, n}\}$ when $p_n \downarrow 0$ as mentioned below (9). Of course, a bound on $R_n(p)$ for *fixed* p is no longer relevant as it was for determining the upper and lower classes for $\{Y_{p,n}\}$. Supposing again that the iid X_i are uniformly distributed on $[0, 1]$, Eicker [8] recently showed that, if $p_n > (\log \log n)/o(n)$, then

$$\limsup_{n \to \infty} [n/2p_n \log \log (np_n)]^{\frac{1}{2}} [S_n(p_n) - p_n] = 1 \quad \text{wp 1.} \quad (12)$$

On the other hand, one of Baxter's results [2] is that, for fixed $c > 0$,

$$\limsup_{n \to \infty} [n(\log \log \log n)/\log \log n] S_n(cn^{-1}) = 1 \quad \text{wp 1.} \quad (13)$$

From the order (wp 1) of R_n^* given in (11), it follows at once from (12) that, if also $n^{\frac{1}{2}} p_n (\log n)^{-1} (\log \log n)^{\frac{1}{2}} \to \infty$,

$$\limsup_{n \to \infty} [n/2 p_n \log \log (n p_n)]^{\frac{1}{2}} [p_n - Y_{p_n, n}] = 1 \quad \text{wp 1.} \tag{14}$$

This is similar to the use of $R_n(p)$ to obtain the LIL for $Y_{p, n}$ as mentioned in §2. On the other hand, no corresponding conclusion about $Y_{c/n, n}$ is derivable from (13) from the estimate (11) on R_n^*; this suggests the problem of finding the order of $R_n(cn^{-1})$ or more generally of $R_n(p_n)$ with $p_n \downarrow 0$. Results on $S_n(p_n)$ and Y_{p_n}, n will appear elsewhere.

If one studies not $\{Y_{p_n, n}\}$ for a single sequence $\{p_n\}$, but rather the whole 'quantile process' $\{Y_{p, n}, 0 \leqslant p \leqslant 1\}$, then the result (11) again yields a LIL from the corresponding analogous result [5] for $\{S_n(t),$ $a \leqslant t \leqslant b\}$ where $[a, b]$ is the interval where $f > 0$. The result is that, for $\lambda_n \uparrow \infty$,

$$P\{ \sup_{0 \leqslant p \leqslant 1} f(\xi_p) |Y_{p, n} - \xi_p| > \lambda_n n^{-\frac{1}{2}} \text{ infinitely often}\}$$

$$= \begin{cases} 1 \Leftrightarrow \infty = \\ 0 \Leftrightarrow \infty > \end{cases} \sum_n n^{-1} \lambda_n^2 e^{2\lambda_n^2}. \tag{15}$$

(An examination of the proof in [5] shows that the same criterion holds for the maximum deviation of one sign rather than maximum absolute deviation, and thus the same holds true in (15).) The proof of (15) develops along familiar lines from the fact [5] that the same criterion on $\{\lambda_n\}$ holds for $\sup_t |S_n(t) - t|$. One first notes, from [5] and (15), that

$$\lambda_n^* = [2^{-1}(1 * \epsilon) \log n]^{\frac{1}{2}}, \quad \text{for} \quad * = + \text{ or } -,$$

gives a sequence in each of the upper and lower classes for

$$\sup_p f(\xi_p) |Y_{p, n} - \xi_p|;$$

and that any $\{\lambda_n\}$ may be taken to lie between those two sequences by replacing λ_n by $(\lambda_n \wedge \lambda_n^+) \vee \lambda_n^-$ (which changes neither the summability class of (15) for $\{\lambda_n\}$, nor the monotonicity of $\{\lambda_n\}$, nor whether it is upper or lower class). Next, one notes from (11) that $R_n^* < \lambda_n^{-1}$ for all large n wp 1, for these restricted $\{\lambda_n\}$. Finally, it is trivial that $\{\lambda_n \pm \lambda_n^{-1}\}$ is monotone for large n and lies in the same summability class of (15) as $\{\lambda_n\}$.

REFERENCES

[1]   R. R. Bahadur. A note on quantiles in large samples. *Ann. Math. Statist.* **37**, 577–80 (1966).

[2]   G. Baxter. An analogue of the LIL. *Proc. Am. Math. Soc.* **6**, 177–81 (1955).

[3]   L. Breiman. *Probability*. Addison–Wesley (1968).

[4]   D. R. Brillinger. An asymptotic representation of the sample df. *Bull. Am. Math. Soc.* **75**, 545–7 (1969).

[5]   K. L. Chung. An estimate concerning the Kolmogoroff limit distribution. *Trans. Am. Math. Soc.* **67**, 36–50 (1949).

[6]   F. Eicker. A new proof of the Bahadur–Kiefer representation of sample quantiles. This volume, pp. 321–42.

[7]   F. Eicker. On the probability that a sample df lies below a polygon. (To appear.)

[8]   F. Eicker. A log log law for double sequences of random variables. (To appear.)

[9]   W. Feller. The general form of the so-called law of the iterated logarithm. *Trans. Am. Math. Soc.* **54**, 373–402 (1943).

[10] W. Feller. *An Introduction to Probability Theory and its Applications*, vol. 1, second edn. (1957); vol. 2, first edn. (1966). New York: John Wiley and Sons.

[11] J. Kiefer. On Bahadur's representation of sample quantiles. *Ann. Math. Statist.* **38**, 1323–42 (1967).

[12] J. Kiefer. Deviations between the sample quantile process and the sample df. This volume, pp. 299–319.

[13] J. Kiefer. On a Brownian bridge approximation to the sample df. (To appear.)

[14] M. Loève. Ranking limit problems. *Proc. third Berkeley Symp.* vol. 2, 177–98 (1956).

[15] W. A. Rosenkrantz. On rates of convergence for the invariance principle. *Trans. Am. Math. Soc.* **129**, 542–52 (1967).

[16] W. A. Rosenkrantz. A rate of convergence for the Von Mises statistic. (To appear.)

[17] A. V. Skorokhod. *Studies in the Theory of Random Processes*. Addison–Wesley (1965).

[18] N. Smirnov. Limit distributions for the terms of a variational series. *Am. Math. Soc. Transl.* **67**, 1–64, Providence (1952); originally in *Trudy mat. Inst. V. A. Steklova* **25**, 60 pp. (1949).

[19] V. Strassen. An invariance principle for the LIL. *Z. Wahrscheinlichkeitstheorie und Verw. Gebiete* **3**, 211–26 (1964).

[20] V. Strassen. Almost sure behavior of sums of independent random variables and martingales. *Proc. fifth Berkeley Symp.* vol. 2, part 1, 315–43 (1966).

[21] L. Weiss. Asymptotic distribution of quantiles in some nonstandard cases. This volume, pp. 343–8.

COMBINATORIAL METHODS IN THE
THEORY OF ORDER STATISTICS†

LAJOS TAKÁCS

1 INTRODUCTION

In the theory of order statistics there are many combinatorial problems connected with the comparisons of theoretical and empirical distribution functions. In this paper we shall show how a simple combinatorial theorem can be used in finding the distributions of many statistics in order statistics.

First we shall formulate our fundamental theorem which is a generalization of the classical ballot theorem. Then we deduce several auxiliary theorems from the fundamental theorem. We shall apply these theorems to find the distributions of several statistics depending on the deviations between a theoretical and an empirical distribution function and between two empirical distribution functions.

The main new results of this paper are some generalizations of a theorem of V. S. Mihalevič [23] and a theorem of I. Vincze [33].

2 A FUNDAMENTAL THEOREM

In this section we shall formulate a generalization of the classical ballot theorem. We shall present this theorem first in a simple combinatorial way and then in a slightly more general form.

Theorem 1. Let $k_1, k_2, ..., k_n$ be nonnegative integers with sum

$$k_1 + k_2 + ... + k_n = k \leq n.$$

Among the $n!$ permutations of $(k_1, k_2, ..., k_n)$ there are exactly $(n-1)!(n-k)$ for which the rth partial sum is less than r for all $r = 1, 2, ..., n$.

Theorem 1 was found in 1960 by the author [26], [27]. The proofs given in [26] and [27] are based on mathematical induction. A direct combinatorial proof can be found in [31].

It might be interesting to mention briefly the history of Theorem 1. If we assume that each k_i $(i = 1, 2, ..., n)$ is either 0 or 2, then Theorem 1 reduces to the classical ballot theorem which was first formulated in 1887 by J. Bertrand [5] and proved in the same year by D. André [3].

† This research was sponsored by the National Science Foundation under Contract No. GP-7847.

It should be noted, however, that this particular case can also be deduced from a result of duration of plays which was found in 1708 by A. De Moivre [9, p. 262] and in a different version in 1718 also by A. De Moivre [10, p. 121]. A. De Moivre did not give proofs of his results. Proofs for De Moivre's results were given only in 1773 by P. S. Laplace [22, pp. 188–93] and in 1776 by J. L. Lagrange [21, pp. 230–8]. See also W. A. Whitworth [36], [37].

If we assume that each k_i $(i = 1, 2, \ldots, n)$ is either 0 or $\mu + 1$ where μ is a positive integer, then Theorem 1 reduces to a generalization of the classical ballot theorem which was formulated in 1887 by É. Barbier [4] and proved in 1924 by A. Aeppli [1]. See also A. Dvoretzky and Th. Motzkin [12], H. D. Grossman [18], S. G. Mohanty and T. V. Narayana [24] and the author [27].

Now we shall formulate and prove Theorem 1 in a slightly more general form which we shall need in this paper.

Theorem 2. Let $v_1, v_2, \ldots, v_n$ be interchangeable random variables taking on nonnegative integers. Set $N_r = v_1 + \ldots + v_r$ for $r = 1, 2, \ldots, n$. Then we have

$$P\{N_r < r \text{ for } r = 1, \ldots, n | N_n = k\} = \begin{cases} 1 - \dfrac{k}{n} & \text{for } k = 0, 1, \ldots, n, \\ 0 & \text{otherwise}, \end{cases} \quad (1)$$

where the conditional probability is defined up to an equivalence.

Proof. If $n = 1$, then (1) is evidently true. Suppose that (1) is true when n is replaced by $n - 1$ $(n = 2, 3, \ldots)$. We shall prove that it is true for n. Hence by mathematical induction it follows that (1) is true for all

$$n = 1, 2, \ldots.$$

If $k \geqq n$, then (1) is obviously true. Let $k < n$. By assumption

$$P\{N_r < r \text{ for } r = 1, \ldots, n-1 | N_{n-1} = j\}$$
$$= \begin{cases} 1 - \dfrac{j}{n-1} & \text{if } j = 0, 1, \ldots, n-1, \\ 0 & \text{otherwise}, \end{cases} \quad (2)$$

where the conditional probability is defined up to an equivalence. Thus by the theorem of total probability

$$P\{N_r < r \text{ for } r = 1, \ldots, n | N_n = k\} = \sum_{j=0}^{n-1} \left(1 - \frac{j}{n-1}\right) P\{N_{n-1} = j | N_n = k\}$$

$$= 1 - \frac{1}{n-1} E\{N_{n-1} | N_n = k\} = 1 - \frac{1}{n-1} \frac{(n-1)k}{n} = 1 - \frac{k}{n} \quad (3)$$

for $k = 0, 1, \ldots, n-1$. This completes the proof of (1).

Finally, we note that by Theorem 1 it follows that

$$\mathbf{P}\{N_r < r \quad \text{for} \quad r = 1, 2, \ldots, n\} = \mathbf{E}\left\{\left[1 - \frac{N_n}{n}\right]^+\right\}, \tag{4}$$

where $[x]^+ = \max(0, x)$.

An example. Here we shall give a simple example for the application of Theorem 1. This example has some useful applications in the theory of order statistics.

Theorem 3. Suppose that n random points are distributed independently and uniformly on the interval $(0, t)$. Let $\chi(u)$ $(0 \leq u \leq t)$ be c times the number of points in the interval $(0, u]$ where c is a positive constant. Then

$$\mathbf{P}\{\chi(u) \leq u \quad \text{for} \quad 0 \leq u \leq t\} = \begin{cases} 1 - \dfrac{nc}{t} & \text{for} \quad 0 \leq nc \leq t, \\ 0 & \text{otherwise.} \end{cases} \tag{5}$$

Proof. The theorem is obviously true for $nc \geq t$. Let $nc < t$. Denote by v_r $(r = 1, 2, \ldots, n)$ the number of random points in the interval $((r - 1)c, rc]$. Set $N_r = v_1 + \ldots + v_r$ for $r = 1, 2, \ldots, n$. Now $v_1, v_2, \ldots, v_n$ are interchangeable random variables taking on nonnegative integers. We have $N_n \leq n$ and N_n has a Bernoulli distribution with parameters n and $p = nc/t$. Thus $\mathbf{E}\{N_n\} = n^2c/t$. By (4) it follows that

$$\mathbf{P}\{\chi(u) \leq u \quad \text{for} \quad 0 \leq u \leq t\} = \mathbf{P}\{N_r < r \quad \text{for} \quad r = 1, 2, \ldots, n\}$$

$$= \mathbf{E}\left\{\left[1 - \frac{N_n}{n}\right]^+\right\} = \mathbf{E}\left\{1 - \frac{N_n}{n}\right\} = 1 - \frac{nc}{t} \tag{6}$$

which was to be proved.

We note that (5) can also be expressed as follows

$$\mathbf{P}\{\chi(u) \leq u \quad \text{for} \quad 0 \leq u \leq t\} = \begin{cases} 1 - \dfrac{\chi(t)}{t} & \text{for} \quad 0 \leqslant \chi(t) \leqslant t, \\ 0 & \text{otherwise.} \end{cases} \tag{7}$$

An identity. Let $v_1, v_2, \ldots, v_n$ be interchangeable random variables taking on nonnegative integers. Let $N_r = v_1 + \ldots + v_r$ for $r = 1, 2, \ldots, n$ and $N_0 = 0$. For any l satisfying the inequalities $1 \leq l < k \leq n$ we have

$$\sum_{i=1}^{n-1} \frac{l(k-l)}{i(n-i)} \mathbf{P}\{N_i = i - l, N_n = n - k\} = \frac{k}{n} \mathbf{P}\{N_n = n - k\}. \tag{8}$$

This can be proved as follows. Define $\rho(k)$ $(k = 0, 1, \ldots, n)$ as the smallest $r = 0, 1, \ldots, n$ for which $r - N_r = k$ if such an r exists. $\rho(k)$ is the first passage time of $r - N_r$ $(r = 0, 1, \ldots, n)$ through k. By Theorem 2 we have

$$\mathbf{P}\{\rho(k) = n\} = \frac{k}{n} \mathbf{P}\{N_n = n - k\} \tag{9}$$

for $1 \leqq k \leqq n$. For

$$
\begin{aligned}
\mathbf{P}\{\rho(k) = n\} &= \mathbf{P}\{r - N_r < k \text{ for } r = 1, \ldots, n-1 \text{ and } n - N_n = k\} \\
&= \mathbf{P}\{N_n - N_r < n - r \text{ for } r = 1, \ldots, n-1 \text{ and } N_n = n - k\} \\
&= \mathbf{P}\{N_i < i \text{ for } i = 1, \ldots, n-1 \text{ and } N_n = n - k\} \\
&= \frac{k}{n} \mathbf{P}\{N_n = n - k\}. \quad (10)
\end{aligned}
$$

If $1 \leqq l < k$, then evidently

$$
\sum_{i=1}^{n} \mathbf{P}\{\rho(l) = i, \rho(k) - \rho(l) = n - i\} = \mathbf{P}\{\rho(k) = n\}. \quad (11)
$$

Since $\rho(k) - \rho(l)$ has the same distribution as $\rho(k-l)$, we get (8) by (9).

3 ANOTHER COMBINATORIAL THEOREM

The applicability of Theorem 2 can be greatly increased if we also make use of the following combinatorial theorem which was found in 1953 by E. S. Andersen [2] and in 1959 by W. Feller [15].

Theorem 4. Let $\xi_1, \xi_2, \ldots, \xi_n$ be interchangeable random variables taking on real values. Define $\zeta_r = \xi_1 + \ldots + \xi_r$ for $r = 1, 2, \ldots, n$ and $\zeta_0 = 0$. Denote by Δ_n the number of positive members in the sequence $\zeta_1, \zeta_2, \ldots, \zeta_n$ and by Δ_n^* the number of nonnegative members in the sequence $\zeta_1, \zeta_2, \ldots, \zeta_n$. Denote by ρ_n the subscript of the first maximal member in the sequence $\zeta_0, \zeta_1, \ldots, \zeta_n$ and by ρ_n^* the subscript of the last maximal member in the sequence $\zeta_0, \zeta_1, \ldots, \zeta_n$. We have

$$
\mathbf{P}\{\Delta_n = j\} = \mathbf{P}\{\rho_n = j\} \quad (12)
$$

and

$$
\mathbf{P}\{\Delta_n^* = j\} = \mathbf{P}\{\rho_n^* = j\} \quad (13)
$$

for $j = 0, 1, \ldots, n$.

Proof. First we note that (12) implies (13) and conversely (13) implies (12). If we apply (12) to the random variables $-\xi_1, -\xi_2, \ldots, -\xi_n$, then it follows that $\mathbf{P}\{n - \Delta_n^* = j\} = \mathbf{P}\{n - \rho_n^* = j\}$ for $j = 0, 1, \ldots, n$ which is exactly (13). If we apply (13) to the random variables $-\xi_1, -\xi_2, \ldots, -\xi_n$, then it follows that $\mathbf{P}\{n - \Delta_n = j\} = \mathbf{P}\{n - \rho_n = j\}$ for $j = 0, 1, \ldots, n$, which is exactly (12).

Next we shall prove (12) and (13) in the particular case when

$$
(\xi_1, \xi_2, \ldots, \xi_n)
$$

is a random permutation of n real numbers $(c_1, c_2, \dots, c_n)$ and every permutation has the same probability. We shall prove (12) and (13) by mathematical induction. If $n = 1$, then $\Delta_1 = \rho_1$ and $\Delta_1^* = \rho_1^*$ and thus (12) and (13) are true. Suppose that both (12) and (13) are true when n is replaced by $n - 1$ ($n = 2, 3, \dots$). We shall prove that they are true for n too. We shall distinguish two cases. If $\zeta_n = c_1 + \dots + c_n \leqq 0$, then the first maximum cannot occur at the nth place because $\zeta_0 = 0$ and thus $\mathbf{P}\{\Delta_n = j\} = \mathbf{P}\{\rho_n = j\}$ for $j = 0, 1, \dots, n$ if (12) is true when n is replaced by $n - 1$. If $\zeta_n = c_1 + \dots + c_n \geqq 0$, then

$$\mathbf{P}\{n - \Delta_n^* = j\} = \mathbf{P}\{n - \rho_n^* = j\} \quad \text{for} \quad j = 0, 1, \dots, n$$

if (13) is true when n is replaced by $n - 1$. This completes the proof of the theorem in the particular case stated above.

The general case follows immediately from the particular case above if we apply it to the $n!$ permutations of $(c_1, c_2, \dots, c_n)$ where $c_1, c_2, \dots, c_n$ are any realizations of the random variables $\xi_1, \xi_2, \dots, \xi_n$.

An example. Here we shall show that if we combine Theorem 2 and Theorem 4, then we obtain the following interesting result.

Theorem 5. Let $v_1, v_2, \dots, v_n$ be interchangeable random variables taking on nonnegative integers. Set $N_r = v_1 + \dots + v_r$ for $r = 1, \dots, n$ and $N_0 = 0$. Denote by Δ_n the number of subscripts $r = 1, 2, \dots, n$ for which $N_r < r$ holds. If $\mathbf{P}\{N_n = n - 1\} > 0$, then we have

$$\mathbf{P}\{\Delta_n = j \mid N_n = n - 1\} = \frac{1}{n} \tag{14}$$

for $j = 1, 2, \dots, n$.

Proof. Δ_n can be interpreted as the number of positive members in the sequence $r - N_r$ ($r = 1, 2, \dots, n$). If we apply Theorem 4 to the random variables $(1 - v_1), (1 - v_2), \dots, (1 - v_n)$ and Theorem 2 to the random variables $v_j, v_{j-1}, \dots, v_1, v_n v_{n-1}, \dots, v_{j+1}$, then we obtain that

$$\mathbf{P}\{\Delta_n = j \mid N_n = n - 1\} = \mathbf{P}\{i - N_i < j - N_j \quad \text{for} \quad 0 \leqq i < j$$
$$\text{and} \qquad\qquad i - N_i \leqq j - N_j \quad \text{for} \quad j \leqq i \leqq n \mid N_n = n - 1\}$$
$$= \mathbf{P}\{N_j - N_i < j - i \quad \text{for} \quad 0 \leqq i < j$$
$$\text{and} \quad N_n - N_i + N_j < n - i + j \quad \text{for} \quad j \leqq i \leqq n \mid N_n = n - 1\} = 1/n$$

for $j = 1, 2, \dots, n$. A direct combinatorial proof for (14) has been given in [28].

Note. If Δ_n^* denotes the number of subscripts $r = 1, 2, \ldots, n$ for which $N_r \leqq r$, then in a similar way as (14) we obtain that

$$\mathbf{P}\{\Delta_n^* = j | N_n = n+1\} = \frac{1}{n} \tag{15}$$

for $j = 0, 1, \ldots, n-1$, provided that $\mathbf{P}\{N_n = n+1\} > 0$.

4 AUXILIARY THEOREMS

In this section we shall prove several theorems which can be used in the theory of order statistics. All these theorems are consequences of Theorem 2 and Theorem 4.

Throughout this section we shall assume that $v_1, v_2, \ldots, v_n$ are interchangeable random variables taking on nonnegative integers. We use the notation $N_r = v_1 + \ldots + v_r$ for $r = 1, 2, \ldots, n$ and $N_0 = 0$.

Let us denote by $\Delta_n^{(c)}$ the number of subscripts $r = 1, 2, \ldots, n$ for which $N_r < r+c$ where $c = 0, \pm 1, \pm 2, \ldots$. In particular, we use the notation $\Delta_n = \Delta_n^{(0)}$ and $\Delta_n^* = \Delta_n^{(1)}$.

In this section we shall determine the distributions of the random variables $\Delta_n^{(c)}$ ($c = 0, \pm 1, \pm 2, \ldots$).

The distribution of Δ_n. The random variable Δ_n denotes the number of subscripts $r = 1, 2, \ldots, n$ for which $N_r < r$.

Theorem 6. We have

$$\mathbf{P}\{\Delta_n = 0\} = 1 - \sum_{i=1}^{n} \frac{1}{i} \mathbf{P}\{N_i = i-1\} \tag{16}$$

and

$$\mathbf{P}\{\Delta_n = j\} = \sum_{l=0}^{j} \left(1 - \frac{l}{j}\right) \left[\mathbf{P}\{N_j = l\} \right.$$

$$\left. - \sum_{i=j+1}^{n} \frac{1}{(i-j)} \mathbf{P}\{N_j = l \text{ and } N_i - N_j = i-j-1\} \right] \tag{17}$$

for $j = 1, 2, \ldots, n$.

Proof. First we shall find $\mathbf{P}\{\Delta_n > 0\}$. If $\Delta_n > 0$, then $N_r = r-1$ for some $r = 1, 2, \ldots, n$. Denote by i the smallest such r. Then

$$\mathbf{P}\{\Delta_n > 0\} = \sum_{i=1}^{n} \mathbf{P}\{N_r \geqq r \text{ for } r = 1, \ldots, i-1 \text{ and } N_i = i-1\}$$

$$= \sum_{i=1}^{n} \mathbf{P}\{N_i - N_r < i-r \text{ for } r = 1, \ldots, i-1 \text{ and } N_i = i-1\}$$

$$= \sum_{i=1}^{n} \frac{1}{i} \mathbf{P}\{N_i = i-1\}, \tag{18}$$

where the last equality follows from Theorem 2 if we apply it to the random variables $v_i, v_{i-1}, ..., v_1$. This proves (16).

Next we shall prove (17). We observe that the random variable Δ_n can be interpreted as the number of positive members in the sequence $r - N_r (r = 1, 2, ..., n)$. Then by Theorem 4 we can write that

$$\mathbf{P}\{\Delta_n = j\} = \mathbf{P}\{i - N_i < j - N_j \quad \text{for} \quad 0 \leq i < j$$

$$\text{and} \quad i - N_j \leq j - N_j \quad \text{for} \quad j \leq i \leq n\}. \quad (19)$$

Hence for $j = 1, 2, ..., n$

$$\mathbf{P}\{\Delta_n = j\} = \sum_{l=0}^{j} \mathbf{P}\{N_j - N_i < j - i \quad \text{for} \quad 0 \leq i < j \,|\, N_j = l\}$$

$$\times \mathbf{P}\{N_j - N_i \leq j - i \quad \text{for} \quad j \leq i \leq n \quad \text{and} \quad N_j = l\} \quad (20)$$

$$= \sum_{l=0}^{j} \left(1 - \frac{l}{j}\right) \left[\mathbf{P}\{N_j = l\} - \sum_{i=j+1}^{n} \frac{1}{(i-j)} \mathbf{P}\{N_j = l \right.$$

$$\left. \text{and} \quad N_i - N_j = i - j - 1\}\right],$$

where we applied Theorem 2 to the random variables $v_j, v_{j-1}, ..., v_1$ and formula (16) to the random variables $v_{j+1}, ..., v_n$. This proves (17).

Note. Let us introduce the notation

$$Q_j(n|k) = Q_j(N_1, N_2, ..., N_n \,|\, N_n = k) \quad (21)$$

for the probability $\mathbf{P}\{\Delta_n = j \,|\, N_n = k\}$. Then by Theorem 6 we obtain that

$$Q_0(n|k) = 1 - \sum_{i=1}^{n} \frac{1}{i} \mathbf{P}\{N_i = i - 1 \,|\, N_n = k\}, \quad (22)$$

$$Q_n(n|k) = \begin{cases} 1 - \dfrac{k}{n} & \text{if} \quad k = 0, 1, ..., n, \\[2mm] 0 & \text{otherwise,} \end{cases} \quad (23)$$

and $\quad Q_j(n|k) = \sum_{l=0}^{j} \mathbf{P}\{N_j = l \,|\, N_n = k\} Q_j(j|l) Q_0(n - j \,|\, k - l) \quad (24)$

for $j = 0, 1, ..., n$.

Particular cases. If we make use of Theorem 5, then we can obtain the following particular cases of Theorem 6. If $k = 0, 1, ..., n - 2$, then

$$Q_j(n|k) = \begin{cases} 0 \quad \text{for} \quad j = 0, 1, ..., n - k - 1, \\[2mm] \displaystyle\sum_{i=n-j+1}^{k+1} \frac{(n-k-1)}{i(n-i)} \mathbf{P}\{N_i = i - 1 \,|\, N_n = k\} \\ \qquad\qquad\qquad \text{for} \quad j = n - k, ..., n - 1, \\[2mm] 1 - \dfrac{k}{n} \quad \text{for} \quad j = n. \end{cases} \quad (25)$$

Furthermore, $\quad Q_j(n|n-1) = \dfrac{1}{n} \quad$ for $\quad j = 1, 2, \ldots, n$ $\qquad$ (26)

and

$$Q_j(n|n) = \begin{cases} 1 - \displaystyle\sum_{i=1}^{n-1} \frac{1}{i} \, P\{N_i = i-1 | N_n = n\} & \text{for} \quad j = 0, \\[4mm] \displaystyle\sum_{i=1}^{n-j} \frac{1}{i(n-i)} \, P\{N_i = i-1 | N_n = n\} & \text{for} \quad j = 1, 2, \ldots, n-1. \end{cases}$$ $\qquad$ (27)

The distribution of $\Delta_n^{(-c)}$. For $c = 0, 1, 2, \ldots$ the random variable $\Delta_n^{(-c)}$ denotes the number of subscripts $r = 1, 2, \ldots, n$ for which $N_r < r - c$.

Theorem 7. If $c = 0, 1, \ldots, n$, then

$$P\{\Delta_n^{(-c)} = 0\} = 1 - \sum_{i=c+1}^{n} \frac{c+1}{i} \, P\{N_i = i-c-1\} \qquad (28)$$

and if $c = 0, 1, \ldots, n-1$ and $j = 1, 2, \ldots, n-c$, then

$$P\{\Delta_n^{(-c)} = j\} = \sum_{l=0}^{j} \left(1 - \frac{l}{j}\right) \left[\sum_{i=j+c}^{n} \frac{c}{(i-j)} \, P\{N_j = l, \, N_{i-j} = i-j-c\} \right.$$
$$\left. - \sum_{i=j+c+1}^{n} \frac{c+1}{(i-j)} \, P\{N_j = l \text{ and } N_i - N_j = i-j-c-1\} \right]. \qquad (29)$$

Proof. If $c = 0$, then Theorem 7 reduces to Theorem 6. First we shall prove (28). We have

$$P\{\Delta_n^{(-c)} = 0\} = P\{N_r \geq r-c \text{ for } r = 1, 2, \ldots, n\}$$
$$= 1 - P\{N_r < r-c \text{ for some } r = 1, 2, \ldots, n\}. \qquad (30)$$

If the event $\{N_r < r-c$ for some $r = 1, 2, \ldots, n\}$ occurs, then there is an r such that $N_r = r-c-1$. Denote by i the smallest such r. Then by Theorem 2 we have

$$P\{\Delta_n^{(-c)} = 0\} = 1 - \sum_{i=c+1}^{n} P\{N_i - N_r < i-r$$
$$\text{for } r = 1, \ldots, i-1 \text{ and } N_i = i-c-1\}$$
$$= 1 - \sum_{i=c+1}^{n} \frac{c+1}{i} \, P\{N_i = i-c-1\} \qquad (31)$$

which proves (28).

Next we shall prove (29). Suppose that $c = 1, 2, \ldots, n-1$ and

$$j = 1, 2, \ldots, n-c.$$

If $\Delta_n^{(-c)} = j$, then there is an $r = 1, 2, \ldots, n$ such that $N_r = r-c$. Denote by s the smallest r with this property Then $N_r > r-c$ for $1 \leq r < s$, $N_s = s-c$

and $N_r < r - c$ for j subscripts $r = s + 1, \ldots, n$. The last condition can be replaced by the following : $N_r - N_s < r - s$ for j subscripts $r = s + 1, \ldots, n$, or equivalently, the first maximum in the sequence

$$(r - N_r) - (s - N_s) \quad (r = s + 1, \ldots, n)$$

occurs at $r = s + j$. If we define $\rho(k)$ $(k = 0, 1, \ldots, n)$ as the smallest $r = 0, 1, \ldots, n$ for which $r - N_r = k$, if such an r exists, then we can write that

$$\mathbf{P}\{\Delta_n^{(-c)} = j\}$$

$$= \sum_{s=c}^{n-j} \sum_{l=1}^{j} \mathbf{P}\{\rho(c) = s, \rho(c + l) - \rho(c) = j, \rho(c + l + 1) - \rho(c + l)$$

$$> n - s - j\} = \sum_{l=1}^{j} \mathbf{P}\{\rho(l) = j, \rho(l + c) - \rho(l) \leq n - j, \rho(l + c + 1)$$

$$-\rho(l) > n - j\}$$

$$= \sum_{l=1}^{j} [\mathbf{P}\{\rho(l) = j, \rho(l + c) - \rho(l) \leq n - j\} - \mathbf{P}\{\rho(l) = j, \rho(l + c + 1)$$

$$-\rho(l) \leq n - j\}]$$

$$= \sum_{l=1}^{j} \frac{l}{j} \left[\sum_{r=c}^{n-j} \frac{c}{h} \mathbf{P}\{N_j = j - l, N_{j+r} - N_j = r - c\} \right.$$

$$\left. - \sum_{r=c+1}^{n-j} \frac{c+1}{r} \mathbf{P}\{N_j = j - l, N_{j+r} - N_j = r - c - 1\} \right] \tag{32}$$

where we used (8) and (9). This proves (29).

The distribution of Δ_n^*. The random variable Δ_n^* denotes the number of subscripts $r = 1, 2, \ldots, n$, for which $N_r \leq r$.

Theorem 8. We have

$$\mathbf{P}\{\Delta_n^* = 0\} = \mathbf{P}\{N_1 > 1\} - \sum_{i=2}^{n} \frac{1}{(i-1)} \mathbf{P}\{N_1 = 0 \quad \text{and} \quad N_i = i\} \tag{33}$$

and $\quad \mathbf{P}\{\Delta_n^* = j\} = \sum_{l=0}^{j} \left[\mathbf{P}\{N_j = l, N_{j+1} > l + 1\} \right.$

$$- \sum_{i=1}^{j-1} \frac{(j + 1 - l)}{j - i} \mathbf{P}\{N_i = i + 1, N_j = l, N_{j+1} > l + 1\} \Big]$$

$$- \sum_{l=0}^{j} \sum_{r=j+2}^{n} \left[\frac{1}{(r - j - 1)} \mathbf{P}\{N_j = l, N_{j+1} = l, N_r = l + r - j\} \right.$$

$$- \sum_{i=1}^{j-1} \frac{(j + 1 - l)}{(j - i)(r - j - 1)} \mathbf{P}\{N_i = i + 1, N_j = l, N_{j+1} = l,$$

$$N_r = l + r - j\} \Big] \tag{34}$$

for $j = 0, 1, \ldots, n - 1$.

Further

$$\mathbf{P}\{\Delta_n^* = n \quad \text{and} \quad N_n = k\} = \mathbf{P}\{N_n = k\} - \sum_{i=1}^{n-1} \frac{n + 1 - k}{n - i} \mathbf{P}\{N_i = i + 1$$

$$\text{and} \quad N_n = k\} \tag{35}$$

for $k = 0, 1, \ldots, n$.

Proof. To prove (33) we can write that

$$\mathbf{P}\{\Delta_n^* = 0\} = \mathbf{P}\{N_r > r \quad \text{for} \quad r = 1, 2, \ldots, n\}$$

$$= \mathbf{P}\{N_1 > 1\} - \mathbf{P}\{N_1 > 1 \quad \text{and} \quad N_r \leqq r \quad \text{for some} \quad r = 2, \ldots, n\}. \quad (36)$$

To find the last probability we take into consideration that there is an $r = 2, \ldots, n$, such that $N_r = r$. Denote by i the smallest such r. Then

$$\mathbf{P}\{\Delta_n^* = 0\} = \mathbf{P}\{N_1 > 1\} - \sum_{i=2}^{n} \mathbf{P}\{N_r > r \quad \text{for} \quad r = 1, \ldots, i-1 \text{ and } N_i = i\}$$

$$= \mathbf{P}\{N_1 > 1\} - \sum_{i=2}^{n} \sum_{s=2}^{i} \frac{(s-1)}{(i-1)} \mathbf{P}\{N_1 = s \quad \text{and} \quad N_i = i\}$$

$$= \mathbf{P}\{N_1 > 1\} - \sum_{i=2}^{n} \frac{1}{i-1} \mathbf{P}\{N_1 = 0 \quad \text{and} \quad N_i = i\}, \quad (37)$$

where we used Theorem 2. This proves (33).

To prove (35) we can write that

$$\mathbf{P}\{\Delta_n^* = n \quad \text{and} \quad N_n = k\} = \mathbf{P}\{N_r < r+1 \quad \text{for} \quad r = 1, \ldots, n$$
$$\text{and} \quad N_n = k\}$$

$$= \mathbf{P}\{N_n = k\} - \mathbf{P}\{N_r \geqq r+1 \quad \text{for some} \quad r = 1, \ldots, n$$
$$\text{and} \quad N_n = k\}. \quad (38)$$

To find the last probability we take into consideration that there is an $r = 1, 2, \ldots, n-1$, such that $N_r = r+1$. Denote by i the largest such r. Then

$$\mathbf{P}\{\Delta_n^* = n \quad \text{and} \quad N_n = k\} = \mathbf{P}\{N_n = k\} - \sum_{i=1}^{n-1} \mathbf{P}\{N_i = i+1, N_r < r+1$$
$$\text{for} \quad i < r \leqq n, N_n = k\}$$

$$= \mathbf{P}\{N_n = k\} - \sum_{i=1}^{n-1} \mathbf{P}\{N_r - N_i < r-i$$
$$\text{for} \quad r = i+1, \ldots, n \quad \text{and} \quad N_i = i+1, N_n = k\}$$

$$= \mathbf{P}\{N_n = k\} - \sum_{i=1}^{n-1} \frac{n+1-k}{n-i} \mathbf{P}\{N_i = i+1,$$
$$N_n = k\}, \quad (39)$$

where we used Theorem 2. This proves (35).

If we use Theorem 4, then we can write for $j = 0, 1, \ldots, n$ that

$$\mathbf{P}\{\Delta_n^* = j\} = \mathbf{P}\{r - N_r \leqq j - N_j \quad \text{for} \quad 0 \leqq r \leqq j \quad \text{and} \quad r - N_r < j - N_j$$
$$\text{for} \quad j < r \leqq n\}$$

$$= \sum_{l=0}^{j} \mathbf{P}\{N_j - N_r \leqq j - r \quad \text{for} \quad 0 \leqq r \leqq j \quad \text{and} \quad N_j = l\}$$
$$\times \mathbf{P}\{N_j - N_r < j - r \quad \text{for} \quad j < r \leqq n | N_j = l\}. \quad (40)$$

In the sum the first factor can be obtained by (35) if we apply it to the random variables $v_j, v_{j-1}, \ldots, v_1$ and the second factor by (33) if we apply it to the random variables $v_{j+1}, \ldots, v_n$. Thus we get (34).

Note. If we introduce the notation

$$Q_j^*(n|k) = Q_j^*(N_1, \ldots, N_n | N_n = k) \tag{41}$$

for the probability $P\{\Delta_n^* = j | N_n = k\}$, then by Theorem 8 we obtain that

$$Q_0^*(n|k) = P\{N_1 > 1 | N_n = k\} - \sum_{i=2}^{n} \frac{1}{i-1} P\{N_1 = 0, N_i = i | N_n = k\} \tag{42}$$

and

$$Q_n^*(n|k) = 1 - \sum_{i=1}^{n-1} \frac{n+1-k}{n-i} P\{N_i = i+1 | N_n = k\} \tag{43}$$

for $k = 0, 1, \ldots, n$ and $Q_n^*(n|k) = 0$ if $k > n$. Furthermore,

$$Q_j^*(n|k) = \sum_{l=0}^{j} P\{N_j = l | N_n = k\} Q_j^*(j|l) Q_0^*(n-j|k-l) \tag{44}$$

if $j = 0, 1, \ldots, n$.

Particular cases. If we make use of (15), then we can obtain the following particular cases of Theorem 8. If $k = 1, 2, \ldots, n$, then

$$Q_j^*(n|k) = \begin{cases} \displaystyle\sum_{i=n-j}^{k-1} \frac{(n+1-k)}{i(n-i)} P\{N_i = i+1 | N_n = k\} & \text{for} \quad n-k < j < n, \\[4mm] 1 - \displaystyle\sum_{i=1}^{k-1} \frac{(n+1-k)}{(n-i)} P\{N_i = i+1 | N_n = k\} & \text{for} \quad j = n. \end{cases} \tag{45}$$

The distribution of $\Delta_n^{(c)}$. For $c = 0, 1, 2, \ldots$ the random variable $\Delta_n^{(c)}$ denotes the number of subscripts $r = 1, 2, \ldots, n$ for which $N_r < r+c$.

Theorem 9. If $c = 1, 2, \ldots$ and $l = 1, 2, \ldots, n+c$, then

$$P\{\Delta_n^{(c)} = j \quad \text{and} \quad N_n = n+c-l\} = \sum_{s=1}^{n-j} \frac{s}{n-j} \left[\sum_{r=l-1}^{j} \frac{l-1}{r} P\{N_{n-j} \right.$$
$$= n-j-s, N_{n-j+r} - N_{n-j} = r-l+1, N_n = n+c-l\}$$
$$\left. - \sum_{r=l}^{j} \frac{l}{r} P\{N_{n-j} = n-j-s, N_{n-j+r} - N_{n-j} = r-l, N_n = n+c-l\} \right] \tag{46}$$

for $j = 0, 1, \ldots, n-1$, and

$$P\{\Delta_n^{(c)} = n \quad \text{and} \quad N_n = n+c-l\} = P\{N_n = n+c-l\}$$
$$- \sum_{i=l}^{n} \frac{l}{i} P\{N_i = i-l \quad \text{and} \quad N_n = n+c-l\}. \tag{47}$$

Further for $c = 1, 2, \ldots$ *and* $j = 0, 1, \ldots, n - 1$

$$\mathbf{P}\{\Delta_n^{(c)} = j \text{ and } N_n = n + c\} = \mathbf{P}\{\Delta_n^* = n - 1 - j \text{ and } N_n = n + c\},$$
(48)

where the right-hand side is given by Theorem 8.

Proof. If $c = 0, 1, 2, \ldots$ and $l = 0, 1, \ldots, n + c$, then

$$\mathbf{P}\{\Delta_n^{(c)} = j \text{ and } N_n = n + c - l\}$$
$$= \mathbf{P}\{N_r < r + c \text{ for } j \text{ subscripts } r = 1, \ldots, n \text{ and } N_n = n + c - l\}$$
$$= \mathbf{P}\{N_n - N_r > n - r - l \text{ for } j \text{ subscripts } r = 1, \ldots, n$$
$$\text{and } N_n = n + c - l\}$$
$$= \mathbf{P}\{N_i < i - l + 1 \text{ for } n - j \text{ subscripts } i = 0, 1, \ldots, n - 1$$
$$\text{and } N_n = n + c - l\}. \quad (49)$$

Accordingly

$$\mathbf{P}\{\Delta_n^{(c)} = j \text{ and } N_n = n + c - l\} = \mathbf{P}\{\Delta_n^{(1-l)} = n - j$$
$$\text{and } N_n = n + c - l\} \quad (50)$$

if $c \geq 1$ and $l \geq 1$ and the right-hand side is given by Theorem 7. Thus we get (46) and (47).

Furthermore

$$\mathbf{P}\{\Delta_n^{(c)} = j \text{ and } N_n = n + c\} = \mathbf{P}\{\Delta_n^{(1)} = n - j - 1 \text{ and } N_n = n + c\}$$
(51)

if $c \geq 1$ and the right-hand side is given by Theorem 8.

5 THE COMPARISON OF A THEORETICAL AND AN EMPIRICAL DISTRIBUTION FUNCTION

Let $\xi_1, \xi_2, \ldots, \xi_m$ be mutually independent random variables having a common distribution function $\mathbf{P}\{\xi_r \leq x\} = F(x)$ $(r = 1, 2, \ldots, n)$. Let $F_m(x)$ be the empirical distribution function of the sample $(\xi_1, \xi_2, \ldots, \xi_m)$ that is, $F_m(x)$ is defined as the number of variables less than or equal to x divided by m.

Denote by $\xi_1^*, \xi_2^*, \ldots, \xi_m^*$ the random variables $\xi_1, \xi_2, \ldots, \xi_m$ arranged in increasing order of magnitude. The random variable ξ_r^* $(r = 1, 2, \ldots, m)$ is called the rth order statistic of the sample $(\xi_1, \xi_2, \ldots, \xi_m)$.

Consider the deviations

$$\delta_m(r) = F_m(\xi_r^*) - F(\xi_r^*) \quad (r = 1, 2, \ldots, m). \quad (52)$$

There are several random variables connected with the sequence $\delta_m(1), \delta_m(2), \ldots, \delta_m(m)$ which have some importance in order statistics.

The most frequently used variable is

$$\delta_m^+ = \max_{1 \le r \le m} \delta_m(r) \tag{53}$$

which can also be expressed in the following equivalent form

$$\delta_m^+ = \sup_{-\infty < x < \infty} [F_m(x) - F(x)]. \tag{54}$$

Denote by ρ_m the number of nonnegative elements among $\delta_m(r)$ $(r = 1, 2, ..., m)$.

Define ρ_m^* as the largest r for which $\delta_m(r)$ $(r = 1, 2, ..., m)$ attains its maximum.

If we assume that $F(x)$ is a continuous distribution function, then the joint distribution of the random variables $\delta_m(r)$ $(r = 1, 2, ..., m)$ does not depend on $F(x)$, and consequently the distributions of the random variables δ_m^+, ρ_m and ρ_m^* are also independent of $F(x)$. In what follows we shall investigate exclusively this case. In this case we may assume, without loss of generality, that

$$F(x) = \begin{cases} 0 & \text{for } x \le 0, \\ x & \text{for } 0 \le x \le 1, \\ 1 & \text{for } x \ge 1. \end{cases} \tag{55}$$

Then $\qquad \delta_m(r) = F_m(\xi_r^*) - \xi_r^* \quad \text{for } r = 1, 2, ..., m. \tag{56}$

In this section we shall assume that $F(x)$ is a continuous distribution function and find the distributions of the random variables δ_m^+, ρ_m and ρ_m^*.

Theorem 10. *If* $k = 1, 2, ..., m$, *then we have*

$$\mathbf{P}\left\{\delta_m^+ \le \frac{k}{m}\right\} = 1 - \sum_{j=1}^{m-k} \frac{k}{m-j} \binom{m}{j+k} \left(\frac{j}{m}\right)^{j+k} \left(1 - \frac{j}{m}\right)^{m-j-k}. \tag{57}$$

Proof. Suppose that $F(x)$ is given by (55). Denote by v_r $(r = 1, 2, ..., m)$ the number of variables $\xi_1, \xi_2, ..., \xi_m$ falling in the interval

$$\left(\frac{r-1}{m}, \frac{r}{m}\right] \quad \text{and define} \quad N_r = v_1 + ... + v_r$$

for $r = 1, 2, ..., m$. Then

$$\mathbf{P}\{N_r = i\} = \binom{m}{i} \left(\frac{r}{m}\right)^i \left(1 - \frac{r}{m}\right)^{m-i} \tag{58}$$

for $r = 1, 2, ..., m$ and $N_m = m$. Now we have

$$\mathbf{P}\left\{\delta_m^+ \le \frac{k}{m}\right\} = \mathbf{P}\{\max_{1 \le r \le m} (N_r - r) < k\} = 1 - \sum_{j=1}^{m-k} \frac{k}{m-j} \mathbf{P}\{N_j = j + k\},$$

$$\tag{59}$$

where the last equality follows from (47). This proves (57).

Theorem 11. *If* $0 < x \leqq 1$, *then we have*

$$\mathbf{P}\{\delta_m^+ \leqq x\} = 1 - \sum_{mx \leqq j \leqq m} \frac{mx}{m+mx-j} \binom{m}{j} \left(\frac{j}{m}-x\right)^j \left(1+x-\frac{j}{m}\right)^{m-j}.$$

(60)

Proof. We shall calculate the probability $\mathbf{P}\{\delta_m^+ > x\}$. Suppose that $F(x)$ is given by (55). If $\delta_m^+ > x$, then for some u $(0 < u < 1)$ the empirical distribution function $F_m(u)$ intersects the line $x+u$ $(0 < u < 1)$. Suppose that the last intersection occurs at $u = v$. Then $F_m(v) = j/m$ for some j $(mx \leqq j \leqq m)$ and $v = (j-mx)/m$. In this case there are j elements of the sample in the interval $(0, v]$ and $m-j$ elements in the interval $(v,1]$. This event has probability

$$\binom{m}{j} v^j (1-v)^{m-j}.$$

(61)

Furthermore if the last intersection occurs at $u = v$, then

$$F_m(u) \leqq u+x \quad \text{for} \quad v \leqq u \leqq 1$$

or $\qquad\qquad F_m(u) - F_m(v) \leqq u-v \quad \text{for} \quad v \leqq u \leqq 1.$

Since $F_m(1) - F_m(v) = 1 - (v+x)$, the latter event has probability

$$\frac{x}{1-v}$$

(62)

by Theorem 3. If we put $v = (j-mx)/m$ into (61) and (62), multiply them, and add the product for $mx \leqq j \leqq m$, then we get $\mathbf{P}\{\delta_m^+ > x\}$. Thus we obtain (60).

In the particular case $x = k/m$, (60) reduces to (57).

The distribution function of δ_m^+ was found in 1944 by N. V. Smirnov [25]. See also A. Wald and J. Wolfowitz [35], Z. W. Birnbaum and F. H. Tingey [7], B. L. van der Waerden [32], A. P. Dempster [11] and M. Dwass [14].

Theorem 12. *We have*

$$\mathbf{P}\{\rho_m = j\} = \frac{1}{m} \sum_{i=1}^{j} \frac{1}{i} \binom{m}{i-1} \left(\frac{i}{m}\right)^{i-1} \left(1-\frac{i}{m}\right)^{m-i}$$

(63)

for $j = 1, 2, \ldots, m$.

Proof. Let us use the same notation as in the proof of Theorem 10. Since $\delta_m(r) \geqslant 0$ if and only if $N_r \geqslant r$, consequently ρ_m is equal to the number of subscripts $r = 1, 2, ..., m$ for which $N_r \geq r$. Since $N_m = m$, by (27) we obtain that

$$P\{\rho_m = j\} = \begin{cases} \sum_{i=1}^{j} \dfrac{1}{i(m-i)} P\{N_i = i-1\} & \text{for} \quad j = 1, 2, ..., m-1, \\ 1 - \sum_{i=1}^{m-1} \dfrac{1}{i} P\{N_i = i-1\} & \text{for} \quad j = m. \end{cases} \tag{64}$$

Hence by (58) we get (63).

The distribution of ρ_m was found in 1958 by P. Cheng [8].

Theorem 13. We have
$$P\{\rho_m^* = j\} = P\{\rho_m = j\} \tag{65}$$
for $j = 1, 2, ..., m$.

Proof. The random variable ρ_m is equal to the number of subscripts $r = 1, 2, ..., m$ for which $\delta_m(r) \geq 0$, that is, $N_r - r \geq 0$, and ρ_m^* is the largest subscript $r = 1, 2, ..., m$ for which $N_r - r$ attains its maximum. By Theorem 4 the position of the last maximum in $N_r - r$ $(r = 0, 1, ..., m)$ has the same distribution as the number of nonnegative elements in $N_r - r$ $(r = 1, 2, ..., m)$. This proves (65).

We note that the random variables $\delta_m(r)$ $(r = 1, 2, ..., m)$ are continuous and with probability 1 there is only a single maximum in the sequence $\delta_m(r)$ $(r = 1, 2, ..., m)$.

The distribution of ρ_m^* was found in 1958 by Z. W. Birnbaum and R. Pyke [6].

We mention briefly two more theorems.

Theorem 14. The random variable

$$\frac{\rho_m^*}{m} - \delta_m^+$$

has a uniform distribution over the interval $(0, 1)$, that is

$$P\left\{\frac{\rho_m^*}{m} - \delta_m^+ \leq x\right\} = x \tag{66}$$

for $0 \leq x \leq 1$.

This theorem was found in 1958 by Z. W. Birnbaum and R. Pyke [6]. For other proofs see M. Dwass [13], N. H. Kuiper [20] and the author [30]. Theorem 14 can easily be proved by using Theorem 3.

Theorem 15. Let

$$G_k(x) = P\left\{\frac{\rho_m^*}{m} - \delta_m^+ \leq \frac{x}{m} \quad \text{and} \quad \rho_m^* = k\right\} \tag{67}$$

for $k = 1, 2, ..., m$. If $x \leq 0$, then $G_k(x) = 0$. If $x \geq k$, then

$$G_k(x) = P\{\rho_m^* = k\}.$$

If $0 < x < k$, *then*

$$\frac{dG_k(x)}{dx} = \binom{m}{k} \frac{(k-x)(m-x)^{m-k-1}}{m^m} \left[kx^{k-1} - \sum_{j=1}^{[x]} \binom{k}{j} j^{j-1}(x-j)^{k-j} \right].$$

(68)

Probability (67) was found in 1958 by Z. W. Birnbaum and R. Pyke [6]. See also [30]. Formula (68) can be proved by using Theorem 3.

6 THE COMPARISON OF TWO EMPIRICAL DISTRIBUTION FUNCTIONS

Let $\xi_1, \xi_2, \ldots, \xi_m, \eta_1, \eta_2, \ldots, \eta_n$ be mutually independent random variables having a common distribution function $F(x)$. Denote by $F_m(x)$ and $G_n(x)$ the empirical distribution functions of the samples $(\xi_1, \xi_2, \ldots, \xi_m)$ and $(\eta_1, \eta_2, \ldots, \eta_n)$ respectively.

Denote by $\eta_1^*, \eta_2^*, \ldots, \eta_n^*$ the random variables $\eta_1, \eta_2, \ldots, \eta_n$ arranged in increasing order of magnitude.

Consider the deviations

$$F_m(\eta_r^*) - G_n(\eta_r^* - 0) \quad (r = 1, 2, \ldots, n).$$

(69)

There are several random variables depending on the deviations (69) which have some importance in order statistics. In what follows we shall consider a few examples.

Let

$$\delta^+(m, n) = \max_{1 \leq r \leq n} [F_m(\eta_r^*) - G_n(\eta_r^* - 0)].$$

(70)

This statistic can be expressed in the following familiar form

$$\delta^+(m, n) = \sup_{-\infty < x < \infty} [F_m(x) - G_n(x)].$$

(71)

Denote by $\gamma_c(m, n)$ the number of subscripts $r = 1, 2, \ldots, n$ for which

$$F_m(\eta_r^*) < G_n(\eta_r^*) - \frac{c}{n},$$

(72)

where $c = 0, \pm 1, \ldots, \pm(n-1)$.

Denote by $\tau(m, n)$ the smallest $r = 1, 2, \ldots, n$ for which

$$F_m(\eta_r^*) - G_n(\eta_r^* - 0)$$

(73)

attains its maximum.

If we suppose that $F(x)$ is a continuous distribution function, then the distributions of the random variables $\delta^+(m, n)$, $\gamma_c(m, n)$ and $\tau(m,n)$ do not depend on $F(x)$.

A mathematical method of finding the distributions of $\delta^+(m,n)$, $\gamma_c(m,n)$ and $\tau(m,n)$.

Let v_r $(r = 1, 2, \ldots, n+1)$ be defined as p times the number of variables $\xi_1, \xi_2, \ldots, \xi_m$ falling in the interval $(\eta_{r-1}^*, \eta_r^*]$ where $\eta_0^* = -\infty$ and $\eta_{n+1}^* = \infty$. Set $N_r = v_1 + \ldots + v_r$ for $r = 1, 2, \ldots, n+1$. Obviously $N_{n+1} = mp$. By using this notation we can write that

$$F_m(\eta_r^*) = \frac{N_r}{mp} \quad \text{and} \quad G_n(\eta_r^*) = \frac{r}{n} \quad \text{for} \quad r = 1, 2, \ldots, n. \quad (74)$$

Accordingly we have

$$\delta^+(m,n) = \max_{1 \le r \le n} \left[\frac{N_r}{mp} - \frac{r-1}{n} \right]. \quad (75)$$

$\gamma_c(m,n)$ is equal to the number of subscripts $r = 1, 2, \ldots, n$ for which

$$\frac{N_r}{mp} < \frac{r-c}{n}. \quad (76)$$

$\tau(m,n)$ is the smallest $r = 1, 2, \ldots, n$ for which

$$\frac{N_r}{mp} - \frac{r-1}{n} \quad (77)$$

attains its maximum.

If we suppose that $F(x)$ is a continuous distribution function, then $v_1, v_2, \ldots, v_{n+1}$ are interchangeable random variables with sum

$$N_{n+1} = v_1 + \ldots + v_{n+1} = mp$$

and we have
$$\mathbf{P}\{N_i = sp\} = \frac{\binom{i+s-1}{s}\binom{m+n-i-s}{m-s}}{\binom{m+n}{m}} \quad (78)$$

for $i = 1, 2, \ldots, n$, and $s = 0, 1, \ldots, m$. This can be proved as follows: Let us imagine that a box contains m black and n white balls and we draw all the $m+n$ balls without replacement. Suppose that every result has the same probability. It can easily be seen that $\mathbf{P}\{N_i = sp\}$ can be interpreted as the probability that we obtain the ith white ball at the $(i+s)$th drawing.

In a similar way we obtain that

$$\mathbf{P}\{N_i = sp, N_{i+j} - N_i = tp\} = \frac{\binom{i+s-1}{s}\binom{j+t-1}{t}\binom{m+n-i-j-s-t}{m-s-t}}{\binom{m+n}{m}} \quad (79)$$

for $1 \leqq i < i+j \leqq n$ and $0 \leqq s+t \leqq m$, and

$$\mathbf{P}\{N_i = sp, N_{i+j} - N_i = tp, N_{i+j+k} - N_{i+j} = up\}$$

$$= \frac{\binom{i+s-1}{s} \binom{j+t-1}{t} \binom{k+u-1}{u} \binom{m+n-i-j-k-s-t-u}{m-s-t-u}}{\binom{m+n}{m}} \qquad (80)$$

for $1 \leqq i < i+j < i+j+k \leqq n$ and $0 \leqq s+t+u \leqq m$.

Note. If $F(x)$ is a continuous distribution function and we know the distribution of the random variable $\gamma_c(m, n)$ for $c \geqq 0$, then we can obtain immediately the distribution of $\gamma_{-c}(m, n)$. As we shall show

$$\mathbf{P}\{\gamma_c(m, n) = j\} = \mathbf{P}\{\gamma_{-c}(m, n) = n-j\} \qquad (81)$$

holds for $j = 0, 1, \ldots, n$ and $c = 0, \pm 1, \pm 2, \ldots$. This follows from the following relations:

$$\mathbf{P}\{\gamma_c(m, n) = j\} = \mathbf{P}\left\{\frac{N_r}{mp} < \frac{r-c}{n} \quad \text{for } j \text{ subscripts } r = 1, 2, \ldots, n\right\}$$

$$= \mathbf{P}\left\{\frac{N_{n+1} - N_r}{mp} > \frac{n-r+c}{n} \quad \text{for } j \text{ subscripts } r = 1, 2, \ldots, n\right\}$$

$$= \mathbf{P}\left\{\frac{N_s}{mp} > \frac{s+c-1}{n} \quad \text{for } j \text{ subscripts } s = 1, 2, \ldots, n\right\}$$

$$= \mathbf{P}\left\{\frac{N_s}{mp} < \frac{s+c}{n} \quad \text{for } n-j \text{ subscripts } s = 1, 2, \ldots, n\right\}$$

$$= \mathbf{P}\{\gamma_{-c}(m, n) = n-j\}. \qquad (82)$$

We shall be able to find easily the distributions of $\delta^+(m, n)$, $\gamma_c(m, n)$ and $\tau(m, n)$ if we assume that $F(x)$ is a continuous distribution function and $n = mp$ where p is a positive integer. In this case $v_1, v_2, \ldots, v_{n+1}$ are interchangeable random variables taking on nonnegative integers and their sum is $N_{n+1} = v_1 + \ldots + v_{n+1} = mp = n$. Furthermore we have

$$\delta^+(m, n) = \frac{1}{n} \max_{1 \leqq r \leqq n} (N_r - r + 1). \qquad (83)$$

$\gamma_c(m, n)$ is equal to the number of subscripts $r = 1, 2, \ldots, n$ for which $N_r < r-c$.

$\tau(m, n)$ is the smallest r for which $N_r - r + 1$ attains its maximum.

In what follows we shall assume that $F(x)$ is a continuous distribution function and that $n = mp$ where p is a positive integer and we shall determine the distributions of the random variables $\delta^+(m, n)$, $\gamma_c(m, n)$ and $\tau(m, n)$.

The distribution of $\delta^+(m, n)$. We suppose that $F(x)$ is a continuous distribution function.

Theorem 16. If $n = mp$ where p is a positive integer, and $c = 0, 1, \ldots, n$, then

$$\mathbf{P}\left\{\delta^+(m, n) \leq \frac{c}{n}\right\} = 1 - \frac{1}{\binom{m+n}{m}} \sum_{(c+1)/p \leq s \leq m}$$

$$\frac{c+1}{n+c+1-sp}\binom{sp+s-c-1}{s}\binom{m+n+c-sp-s}{m-s} \tag{84}$$

and, in particular

$$\mathbf{P}\left\{\delta^+(m, m) \leq \frac{c}{m}\right\} = 1 - \frac{\binom{2m}{m+1+c}}{\binom{2m}{m}} \tag{85}$$

for $c = 0, 1, \ldots, m$.

Proof. In this case $\delta^+(m, n)$ is given by (83) and $N_{n+1} = n$. Hence

$$\mathbf{P}\left\{\delta^+(m, n) \leq \frac{c}{n}\right\} = \mathbf{P}\{N_r < r+c \quad \text{for} \quad r = 1, \ldots, n+1\}$$

$$= 1 - \sum_{j=1}^{n} \frac{c+1}{n+1-j}\mathbf{P}\{N_j = j+c\} \tag{86}$$

for $c = 0, 1, \ldots, n$ which follows from (47). By (78) we obtain (84). If, in particular, $p = 1$, then (84) reduces to (85).

The distribution of the random variable $\delta^+(m, n)$ for $n = m$ was found in 1951 by B. V. Gnedenko and V. S. Korolyuk [17] and for $n = mp$ where p is a positive integer in 1955 by V. S. Korolyuk [19]. See also the author [29].

Note. If we suppose that $c = [nx]$, where $0 < x \leq 1$ and $n = mp$, then by (84) we obtain that

$$\lim_{p \to \infty} \mathbf{P}\left\{\delta^+(m, n) \leq \frac{c}{n}\right\} = \mathbf{P}\{\delta_m^+ \leq x\}$$

$$= 1 - \sum_{mx \leq j \leq m} \frac{mx}{m+mx-j}\binom{m}{j}\left(\frac{j}{m}-x\right)^j\left(1+x-\frac{j}{m}\right)^{m-j} \tag{87}$$

which is in agreement with (60).

The distribution of $\gamma_c(m, n)$. We suppose that $F(x)$ is a continuous distribution function.

Theorem 17. *Let* $n = mp$ *where* p *is a positive integer. We have*

$$\mathbf{P}\{\gamma_0(m,n) = j\} = \frac{1}{n+1} \tag{88}$$

for $j = 0, 1, 2, \ldots, n$. *If* $c = 1, 2, \ldots, n$, *then*

$$\mathbf{P}\{\gamma_c(m,n) = 0\} = 1 - \frac{1}{\binom{m+n}{m}} \sum_{(c+1)/p \leq s \leq m}$$

$$\frac{c+1}{n+c+1-sp} \binom{sp+s-c-1}{s} \binom{m+n+c-sp-s}{m-s} \tag{89}$$

and

$$\mathbf{P}\{\gamma_c(m,n) = j\} = \frac{1}{\binom{m+n}{m}} \sum_{s=0}^{[(j-1)/p]} \left(1 - \frac{sp}{j}\right) \left[\sum_{t=0}^{[(n-j-c)/p]} \left(\frac{c}{c+tp}\right) \binom{j+s-1}{s}\right.$$

$$\times \binom{c+tp+t-1}{t} \binom{m+n-j-s-c-tp-t}{m-s-t}$$

$$- \sum_{t=0}^{[(n-j-c-1)/p]} \frac{c+1}{c+1+tp} \binom{j+s-1}{s} \binom{c+tp+t}{t}$$

$$\left. \times \binom{m+n-j-s-c-tp-t-1}{m-s-t}\right] \tag{90}$$

for $j = 1, 2, \ldots, n-c$.

If, in particular, $p = 1$, *then* (89) *reduces to*

$$\mathbf{P}\{\gamma_c(m,m) = 0\} = 1 - \frac{\binom{2m}{m+1+c}}{\binom{2m}{m}} \tag{91}$$

for $c = 1, 2, \ldots, m$ *and* (90) *becomes*

$$\mathbf{P}\{\gamma_c(m,m) = j\} = \frac{1}{\binom{2m}{m}} \sum_{i=j}^{m-c} \frac{c}{(i+1)(m-i)} \binom{2i}{i} \binom{2m-2i}{m+c-i} \tag{92}$$

for $c = 1, 2, \ldots, m-1$ *and* $j = 1, 2, \ldots, m-c$.

Proof. By definition

$$\mathbf{P}\{\gamma_c(m,n) = j\} = \mathbf{P}\{N_r < r-c \quad \text{for } j \text{ subscripts} \quad r = 1, 2, \ldots, n\}. \tag{93}$$

If $c = 0$ and we take into consideration that $N_{n+1} = n$, then by (14) we obtain (88).

If $c = 1, 2, \ldots, n$ and $j = 0$, then

$$\mathbf{P}\{\gamma_c(m,n) = 0\} = \mathbf{P}\left\{\delta^+(m,n) \leq \frac{c}{n}\right\} \tag{94}$$

and thus (89) and (91) follow from (84) and (85) respectively.

If $c > 0$ and $j > 0$, then by (32) we have

$$\mathbf{P}\{\gamma_c(m,n) = j\} = \sum_{l=1}^{j} \frac{l}{j} \left[\sum_{r=c}^{n-j} \frac{c}{r} \mathbf{P}\{N_j = j-l,\ N_{j+r} - N_j = r-c\} \right.$$

$$\left. - \sum_{r=c+1}^{n-j} \frac{c+1}{r}\, \mathbf{P}\{N_j = j-l, N_{j+r} - N_j = r-c-1\} \right] \qquad (95)$$

and the probabilities on the right-hand side are given by (78) and (79). This proves (90). If, in particular, $p = 1$, then (90) reduces to (92).

The distribution of the random variable $\gamma_c(m, n)$ in the particular case $m = n$ was found in 1952 by V. S. Mihalević [23].

The distribution of $\tau(m, n)$. We shall find the joint distribution of $\delta^+(m,n)$ and $\tau(m, n)$ in the case where $F(x)$ is a continuous distribution function and $n = mp$ where p is a positive integer.

Theorem 18. If $p \geqq 1, k = 0, 1, \ldots, n, j = 1, 2, \ldots, n-k+1$ and

$$j + k = tp + 1 \qquad (t = 0, 1, 2, \ldots),$$

then we have

$$\mathbf{P}\left\{\delta^+(m,n) = \frac{k}{n}, \tau(m,n) = j\right\} = \frac{1}{\binom{m+n}{n}}$$

$$\times \left[\binom{j+t-\alpha(p)}{j-1} - \sum_{2 \leqq sp \leqq j-1} \frac{1}{sp-1} \binom{sp+s-2}{s} \binom{j+t-sp-s-1}{t-s} \right]$$

$$\times \left[\binom{m+n-j-t}{m-t} - \sum_{2 \leqq sp \leqq n+1-j} \frac{k+1}{n+2-j-sp} \binom{sp+s-2}{s} \right.$$

$$\times \left. \binom{m+n-j-t-sp-s+1}{m-t-s} \right], \qquad (96)$$

where $\alpha(p) = 3$ if $p = 1$ and $\alpha(p) = 2$ if $p > 1$.

If, in particular, $k = 1$ and $j = 2, 3, \ldots, n$ where $j = tp$, then

$$\mathbf{P}\left\{\delta^+(m,n) = \frac{1}{n}, \tau(m,n) = j\right\} = \frac{1}{(j-1)} \frac{\binom{j+t-2}{t}}{\binom{m+n}{m}}$$

$$\times \left[\binom{m+n-j-t}{m-t} - \sum_{2 \leqq sp \leqq n+1-j} \frac{2}{n+2-j-sp} \binom{sp+s-2}{s} \right.$$

$$\times \left. \binom{m+n-j-t-sp-s+1}{m-t-s} \right] \qquad (97)$$

and $\quad \mathbf{P}\{\delta^+(m,n) = 0, \tau(m,n) = 1\} = \dfrac{1}{\dbinom{m+n}{m}}\left[\dbinom{m+n-1}{m}\right.$

$$\left. - \sum_{2 \le sp \le n} \frac{1}{n+1-sp}\dbinom{sp+s-2}{s}\dbinom{m+n-sp-s}{m-s}\right]. \qquad (98)$$

If, in particular, $p = 1$, then we have

$$\mathbf{P}\left\{\delta^+(m,m) = \frac{k}{m}, \tau(m,m) = j\right\} = \frac{k(k+1)}{(k+2j-2)(m+2-j)}$$

$$\times \dbinom{k+2j-2}{j-1}\dbinom{2m+2-2j-k}{m+1-j} \qquad (99)$$

for $k = 1, 2, ..., m$ and $j = 1, 2, ..., m+1-k$, and

$$\mathbf{P}\{\delta^+(m,m) = 0, \tau(m,m) = 1\} = \frac{1}{m+1}. \qquad (100)$$

Proof. We can write that

$$\mathbf{P}\left\{\delta^+(m,n) = \frac{k}{n}, \tau(m,n) = j\right\} = \mathbf{P}\{N_r - r + 1 < N_j - j + 1 = k$$

$$\text{for}\quad 1 \le r < j \quad\text{and}\quad N_r - r + 1 \le N_j - j + 1\quad\text{for}\quad j \le r \le n\}$$

$$= \mathbf{P}\{N_j - N_r > j - r \quad\text{for}\quad 1 \le r < j \quad\text{and}\quad N_j = j + k - 1\}$$

$$\times \mathbf{P}\{N_r - N_j \le r - j \quad\text{for}\quad j \le r \le n | N_j = j + k - 1\}. \qquad (101)$$

By Theorem 8, the first factor on the right-hand side of (101) is

$$\mathbf{P}\{N_1 = k\} \quad\text{if}\quad j = 1$$

and

$$\mathbf{P}\{N_1 > 1, N_j = j + k - 1\} - \sum_{i=2}^{j-1}\frac{1}{(i-1)}\mathbf{P}\{N_1 = 0, N_i = i, N_j = j + k - 1\}$$

$$\qquad (102)$$

if $j = 2, ..., n+1-k$, and 0 if $j > n+1-k$. We note that if $k = 1$, then (102) reduces to

$$\frac{1}{(j-1)}\mathbf{P}\{N_1 = 0, N_j = j\} \qquad (103)$$

which follows from (27).

If we take into consideration that $N_{n+1} = n$, then by (35) we obtain that the second factor on the right-hand side of (101) is

$$1 - \sum_{i=1}^{n-j}\frac{k+1}{n+1-j-i}\mathbf{P}\{N_{j+i} - N_j = i + 1 | N_j = j + k - 1\} \qquad (104)$$

for $j = 1, 2, ..., n+1-k$.

We can easily see that (101) is 0 unless

$$k = 0, 1, 2, \ldots, n, \quad j = 1, 2, \ldots, n+1-k \quad \text{and} \quad j+k = tp+1$$

where $t = 0, 1, 2, \ldots$. If we use formulas (78), (79) and (80), then we obtain (96), (97) and (98) and the particular cases (99) and (100).

The joint distribution of the random variables $\delta^+(m, n)$ and $\tau(m, n)$ in the particular case $m = n$ was found in 1957 by I. Vincze [33], [34].

Some limit theorems. Now we shall find the limiting distributions of the random variables $\delta^+(m, n)$ and $\tau(m, n)/n$ if $n = mp$ and $p \to \infty$.

Theorem 19. *If* $n = mp$, *then*

$$\lim_{p \to \infty} \mathbf{P} \left\{ \frac{\tau(m, n)}{n} \leqq x \right\} = x \tag{105}$$

for $0 \leqq x \leqq 1$ *and*

$$\lim_{p \to \infty} \mathbf{P} \left\{ \delta^+(m, n) \leqq x, \frac{\tau(m, n)}{n} \leqq y \right\} = \sum_{k=1}^{[m(x+y)]} [G_k(my) - G_k(k - mx)] \tag{106}$$

for $0 \leqq x, 0 \leqq y, x+y \leqq 1$, *where* $G_k(x)$ *is given in Theorem* 15.

Proof. Without loss of generality we may assume that $F(x)$ is given by (55). Since by a theorem of V. Glivenko [16] it follows that

$$\mathbf{P}\{ \lim_{n \to \infty} \sup_{-\infty < x < \infty} |G_n(x) - F(x)| = 0 \} = 1, \tag{107}$$

we can conclude that if $n = mp$, then

$$\mathbf{P}\{ \lim_{p \to \infty} \delta^+(m, n) = \delta_m^+ \} = 1, \tag{108}$$

where δ_m^+ is defined by (54) and

$$\mathbf{P} \left\{ \lim_{p \to \infty} \frac{N_{\tau(m, n)}}{p} = \rho_m^* \right\} = 1, \tag{109}$$

where ρ_m^* is defined after formula (54).

Since by (75)

$$\delta^+(m, n) = \frac{1}{n} [N_{\tau(m, n)} - \tau(m, n) + 1], \tag{110}$$

it follows that
$$\mathbf{P} \left\{ \lim_{p \to \infty} \frac{\tau(m, n)}{n} = \frac{\rho_m^*}{m} - \delta_m^+ \right\} = 1. \tag{111}$$

382   LAJOS TAKÁCS

Accordingly we have

$$\lim_{p\to\infty} \mathbf{P}\left\{\frac{\tau(m,n)}{n} \leq x\right\} = \mathbf{P}\left\{\frac{\rho_m^*}{m} - \delta_m^+ \leq x\right\} \qquad (112)$$

and the right-hand side is given by (66). This proves (105).
By (108) and (111) we have

$$\lim_{p\to\infty} \mathbf{P}\left\{\delta^+(m,n) \leq x, \frac{\tau(m,n)}{n} \leq y\right\} = \mathbf{P}\left\{\delta_m^+ \leq x, \frac{\rho_m^*}{m} - \delta_m^+ \leq y\right\} \qquad (113)$$

and the right-hand side can be obtained by Theorem 15. Thus we get
(106).

Note. If we do not make the assumption that $n = mp$, then (105) and (106)
hold unchangeably if $n \to \infty$.

REFERENCES

[1] A. Aeppli. Zur Theorie verketteter Wahrscheinlichkeiten. Markoffsche
     Ketten höherer Ordnung. *Diss. Eidg. Tech. Hochschule.* Zürich, 1924.
[2] E. S. Andersen. On sums of symmetrically dependent random variables.
     Skand. Aktuarietids. 36 (1953) 123–38.
[3] D. André. Solution directe du problème résolu par M. Bertrand. *C.R.
     Acad. Sci. Paris* 105 (1887) 436–7.
[4] É. Barbier. Generalisation du problème résolu par M. J. Bertrand.
     C.R. Acad. Sci. Paris 105 (1887) 407 and 440 (errata).
[5] J. Bertrand. Solution d'un problème. *C.R. Acad. Sci. Paris* 105 (1887)
     369.
[6] Z. W. Birnbaum and R. Pyke. On some distributions related to the
     statistic D_n^+. *Ann. Math. Statist.* 29 (1958) 179–87.
[7] Z. W. Birnbaum and F. H. Tingey. One sided confidence contours for
     probability distribution functions. *Ann. Math. Statist.* 22 (1951)
     592–6.
[8] P. Cheng. Non-negative jump points of an empirical distribution
     function relative to a theoretical distribution function. (Chinese.)
     Acta Math. Sin. 5 (1955) 347–68. [English translation: Selected
     Translations in Mathematical Statistics and Probability. *IMS and
     AMS*, 4 (1963) 17–38.]
[9] A. De Moivre. De mensura sortis, seu, de probabilitate eventuum in
     ludis a casu fortuito pendentibus. *Philos. Trans.* (London) 27 (1711)
     213–64.
[10] A. De Moivre. *The Doctrine of Chances: or, A Method of Calculating the
     Probability of Events in Play.* London, 1718.
[11] A. Dempster. Generalized D_n^+ statistics. *Ann. Math. Statist.* 30 (1959)
     593–7.
[12] A. Dvoretzky and Th. Motzkin. A problem of arrangements. *Duke
     Math. J.* 14 (1947) 305–13.

[13] M. Dwass. On several statistics related to empirical distribution functions. *Ann. Math. Statist.* **29** (1958) 188–91.

[14] M. Dwass. The distribution of a generalized D_n^+ statistic. *Ann. Math. Statist.* **30** (1959) 1024–8.

[15] W. Feller. On combinatorial methods in fluctuation theory. Probability and Statistics. *The Harald Cramér Volume.* Edited by U. Grenander. Stockholm: Almqvist and Wiksell, 1959, pp. 75–91.

[16] V. Glivenko. Sulla determinazione empirica delle leggi di probabilitá. *G. dell'Istituto Ital. Degli Attuari* **4** (1933) 92–9.

[17] B. V. Gnedenko and V. S. Korolyuk. On the maximum discrepancy between two empirical distribution functions. (Russian.) *Dokl. Akad. Nauk SSSR* **80** (1951) 525–8. [English translation: Selected Translations in Mathematical Statistics and Probability. *IMS and AMS,* **1** (1961) 13–16.]

[18] H. D. Grossman. Another extension of the ballot problem. *Scr. Math.* **16** (1950) 120–4.

[19] V. S. Korolyuk. On the discrepancy of empirical distribution functions for the case of two independent samples. *Izv. Akad. Nauk SSSR Ser. Math.* **19** (1955) 81–96. [English translation: Selected Translations in Mathematical Statistics and Probability. *IMS and AMS,* **4** (1963) 105–21.]

[20] N. H. Kuiper. Alternative proof of a theorem of Birnbaum and Pyke. *Ann. Math. Statist.* **30** (1959) 251–2.

[21] J. L. Lagrange. Recherches sur les suites récurrentes dont les termes varient de plusieurs manières différentes, ou sur l'intégration des équations linéaires aux différences finies et partielles; et sur l'usage de cas équations dans la théorie des hasards. *Nouveaux Mémoires de l'Académie Royale des Sciences et Belles-Lettres de Berlin,* année 1775 (1777) pp. 183–272. [*Oeuvres de Lagrange,* **4**, pp. 151–251. Paris: Gauthier-Villars, 1869.]

[22] P. S. Laplace. Recherches sur l'intégration des équations différentielles aux différences finies et sur leur usage dans la théorie des hasards. *Mémoires de l'Académie Royale des Sciences de Paris,* année 1773, **7** (1776). [*Oeuvres complètes de Laplace,* **8** (1891) 69–197.]

[23] V. S. Mihalevič. On the mutual disposition of two empirical distribution functions. (Russian.) *Dokl. Akad. Nauk SSSR* **85** (1952) 485–8. [English translation: Selected Translations in Mathematical Statistics and Probability. *IMS and AMS,* **1** (1961) 63–7.]

[24] S. G. Mohanty and T. V. Narayana. Some properties of compositions and their application to probability and statistics. *Biometr. Z.* **3** (1961) 252–8 and **5** (1963) 8–18.

[25] N. V. Smirnov. Approximate laws of distribution of random variables from empirical data. (Russian.) *Usp. Mat. Nauk* **10** (1944) 179–206.

[26] L. Takács. The probability law of the busy period for two types of queuing processes. *Operations Res.* **9** (1961) 402–7.

[27] L. Takács. A generalization of the ballot problem and its application in the theory of queues. *J. Am. Statist. Ass.* **57** (1962) 327–37.

[28] L. Takács. Ballot problems. *Z. Wahrscheinlichkeitstheorie* **2** (1963) 118–21.

[29] L. Takács. An application of a ballot theorem in order statistics. *Ann. Math. Statist.* **35** (1964) 1356–8.

[30] L. Takács. The distributions of some statistics depending on the deviations between empirical and theoretical distribution functions. *Sankhyā, Ser. A,* **27** (1965) 93–100.

[31] L. Takács. *Combinatorial Methods in the Theory of Stochastic Processes.* New York: John Wiley, 1967.

[32] B. L. Van der Waerden. Testing a distribution function. *Indag. Math.* **15** (1953) 201–7.

[33] I. Vincze. Einige zweidimensionale Verteilungs- und Grenzverteilungssätze in der Theorie der geordneten Stichproben. *Publ. Math. Inst. Hungar. acad. Sci.* **2** (1957) 183–209.

[34] I. Vincze. On some joint distributions and joint limiting distributions in the theory of order statistics, II. *Publ. Math. Inst. Hungar. Acad. Sci.* **4** (1959) 29–47.

[35] A. Wald and J. Wolfowitz. Confidence limits for continuous distribution functions. *Ann. Math. Statist.* **10** (1939) 105–18.

[36] W. A. Whitworth. Arrangements of *m* things of one sort and *n* things of another sort, under certain conditions of priority. *Mess. Math.* **8** (1879) 105–14.

[37] W. A. Whitworth. *Choice and Chance.* Fourth edition. Deighton Bell, Cambridge, 1886.

ON KOLMOGOROV–SMIRNOV
TYPE DISTRIBUTION THEOREMS

I. VINCZE

0 INTRODUCTION

0.1. In the following, two related topics will be considered: the first part (§1 and §2) of this paper deals with the distribution of the Kolmogorov–Smirnov two-sample statistics in the case of discontinuous distribution functions. In the second part (§3) the proofs of the distribution theorem given by Birnbaum and Tingey, and independently by Smirnov will be investigated.

0.2. In his paper [6] Schmid has given the limiting distribution of the one-sided and two-sided Kolmogorov–Smirnov statistics, that is, of

$$D_n^+ = \sup_{-\infty < x < \infty} (F_n(x) - F(x)),$$

$$D_n = \sup_{-\infty < x < \infty} |F_n(x) - F(x)|$$

for discontinuous $F(x)$. In §1 the distribution laws of the corresponding two-sample statistics $D_{n,n}^+$ and $D_{n,n}$ will be determined in the case of equal sample sizes. This gives the Gnedenko–Koroljuk [3] distributions for discontinuous random variables; letting n tend to infinity we obtain an elementary and simple method of determining the limiting distributions; however, we come to a different form from that given by Schmid. Although our formula has perhaps a slightly more complicated structure the domain of integration is identical for each term and at the same time much simpler; therefore it seems to be more appropriate for theoretical investigations. Furthermore, our treatment admits a certain generalization, making possible the determination of the probability of the event.

$$\{y_1(x) < F_n(x) - F(x) < y_2(x), \ -\infty < x < \infty\},$$

where $y_i(x)$ $(y_1(\pm\infty) < 0 < y_2(\pm\infty))$ are step functions.

0.3. In his book [8] Takács applies his method to the determination of probabilities of events related to certain stochastic processes; in particular, he derives with its help many one-sided one- and two-sample statistics. With this method proofs of the corresponding distribution laws are obtained in a very elementary way and in simple form. The present author wishes to call attention to the fact that by means of this method the Smirnov–Birnbaum–Tingey [7, 1] distribution, for example, may be obtained in an elementary way, appropriate for teaching at the very beginning of probabilistic studies; this circumstance is not obvious from the works of Takács. His method is based on the so-called 'ballot lemma'. The author wishes to point out that a certain generalization of the ballot lemma, proved by G. Tusnády and K. Sarkadi, admits a further simplification of the proofs of such types of distribution theorems. A more general ballot lemma due to E. Csáki will be mentioned too; this is also a generalization of a result of Nef [5]. The last part of the paper does not contain the proofs of all theorems mentioned therein.

1   THE DISTRIBUTION OF $D_{n,n}^{+}$ AND $D_{n,n}$ FOR DISCONTINUOUS DISTRIBUTION FUNCTIONS

1.1. Let X and Y be independent random variables with distribution functions $F(x)$ and $G(x)$ respectively. In the present paper the validity of the null hypothesis
$$H_0 : F(x) \equiv G(x)$$
will be assumed throughout. Let $F(x)$ have jumps at
$$x_1, x_2, \ldots, x_r \quad (-\infty < x_1 < x_2 < \ldots < x_r < \infty)$$
and be continuous otherwise. For the left, continuous distribution function $F(x)$ the following relations hold with $x_0 = -\infty$, $x_{r+1} = \infty$:
$$\left.\begin{aligned} F(x_i) - F(x_{i-1} + 0) &= p_i \quad (i = 1, 2, \ldots r+1), \\ F(x_i + 0) - F(x_i) &= q_i \quad (i = 1, 2, \ldots, r), \end{aligned}\right\} \tag{1}$$
and
$$\sum_{i=1}^{r+1} p_i + \sum_{i=1}^{r} q_i = 1.$$

Let us consider a sample $X_1, X_2, \ldots, X_n$ of size n on X and $Y_1, Y_2, \ldots, Y_n$ of the same size on Y. The corresponding empirical distribution functions —defined in the usual way—will be denoted by $F_n(x)$ and $G_n(x)$ respectively. The one- and two-sided statistics
$$D_{n,n}^{+} = \sup_{-\infty < x < \infty} (F_n(x) - G_n(x)),$$
$$D_{n,n} = \sup_{-\infty < x < \infty} |F_n(x) - G_n(x)|$$
will be considered.

The event $\{D_{n,n}^{+} < k/n\}$, k being a non-negative integer, means that the random process
$$\zeta_n(x) = F_n(x) - G_n(x) \quad (-\infty < x < \infty)$$

lies below height k/n throughout. Assume that the X_i's and Y_j's lie on the real axis in such a way that out of the X_i's ν_i will be found inside the interval $(x_{i-1}, x_i), i = 1, 2, \ldots, r+1$, while the event $X_n = x_i$ occurs α_i times $(i = 1, 2, \ldots, r)$, with

$$\sum_{i=1}^{r+1} \nu_i + \sum_{i=1}^{r} \alpha_i = n.$$

The corresponding quantities for the second sample will be denoted by μ_i and β_i for which

$$\sum_{i=1}^{r+1} \mu_i + \sum_{i=1}^{r} \beta_i = n$$

holds. Obviously

$$n\zeta_n(x_i) = \sum_{j=1}^{i} \nu_j - \sum_{j=1}^{i} \mu_j + \sum_{j=1}^{i-1} \alpha_j - \sum_{j=1}^{i-1} \beta_j$$

and

$$n\zeta_n(x_i+0) = \sum_{j=1}^{i} \nu_j - \sum_{j=1}^{i} \mu_j + \sum_{j=1}^{i} \alpha_j - \sum_{j=1}^{i} \beta_j.$$

The following notations will be used

$$\left. \begin{array}{l} n_0 = 0, \quad \sum_{j=1}^{i} \nu_j = n_i \quad (i = 1, 2, \ldots, r+1), \\[2ex] m_0 = 0, \quad \sum_{j=1}^{i} \mu_j = m_i \quad (i = 1, 2, \ldots, r+1), \\[2ex] a_0 = 0, \quad \sum_{j=1}^{i} \alpha_j = a_i \quad (i = 1, 2, \ldots, r), \\[2ex] b_0 = 0, \quad \sum_{j=1}^{i} \beta_j = b_i \quad (i = 1, 2, \ldots, r), \\[2ex] n_{r+1} + a_r = n, \\[1ex] m_{r+1} + b_r = n, \end{array} \right\} \qquad (2)$$

furthermore
$$\{\nu\} = (\nu_1, \nu_2, \ldots, \nu_{r+1}),$$
$$\{\mu\} = (\mu_1, \mu_2, \ldots, \mu_{r+1}),$$
$$\{\alpha\} = (\alpha_1, \alpha_2, \ldots, \alpha_r),$$
$$\{\beta\} = (\beta_1, \beta_2, \ldots, \beta_r).$$

The events $\qquad A_k^{(i)} = \{n\zeta_n(x) < k, \quad x_{i-1} < x \leqslant x_i\}$

for $i = 1, 2, \ldots, r+1$ will be considered under the condition

$$B_{\nu, \mu}^{\alpha, \beta} = \{\{\nu\}, \{\mu\}, \{\alpha\}, \{\beta\}\}.$$

But under the stated condition the events $A_k^{(i)}$ for $i \dots 1, 2, \dots, r+1$ are independent. Consequently the relation

$$\mathbf{P}\left(D_{n,n}^+ < \frac{k}{n}\right) = \sum_{\{v\}} \sum_{\{\mu\}} \sum_{\{\alpha\}} \sum_{\{\beta\}} P\left(D_{n,n}^+ < \frac{k}{n}\middle| B_{v,\mu}^{\alpha,\beta}\right) P(B_{v,\mu}^{\alpha,\beta})$$

$$= \sum_{\{v\}} \sum_{\{\mu\}} \sum_{\{\alpha\}} \sum_{\{\beta\}} \prod_{i=1}^{r+1} P(A_k^{(i)}|B_{v,\mu}^{\alpha,\beta}) P(B_{v,\mu}^{\alpha,\beta})$$

must hold. The summation is to be taken for all those systems of the indices, for which the following conditions are fulfilled:

$$\left.\begin{array}{l} v_i \geqslant 0, \quad \mu_i \geqslant 0 \quad (i = 1, 2, \dots, r+1), \\[4pt] \alpha_i \geqslant 0, \quad \beta_i \geqslant 0 \quad (i = 1, 2, \dots, r), \quad \text{integers} \\[4pt] \displaystyle\sum_{i=1}^{r+1} v_i + \sum_{i=1}^{r} \alpha_i = n, \quad \sum_{i=1}^{r+1} \mu_i + \sum_{i=1}^{r+1} \beta_i = n, \end{array}\right\} \tag{3}$$

furthermore

$$n\zeta_n(x_{i-1}+0) = n_{i-1} - m_{i-1} + a_{i-1} - b_{i-1} < k,$$

$$n\zeta_n(x_i) = n_i - m_i + a_{i-1} - b_{i-1} < k,$$

$$(i = 1, 2, \dots, r+1).$$

The case of the event $B_{v,\mu}^{\alpha,\beta}$ leads to the multinomial distribution; the two samples being independent, we have

$$P(B_{v,\mu}^{\alpha,\beta}) = \frac{n!}{\displaystyle\prod_{i=1}^{r+1} v_i! \prod_{i=1}^{r} \alpha_i!} \frac{n!}{\displaystyle\prod_{i=1}^{r+1} \mu_i! \prod_{i=1}^{r} \beta_i!} \prod_{i=1}^{r+1} p_i^{v_i+\mu_i} \prod_{i=1}^{r} q_i^{\alpha_i+\beta_i}.$$

Finally the probabilities $\qquad P(A_k^{(i)}|B_{v,\mu}^{\alpha,\beta})$

are to be determined. Under the given condition there are $v_i X_j$'s and $\mu_i Y_l$'s in the interval $(X_{i-1} < X \leqslant X_i)$ all being independent and identically distributed. Consequently each array of the X_j's and Y_l's is of the same probability

$$\binom{v_i + \mu_i}{v_i}^{-1}.$$

The number of arrays for which the event $A_k^{(i)}$ takes place can be obtained by well-known random walk methods: the number of paths is to be determined for a simple random walk starting at height

$$s_0 = n_{i-1} - m_{i-1} + a_{i-1} - b_{i-1}$$

and arriving after $v_i + \mu_i$ steps at height $s_{v_i+\mu_i} = n_i - m_i + a_{i-1} - b_{i-1}$ without ever before reaching height k

$$s_j < k \quad (j = 0, 1, 2, \dots, v_i + \mu_i).$$

The use of the reflection principle results in the formula:

$$P(A_k^{(i)} | B_{\nu,\mu}^{\alpha,\beta}) = 1 - \frac{\binom{\nu_i + \mu_i}{n_i - m_{i-1} + a_{i-1} - b_{i-1} - k}}{\binom{\nu_i + \mu_i}{\nu_i}} \qquad (i = 1, 2, \ldots, r+1).$$

We are now in the position to formulate the following theorem, the proof of which requires but a complementary remark for the two-sided case:

Theorem 1. Having two samples of the same size n from populations with common discontinuous distribution functions given in (1) and denoting by $F_n(x)$ and $G_n(x)$ the corresponding empirical distribution functions, for the distributions of the maximal deviations

$$D_{n,n}^+ = \sup_{-\infty < x < \infty} (F_n(x) - G_n(x)),$$

and

$$D_{n,n} = \sup_{-\infty < x < \infty} |F_n(x) - G_n(x)|$$

we have

$$D\left(D_{n,m}^+ < \frac{k}{n}\right) = \sum_{\{\nu\}} \sum_{\{\mu\}} \sum_{\{\alpha\}} \sum_{\{\beta\}} \prod_{i=1}^{r+1} \left(1 - \frac{\binom{\nu_i + \mu_i}{n_i - m_{i-1} + a_{i-1} - b_{i-1} - k}}{\binom{\nu_i + \mu_i}{\nu_i}}\right)$$

$$\times \frac{(n!)^2 \prod\limits_{i=1}^{r+1} p_i^{\nu_i + \mu_i} \prod\limits_{i=1}^{r} q_i^{\alpha_i + \beta_i}}{\prod\limits_{i=1}^{r+1} (\nu_i! \mu_i!) \prod\limits_{i=1}^{r} (\alpha_i! \beta_i!)}, \qquad (4)$$

$k = 0, 1, 2, \ldots, n$, *where the notation (2) is used; the summation is to be taken over the lattice points of the 4r-dimensional space with the restrictions given in (3). For the two-sided case we have*

$$P\left(D_{n,n} < \frac{k}{n}\right) = \sum_{\{\nu\}} \sum_{\{\mu\}} \sum_{\{\alpha\}} \sum_{\{\beta\}} \prod_{i=1}^{r+1} \left\{ \sum_{\gamma = -\infty}^{\infty} \left[\binom{\nu_i + \mu_i}{\nu_i + 2\gamma k} \right. \right.$$

$$\left. \left. - \binom{\nu_i + \mu_i}{n_i - m_{i-1} + a_{i-1} - b_{i-1} + (2\gamma + 1) k} \right] \frac{1}{\binom{\nu_i + \mu_i}{\nu_i}} \right\}$$

$$\times \frac{(n!)^2 \prod\limits_{i=1}^{r+1} p_i^{\nu_i + \mu_i} \prod\limits_{i=1}^{r} q_i^{\alpha_i + \beta_i}}{\prod\limits_{i=1}^{r+1} (\nu_i! \mu_i!) \prod\limits_{i=1}^{r} (\alpha_i! \beta_i!)} \qquad (k = 2, 3, \ldots, n), \qquad (5)$$

where for the summation the domain given in (3) is to be modified by replacing the last two rows by

$$-k < n_{i-1} - m_{i-1} + a_{i-1} - b_{i-1} < k \quad (i = 1, 2, ..., r+1)$$

and $\quad -k < n_i - m_i + a_{i-1} - b_{i-1} < k \quad (i = 1, 2, ..., r+1).$

In order to complete the proof we have to take into account only that the probability of the event

$$\{-k < n\zeta_n(x) < k, x_{i-1} < x \leqslant x_i\}$$

under condition $B_{\nu, \mu}^{\alpha, \beta}$ can be determined by making use of a random walk with two barriers. Applying a formula of Ellis (see e.g. [4]), exactly appropriate for the present case we come to our above result.

1.2. A slight modification in these considerations—more precisely, a certain change in the combinatorial formulae derived with the aid of the random walk model—allow us to obtain the probabilities of the following events:

$$\{k_{i1} < n(F_n(x) - G_n(x)) < k_{i2}, x_{i-1} < x \leqslant x_i\} \quad (i = 1, 2, ..., r+1)$$

with

$$k_{i2} - k_{2i} \geqslant 3 \quad (i = 2, 3, ..., r, k_{11} < -1, k_{21} > 1, k_{r+1,1} < -1, k_{r+1,2} > 1)$$

and the intervals $(k_{i-1,1}, k_{i-1,2})$ and $(k_{i,1}, k_{i,2})$ having a common part at least of lengths 2 for $i = 2, 3, ..., r$. We do not go into details here. The same applies for the limiting distributions given in the following section.

2   LIMITING DISTRIBUTION LAWS

We turn now to the proof of the following theorem:

Theorem 2. Using the notations given in §1, the following limiting relations are valid: $(y \geqslant 0)$

$$\lim_{n \to \infty} P\left(\sqrt{\frac{n}{2}} D_{n,n}^+ < y \right)$$

$$= \frac{1}{(2\pi)^r \sqrt{p_{r+1}}} \int_{G^+} \cdots \int \prod_{i=1}^{r+1} \left(1 - \exp\left[-\frac{2}{p_i} (y - S_{i-1} - T_{i-1})(y - S_i - T_i) \right] \right)$$

$$\times \exp\left[-\frac{1}{2} \sum_{i=1}^{r} (u_i^2 + w_i^{-2}) - \frac{1}{2p_{r+1}} (S_r + T_r)^2 \right] \prod_{i=1}^{r} du_i dw_i$$

where $\qquad S_0 = 0, \quad S_i = \sum_{j=1}^{i} u_j \sqrt{p_j} \quad (i = 1, 2, ..., r),$

$$T_0 = 0, \quad T_i = \sum_{j=1}^{i} w_j \sqrt{q_j} \quad (i = 1, 2, ..., r),$$

and for the domain of integration we have

$$G^+ = \{S_{i-1}+T_{i-1} < y, S_i+T_{i-1} < y;\ (i = 1, 2, ..., r)\}.$$

Furthermore—with $y > 0$—

$$\lim_{n\to\infty} P\left(\sqrt{\frac{n}{2}} D_{n,n} < y\right)$$

$$= \frac{1}{(2\pi)^r \sqrt{p_{r+1}}} \int \cdots \int_G \prod_{i=1}^{r+1} \left\{ \sum_{\gamma=-\infty}^{\infty} \left(\exp\left[-\frac{2}{p_i} 2\gamma y(2\gamma y - u_i\sqrt{p_i}) \right] \right.\right.$$

$$\left.\left. - \exp\left[-\frac{2}{p_i} [(2\gamma+1)y + S_i + T_{i-1}][(2\gamma+1)y + S_{i-1} + T_{i-1}] \right] \right) \right\}$$

$$\times \exp\left[-\frac{1}{2} \sum_{i-1}^{r} (u_i^2+w_i^2) = \frac{1}{2p_{r+1}} (T_r+S_r)^2 \right] \prod_{i-1}^{r} du_i\, dw_i$$

where S_i and T_i have the same meaning as above, while for the integration domain we have

$$G = \{-y < S_{i-1}+T_{i-1} < y,\ -y < S_i+T_{i-1} < y;\ (i = 1, 2, ..., r)\}.$$

Proof. From formula (3) our first statement concerning $\sqrt{\frac{1}{2}n} D_{n,n}^+$ will be derived, making use of Stirling's formula, of certain integral transformations and of the following simple asymptotic relation, the proof of which is also a straightforward application of Stirling's theorem, and which will be omitted here: Let $a > 0$, b, c, d, real; then for sufficiently large n the asymptotic relation

$$\frac{\binom{2an+b\sqrt{n}}{an+c\sqrt{n}}}{\binom{2an+b\sqrt{n}}{an+d\sqrt{n}}} \sim e^{-c^2+d^2+bc-bd} \tag{6}$$

holds.

Turning to relation (3) of Theorem 1 the continuous variables

$$u_i, v_i \quad (i = 1, 2, ..., r+1), \quad w_i, z_i \quad (i = 1, 2, ..., r)$$

will be introduced in the following way:

$$\left.\begin{array}{l} v_i \sim np_i + u_i\sqrt{(np_i)}, \quad \alpha_i \sim nq_i + w_i\sqrt{(nq_i)}, \\ \mu_i \sim np_i + v_i\sqrt{(np_i)}, \quad \beta_i \sim nq_i + z_i\sqrt{(nq_i)}. \end{array}\right\} \tag{7}$$

Taking into account the relations

$$\sum_{i=1}^{r+1} v_i + \sum_{i=1}^{r} \alpha_i = \sum_{i=1}^{r+1} \mu_i + \sum_{i=1}^{r} \beta_i = n,$$

we obtain $\quad \sum_{i=1}^{r+1} u_i\sqrt{p_i} + \sum_{i=1}^{r} w_i\sqrt{q_i} = \sum_{i=1}^{r+1} v_i\sqrt{p_i} + \sum_{i=1}^{r} z_i\sqrt{q_i} = 0. \tag{8}$

Consequently u_{r+1} can be expressed in terms of the other $2r$ variables; a similar statement holds for v_{r+1}. Furthermore, from the above relations we have

$$du_i = dv_i \sim \frac{1}{\sqrt{p_i}}\frac{1}{\sqrt{n}} \quad (i = 1, 2, \ldots, r),$$

$$dw_i = dz_i \sim \frac{1}{\sqrt{q_i}}\frac{1}{\sqrt{n}} \quad (i = 1, 2, \ldots, r).$$

Making use of the Stirling formula and the above substitutions we obtain

$$\frac{p_i^{\nu_i}}{\nu_i!} \sim \left(e\frac{p_i}{\nu_i}\right)^{\nu_i}\frac{1}{\sqrt{(2\pi\nu_i)}} \sim \left(\frac{e}{n}\right)^{\nu_i}\left(1+\frac{u_i\sqrt{p_i}}{\sqrt{n}}\right)^{-np_i-u_i\sqrt{np_i}}$$

$$\times\frac{1}{\sqrt{(2\pi np_i)}} \sim \frac{du_i}{\sqrt{(2\pi)}}\left(\frac{e}{n}\right)^{\nu_i}\exp\left[-(np_i+u_i\sqrt{(np_i)})\log\left(1+\frac{u_i\sqrt{p_i}}{\sqrt{n}}\right)\right]$$

$$\sim \left(\frac{e}{n}\right)^{\nu_i}\frac{1}{\sqrt{(2\pi)}}\exp\left[-u_i\sqrt{(np_i)}-\tfrac{1}{2}u_i^2\right]du_i.$$

Applying similar transformations for

$$\frac{q_i^{\mu_i}}{\mu_i!}, \quad \frac{q_i^{\alpha_i}}{\alpha_i!} \quad \text{and} \quad \frac{q_i^{\beta_i}}{\beta_i!}$$

and taking into account (8) the following relation will be obtained:

$$\frac{(n!)^2\prod\limits_{i=1}^{r+1}p_i^{\nu_i+\mu_i}\prod\limits_{i=1}^{r}q_i^{\alpha_i+\beta_i}}{\prod\limits_{i=1}^{r+1}(\nu_i!\,\mu_i)!\prod\limits_{i=1}^{r}(\alpha_i!\,\beta_i!)} \sim \frac{1}{(2\pi)^{4r}\,p_{r+1}}$$

$$\times\exp\left[-\frac{1}{2}\sum_{i=1}^{r+1}(u_i^2+v_i^2)-\frac{1}{2}\sum_{i=1}^{r}(w_i^2+z_i^2)\right]\prod_1^{r}du_i\,dv_i\,dw_i\,dz_i. \tag{9}$$

Turning to the 'random walk' term and replacing the corresponding quantities from (7) we obtain

$$H_i = \frac{\dbinom{\nu_i+\mu_i}{n_i-m_{i-1}+a_{i-1}-b_{i-1}-k}}{\dbinom{\nu_i+\mu_i}{\nu_i}}$$

$$\sim \frac{\dbinom{2np_i+(u_i+v_i)\sqrt{(p_in)}}{np_i+\left[u_i\sqrt{(p_i+\sum\limits_{j=1}^{i-1}(u_j-v_j)\sqrt{p_j}+\sum\limits_{j=1}^{i-1}(w_j-z_j)\sqrt{q_j}-y\sqrt{2}\right]\sqrt{n}}}{\dbinom{2np_i+(u_i+v_i)\sqrt{(p_in)}}{np_i+u_i\sqrt{(p_in)}}}.$$

Making use of the notation

$$S_0 = 0, \quad S_i = \sum_{j=1}^{i} (u_j - v_j) \sqrt{p_j} \quad (i = 1, 2, \ldots, r+1),$$

$$T_0 = 0, \quad T_i = \sum_{j=1}^{i} (w_j - z_j) \sqrt{q_j} \quad (i = 1, 2, \ldots, r),$$

and applying relation (6) after a slight calculation we obtain for H_i

$$H_i \sim \exp\left[-\frac{1}{p_i} (y\sqrt{2} - S_{i-1} - T_{i-1})(y\sqrt{2} - S_i - T_{i-1}) \right]. \tag{10}$$

The product of the right sides in (9) and (10) yields the required asymptotic expression for $P(\sqrt{\tfrac{1}{2}n}\, D_{n,n}^+ < y)$ if a $4r$-dimensional integral is taken with domain of integration corresponding to the domain of summation in (3):

$$
\left.
\begin{aligned}
&\frac{1}{(2\pi)^{2r} p_{r+1}} \int \ldots \int_{G_0^+} \prod_{i=1}^{r+1} \\
&\times \left(1 - \exp\left[-\frac{1}{p_i} (y\sqrt{(2 - S_{i-1})} - T_{i-1})(y\sqrt{(2 - S_i)} - T_{i-1}) \right] \right) \\
&\times \exp\left[-\frac{1}{2} \sum_{1}^{r+1} (u_i^2 + v_i^2) - \frac{1}{2} \sum_{i=1}^{r} (w_i^2 + z_i^2) \right] \prod_{i=1}^{r} du_i\, dv_i\, dw_i\, dz_i, \\
&G_0^+ = \{ S_{i-1} + T_{i-1} < y\sqrt{2},\, S_i + T_{i-1} < y\sqrt{2};\, (i = 1, 2, \ldots, r+1) \}.
\end{aligned}
\right\} \tag{11}
$$

Let us now apply the following orthonormal transformation to the $4r$-dimensional space:

$$
\left.
\begin{aligned}
u_i' &= \frac{u_i - v_i}{\sqrt{2}}, \quad v_i' = \frac{u_i + v_i}{\sqrt{2}}, \\
w_i' &= \frac{w_i - z_i}{\sqrt{2}}, \quad z_i' = \frac{w_i + z_i}{\sqrt{2}},
\end{aligned}
\right\} \quad (i = 1, 2, \ldots, r).
$$

Then $\qquad \prod_{i=1}^{r} du_i\, dv_i\, dw_i\, dz_i = \prod_{i=1}^{r} du_i'\, dv_i'\, dw_i'\, dz_i'$

holds, furthermore

$$\sum_{i=1}^{r} (u_i'^2 + v_i'^2) = \sum_{i=1}^{r} (u_i^2 + v_i^2), \quad \sum_{i=1}^{r} (w_i'^2 + z_i'^2) = \sum_{i=1}^{r} (w_i^2 + z_i^2)$$

and $\qquad S_i = \sum_{j=1}^{i} (u_j - v_j)\sqrt{p_j} = \sqrt{2} \sum_{j=1}^{i} u_j' \sqrt{p_j} = \sqrt{2}\, S_i',$

$$T_i = \sum_{j=1}^{i} (w_j - z_j)\sqrt{q_j} = \sqrt{2} \sum_{j=1}^{i} w_j' \sqrt{q_j} = \sqrt{2}\, T_i'$$

finally

$$p_{r+1}(u_{r+1}^2 + v_{r+1}^2) = \left[\sum_{i=1}^{r} (u_i\sqrt{p_i} + w_i\sqrt{q_i})\right]^2 + \left[\sum_{i=1}^{r} (v_i\sqrt{p_i} + z_i\sqrt{q_i})\right]^2$$

$$= 2^{-1}\left[S_r' + T_r' + \sum_{i=1}^{r} (v_i'\sqrt{p_i} + z_i'\sqrt{q_i})\right]^2 + 2^{-1}\left[S_r' + T_r' - \sum_{i=1}^{r} (v_i'\sqrt{p_i} + z_i'\sqrt{q_i})\right]^2$$

$$= (S_r' + T_r')^2 + \left[\sum_{i=1}^{r} (v_i'\sqrt{p_i} + z_i'\sqrt{q_i})\right]^2.$$

Making use of the above relations, the limiting distribution function of $\sqrt{\tfrac{1}{2}n}\,D_{n,n}^+$ will take on the form

$$\frac{1}{(2\pi)^{2r}p_{r+1}} \int_{G_1^+} \ldots \int \prod_{i=1}^{r+1} \left(1 - \exp\left[-\frac{2}{p_i}(y - S_{i-1}' - T_{i-1}')(y - S_i' - T_{i-1}')\right]\right)$$

$$\times \exp\left\{\left[-\frac{1}{2}\sum_1^r (u_i'^2 + v_i'^2 + w_i'^2 + z_i'^2)\right] - \frac{1}{2p_{r+1}}\left[\sum_1^r (v_i'\sqrt{p_i} + z_i'\sqrt{q_i})\right]^2\right.$$

$$\left. - \frac{1}{2p_{r+1}}(S_r' + T_r')^2\right\} \prod_1^r du_i' dv_i' dw_i' dz_i'. \quad (12)$$

As the variables v_i', z_i' $(i = 1, 2, \ldots, r)$ are orthogonal to $u_i' - s$ and $w_i' - s$, the domain of integration does not depend on them in the sense:

$$G_1^+ = \begin{cases} -\infty < v_i' < \infty, & -\infty < z_i' < \infty \quad (i = 1, 2, \ldots, r), \\ S_{i-1}' + T_{i-1}' < y, & S_{i+}' T_{i-1}' < y \quad (i = 1, 2, \ldots, r). \end{cases}$$

On the other hand these variables can be separated; denoting by R_{2r} the Euclidean $2r$-space, the part of the integral in which they occur has the form

$$\frac{1}{(2\pi)^r \sqrt{p_{r+1}}} \int_{R_{2r}} \ldots \int \exp\left\{\left[-\frac{1}{2}\sum_1^r (v_i'^2 + z_i'^2)\right]\right.$$

$$\left. - \frac{1}{2p_{r+1}}\left[\sum_1^r (v_i'\sqrt{p_i} + z_i'\sqrt{q_i})\right]^2\right\} \prod_1^r dv_i' dz_i'.$$

Using an orthonormal transformation with the first variable

$$t = \frac{1}{1 - p_{r+1}}\left[\sum_1^r (v_i'\sqrt{p_i} + \sum_1^r z_i'\sqrt{q_i})\right],$$

$t_2, t_3, \ldots, t_{2r}$ being arbitrary, with

$$\sum_1^{2r} t_i^2 = \sum_1^r (v_i'^2 + z_i'^2), \quad \prod_1^{2r} dt_i = \prod_1^r dv_i' dz_i'.$$

The above integral takes on the form and the value

$$\frac{1}{(2\pi)^r \sqrt{p_{r+1}}} \int \ldots \int_{R_{2r}} \exp\left[-\frac{1}{2} \sum_1^{2r} t_i^2 - \frac{1-p_{r+1}}{2p_{r+1}} t_1^2 \right] \prod_1^{2r} dt_i$$

$$= \frac{1}{\sqrt{2\pi}\sqrt{p_{r+1}}} \int_{-\infty}^{\infty} \exp\left[-\frac{1}{2} \frac{t_1^2}{p_{r+1}} \right] dt_1 = 1.$$

Having only the remainder part of the integral given in (12) we come to the proof of Theorem 2 in the one-sided case. The two-sided case differs only by the determination of the limiting form of the terms of the infinite sum in (5); here again these terms do not depend on the v_i''s and z_i'''s, consequently the same procedure applies for this case as well.

2.2. Theorems 1 and 2 have a further interpretation: Let us assume that the distribution function $F(x)\,(\equiv G(x))$ is continuous, the events

$$\left\{ F_n(x) - G_n(x) < \frac{k}{n} \right\}, \quad \left\{ -\frac{k}{n} < F_n(x) - G_n(x) < \frac{k}{n} \right\}$$

are considered in consecutive intervals (x_{2i}, x_{2i+1}) only with

$$F(x_{2i+1}) - F(x_{2i}) = p_{i+1}, \quad x_0 = -\infty, \quad x_{2r+1} = \infty$$

and the process $\zeta_n(x)$ may be arbitrary within the intermediate intervals (x_{2i-1}, x_{2i}) with

$$F(x_{2i}) - F(x_{2i-1}) = q_i \quad (i = 1, 2, \ldots, r).$$

Then our distribution theorems give the probabilities for the corresponding events. The limiting distribution theorems have analogous interpretations for the continuous Gaussian stochastic process $\zeta(x)$ which satisfies the conditions

$$E(\zeta(x)) = 0 \quad (-\infty < x < \infty),$$

$$D^2(\zeta(x)) = F(x)(1 - F(x)) \quad (-\infty < x < \infty),$$

$$E[\zeta(x)\,\zeta(x')] = F(x)(1 - F(x')) \quad (-\infty < x < x' < \infty),$$

$\{\zeta(x^{(1)}), \zeta(x^{(2)}), \ldots, \zeta(x^{(s)})\}$ is distributed according to the s-dimensional Gaussian law with the above parameters for any $s = 1, 2, \ldots$ and for any real vector $(x^{(1)}, x^{(2)}, \ldots, x^{(s)})$. (See also (9).)

3 THE PROOF OF THE SMIRNOV–BIRNBAUM–TINGEY THEOREM AND OF RELATED DISTRIBUTION LAWS BY APPLYING THE BALLOT LEMMA

3.1. As mentioned in 0.3, the ballot lemma enables us to give elementary and very simple proofs for many theorems concerning the one-sided deviation

$$D_n^+ = \sup_{-\infty < x < \infty} (F_n(x) - F(x))$$

and for related statistics where $F(x)$ denotes a continuous distribution function and $F_n(x)$ is the empirical distribution function belonging to a sample of size n taken from a population having distribution function $F(x)$. The method was introduced and carried out for several special cases by L. Takács (see [8], pp. 162–87). In his investigations the simplicity and elegance of the proofs are slightly impaired by the formulation of rather general theorems and their application to different special cases. Our intention is now to point out the simplicity of the proof when directly applying the ballot lemma and to show that an elementary modification of this lemma leads to further simplifications.

We turn now to the formulation of the ballot lemma.

Ballot lemma (L. Takács): *Let $k_1, k_2, \dots, k_n$ be non-negative integers with $k_1 + k_2 + \dots + k_n = k \leqslant n$. Among the n cyclic permutations of $(k_1, k_2, \dots, k_n)$ there are exactly $n - k$ for which the sum of the first r elements is less than, r for all $r = 1, 2, \dots, n$.*

For the proof we have to consider only the plot of the $2n$ points $(i; k_1 + k_2 + \dots + k_i)$, $i = 1, 2, \dots, 2n$, with $k_{j+n} = k_j$, in a co-ordinate system (Fig. 1). A glance at this figure makes the proof obvious: regarding the index i^* of the last maximum of the function $k_1 + k_2 + \dots + k_i - i$, it is clear that preceding this, no 'good' starting index can exist, i.e. no index for which the consecutive partial sums satisfy the property required by the lemma; the same is true for indices large than or equal to $i^* + n$. But let us consider the straight lines of slope 45° through the $(n - k)$ points with co-ordinates:

$$\left(i^* + n, \sum_{i=1}^{i^*} k_i - j \right) \quad (j = n, n - 1, \dots, n - k + 1).$$

These intersect our configuration in points belonging to certain abscissas; let the last indices of intersections be $j_1, j_2, \dots, j_{n-k}$ for each of these lines. Then the indices $j_h + 1, h = 1, 2, \dots, n - k$ will be 'good' starting indices. This is the essential idea of the proof given by Takács but more illustratively formulated than by him, it applies for his continuous version of the lemma as well (see [8], p. 1).

This lemma, together with the binomial, or multinomial law, provides tools for determining the Smirnov–Birnbaum–Tingey distribution. We turn now to an extension of this lemma, formulating the extended ballot lemma.

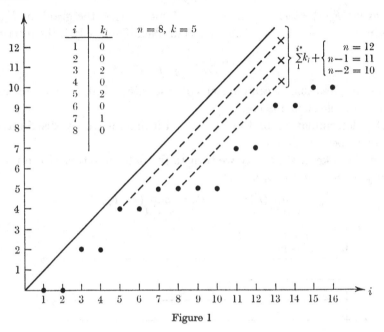

Figure 1

Extended ballot lemma (G. Tusnády). *Let $A_0, A_1, A_2, ..., A_n$ be a complete system of events, for $P(A_0) = q$ and $P(A_j) = p, j = 1, 2, ..., n$ holds. Making n independent observations, let ν_i be the number of cases in which A_i occurred. Then*

$$P\left(\sum_{i=1}^{j} \nu_i < j : j = 1, 2, ..., n \right) = q.$$

This lemma can be obtained from the ballot lemma given above, but direct, elementary proofs were also given by G. Tusnády and later by K. Sarkadi.

Sarkadi utilized the fact that this lemma is equivalent to the following, known theorem proved first by Daniels [2].

If $F_n(x)$ denotes the empirical distribution function corresponding to a sample of size n taken on a random variable distributed according to the uniform law in (0, 1), then

$$P\left(\frac{F_n(x) - F(x)}{F(x)} < y, 0 \leqslant x \leqslant 1 \right) = \frac{y}{y + 1} \quad (0 \leqslant y \leqslant \infty).$$

It is very easy to show that the two statements are equivalent: considering the straight line passing through the points $(0, 0)$ and

$$\left(u = \frac{y}{1+y}, \quad 1\right)$$

the event $F_n(x) < x(1+y)$ for $0 \leqslant x \leqslant 1$ means that the graph of $F_n(x)$ does not intersect this line; but this is fulfilled if and only if the interval

$$\left(0, v = \frac{1-u}{n}\right)$$

containing less than 1 sample element, the interval $(0, 2v)$ contains less than 2 sample elements etc.

The determination of the Smirnov–Birnbaum–Tingey distribution law now goes as follows:

For the sake of simplicity we turn to the determination of the probability of the event $(0 \leqslant y < 1)$

$$\{D_n^- = \sup_{0 \leqslant x \leqslant 1} (F(x) - F_n(x)) = \sup_{0 \leqslant x \leqslant 1} (x - F_n(x)) > y\}.$$

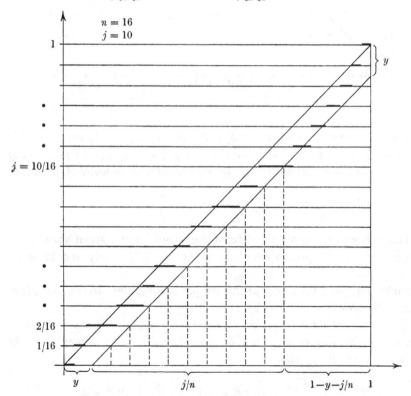

Figure 2

Suppose that the graph of $F_n(x)$ intersects the straight line

$$Y = x + y \quad (0 \leqslant x \leqslant 1, \quad 0 < y < 1)$$

at the jth level, i.e. at the height

$$Y = \frac{j}{n},$$

but not below it. This event takes place if and only if the following two independent events occur: (a) the interval

$$\left(y + \frac{j}{n}, \quad 1\right)$$

contains exactly $n - j$ sample elements, (b) the interval

$$\left(y + \frac{j-1}{n}, \quad y + \frac{j}{n}\right)$$

contains less than 1 sample element, the interval

$$\left(y + \frac{j-2}{n}, \quad y + \frac{j-1}{n}\right)$$

contains less than two sample elements etc. Having now the complete system of events: $A_i = \Big\{$a random point distributed uniformly in the interval

$$\left(0, y + \frac{j}{n}\right)$$

is contained in the interval

$$\left(y + \frac{i-1}{n}, \quad y + \frac{j}{n}\right)\Big\} \quad (i = 1, 2, ..., n)$$

and $A_0 = \Big\{$a random point distributed uniformly in the interval

$$\left(0, y + \frac{j}{n}\right) \quad \text{lies in} \quad (0, y)\Big\}$$

with

$$q = \frac{y}{y + \dfrac{j}{n}}, \quad p = \frac{1-q}{j}$$

the extended ballot lemma can be applied and it yields the value

$$\frac{y}{y + \dfrac{j}{n}}$$

for the probability of the event given in (b). The simultaneous occurrence of (a) and (b) has now the probability

$$\binom{n}{n-j}\left(1-y-\frac{j}{n}\right)^{n-j}\left(y+\frac{j}{n}\right)^{j}\frac{y}{y+\dfrac{j}{n}}.$$

This leads to the known distribution theorem:

$$P(D_n^- < y) = 1 - y \sum_{j=0}^{[n(1-y)]} \binom{n}{j}\left(y+\frac{j}{n}\right)^{j-1}\left(1-y-\frac{j}{n}\right)^{n-j} \quad (0 < y < 1).$$

Finally we mention the following theorem which is closely related to results of Nef [5]:

Generalized ballot lemma (E. Csáki): *Making use of the notation given in the extended ballot lemma let us denote by* λ *the number of indices* j *for which*

$$\sum_{i=1}^{j} r_i = j \text{ is fulfilled. Then}$$

$$P(\lambda \geqslant l) = l!\binom{n}{l}p^l \quad (l = 0, 1, 2, \ldots, n).$$

This lemma can be formulated also in the following way:

Let $F_n(x)$ *be the empirical distribution function belonging to a sample of size* n *on a random variable distributed uniformly in* $(0, 1)$, *let us denote by* $\overline{\lambda}$ *the number of (horizontal) intersections of the graph of* $F_n(x)$ *with the straight line*

$$y = \frac{1}{np}x \quad (np \leqslant 1),$$

then $$P(\overline{\lambda} \geqslant l) = l!\binom{n}{l}p^l \quad (l = 0, 1, 2, \ldots, n).$$

For $p = 1/n$ this result agrees with a theorem of Nef [5].

The proofs of the statements given in the present § 3 will be published by the authors cited.

ACKNOWLEDGEMENT

The author is very indebted to Professor H. A. David for his valuable comments.

REFERENCES

[1] Birnbaum, Z. W. and F. H. Tingey. One sided confidence contours for probability distribution functions. *Ann. Math. Statist.* **22** (1951) 592–6.
[2] Daniels, H. E. The statistical theory of the strengths of bundles of threads. I. *Proc. Roy. Soc.* A **183** (1945) 405–35.

[3] Gnedenko, B. V. and V. S. Koroljuk. On the maximum discrepancy between two empirical distribution functions. *IMS* and *AMS*. Selected Translations in Mathematical Statistics and Probability 1 (1961) 13–16.

[4] Jordan, Ch. Fejezetek a klasszikus valószinüségszámitásból. Budapest, 1956. (To appear in English.)

[5] Nef, W. Über die Differenz zwischen theoretischer und empirischer Verteilungsfunktion. *Zeitschr. für Wahrscheinlichkeitstheorie* 3 (1964) 154–62.

[6] Schmid, P. On the Kolmogorov and Smirnov limit theorems for discontinuous distribution functions. *Ann. Math. Statist.* 29 (1958) 1011–27.

[7] Smirnov, N. V. Approximate laws of distribution of random variables from empirical data. (Russian.) *Uspehi Mat. Nauk* 10 (1944) 179–206.

[8] Takács L. *Combinatorial methods in the theory of stochastic processes.* New York: John Wiley and Sons, Inc. 1967.

[9] Noether, G. E. Note on the Kolmogorov Statistics in the Discrete Case. *Metrika* 7 (1963) 115–16.

COMMENTS ON THE PAPERS BY
TAKÁCS AND VINCZE

H. A. DAVID

When agreeing to discuss the two papers just delivered, I did so on the strength of the prominence with which the words 'theory of order statistics' appeared in their titles.† I now find myself trapped by terminology. In Dr Takács' case, there admittedly was a warning in the form of his book but I was wholly unprepared, although perhaps I should not have been, to find Dr Vincze's paper also to be largely concerned with a treatment of Kolmogorov–Smirnov and related statistics. Indeed, Dr Vincze's paper contains no further mention of order statistics after the title! Is it too late to persuade the authors, especially Dr Vincze, to change the *titles* of their papers? Calculations of the single-sample Kolmogorov statistic δ_m^+ requires knowledge of the order statistics only up to a monotonic increasing transformation; the two-sample statistic $\delta^+(m, n)$ requires even less and is just a rank-order statistic. I do believe it is useful to make the distinction between order statistics, which involve the actual observations, and rank-order statistics, which depend on ranks only.

Dr Takács has prepared an elegant and polished paper consisting mainly of a masterly review of an area to which he has contributed significantly. It is impressive to see how much can be accomplished by an ingenious sustained use of essentially elementary combinatorial methods. For the purposes of the present paper Dr Takács has left his remarkable Theorem 1 in less general form than proved in his book and, in fact, stated by Dr Vincze as the ballot lemma (p. 396). From permutations of numbers it is a natural step to interchangeable (= exchangeable) variates. We have here incidentally an illustration of a fact which is well known but deserves to be still better known: much of the null distribution theory of nonparametric statistics requires only exchangeability of the underlying variates rather than independence, because what is needed usually is merely that all orderings of the variates be equiprobable. Thus δ_m^+ is completely robust 'against' exchangeability and $\delta^+(m, n)$ against exchangeability of all $m + n$ variates (unfortunately not against exchangeability among the ξ's and the η's separately).

† Dr Vincze's paper was originally entitled 'On some distribution laws in the theory of order statistics'.

Whereas Dr Takács considers one-sided one- and two-sample KS and related statistics, extending some results from the case $m = n$ to n an integral multiple of m, Dr Vincze deals with one- and two-sided two-sample statistics for $m = n$ when the common parent distribution $F(x)$ is discontinuous. If $F(x)$ has jumps at (see the figure)

$$x_1, x_2, \ldots, x_r \ (-\infty = x_0 < x_1 < \ldots < x_r < x_{r+1} = \infty)$$

with $\nu_i X$'s and $\mu_i Y$'s in (x_{i-1}, x_i) and $\alpha_i X$'s and $\beta_i Y$'s at x_i, then Dr Vincze points out that conditionally on $\{\boldsymbol{\nu}, \boldsymbol{\mu}, \boldsymbol{\alpha}, \boldsymbol{\beta}\}$ independent random walks are executed within the intervals (x_{i-1}, x_i). Well-known theory using the reflection principle may be applied and his Theorem 1 follows on unconditionalizing. These exact small-sample results as well as the corresponding asymptotic results of Theorem 2 depend on $F(x)$ through the $p_i = \Pr\{x_{i-1} < X < x_i\}$ and $q_i = \Pr\{X = x_i\}$ $(i = 1, 2, \ldots, r+1)$. Since standard (i.e. continuous theory) tests of significance with $D_{n,n}^+$ and $D_{n,n}$ are conservative when $F(x)$ is discontinuous, the means have been provided to answer the question of just how conservative.

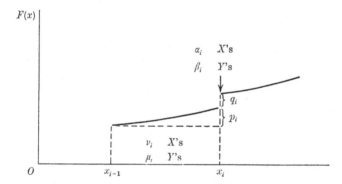

The final section of Dr Vincze's paper gives some alternative proofs of known ballot theorems and indicates extensions.

Clearly the two papers complement each other to an unusual degree. They are important contributions to nonparametric inference—rather than to order statistics.

PITMAN EFFICIENCIES OF TESTS
BASED ON SPACINGS

J. SETHURAMAN AND J. S. RAO

1 INTRODUCTION

Let $X_1, X_2, \ldots, X_{n-1}$ be $(n-1)$ independently and identically distributed random variables $(n > 2)$ with a common distribution function (df) $G(x)$. In this paper we are concerned with the Pitman efficiencies of tests based on spacings (see (1) below) in the goodness-of-fit problem, i.e. the problem of testing $G(x)$ is equal to a specified continuous df $G_0(x)$. In this case, the simple probability transformation on the random variables would permit us to equate $G_0(x)$ with the uniform df on $[0, 1]$. We assume that this is done, and so the null hypothesis states $G(x)$ = the uniform df on $[0, 1]$.

Let $X_1' \leqslant X_2' \leqslant \ldots \leqslant X_{(n-1)}'$ be the order statistics. The sample spacings $\{D_1, D_2, \ldots, D_n\}$ are defined by

$$D_i = X_i' - X_{i-1}' \quad (i = 1, \ldots, n), \qquad (1)$$

where we put $X_0' = 0$, $X_n' = 1$.

Some of the spacings tests for the goodness-of-fit proposed in the literature are those based on $V_r(n)$, $U(n)$ and $L(n)$ defined below. In each case the hypothesis is rejected when the absolute value of the statistic is large. The statistics

$$V_r(n) = \sum_1^n (nD_i)^r / n \quad (r > -\tfrac{1}{2})$$

were suggested by Kimball (1950)—of which the statistic suggested by Greenwood (1946) is the case with $r = 2$. Kendall in the discussion on Greenwood (1946) suggested the statistic

$$U(n) = \sum_1^n |nD_i - 1| / n$$

which was later studied by Sherman (1950). Darling (1953) proposed the statistic

$$L(n) = \sum_1^n \log (nD_i) / n.$$

In this paper we present a unified treatment of the computation of Pitman (asymptotic relative) efficiencies of such tests. Though it is known that the efficiency of any test symmetric in the spacings is zero relative to the Kolmogorov–Smirnov test (as shown for instance by Cibisov (1961))

[405]

it would be useful to know about the efficiency of one symmetric spacings test relative to another. Table 1 in § 3 gives the efficacies of the tests based on $V_r(n)$, $U(n)$ and $L(n)$ from which the relative efficiencies can be computed. Finally it is shown that among a large class of symmetric tests based on spacings the test based on $V_2(n)$ has maximum efficacy.

We now describe the alternative hypothesis. Being concerned with Pitman efficiencies, we will specify the alternative hypothesis by a df depending on n and converging to the null hypothesis. Under the alternative hypothesis,

$$G_n(x) = x + L_n(x)/n^\delta \quad (x \in [0, 1]), \tag{2}$$

where $L_n(0) = L_n(1) = 0$ and $\delta \geqslant \frac{1}{4}$. We further assume that $L_n(x)$ is twice differentiable on $[0, 1]$ and there is a function $L(x)$ which is twice continuously differentiable and

$$L(0) = L(1) = 0,$$

$$n^{\delta^*} \sup_{0 \leqslant x \leqslant 1} |L_n(x) - L(x)| = o(1),$$

$$n^{\delta^*} \sup_{0 \leqslant x \leqslant 1} |L_n'(x) - l(x)| = o(1),$$

$$n^{\delta^*} \sup_{0 \leqslant x \leqslant 1} |L_n''(x) - l'(x)| = o(1),$$

where $l(x)$ and $l'(x)$ are the first and second derivatives of $L(x)$ and $\delta^* = \max(0, \frac{1}{2} - \delta)$.

These sequence of alternatives are 'smooth' in a certain sense and have been considered before in this problem, for instance, see Cibisov (1961), Weiss (1965). These alternatives have carrier $[0, 1]$ just as the hypothesis. (The carrier of a distribution is the smallest closed set with probability 1.) This seems to be a natural condition on the alternatives if the test is to be based on spacings since the spacings, as defined in (1), do not make allowances for sub-intervals of $[0, 1]$ with probability zero and the intervals $(-\infty, 0)$ and $(1, \infty)$, if they have positive probability. Weiss (1962) has constructed an interesting example of a sequence of alternatives with non-constant carrier having limiting power 1 under a spacings test but which have limiting power equal to test size under the Kolmogorov–Smirnov test. We discuss this example again in § 3.

Previous approaches to the computation of Pitman efficiencies mostly used fixed alternatives. In this case, the limiting distributions of test statistics based on spacings become difficult to obtain. Some special cases have been treated in Weiss (1957), Pyke (1965), Jackson (1967). This still does not give Pitman efficiencies since the usual method of differentiating the asymptotic mean, etc., cannot in general be justified. Cibisov (1961) and Weiss (1965) met with success using alternatives of the form (2). Theorems 1 and 2 in § 2 give the limiting distribution of the empirical df of the 'normalized' and 'modified' spacings, where these terms are defined. In § 3 we show how to obtain Pitman efficiencies of tests based on spacings, including many of those proposed in the literature. One can also consider the problem of testing for uniformity of a distribution on a circle. In § 3 we remark that the asymptotic theory of circular spacing tests is the same as that of linear spacing tests. Thus Pitman efficiencies of circular spacings tests are the same as those of the corresponding linear spacings tests.

2 LIMITING DISTRIBUTION OF THE EMPIRICAL DF OF 'NORMALIZED' AND 'MODIFIED' SPACINGS

Under the null hypothesis, $E(D_i) = 1/n$ for all i. We will therefore call $\{nD_i, i = 1, ..., n\}$ 'normalized' spacings. Let

$$\{h_{n1}, h_{n2}, ..., h_{nn}\} \quad (n = 2, 3, ...),$$

be a triangular array of positive numbers. We shall call

$$\{nD_1/h_{n1}, nD_2/h_{n2}, ..., nD_n/h_{nn}\}$$

'modified' or 'adjusted' spacings. For example, one way of adjusting the spacings is to divide them by their expectations under some arbitrary df. This might help to enlarge the class of statistics based on spacings and may be likened to the method of using Normal scores and other scores in rank statistics.

We now define the empirical df's $H_n(x)$ and $H_n^*(x)$ of the normalized and modified spacings, respectively. Let

$$H_n(x) = \sum_1^n I(nD_i; x)/n \quad (x \geqslant 0) \tag{3}$$

and

$$H_n^*(x) = \sum_1^n I(nD_i/h_{ni}; x)/n \quad (x \geqslant 0), \tag{4}$$

where

$$I(z; x) = \begin{cases} 1 & z \leqslant x \\ 0 & z > x \end{cases}. \tag{5}$$

When we deal with modified spacings we shall always assume that there exists a continuous function $h(p)$ on $(0, 1)$ such that

$$\max_{1 \leqslant i \leqslant n} \sqrt{n} \, |h(i/(n+1)) - h_{ni}| = o(1)$$

and that, for each x,

$$\int_0^1 (1 - e^{-xh(p)})\, dp < \infty, \quad \int_0^1 e^{-xh(p)} l(p)\, h(p)\, dp < \infty,$$

$$\int_0^1 L(p)\, l'(p)\, h(p)\, e^{-xh(p)}\, dp < \infty \quad \text{and} \quad \int_0^1 h^2(p)\, l^2(p)\, e^{-xh(p)}\, dp < \infty.$$

Call this condition (M).

Define
$$\zeta_n(x) = \sqrt{n}(H_n(x) - F_n(x)) \quad (x \geqslant 0) \tag{6}$$

and
$$\zeta_n^*(x) = \sqrt{n}(H_n^*(x) - F_n^*(x)) \quad (x \geqslant 0), \tag{7}$$

where
$$F_n(x) = 1 - e^{-x} + \left(\int_0^1 l^2(p)\, dp\right) e^{-x}(x - x^2/2)/n^{2\delta} \tag{8}$$

and

$$F_n^*(x) = \int_0^1 (1 - e^{-xh(p)})\, dp + \left(\int_0^1 x e^{-xh(p)} l(p)\, h(p)\, dp\right)\Big/ n^\delta$$

$$+ \left(\int_0^1 [-xL(p)\, l'(p)\, h(p)\, e^{-xh(p)} - x^2 h^2(p)\, l^2(p)\, e^{-xh(p)}/2]\, dp\right)\Big/ n^{2\delta}. \tag{9}$$

Note that
$$\lim_{x \to \infty} \zeta_n(x) = 0 \quad \text{and} \quad \lim_{x \to \infty} \zeta_n^*(x) = 0.$$

We therefore put $\zeta_n(\infty) = \zeta_n^*(\infty) = 0$. Then, it is easy to see that

$$\{\zeta_n(x), 0 \leqslant x \leqslant \infty\} \quad \text{and} \quad \{\zeta_n^*(x), 0 \leqslant x \leqslant \infty\}$$

are measurable and are stochastic processes in $D[0, \infty]$ endowed with the Skorohod topology. The original Skorohod topology was for $D[0, 1]$, but is applicable to any $D(K)$, when there is a strictly increasing continuous transformation of $[0, 1]$ onto K.

We now state two theorems.

Theorem 1. *Under the alternatives in* (2), *the processes* $\{\zeta_n(x), 0 \leqslant x \leqslant \infty\}$ *converge weakly to a Gaussian process* $\{\zeta(x), 0 \leqslant x \leqslant \infty\}$ *in* $D[0, \infty]$ *having mean function zero and covariance function*

$$K(x, y) = e^{-y}(1 - e^{-x} - xy\, e^{-x}) \quad \text{for } 0 \leqslant x \leqslant y \leqslant \infty. \tag{10}$$

Theorem 2. *Under the alternatives* (2) *and under condition* (M), *the process* $\{\zeta_n^*(x), 0 \leqslant x \leqslant \infty\}$ *converges weakly to the Gaussian process*

$$\{\zeta^*(x),\ 0 \leqslant x \leqslant \infty\}$$

in $D[0, \infty]$ *with mean function zero and covariance function*

$$K^*(x, y) = \int_0^1 e^{-yh(p)}(1 - e^{-xh(p)})\, dp$$

$$- xy\left(\int_0^1 h(p)\, e^{-xh(p)}\, dp\right)\left(\int_0^1 h(p)\, e^{-yh(p)}\, dp\right) \quad \text{for } 0 \leqslant x \leqslant y \leqslant \infty. \tag{11}$$

Theorem 2 implies Theorem 1. Theorem 2 is no more difficult to prove than Theorem 1. The proof of these theorems are involved and will not be given here. They use the well-known transformation from general spacings to the uniform spacings and from the uniform spacings to ratios of exponential random variables to their sum. See for instance Pyke (1965), Weiss (1962). After this stage, we have to deal with the empirical distribution function of random variables with random perturbations and a random scale factor, i.e. with something like

$$\sum_{1}^{n} I(W_i/\theta_{ni} \, W_n^*; x)/n, \qquad (12)$$

where $W_1, \ldots, W_n$ are independently and identically distributed random variables, θ_{ni} are random variables approximately equal to $\theta(i/(n+1))$, where $\theta(p)$ is a well behaved function and W_n^* is a random variable with asymptotic normal distribution with mean 1 and variance $1/\sqrt{n}$. In a subsequent paper, Sethuraman and J. S. Rao (1969), we establish weak convergence of such empirical df's under more general conditions than required here. Since it seems to us that such results are of sufficient independent interest, we have relegated the proof of Theorems 1 and 2 to our forthcoming paper.

Theorems 1 and 2 allow us to obtain the limiting distributions of a host of functionals of $\zeta_n(x)$ and $\zeta_n^*(x)$, by just invoking the invariance principle. We do this in the next section and compute the Pitman efficiencies of several tests.

3 PITMAN EFFICIENCIES OF TESTS BASED ON SPACINGS

The Pitman asymptotic relative efficiency (ARE) of a test relative to another test is defined to be the limit of the inverse ratio of sample sizes required to obtain the same limiting power at a sequence of alternatives converging to the null hypothesis. This limiting power should be a value between the limiting test size, α, and the maximum power, 1. If the limiting power of a test at a sequence of alternatives is α, then its ARE with respect to any other test with the same test size and with limiting power greater than α, is zero. On the other hand, if the limiting power of a test at a sequence of alternatives converges to a number in the open interval $(\alpha, 1)$, then a measure of rate of convergence, called 'efficacy' can be computed. Under certain standard regularity assumptions (see e.g. Fraser (1957)), which include a condition about the nature of the alternative, asymptotic normal distribution of the test statistic under the sequence of alternatives, etc., this efficacy is given by

$$\text{efficacy} = \mu^2/\sigma^2. \tag{13}$$

Here μ and σ^2 are the mean and variance of the limiting normal distribution under the sequence of alternatives when the test statistic has been normalized to have a limiting standard normal distribution under the hypothesis. In such a situation, the ARE of one test with respect to another is simply the ratio of their efficacies.

We first consider tests statistics based on the normalized spacings. Let $m(x)$ be a function on $(0, \infty)$ and let

$$T_n = \sum_1^n m(nD_i)/n = \int_0^\infty m(x)\, dH_n(x). \tag{14}$$

Let $m(x)$ be absolutely continuous and of bounded variation in the interval $(\epsilon, 1/\epsilon)$ for each $\epsilon > 0$, and let

$$\int_0^1 m'(x)\sqrt{((1-e^{-x})\log\log(1-e^{-x})^{-1})}\, dx < \infty \tag{15}$$

and $$\int_0^\infty m'(x)\sqrt{(e^{-x}\log x)}\, dx < \infty. \tag{16}$$

Further let $$\int_0^\infty m(x)\, dF_n(x) < \infty \tag{17}$$

and $$\int_0^\infty \int_0^\infty m'(x)\, m'(y)\, K(x,y)\, dx\, dy < \infty, \tag{18}$$

where $F_n(x)$ and $K(x,y)$ are as in (8) and (10).

Conditions (15) through (18) are sufficient conditions for

$$\int_0^\infty m(x)\, dy\,(x), y(x) \in D$$

to be a functional on $D[0, \infty]$ which is continuous with probability 1 under $\{\zeta(x), 0 \leqslant x \leqslant \infty\}$. Conditions (15) and (16) are smoothness conditions on $m(x)$ based on the growth rate of $\zeta(x)$ at $x = 0$ and $x = \infty$ given by the law of the iterated logarithm while (17) and (18) make the mean and variance of T_n finite. Some examples of functions $m(x)$ satisfying these conditions are

$$m(x) = x^r \quad (r > -\tfrac{1}{2}),$$
$$m(x) = |x-1|/2, \tag{19}$$
$$m(x) = \log x.$$

T_n will be called a regular spacings statistic when $m(x)$ satisfies (15) through (18).

The following is easily deduced from Theorem 1.

Theorem 3. *Let* $m(x)$ *satisfy conditions* (15) *through* (18). *Let*

$$S_n = \sqrt{n} \left(\int_0^\infty m(x) \, dH_n(x) - \int_0^\infty m(x) \, dF_n(x) \right).$$ (20)

Under the sequence of alternatives (2), S_n *has a limiting normal distribution with mean* 0 *and variance*

$$\int_0^\infty \int_0^\infty m'(x) \, m'(y) \, K(x,y) \, dx \, dy < \infty.$$

Conditions (15) through (18) can be relaxed in Theorem 3. However, since the test statistics that we deal with are covered by the examples of (19) we will not try to do this here.

By expanding the mean $\int_0^\infty m(x) \, dF_n(x)$ of T_n under the null and alternative hypothesis we readily obtain that

$$\text{efficacy of } T_n \quad \begin{cases} 0 & \text{if } \delta > \tfrac{1}{4}, \\[2ex] \dfrac{\left(\int_0^1 l^2(p) \, dp \right)^2 \left(\int_0^\infty m'(x) \, e^{-x}(x - x^2/2) \, dx \right)^2}{\displaystyle\int_0^\infty \int_0^\infty m'(x) \, m'(y) \, K(x,y) \, dx \, dy} & \text{if } \delta = \tfrac{1}{4}. \end{cases}$$

(21)

It thus turns out that tests which are symmetric in the sample spacings cannot discriminate alternatives if they are at a distance of order $n^{-\delta}$ from the hypothesis and $\delta > \tfrac{1}{4}$. This is a disturbing feature. It was first demonstrated by Cibisov (1961) where he computed the likelihood ratio of the ordered sample spacings. However, we know that there exist tests which can discriminate alternatives at a distance of order $n^{-\frac{1}{2}}$ from the hypothesis. The Kolmogorov–Smirnov test is an example. This poor showing of tests symmetrically based on the spacings may be somewhat explained as follows. The sample spacings form a sufficient statistic since they are equivalent to the order statistics. Under the null hypothesis they form an interchangeable collection of random variables. However, limiting oneself to symmetric functions of the spacings entails loss of information, since the original order statistics cannot be recovered now.

Formula (21) enables us to compute the efficacies of tests symmetrically based on spacings. We present below a table giving the efficacies of the tests $V_r(n)$, $L(n)$ and $U(n)$.

Table 1. *Efficacies of some tests based on spacings*

Test‾statistic	Efficacy$\Big/ \left(\int_0^1 l^2(p)\,dp \right)^2$
$V_r(n): r = 0 \cdot 0$	0·0000
$r = 0 \cdot 5$	0·6760
$r = 1 \cdot 0$	0·0000
$r = 1 \cdot 5$	0·9700
$r = 2 \cdot 0$	1·0000
$r = 2 \cdot 5$	0·9728
$r = 3 \cdot 0$	0·9000
$r = 3 \cdot 5$	0·7976
$r = 4 \cdot 0$	0·6792
$U(n):$	0·5726
$L(n):$	0·3876

It is interesting to note that the efficacies above depend on the alternative only through the multiplying constant, $\left(\int_0^1 l^2(p)\,dp \right)^2$. Thus the efficiencies become independent of the alternative hypothesis. The table also indicates that the Greenwood statistic $V_2(n)$ has maximum efficacy in the limited comparison that was made. Let $T_n = \int_0^\infty m(x)\,dH_n(x)$ be a regular spacings test. It has maximum efficacy among all regular spacings test if and only if

$$\frac{\left(\int_0^\infty m'(x)\,e^{-x}(x - x^2/2)\,dx \right)^2}{\int_0^\infty \int_0^\infty m'(x)\,m'(y)\,K(x,y)\,dx\,dy}$$

is a maximum. This condition is equivalent to the condition that there is a constant λ such that

$$\int_0^\infty m'(y)\,K(x,y)\,dy = \lambda e^{-x}(x - x^2/2).$$

With $m(x) = x^2, m'(x) = 2x$, it is easy to verify the above condition. Thus the Greenwood statistic $V_2(n)$ has maximum efficacy among all regular spacings statistics.

Using Theorem 2 we can give the efficacies of tests of the form $\sum_1^n m(nD_i/h_{ni})/n$. We shall not give any general theorems on and computation of efficacies of such tests symmetrically based on modified spacings, since it is not clear which kind of modification to choose. However, we will broadly indicate what possibilities occur when one uses modified spacings. Put $h_{ni} = n/(n-i+1)$. Then the modified spacings become $\{nD_1, (n-1)D_2, ..., D_n\}$. Consider a test based on the mean, $M(n)$, of these modified spacings. In large samples, it is essentially a test based on the mean of $X_1, ..., X_n$. Either using Theorem 2, or directly, we can show that the efficacy of $M(n)$ is

$$12 \left(\int_0^1 l(p)\, p\, dp \right)^2 \quad \text{if} \quad \int_0^1 p l(p)\, dp \neq 0 \quad \text{and} \quad \delta = \tfrac{1}{2},$$

$$0 \quad \text{if} \quad \int_0^1 p l(p)\, dp = 0 \quad \text{and} \quad \delta = \tfrac{1}{4}.$$

Thus, this test has efficiency ∞, at a sequence of alternatives with $\delta = \tfrac{1}{4}$, relative to a test based symmetrically on the spacings, say $V_2(n)$, when $\int_0^1 p l(p)\, dp \neq 0$, but has efficiency 0 when $\int_0^1 p l(p)\, dp = 0$. Thus at least for this test $M(n)$, there is a sequence of alternatives at which it does not fare as well as tests based symmetrically on the spacings. Whether this is a general phenomenon or not is not known. It would be interesting to investigate this problem.

It is known that the Kolmogorov–Smirnov statistic discriminates alternatives of the form (2) with $\delta = \tfrac{1}{2}$ or more generally, alternatives with df $G_n(x)$ with $\sqrt{n} \sup_x |G_n(x) - x| \to a$ constant $\neq 0$. Thus the Kolmogorov–Smirnov test has efficiency ∞ with respect to any test which is symmetric in the spacings. (See for instance Cibisov (1961).) Weiss (1962) has given an example of a sequence of alternatives, namely the uniform distribution on $\left[\dfrac{1}{\log n}, 1 \right]$, at which the Kolmogorov–Smirnov test has efficiency 0 compared to the symmetric spacings test based on

$$D(n) = \max_i D_i.$$

The alternatives here do not have a constant carrier and the definition of the first sample spacing D_1 as X_1' inflates it considerably when the alternative is true. This accounts for the better performance of the test based on $D(n)$. As we described in § 1, we feel that all distributions considered must have the same carrier if spacings are to retain their meaning and do not get inflated as in this case.

The results established here apply with equal force to the goodness-of-fit problems on the circle. Consider n random variables $X_1, X_2, ..., X_n$ distributed independently and identically on a circle with unit circumference. The null hypothesis is one which states that the distribution is uniform on the circle. Ordering the observations as $X_1' \leqslant X_2' \leqslant \ldots \leqslant X_n'$, the sample spacings ($n$ in number), which are the arc-lengths between the successive sample observations, may be defined by

$$D_i = X_i' - X_{i-1}' \quad (i = 1, \ldots, n),$$

where we put $X_0' = X_n' - 1$. Under the null hypothesis, the distribution of these spacings is the same as those from a sample of size $(n-1)$ from the uniform distribution on the interval $[0, 1]$. Under the alternative, we can choose and fix an arbitrary point on the circle as the zero-direction and cut open the circle at that point to get the line segment $[0, 1]$. Now the $(n-1)$ circular spacings, which do not contain the cut off point, will have the same distribution as $(n-1)$ linear spacings on $[0, 1]$, not containing 0 and 1, while the nth circular spacing, containing the cut off point, will have the same distribution as the sum of the remaining two linear spacings. It is easy to see therefore that the limiting distribution of the empirical distribution function of the normalized or modified spacings is the same in the circular and linear cases. Hence all our statements regarding the ARE's of spacings tests made above hold for the circular case. It is rather fortunate that the circular case, which was our initial interest for this series of investigations on efficiencies, fits into the linear case in the asymptotic theory, and does not create difficult problems as it often does in the case of finite sample sizes.

ACKNOWLEDGEMENTS

The authors wish to thank Professor C. R. Rao for the keen interest he has shown in the work and Professors I. R. Savage and J. K. Ghosh for having gone through an earlier draft of the paper.

REFERENCES

Cibisov, D. M. (1961). On the tests of fit based on sample spacings. *Theoria vero. i. Prim.* **6**, 354–8, *Theory Probab. Applic.* **6**, 325–9.

Darling, D. A. (1953). On a class of problems related to the random division of an interval. *Ann. Math. Statist.* **24**, 239–53.

Fraser, D. A. S. (1957). *Nonparametric Methods in Statistics.* New York: John Wiley and Sons.

Greenwood, Major (1946). The statistical study of infectious diseases. *J. R. Statist. Soc.* **109**, 85–103.

Jackson, O. A. Y. (1967). An analysis of departures from the exponential distribution. *J. R. Statist. Soc.* (Ser. B), **29**, 540–9.

Kimball, B. F. (1950). On the asymptotic distribution of the sum of powers of unit frequency differences. *Ann. Math. Statist.* **21**, 263–71.

Pyke, R. (1965). Spacings. *J. R. Statist. Soc.* (Ser. B), **27**, 395–449.

Sethuraman, J. and Rao, J. S. (1969). Weak convergence of empirical distribution functions of random variables subject to perturbations and scale factors (in preparation).

Sherman, B. (1950). A random variable related to the spacing of sample values. *Ann. Math. Statist.* **21**, 339–61.

Weiss, L. (1957). The asymptotic power of certain tests of fit based on sample spacings. *Ann. Math. Statist.* **28**, 783–6.

Weiss, L. (1962). Review of Cibisov (1961). *Mathl Rev.* **24**, No. A, 1776.

Weiss, L. (1965). On asymptotic sampling theory for distributions approaching the uniform distribution. *Z. Wahrscheinlichkeitstheorie Verw. Geb.* **4**, 217–21.

ORDER STATISTICS OF A SAMPLE AND
OF AN EXTENDED SAMPLE†

M. M. SIDDIQUI

1 INTRODUCTION

For any numbers $a_1, a_2, \ldots$ and $n = 1, 2, \ldots$, let

$$a_{1,n} \leqslant a_{2,n} \leqslant \ldots \leqslant a_{n,n}$$

denote the ordered values of $a_1, \ldots, a_n$ arranged in the ascending order. Let $X_i, i = 1, 2, \ldots$, be a sequence of mutually independent random variables. Our objective is to obtain the joint distribution of

$$(X_{k,n}, X_{k+s, n+m})$$

for any $k, s, 1 \leqslant k \leqslant n, 1 \leqslant k+s \leqslant n+m$, under the following alternative hypotheses:

$H_0 = X_i, i = 1, \ldots, n+m$ are identically distributed with a common df F,

$H_1 = X_i, i = 1, \ldots, n$ are identically distributed with a common df F, and

$X_i, i = n+1, \ldots, n+m$ are identically distributed with a common df G.

The probability measures under H_0 and H_1 will be denoted by P_0 and P_1 respectively.

2 PROBABILITY THAT $X_{k,n} = X_{k+j, n+m}$

Clearly $\quad X_{k,n} = X_{k+j, n+m} \quad$ for some $j = 0, 1, \ldots, m$.

In fact, given $X_{k,n} = x$, $X_{k+j, n+m} = X_{k,n}$ if exactly j of $X_{n+1}, \ldots, X_{n+m}$ are less than x. Hence,

$$P_1(X_{k,n} = X_{k+j, n+m}) = \int_{-\infty}^{\infty} P_1(X_{k+j, n+m} = x \mid X_{k,n} = x)$$

$$\times P_1(x \leqslant X_{k,n} \leqslant x+dx) = n \binom{n-1}{k-1} \binom{m}{j}$$

$$\times \int_{-\infty}^{\infty} F^{k-1}(x) [1 - F(x)]^{n-k} G^j(x) [1 - G(x)]^{m-j} dF(x). \quad (2.1)$$

† This research was partly supported by the National Science Foundation under Grant No. GP 8835.

14   [417]   PNT

This integral cannot be evaluated unless G and F are specified. However, the probability of this event under the null hypothesis can be readily evaluated. Let

$$p_{k,\,k+j} = P_0(X_{k,\,n} = X_{k+j,\,n+m}),$$

then

$$p_{k,\,k+j} = n\binom{n-1}{k-1}\binom{m}{j}\int_0^1 F^{k+j-1}(1-F)^{n+m-k-j}\,dF$$

$$= \binom{m+n}{n}^{-1}\binom{k+j-1}{k-1}\binom{m+n-k-j}{n-k}. \tag{2.2}$$

The probability $p_{k,\,k+j}$ can also be looked upon as the probability that the kth largest among the first n observations is exceeded $m-j$ times in the following m observations. Gumbel [1] has studied the distribution of 'exceedances' in detail and in his notation ([1], p. 58)

$$p_{k,\,k+j} = w(n, k, m, m-j).$$

For additional references on this subject one may consult the bibliography given by Gumbel.

By choosing a k, one can construct a simple test for H_0 against the following alternatives:

$$H_{11}: \{G(x) < F(x),\ 0 < F(x) < 1\},$$

$$H_{12}: \{G(x) > F(x),\ 0 < F(x) < 1\},$$

$$H_1: H_{11} \cup H_{12}.$$

Consider for example H_{11} against H_0. Clearly, when $G < F$,

$$G^j(1-G)^{m-j} > F^j(1-F)^{m-j}$$

for small values of j, in fact, for

$$j < \frac{n[\log(1-G) - \log(1-F)]}{\log(1-G) - \log(1-F) - \log G + \log F}.$$

Thus, given α, $0 < \alpha < 1$, the α-level critical region for rejecting H_0 will be of the form $j \leqslant j(\alpha)$, where $j(\alpha)$ is the largest $\bar{r}$ such that

$$\sum_{j=0}^{\bar{r}} p_{k,\,k+j} \leqslant \alpha.$$

For a two-sided alternative H_1 the critical region will be $\{j \leqslant j_1$ or $j \geqslant j_2\}$, where one can choose (j_1, j_2) to be that number pair (r_1, r_2) for which $r_2 - r_1$ is shortest while

$$\sum_{j=r_1}^{r_2} p_{k,\,k+j} \geqslant 1-\alpha.$$

An easy evaluation gives,

$$\frac{p_{k,\,k+j+1}}{p_{k,\,k+j}} = 1 + \frac{(m+n)(k-1)-(n-1)(k+j)}{(m+n-k-j)(j+1)} \tag{2.3}$$

so that, with $(k-1)/(n-1) = p$,

$$p_{k,\,k+j+1} \gtreqless p_{k,\,k+j} \quad \text{as} \quad \frac{k+j}{m+n} \gtreqless p. \tag{2.4}$$

If $(m+n)p$ is not an integer, let $k+r$ be the smallest integer greater than $(m+n)p$. We then have,

$$p_{k,\,k} < p_{k,\,k+1} < \cdots < p_{k,\,k+r-1} < p_{k,\,k+r} > p_{k,\,k+r+1} > \cdots > p_{k,\,k+m}$$

so that the accumulation of probabilities can start with $p_{k,\,k+r}$ going on to $p_{k,\,k+r-1}$ or $p_{k,\,k+r+1}$, whichever be the larger, and continue selecting the largest of the unselected terms until the sum is equal to $1-\alpha$ or just exceeds it.

The following estimates of the tail probabilities may be useful.

Let $1 \leqslant s < r$. Then for any j, $0 \leqslant j \leqslant s-1$,

$$\frac{p_{k,\,k+j+1}}{p_{k,\,k+j}} > 1 + \left(p - \frac{k+s}{m+n}\right)\frac{(m+n)(n-1)}{(s+1)(m+n-k)} = 1 + \delta,$$

where $\delta > 0$. Then

$$p_{k,\,k+j} = p_{k,\,k+s} \prod_{i=1}^{s-j} \frac{p_{k,\,k+j+i-1}}{p_{k,\,k+j+i}} < p_{k,\,k+s}(1+\delta)^{-(s-j)} \tag{2.5}$$

and

$$\sum_{j=0}^{s} p_{k,\,k+j} < \frac{1+\delta}{\delta} p_{k,\,k+s}. \tag{2.6}$$

Similarly, for $r < s \leqslant m$, $s+1 \leqslant j \leqslant m$, one can find a $\tau > 0$ such that

$$\left.\begin{array}{l} p_{k,\,k+j} < (1-\tau)^{j-s} p_{k,\,k+s} \\[2mm] \displaystyle\sum_{j=s}^{m} p_{k,\,k+j} < \frac{1}{\tau} p_{k,\,k+s}. \end{array}\right\} \tag{2.7}$$

For an asymptotic approximation, which is due to Wilks, when $m \to \infty$, $s \to \infty$ and n remains finite, see Gumbel ([1], p. 65).

3 JOINT DISTRIBUTION OF $(X_{k,\,n}, X_{k+s,\,n+m})$.

In this section we consider the joint distribution of $X_{k,\,n}$ and $X_{k+s,\,n+m}$, $0 \leqslant s \leqslant m$, when H_0 is true, i.e. when $X_i, i = 1, \ldots, m+n$ are iid (independent and identically distributed). Since $U_j = F(X_j)$ are independent uniform [0, 1] random variables, it is convenient to begin with the joint distribution of $(U_{k,\,n}, U_{k+s,\,n+m})$.

For $0 \leqslant x \leqslant 1$, $0 \leqslant y \leqslant 1$, let

$$B_{x,y} = \{x \leqslant U_{k,n} \leqslant x+dx, \, y \leqslant U_{k+s,\,n+m} \leqslant y+dy\},$$

$$A_{k,r} = \{U_{k,n} = U_{r,\,n+m}\}.$$

We then have
$$P(B_{x,y}) = \sum_{j=0}^{m} P(B_{x,y} \cap A_{k,\,k+j}). \tag{3.1}$$

Clearly, for any (x,y) some of the sets $B_{x,y} \cap A_{k,\,k+j}$ are empty. In fact, we evaluate (3.1) separately for three cases: (1) $y = x$, (2) $y > x$, (3) $y < x$.

Case 1: $y = x$. In this case the only non-zero term in the r.h.s. of (3.1) corresponds to $j = s$, and we have

$$P(B_{x,x}) = n\binom{n-1}{k-1}\binom{m}{s} x^{k+s-1}(1-x)^{n+m-k-s}\,dx. \tag{3.2}$$

Case 2: $y > x$. The non-zero terms in the r.h.s. of (3.1) correspond to $j = 0, 1, ..., s-1$. However, if we adopt the convention that

$$\binom{a}{b} = 0, \quad \text{if } b > a \text{ or } b < 0,$$

we need not place any restriction on j.

By a straightforward combinatorial argument, we get

$$P(B_{x,y} \cap A_{k,\,k+j}) = n(n+m-k-s+1)\binom{n-1}{k-1}\binom{m}{j}\binom{n+m-k-j}{s-j-1}$$
$$\times x^{k+j-1}\,(y-x)^{s-j-1}\,(1-y)^{n+m-s-k}\,dx\,dy. \tag{3.3}$$

Case 3: $y < x$. In this case,

$$P(B_{x,y} \cap A_{k,\,k+j})$$
$$= n(k+s)\binom{n-1}{k-1}\binom{m}{j}\binom{k+j-1}{j-s-1}(1-x)^{n+m-k-j}\,(x-y)^{j-s-1}\,y^{s+k-1}\,dx\,dy,$$
$$\tag{3.4}$$

and the positive terms correspond to $j = s+1, ..., m$.

3.1. The moments.

It is known that, for $j \geqslant k$,

$$\left.\begin{aligned} EU_{k,n} &= \frac{k}{n+1}, \\[2mm] \operatorname{cov}(U_{k,n}, U_{j,n}) &= \frac{1}{(n+2)}\frac{k}{(n+1)}\left(1 - \frac{j}{n+1}\right). \end{aligned}\right\} \tag{3.5}$$

Now,

$$\operatorname{cov}(U_{k,n}, U_{k+s,m+n}) = \sum_{j=0}^{m} P(U_{k,n} = U_{k+j,n+m}) \operatorname{cov}(U_{k+j,n+m}, U_{k+s,n+m})$$

$$= \sum_{j=0}^{s} p_{k,k+j} \frac{1}{(m+n+2)} \frac{k+j}{(m+n+1)} \left(1 - \frac{k+s}{m+n+1}\right)$$

$$+ \sum_{j=s+1}^{m} p_{k,k+j} \frac{1}{(m+n+2)} \frac{k+s}{(m+n+1)} \left(1 - \frac{k+j}{m+n+1}\right),$$

and $\displaystyle\sum_{j=0}^{m} p_{k,k+j} = 1, \quad \sum_{j=0}^{m} p_{k,k+j} \frac{k+j}{m+n+1} = EU_{k,n} = \frac{k}{n+1}.$

Hence

$$\operatorname{cov}(U_{k,n}, U_{k+s,m+n}) = \frac{1}{(m+n+2)}$$

$$\times \left\{ \frac{k+s}{(m+n+1)} \left(1 - \frac{k}{n+1}\right) - \frac{1}{m+n+1} \sum_{j=0}^{s-1} (s-j) p_{k,k+j} \right\}. \quad (3.6)$$

This can also be written as

$$\operatorname{cov}(U_{k,n}, U_{k+s,m+n}) = \frac{1}{(m+n+2)}$$

$$\times \left\{ \frac{k}{n+1} \left(1 - \frac{k+s}{m+n+1}\right) - \frac{1}{m+n+1} \sum_{j=s+1}^{m} (j-s) p_{k,k+j} \right\}. \quad (3.7)$$

3.2. Asymptotic distribution

Set
$$k(n+1)^{-1} = p_1, \quad (k+s)(m+n)^{-1} = p_2.$$

If $p_2 < p_1$, we can use the estimate (2.5) for $p_{k,k+j}$, and if $p_2 > p_1$, the estimate (2.7), in (3.6) and (3.7) respectively. Thus for large $n+m$

$$\operatorname{cov}(U_{k,n}, U_{k+s,m+n}) = (n+m+2)^{-1} p_2(1-p_1) + O((n+m)^{-2}) \quad \text{(if } p_2 < p_1),$$

$$= (n+m+2)^{-1} p_1(1-p_2) + O((n+m)^{-2}) \quad \text{(if } p_1 < p_2).$$

By continuity the result holds also for $p_1 = p_2$. It is not difficult to show that as $n \to \infty$, $m \to \infty$, while $k(n+1)^{-1} \to p_1$ and $(k+s)(n+m+1)^{-1} \to p_2$, the asymptotic distribution of $(U_{k,n}, U_{k+s,m+n})$ will be normal.

Remarks. (1) Large sample results for the joint distribution of $(X_{k,n}, X_{k+s,m+n})$ are obtained in the usual way with the additional assumptions that $f(F^{-1}(p_1))$ and $f(F^{-1}(p_2))$ be positive. Actually, an easier proof of the asymptotic result can be constructed through the representation of $X_{k,n}$ in terms of the empirical distribution function.

(2) One can evaluate

$$P(x \leqslant U_{k,n} \leqslant x+dx, \ y \leqslant U_{k-s,m+n} \leqslant y+dy),$$

when $1 \leqslant s \leqslant k-1$, in a similar fashion. In fact, since $U_{k-s,m+n} < U_{k,n}$, the cases, '$y = x$' and '$y > x$' do not arise. We have, for $y < x$,

$$P(x \leqslant U_{k,n} \leqslant x+dx, \ y \leqslant U_{k-s,m+n} \leqslant y+dy)$$

$$= (k-s)\,k \sum_j \binom{m}{j} \binom{k+j-1}{k-s} (1-x)^{n+m-k-j} (x-y)^{s+j-1} y^{k-s-1} \, dy \, dx.$$

4 JOINT DISTRIBUTION OF $(X_{k,n}, X_{k+s,n+m})$ UNDER THE ALTERNATIVE H_1

The partitioning of the event

$$B_{x,y} = \{x \leqslant X_{k,n} \leqslant x+dx, \ y \leqslant X_{k+s,m+n} \leqslant y+dy\},$$

by

$$A_{k,k+j} = \{X_{k,n} = X_{k+j,m+n}\} \quad (j = 0, 1, \ldots, m),$$

is not very helpful in evaluating $P_1(B_{x.y})$, under the alternative H_1. An alternative partitioning is needed. Again, we consider the three cases separately:

$$(1) \ y = x, \quad (2) \ y > x, \quad (3) \ y < x.$$

Case 1: $y = x$. Clearly,

$$P_1(U_{k+s,m+n} = x \,|\, U_{k,n} = x) = \binom{m}{s} G^s(x) \, [1 - G(x)]^{m-s},$$

so that

$$P_1(B_{x.x}) = n \binom{n-1}{k-1} \binom{m}{s} G^s(x) \, [1 - G(x)]^{m-s} \, F^{k-1}(x) \, [1 - F(x)]^{n-k} \, dF(x).$$

$$(4.1)$$

Case 2: $y > x$. Introduce the following partitioning events:

$A_{i1} = \{$Exactly i of $X_1, \ldots, X_n$ between x and y, and one of these in $(y, y+dy)\}$,

$A_{i2} = \{$Exactly i of $X_1, \ldots, X_n$ between x and y, and none of them in $(y, y+dy)\}$.

Then,

$$P_1(B_{x,y} \cap A_{i1}) = \frac{n!}{(k-1)!\,i!\,(n-k-i-1)!} \binom{m}{s-i-1} G^{s-i-1}(y)$$

$$\times [1 - G(y)]^{m-s+i+1} \, F^{k-1}(x) \, [F(y) - F(x)]^i \, [1 - F(y)]^{n-k-i-1} \, dF(x) \, dF(y),$$

$$P_1(B_{x,y} \cap A_{i,2}) = \frac{n!}{(k-1)!\,i!\,(n-k-i)!} \, m \binom{m-1}{s-i-1} G^{s-i-1}(y)$$

$$\times [1 - G(y)]^{m-s+i} \, F^{k-1}(x) \, [F(y) - F(x)]^i \, [1 - F(y)]^{n-k-i} \, dF(x) \, dG(y).$$

Thus $\qquad P_1(B_{x,y}) = \sum_i P_1(B_{x,y} \cap A_{i1}) + \sum_i P_1(B_{x,y} \cap A_{i2}),$

where the summation on i need not be specified due to our convention for the combinatorial symbol $\binom{a}{b}$.

Case 3: $y < x$. We now have

$$P_1(B_{x,y} \cap A_{i1}) = \frac{n!}{(k-i-2)!\,i!\,(n-k)!} \binom{m}{s+i+1} G^{s+i+1}(y)$$

$$\times [1 - G(y)]^{m-s-i-1} F^{k-i-2}(y) [F(x) - F(y)]^i [1 - F(x)]^{n-k} dF(y)\,dF(x),$$

$$P_1(B_{x,y} \cap A_{i,2}) = \frac{n!}{(k-i-1)!\,i!\,(n-k)!}\, m \binom{m-1}{s+i} G^{s+i}(y)$$

$$\times [1 - G(y)]^{m-s-i-1} F^{k-i-1}(x) [F(x) - F(y)]^i [1 - F(x)]^{n-k} dG(y)\,dF(x),$$

and $\qquad P_1(B_{x,y}) = \sum_i P(B_{x,y} \cap A_{i,1}) + \sum_i P(B_{x,y} \cap A_{i,2}).$

REFERENCE

[1] E. J. Gumbel. *Statistics of Extremes*. New York: Columbia University Press, 1958.

PART 4

GENERAL THEORY

ON A STATISTIC SIMILAR
TO STUDENT'S t

1 THE PROBLEM OF A STUDENTIZED CHEBYSHEV INEQUALITY

When a random variable X has expectation μ and variance σ^2, and $X_1, X_2, \ldots, X_n$ is a random sample of X and $\bar{X} = \sum_{j=1}^{n} X_j/n$ the sample mean, then Chebyshev's inequality yields

$$P\{|\bar{X} - \mu| > \lambda\sigma\} \leqslant \frac{1}{n\lambda^2} \quad \text{for every} \quad \lambda > 0. \tag{1.1}$$

When X has a normal distribution $N(\mu, \sigma^2)$, then one has from Student's theorem

$$P\{|\bar{X} - \mu| > \lambda/\sqrt{[s_n^2/(n-1)]}\} = \frac{2\Gamma(\frac{1}{2}n)}{\Gamma(\frac{1}{2}n - \frac{1}{2})\sqrt{[\pi(n-1)]}} \int_{\lambda\sqrt{n}}^{+\infty} \left(1 + \frac{t^2}{n-1}\right)^{-\frac{1}{2}n} dt. \tag{1.2}$$

In (1.1) $\bar{X}$ is an estimate of a location parameter (expectation) and σ is the known value of a scale parameter (standard deviation). Chebyshev's inequality provides in (1.1), for a very general class of distributions of X, an upper bound on the probability of the event $|\bar{X} - \mu| > \lambda\sigma$.

If the additional assumption is made that X has a normal distribution, then σ can be replaced by its estimate $\hat{\sigma} = \sqrt{[s_n^2/(n-1)]}$ and (1.2) provides the exact probability of the event $|\bar{X} - \mu| > \lambda\hat{\sigma}$, even though σ is not known.

For many years statisticians have from time to time given some thought to a question which may be called the 'problem of a studentized Chebyshev inequality', and which can be stated as follows: is there a sequence of functions $\Psi_n(\lambda)$, decreasing to 0 as $\lambda \to +\infty$, such that

$$P\{|\bar{X} - \mu| > \lambda\sqrt{[s_n^2/(n-1)]}\} \leqslant \Psi_n(\lambda) \tag{1.3}$$

no matter what probability distribution X may have? To the author's knowledge, no answer has been given to this question.

In the following section a particular answer is given to the following narrower question:

† Research supported by the Office of Naval Research.

[427]

Let $\mathscr{F}$ be a two-parametric family of probability distributions, with a location parameter μ and scale parameter σ (not necessarily the standard deviation). Making suitable assumptions on $\mathscr{F}$ and using suitable estimates $\hat{\mu}$ and $\hat{\sigma}$ for μ and σ, can one obtain a sequence of functions $\Psi_n(\lambda)$, decreasing to 0 as $\lambda \to +\infty$, such that

$$P\{|\hat{\mu} - \mu| > \lambda \hat{\sigma}\} \leqslant \Psi_n(\lambda)$$

when X has a distribution in $\mathscr{F}$?

2 A STUDENT-TYPE STATISTIC BASED ON ORDER STATISTICS

2.1. Definitions

Let $X_{(1)} \leqslant X_{(2)} \leqslant \ldots \leqslant X_{(2m+1)}$ be the order statistics for a sample of odd size $2m+1$ of a random variable X. We consider the following three order statistics

$$U = X_{(m+1-r)}, \quad V = X_{(m+1)}, \quad W = X_{(m+1+r)} \tag{2.1.1}$$

for some integer r, $1 \leqslant r \leqslant m$, so that V is the sample median, and $W - U$ an inter-quantile range between two sample-quantiles. The statistic we will consider is

$$S = \frac{V - \mu}{W - U}. \tag{2.1.2}$$

A class of similar statistics has been considered in [1].

From now on it will always be assumed that X has a 'bell-shaped' probability density $f(x)$, i.e. that

$$f(\mu - x) = f(\mu + x) \quad \text{for} \quad x \geqslant 0 \tag{2.1.3}$$

and $\qquad f(\mu + x)$ is non-increasing for $x \geqslant 0$, $\tag{2.1.4}$

and $\mathscr{F}_B$ will denote the family of distributions with densities satisfying (2.1.3) and (2.1.4) for some median μ.

2.2. An inequality

If X has a probability distribution in $\mathscr{F}_B$, then

$$P\{|S| > \lambda\} \leqslant \binom{2m+1}{m-r}\binom{2r}{r}[\lambda(\lambda-1)]^{-r}2^{-(m+r)} \quad \text{for} \quad \lambda > 1. \tag{2.2.1}$$

Proof. Clearly the value of S remains unchanged under linear transformations of X, i.e. S is independent of the location parameter and the scale parameter. To study the probability distribution of S we shall, therefore, without loss of generality, assume

$$\mu = 0 \qquad (2.2.2)$$

and we may also choose arbitrarily the scale parameter. Our random variable X is thus assumed to have a probability density $f(x)$ which has its maximum at 0 and decreases symmetrically about 0 as $|x| \to \infty$, and the corresponding cumulative probability function $F(x)$ is convex for $x < 0$, concave for $x > 0$ and $F(0) = \frac{1}{2}$.

The joint probability density of U, V, W is

$$g(u, v, w) = \frac{(2m+1)!}{[(m-r)!\,(r-1)!]^2} f(u) f(v) f(w)\, F^{m-r}(u)\, [F(v) - F(u)]^{r-1}$$

$$\times\, [F(w) - F(v)]^{r-1} [1 - F(w)]^{m-r} \quad \text{for} \quad u < v < w$$

$$(2.2.3)$$

and zero elsewhere.

Writing

$$P\{S > \lambda\} = P(\lambda)$$

we have according to (2.2.3), for $\lambda > 1$,

$$P(\lambda) = \int_{v=0}^{+\infty} \int_{u=(1-1/\lambda)v}^{v} \int_{w=v}^{u+v/\lambda} g(u, v, w)\, dw\, du\, dv$$

$$= K(m, r)\, I(m, r, \lambda) \quad (2.2.4)$$

where

$$K(m, r) = \frac{(2m+1)!}{[(m-r)!\,(r-1)!]^2} \qquad (2.2.5)$$

and

$$I(m, r, \lambda) = \int_{v=0}^{+\infty} f(v) \int_{u=(1-1/\lambda)v}^{v} f(u)\, F^{m-r}(u)\, [F(v) - F(u)]^{r-1}$$

$$\times \int_{w=v}^{u+v/\lambda} f(w)\, [F(w) - F(v)]^{r-1} [1 - F(w)]^{m-r}\, dw\, du\, dv. \quad (2.2.6)$$

We introduce the abbreviations

$$I(u, v) = \int_{w=v}^{u+v/\lambda} f(w)\, [F(w) - F(v)]^{r-1} [1 - F(w)]^{m-r}\, dw \qquad (2.2.7)$$

and

$$I(v) = \int_{u=(1-1/\lambda)v}^{v} f(u)\, F^{m-r}(u)\, [F(v) - F(u)]^{r-1} I(u, v)\, du \qquad (2.2.8)$$

so that (2.2.6) becomes

$$I(m, r, \lambda) = \int_{v=0}^{+\infty} f(v)\, I(v)\, dv. \qquad (2.2.9)$$

We note the following inequalities which follow from the assumption that $f(x)$ is bell-shaped about $\mu = 0$.

For $0 \leqslant u \leqslant v \leqslant w$,

$$f(w) = \frac{F(w) - F(v)}{w - v} \leqslant f(v) \leqslant \frac{F(v) - F(u)}{v - u} \leqslant f(u) \leqslant \frac{F(u) - \frac{1}{2}}{u - \frac{1}{2}} \leqslant f(0)$$

(2.2.10)

and, the special case obtained from (2.2.10) by setting $v = 0$, $w = x$,

$$x f(x) \leqslant F(x) - \tfrac{1}{2}.$$ (2.2.11)

From (2.27) and (2.2.10) follows

$$I(u, v) \leqslant [1 - F(v)]^{m-r} \int_{w=v}^{u+v/\lambda} f(w) [F(w) - F(v)]^{r-1} dw$$

$$= [1 - F(v)]^{m-r} \frac{1}{r} \left[F\left(u + \frac{v}{\lambda} \right) - F(v) \right]^r$$

$$\leqslant [1 - F(v)]^{m-r} \frac{1}{r} f^r(v) \left(u + \frac{v}{\lambda} - v \right)^r$$

and from (2.2.8)

$$I(v) \leqslant \frac{1}{r} [1 - F(v)]^{m-r} f^r(v) \int_{u=(1-1/\lambda)v}^{v} f(u) F^{m-r}(u) [F(v) - F(u)]^{r-1}$$

$$\times \left[u - v \left(1 - \frac{1}{\lambda} \right) \right]^r du$$

$$= \frac{1}{r_1} [1 - F(v)]^{m-r} F^{m-1}(v) f^r(v) \int_{u=(1-1/\lambda)v}^{v} f(u) \left[\frac{F(u)}{F(v)} \right]^{m-r}$$

$$\times \left[1 - \frac{F(u)}{F(v)} \right]^{r-1} \left[u - v \left(1 - \frac{1}{\lambda} \right) \right]^r du$$

$$\leqslant \frac{1}{r} [1 - F(v)]^{m-r} F^{m-1}(v) f^r(v) \left(\frac{v}{\lambda} \right)^r \int_{u=(1-1/\lambda)v}^{v} \left[\frac{F(u)}{F(v)} \right]^{m-r}$$

$$\times \left[1 - \frac{F(u)}{F(v)} \right]^{r-1} f(u) \, du$$

$$= \frac{1}{r} [1 - F(v)]^{m-r} F^{m-1}(v) f^r(v) \left(\frac{v}{\lambda} \right)^r \int_{z=\frac{F(v-v/\lambda)}{F(v)}}^{1} z^{m-r} (1-z)^{r-1} F(v) \, dz$$

$$= \frac{1}{r} F^m(v) [1 - F(v)]^{m-r} [v f(v)]^r \lambda^{-r} \int_{z=\frac{F(v-v/\lambda)}{F(v)}}^{1} z^{m-r} (1-z)^{r-1} \, dz$$

$$\leqslant \frac{1}{r} F^m(v) [1 - F(v)]^{m-r} [v f(v)]^r \lambda^{-r} \int_{z=\frac{F(v-v/\lambda)}{F(v)}}^{1} (1-z)^{r-1} \, dz$$

$$= \frac{1}{r^2} F^m(v) [1 - F(v)]^{m-r} [v f(v)]^r \lambda^{-r} \left[1 - \frac{F(v - v/\lambda)}{F(v)} \right]^r$$

$$= \frac{1}{r^2} F^{m-r}(v) [1 - F(v)]^{m-r} [v f(v)]^r \lambda^{-r} [F(v) - F(v - v/\lambda)]^r.$$

Using this bound on $I(v)$ in (2.9) one has

$$I(m,r,\lambda) \leqslant \frac{1}{r^2}\lambda^{-r}\int_{v=0}^{+\infty} F^{m-r}(v)\,[1-F(v)]^{m-r}\,[vf(v)]^r$$
$$\times [F(v)-F(v-v/\lambda)]^r f(v)\,dv,$$

and making use of (2.2.10) and (2.2.11)

$$I(m,r,\lambda) \leqslant r^{-2}\lambda^{-r}\int_{v=0}^{+\infty} F^{m-r}(v)\,[1-F(v)]^{m-r}\,[F(v)-\tfrac{1}{2}]^r$$
$$\times f^r\left(v-\frac{v}{\lambda}\right)\left(\frac{v}{\lambda}\right)^r f(v)\,dv$$
$$= r^{-2}\lambda^{-2r}\int_{v=0}^{+\infty} F^{m-r}(v)\,[1-F(v)]^{m-r}\,[F(v)-\tfrac{1}{2}]^r\left[f\left(v-\frac{v}{\lambda}\right)\right.$$
$$\times\left.\left(v-\frac{v}{\lambda}\right)\right]^r\left(\frac{\lambda}{\lambda-1}\right)^r f(v)\,dv$$
$$\leqslant r^{-2}\lambda^{-2r}\left(\frac{\lambda}{\lambda-1}\right)^r\int_{v=0}^{+\infty} F^{m-r}(v)\,[1-F(v)]^{m-r}\,[F(v)-\tfrac{1}{2}]^r$$
$$\times\left[F\left(v-\frac{v}{\lambda}\right)-\tfrac{1}{2}\right]^r f(v)\,dv$$
$$\leqslant r^{-2}\,[\lambda(\lambda-1)]^{-r}\int_{v=0}^{+\infty} F^{m-r}(v)\,[1-F(v)]^{m-r}\,[F(v)-\tfrac{1}{2}]^{2r}f(v)\,dv$$
$$= r^{-2}[\lambda(\lambda-1)]^{-r}\int_{z=\frac{1}{2}}^{1} z^{m-r}(1-z)^{m-r}\,(z-\tfrac{1}{2})^{2r}\,dz$$
$$< r^{-2}\,[\lambda(\lambda-1)]^{-1}\int_{z=\frac{1}{2}}^{1}(1-z)^{m-r}\,(z-\tfrac{1}{2})^{2r}\,dz$$
$$= r^{-2}\,[\lambda(\lambda-1)]^{-r}\int_{s=0}^{1}(\tfrac{1}{2}-\tfrac{1}{2}s)^{m-r}\,(\tfrac{1}{2}s)^{2r}\,\tfrac{1}{2}ds$$
$$= r^{-2}[\lambda(\lambda-1)]^{-r}\,2^{-(m+r+1)}\,B(m-r+1,2r+1)$$
$$= r^{-2}\,[\lambda(\lambda-1)]^{-r}\,2^{-(m+r+1)}\,\frac{(m-r)!\,(2r)!}{(m+r+1)!}. \tag{2.2.12}$$

Combining (2.2.4), (2.2.5) and the upper bound (2.2.12) for $I(m,r,\lambda)$, one obtains (2.2.1).

3 PROPERTIES OF THE STATISTIC S AND RELATED STATISTICS

3.1. The statistic S is, as has been noted before, invariant under linear transformations of the random variable i.e. is independent of the location- and scale-parameters. We may therefore choose any particular family of bell-shaped probability distributions with the two parameters, median $= \mu$ and scale parameter $= \gamma$, and conclude that the probability distribution of S for given m, r will be the same for every random variable X in this family. In particular we may choose the family of normal distributions, tabulate the probability distributions of S for a reasonable range of r and m, and use the statistic S in the same situations in which Student's t is used.

3.2. Even under the assumption of normally distributed X, for problems which are usually dealt with by using Student's t, the S-statistic may have the advantage that it involves very little computation and may therefore lend itself to routine use by mathematically untrained personnel. There are, however, situations in which, even under the assumption of normality, the S-statistic can be used while t could not. A well-known practically important situation of this kind arises when $n = 2m + 1$ observations are made but some of the smallest and some of the largest are 'censored' out. Then as long as the values of the sample median $X_{(m+1)}$ and of two order-statistics $X_{(m+1-r)}$, $X_{(m+1+r)}$ are known, the S-statistic can be used to estimate μ or to test hypotheses about μ, while t cannot be calculated.

3.3. For a specific two-parametric family of distributions, such as the normal distributions, one can use for U and W in the definition of S any two order statistics $X_{(m+1-r)}, X_{(m+1+r')}$ with r and r' not necessarily equal. This may be of practical importance e.g. when there is only censoring from above. The numerical computation of the distribution of S in this case is not prohibitive.

3.4. A numerical computation of the probability distributions of S, as described in § 3.2, under assumption of normality is being completed. A study is also being carried out to obtain the limiting distributions of S.

3.5. If one-sided conclusions (one-sided confidence bounds or tests against one-sided alternatives) about μ are desired, then the simpler analogue to S may be used

$$S' = \frac{V - \mu}{W - V}.$$

An inequality such as (2.2.1) can be derived for S' under the assumption that X has a bell-shaped distribution, and within a family of two-parametric distributions for X with location- and scale-parameters the statistic S' obviously has for each m and r a probability distribution independent of both parameters. Some numerical calculations have been carried out for the probability distributions of S' for finite sample sizes under the assumption of a normal distribution for X.

REFERENCE

[1]   F. N. David and N. L. Johnson. Some tests of significance with ordered variables. *J. R. Statist. Soc.* p. **18** (1956) 1–20.

ASYMPTOTIC DISTRIBUTIONS
OF SOME STATISTICS BASED ON THE
BIVARIATE SAMPLE DISTRIBUTION
FUNCTION

J. DURBIN

1 INTRODUCTION

Suppose we have a sample of n independent bivariate observations and wish to test the hypothesis that they come from a distribution with specified continuous distribution function (df) $F(x, y)$. On making the transformation $s = F(x)$, $t = F(y|x)$ where $F(x)$ is the marginal df of x and $F(y|x)$ is the conditional df of y given x, the hypothetical distribution is transformed to the uniform distribution on the square $0 \leqslant s, t \leqslant 1$. The problem has therefore been reduced to that of testing for uniformity on the unit square. However, the reduction is not unique since (x, y) can be transformed arbitrarily before transforming to the uniform distribution by means of the marginal and conditional df's. Transformations of this kind have been studied by Rosenblatt (1952a).

Let $F_n(s, t)$ denote the sample df defined to equal r/n when exactly r of the observed pairs (s_i, t_i) satisfy the relations $s_i \leqslant s, t_i \leqslant t$ for $i = 1, ..., n$ and for $0 \leqslant s, t \leqslant 1$. This paper considers the asymptotic distributions of a number of statistics based on $F_n(s, t)$.

In §2 the random process $F_n(s, t) - st$ is approximated by a two-dimensional Brownian process 'tied down' to equal zero at $(1, 1)$. This approximation enables the asymptotic characteristic function of the two-dimensional Cramér–von Mises statistic to be determined by using the properties of regression residuals. The characteristic function is inverted numerically. Asymptotic distributions of statistics such as

$$\sqrt{n} \max_{0 \leqslant s \leqslant 1} \left[\int_0^1 F_n(s, t) \, dt - E \left\{ \int_0^1 F_n(s, t) \, dt \right\} \right]$$

and

$$\int_0^1 \int_0^1 F_n(s, t) \, ds \, dt$$

are given in §6.

In case the reader might regard the basic problem considered in this paper as rather artificial in view of the element of arbitrariness in the transformation to the unit square, I might refer to the fact that the practi-

[435]

cal importance of statistics based on the univariate sample distribution function comes not only from their use for goodness-of-fit tests but also from their application in a wide variety of other situations where cumulative-sum techniques are found to be useful, e.g. in tests of the Poisson hypothesis, tests of serial independence in time series analysis, tests of constancy over time of regression models and tests of randomness of points on a line. My main incentive in examining the two-dimensional case was the hope that some results might be obtained with a wider range of applicability than to tests of goodness-of-fit even though, for convenience of exposition, the discussion is conducted in goodness-of-fit terms.

2   APPROXIMATION BY A TWO-DIMENSIONAL NORMAL PROCESS

Consider the process

$$u_n(s, t) = \sqrt{n}\{F_n(s, t) - st\} \quad (0 \leqslant s, t \leqslant 1). \tag{1}$$

The mean and covariance functions are

$$\left.\begin{aligned}
E\{u_n(s, t)\} &= 0, \\
E\{u_n(s_1, t_1) u_n(s_2, t_2)\} &= \min(s_1, s_2)\min(t_1, t_2) - s_1 s_2 t_1 t_2 \\
&\qquad (0 \leqslant s_1, s_2, t_1, t_2 \leqslant 1)
\end{aligned}\right\} \tag{2}$$

and the joint distribution of the values of $u_n(s, t)$ at any finite number of points (s, t) is asymptotically normal. These results follow from standard multinomial theory (cf. Rosenblatt, 1952b).

Let $b(s, t)$ be the normal process with mean zero and covariance function

$$E\{b(s_1, t_1) b(s_2, t_2)\} = \min(s_1, s_2)\min(t_1, t_2) \quad (0 \leqslant s_1, t_1, t_2 \leqslant 1), \tag{3}$$

and let $u(s, t)$ be the residual from the regression of $b(s, t)$ on $b(1, 1)$. The regression coefficient is

$$E\{b(s, t) b(1, 1)\}/E\{b^2(1, 1)\} = st,$$

so $u(s, t) = b(s, t) - st\, b(1, 1)$ which is normally distributed with mean zero and covariance function

$$E\{u(s_1, t_1) u(s_2, t_2)\} = E[\{b(s_1, t_1) - s_1 t_1 b(1, 1)\}\{b(s_2, t_2) - s_2 t_2 b(1, 1)\}]$$

$$= \min(s_1, s_2)\min(t_1, t_2) - s_1 s_2 t_1 t_2,$$

which is the same as (2). Our technique is to obtain asymptotic distributions by approximating $u_n(s, t)$ by $u(s, t)$.

The process $b(s, t)$ has been referred to by Pyke (1968) and some of its properties have been studied by Kuelbs (1968).

We now obtain a representation of $b(s, t)$ in terms of the eigenvalues λ_{jk} and eigenfunctions $\phi_{jk}(s, t)$ of the covariance function (3). These are obtained as the eigen-solutions of the integral equation

$$\int_0^1 \int_0^1 \min(s, \sigma) \min(t, \tau) \phi(\sigma, \tau) \, d\sigma \, d\tau = \lambda \phi(s, t). \tag{4}$$

It is well known that the eigen-solutions of the one-dimensional integral equation

$$\int_0^1 \min(s, \sigma) \phi(\sigma) \, d\sigma = \lambda \phi(s) \tag{5}$$

are given by

$$\phi_j(s) = \sqrt{2} \sin\left(j - \tfrac{1}{2}\right)\pi s, \quad \lambda_j^{-1} = \left(j - \tfrac{1}{2}\right)^2 \pi^2 \quad (j = 1, 2, \ldots).$$

Writing the left-hand side of (4) as the product of two integrals of the form of the left-hand side of (5) we see that the eigen-solutions of (4) are given by

$$\phi_{jk}(s, t) = 2 \sin\left(j - \tfrac{1}{2}\right)\pi s \sin\left(k - \tfrac{1}{2}\right)\pi t, \quad \lambda_{jk}^{-1} = \left(j - \tfrac{1}{2}\right)^2 \left(k - \tfrac{1}{2}\right)^2 \pi^4$$

$$(j, k = 1, 2, \ldots).$$

It follows that $b(s, t)$ has the representation

$$b(s, t) = 2 \sum_{j, k=1}^{\infty} \frac{z_{jk} \sin\left(j - \tfrac{1}{2}\right)\pi s \sin\left(k - \tfrac{1}{2}\right)\pi t}{\left(j - \tfrac{1}{2}\right)\left(k - \tfrac{1}{2}\right)\pi^2} \quad (0 \leqslant s, t \leqslant 1), \tag{6}$$

where $\{z_{jk}\}$ is a double sequence of independent $N(0, 1)$ variables. A convenient treatment of this type of representation is given by Rosenblatt (1962, pp. 185–95). The papers by Kuelbs (1968) and Blum, Kiefer and Rosenblatt (1962) should also be consulted.

3   THE ASYMPTOTIC CHARACTERISTIC FUNCTION OF THE CRAMÉR–VON MISES STATISTIC

Rosenblatt (1952) has shown that the asymptotic distribution of the Cramér–von Mises statistic

$$W_n^2 = \int_0^1 \int_0^1 u_n^2(s, t) \, ds \, dt$$

is the same as the distribution of the random variable

$$W^2 = \int_0^1 \int_0^1 u^2(s, t) \, ds \, dt. \tag{7}$$

It was shown in §2 that $u(s, t)$ can be represented as the residual from regression of $b(s, t)$ on $b(1, 1)$. By the properties of normal regression any function of the residuals $u(s, t)$ from regression of $b(s, t)$ on $b(1, 1)$ has the same distribution as the same function of the $b(s, t)$ conditional on $b(1, 1) = 0$. The distribution of W^2 is therefore the same as that of

$$Q = \int_0^1 \int_0^1 b^2(s, t)\, ds\, dt \tag{8}$$

conditional on $b(1, 1) = 0$.

Substituting from (6) we have

$$Q = 4 \sum_{j, k=1}^{\infty} \sum_{p, q=1}^{\infty} \frac{z_{jk} z_{pq}}{(j - \frac{1}{2})(k - \frac{1}{2})(p - \frac{1}{2})(q - \frac{1}{2}) \pi^4}$$

$$\times \int_0^1 \int_0^1 \sin(j - \tfrac{1}{2}) \pi s \sin(k - \tfrac{1}{2}) \pi t \sin(p - \tfrac{1}{2}) \pi s \sin(q - \tfrac{1}{2}) \pi t\, ds\, dt$$

$$= \sum_{j, k=1}^{\infty} \frac{z_{jk}^2}{(j - \frac{1}{2})^2 (k - \frac{1}{2})^2 \pi^4} \tag{9}$$

in virtue of the relations

$$\int_0^1 \sin(j - \tfrac{1}{2}) \pi s \sin(p - \tfrac{1}{2}) s\, ds = 0\ (p \neq j) \quad \text{and} \quad = \tfrac{1}{2}\ (p = j).$$

We have to find the distribution of this subject to

$$b(1, 1) = 2 \sum_{j, k=1}^{\infty} \frac{(-1)^{j+k} z_{jk}}{(j - \frac{1}{2})(k - \frac{1}{2}) \pi^2} = 0. \tag{10}$$

Lemma. Suppose that $x_1, x_2, \ldots, x_m$ *are independent* $N(0, 1)$ *variables. The characteristic function of* $q = \sum_{j=1}^{m} \lambda_j x_j^2$ *conditional on* $\sum_{j=1}^{m} x_j l_j = 0$, *where* $\sum_{j=1}^{m} l_j^2 = 1$, *is*

$$\phi(t) = \left\{ \prod_{j=1}^{m} (1 - 2\lambda_j it) \sum_{j=1}^{m} \frac{l_j^2}{1 - 2\lambda_j it} \right\}^{-\frac{1}{2}}. \tag{11}$$

This lemma and its proof are essentially contained in some results of Durbin and Watson (1950, p. 414). It is a special case of a more general result of Bateman (1949).

Since

$$\sum_{j, k=1}^{\infty} \frac{1}{(j - \frac{1}{2})^2 (k - \frac{1}{2})^2} = \left\{ \sum_{j=1}^{\infty} \frac{1}{(j - \frac{1}{2})^2} \right\}^2 = \frac{\pi^4}{4}$$

(Abramovitz and Stegun, 1964, 23.2.28), on applying the lemma to the distribution of

$$\sum_{j, k=1}^{h} \frac{z_{jk}^2}{(j - \ ^? (k - \frac{1}{2})^2 \pi^4} \quad \text{subject to} \quad \sum_{j, k=1}^{h} \frac{(-1)^{j+k} z_{jk}}{(j - \frac{1}{2})(k - \frac{1}{2}) \pi^2} = 0$$

and letting $h \to \infty$, we obtain for the characteristic function of W^2

$$\phi(t) = \left\{ \prod_{j,\,k=1}^{\infty} \left(1 - \frac{2it}{(j-\frac{1}{2})^2 (k-\frac{1}{2})^2 \pi^4} \right) \right.$$

$$\left. \times \sum_{j,\,k=1}^{\infty} \frac{4}{(j-\frac{1}{2})^2 (k-\frac{1}{2})^2 \pi^4} \frac{1}{1 - (2it/(j-\frac{1}{2})^2 (k-\frac{1}{2})^2 \pi^4)} \right\}^{-\frac{1}{2}}. \qquad (12)$$

The infinite product for the cosine (Abramovitz and Stegun, 1964, 4.3.90) gives

$$\cos\sqrt{\theta} = \prod_{j=1}^{\infty} \left(1 - \frac{\theta}{(j-\frac{1}{2})^2 \pi^2} \right). \qquad (13)$$

Differentiating this we obtain

$$\frac{\sin\sqrt{\theta}}{\sqrt{\theta}} = \prod_{j=1}^{\infty} \left(1 - \frac{\theta}{(j-\frac{1}{2})^2 \pi^2} \right) \sum_{j=1}^{\infty} \frac{2}{(j-\frac{1}{2})^2 \pi^2} \frac{1}{1 - (\theta/(j-\frac{1}{2})^2 \pi^2)}. \qquad (14)$$

On differentiating the function

$$\psi(\theta) = -4 \prod_{j=1}^{\infty} \cos\left\{ \frac{\sqrt{\theta}}{(j-\frac{1}{2})\pi} \right\}$$

and using (13) and (14) we therefore find that (12) can be written in the form

$$\frac{1}{\phi^2(t)} = \frac{d\psi}{d\theta}\bigg|_{\theta=2it}. \qquad (15)$$

This gives a nice generalisation of the result for the univariate Cramér-von Mises statistic for which (15) holds with $\psi(\theta) = -2\cos\sqrt{\theta}$.

Added in proof.

Since this article went to press the following article has appeared.

Dugué, D. (1969). Characteristic functions of random variables connected with Brownian motion and of the von Mises multidimensional ω_n^2. *Multivariate Analysis*, vol. 2. Edited by P. R. Krishnaiah. New York: Academic Press.

This article gives the characteristic function of W^2 in essentially the form (15) above and contains references to related work by the author. I was unaware of this previous work by Professor Dugué during my own investigations.

4   NUMERICAL INVERSION OF THE CHARACTERISTIC FUNCTION OF W^2

The most practical way of obtaining numerical values for the distribution function of W^2 seems to be by numerical inversion of $\phi(t)$. However, this is difficult to do directly since the infinite sums and products in (12) and

(15) do not converge sufficiently rapidly. The following slightly modified method was therefore used.

We note that W^2 may be written in the form

$$W^2 = \sum_{r=1}^{\infty} \nu_r \xi_r^2, \quad \text{where} \quad \nu_1 \geqslant \nu_2 \geqslant \dots$$

are the reciprocals of the eigenvalues of the covariance function (3) and $\xi_1, \xi_2, \dots$ are independent $N(0, 1)$. In terms of the ν_j's the characteristic function of W^2 is

$$\phi(t) = \prod_{r=1}^{\infty} (1 - 2\nu_r it)^{-\frac{1}{2}}.$$

On equating with (12) it follows that $\nu_1, \nu_2, \dots$ are the roots of the equation

$$\prod_{j, k=1}^{\infty} (x - \lambda_{jk}) \sum_{j, k=1}^{\infty} \frac{\lambda_{jk}}{x - \lambda_{jk}} = 0, \tag{16}$$

where $\lambda_{jk} = \{(j - \frac{1}{2})^2 (k - \frac{1}{2})^2 \pi^4\}^{-1}$. For $j \neq k$, $x - \lambda_{jk}$ is a factor of (16) since $\lambda_{jk} = \lambda_{kj}$. For other roots the term $\prod_{j, k=1}^{\infty} (x - \lambda_{jk})$ may be dropped from (16) and the equation written in the form

$$\left\{ \sum_{j=1}^{h} \sum_{k=1}^{\infty} + \sum_{j=1}^{\infty} \sum_{k=1}^{h} - \sum_{j=1}^{h} \sum_{k=1}^{h} + \sum_{j=h+1}^{\infty} \sum_{k=h+1}^{\infty} \right\} \frac{\lambda_{jk}}{x - \lambda_{jk}} = 0, \tag{17}$$

where h is a suitably chosen integer.

From (13) and (14) we have

$$\sum_{k=1}^{\infty} \frac{1}{(k - \frac{1}{2})^2 \pi^2} \frac{1}{1 - (\theta/(k - \frac{1}{2})^2 \pi^2)} = \frac{\tan \sqrt{\theta}}{2\sqrt{\theta}},$$

from which the first two terms of (17) are

$$\sum_{j=1}^{h} \sum_{k=1}^{\infty} = \sum_{j=1}^{\infty} \sum_{k=1}^{h} = \frac{1}{2\sqrt{x}} \sum_{j=1}^{h} \frac{1}{(j - \frac{1}{2})\pi} \tan \{x^{\frac{1}{2}}(j - \frac{1}{2}) \pi\}^{-1}.$$

The third term is evaluated by direct summation and the fourth is written in the form

$$\frac{1}{\pi^4 x} \sum_{j=h+1}^{\infty} \sum_{k=h+1}^{\infty} \sum_{r=0}^{\infty} \frac{1}{(j - \frac{1}{2})^{2r+2} (k - \frac{1}{2})^{2r+2} x^r \pi^{4r}} = \sum_{r=1}^{\infty} a_r y^r,$$

where $\quad y = (\pi^4 x)^{-1} \quad$ and $\quad a_r = \left\{ \sum_{j=h+1}^{\infty} \frac{1}{(j - \frac{1}{2})^{2r}} \right\}^2$

which is evaluated using formula 23.2.20 and Table 23.3 of Abramowitz and Stegun (1964). For the value of h chosen, namely 10, it was found that the series converged very rapidly.

Similar methods were used to obtain a computable form for the derivative of the left-hand side of (17) and the equation was solved by Newton's method. In this way the 30 largest roots $\nu_1, \ldots, \nu_{30}$ of (16) were calculated. W^2 was then approximated by $W_0^2 = \sum_1^{31} \nu_j \xi_j^2$ where $\xi_1, \ldots, \xi_{30}$ are independent $N(0, 1)$ and ξ_{31}^2 is a χ^2 variable with h_{31} degrees of freedom. The values of ν_{31}, h_{31} were chosen so that W_0^2 and W^2 have exactly the same mean and variance. The values obtained are listed below.

$$\nu_1 = 0{\cdot}06323661 \qquad\qquad \nu_{14} = 0{\cdot}00088869$$
$$\nu_2 = 0{\cdot}01825064 \qquad \nu_{15} = \nu_{16} = \nu_{17} = 0{\cdot}00073003$$
$$\nu_3 = 0{\cdot}01136115 \qquad\qquad \nu_{18} = 0{\cdot}00063364$$
$$\nu_4 = 0{\cdot}00657023 \qquad\qquad \nu_{19} = 0{\cdot}00056836$$
$$\nu_5 = 0{\cdot}00491204 \qquad\qquad \nu_{20} = 0{\cdot}00051839$$
$$\nu_6 = 0{\cdot}00335216 \qquad\qquad \nu_{21} = 0{\cdot}00045500$$
$$\nu_7 = 0{\cdot}00278038 \qquad\qquad \nu_{22} = 0{\cdot}00042790$$
$$\nu_8 = \nu_9 = 0{\cdot}00202785 \qquad \nu_{23} = \nu_{24} = \nu_{25} = 0{\cdot}00037246$$
$$\nu_{10} = 0{\cdot}00165297 \qquad\qquad \nu_{26} = 0{\cdot}00033749$$
$$\nu_{11} = 0{\cdot}00135749 \qquad\qquad \nu_{27} = 0{\cdot}00031050$$
$$\nu_{12} = 0{\cdot}00118583 \qquad\qquad \nu_{28} = 0{\cdot}00029273$$
$$\nu_{13} = 0{\cdot}00097193 \qquad \nu_{29} = \nu_{30} = 0{\cdot}00026281$$

$$\nu_{31} = 0{\cdot}00007639 \qquad\qquad h_{31} = 143{\cdot}155383$$

The distribution function of W_0^2 was obtained by Imhof's (1961) method, i.e. inversion of the characteristic function by numerical integration using the formula

$$\Pr\left(W_0^2 \leqslant x\right) = \tfrac{1}{2} - \frac{1}{\pi} \int_0^\infty \frac{\sin \theta(u)}{u\rho(u)}\, du, \tag{18}$$

where
$$\theta = \tfrac{1}{2} \sum_{r=1}^{31} h_r \tan^{-1}(\nu_r u),$$

$$\rho(u) = \prod_1^{31} \left(1 + \nu_r^2 u^2\right)$$

with $h_r = 1$ for $r = 1, \ldots, 30$. It can be shown by a slight variant of Imhof's treatment that the truncation error due to replacing ∞ in (18) by U is bounded above by

$$\left\{ \pi \max k U^k \prod_{r=1}^j \nu_r^{\frac{1}{2}h_r} \right\}^{-1} \quad \text{where} \quad k = \frac{1}{2} \sum_{r=1}^j h_r.$$

$U = 2100$ was taken and this gave a bound $< 10^{-5}$. The integrals were evaluated by the trapezoidal rule, using a variable length of sub-interval chosen to make the increments $\{u\rho(u)\}^{-1}\,du$ approximately constant. Lengths of sub-intervals were halved in successive evaluations until the change from one evaluation to the next was less than 10^{-5}. The results are given in Table 1. Lagrangian interpolation was used to obtain the 10, 5, and 1% significance points.

Table 1. $Pr(W^2 \leqslant x)$

x	Pr $(W^2 \leqslant x)$	x	Pr $(W^2 \leqslant x)$	x	Pr $(W^2 \leqslant x)$
0·01	0·00000	0·36	0·96370	0·71	0·99838
0·02	0·00000	0·37	0·96693	0·72	0·99851
0·03	0·00140	0·38	0·96986	0·73	0·99864
0·04	0·01581	0·39	0·97253	0·74	0·99875
0·05	0·05658	0·40	0·97494	0·75	0·99885
0·06	0·12126	0·41	0·97714	0·76	0·99894
0·07	0·19900	0·42	0·97914	0·77	0·99903
0·08	0·28003	0·43	0·98095	0·78	0·99911
0·09	0·35819	0·44	0·98260	0·79	0·99918
0·10	0·43036	0·45	0·98411	0·80	0·99925
0·11	0·49535	0·46	0·98548	0·81	0·99931
0·12	0·55304	0·47	0·98673	0·82	0·99937
0·13	0·60386	0·48	0·98787	0·83	0·99942
0·14	0·64844	0·49	0·98890	0·84	0·99947
0·15	0·68751	0·50	0·98985	0·85	0·99951
0·16	0·72173	0·51	0·99071	0·86	0·99955
0·17	0·75176	0·52	0·99150	0·87	0·99959
0·18	0·77815	0·53	0·99222	0·88	0·99962
0·19	0·80138	0·54	0·99288	0·89	0·99965
0·20	0·82188	0·55	0·99348	0·90	0·99968
0·21	0·84002	0·56	0·99403	0·91	0·99971
0·22	0·85610	0·57	0·99453	0·92	0·99973
0·23	0·87039	0·58	0·99499	0·93	0·99975
0·24	0·88311	0·59	0·99541	0·94	0·99977
0·25	0·89445	0·60	0·99580	0·95	0·99979
0·26	0·90460	0·61	0·99615	0·96	0·99981
0·27	0·91368	0·62	0·99647	0·97	0·99982
0·28	0·92183	0·63	0·99676	0·98	0·99984
0·29	0·92914	0·64	0·99703	0·99	0·99985
0·30	0·93572	0·65	0·99728	1·00	0·99986
0·31	0·94165	0·66	0·99750		
0·32	0·94700	0·67	0·99771		
0·33	0·95182	0·68	0·99790		
0·34	0·95618	0·69	0·99807		
0·35	0·96012	0·70	0·99823		

Significance values: 10%, 0·25533; 5%, 0·32611; 1%, 0·50166.

The distribution function was recalculated at a sample grid of points by the same method but using the 60 largest roots of (16) instead of the 30 largest. The results were identical to the number of places printed out, namely six. As a general check on the method it was used to recalculate the df of the statistic B of Blum, Kiefer and Rosenblatt (1961), which has a similar form to W^2, at a sample of 16 points. The results agree with the values given in Blum, Kiefer and Rosenblatt's Table 1 to one unit in the fifth decimal place in 15 cases and to two units in the remaining case.

5   COMPUTATION OF W_n^2

Consider the practical evaluation of the integral

$$W_n^2 = n \int_0^1 \int_0^1 \{F_n(s,t) - st\}^2 \, ds \, dt$$

for a particular sample of observations. For given s, denote the values of t for points whose s co-ordinate is less than this by $t_{1s} \leqslant t_{2s} \leqslant \ldots \leqslant t_{rs}$. Then

$$n \int_0^1 \{F_n(s,t) - st\}^2 \, dt = n \sum_{j=0}^r \int_{t_{js}}^{t_{(j+1)s}} \left(\frac{j}{n} - st\right)^2 dt \quad (t_{0s} = 0, t_{(r+1)s} = 1)$$

$$= n \sum_{j=0}^r \left[\frac{j^2}{n^2} (t_{(j+1)s} - t_{js}) - \frac{j}{n} s(t_{(j+1)s}^2 - t_{js}^2) \right.$$
$$\left. + \frac{s^2}{3} (t_{(j+1)s}^3 - t_{js}^3) \right]$$

$$= n \left[-\frac{1}{n^2} \sum_{j=1}^r (2j-1) t_{js} + \frac{r^2}{n^2} + \frac{s}{n} \sum_{j=1}^r t_{js}^2 - \frac{rs}{n} + \frac{s^2}{3} \right].$$

Denoting the n observed values of s by $s_1 \leqslant s_2 \leqslant \ldots \leqslant s_n$ we therefore have

$$W_n^2 = n \sum_{r=0}^n \left[\left\{ -\frac{1}{n^2} \sum_{j=1}^r (2j-1) t_{jr} + \frac{r^2}{n^2} \right\} (s_{r+1} - s_r) \right.$$
$$\left. + \frac{1}{2} \left(\frac{1}{n} \sum_{j=1}^r t_{jr}^2 - \frac{r}{n} \right) (s_{r+1}^2 - s_r^2) + \tfrac{1}{9}(s_{r+1}^3 - s_r^3) \right] \quad (s = 0, s_{n+1} = 1),$$

where t_{jr} denotes an observed value of t which is jth in order of magnitude among those whose corresponding s co-ordinates are less than s_{r+1}.
The only term in this expression giving rise to difficulty is

$$\sum_{r=0}^n \sum_{j=1}^r jt_{jr}(s_{r+1} - s_r).$$

Consider the contribution to this term of a particular observed value of t. As r increases, suppose this value first comes in as the contribution from the point $(s_{r_1}, t_{j_1 r_1})$ and suppose that the next point coming into this term whose t co-ordinate is less than $t_{j_1 r_1}$ has s co-ordinate equal to s_{r_2}. The contribution of $t_{j_1 r_1}$ to the sum from $r = r_1$ to $r = r_2$ is then

$$j_1 t_{j_1 r_1}(s_{r_1+1} - s_{r_1}) + j_1 t_{j_1 r_1}(s_{r_1+2} - s_{r_1+1}) + \ldots + j_1 t_{j_1 r_1}(s_{r_2} - s_{r_2-1})$$

$$+ (j_1 + 1) t_{j_1 r_1}(s_{r_2+1} - s_{r_2})$$

$$= -j_1 t_{j_1 r_1} s_{r_1} - t_{j_1 r_1} s_{r_2} + (j_1 + 1) t_{j_1 r_1} s_{r_2+1}.$$

Noting that $nF_n(s_{r_1}, t_{j_1 r_1}) = j_1$ and $s_{n+1} = 1$ we obtain for the contribution of $t_{j_1 r_1}$ to the overall sum $\sum_{r=0}^{n} \sum_{j=1}^{r} jt_{jr}(s_{r+1} - s_r)$ the result

$$- nF_n(s_{r_1}, t_{j_1 r_1}) s_{r_1} t_{j_1 r_1} - \sum_{r}' t_{j_1 r_1} s_r + T(t_{j_1 r_1}) t_{j_1 r_1},$$

where $T(t_{j_1 r_1})$ is the rank of $t_{j_1 r_1}$ among the whole set of n values of t, and $\sum_{r}'$ denotes summation over those values of $r > r_1$ for which the t co-ordinate is less than $t_{j_1 r_1}$. Taking all t values we therefore have

$$- \frac{2}{n^2} \sum_{r=0}^{n} \sum_{j=1}^{r} jt_{jr}(s_{r+1} - s_r) = \frac{2}{n} \sum_{i=1}^{n} F_n(s_i, t_i) s_i t_i + \frac{2}{n^2} \sum_{i=1}^{n} R_i t_i - \frac{2}{n^2} \sum_{i=1}^{n} T_i t_i,$$

where R_i is the sum of all s values for points (s, t) in the region $s > s_i$, $t < t_i$ and T_i is the rank of the t co-ordinate of the point (s_i, t_i) in the whole set of n values of t.

The other terms give

$$\frac{1}{n^2} \sum_{r=0}^{n} \sum_{j=1}^{r} t_{jr}(s_{r+1} - s_r) = \frac{1}{n^2} \sum_{i=1}^{n} (1 - s_i) t_i,$$

$$\sum_{r=0}^{n} \frac{r^2}{n^2}(s_{r+1} - s_r) = \frac{1}{n^2}[s_2 - s_1 + 4(s_3 - s_2) + \ldots + n^2(1 - s_n)]$$

$$= 1 - \frac{1}{n^2} \sum_{i=1}^{n} (2i - 1) s_i,$$

$$\frac{1}{2n} \sum_{r=0}^{n} \sum_{j=1}^{r} t_{jr}^2(s_{r+1}^2 - s_r^2) = \frac{1}{2n} \sum_{i=1}^{n} (1 - s_i^2) t_i^2,$$

$$- \frac{1}{2n} \sum_{r=0}^{n} r(s_{r+1}^2 - s_r^2) = \frac{1}{2n} \sum_{i=1}^{n} s_i^2 - \frac{1}{2},$$

$$\frac{1}{9} \sum_{r=0}^{n} (s_{r+1}^3 - s_r^3) = \frac{1}{9}.$$

After some elementary but tedious reductions we obtain

$$W_n^2 = 2 \sum_{i=1}^n F_n(s_i, t_i)\, s_i t_i + \frac{2}{n} \sum_{i=1}^n R_i t_i - \frac{1}{2} \sum_{i=1}^n s_i^2 t_i^2 - \frac{1}{n} \sum_{i=1}^n s_i t_i$$

$$+ \frac{1}{2} \sum_{i=1}^n (s_i^2 + t_i^2) + \frac{1}{n} \sum_{i=1}^n (s_i + t_i) - \frac{2}{n} \sum_{i=1}^n (S_i s_i + T_i t_i) + \tfrac{11}{18} n, \quad (19)$$

where as before R_i is the sum of the s co-ordinates of points (s_j, t_j) for which $s_j > s_i$ and $t_j < t_i$, S_i is the rank of s_i and T_i is the rank of t_i in the sets of n values of s and t respectively.

Further manipulation yields the alternative form

$$W_n^2 = W_{ns}^2 + W_{nt}^2 - \sum_{i=1}^n \left(\frac{r_i - \frac{1}{2}}{n} - s_i t_i \right)^2 + \frac{1}{2} \sum_{i=1}^n s_i^2 t_i^2 - \frac{1}{2} \sum_{i=1}^n (s_i^2 + t_i^2)$$

$$+ \frac{2}{n} \sum_{i=1}^n R_i t_i + \frac{1}{n^2} \sum_{i=1}^n (r_i - \tfrac{1}{2})^2 - \tfrac{1}{18} n, \quad (20)$$

where W_{ns}^2 is the univariate Cramér–von Mises statistic for the s values, i.e.

$$W_{ns}^2 = \sum_{i=1}^n \left(\frac{S_i - \frac{1}{2}}{n} - s_i \right)^2 + \frac{1}{12n},$$

W_{nt}^2 is the same statistic calculated from the t values and $r_i = n F_n(s_i, t_i)$.

The complexity of these formulae suggests that it might be preferable to use $dF_n(s, t)$ in place of $ds\,dt$ in the expression for W_n^2. This gives the statistic

$$\overline{\overline{W}}_n^2 = n \int_0^1 \int_0^1 \{ F_n(s, t) - st \}^2\, dF_n(s, t).$$

Since $dF_n(s, t) = n^{-1}$ at observation points (s_i, t_i) and $= 0$ elsewhere we obtain

$$\overline{\overline{W}}_n^2 = \sum_{i=1}^n \left(\frac{r_i}{n} - s_i t_i \right)^2,$$

where $r_i = n F_n(s_i, t_i)$ as before. By analogy with the univariate case and in an attempt to introduce a form of continuity correction it would seem preferable to use instead of $\overline{\overline{W}}_n^2$ the statistic

$$\overline{W}_n^2 = \sum_{i=1}^n \left(\frac{r_i - \frac{1}{2}}{n} - s_i t_i \right)^2 + c_n,$$

where c_n is chosen to make the mean of $\overline{W}_n^2$ the same as that of W_n^2. By straight-forward algebra the value of c_n is found to be $1/(36n)$. We therefore obtain finally the statistic

$$\overline{W}_n^2 = \sum_{i=1}^n \left(\frac{r_i - \frac{1}{2}}{n} - s_i t_i \right)^2 + \frac{1}{36n}. \quad (21)$$

By arguments similar to those of Kiefer (1959, §2) and Kiefer and Wolfowitz (1958) it follows that $\overline{W}_n^2$ has the same asymptotic distribution as W_n^2.

6 SOME FURTHER STATISTICS BASED ON $F_n(s,t)$

Let $G_n(s)$ denote the mean of $F_n(s,t)$ for given s, i.e. $G_n(s) = \int_0^1 F_n(s,t)\,dt$.
In view of the mathematical intractability of the Kolmogorov–Smirnov statistics, typified by $D_n^+ = \max_{s,t}\{F_n(s,t)-st\}$, in the two-dimensional case, the following statistics seem worth considering:

$$G_n^+ = \max_{0\leqslant s\leqslant 1}\sqrt{n}\{G_n(s)-\tfrac12 s\},$$

$$G_n^- = \max_{0\leqslant s\leqslant 1}\sqrt{n}\{\tfrac12 s - G_n(s)\},$$

$$G_n = \max(G_n^+, G_n^-).$$

The term $\tfrac12 s$ comes from the fact that $E\{G_n(s)\} = \tfrac12 s$. A similar set of statistics is obtained by considering the mean of $F_n(s,t)$ for given t. The asymptotic distributions of these statistics will be sketched out briefly.

Suppose that exactly r of the n points (s_i, t_i) have their s co-ordinate $\leqslant s$ and that the t co-ordinates of these r points are $t_{r1} \leqslant t_{r2} \leqslant \ldots \leqslant t_{rr}$. Then

$$G_n(s) = \frac{1}{n}(t_{r2}-t_{r1}) + \frac{2}{n}(t_{r3}-t_{r2}) + \ldots + \frac{r}{n}(1-t_{rr}) = \frac{1}{n}\left(r - \sum_{j=1}^r t_{rj}\right).$$

Thus, $E\{G_n(s)|r\} = \frac{1}{n}(r - \tfrac12 r) = \frac{1}{2}\frac{r}{n}$,   so $E\{G_n(s)\} = \tfrac12 s$

as stated.

Given $s' > s$ suppose exactly r' points have s co-ordinate $\leqslant s'$. Then

$$E\{G_n(s)\,G_n(s')|r,r'\} = \frac{1}{n^2}E\left\{rr' - r\sum_{j=1}^{r'} t_{r'j} - r'\sum_{j=1}^{r} t_{rj} + \sum_{j=1}^{r} t_{rj}\sum_{j=1}^{r'} t_{r'j}\bigg|r,r'\right\}$$

$$= \frac{1}{n^2}E(rr' - \tfrac12 rr' - \tfrac12 rr' + \tfrac14 rr' + \tfrac{1}{12}r)$$

since $E\left\{\sum_{j=1}^{r} t_{rj}\bigg|r,r',t_{r'1},\ldots,t_{r'r'}\right\} = \frac{r}{r'}\sum_{j=1}^{r'} t_{r'j}$

and $E\left\{\left(\sum_{j=1}^{r'} t_{r'j}\right)^2\bigg|r,r'\right\} = \tfrac14 r'^2 + \tfrac{1}{12}r'.$

Thus
$$E\{G_n(s)\,G_n(s')\} = \frac{1}{n^2}\left[\frac{1}{4}\frac{s}{s'}\{n^2s'^2 + ns'(1-s')\} + \tfrac{1}{12}ns\right]$$

$$= \tfrac{1}{4}ss' + \frac{1}{12n}\,s(4-3s'),$$

from which we have

$$nC\{G_n(s),\,G_n(s')\} = \tfrac{1}{12}s(4-3s'). \tag{22}$$

Let $g(s)$ be the normal process with mean zero and covariance function given by the right-hand side of (22). It follows from the results of Dudley (1966) that G_n^+, G_n^- and G_n are distributed asymptotically as

$$G^+ = \max_{0\leqslant s\leqslant 1} g(s), \quad G^- = \min_{0\leqslant s\leqslant 1} g(s) \quad \text{and} \quad G = \max\,(G^+, G^-).$$

The form of the covariance function (22) then enables us to obtain the required distributions by Doob's method (Doob, 1949).

Putting
$$\xi(t) = (\tfrac{1}{4} + 9t)\,g\left\{\frac{4t}{\frac{1}{12} + 3t}\right\} \quad (0 \leqslant t \leqslant \tfrac{1}{12}),$$

we find that $\xi(t)$ has zero mean and covariance function

$$E\{\xi(t_1)\,\xi(t_2)\} = \min\,(t_1, t_2),$$

so $\xi(t)$ is a Brownian process. Also, $g(s) \leqslant a$ for $0 \leqslant s \leqslant 1$ if $\xi(t)/(\tfrac{1}{4} + 9t) < a$, i.e. if $\xi(t) < \tfrac{1}{4}a + 9at$ for $0 \leqslant t \leqslant \tfrac{1}{12}$. By standard reflection techniques (cf. Cox and Miller, 1965, Ex. 5.1, p. 220) we find that the probability that a Brownian path crosses the line $y = b + ct$ before $t = t_0$ and passes between the points (t_0, y_0), $(t_0, y_0 + dy)$ is

$$e^{-2bc}\frac{1}{\sqrt{2\pi t_0}}\exp - \left[\frac{(y_0 - 2b)^2}{2t_0}\right] dy \quad \text{for} \quad y_0 < b + ct_0$$

and
$$\frac{1}{\sqrt{2\pi t_0}}\exp - \left[\frac{y_0^2}{2t_0}\right] dy \quad \text{for} \quad y_0 \geqslant b + ct_0.$$

The probability that the line is not crossed is therefore

$$\Phi\left(\frac{b + ct_0}{\sqrt{t_0}}\right) - e^{-2bc}\,\Phi\left(\frac{-b + ct_0}{\sqrt{t_0}}\right),$$

where $\Phi(x)$ is the probability that an $N(0, 1)$ variable $\leqslant x$. Putting $b = \tfrac{1}{4}a$, $c = 9a$ and $t_0 = \tfrac{1}{12}$ we have

$$\Pr\,(G_n^+ \leqslant a) = \Phi(2\sqrt{3}a) - e^{-\frac{9}{4}a^2}\Phi(\sqrt{3}a). \tag{23}$$

By symmetry, G^- has the same distribution.

By Doob's (1949) reflection technique, the probability of crossing the line $y = b + ct$, then the line $y = -b - ct$, then $y = b + ct, \dots$ and so on (k crossings) at least once in $(0, \frac{1}{12})$ is the same as the probability of crossing the line $y = kb + kct$, i.e. $H(ka)$ where $1 - H(a)$ denotes the probability (23). But the event, the path crosses either line $y = \pm (b + ct)$, is the same as the event

$$\sum_{k=1}^{\infty} (-1)^{k-1} E_k,$$

where E_k is the event, the path crosses $y = b + ct$, then $y = -b - ct$, then $y = b + ct, \dots$ (k crossings) at least once or crosses $y = -b - ct$ then $y = b + ct$, then $y = -b - ct, \dots$ (k crossings) at least once. Thus

$$\Pr (G \leqslant a) = 1 - 2 \sum_{k=1}^{\infty} (-1)^{k-1} H(ka). \tag{24}$$

A further statistic which merits consideration is the mean $\bar{F}_n$ of $F_n(s, t)$ over the entire unit square, i.e.

$$\bar{F}_n = \int_0^1 \int_0^1 F_n(s, t) \, ds \, dt = \int_0^1 G_n(s) \, dt.$$

As before, $\qquad G_n(s) = \frac{1}{n} \left(r - \sum_{j=1}^{r} t_{rj} \right) \quad \text{for} \quad s_r \leqslant s < s_{r+1}.$

Thus $\qquad \bar{F}_n = \frac{1}{n} \sum_{r=0}^{n} r(s_{r+1} - s_r) - \frac{1}{n} \sum_{r=0}^{n} \sum_{j=1}^{r} t_{rj}(s_{r+1} - s_r). \tag{25}$

Let the t co-ordinates of the n points be numbered according to the suffixes of the s co-ordinates, i.e. the n points are denoted by $(s_i, t_i), i = 1, \dots, n$ where $s_1 \leqslant s_2 \leqslant \dots \leqslant s_n$. The contribution to the last term of (25) from t_j is then

$$-\frac{1}{n} t_j(s_{j+1} - s_j + s_{j+2} - s_{j+1} + \dots + 1 - s_n) = -\frac{1}{n} t_j + \frac{1}{n} s_j t_j.$$

Substituting in (25) and evaluating the sum we have

$$\bar{F}_n = 1 - \frac{1}{n} \sum_{j=1}^{n} s_j - \frac{1}{n} \sum_{j=1}^{n} t_j + \frac{1}{n} \sum_{j=1}^{n} s_j t_j$$

$$= \frac{1}{n} \sum_{j=1}^{n} (1 - s_j)(1 - t_j).$$

Since s_j, t_j are independent uniform $(0, 1)$ variables so are $1 - s_j, 1 - t_j$. Thus $\bar{F}_n$ is distributed as the mean of n independent random variables each of which is distributed as the product of two independent uniform $(0, 1)$ variables. We have $E(st) = \frac{1}{4}$ and $V(st) = \frac{1}{9} - \frac{1}{16} = \frac{7}{144}$. It follows that $\sqrt{n}(\bar{F}_n - \frac{1}{4})$ is asymptotically normal with mean zero and variance $\frac{7}{144}$.

The tests in this paper refer to a fully specified null hypothesis. Where there are nuisance parameters present these can sometimes be eliminated by the randomization device suggested in Durbin (1961).

ACKNOWLEDGEMENTS

The calculations for this paper were done by Susannah Brown and Roger R. Heard. I am extremely grateful for this assistance. I am indebted to D. R. Brillinger, H. A. David and R. Pyke for some useful references.

REFERENCES

Abramowitz, M. and Stegun, I. A. (1964). *Handbook of mathematical functions. National Bureau of Standards, U.S.A.*

Bateman, G. I. (1949). The characteristic function of a weighted sum of noncentral squares of normal variates subject to s linear restraints. *Biometrika*, **36**, 460.

Blum, J. R., Kiefer, J. and Rosenblatt, M. (1962). Distribution-free tests of independence based on the sample distribution function. *Ann. Math. Statist.* **32**, 485.

Cox, D. R. and Miller, H. D. (1965). *The theory of stochastic processes*. Methuen, London.

Doob, J. L. (1949). Heuristic approach to the Kolmogorov–Smirnov theorems. *Ann. Math. Statist.* **20**, 277.

Dudley, R. M. (1966). Weak convergence of probabilities on nonseparable metric spaces and empirical measures on Euclidean spaces. *Illinois J. Math.* **10**, 109.

Durbin, J. (1961). Some methods of constructing exact tests. *Biometrika*, **48**, 41.

Durbin, J. and Watson, G. S. (1950). Testing for serial correlation in least-squares regression I. *Biometrika*, **37**, 409.

Imhof, J. P. (1961). Computing the distribution of quadratic forms in normal variables. *Biometrika*, **48**, 419.

Kiefer, J. (1959). K-sample analogues of the Kolmogorov–Smirnov and Cramer–von Mises tests. *Ann. Math. Statist.* **30**, 420.

Kiefer, J. and Wolfowitz, J. (1958). On the deviations of the empiric distribution function of vector chance variables. *Trans. Am. Math. Soc.* **87**, 173.

Kuelbs, J. (1968). The invariance principle for a lattice of random variables. *Ann. Math. Statist.* **39**, 382.

Pyke, R. (1968). The weak convergence of the empirical process with random sample size. *Proc. Camb. Phil. Soc.* **64**, 155.

Rosenblatt, M. (1952a). Remarks on a multivariate transformation. *Ann. Math. Statist.* **23**, 470.

Rosenblatt, M. (1952b). Limit theorems associated with variants of the von Mises statistic. *Ann. Math. Statist.* **23**, 617.

Rosenblatt, M. (1962). *Random processes*. New York: Oxford University Press.

DISCUSSION ON DURBIN'S PAPER

OSCAR KEMPTHORNE

I am very pleased to see the developments of Dr Durbin. The testing of goodness-of-fit is, I think, a critical aspect of data analysis and interpretation. The simple case of a random sample from a univariate distribution has been worked extensively and the results are informative. They seem to indicate that there is an indefinitely large number of possible tests, each of which is most sensitive with regard to its own noncentrality, but each of which is relatively insensitive to other types of deviation from the hypothesized distribution.

From the viewpoint of tests of significance (rather than accept-reject rules of hypotheses testing theory), the problem in the univariate case is that each sample of size N can be represented as a point in N-dimensional Euclidean space, and a test of significance is really just an ordering of the possible points. The choice of an ordering is simple only if there is a completely specified distribution D_0 and a completely specified alternative distribution D_1, say, because then an ordering which is difficult to improve on is the ordering:

$x_1, x_2, ..., x_N$ is more significant than $x_1', x_2', ..., x_N'$ if

$$\frac{P(x_1, x_2, ..., x_N; D_1)}{P(x_1, x_2, ..., x_N; D_0)} > \frac{P(x_1', x_2', ..., x_N'; D_1)}{P(x_1', x_2', ..., x_N'; D_0)}.$$

Indeed, it seems unquestionable that this ordering cannot be improved on.

This solution is not, of course, restricted to the univariate case, and goes over directly to the multivariate case if the x_i are interpreted as vectors. The extension to a case of a family of alternative distributions depends critically on some monotone likelihood ratio property.

There has been extensive work on Cramer–von Mises statistic and the Kolmogorov–Smirnov statistic. These statistics have interest in contexts quite different from that of goodness-of-fit. But as a user of statistical methods I have acquired a strong impression that this type of statistic is rather insensitive for the assaying of goodness-of-fit. I found the results of Shapiro and Wilk (*Biometrika*, 1965) rather convincing. I wonder if there really is any case in view of what I say above for the use of the KS statistic for goodness-of-fit. Furthermore, the use of maximum discrepancy of the sample cdf seems very questionable in view of the fact that

[450]

the variance of actual discrepancy, which is, of course, binomial varies so much over the total range. Hence the discrepancies have quite unequal precisions and the mere determination of which is largest without taking account of the unequal precisions in some way seems of very doubtful efficacy. The statistic W_n^2 seems to treat all $[F_n(s, t) - st]$ on an equal basis and seems to suffer from the same defect. The same seems to apply to the other statistics discussed in §6.

The arbitrariness of the initial transformation of Dr Durbin does, I believe, make the development somewhat tangential. I wonder how sensitive the tests are to this initial choice. Can this initial arbitrariness be exploited to yield sensitivity in different 'directions'?

However, the distribution of these statistics will surely be useful in contexts, perhaps, such as Wiener process theory, other than goodness-of-fit; and I commend Dr Durbin for his derivations which I found very interesting.

As regards testing of goodness-of-fit, it seems that those of us, including myself, who regarded the idea as crucial, must clarify what we want before we can claim to have made any real progress in data interpretation.

I have a final rather minor point. I believe the idea of Dr Durbin (*Biometrika*, 1961) with regard to nuisance parameters is of very questionable utility but obviously differences of opinion are possible and should be tolerated.

STUDENTIZING ROBUST ESTIMATES

PETER J. HUBER

1 INTRODUCTION

In most estimation contexts, the experimenter will not be satisfied with the mere value of some estimate T of the unknown θ; he will also want an indication of its accuracy. Sometimes, a confidence interval will be appropriate, but often, especially if the estimate is to be processed further and to be combined with other estimates, some kind of estimated standard deviation $s(T)$ will be more convenient.

While a lot of work has been spent on determining the limiting distributions and limiting efficiencies of robust estimates, relatively little seems to be known (or has found its way into print) about the $s(T)$ appropriate for a given estimator T and about the Studentized ratio $(T - \theta)/s(T)$. It is hoped that this preliminary and heuristic study will help to clarify some of the issues and to encourage some further work in this important area. After some hesitations I decided to put forth even rather conjectural results, since they might serve as guideposts for badly needed empirical sampling studies.

The general philosophy behind the choice of numerator $T - \theta$ and denominator $s(T)$ has been outlined by Tukey and McLaughlin [14]:

choose the T in the numerator to achieve a high robustness of performance;

then match it with a denominator $s(T)$ to achieve a high robustness of validity over a broad range of distributions.

For small sample sizes both requirements may be interpreted in several ways; but for large sample sizes, the second requirement simply means that the square of the denominator should be a consistent estimate of the asymptotic variance of the numerator, if both are suitably normed.

After one has chosen a pair T, $s(T)$, one would like to know whether, for which distributions and how fast the law of $(T - \theta)/s(T)$ does approximate the standard normal law. Can one approximate it more accurately by a t-distribution with the appropriate number of degrees of freedom, if the sample is moderately large, and what is this appropriate number? Although there may be better direct methods for finding confidence intervals (cf. §3), one should at least know whether and when it is reasonable to act as if T were normal and $s(T)$ were its true standard deviation.

[453]

PETER J. HUBER

That is, if we separate problems, we should know (i) whether and how fast $n^{\frac{1}{2}}(T-\theta)$ approaches a normal law; (ii) whether and how fast $ns(T)^2$ approaches a constant value, and whether this value coincides with the asymptotic variance of $n^{\frac{1}{2}}(T-\theta)$; (iii) whether numerator and denominator can approximately be treated as if they were stochastically independent.

These problems present severe technical difficulties on the theoretical side, since limiting distributions alone will not quite suffice and one has to consider also some higher order terms in the asymptotic expansions. A purely empirical approach by Monte Carlo techniques is out of the question because of the sheer magnitude of the task. On the other hand, since this type of asymptotic theory is directed toward intermediate sample sizes, say for n between 10 or 20 and 50, one urgently needs some empirical spot checks whether the theory makes any sense in the range for which it is primarily intended.

Actually, small sample investigations (cf. [7] and Fig. 1 in [12]) and some attempts at asymptotic expansions suggest that the numerator typically approximates the normal law rather fast, at least if the true underlying distribution of the observations is symmetric and has not too heavy tails (cf. [5], Ch. 8, and [13] for the principal ideas).

If we may assume (iii), this would mean that the distribution of the Studentized ratios depends on the sample size n mainly through the distribution of the denominator, as in the classical case.

It remains to be seen how accurate this hypothesis is, but in any case it seemed worthwhile to have a closer look at the asymptotic behavior of the denominator $s(T)$, i.e. to attack problem (ii) first.

2   THE CLASSICAL CASE

If $X_1, ..., X_n$ are independent identically distributed observations from $\mathcal{N}(\theta, \sigma^2)$, the classical estimates

$$\bar{X} = \frac{1}{n}(X_1 + X_2 + ... + X_n), \qquad s(\bar{X})^2 = \frac{1}{n(n-1)}\Sigma(X_i - \bar{X})^2$$

are stochastically independent, the former is normal $\mathcal{N}(\theta, \sigma^2/n)$, the latter is (apart from a constant factor) chi-square with $n-1$ degrees of freedom, and $(\bar{X} - \theta)/s(\bar{X})$ is asymptotically normal $\mathcal{N}(0,1)$. The limiting value of $ns(\bar{X})^2$ is σ^2, and for large n, the quotient

$$\frac{ns(\bar{X})^2}{\sigma^2}$$

is approximately normal $\mathcal{N}(1, 2/n)$.

We shall try to find out by which expression the variance $2/n$ has to be replaced for other estimators and distributions. This will give (i) an indication how fast $ns(T)^2$ converges, and (ii) it may suggest the appropriate number of degrees of freedom, if one wants to approximate the distribution of $s(T)^2$ by means of a chi-square distribution. In the next three sections we will discuss typical representatives of the three main types of robust estimators.

3   ESTIMATES DERIVED FROM RANK TESTS

Because of the origin of these estimates, confidence intervals can be found in a straightforward and natural fashion with the aid of the underlying tests (cf. [11]).

Consider the case of the Hodges–Lehmann estimate T (i.e. the sample median of the pairwise means $1/2(X_i + X_j)$), which derives from the Wilcoxon test. As usual, one assumes that the true underlying distribution has a sufficiently smooth symmetric density f satisfying

$$\int f(x)^2\, dx < \infty.$$

Confidence intervals for T can be obtained from Wilcoxon's one-sample test; if one divides the length of such a confidence interval by the corresponding interquantile distance of the standard normal distribution, one obtains an $s(T)$ such that $ns(T)^2$ is a consistent estimate of the asymptotic variance $\sigma_{HL}^2 = 1/(12(\int f^2 dx)^2)$ of $n^{\frac{1}{2}}(T - \theta)$ (Theorem 1' of [11]). To my knowledge, the speed with which $ns(T)^2$ approaches its limiting value has not been investigated so far. I conjecture that

$$\frac{ns(T)^2}{\sigma_{HL}^2}$$

is approximately normal $\mathcal{N}(1, C/n)$ with

$$C = 16\left[\frac{\int f^3\, dx}{(\int f^2\, dx)^2} - 1\right].$$

If we compare this with the preceding section, we might say that asymptotically, $s(T)^2$ has the equivalent of $(2/C)n$ degrees of freedom (for the normal and for the double-exponential density f one would have

$$2/C = 0\cdot808 \quad \text{and} \quad 2/C = 0\cdot375$$

respectively).

The base for this conjecture is as follows. Let ξ_α be the upper α-quantile of the normal distribution ($\Phi(-\xi_\alpha) = \alpha$). Then (θ^*, θ^{**}) is a confidence interval with asymptotic level $1-2\alpha$, where the boundaries θ^*, θ^{**} are determined by

$$\frac{\sqrt{12}}{n^{\frac{3}{2}}} \Sigma \left[I\left(\frac{X_i+X_j}{2} > \theta^*\right) - \tfrac{1}{2} \right] = \xi_\alpha$$

$$\frac{\sqrt{12}}{n^{\frac{3}{2}}} \Sigma \left[I\left(\frac{X_i+X_j}{2} > \theta^{**}\right) - \tfrac{1}{2} \right] = -\xi_\alpha.$$

Here, $I(A)$ is the indicator of the set A, and the sums are taken over the set of all pairs $i < j$.

The difference of these two equations is

$$\frac{\sqrt{12}}{n^{\frac{3}{2}}} \Sigma I\left(\theta^* < \frac{X_i+X}{2} < \theta^{**}\right) = 2\xi_\alpha.$$

We know that

$$n^{\frac{1}{2}} s(T) = \frac{n^{\frac{1}{2}}(\theta^{**} - \theta^*)}{2\xi_\alpha} \to \sigma_{HL}$$

and that θ^*, θ^{**} are of the stochastic order $O(n^{-\frac{1}{2}})$, if the X_i have the smooth symmetric density $f(x)$.

If t and Δ are fixed, non-stochastic quantities, one obtains by projection (Hájek [6], p. 335) that

$$U = \frac{\sqrt{12}}{n^{\frac{3}{2}}} \Sigma I\left(n^{-\frac{1}{2}}(t-\Delta) < \frac{X_i+X_j}{2} < n^{-\frac{1}{2}}(t+\Delta)\right)$$

is asymptotically equal to

$$\frac{2\Delta}{\sigma_{HL}} + \frac{\sqrt{(192)}\,\Delta}{n} \Sigma_i \left(f(X_i) - \int f^2\, dx \right).$$

In other words, we have asymptotically

$$U = \frac{2\Delta}{\sigma_{HL}}(1+Z),$$

where

$$Z = \frac{2}{n \int f^2\, dx} \Sigma \left(f(X_i) - \int f^2\, dx \right)$$

is asymptotically normal

$$\mathscr{N}\left(0, \frac{4}{n} \left[\frac{\int f^3\, dx}{\left(\int f^2\, dx\right)^2} - 1 \right] \right).$$

One now argues heuristically that the same relation should also hold with the stochastic quantities $n^{\frac{1}{2}}(\theta^* + \theta^{**})$ and $n^{\frac{1}{2}}(\theta^{**} - \theta^*)$ in place of

$2t$ and 2Δ respectively. Then we would obtain

$$\frac{n^{\frac{1}{2}}(\theta^{**}-\theta^{*})}{\sigma_{HL}}(1+Z) = 2\xi_{\alpha} \quad \text{or} \quad \frac{n^{\frac{1}{2}}s(T)}{\sigma_{HL}}(1+Z) = 1.$$

This would mean that $(n^{\frac{1}{2}}s(T)/\sigma_{HL})$ has the same asymptotic distribution as $1/(1+Z) \approx 1-Z$.

4 LINEAR COMBINATION OF ORDER STATISTICS

It is known by now that a suitably trimmed mean is an asymptotically minimax estimate of location for symmetrically ϵ-contaminated normal distributions ([1], [8], [10]). This makes the use of the trimmed mean even more attractive than before (cf. [14]).

Given n-ordered observations

$$X_1 \leqslant X_2 \leqslant \dots \leqslant X_n,$$

the *g-times trimmed sample* is the sample of size $h = n - 2g$ obtained by omitting the g smallest and the g largest observations. The *g-times Winsorized sample* is obtained by replacing the g smallest values by the nearest other X, namely X_{g+1}, and replacing the g largest values by the nearest other X, namely X_{n-g}.

We shall use below the g-times trimmed mean:

$$X_{Tg} = \frac{1}{n-2g}(X_{g+1}+X_{g+2}+\dots+X_{n-g})$$

and the g-times Winsorized sum of squared deviations from X_{Tg}:

$$SSD_{Wg} = g(X_{g+1}-X_{Tg})^2+(X_{g+1}-X_{Tg})^2$$
$$+(X_{g+2}-X_{Tg})^2+\dots+(X_{n-g}-X_{Tg})^2+g(X_{n-g}-X_{Tg})^2.$$

Instead of 'g-times trimmed (Winsorized)' we shall also say 'α-trimmed (Winsorized)', where $\alpha = g/n$ is the fraction of observations affected by trimming or Winsorizing on each side.

Tukey and McLaughlin [14] proposed, among other possibilities, to Studentize the trimmed mean X_{Tg} by dividing it by

$$s(X_{Tg}) = \sqrt{\left(\frac{SSD_{Wg}}{h(h-1)}\right)},$$

where $h = n - 2g$ is the number of observations left unaffected by trimming or Winsorizing. They were led to this curious pairing, as it seems, through 'trial and consideration' involving *small samples* $(n \leqslant 20)$.

They failed to notice that an even better case can be made for this choice of the denominator, if one also takes *large sample* theory into account.

The point is that $ns(X_{Tg})^2$ is a consistent estimate of the asymptotic variance of $n^{\frac{1}{2}}(X_{Tg} - \theta)$, whenever the true underlying distribution F is symmetric, continuous and strictly increasing at the points $\pm\xi_\alpha$, where $F(-\xi_\alpha) = \alpha$ is the limiting fraction g/n. So the quotient

$$\frac{X_{Tg} - \theta}{s(X_{Tg})}$$

is asymptotically normal $\mathcal{N}(0, 1)$ for all these F.

This can be seen as follows: $n^{\frac{1}{2}}(X_{Tg} - \theta)$ has asymptotic mean 0 and asymptotic variance

$$\sigma(\alpha)^2 = \frac{\displaystyle\int_{-\xi_\alpha}^{\xi_\alpha} x^2 F(dx) + 2\alpha\xi_\alpha^2}{(1 - 2\alpha)^2},$$

cf. [1].

A straightforward application of the Glivenko–Cantelli theorem shows that

$$\frac{SSD_{Wg}}{n} \to \int_{-\xi_\alpha}^{\xi_\alpha} x^2 F(dx) + 2\alpha\xi_\alpha^2,$$

hence $\qquad\qquad\qquad ns(X_{Tg})^2 \to \sigma(\alpha)^2.$

How fast is this convergence? I assert that

$$n^{\frac{1}{2}}\left(\frac{SSD_{Wg}}{n} - (1 - 2\alpha)^2 \sigma(\alpha)^2\right)$$

is asymptotically normal with asymptotic mean 0 and asymptotic variance

$$\tau(\alpha)^2 = \int_{-\xi_\alpha}^{\xi_\alpha} x^4\, dF + 2\alpha\left(\xi_\alpha^2 + \frac{2\alpha\xi_\alpha}{f(\xi_\alpha)}\right)^2 - \left[\int_{-\xi_\alpha}^{\xi_\alpha} x^2\, dF + 2\alpha\left(\xi_\alpha^2 + \frac{2\alpha\xi_\alpha}{f(\xi_\alpha)}\right)\right]^2,$$

where f is the density of F, which is assumed to exist and to be continuous near $\pm\xi_\alpha$.

This can be proved in a similar way as Bickel [1] proved the asymptotic normality of the trimmed and Winsorized means; a conceptually even simpler proof is possible with the invariance principle of Doob [4] and Donsker [3]. Hence

$$\frac{ns(X_{Tg})^2}{\sigma(\alpha)^2}$$

is approximately normal $\mathcal{N}(1, C/n)$ with

$$C = \frac{\tau(\alpha)^2}{(1-2\alpha)^4 \sigma(\alpha)^4}.$$

If we compare this asymptotic expression with the classical one (§2), we may say that, asymptotically, $s(X_{T_g})^2$ has the equivalent of $(2/C)n$ degrees of freedom.

Small sample experiments suggested (see [14]) that a t-distribution with $h-1$, or perhaps slightly less, degrees of freedom gives a good approximation to the distribution of $(X_{T_g} - \theta)/s(X_{T_g})$. This agrees with the above asymptotic results: the factor $2/C$ is slightly smaller than $(1-2\alpha) = h/n$ if the true underlying distribution is normal. But note that it is much smaller for longer-tailed distributions, for instance for the double-exponential one (Table 1).

Table 1

α	Normal case		Double exponential case	
	$\sigma(\alpha)^2$	$2/C$	$\sigma(\alpha)^2$	$2/C$
0	1·000	1·000	2·000	0·400
0·01	1·004	0·967	1·878	0·424
0·05	1·026	0·847	1·654	0·418
0·10	1·060	0·711	1·494	0·383
0·25	1·195	0·372	1·227	0·248

By the way, the formula for $\tau(\alpha)^2$ shows once more that Winsorizing is very sensitive to a vanishing density at the points $\pm \xi_\alpha$, so some care is needed if one works with grouped data. For instance, one might de-group the data by spreading the observations evenly over the critical intervals.

Note that the seemingly logical matches between the trimmed mean and the trimmed e.s.d. (estimated standard deviation), or between the Winsorized mean and the Winsorized e.s.d. [2] do not fit too well: the asymptotic variance of $(T - \theta)/s(T)$ will be different from 1 for most underlying distributions, and therefore the robustness of *validity* will be low for large sample sizes and extreme confidence levels. Nevertheless, matching a trimmed mean with a (possibly less) trimmed e.s.d. might be preferable to the Winsorized e.s.d. in certain small sample cases: the trimmed e.s.d., although biased, is more robust as an estimate of scale, and so the quotient might have better robustness of *performance*.

5   MAXIMUM LIKELIHOOD TYPE ESTIMATES

Some years ago I proposed to estimate location θ and scale σ by solving the simultaneous equations

$$\Sigma \psi \left(\frac{X_i - T}{S}\right) = 0, \quad \Sigma \chi \left(\frac{X_i - T}{S}\right) = 0$$

for T and S. Here, ψ is an odd, χ an even function, both being monotone increasing for positive arguments. More precisely, I had suggested choosing

$$\psi(x) = \min(k, \max(-k, x)), \quad \chi(x) = \psi(x)^2 - \beta,$$

with $\beta = E_\Phi(\psi(x)^2)$ ('proposal 2' of [8]).

These estimates are obviously translation and scale invariant; they have limiting values (are 'consistent') and are asymptotically normal under very general conditions [9]. If the true underlying distribution is symmetric around 0, then $\lim T = 0$; for simplicity, we shall assume that $\lim S = 1$.

The following calculations are of a heuristic nature, but they can easily be made rigorous by spelling out regularity conditions on ψ, χ and f, and by estimating remainder terms.

A Taylor expansion of the defining equations gives to a first-order approximation

$$n^{\frac{1}{2}}T = n^{-\frac{1}{2}} \Sigma \psi(X_i)/E(\psi'(X)), \quad n^{\frac{1}{2}}(S-1) = n^{-\frac{1}{2}} \Sigma \chi(X_i)/E(X\chi'(X)).$$

In particular, $n^{\frac{1}{2}}T$ is asymptotically normal with mean 0 and variance

$$\sigma(T)^2 = \frac{E(\psi(X)^2)}{(E\psi'(X))^2}.$$

Hence (cf. [8], p. 97),

$$ns(T)^2 = \frac{n}{n-1} \frac{\frac{1}{n}\Sigma \psi \left(\frac{X_i - T}{S}\right)^2}{\left[\frac{1}{n}\Sigma \psi' \left(\frac{X_i - T}{S}\right)\right]^2} S^2$$

would define a reasonable 'estimated standard deviation' $s(T)$ for T. We shall now expand the right-hand side of this definition into an asymptotic series. We have, with $\Delta S = S - 1$,

$$\Sigma \psi \left(\frac{X_i - T}{S}\right)^2 = nE(\psi^2) + \Sigma(\psi(X_i)^2 - E(\psi^2)) - n2E(X\psi'\psi)\Delta S + \ldots,$$

$$\Sigma \psi' \left(\frac{X_i - T}{S}\right) = nE(\psi') + \Sigma(\psi'(X_i) - E(\psi')) - nE(X\psi'')\Delta S + \ldots,$$

$$S^2 = 1 + 2\Delta S + \ldots.$$

If we put for short

$$U_i = \frac{1}{E(\psi^2)} (\psi(X_i)^2 - E(\psi^2)),$$

$$V_i = \frac{1}{E(\psi')} (\psi'(X_i) - E(\psi')),$$

$$W_i = \frac{2}{E(X\chi')} \left[1 + \frac{E(X\psi'')}{E(\psi')} - \frac{E(X\psi'\psi)}{E(\psi^2)} \right] \chi(X_i),$$

we may write

$$ns(T)^2 = \frac{n}{n-1} \sigma(T)^2 \left(1 + \frac{1}{n} \Sigma (U_i - 2V_i + W_i) + ... \right).$$

So, if we put $C = \text{var}(U_i - 2V_i + W_i),$

we find that $(ns(T)^2/\sigma(T)^2)$ is approximately normal $\mathcal{N}(1, C/n)$.

Again, if we compare this asymptotic expression with the classical one (§2), we may say that asymptotically, $s(T)^2$ has the equivalent of $2/C$ degrees of freedom.

For 'proposal 2' one obtains

$$C = A^2 \frac{m_4}{(m_2)^2} + 2\alpha B^2 - (A + 2\alpha B)^2$$

with

$$m_p = \int_{-k}^{+k} x^p f(x)\, dx,$$

$$\alpha = F(-k),$$

$$A = 1 - \frac{2kf(k)}{1-2\alpha},$$

$$B = A \frac{k^2}{m_2} + \frac{2}{1-2\alpha}.$$

Table 2 contains some numerical results; to facilitate comparison, the values of k were adjusted to give the same values of α as in Table 1 (for $\alpha = F(-k)$, the trimmed mean and T of 'proposal 2' have the same asymptotic variance).

Table 2

	Normal case			Double exponential case		
α	k	$\sigma(T)^2$	$2/C$	k	$\sigma(T)^2$	$2/C$
0	∞	1·000	1·000	∞	2·000	0·400
0·01	2·326	1·004	0·967	3·912	1·878	0·444
0·05	1·645	1·026	0·846	2·303	1·654	0·432
0·10	1·282	1·060	0·710	1·609	1·494	0·392
0·25	0·674	1·195	0·372	0·693	1·227	0·249

It would seem that the procedures of §§4 and 5 have almost equivalent properties. Because of their respective ease of computation, one would probably favor the trimmed mean for estimating a single location parameter, but (the obvious generalization of) 'proposal 2' for more general regression problems.

I would like to draw attention to the fact that C contains a term $E(X\psi'')$, which equals $-2kf(k)$ for 'proposal 2'. Hence, there will be trouble if the density becomes large at the points $\pm k$, so again some care is needed if one works with grouped data (although for a reason opposite to that of §4), and it may be advisable to smooth ψ near $\pm k$.

6 CONCLUSIONS

1. The trimmed mean, scaled by the Winsorized e.s.d., has both excellent small sample and excellent large sample properties; moreover, it is easy to compute. So it can strongly be recommended for practical use. Some further investigations of its small and intermediate sample size behavior are certainly desirable.

The proper trimming rate α will have to be determined empirically, e.g. with the aid of formula (23) of [14], using actual observational data. If one uses the conservative approach, i.e. if one determines α on the basis of previous samples, then, in my experience, sample sizes $\geqslant 1000$ seem to be needed for half-way stable and trustworthy results. For so-called good data, trimming rates between 0·05 and 0·10 turned out to be optimal, in fact, some empirical distributions suggested that a suitably trimmed mean would be an approximately efficient estimate. It does not seem to matter much whether one takes $\alpha = 0·05$ or $\alpha = 0·10$, but $\alpha = 0·25$ (Tukey's 'midmean') is usually too high and will also give a poor Winsorized e.s.d. for smaller samples.

2. The trimmed mean and the estimate T considered in §5 behave well even if the underlying distribution fails to have a density, but the corresponding estimates $s(T)$ do not.

This raises an interesting problem. Since the choice of the functional θ of F, which is to be estimated consistently, practically forces the choice of both the estimator T and of $s(T)$ upon us, one wonders how one has to choose this functional such that both T and $s(T)$ have optimal robustness properties.

REFERENCES

[1] Bickel, P. J. (1965). On some robust estimates of location. *Ann. Math. Statist.* **36**, 847–58.

[2] Dixon, W. J. and Tukey, J. W. (1968). Approximate behavior of the distribution of Winsorized t (Trimming/Winsorization 2). *Technometrics*, **10**, 83–98.

[3] Donsker, M. D. (1952). Justification and extension of Doob's heuristic approach to the Kolmogorov–Smirnov theorems. *Ann. Math. Statist.* **23**, 277.

[4] Doob, J. L. (1949). Heuristic approach to the Kolmogorov–Smirnov theorems. *Ann. Math. Statist.* **20**, 393–403.

[5] Gnedenko, B. V. and Kolmogorov, A. N. *Limiting Distributions for Sums of Independent Random Variables.* Cambridge, Mass.: Addison–Wesley, 1954.

[6] Hájek, J. (1968). Asymptotic normality of simple linear rank statistics under alternatives. *Ann. Math. Statist.* **39**, 325–46.

[7] Hodges, J. L. (1967). Efficiency in normal samples and tolerance of extreme values for some estimates of location. *Proc. Fifth Berkeley Symp. Math. Statist. and Probability*, Vol. 1, 163–86.

[8] Huber, P. J. (1964). Robust estimation of a location parameter. *Ann. Math. Statist.* **35**, 73–101.

[9] Huber, P. J. (1967). The behavior of maximum likelihood estimates under nonstandard conditions. *Proc. Fifth Berkeley Symp. Math. Statist. and Probability*, Vol. 1, 221–33.

[10] Huber, P. J. (1968). Robust confidence limits. *Z. f. Wahrscheinlichkeitstheorie* **10**, 269–78.

[11] Lehmann, E. L. (1963). Nonparametric confidence intervals for a shift parameter. *Ann. Math. Statist.* **34**, 1507–12.

[12] Leone, F. C., T. Jayachandran and S. Eisenstat (1967). A study of robust estimators. *Technometrics* **9**, 652–60.

[13] Tukey, J. W. (1960). A survey of sampling from contaminated distributions. In *Contributions to Probability and Statistics* (ed. Olkin), p. 448–85. Stanford Univ. Press.

[14] Tukey, J. W. and McLaughlin, D. H. (1963). Less vulnerable confidence and significance procedures for location based on a single sample. *Sankhyā*, Ser. A **25**, 331–52.

CHEBYSHEV BOUNDS FOR RISKS AND ERROR PROBABILITIES IN SOME CLASSIFICATION PROBLEMS

ALBERT W. MARSHALL AND INGRAM OLKIN

1 INTRODUCTION

In problems of classifying items that may belong to one of several classes, it is often the case that each item is characterized by a variable Y. The actual class to which the item belongs is determined by which of several sets contains Y. When the observation of Y is costly, inconvenient or impossible, it may be necessary or desirable to classify an item on the basis of an observation X correlated with Y. Then, a procedure making optimal use (in the sense of minimizing the risk, i.e. expected loss) of the variable X may be desired. The determination of such procedures was considered by Marshall and Olkin (1968) under the assumption that the conditional distribution $F(y|x)$ of Y given X is known. Examples studied there in detail with various loss functions were mostly for the case that $F(y|x)$ is normal; certain discrete distributions were also considered.

This paper is concerned with the case that means, variances and the covariance of X and Y are known, but that otherwise $F(x|y)$ is unknown. With this incomplete information, it is not possible to compute risks for given procedures or to determine minimum risk procedures. However, for any given procedure, a 'Chebyshev' bound for the risk is often obtainable. Our purpose here is to obtain such results for several important loss functions. A discussion of the various loss functions is given in § 2 together with bounds for the corresponding risks. Some Chebyshev bounds for probabilities of misclassification are given in § 3, but derivations are deferred to § 4. The discussion there may aid the reader who encounters a loss function not treated in § 2.

2 BOUNDS FOR THE RISK

2.1. Constant Loss I

Suppose an item is satisfactory if $|Y| \leqslant \delta$ and unsatisfactory if $|Y| > \delta$. It is to be classified as satisfactory or unsatisfactory according as $|X| \leqslant \epsilon$ or $|X| > \epsilon$. Suppose further that the loss is 0 if a satisfactory item is correctly classified (i.e. $|X| \leqslant \epsilon$, $|Y| \leqslant \delta$), the loss is λ_1 if an item is

[465]

classified unsatisfactory (i.e. $|X| > \epsilon$), and the loss is $\lambda_2 > \lambda_1$ if an unsatisfactory item is classified satisfactory (i.e. $|X| \leqslant \epsilon, |Y| > \delta$), as in Figure 1.

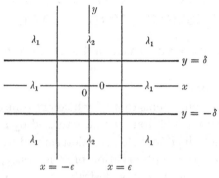

Figure 1

If $EX = EY = 0$, $\mathrm{var}(X) = \epsilon^2 \psi_{11}$, $\mathrm{var}(Y) = \delta^2 \psi_{22}$, $\mathrm{cov}(X, Y) = \epsilon\delta\psi_{12}$, then the risk, R, is bounded above as follows:

$$R \leqslant \min(\beta_1, \beta_2, ..., \beta_5), \tag{2.1}$$

where

$$\beta_1 = \tfrac{1}{2}\{\lambda_1\psi_{11} + \lambda_2\psi_{22} + [(\lambda_1\psi_{11} + \lambda_2\psi_{22})^2 - 4\lambda_1\lambda_2\psi_{12}^2]^{\frac{1}{2}}\};$$

$$\beta_2 = \lambda_2\psi_{11} - (\lambda_2 - \lambda_1)\frac{(\psi_{11} - |\psi_{12}|)^2}{\psi_{11} + \psi_{22} - 2|\psi_{12}|},$$

if $\lambda_1(\psi_{11} - |\psi_{12}|) + \lambda_2(\psi_{22} - |\psi_{12}|) \leqslant 0$; $\beta_2 = \infty$, otherwise;

$$\beta_3 = \lambda_1 + (\lambda_2 - \lambda_1)\frac{\psi_{11}\psi_{22} - \psi_{12}^2}{\psi_{11} + \psi_{22} - 2|\psi_{12}|},$$

if $\psi_{22} \leqslant |\psi_{12}| \leqslant \psi_{11}$; $\beta_3 = \infty$, otherwise;

$$\beta_4 = \lambda_1 + \psi_{22}(\lambda_2 - \lambda_1);$$

$$\beta_5 = \lambda_2.$$

2.2 Constant Loss II

In contrast with the previous example, it sometimes happens that unsatisfactory items ($|Y| > \delta$) are to be distinguished as 'too small' ($Y < -\delta$) or 'too large' ($Y > \delta$). The item is classified as 'too small' if $x < -\epsilon$ and as 'too large' if $x > \epsilon$. Suppose the losses are as in Figure 2, where $\lambda_2 \geqslant \lambda_1$.

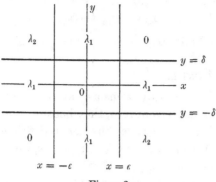

Figure 2

If $EX = EY = 0$, $\text{var}(X) = \epsilon^2\psi_{11}$, $\text{var}(Y) = \delta^2\psi_{22}$, $\text{cov}(X, Y) = \epsilon\delta\psi_{12}$, then the risk, R, is bounded above as follows:

$$R \leqslant \min(\beta_1, \beta_2, ..., \beta_6), \qquad (2.2)$$

where $\beta_1 = \frac{1}{2}\lambda_1\{\psi_{11} + \psi_{22} + [(\psi_{11} + \psi_{22})^2 - 4\psi_{12}^2]^{\frac{1}{2}}\}$,

if $(\lambda_2 - \lambda_1)^2 (\psi_{11} + \psi_{22} - 2\psi_{12}) \leqslant \lambda_1^2(\psi_{11} + \psi_{22} + 2\psi_{12})$; $\beta_1 = \infty$, otherwise;

$$\beta_2 = \frac{(\psi_{11}\psi_{12} - \psi_{12}^2)\lambda_2 + \lambda_1[\max(\psi_{11}, \psi_{22}) + \psi_{12}]^2}{\psi_{11} + \psi_{22} + 2\psi_{12}},$$

if $\lambda_1(\psi_{11} + \psi_{22} + 2\psi_{12}) \leqslant (\lambda_2 - \lambda_1)|\psi_{11} - \psi_{22}|$; $\beta_2 = \infty$, otherwise;

$$\beta_3 = \frac{\lambda_2}{4(\lambda_2 - \lambda_1)} [\lambda_2(\psi_{11} + \psi_{22} - 2\psi_{12}) + 4\lambda_1\psi_{12}];$$

$$\beta_4 = \lambda_1 + (\lambda_2 - \lambda_1) \frac{\psi_{11}\psi_{12} - \psi_{12}^2}{\psi_{11} + \psi_{22} + 2\psi_{12}},$$

if $-\psi_{12} \leqslant \min(\psi_{11}, \psi_{22})$; $\beta_4 = \infty$, otherwise;

$$\beta_5 = \lambda_1 + (\lambda_2 - \lambda_1)\min(\psi_{11}, \psi_{12});$$

$$\beta_6 = \lambda_2.$$

2.3. Quadratic Loss I

In some situations the distinction between 'satisfactory' and 'unsatisfactory' is ill-defined. Rather, an item becomes less and less satisfactory as Y deviates from some optimal or specified value δ. Then, the loss $\lambda_2(y)$ incurred by accepting an item ($|X| \leqslant \epsilon$) may be a continuous function, increasing as Y deviates from δ. However, the loss in rejecting an item ($|X| > \epsilon$) may still be a constant, λ, as in §2.1.

If, in particular, $\delta = 0$, $\lambda_2(y) = y^2$, $EX = EY = 0$, $\text{var}(X) = \epsilon^2 \psi_{11}$, $\text{var}(Y) = \psi_{22}$, and $\text{cov}(X, Y) = \epsilon \psi_{12}$, then

$$R \leqslant \tfrac{1}{2}[\psi_{22} + \lambda \min(\psi_{11}, 2 - \psi_{11}) + \{(\lambda \psi_{11} + \psi_{22})^2 - 4\lambda \psi_{12}^2\}^{\frac{1}{2}}]. \qquad (2.3)$$

2.4. Quadratic Loss II

It may be desirable to classify items as 'deficient' or 'adequate' when, as Y increases (decreases) from zero, they become more and more useful (detrimental). If $X < \epsilon$, we may say the item is deficient, and suffer a loss $\lambda_1(y) = y^2$ for $y \geqslant 0$, $\lambda_1(y) = 0$ for $y \leqslant 0$. If $X \geqslant \epsilon$, we may say the item is adequate, and suffer a loss $\lambda_2(y) = y^2$ for $y \leqslant 0$, $\lambda_2(y) = 0$ for $y \geqslant 0$.

If $EX = EY = 0$, $\text{var}(X) = \psi_{11}$, $\text{var}(Y) = \psi_{22}$, $\text{cov}(X, Y) = \psi_{12}$, then

$$\left.\begin{array}{ll} R \leqslant \psi_{22} - \psi_{12}^2/(\psi_{11} + \epsilon^2) & \text{if } \psi_{22} \geqslant 0, \\[2mm] R \leqslant \psi_{22}, & \text{if } \psi_{12} \leqslant 0. \end{array}\right\} \qquad (2.4)$$

3 BOUNDS FOR ERROR PROBABILITIES

In §§ 2.1 and 2.2, an item is in a state i according as $Y \in S_i$, but must be classified as being in state i if $X \in A_i$. Events such as $\{X \in A_i, Y \in S_j\}$ and $\{X \notin A_i, Y \in S_i\}$ are of interest, and we bound their probabilities when only the first and second moments are known.

As in § 2.1 (Figure 1), suppose an item is satisfactory or unsatisfactory according as $|Y| \leqslant \delta$ or $|Y| > \delta$ and is classified as satisfactory or unsatisfactory according as $|X| \leqslant \epsilon$ or $|X| > \epsilon$. If

$$EX = EY = 0, \quad \text{var}(X) = \epsilon^2 \psi_{11}, \quad \text{var}(Y) = \delta^2 \psi_{22}, \quad \text{cov}(X, Y) = \epsilon \delta \psi_{12},$$

an upper bound for the probability of misclassifying an unsatisfactory item is given by

$$P\{|X| \leqslant \epsilon, |Y| > \delta\} \leqslant \min(\psi_{22}, p, 1), \qquad (3.1)$$

where

$$p = (\psi_{11}\psi_{22} - \psi_{12}^2)/(\psi_{11} + \psi_{22} - 2|\psi_{12}|) \quad \text{if} \quad \psi_{22} \leqslant |\psi_{12}|,$$
$$p = \infty, \text{ otherwise.}$$

Similarly, an upper bound for the probability of misclassifying a satisfactory item is obtained by interchanging ψ_{11} and ψ_{22} above. Formula (3.1) can be obtained from (2.1) with $\lambda_1 = 0$, $\lambda_2 = 1$. It can also be obtained via Theorem 3.1 of Marshall and Olkin (1960). According to that theorem, equality can be attained in (3.1).

In the context of §§ 2.1 or 2.2, a lower bound for the probability that an item conforms ($|Y| \leqslant \delta$) and is so classified ($|X| \leqslant \epsilon$) is given by

$$P\{|X| \leqslant \epsilon, |Y| \leqslant \delta\} \geqslant 1 - \tfrac{1}{2}\{\psi_{11} + \psi_{22} + [(\psi_{11} + \psi_{22})^2 - 4\psi_{12}^2]^{\frac{1}{2}}\}. \qquad (3.2)$$

This result was obtained by Lal (1955), and follows from (2.1) with $\lambda_1 = \lambda_2 = 1$.

A bound for the probability of correctly classifying an item as unsatisfactory is given by

$$P\{|X| > \epsilon, |Y| > \delta\} \leqslant \min(\psi_{11}, \psi_{22}). \tag{3.3}$$

This result is elementary, but may be obtained from Theorem 4.1 of Marshall and Olkin (1960).

In the context of § 2.2, we may be concerned with the probability of classifying as satisfactory an item which is 'too big'. An upper bound for this probability is given by

$$P\{|X| \leqslant \epsilon, Y > \delta\} \leqslant \min[\psi_{22}/(1 + \psi_{22}), p/(1 + p), \tfrac{1}{2}], \tag{3.4}$$

where p is as in (3.1). Similarly, an upper bound for the probability of classifying a satisfactory item as 'too big' is obtained by interchanging ψ_{11} and ψ_{22} in (3.4). Inequality (3.4) follows from (3.1) and a relation between 'one-sided' and 'two-sided' bounds given by Theorem 3.1 of Marshall and Olkin (1960).

A bound for the probability of classifying an unsatisfactory item as too big is

$$P\{X > \epsilon, |Y| > \delta\} \leqslant \min\left(\frac{\psi_{11}}{1 + \psi_{11}}, \psi_{22}\right), \tag{3.5}$$

and follows from

$$P\{X > \epsilon, |Y| > \delta\} \leqslant P\{X > \epsilon\} \leqslant \psi_{11}/(1 + \psi_{11}),$$

$$P\{X > \epsilon, |Y| > \delta\} \leqslant P\{|Y| > \delta\} \leqslant \psi_{22}.$$

When intervals for Y determine the state of an item and corresponding intervals for X determine how it will be classified, we may encounter regions of the form $\{\epsilon \leqslant X \leqslant \epsilon c_1, \delta \leqslant Y \leqslant \delta c_2\}$, where $\epsilon, \delta > 0$, $c_1, c_2 > 1$, and $EX = EY = 0$. For example, with $c_1 = c_2 = \infty$, this event may represent the correct classification of an item which is 'too big'.

If $EX = EY = 0$, $\operatorname{var} X = \epsilon^2 \psi_{11}$, $\operatorname{var} Y = \delta^2 \psi_{22}$, $\operatorname{cov}(X, Y) = \epsilon \delta \psi_{12}$, then

$$P\{\epsilon \leqslant X \leqslant \epsilon c_1, \delta \leqslant Y \leqslant \delta c_2\} \leqslant p, \tag{3.6}$$

where $\quad p = \psi_{11}/(1 + \psi_{11}) \quad$ if $\psi_{11} \leqslant \psi_{12} \leqslant c_2 \psi_{11}$,

$\quad\quad\quad\quad p = \psi_{22}/(1 + \psi_{22}) \quad$ if $\psi_{22} \leqslant \psi_{12} \leqslant c_1 \psi_{22}$,

$$p = \frac{\psi_{11}\psi_{22} - \psi_{12}^2}{(\psi_{11}\psi_{22} - \psi_{12}^2) + (\alpha_1^2 \psi_{11} - 2\alpha_1\alpha_2\psi_{12} + \alpha_2^2 \psi_{22})},$$

where $(\alpha_1, \alpha_2) = (1, 1)$ if $\psi_{12} \leqslant \min(\psi_{11}, \psi_{22})$,

$(\alpha_1, \alpha_2) = (c_2, 1)$ if $c_2 \psi_{11} \leqslant \psi_{12}$,

and $(\alpha_1, \alpha_2) = (1, c_1)$ if $c_1 \psi_{22} \leqslant \psi_{12}$.

The proof of this result is outlined in §4.

4 CHEBYSHEV BOUNDS

When a measure P is known to satisfy

$$\int f_i(z)\, dP = \nu_i \quad (i = 1, 2, ..., n), \tag{4.1}$$

but is otherwise unknown, the integral $\int \tau(z)\, dP(z)$ can be computed only for very special functions τ. However, there is a standard method for obtaining a bound on such an integral which is often useful. If

$$\mathscr{A} = \left\{ a = (a_1, ..., a_n) : \sum_1^n a_i f_i(z) \geqslant \tau(z) \quad \text{for all } z \right\}$$

is not empty, then for any $a \in \mathscr{A}$,

$$\int \tau(z)\, dP(z) \leqslant \int \Sigma a_i f_i(z)\, dP(z) = \Sigma a_i \nu_i,$$

and moreover $$\int \tau(z)\, dP(z) \leqslant \inf_{a \in \mathscr{A}} \Sigma a_i \nu_i.$$

Under quite general conditions such an inequality is sharp in the sense that for every $\epsilon > 0$, and every possible value of $(\nu_1, ..., \nu_n)$, there exists a measure P satisfying (4.1) such that

$$\int \tau(z)\, dP(z) \geqslant \inf_{a \in \mathscr{A}} \Sigma a_i \nu_i - \epsilon.$$

Often it is possible to achieve equality, i.e. to find P satisfying (4.1) such that

$$\int \tau(z)\, dP(z) = \inf_{a \in \mathscr{A}} \Sigma a_i \nu_i,$$

and in order to prove that the infimum has indeed been found, it is sufficient to exhibit the measure achieving equality.

This general method and the sharpness of the inequalities it yields has been discussed by various authors; see, e.g. Isii (1959), Kemperman (1965), or Karlin and Studden (1967, p. 472).

In the following proofs, we often give only the function $f(z) = \Sigma a_i^* f(z_i)$ which satisfies $\inf_{a \in \mathscr{A}} \Sigma a_i \nu_i = \Sigma a_i^* \nu_i$, with $a^* \in \mathscr{A}$.

Ordinarily, the extremal vector a^* depends upon the ν_i; moreover, it may take different functional forms, say $a^{*(1)}, \ldots, a^{*(k)}$ for $(\nu_1, \ldots, \nu_n)$ in corresponding sets $\omega_1, \ldots, \omega_k$. Often the bound $\Sigma_i a_i^{*(j)} \nu_i$ is valid over a set larger than
$$\omega_j = \{\nu = (\nu_1, \ldots, \nu_n) : a^*(\nu) = a^{*(j)}(\nu)\}.$$

These larger sets are sometimes not difficult to describe, whereas the sets $\omega_1, \ldots, \omega_k$ may have complicated forms. In such cases, it is most convenient to express the bound as a minimum of several valid bounds, as in §§ 2.1 and 2.2.

In each of the cases discussed below, there is a convenient normalization of X and Y which yields variables U, V. With $z = (u, v)$, the function $f(z) = \Sigma a_i^* f_i(z)$ is in these cases of the form
$$f(u, v) = a_{11} u^2 + 2a_{12} uv + a_{22} v^2 + c. \tag{4.2}$$

Discussion on the bound for §2.1

It is convenient to introduce the notation $U = X/\epsilon$, $V = Y/\delta$.

For the bound β_1, take in (4.2)
$$a_{11} = b_{22}/|B|, \quad a_{22} = b_{11}/|B|, \quad a_{12} = -b_{12}/|B|,$$
where
$$|B| = b_{11} b_{22} - b_{12}^2,$$
$$b_{11} = 1/\lambda_1, \quad b_{22} = 1/\lambda_2,$$
$$b_{12} = \frac{(\lambda_1 \psi_{11} + \lambda_2 \psi_{22}) - [(\lambda_1 \psi_{11} + \lambda_2 \psi_{22})^2 - 4\lambda_1 \lambda_2 \psi_{12}^2]^{\frac{1}{2}}}{2\lambda_1 \lambda_2 \psi_{12}}.$$
and $c = 0$.

If
$$\psi_{12}^4 (\lambda_1 + \lambda_2)^2 - \psi_{12}^2 [(\psi_{11} + \psi_{22})(\lambda_1^2 \psi_{11} + \lambda_2^2 \psi_{22}) + 4\lambda_1 \lambda_2 (\psi_{11} + \psi_{22} - 1)]$$
$$+ (\psi_{11} + \psi_{22} - 1)(\lambda_1 \psi_{11} + \lambda_2 \psi_{22})^2 \leqslant 0,$$
and $\lambda_1 (\psi_{11} - |\psi_{12}|) + \lambda_2 (\psi_{22} - |\psi_{12}|) \geqslant 0$, then equality is attained by the distribution (for U, V) which places probability $p_1/2$ at the points $\pm (b_{12}/b_{22}, 1)$, probability $p_2/2$ at $\pm (1, b_{12}/b_{11})$, and probability $1 - p_1 - p_2$ at $(0, 0)$, where
$$p_1 = b_{22}(b_{11} \psi_{22} - b_{12} \psi_{12})/|B|, \quad p_2 = b_{11}(b_{22} \psi_{11} - b_{12} \psi_{12})/|B|.$$
Remark. The choice of b_{11} and b_{22} insures that $\min_u f(u, \delta) = \lambda_2$ and $\min_v f(\epsilon, v) = \lambda_1$. When $\min_u f(u, \delta)$ occurs at u^* satisfying $u^* > \epsilon$ (or

$u^* < -\epsilon$) the constraint $\min_u f(u, \delta) \geqslant \lambda_2$ need not be satisfied, but can be replaced by $f(\epsilon, \delta) = \lambda_2$ (or by $f(-\epsilon, \delta) = \lambda_2$). In this case the bound β_2 is obtained, but of course it is valid only when $|u^*| > \epsilon$; this is the origin of the constraint given for the validity of β_2.

For the bound β_2, take

$$a_{11} = \lambda_2 + (2\alpha - 1) a_{22}, \quad a_{12} = -\alpha a_{22}(\mathrm{sign}\, \psi_{12}),$$

$$a_{22} = (\lambda_2 - \lambda_1)/(1-\alpha)^2, \quad c = 0,$$

where

$$\alpha = (|\psi_{12}| - \psi_{22})/(\psi_{11} - |\psi_{12}|).$$

The bound obtained from this function is valid whenever

$$\lambda_1(\psi_{11} - |\psi_{12}|) + \lambda_2(\psi_{22} - |\psi_{12}|) \leqslant 0.$$

If, in addition, $\psi_{11} \leqslant 1$, then equality is attained by the distribution (for U, V) which places probability $p_1/2$ at the points $\pm (\mathrm{sign}\, \psi_{12}, 1)$ probability $p_2/2$ at the points $\pm (1, -a_{12}/a_{22})$, and probability $1 - p_1 - p_2$ at $(0, 0)$, where

$$p_1 = \frac{\psi_{11}\psi_{22} - \psi_{12}^2}{\psi_{11} + \psi_{22} - 2|\psi_{12}|}, \quad p_2 = \psi_{11} - p_1.$$

For the bound β_3, take

$$a_{11} = a_1^2, \quad a_{22} = a_2^2, \quad a_{12} = -a_1 a_2(\mathrm{sign}\, \psi_{12}), \quad c = \lambda_1,$$

where

$$a_1 = \frac{(|\psi_{12}| - \psi_{22})\sqrt{(\lambda_2 - \lambda_1)}}{\psi_{11} + \psi_{22} - 2|\psi_{12}|}, \quad a_2 = a_1 + \sqrt{(\lambda_2 - \lambda_1)}.$$

The bound β_3 is valid whenever $\psi_{11} \geqslant |\psi_{12}| \geqslant \psi_{22}$. If, in addition,

$$\psi_{11} \geqslant 1, \quad \psi_{11} + \psi_{22} - 2|\psi_{12}| \geqslant \psi_{11}\psi_{22} - \psi_{12}^2,$$

then equality may be attained by any distribution (for U, V) which places probability $p_1/2$ at the points $\pm (1, 1), p_2/2$ at the points $\pm (\gamma_1, \gamma_2)$ and probability $(1 - p_1 - p_2)/2$ at the points $\pm (1, \gamma_2/\gamma_1)$, where

$$p_1 = \frac{\psi_{11}\psi_{22} - \psi_{12}^2}{\psi_{11} + \psi_{22} - 2|\psi_{12}|}, \quad p_2 = \frac{\psi_{11} - 1}{\gamma_1^2 - 1}, \quad \gamma_2 = \gamma_1 \frac{|\psi_{12}| - \psi_{22}}{\psi_{11} - |\psi_{12}|},$$

and γ_1 satisfies

$$\gamma_1^2 \geqslant \frac{(\psi_{11} - |\psi_{12}|)^2}{\psi_{11} + \psi_{22} - 2|\psi_{12}| - \psi_{11}\psi_{22} + \psi_{12}^2}.$$

For the bound β_4, take $a_{11} = a_{12} = 0$, $a_{22} = \lambda_2 - \lambda_1$, $c = \lambda_1$. If

$$\psi_{11} \geqslant 1 > \psi_{22} \geqslant |\psi_{12}|,$$

then equality is attained by any distribution for (U, V) which places probability $(\psi_{22} + \psi_{12})/4$ at the points $\pm (1, 1)$, probability $(\psi_{22} - \psi_{12})/4$ at the points $\pm (1, -1)$, probability $(\psi_{11} - 1)/2(\gamma^2 - 1)$ at the points $\pm (\gamma, 0)$, and probability $[\gamma^2(1 - \psi_{22}) + (\psi_{22} - \psi_{11})]/2(\gamma^2 - 1)$ at the points $\pm (1, 0)$, where $\gamma^2 \geqslant (\psi_{11} - \psi_{22})/(1 - \psi_{22})$. If

$$1 \geqslant \psi_{11}, \psi_{22} \quad \text{and} \quad \psi_{11}\psi_{22} - \psi_{12}^2 \geqslant \psi_{22}(1 - \psi_{22}),$$

equality is attained by the distribution for (U, V) which places probability $p_1/2$ at the points $\pm (\alpha, 1)$, probability $p_2/2$ at $\pm (\alpha, -1)$, and probability $(1 - p_1 - p_2)/2$ at the points $\pm (1, 0)$, where

$$p_1 = (\alpha\psi_{22} + \psi_{12})/2\alpha, \quad p_2 = (\alpha\psi_{22} - \psi_{12})/2\alpha,$$

and
$$\alpha^2 = (\psi_{11} + \psi_{22} - 1)/\psi_{22}.$$

For the bound β_5, take $a_{11} = a_{12} = a_{22} = 0$, $c = \lambda_2$. If $\psi_{11} \leqslant 1 \leqslant \psi_{22}$, equality may be attained by the distribution for (U, V) which places probability $p/2$ at the points $\pm (\sqrt{\psi_{11}}, \sqrt{\psi_{22}})$, and probability $(1 - p)/2$ at the points $\pm (\sqrt{\psi_{11}}, -\sqrt{\psi_{22}})$, where $p = \frac{1}{2}[1 + \psi_{12}/\sqrt{(\psi_{11}\psi_{22})}]$. This bound can be approximated in various other circumstances.

Discussion of §2.2

For the bound β_1, take

$$a_{11} = b_{22}/|B|, \quad a_{22} = b_{11}/|B|, \quad a_{12} = -b_{12}/|B|, \quad c = 0,$$

where
$$|B| = b_{11}b_{22} - b_{12}^2, \quad b_{11} = b_{22} = 1/\lambda_1,$$

$$b_{12} = \{\psi_{11} + \psi_{22} - [(\psi_{11} + \psi_{22})^2 - 4\psi_{12}^2]^{\frac{1}{2}}\}/2\lambda_1\psi_{12}.$$

The bound is valid if $(\lambda_2 - \lambda_1)^2 (\psi_{11} + \psi_{22} - 2\psi_{12}) \leqslant \lambda_1^2(\psi_{11} + \psi_{22} + 2\psi_{12})$. If, in addition, $\psi_{11} + \psi_{22} \leqslant \psi_{12}^2 + 1$, then equality is attained by the distribution for $(U, V) = (X/\epsilon, Y/\delta)$ which places probability $p_1/2$ at the points $\pm (b_{12}/b_{22}, 1)$, probability $p_2/2$ at the points $\pm (1, b_{12}/b_{11})$, and probability $1 - p_1 - p_2$ at $(0, 0)$, where

$$p_1 = b_{22}(b_{11}\psi_{22} - b_{12}\psi_{12})/|B|, \quad p_2 = b_{11}(b_{22}\psi_{11} - b_{12}\psi_{12})/|B|.$$

For the bound β_2 with $\psi_{11} \geqslant \psi_{22}$, take

$$a_{11} = \lambda_1 + m^2 a_{22}, \quad a_{12} = -ma_{22}, \quad a_{22} = (\lambda_2 - \lambda_1)/(1 + m)^2, \quad c = 0,$$

where
$$m = \frac{\psi_{12} + \min (\psi_{11}, \psi_{22})}{\psi_{12} + \max (\psi_{11}, \psi_{22})}.$$

If $\psi_{11} \leqslant \psi_{22}$, interchange a_{11} and a_{22}. This bound is valid whenever

$$\frac{\lambda_1}{\lambda_2 - \lambda_1} \leqslant \frac{|\psi_{11} - \psi_{22}|}{\psi_{11} + \psi_{22} + 2\psi_{12}}.$$

If in addition, $\max(\psi_{11}, \psi_{22}) \leqslant 1$, then equality is attained for the distribution which places probability $p_1/2$ at the points $\pm(1, -1)$, probability $p_2/2$ at the points $\pm(\alpha_1, \alpha_2)$, and probability $1 - p_1 - p_2$ at the point $(0, 0)$, where

$$p_1 = (\psi_{11}\psi_{22} - \psi_{12}^2)/(\psi_{11} + \psi_{22} + 2\psi_{12}), \quad p_2 = \psi_{11} - p_1,$$

$(\alpha_1, \alpha_2) = (1, m)$ if $\psi_{22} \leqslant \psi_{11}$ and $(\alpha_1, \alpha_2) = (m, 1)$ if $\psi_{11} \leqslant \psi_{22}$.

For the bound β_3, take

$$a_{11} = a_{22} = \frac{\lambda_2^2}{4(\lambda_2 - \lambda_1)}, \quad a_{12} = \frac{\lambda_2(2\lambda_1 - \lambda_2)}{4(\lambda_2 - \lambda_1)}, \quad c = 0.$$

If

$$\left(\frac{\psi_{11} + \psi_{22} + 2\psi_{12}}{\psi_{11} + \psi_{22} - 2\psi_{12}}\right)^{\frac{1}{2}} \leqslant \frac{\lambda_2 - \lambda_1}{\lambda_1} \leqslant \min\left[\frac{\psi_{11} + \psi_{22} + 2\psi_{12}}{|\psi_{11} - \psi_{22}|}, \frac{2 - \psi_{11} - \psi_{22}}{\psi_{11} + \psi_{22} + 2\psi_{12}}\right],$$

then equality is attained for the distribution for (U, V) which places probability $p_1/2$ at the points $\pm(1, -1)$, probability $p_2/2$ at the points $\pm(1, \alpha)$, probability $p_3/2$ at the points $\pm(\alpha, 1)$, and probability $1 - p_1 - p_2 - p_3$ at $(0, 0)$, where $\alpha = (\lambda_2 - 2\lambda_1)/\lambda_2$,

$$p_1 = \frac{\alpha}{(1 + \alpha)^2}(\psi_{11} + \psi_{22} + 2\psi_{12}) - \psi_{12},$$

$$p_2 = \frac{(\psi_{11} + \psi_{12}) - \alpha(\psi_{22} + \psi_{12})}{(1 + \alpha)(1 - \alpha^2)},$$

$$p_3 = \frac{(\psi_{22} + \psi_{12}) - \alpha(\psi_{11} + \psi_{12})}{(1 + \alpha)(1 - \alpha^2)}.$$

For the bound β_4, take

$$a_{11} = t^2 a_{22}, \quad a_{12} = -t a_{22}, \quad a_{22} = (\lambda_2 - \lambda_1)/(1 + t)^2, \quad c = \lambda_1,$$

where $t = (\psi_{22} + \psi_{12})/(\psi_{11} + \psi_{12})$. If

$$-\psi_{12} \leqslant \min(\psi_{11}, \psi_{22}) \leqslant 1 \leqslant \max(\psi_{11}, \psi_{22}),$$

then equality is attained by the distribution for (U, V) which places probability $p_1/2$ at the points $\pm(-1, 1)$, probability $p_2/2$ at the points $\pm(1/t, 1)$, and probability $(1 - p_1 - p_2)/2$ at the points $\pm(1, t)$, where

$$p_1 = \frac{\psi_{11}\psi_{22} - \psi_{12}^2}{\psi_{11} + \psi_{22} + 2\psi_{12}}, \quad p_2 = \frac{t^2(1 - \psi_{11})}{t^2 - 1}.$$

If $\min(\psi_{11}, \psi_{22}) \geqslant 1, \quad \min(\psi_{11}, \psi_{22}) \geqslant -\psi_{12},$

and $\psi_{11}\psi_{22} - \psi_{12}^2 \leqslant \psi_{11} + \psi_{22} + 2\psi_{12},$

then the bound is approached by the distribution for (U, V) which places probability $p_1/2$ at the points $\pm(-1, 1)$, probability $p_2/2$ at the points $\pm(\gamma_1, \gamma_2)$, and probability $(1 - p_1 - p_2)/2$ at the points $\pm\alpha(\gamma_1, \gamma_2)$, where $(\gamma_1, \gamma_2) = (1, t)$ if $\psi_{22} \geqslant \psi_{11}$, $(\gamma_1, \gamma_2) = (t, 1)$ if $\psi_{22} \leqslant \psi_{11}$,

$$p_1 = \frac{\psi_{11}\psi_{22} - \psi_{12}^2}{\psi_{11} + \psi_{22} + 2\psi_{12}}, \quad p_2 = \frac{\alpha^2 - \min(\psi_{11}, \psi_{22})}{\alpha^2 - 1} - p_1, \quad \text{and} \quad \alpha^2 \to \infty.$$

For the bound β_5, with $\psi_{11} \leqslant \psi_{22}$, take

$$a_{11} = \lambda_2 - \lambda_1, \quad a_{12} = a_{22} = 0, \quad c = \lambda_1;$$

if $\psi_{11} \geqslant \psi_{22}$, interchange a_{11} and a_{22}. If

$$\psi_{22} \geqslant 1 \geqslant \psi_{11}, \quad -\psi_{12} \geqslant \psi_{11}, \quad \psi_{11}\psi_{22} - \psi_{12}^2 \geqslant \psi_{11}(1 - \psi_{11}),$$

then equality is attained for the distribution which places probability $\psi_{11}/2$ at the points $\pm(1, \psi_{12}/\psi_{11})$, probability $(1 - \psi_{11})/2$ at the points $\pm(0, \gamma)$, where $\gamma^2 = (\psi_{11}\psi_{22} - \psi_{12}^2)/[\psi_{11}(1 - \psi_{11})]$. A symmetric case is obtained by interchanging U and V, ψ_{11} and ψ_{22}.

For the bound β_6, take $a_{11} = a_{12} = a_{22} = 0, c = \lambda_2$. If $1 \leqslant \sqrt{\psi_{11}} \leqslant -\psi_{12}$, or $-\psi_{11}/\psi_{12} \geqslant 1$, $\psi_{11} \geqslant 1$, or $\psi_{11} \geqslant 1$, $\psi_{11}\psi_{22} - \psi_{12}^2 \geqslant \psi_{11} + \psi_{22} + 2\psi_{12}$, then the bound can be approached by distributions for (U, V) which place probability $p/2$ at the points $\pm(-\alpha^*, \gamma^*)$ and probability $(1 - p)/2$ at the points $\pm(\alpha, \gamma)$, where $\alpha^* \geqslant 1$, $\gamma^* \geqslant 1$, and $p \to 1$.

Discussion of §2.3

The function $f(u, v)$ is determined by the choice

$$a_{11} = \lambda\tau - c, \quad a_{12} = -[\lambda\tau(\tau - 1)]^{\frac{1}{2}}(\text{sign } \psi_{12}), \quad a_{22} = \tau,$$

where $\tau = \frac{1}{2}\{1 + (\lambda\psi_{11} + \psi_{22})/[(\lambda\psi_{11} + \psi_{22})^2 - 4\lambda\psi_{12}^2]^{\frac{1}{2}}\}$,

and $c = 0$ if $\psi_{11} \leqslant 1$, $c = \lambda$ if $\psi_{11} > 1$. Here, $(U, V) = (X/\epsilon, Y)$. Equality cannot be attained in (2.3), but the bound can be approached arbitrarily closely. If $\psi_{11} \leqslant 1$ and the loss function is altered to be λ when $X = \pm\epsilon$, $Y \leqslant \epsilon$, then the inequality still holds, and equality is attained by the distribution for (U, V) which places probability $p_1/2$ at the points

$$\pm(1, [\lambda\tau/(\tau - 1)]^{\frac{1}{2}}(\text{sign } \psi_{12})),$$

probability $p_2/2$ at $\pm(1, [\lambda(\tau - 1)/\tau]^{\frac{1}{2}}(\text{sign } \psi_{12}))$

and probability $1 - p_1 - p_2$ at $(0, 0)$, where

$$p_1 = [\psi_{22}\tau(\tau - 1) - \psi_{11}\lambda(\tau - 1)^2]/\lambda(2\tau - 1), \quad p_2 = \psi_{11} - p_1.$$

If $\psi_{11} > 1$, then the bound can be approached as $\alpha \to \infty$ by a distribution which satisfies the moment conditions and concentrates on points of the form $\pm a(1, [\lambda\tau/(\tau-1)]^{\frac{1}{2}})$, $\pm \alpha(1, [\lambda(\tau-1)/\tau]^{\frac{1}{2}})$.

Discussion of §2.4

The function $f(X, Y)$ is determined by the choice

(i) $a_{11} = \psi_{12}^2/(\psi_{11}+\epsilon^2)^2$, $a_{12} = -\psi_{12}/(\psi_{11}+\epsilon^2)$, $a_{22} = 1$, $c = 0$,

$$\text{if } \psi_{12} \geqslant 0,$$

(ii) $a_{11} = a_{12} = 0$, $a_{22} = 1$ $(c = 0)$, if $\psi_{12} \leqslant 0$.

When $\epsilon = 0$, equality is obtained by any distribution which places probability $p/2$ at the points $\pm(0, \alpha)$ and probability $(1-p)/2$ at $\pm\beta(\psi_{11}, \psi_{12})$, where

$$\beta^2 = \alpha^2/(\alpha^2\psi_{11} - \psi_{11}\psi_{22} + \psi_{12}^2), \quad p = (\psi_{11}\psi_{22} - \psi_{12}^2)/\psi_{11}\alpha^2,$$

and

$$\alpha^2 \geqslant (\psi_{11}\psi_{22} - \psi_{12}^2)/\psi_{11}.$$

Discussion of (3.6)

With $(U, V) = (X/\epsilon, Y/\delta)$, (3.6) is obtained from the function

$$f(u, v) = (u + \psi_{11})^2/(1 + \psi_{11})^2 \quad \text{if } \psi_{11} \leqslant \psi_{12} \leqslant c_2\psi_{11},$$

$$f(u, v) = (v + \psi_{22})^2/(1 + \psi_{22})^2 \quad \text{if } \psi_{22} \leqslant \psi_{12} \leqslant c_1\psi_{22},$$

$$f(u, v) = [(x, y)\Psi^{-1}\alpha' + 1]^2/(\alpha\Psi^{-1}\alpha' + 1)^2,$$

where $\Psi = (\psi_{ij})$ $(i, j = 1, 2)$,

$$\alpha = (1, 1) \quad \text{if } \psi_{12} \leqslant \min(\psi_{11}, \psi_{22}),$$

$$\alpha = (c_2, 1) \quad \text{if } c_2\psi_{11} \leqslant \psi_{12},$$

and $\alpha = (1, c_1) \quad \text{if } c_1\psi_{22} \leqslant \psi_{12}.$

It follows from Theorem 3.1 of Marshall and Olkin (1960) that equality can be attained in (3.6).

REFERENCES

Isii, K. (1959). On a method for generalizations of Tchebycheff's inequality. *Ann. Inst. Statist. Math.* **10**, 65–88.

Karlin, S. J. and Studden, W. J. (1967). *Tchebycheff systems: with applications in analysis and statistics.* New York: John Wiley and Sons.

Kemperman, J. H. B. (1965). On the sharpness of Tchebycheff type inequalities, I, II, III. *Indag. Math.* **27**, 554–71; 572–87; 588–601.

Lal, D. N. (1955). A note on a form of Tchebycheff's inequality for two or more variables. *Sankhyā*, **15**, 317–20.

Marshall, A. W. and Olkin, I. (1960). Multivariate Chebyshev inequalities. *Ann. Math. Statist.* **31**, 1001–14.

Marshall, A. W. and Olkin, I. (1968). A general approach to some screening and classification problems. *J. Roy. Statist. Soc.*, Ser. B, **30**, 407–43.

AN ASYMPTOTIC MINIMAX
PROPERTY OF CERTAIN LIKELIHOOD
RATIO TESTS

KENNETH S. MOUNT AND H. T. DAVID

We shall consider (Theorem 1) an asymptotic minimax property of certain likelihood ratio procedures for exponential families. The hypotheses involved typically will differ with respect to features other than location or scale; in this sense our remarks pertain to goodness-of-fit; for this reason, too, the exact-minimax status of likelihood ratio procedures for the sorts of situations discussed below would seem to be hard to ascertain.

In [8], it is shown that, for general multivariate normal hypotheses H_0, rather arbitrary tests of $(H_0) v.$ (not H_0) can be improved upon in a certain asymptotic sense by suitable likelihood ratio tests, provided test size is made to decrease appropriately with n. A different point of view is adopted in [6]; here it is shown that, for testing $H_0: \mu = 0 v.$ a certain H_λ 'separated' from H_0, and increasingly so as $\lambda \to \infty$, the likelihood ratio test is not 'asymptotically logarithmically minimax'; this in a sense involving fixed test size and sample size, and increasing λ. Efron and Truax[5] treat the testing of a simple hypothesis (H_0) concerning an exponential family $v.$ (not H_0). As in [8], Efron and Truax deal with increasing n, and with the comparison (asymptotically favorable to the likelihood ratio test in a certain sense) of fairly general tests with suitably chosen likelihood ratio tests; test size in [5] is held fixed.

In this note, we consider hypotheses H_0 and H_1 concerning an exponential family that are 'separated' as were those in [6]; we show that a certain likelihood ratio test is 'asymptotically logarithmically minimax' in this situation; our meaning here is reminiscent of that in [7] and [6], but differs from the latter in that, here, as in [5] and [8], it is sample size that increases, while H_0 and H_1 are fixed; in addition, rather than fixing test size, we consider, symmetrically, a risk that sums the errors of both kinds, with a corresponding adaptation of the minimax operator; finally, our (less demanding) limit is taken 'first'.

That the likelihood ratio test is 'asymptotically logarithmically minimax' in our sense is based, essentially on the following fact: let $P_\theta(a)$ be the (dominating) exponential term of the probability of a large deviation of $\bar{x}_n$ to neighborhoods of a, when sampling from an exponential density

$\exp(\theta.x - \psi(\theta) + g(x))$; also, let $\Pi_\theta(a)$ be the density of $\bar{x}_n$, evaluated at a. Then

$$\ln P_\theta(a) = \ln \Pi_\theta(a) + \delta(a). \tag{1}$$

Hence, at least to first order, it is immediate to compute ratios of probabilities of large deviations into certain regions specified by ratios of likelihoods.

Consider an exponential family with density $p_\theta(x)$ equal to

$$\exp(\theta.x - \psi(\theta) + g(x)); \quad x \in X \subset E_k, \theta \in \Theta \subset E_k$$

with respect to Lebesgue measure, such that

$$\text{the natural parameter space } \Theta \text{ is open.} \tag{2}$$

Suppose as well that,

for all $\theta \in \Theta$, there is an n such that the n-fold convolution of $p_\theta(x)$
    is bounded in any of the senses prescribed by condition (I) of [2].   (3)

Then, specialization of arguments in [2], [5] and [10] yields the following facts, analogous, in part, to the properties given for $J(\dot{p}, p)$ in [8].

(F 1) The cumulant generating function of $p_\theta(x)$ is $\psi(\theta + t) - \psi(\theta)$, $\theta + t \in \Theta$.

(F 2) $\psi(\theta)$ has derivatives of all orders for $\theta \in \Theta$.

(F 3) Let Ω be the interior of the convex hull of X. Consider the map $f(\theta) = a$ from Θ into E_k, defined by

$$a = \operatorname{grad} \psi(s)|_\theta.$$

Then f is a continuous 1–1 map from Θ onto Ω.   (4)

(F 4) Define $J(a, \theta) \equiv a.[f^{-1}(a) - \theta] + \psi(\theta) - \psi(f^{-1}(a))$. Then $J(a, \theta) \geqslant 0$, and $= 0$ iff $a = f(\theta)$; further, $J(a, \theta)$ is continuous in a, and also in θ, for all (a, θ) in $\Omega \times \Theta$.

(F 5) Define sets $A_{\theta, J} \subset \Omega$ by $A_{\theta, J} = \{a : J(a, \theta) \leqslant J\}$. Then, for all $\theta \in \Theta$ and all $J \in [0, \infty)$, $A_{\theta, J}$ is convex.

(F 6) Consider a region D in Ω with boundary $\mathscr{D}$ relative to Ω, and define

$$J''(D, \theta) \equiv \inf\{J : \mu(D \cap A_{\theta, J}) > 0\},$$

where μ is Lebesgue measure. Also define

$$J'(D, \theta) \equiv \inf\{J(a, \theta) : a \in D\},$$

$$J(D, \theta) \equiv \inf\{J(a, \theta) : a \in \mathscr{D}\}.$$

Suppose that p_θ and D are such that, for all $\theta \in \Theta$,

$$J''(D, \theta) = J'(D, \theta), \quad \text{and} \quad A_{\theta, J'(D, \theta)} \text{ is bounded.} \tag{5}$$

Also suppose that, for all $\theta \in \Theta$, the order l_θ of contact of D with the regions $A_{\theta, J}$ is defined in the sense of [2]. Then, if $J'(D, \theta) > 0$, $J'(D, \theta) = J(D, \theta)$ in view of the compactness of the boundary of $A_{\theta, J'(D, \theta)}$ guaranteed by the convexity and boundedness of $A_{\theta, J'(D, \theta)}$, and

$$\Pr\{\bar{x}_n \in D | \theta\} \sim K_\theta n^{k/2-1-l_\theta} e^{-nJ(D, \theta)}. \tag{6}$$

We now proceed to the asymptotic minimax property given in Theorem 1, giving along the way the additional assumptions needed. The first of these is that

there is an n_0 such that, for all $\theta \in \Theta$ and all $n \geqslant n_0$, the support of the distribution of $\bar{x}_n$ is Ω.   (7)

Consider now two hypotheses (i.e., regions) H_0 and H_1 in Θ such that

the distance between $f(H_0)$ and $f(H_1)$ exceeds zero.   (8)

For $n \geqslant n_0$, let $\Pi_{\theta, n}(a)$ be the density of $\bar{x}_n$ at $a \in \Omega$ under $p_\theta(x)$, and consider the likelihood ratio test that accepts $H_1(\text{resp. } H_0)$ when

$$l_n \equiv \frac{\sup\limits_{\theta \in H_1} \Pi_{\theta, n}(a)}{\sup\limits_{\theta \in H_0} \Pi_{\theta, n}(a)} \geqslant 1(\text{resp.} < 1). \tag{9}$$

Let L (resp. $\Omega - L$) be the region in Ω leading to acceptance of $H_1(\text{resp. } H_0)$, and let $\mathcal{L}_1(\text{resp. } \mathcal{L}_0)$ be the boundary of $L(\text{resp. } \Omega - L)$ relative to Ω; suppose that
$$\mathcal{L} \equiv \mathcal{L}_1 = \mathcal{L}_0 = \{a \in \Omega: l_n = 1\}. \tag{10}$$

We can now state

Lemma 1. *For* $n \geqslant n_0$, (i) $\mathcal{L}$ *does not depend on* n, *and* (ii) L, $\Omega - L$ *and* $\mathcal{L}$ *are characterized, respectively, by*

$$\frac{\inf\limits_{\theta \in H_1} J(a, \theta)}{\inf\limits_{\theta \in H_0} J(a, \theta)} \leqslant, >, = 1. \tag{11}$$

Proof. (i) and (ii) follow from (7), (9), (10) and the definition of $J(a, \theta)$ in (F 4).

We now define
$$J_i(a) \equiv \inf\limits_{\theta \in H_i} J(a, \theta) \tag{12}$$

and assume that, on $\mathcal{L}$ (where, by Lemma 1, $J_0(a) = J_1(a) \equiv J(a)$), $\inf J_0(a)$ is achieved, say at a^*:

$$\inf\limits_{a \in \mathcal{L}} J_1(a) \equiv J_1(a^*) \equiv J(a^*) \equiv J_0(a^*) \equiv \inf\limits_{a \in \mathcal{L}} J_0(a). \tag{13}$$

We assume as well that $\inf_{\theta \in H_i} J(a^*, \theta)$ is achieved, say at θ_i^*:

$$\inf_{\theta \in H_i} J(a^*, \theta) \equiv J(a^*, \theta_i^*). \tag{14}$$

(Note that, in view of (F 4), (14) is guaranteed when H_i is compact.)

Lemma 2. Consider a point set $\mathscr{D}$ in Ω with the following property:

If $\mathscr{E}$ is any connected set with $f(H_i) \cap \mathscr{E} \neq \phi, i: 0, 1$, then $\mathscr{D} \cap \mathscr{E} \neq \phi$. (15)

Then

$$\inf_{\substack{a \in \mathscr{D} \\ \theta \in H_0 \cup H_1}} J(a, \theta) \leqslant \inf_{\substack{a \in \mathscr{L} \\ \theta \in H_0 \cup H_1}} J(a, \theta) = J(a^*, \theta_1) = J(a^*, \theta_0) = J(a^*) > 0. \tag{16}$$

Proof. With regard to the equalities, and last inequality,

$$\inf_{\substack{a \in \mathscr{L} \\ \theta \in H_0 \cup H_1}} J(a, \theta) = \inf_{a \in \mathscr{L}} [\inf_{\theta \in H_0 \cup H_1} J(a, \theta)] = \inf_{a \in \mathscr{L}} [\min_{i: 0, 1} (\inf_{\theta \in H_i} J(a, \theta))]$$

$$\overset{\text{Lem 1}}{=} \inf_{a \in \mathscr{L}} [\inf_{\theta \in H_0} J(a, \theta)] \overset{\text{Lem 1}}{=} \inf_{a \in \mathscr{L}} [\inf_{\theta \in H_1} J(a, \theta)] \overset{(12)}{\equiv} \inf_{a \in \mathscr{L}} J_0(a)$$

$$\overset{(13)}{\equiv} \inf_{a \in \mathscr{L}} J_1(a) \overset{(13)}{\equiv} J_0(a^*) \overset{(13)}{\equiv} J(a^*) \overset{(13)}{\equiv} J_1(a^*) \overset{(12)}{\equiv} \inf_{\theta \in H_0} J(a^*, \theta)$$

$$\overset{(12)}{\equiv} \inf_{\theta \in H_1} J(a^*, \theta) \overset{(14)}{\equiv} J(a^*, \theta_0^*) \overset{(14)}{=} J(a^*, \theta_1^*) \overset{(8)(F4)}{>} 0,$$

where facts validating the various equalities are referenced above the signs.

With regard to the first inequality in (16), define

$$B_i \equiv \{a: J_i(a) \leqslant J(a^*)\}.$$

Since $[J(a, \theta_i^*) \leqslant J(a^*)]$ implies

$$[J_i(a) \equiv \inf_{\theta \in H_i} J(a, \theta) \leqslant J(a_1, \theta_i^*) \leqslant J(a^*)],$$

we have that $\quad\quad\quad\quad A_{\theta_i^*, J(a^*)} \subset B_i,$ (17)

where $A_{\theta, J}$ is as defined in (F 5). Also, since, in validating the equalities, we pointed out that $J(a^*, \theta_i^*) = J(a^*)$,

$$a^* \in A_{\theta_i^*, J(a^*)} \tag{18}$$

and, in view of (F 4), $\quad\quad f(\theta_i^*) \in A_{\theta_i^*, J(a^*)}.$ (19)

But, in view of (F 5), $\quad\quad A_{\theta_i^*, J(a^*)}$ is convex. (20)

Hence, in view of (18), (19) and (20), $\mathscr{A} \equiv A_{\theta_0^*, J(a^*)} \cup A_{\theta_1^*, J(a^*)}$ is a connected set having a non-null intersection with $f(H_0)$ and $f(H_1)$, so that, in view of (15),

$$\mathscr{D} \cap \mathscr{A} \neq \phi.$$

Hence suppose, without loss of generality, that there is a point $\tilde{a}$ with

$$\tilde{a} \in \mathscr{D} \cap A_{\theta_0^*, J(a^*)}.$$

Then, in view of (17), $\tilde{a} \in \mathscr{D} \cap B_0$,

so that

$$\inf_{\substack{a \in \mathscr{D} \\ \theta \in H_0 \cup H_1}} J(a, \theta) = \min_{i: 0, 1} [\inf_{a \in \mathscr{D}} J_i(a)] \leqslant \min_{i: 0, 1} J_i(\tilde{a}) \leqslant J_0(\tilde{a}) \leqslant J(a^*),$$

where the first inequality holds because $\tilde{a} \in \mathscr{D}$, and the third because $\tilde{a} \in B_0$.

Lemma 3. Consider a point set $\mathscr{D}$ in Ω which, in addition to (15), *has the property that*

for $(\theta_0, \theta_1) \in H_0 \times H_1$, the function $\min_{i: 0, 1} [\inf_{a \in \mathscr{D}} J(a, \theta_i)]$

achieves its infimum, say at $(\theta_0(\mathscr{D}), \theta_1(\mathscr{D}))$. (21)

Then $(\theta_0(\mathscr{D}), \theta_1(\mathscr{D}))$ also has the property that, for $(\theta_0, \theta_1) \in H_0 \times H_1$,

$$\min_{i: 0, 1} [\inf_{a \in \mathscr{D}} J(a, \theta_i(\mathscr{D}))] \leqslant \min_{i: 0, 1} [\inf_{a \in \mathscr{L}} J(a, \theta_i)].$$

Proof. Rearranging inf's in the conclusion of Lemma 2, we find that

$$\inf_{(\theta_0, \theta_1) \in H_0 \times H_1} [\min_{i: 0, 1} [\inf_{a \in \mathscr{D}} J(a, \theta_i)]] \leqslant \inf_{(\theta_0, \theta_1) \in H_0 \times H_1} [\min_{i: 0, 1} [\inf_{a \in \mathscr{L}} J(a, \theta_i)]],$$

from which the conclusion follows by (21).

Theorem 1. In testing the hypothesis H_0 v. H_1 for the exponential density $p_\theta(x_1, \ldots, x_n)$, consider any $\bar{x}_n$-critical region D in Ω with $\mathscr{D}$ the boundary of D relative to Ω. Suppose that, in addition to properties (15) *and* (21), *D has the further property that*

for $(\theta_0, \theta_1) \in H_0 \times H_1$, $J'(D, \theta_0) > 0$ and $J'(\Omega - D, \theta_1) > 0$. (22)

Suppose as well that conditions (2), (3), (7), (8), (10), (13) *and* (14) *are satisfied, as well as condition* (5) *for both D and $\Omega - D$.*

Then there is a $(\hat{\theta}_0, \hat{\theta}_1)$ (depending on D), such that, for any

$$(\theta_0, \theta_1) \in H_0 \times H_1,$$

$$\lim_{n \to \infty} r_n \leqslant 1,$$

where $$r_n = \frac{\ln\left[\alpha_n(D,\theta_0)+\beta_n(D,\theta_1)\right]}{\ln\left[\alpha_n(L,\theta_0)+\beta_n(L,\theta_1)\right]},$$

$$\alpha_n(D,\theta_0) = P\{\bar{x}_n \in D | \theta_0\},$$

$$\beta_n(D,\theta_1) = P\{\bar{x}_n \in (\Omega - D) | \theta_1\}.$$

Proof. For any $(\theta_0',\theta_1') \in H_0 \times H_1$, consider the quantities $J(D,\theta_0')$ and $J(\Omega-D,\theta_1') = J(D,\theta_1')$, as defined in (F 6). Without loss of generality, assume that
$$J(D,\theta_0') \leqslant J(D,\theta_1'). \tag{23}$$

Then it is clear, in view of (22) and (6), that there is a constant c such that $\ln\left[\alpha_n(D,\theta_0')+\beta_n(D,\theta_1')\right]$ eventually is bounded by $-nJ(D,\theta_0') \pm c\log n$, i.e. in view of (23), by $-n(\min_{i:\,0,1} J(D,\theta_i')) \pm (c)(\log n)$. The same argument applied to L shows that, if (θ_0,θ_1) is some other point in $H_0 \times H_1$, then r_n eventually is bounded by

$$\frac{\min_{i:\,0,1} J(D,\theta_i')}{\min_{i:\,0,1} J(L,\theta_i)} \pm (d)\left(\frac{\log n}{n}\right) \tag{24}$$

and the desired result, with $\hat{\theta}_i$ in fact equal to $\theta_i(\mathscr{D})$, follows from (24) and Lemma 3, upon setting $\theta_i' = \theta_i(\mathscr{D})$.

REFERENCES

[1] Bahadur, R. R. and Rao, R. R. On deviations of the sample mean. *Ann. Math. Statist.* **31**, 1015–27 (1960).

[2] Borovkov, A. A. and Rogozin, B. A. On the central limit theorem in the multidimensional case (in Russian). *Teor. Veroyat. Primen.* **10**, 61–9 (1965).

[3] Chernoff, H. A measure of asymptotic efficiency for tests of a hypothesis based on the sum of observations. *Ann. Math. Statist.* **23**, 493–507 (1952).

[4] Chernoff, H. Discussion of a paper by Wassily Hoeffding. *Ann. Math. Statist.* **36**, 405–7 (1965).

[5] Efron, B. and Truax, D. Large deviations theory in exponential families. *Ann. Math. Statist.* **39**, 1402–24 (1968).

[6] Giri, N. Locally and asymptotically minimax tests of a multivariate problem. *Ann. Math. Statist.* **39**, 171–8 (1968).

[7] Giri, N. and Kiefer, J. Local and asymptotic minimax properties of multivariate tests. *Ann. Math. Statist.* **35**, 21–35 (1964).

[8] Herr, D. G. Asymptotically optimal tests for multivariate normal distributions. *Ann. Math. Statist.* **38**, 1829–44 (1967).

[9] Hoeffding, W. Asymptotically optimal tests for multinomial distributions. *Ann. Math. Statist.* **36**, 369–401 (1965).

[10]   Lehman, E. L. *Testing Statistical Hypotheses*. New York: John Wiley and Sons, Inc. (1959).

[11]   Mount, Kenneth S. Minimax properties of likelihood ratio tests related to goodness-of-fit. Unpublished Ph.D. thesis. Iowa State University (1969).

[12]   Sanov, I. N. On the probability of large deviations of random variables. *Sel. Trans. Math. Statist. and Probab.* 1, 213–44 (1961).

[13]   Savage, L. J. An asymptotic connection between the minimax and maximum likelihood rules. Unpublished paper. Mimeo (1952).

[14]   Sethuraman, J. On the probabilities of large deviations of families of sample means. *Ann. Math. Statist.* 35, 1304–16 (1964).

DISCUSSION ON MOUNT
AND DAVID'S PAPER

I can make no claims to having followed the whole of the mathematical development and my questions are natural to a user of mathematical statistics rather than to a developer of the purely mathematical ideas.

What is the relation of the material to goodness-of-fit tests? Theorem 1 seems to me to relate to a pure decision problem as to whether the exponential parameter θ lies in a region H_0 or in a region H_1 of the parameter space. The goodness-of-fit problem is typically, I think, the forming of an opinion or making a decision as to whether a set of data can be regarded as a random sample from a completely or partly specified distribution.

I am a significance tester rather than an accept–reject person, so the problem of testing of goodness of fit reduces, for me, to the determination of an ordering of the possible samples and the calculation of the probability of a sample as significant as or more significant than the actual data with regard to the ordering. Thus the resulting number is the significance level and is a measure of strength of evidence against the hypothesized model.

In the case of the exponential family there exists a sufficient statistic of the dimensionality of the parameter and it is plausible to regard a test of goodness of fit of the particular family, as an ordering of the possible observations conditional on the realized value of the sufficient statistic. Just what ordering should be used is very obscure. One commonly used is ordering by conditional probability, but it seems clear that there is no ordering best for all purposes.

It seems clear also that there are problems which can be formulated as decision problems (or accept–reject problems) which cannot be formulated in terms of significance tests. Suppose, for example, we wish to form a measure of strength of evidence against H_0 with respect to the alternative H_1, where

H_0 is $\{x_1, x_2 \ldots, x_N$ is a R.S. from $N(0, 1)$ or Cauchy $(0, 2)$ say$\}$,

H_1 is $\{x_1, x_2, \ldots, x_N$ is a R.S. from D_1 or $D_2 \ldots$ or D_k which are totally specified$\}$.

It is difficult for one to envisage this as being approachable directly by a significance test. But the situation can be approached as a problem of formulating an accept–reject rule.

It would be foolish to say that I cannot imagine a background of information which would tell me that the 'true state of nature' is either H_0 or H_1. But even if I did know this, I would want to have evidence on whether the distribution is $N(0, 1)$ or is Cauchy $(0, 2)$ or D_1 or $D_2, \ldots$ or D_R.

It is quite different, I believe, to envisage the data being from $N(\mu, 1)$ with μ being unspecified and to ask for a goodness-of-fit test of this composite model.

It is quite different also to envisage the data as having a distribution with density
$$f(x) = (1 - \theta_1 - \theta_2)f_0(x) + \theta_1 f_1(x) + \theta_2 f_2(x)$$

and to ask for a goodness-of-fit test of this with θ_1, θ_2 unspecified. To do so would be very difficult for me because I would like a 2-dimensional statistic for θ_1, θ_2 and this would probably not exist. We can of course take $(\theta_1, \theta_2) = (0, 0)$ to be an interesting null hypothesis and seek a test of this. But we shall find in general that a test sensitive with regard to θ_1 differing from zero is quite insensitive with regard to θ_2 differing from zero. The way out of this dilemma in testing hypothesis theory has been to choose the test with maximum average power or with maximum minimum power or something else. However, from the viewpoint of data evaluation, it seems that one should use many tests and not use just one (as for instance in analysis of variance in which one may use the F test and the range test) to give one insight in different directions in the parameter space. If one is in the business of accepting or rejecting lots, of course, one has to have a single decision rule.

These above views lead one to ask the following questions. What is the role of asymptotics? I understand the role of asymptotics in giving an approximation to a distribution which is difficult to obtain. I understand the role of asymptotics associated with maximum likelihood, which in effect lead to the idea that the maximum likelihood statistic may be asymptotically sufficient, and one therefore thinks that the maximum likelihood statistic is 'nearly' sufficient in finite samples. But I have less understanding of asymptotics of test theory in which sample size tends to infinity and some parameter or function of parameters tends to zero. I have little feel for asymptotics in which sample size tends to infinity and the parameter values remain unchanged. If it should turn out, as discussed in Dr Mount's Ph.D. thesis, that the nature of the decision rule best from a certain viewpoint changes with sample size, what is the relevance of the asymptotic theory? What I am really saying is that I can see relevance to asymptotics as $N \to \infty$, or even as $N \to \infty$ and some $\lambda \to 0$ as regards distribution theory but not in test theory.

Next I wonder about the minimax idea and how relevant it is to the uses of statistics, even in the sphere of acceptance sampling, which is the exemplar field for the application of accept–reject rules. I find that I use a kind of minimax in some situations, but these are cases in which the performance of the rule for the *individual* case is important. It seems clear that, in general, minimaxing will be poor on the average.

I also wonder about the use of a risk which is the sum of the errors of the two kinds. How would things work out if one wished to work with $\alpha + 2\beta$, for instance?

It is commonly said that the exponential class of models covers most of the models used in applications. This is true perhaps because we cannot do better though in fact it is not true. For instance, the multinomial (p_1, p_2, p_3), with $p_1 = f_1(\theta)$, $p_2 = f_2(\theta)$, $p_3 = f_3(\theta)$, and θ a scalar, is not in the class, and this surely is a common case.

I would have liked to have seen a more explicit description of the relation of the material to 'classical' likelihood theory of the exponential family. For instance, the equation $a = \operatorname{grad} \psi(s)/\theta$ is essentially the equation for maximum likelihood estimation, $\bar{x}_n = \operatorname{grad} \psi(\theta)$.

PART 5

RANKING AND SELECTION
PROCEDURES

ON SOME CLASSES OF SELECTION
PROCEDURES BASED ON RANKS†

SHANTI S. GUPTA AND GARY C. MCDONALD

1 INTRODUCTION AND SUMMARY

The shortcomings of the classical tests of homogeneity, i.e. testing the hypothesis of equality of parameters, have long been known. Given k populations and from each population a fixed number of observations whose distribution depends on a parameter θ_i, concluding that all θ_i are not equal may not be sufficient. Often the experimenter is interested in ascertaining which population is associated with the largest (or smallest) θ, which populations possess the t largest (or smallest) θ, etc. Suppose the experimenter is interested in identifying which one of the k populations possesses the largest θ, the so-called 'best' population. The parameter θ may be, for example, the mean, the variance, some quantile, or some function of these quantities. Basically, there have been two approaches to ranking and selection problems, the 'indifference zone' approach and the 'subset selection' approach. In the first a single population is chosen and is guaranteed to be the best with probability P^* whenever a certain indifference zone condition holds. For example, in case the populations have normal distributions with a common known variance and unknown means the experimenter may be interested in guaranteeing this probability to be at least P^* whenever the two largest means are separated by a distance greater than d^*. This formulation is due to Bechhofer [5]. The second approach requires no specifications of the parameter space. However, a single population is not necessarily chosen; rather a subset of the given k populations is selected which is guaranteed to contain the best population with probability P^*, the basic probability requirement in these procedures. In this sense the number of populations in the selected subset is a random variable. This formulation is due to Gupta [7, 10].

In the past ten years many papers have appeared on both formulations of the selection problem. As can be expected, most of this research has been devoted to rules which assume a specific distributional form of the

† This research was supported in part by the Office of Naval Research Contract NONR-1100(26) and the Aerospace Research Laboratories Contract AF 33(615)67C1244 at Purdue University. Reproduction in whole or part is permitted for any purposes of the United States Government.

underlying observations; e.g. normal, binomial, multinomial, etc. Barlow
and Gupta [2] and Barlow, Gupta and Panchapakesan [3] have considered the problem of selecting a subset containing the largest (smallest)
quantile of a given order and a subset containing the largest (smallest)
mean. They assume the observations from each population have a distribution which belongs to certain restricted families, e.g. IFR (Increasing
Failure Rate) distributions, IFRA distributions, etc. Distribution-free
selection procedures, most of which are based on joint ranks of the
observations, have been studied by Lehmann [13]; Patterson [18];
Dudewicz [6]; Rizvi and Sobel [22]; Bartlett and Govindarajulu [4];
Puri and Puri [20, 21]; and McDonald [15].

The present paper deals with three classes of nonrandomized distribution-free ranking and selection procedures under the subset selection
formulation. The main problem is to select a subset of k given populations
which contains the 'best' population with probability of at least P^*.
The random variables associated with a fixed population are assumed to
be independent identically distributed with a continuous distribution
function depending on a scalar parameter. This parameter is assumed to
stochastically order the k distribution functions, and the 'best' population is the stochastically largest (smallest) population. The procedures
presented depend on the individual observations of a given population
only through their ranks in the combined sample. In other words, one is
not required to have at hand the actual observations from each population; it suffices to have these ranks, which in some preference-type tests
or lost data problems may be the only information available to an
experimenter.

In §2 the problem is formally stated and the three classes of rules are
defined. In §3 the probability of making a correct selection, i.e. selecting
the population with the largest parameter, using these rules is investigated. For all these classes this probability is shown to be a nondecreasing
function in the largest parameter. For one of the classes this probability
is further shown to be nonincreasing in all parameters but the largest.
For the other two classes of rules, §4 provides bounds on the probability
of correct selection, which in turn provides conservative bounds on the
constants needed for the actual implementation of these rules. Section 5
presents exact expressions for the means, variances and covariances of
the statistics upon which our selection rules are based. In §6 some distribution theory is presented which arises from consideration of one selection procedure based on the rank sums of each population. Section 7
discusses some properties of these selection rules, e.g. local optimality

and monotonicity, and makes some comparison between rules of the three classes in terms of the expected number of populations included in the selected subset.

2 FORMULATION OF THE PROBLEM AND THREE CLASSES OF RULES

Let $\pi_1, \ldots, \pi_k$ be $k \, (\geq 2)$ independent populations. The associated random variables $X_{ij}, j = 1, \ldots, n_i; \, i = 1, \ldots, k$, are assumed independent and to have a continuous distribution $F_{\theta_i}(x)$, where θ_i belong to some interval Θ on the real line. Suppose $F_\theta(x)$ is a stochastically increasing (SI) family of distributions, i.e. if θ_1 is less than θ_2, then $F_{\theta_1}(x)$ and $F_{\theta_2}(x)$ are distinct and $F_{\theta_2}(x) \leq F_{\theta_1}(x)$ for all x. Examples of such families of distributions are: (1) any location parameter family, i.e. $F_\theta(x) = F(x - \theta)$; (2) any scale parameter family, i.e. $F_\theta(x) = F(x/\theta), x > 0, \theta > 0$; (3) any family of distribution functions whose densities possess the monotone likelihood ratio (or TP_2) property. Let R_{ij} denote the rank of the observation x_{ij} in the combined sample; i.e. if there are exactly r observations less than x_{ij} then $R_{ij} = r + 1$. These ranks are well-defined with probability one, since the random variables are assumed to have a continuous distribution. Let $Z(1) \leq Z(2) \leq \ldots \leq Z(N)$ denote an ordered sample of size $N = \sum_{i=1}^{k} n_i$ from any continuous distribution G, such that

$$-\infty < a(r) \equiv E[Z(r)|G] < \infty \quad (r = 1, \ldots,, N).$$

With each of the random variables X_{ij} associate the number $a(R_{ij})$ and define

$$H_i = n_i^{-1} \sum_{j=1}^{n_i} a(R_{ij}) \quad (i = 1, \ldots, k). \tag{2.1}$$

Using the quantities H_i, we wish to define procedures for selecting a subset of the k populations. Letting $\theta_{[i]}$ denote the ith smallest unknown parameter, we have

$$F_{\theta_{[1]}}(x) \geq F_{\theta_{[2]}}(x) \geq \ldots \geq F_{\theta_{[k]}}(x) \quad (\forall x). \tag{2.2}$$

The population whose associated random variables have the distribution $F_{\theta_{[k]}}(x)$ will be called the best population. In case several populations possess the largest parameter value $\theta_{[k]}$, one of them is tagged at random and called the best. A 'Correct Selection' (CS) is said to occur if and only if the best population is included in the selected subset. In the usual subset selection problem one wishes to select a subset such that the probability is at least equal to a preassigned constant P^* $(1/k < P^* < 1)$

that the selected subset includes the best population. Mathematically, for a given selection rule R,

$$\inf_{\Omega} P(\text{CS}|R) \geq P^*, \tag{2.3}$$

where $\qquad \Omega = \{\boldsymbol{\theta} = (\theta_1, \ldots, \theta_k): \theta_i \in \Theta, i = 1, 2, \ldots, k\}. \tag{2.4}$

The following three classes of selection procedures, which choose a subset of the k given populations, and which depend on the given distribution G, will be considered:

$$R_1(G): \text{select } \pi_i \text{ iff } H_i \geq \max_{1 \leq j \leq k} H_j - d \quad (i = 1, \ldots, k, d \geq 0), \tag{2.5}$$

$$R_2(G): \text{select } \pi_i \text{ iff } H_i \geq c^{-1} \max_{1 \leq j \leq k} H_j \quad (i = 1, \ldots, k, c \geq 1), \tag{2.6}$$

$$R_3(G): \text{select } \pi_i \text{ iff } H_i \geq D \quad (i = 1, \ldots, k, -\infty < D < \infty). \tag{2.7}$$

It should be noted that rules $R_1(G)$, $R_2(G)$, and $R_3(G)$ are equivalent if $k = 2$. The procedures $R_1(G)$ (and their randomized analogs) have been suggested by Bartlett and Govindarajulu [4] for continuous distributions differing by a location parameter. The procedure $R_2(G)$ will be studied in this paper *only* for the case where $H_i \geq 0$ for all i. The constants d and c are usually chosen to be as small as possible, D as large as possible, while satisfying the probability requirement (2.3). The number of populations included in the selected subset is a random variable which takes values 1 to k inclusive for rules $R_1(G)$ and $R_2(G)$. The subset chosen by rule $R_3(G)$, however, could possibly be empty.

Another class of selection rules which includes $R_1(G)$ and $R_3(G)$ as special cases, and depends on an index t ($1 \leq t \leq \infty$), can be defined as follows when H_i are nonnegative:

$$R(G): \text{select } \pi_i \text{ iff } H_i \geq \left(\frac{1}{k-1} \sum_{\substack{j=1 \\ j \neq i}}^{k} H_j^t \right)^{1/t} - d_t \quad (i = 1, \ldots, k; d_t \geq 0). \tag{2.8}$$

For $t = 1$, this rule reduces to a rule of the form $R_3(G)$ since the sum of all the H_j is constant, and for $t = \infty$, $R(G)$ reduces to a rule of the type $R_1(G)$.

Let $\pi_{(i)}$ be the population associated with $\theta_{[i]}$, the ith smallest θ_i. Then the probability of making a correct selection using the procedures $R_i(G)$, $i = 1, 2, 3$, is given, respectively, by

$$P(\text{CS}|R_i(G)) = \begin{cases} P(H_{(k)} \geq \max_{1 \leq j \leq k} H_{(j)} - d) & (i = 1), \\ P(H_{(k)} \geq c^{-1} \max_{1 \leq j \leq k} H_{(j)}) & (i = 2), \\ P(H_{(k)} \geq D) & (i = 3). \end{cases} \tag{2.9}$$

The corresponding rules for choosing a subset of the k populations which contains the population with the smallest parameter, $\pi_{(1)}$, are:

$$R_1'(G): \text{select } \pi_i \text{ iff } H_i \leqslant \min_{1 \leqslant j \leqslant k} H_j + d' \quad (i = 1, \ldots, k; d' \geqslant 0), \quad (2.10)$$

$$R_2'(G): \text{select } \pi_i \text{ iff } H_i \leqslant c' \min_{1 \leqslant j \leqslant k} H_j \quad (i = 1, \ldots, k; c' \geqslant 1), \quad (2.11)$$

$$R_3'(G): \text{select } \pi_i \text{ iff } H_i \leqslant D' \quad (i = 1, \ldots, k; \; -\infty < D' < \infty). \quad (2.12)$$

The constants d', c' and D' are obtained as before. No more consideration will be given to these three rules; results and methods developed for $R_1(G)$, $R_2(G)$ and $R_3(G)$ will have an obvious analog for $R_1'(G)$, $R_2'(G)$ and $R_3'(G)$, respectively.

3 THE INFIMUM OF THE PROBABILITY OF A CORRECT SELECTION

We start with a lemma, which is essentially the same as Lemma 4.2 in Mahamunulu [14] and Lemma 2.1 in Alam and Rizvi [1] both being a generalization of a result of Lehmann [12, p. 112, no. 11] for more than one dimension. We state our version without proof.

Lemma 3.1. Let $\mathbf{X} = (X_{11}, \ldots, X_{1n_1}, \ldots, X_{k1}, \ldots, X_{kn_k})$ be a vector valued random variable of $\sum_{i=1}^{k} n_i (\geqslant 1)$ independent components with X_{ij} having the distribution $F_{\theta_i}(x), j = 1, \ldots, n_i; i = 1, \ldots, k$. Suppose $F_\theta(x)$ is a SI family of distributions. Let Ψ be a function of $x_{11}, \ldots, x_{1n_1}, \ldots, x_{k1}, \ldots, x_{kn_k}$ which, for any fixed i, is a nondecreasing (nonincreasing) function of $x_{i1}, \ldots, x_{in_i}$ when the other components of $\mathbf{x}$ are held fixed. Then $E_\theta[\Psi(\mathbf{X})]$ is a nondecreasing (nonincreasing) function of θ_i.

Theorem 3.1. For rules $R_i(G)$, $i = 1, 2, 3, p_s(R_i(G))$, the probability of including the population $\pi_{(s)}$ in the selected subset is nondecreasing in $\theta_{[s]}$ and, hence,

$$\inf_{\Omega} p_s(R_i(G)) = \inf_{\Omega_s} p_s(R_i(G)) \quad (s = 1, \ldots, k), \quad (3.1)$$

where
$$\Omega_s = \{\boldsymbol{\theta} \in \Omega: \theta_{[s]} = \theta_{[s-1]}\} \quad (3.2)$$

and $\theta_{[0]}$ is the least admissible value of θ.

Proof. We will prove it for the rule $R_1(G)$. Let

$$\Psi_s(\mathbf{X}) = \begin{cases} 1 & \text{if } H_{(s)} \geqslant \max_{j \neq s} H_{(j)} - d, \\ 0 & \text{otherwise.} \end{cases} \quad (3.3)$$

Let $R_{(i)j}$ be the rank of $X_{(i)j}, j = 1, \ldots, n_{(i)}$, and consider an observation $x_{(s)l}$ for some fixed l, $1 \leqslant l \leqslant n_{(s)}$. As $x_{(s)l}$ increases and the other observations remain fixed, either:

(1) $x_{(s)l}$ surpasses first an $x_{(m)j}$, $m \neq s$, so $R_{(s)l}$ increases by 1 and $R_{(m)j}$ decreases by 1; or

(2) $x_{(s)l}$ surpasses first an $x_{(s)j}$, $j \neq l$, so $R_{(s)l}$ increases by 1 and $R_{(s)j}$ decreases by 1; or

(3) $x_{(s)l}$ does not surpass any other observation, so all ranks remain the same.

In all three cases, $H_{(s)}$ is nondecreasing and $H_{(j)}$, $j \neq s$, is nonincreasing and hence so is $\max\limits_{j \neq s} H_{(j)}$. Therefore, $\Psi_s(\mathbf{x})$ is a nondecreasing function of $x_{(s)j}$, $j = 1, \ldots, n_{(s)}$. By Lemma 3.1, $E_\theta[\Psi_s(\mathbf{X})] = p_s(R_1(G))$ is a nondecreasing function of $\theta_{[s]}$. A similar argument proves the result for $R_2(G)$ and $R_3(G)$.

In particular, for $s = k$, Equation (3.1) can be written as

$$\inf_\Omega P(\mathrm{CS}|R_i(G)) = \inf_{\Omega_k} P(\mathrm{CS}|R_i(G)). \qquad (3.4)$$

Remark. If H_i in (2.1) is redefined to be

$$H_i^* = n_i^{-1} \sum_{j=1}^{n_i} Z(R_{ij})$$

and rules $R_1^*(G)$, $R_2^*(G)$ and $R_3^*(G)$ are defined by (2.5), (2.6) and (2.7) with H_i replaced by H_i^*, $i = 1, \ldots, k$, then Theorem 3.1 holds with $R_i(G)$ replaced by $R_i^*(G)$. Thus, Theorem 3.1 is valid for randomized, as well as nonrandomized, selection procedures.

In the case of $R_3(G)$ we can say more on the infimum of the probability of a correct selection.

Theorem 3.2. For the procedure $R_3(G)$,

$$\inf_\Omega P(\mathrm{CS}|R_3(G)) = \inf_{\Omega_0} P(\mathrm{CS}|R_3(G)), \qquad (3.5)$$

where
$$\Omega_0 = \{\boldsymbol{\theta} \in \Omega: \theta_{[1]} = \ldots = \theta_{[k]}\}. \qquad (3.6)$$

Proof. Let
$$\Psi(\mathbf{X}) = \begin{cases} 1 & \text{if } H_{(k)} \geqslant D \\ 0 & \text{otherwise.} \end{cases}$$

By an argument similar to the one employed in the proof of Theorem 3.1, we have that $E_\theta[\Psi(\mathbf{X})]$ is a nondecreasing function of $\theta_{[k]}$ and a nonincreasing function of $\theta_{[j]}$, $j = 1, \ldots, k-1$. This completes the proof.

SELECTION PROCEDURES BASED ON RANKS **497**

For $\boldsymbol{\theta} \in \Omega_0$, the quantity $P_{\boldsymbol{\theta}}(\mathrm{CS}|R_i(G))$, $i = 1, 2, 3$ is independent of the common underlying distribution, $F_{\theta}(x)$. In other words, the distribution of the statistics $\max_{1 \leqslant j \leqslant k} H_j - H_i$ or H_i, or any other statistics involving H_i does not depend on $F_{\theta}(x)$. It is in this sense that the procedures of this paper are distribution-free.

From Theorem 3.1, if $k = 2$, the probability of a correct selection using either rule $R_1(G)$ or $R_2(G)$ is minimized when the two populations are identically distributed. The same result is true in a slippage configuration, i.e. if $\theta_{[1]} = \ldots = \theta_{[k-1]}$ then the probability of a correct selection is minimized when $\theta_{[1]} = \theta_{[2]} = \ldots = \theta_{[k]}$.

It should be pointed out that a theorem similar to Theorem 3.2 involving $R_1(G)$ does not hold in general. This fact is established by means of a counterexample constructed by Rizvi and Woodworth[23] using distributions having two finite disjoint intervals for their supports and lacking the MLR property. McDonald[15] uses the same type of distributions to show that for $k = 3$, $P(\mathrm{CS}|R_1(G))$ need not be monotonic in $\theta_{[2]}$. The main difficulty arises out of the fact that the statistics $H_{(i)}$ are not independent.

We will next obtain bounds on $P(\mathrm{CS}|R_s(G))$, $s = 1, 2$, before investigating further the rule $R_3(G)$.

4 BOUNDS ON $P(\mathrm{CS}|R_i(G))$, $i = 1, 2$

We will assume that $n_i = n, i = 1, \ldots, k$. First consider rule $R_1(G)$. We have

$$H_i = n^{-1} \sum_{j=1}^{n} a(R_{ij}) \quad (i = 1, \ldots, k). \tag{4.1}$$

It is easy to see that

$$(k-1)^{-1} \sum_{j=1}^{k-1} H_{(j)} \leqslant \max_{1 \leqslant j \leqslant k-1} H_{(j)} \leqslant n^{-1} \sum_{r=N-n+1}^{N} a(r). \tag{4.2}$$

Using the inequalities (4.2) in the relation

$$P(\mathrm{CS}|R_1(G)) = P(H_{(k)} \geqslant \max_{1 \leqslant j \leqslant k-1} H_{(j)} - d),$$

we obtain

$$P\left(H_{(k)} \geqslant n^{-1} \sum_{r=N-n+1}^{N} a(r) - d\right) \leqslant P(\mathrm{CS}|R_1(G))$$

$$\leqslant P\left(H_{(k)} \geqslant (k-1)^{-1} \sum_{j=1}^{k-1} H_{(j)} - d\right). \tag{4.3}$$

Letting $\sum\limits_{r=1}^{N} a(r) = A$, it follows that $\sum\limits_{j=1}^{k} H_{(j)} = A/n$, a constant. Using this relation in (4.3), and defining

$$u(d, k, n) = [A - nd(k-1)]/nk, \tag{4.4}$$

$$v(d, k, n) = n^{-1} \sum_{r=N-n+1}^{N} a(r) - d, \tag{4.5}$$

we have
$$P(H_{(k)} \geqslant v) \leqslant P(\mathrm{CS} \,|\, R_1(G)) \leqslant P(H_{(k)} \geqslant u), \tag{4.6}$$
and hence

$$\inf_{\Omega} P(H_{(k)} \geqslant v) \leqslant \inf_{\Omega} P(\mathrm{CS} \,|\, R_1(G)) \leqslant \inf_{\Omega} P(H_{(k)} \geqslant u). \tag{4.7}$$

For the rule $R_2(G)$, we get a corresponding expression

$$\inf_{\Omega} P(H_{(k)} \geqslant v') \leqslant \inf_{\Omega} P(\mathrm{CS} \,|\, R_2(G)) \leqslant \inf_{\Omega} P(H_{(k)} \geqslant u'), \tag{4.8}$$

where
$$u'(d, k, n) = n^{-1} A [1 + c(k-1)]^{-1} \tag{4.9}$$

and
$$v'(d, k, n) = (nc)^{-1} \sum_{r=N-n+1}^{N} a(r). \tag{4.10}$$

From Theorem 3.2, we know that the infima over Ω of expressions of the form $P(H_{(k)} \geqslant K)$ are attained when $\theta_{[1]} = \ldots = \theta_{[k]}$.

For the particular case where $a(r) = r$, we have

$$nH_i = \sum_{j=1}^{n} R_{ij} = T_i \quad \text{say.} \tag{4.11}$$

The T_i are the rank-sum statistics, and in this case we denote the selection rules $R_j(G)$ by simply R_j. The expressions given above take the form

$$A = N(N+1)/2, \tag{4.12}$$

and
$$\sum_{r=N-n+1}^{N} a(r) = n(2N-n+1)/2. \tag{4.13}$$

Thus, equations (4.4), (4.5), (4.9) and (4.10) reduce to

$$u(d, k, n) = (N+1)/2 - d(k-1)/k, \tag{4.14}$$

$$v(d, k, n) = (2N-n+1)/2 - d, \tag{4.15}$$

$$u'(d, k, n) = k(N+1)/[2 + 2c(k-1)], \tag{4.16}$$

$$v'(d, k, n) = (2N-n+1)/2c. \tag{4.17}$$

In the special case $a(r) = r$ a more useful form of the lower bound appearing in (4.7) is given in the next theorem.

Theorem 4.1. *If U is the Mann–Whitney statistic associated with samples of size n and $(k-1)n$ taken from identically distributed populations, then in the case where $a(r) = r$,*

$$\inf_{\Omega} P(\mathrm{CS}|R_1) \geqslant P(U \leqslant nd). \tag{4.18}$$

Proof. We first recall that the Mann–Whitney U statistic, calculated from the samples $x_1, \ldots, x_p$ and $y_1, \ldots, y_q$ of sizes p and q from two independent populations, is the number of times an x_i precedes a y_j. If T_x denotes the rank-sum of the x's in the combined sample, then U and T_x are related by

$$U + T_x = pq + p(p+1)/2. \tag{4.19}$$

In our present case with samples from k populations, we need to evaluate $P(H_{(k)} \geqslant v)$ when all the populations are identical. Considering whether the observations came from the $\pi_{(k)}$ or any one of the rest, we have from (4.7), (4.14) and (4.19) with $p = n$ and $q = (k-1)n$,

$$\inf_{\Omega} P(\mathrm{CS}|R_1) \geqslant P(T_{(k)} \geqslant nv)$$

$$= P(U \leqslant n^2(k-1) + n(n+1)/2 - nv)$$

$$= P(U \leqslant nd). \tag{4.20}$$

A similar theorem holds for rule $R_2(G)$.

Since $\sum_{j=1}^{k} H_{(j)} = A/n$, we see that

$$\max_{1 \leqslant j \leqslant k} H_j \geqslant A/nk. \tag{4.21}$$

Hence, a sufficient, but not necessary, condition for the selection rule $R_3(G)$ to select a nonempty subset is that P^* be sufficiently large so that

$$D \leqslant A/N. \tag{4.22}$$

For large n, this sufficiency condition for rule $R_3(G)$ is satisfied if $P^* > \frac{1}{2}$. For rule R_3, i.e. when $a(r) = r$, the condition (4.22) is $D \leqslant (N+1)/2$. As an example, with $k = 3$, $n = 5$ the sufficient condition $D \leqslant 8$ is satisfied for $P^* \geqslant 0.523$ and for such values a nonempty subset will be selected.

The evaluation of the constants $D = D(k, n, P^*)$ for the rule R_3 can be effected as follows:

$$P^* \leqslant P(T_i \geqslant Dn) = P(U \leqslant n^2(k - \tfrac{1}{2}) - n(D - \tfrac{1}{2})), \tag{4.23}$$

using (4.19) and considering all populations identically distributed. Hence, D is the largest integer satisfying the inequality (4.23). The Mann–Whitney U-statistic has been well-tabulated by Milton [16] and others.

5   MOMENTS OF THE H_i

In this section we will derive the means, variances and covariances of the H_i assuming the independent random variables X_{ij} have the continuous distribution $F_i(x)$, $j = 1, ..., n_i$; $i = 1, ..., k$. Let $p^{(i)}_{j_1, ..., j_{n_i}}$ be the probability that the n_i observations from the population π_i have ranks $j_1, ..., j_{n_i}$ in the combined sample. Then,

$$E(H_i) = n_i^{-1} \sum_{\{j_1, ..., j_{n_i}\}} [a(j_1) + ... + a(j_{n_i})] p^{(i)}_{j_1, ..., j_{n_i}} \quad (i = 1, ..., k), \quad (5.1)$$

where the summation is over all possible subsets of size n_i in the set of integers 1 through N. Alternatively,

$$E(H_i) = n_i^{-1} \sum_{l=1}^{N} a(l) p_l^{(i)} \quad (i = 1, ..., k), \quad (5.2)$$

where $p_l^{(i)}$ is the probability that any one observation from π_i has rank l in the combined sample.

Let $p_{l,m}^{(i)}$ be the probability that any two of the observations from π_i have ranks l and m in the combined sample. Then,

$$E(H_i^2) = n_i^{-2} \sum_{\{j_1, ..., j_{n_i}\}} [a(j_1) + ... + a(j_{n_i})]^2 p^{(i)}_{j_1, ..., j_{n_i}}. \quad (5.3)$$

Alternatively,

$$E(H_i^2) = n_i^{-2} \left[\sum_{l=1}^{N} (a(l))^2 p_l^{(i)} + 2 \sum_{l=1}^{N} \sum_{m=l+1}^{N} a(l) a(m) p_{l,m}^{(i)} \right]. \quad (5.4)$$

Hence,

$$n_i^2 \operatorname{var}(H_i) = \sum_{l=1}^{N} (a(l))^2 p_l^{(i)} + 2 \sum_{l=1}^{N} \sum_{m=l+1}^{N} a(l) a(m) p_{l,m}^{(i)} - \left(\sum_{l=1}^{N} a(l) p_l^{(i)} \right)^2$$

$$= \sum_{l=1}^{N} (a(l))^2 p_l^{(i)} (1 - p_l^{(i)}) + 2 \sum_{l=1}^{N} \sum_{m=l+1}^{N} a(l) a(m) (p_{l,m}^{(i)} - p_l^{(i)} p_m^{(i)}). \quad (5.5)$$

In a similar manner one can show that for $i \neq j$,

$$n_i n_j \operatorname{cov}(H_i, H_j) = \sum_{l=1}^{N} \sum_{m=1}^{N} a(l) a(m) (p_{l,m}^{(i,j)} - p_l^{(i)} p_m^{(j)}), \quad (5.6)$$

where $p_{l,m}^{(i,j)}$ is the probability that one observation from population π_i has rank l and one observation from π_j has rank m. Note that $p_{l,l}^{(i,j)} = 0$, $i \neq j, l = 1, ..., N$.

As we see above, the computation of these moments depends upon the evaluation of $p_l^{(i)}$, $p_{l,m}^{(i)}$ and $p_{l,m}^{(i,j)}$. To evaluate $p_l^{(i)}$, choose one of the observations from π_i to have rank l. Ranks $1, 2, ..., l-1$ are then assumed by $l-1$ of the remaining $N-1$ observations. These $l-1$ observations consist of r_j observations from $\pi_j, j = 1, \ldots, k$, subject to the conditions

$$B: \begin{cases} 0 \leqslant r_i \leqslant n_i - 1, \\ 0 \leqslant r_j \leqslant n_j, \\ r_1 + \ldots + r_k = l - 1, \end{cases} \quad (j = 1, \ldots, k; j \neq i). \qquad (5.7)$$

Thus, $\quad p_l^{(i)} = \sum_B n_i \binom{n_i - 1}{r_i} \int \left\{ \prod_{\substack{j=1 \\ j \neq i}}^{k} \binom{n_j}{r_j} F_j^{r_j} \bar{F}_j^{n_j - r_j} \right\} F_i^{r_i} \bar{F}_i^{n_i - r_i - 1} dF_i, \qquad (5.8)$

where $F_j \equiv F_j(x)$, $\bar{F}_j \equiv 1 - F_j(x), j = 1, \ldots, k$. Taking the summation inside the integral yields

$$p_l^{(i)} = n_i \int A_l^{(i)}(x) \, dF_i(x) \quad (i = 1, \ldots, k; l = 1, \ldots, N), \qquad (5.9)$$

where $\quad A_l^{(i)}(x) = \sum_B \binom{n_i - 1}{r_i} F_i^{r_i} \bar{F}_i^{n_i - r_i - 1} \left\{ \prod_{\substack{j=1 \\ j \neq i}}^{k} \binom{n_j}{r_j} F_j^{r_j} \bar{F}_j^{n_j - r_j} \right\}. \qquad (5.10)$

In a similar fashion we can obtain expressions for the probabilities $p_{l,m}^{(i)}$ and $p_{l,m}^{(i,j)}$ in terms of the given distributions and sample sizes. In the special cases where $F_i(x) = F(x)$ and $n_i = n, i = 1, \ldots, k$, we have:

$$E(H_i) = N^{-1} \sum_{l=1}^{N} a(l) = A/N, \qquad (5.11)$$

$$n^2 \text{var} (H_i) = k^{-2}(k-1) \sum_{l=1}^{N} (a(l))^2$$
$$- 2(k-1) k^{-2}(N-1)^{-1} \sum_{l=1}^{N} \sum_{m=l+1}^{N} a(l) a(m), \qquad (5.12)$$

$$n^2 \text{cov} (H_i, H_j) = k^{-1}(N-1)^{-1} \sum_{l=1}^{N} \sum_{\substack{m=1 \\ l \neq m}}^{N} a(l) a(m) - k^{-2}A^2 \quad (i \neq j). \qquad (5.13)$$

If, in addition, $a(l) = l, l = 1, \ldots, N$, then $nH_i = T_i$, and, hence,

$$E(T_i) = n(N+1)/2, \qquad (5.14)$$

$$\text{var} (T_i) = n^2(k-1)(N+1)/12, \qquad (5.15)$$

$$\text{cov} (T_i, T_j) = -n^2(N+1)/12 \quad (i \neq j), \qquad (5.16)$$

which agree with the known expressions for this special case.

Asymptotic forms for the moments of H_i have been given by Puri [19].

6 THE EXACT AND ASYMPTOTIC DISTRIBUTION OF $\max\limits_{1 \leqslant j \leqslant k} T_j - T_i$ FOR IDENTICALLY DISTRIBUTED POPULATIONS

In this section the random variables $X_{ij}, j = 1, \ldots, n_i; i = 1, \ldots, k$, are assumed independent identically distributed with a continuous distribution $F(x)$. In this case the H_i are exchangeable random variables if $n_i = n, i = 1, \ldots, k$. It should be noted that in a slippage-type configuration (see § 3), the constants required to implement rules $R_i(G)$, $i = 1, 2, 3$, are determined from the basic probability requirement $P(\text{CS}\,|\,R_i(G)) \geqslant P^*$ calculated with identically distributed populations. But the exact distributions of the relevant statistics, e.g. $\max\limits_{1 \leqslant j \leqslant k} H_j - H_1$, are not known for the general scores $a(R_{ij})$. However, in the case $a(R_{ij}) = R_{ij}$ the procedures $R_i(G)$ reduce to the rank sum procedures R_i, $i = 1, 2, 3$. The distribution of the statistic $\max\limits_{1 \leqslant j \leqslant k} T_j - T_1$, both exact and asymptotic, is somewhat easier to obtain than the corresponding distribution of the statistic $\max\limits_{1 \leqslant j \leqslant k} T_j/T_1$. For some results concerning the latter statistic, see McDonald [15]. Our concern here will be the former which is tantamount to considering rule R_1. Corresponding to rule R_3 is the statistic T_1, the distribution of which has been well-treated elsewhere.

For $k = 2$, the rules R_1, R_2 and R_3 are all equivalent. The constants required to implement these rules are obtained in a manner as described at the end of § 4. Some of these values are given in Table 2 where they are compared with asymptotic solutions.

Now suppose $k = 3$ and that we have n_i observations from the ith population. The quantities T_i can be obtained if each observation in the ordered sample is replaced by an i if it came from the ith population. Now one has only to consider a sequence of length Σn_i consisting of n_1 1's, n_2 2's, and n_3 3's. Since the random variables are identically distributed, each of the $\dbinom{\Sigma n_i}{n_1, n_2, n_3}$ different sequences are equally likely. Hence, to find $P[T_1 \geqslant \max\limits_{1 \leqslant j \leqslant 3} T_j - m]$, it suffices to count the number of sequences which possess the attribute $[T_2 - T_1 \leqslant m, T_3 - T_1 \leqslant m]$. The recursion formula presented here is of the same type as that given by Odeh [17] in tabulating the distribution of the maximum rank sum. Let

$$S = n_1 + n_2 + n_3, \tag{6.1}$$

and define

$$N(n_1, n_2, n_3 | m_2, m_3)$$

= number of sequences in which $T_2 - T_1 \leqslant m_2$ and $T_3 - T_1 \leqslant m_3$.   (6.2)

The following symmetry holds:

$$N(n_1, n_2, n_3 | m_2, m_3) = N(n_1, n_3, n_2 | m_3, m_2).$$ (6.3)

Then, by conditioning on the parent population of the last element in a sequence, the following recursion formula is obtained:

$$N(n_1, n_2, n_3 | m_2, m_3) = N(n_1 - 1, n_2, n_3 | m_2 + S, m_3 + S)$$

$$+ N(n_1, n_2 - 1, n_3 | m_2 - S, m_3) + N(n_1, n_2, n_3 - 1 | m_2, m_3 - S),$$ (6.4)

with the boundary conditions:

(1) If for any $i \geqslant 2$, $m_i < [n_i(n_i + 1) - n_1(1 + 2S - n_1)]/2$, then

$$N(n_1, n_2, n_3 | m_2, m_3) = 0.$$

(2) If for every $i \geqslant 2$, $m_i \geqslant [n_i(1 + 2S - n_i) - n_1(n_1 + 1)]/2$, then

$$N(n_1, n_2, n_3 | m_2, m_3) = \binom{S}{n_1, n_2, n_3}.$$

(3) $N(0, n_2, n_3 | m_2, m_3)$ = number of sequences of n_2 2's and n_3 3's such that $S(S+1)/2 - m_3 \leqslant T_2 \leqslant m_2$, so

(a) if $S(S+1)/2 - m_3 > m_2$, $N(0, n_2, n_3 | m_2, m_3) = 0$,

(b) if $m_2 < n_2(n_2 + 1)/2$, $N(0, n_2, n_3 | m_2, m_3) = 0$,

(c) if $m_3 < n_3(n_3 + 1)/2$, $N(0, n_2, n_3 | m_2, m_3) = 0$,

(d) if (a) through (c) do not hold, the term can be evaluated from a Mann–Whitney table.

(4) $N(n_1, 0, n_3 | m_2, m_3)$ = number of sequences of n_1 1's and n_3 3's such that $T_1 \geqslant \max\{-m_2, L(S(S+1)/4 - m_3/2)\} \equiv M$, where $L(x)$ is the smallest integer not less than x, so

(a) if $M > n_1(n_1 + 2n_3 + 1)/2$, $N(n_1, 0, n_3 | m_2, m_3) = 0$,

(b) if $M \leqslant n_1(n_1 + 1)/2$, $N(n_1, 0, n_3 | m_2, m_3) = \binom{S}{n_1}$,

(c) if (a) and (b) fail to hold, the term can be evaluated from a Mann–Whitney table.

(5) $N(n_1, n_2, 0 | m_2, m_3) = N(n_1, 0, n_2 | m_3, m_2)$, so condition (4) applies.

It follows from (6.3) that at an 'equal n_i, equal m_i stage', equation (6.4) can be written as

$$N(n, n, n | m, m) = N(n - 1, n, n | m + 3n, m + 3n)$$

$$+ 2N(n, n - 1, n | m - 3n, m).$$ (6.5)

In order to get $P[T_1 \geqslant \max\limits_{1 \leqslant j \leqslant 3} T_j - m]$ for values of $m \geqslant 0$, one uses the following relation:

$$P[T_1 \geqslant \max_{1 \leqslant j \leqslant 3} T_j - m] = N(n_1, n_2, n_3 | m, m)\left(\frac{S}{n_1, n_2, n_3}\right)^{-1}. \quad (6.6)$$

A recursion formula similar to (6.4) can be written for an arbitrary number of populations. The quantity $N(n, n, n | m, m)$ was computed for $n = 2, 3, 4, 5$; $m = 0, 1, \dots, 2n^2$. Using (6.6), $P[T_1 \geqslant \max\limits_{1 \leqslant j \leqslant 3} T_j - m]$ was then obtained to five decimal places, the fifth being rounded. These computations are given in Table 1.

Asymptotically, we have the following theorem as a special case of a more general result applying to the statistics H_i with populations not necessarily identically distributed. The proof follows directly from Puri [19] and is omitted.

Theorem 6.1. *Let* $X_{ij}, j = 1, \dots, n$; $i = 1, \dots, k$, *be independent identically distributed random variables with a continuous distribution function. Then*

$$P[T_k \geqslant \max_{1 \leqslant j \leqslant k} T_j - m] \approx \int_{-\infty}^{\infty} [\Phi(x + m/z)]^{k-1} \phi(x) \, dx \quad (m \geqslant 0), \quad (6.7)$$

where $\Phi(\cdot)$ *and* $\phi(\cdot)$ *are the cumulative distribution function and density of a standard normal random variable, respectively, and*

$$z = z(n, k) = n[k(nk + 1)/12]^{\frac{1}{2}}. \quad (6.8)$$

Integrals of the type appearing on the right-hand side of equation (6.7) have been considered by Gupta [8]. Table 1 in [8] gives h values satisfying the equation

$$\int_{-\infty}^{\infty} [\Phi(x + h\sqrt{2})]^{k-1} \phi(x) \, dx = P^* \quad (6.9)$$

for $P^* = 0 \cdot 99$, $0 \cdot 975$, $0 \cdot 95$, $0 \cdot 90$, $0 \cdot 75$ and $k = 2(1)51$. If $\tilde{m}$ denotes the value of m based on the normal approximation, then from (6.9) one obtains

$$\tilde{m} = hn[k(nk + 1)/6]^{\frac{1}{2}}, \quad (6.10)$$

h being the entry of Table 1 of [8] corresponding to the given P^* and k.

Remarks. (1) By using (6.10) one can obtain an asymptotic value of m (and, hence, d) in rule R_1 when a slippage configuration in Ω exists (as shown in § 3) for $k = 2(1)51$ and for any common sample size n, n large. (2) In general $\tilde{m}$ will not be an integer. So for the solution the smallest integer not less than $\tilde{m}$, $L(\tilde{m})$ should be taken. This method was used to calculate an asymptotic value of m for $k = 2, 3$; $n = 2(1)25$ and $P^* = 0 \cdot 99$, $0 \cdot 975$,

Table 1. For a given n, the left column is the value $N(n, n, n | m, m)$, i.e. the number of sequences in which $\max\limits_{1 \leqslant j \leqslant 3} T_j - T_1 \leqslant m$; the right column is the quantity $P(T_1 \geqslant \max\limits_{1 \leqslant j \leqslant 3} T_j - m)$. The rank sums $T_i, i = 1, 2, 3$ are based on random variables $X_{ij}, j = 1, \ldots, n; i = 1, 2, 3$, which are independent identically distributed

	$n = 2$		$n = 3$		$n = 4$		$n = 5$	
$m = 0$	38	0·42222	600	0·35714	12,268	0·35405	262,686	0·34712
1	44	0·48889	702	0·41786	13,500	0·38961	283,426	0·37453
2	54	0·60000	808	0·48095	14,958	0·43169	305,560	0·40378
3	62	0·68889	912	0·54286	16,322	0·47105	327,738	0·43308
4	70	0·77778	1,004	0·59762	17,734	0·51180	349,236	0·46149
5	76	0·84444	1,112	0·66190	19,162	0·55302	372,410	0·49211
6	84	0·93333	1,206	0·71786	20,588	0·59417	394,770	0·52166
7	88	0·97778	1,294	0·77024	21,884	0·63157	416,774	0·55074
8	90	1·00000	1,374	0·81786	23,274	0·67169	439,432	0·58068
9			1,438	0·85595	24,500	0·70707	461,534	0·60988
10			1,490	0·88690	25,708	0·74193	482,468	0·63755
11			1,544	0·91905	26,846	0·77478	504,104	0·66614
12			1,584	0·94286	27,956	0·80681	524,454	0·69303
13			1,618	0·96309	28,906	0·83423	543,924	0·71876
14			1,644	0·97857	29,842	0·86124	563,152	0·74417
15			1,664	0·99048	30,616	0·88358	581,280	0·76812
16			1,674	0·99643	31,336	0·90436	598,016	0·79024
17			1,678	0·99881	31,952	0·92214	614,640	0·81220
18			1,680	1·00000	32,496	0·93784	629,818	0·83226
19					32,936	0·95053	643,940	0·85092
20					33,338	0·96214	657,292	0·86857
21					33,644	0·97097	669,588	0·88481
22					33,912	0·97870	680,548	0·89929
23					34,126	0·98488	690,952	0·91304
24					34,296	0·98978	700,152	0·92520
25					34,424	0·99348	708,408	0·93611
26					34,518	0·99619	715,862	0·94596
27					34,574	0·99781	722,454	0·95467
28					34,614	0·99896	728,098	0·96213
29					34,634	0·99954	733,120	0·96877
30					34,644	0·99983	737,408	0·97443
31					34,648	0·99994	741,054	0·97925
32					34,650	1·00000	744,192	0·98339
33							746,838	0·98689
34							749,008	0·98976
35							750,826	0·99216
36							752,296	0·99411
37							753,464	0·99565
38							754,402	0·99689
39							755,130	0·99785
40							755,660	0·99855
41							756,048	0·99906
42							756,316	0·99942
43							756,496	0·99966
44							756,608	0·99980
45							756,680	0·99989
46							756,720	0·99995
47							756,740	0·99998
48							756,750	0·99999
49							756,754	0·99999
50							756,756	1·00000

0·95, 0·90, 0·75. These results are presented in Table 2. Exact m values are given in parentheses where they are known. In most cases where the asymptotic value and exact value do not agree, the asymptotic value is larger and, hence, a conservative constant for the rule R_1. From the values given in this table, it is seen that $-1 \leqslant L(\tilde{m}) - m \leqslant 3$ for $k = 2$, and $0 \leqslant L(\tilde{m}) - m \leqslant 3$ for $k = 3$.

7 REMARKS ON THE PROPERTIES OF THE SELECTION RULES

Expected size of the selected subset

All the selection rules discussed in this paper select a subset of size S, where S is an integer-valued random variable. Since $R_1(G)$ and $R_2(G)$

Table 2. *For given values of k, n, P^*, this table gives the smallest integer m based on asymptotic theory which satisfies $P[T_k \geqslant \max\limits_{1 \leqslant j \leqslant k} T_j - m] \geqslant P^*$. The rank sums $T_i, i = 1, ..., k$, are based on random variables $X_{ij}, j = 1, ..., n$; $i = 1, ..., k$, which are independent identically distributed. Exact m values, where known, are given in parentheses*

	P^*				
n	0·99	0·975	0·95	0·90	0·75
			$k = 2$		
2	7	6	5	4	2 (2)
3	11	9	8 (7)	6 (5)	4 (3)
4	17	14 (14)	12 (12)	9 (8)	5 (4)
5	23 (21)	19 (19)	16 (15)	13 (13)	7 (7)
6	30 (28)	25 (24)	21 (20)	17 (16)	9 (8)
7	37 (35)	31 (31)	26 (25)	21 (21)	11 (11)
8	45 (44)	38 (36)	32 (32)	25 (24)	13 (14)
9	53 (51)	45 (45)	38 (37)	30 (29)	16 (15)
10	62 (60)	52 (52)	44 (44)	34 (34)	18 (18)
11	71 (69)	60 (59)	51 (51)	40 (39)	21
12	81 (80)	68 (68)	57 (58)	45 (44)	24
13	91 (89)	77 (77)	65 (65)	50 (51)	27
14	102 (100)	86 (84)	72 (72)	56 (56)	30
15	113 (111)	95 (95)	80 (79)	62 (63)	33
16	124 (122)	105 (104)	88 (88)	69 (68)	36
17	136 (133)	114 (113)	96 (95)	75 (75)	40
18	148 (146)	124 (124)	104 (104)	82 (82)	43
19	160 (157)	135 (133)	113 (113)	88 (89)	47
20	172 (170)	145 (144)	122 (122)	95 (96)	50
21	185	156	131	102	54
22	199	168	141	110	58
23	212	179	150	117	62
24	226	191	160	125	66
25	240	203	170	133	70

Table 2 (*cont.*)

		$k = 3$			
2	10 (8)	9 (7)	8 (7)	6 (6)	4 (4)
3	18 (15)	15 (14)	13 (13)	11 (11)	7 (7)
4	27 (25)	23 (22)	20 (19)	17 (16)	11 (11)
5	37 (35)	32 (31)	28 (27)	23 (23)	15 (15)
6	48	41	36	30	19
7	60	52	45	37	24
8	73	63	55	45	29
9	87	75	65	54	35
10	101	88	76	63	40
11	117	101	87	72	46
12	133	115	99	82	53
13	149	129	112	92	59
14	167	144	125	103	66
15	185	160	138	114	73
16	203	176	152	125	81
17	222	192	167	137	88
18	242	209	181	149	96
19	262	227	197	162	104
20	283	245	212	175	112
21	304	263	228	188	121
22	326	282	244	201	130
23	349	301	261	215	138
24	371	321	278	229	148
25	395	341	296	244	157

select non-empty subsets, S in these cases takes values 1 through k. As pointed out in § 4, $R_3(G)$ under certain conditions will select a non-empty subset; but generally for $R_3(G)$, S takes values 0 through k. For all these rules:

$$E(S) \equiv E(S|R_i(G)) = \sum_{j=1}^{k} p_j(R_i(G))$$

$$= P(\text{CS}|R_i(G)) + \sum_{j=1}^{k-1} p_j(R_i(G)), \qquad (7.1)$$

where $p_j(R_i(G))$ is as defined in Theorem 3.1. In general, it is difficult to obtain the exact expressions for $E(S)$. But asymptotic expressions can be obtained. We consider R_1 and R_3. Assuming $n_i = n$, for large n, the distribution of $\mathbf{T}' = (T_1, ..., T_k)$ is approximately a multivariate normal distribution with mean vector $\boldsymbol{\mu}'_T = (\mu_1, ..., \mu_k)$ and variance-covariance matrix $\Sigma_T = (\sigma_{ij})$, where

$$\mu_i = E(T_i), \quad \sigma_i^2 = \text{var}(T_i) \quad \text{and} \quad \sigma_{ij} = \text{cov}(T_i, T_j); \ i,j = 1, ..., k; \ i \neq j.$$

Let $$\mathbf{W} = A\mathbf{T}, \qquad (7.2)$$

where A is a $(k-1) \times k$ matrix given by

$$A = \begin{pmatrix} 1 & 0 & 0 & \dots & 0 & -1 \\ 0 & 1 & 0 & \dots & 0 & -1 \\ \vdots & \vdots & & & \vdots & \vdots \\ 0 & 0 & 0 & \dots & 1 & -1 \end{pmatrix}. \tag{7.3}$$

Thus $\qquad W_i = T_i - T_k \quad (i = 1, \dots, k-1). \tag{7.4}$

Then, for large n, $\mathbf{W'} = (W_1, \dots, W_{k-1})$ is approximately distributed as a multivariate normal random vector with mean vector given by $\mu_W = A\,\mu_T$ and the variance-covariance matrix $\Sigma_W = A\,\Sigma_T\,A'$.

Now, for $\nu = 1, \dots, k$, we define

$$\mathbf{W}^\nu = A_\nu\,\mathbf{T}, \tag{7.5}$$

where A_ν is the $(k-1) \times k$ matrix obtained from matrix A defined in (7.3) by moving column j to column $j+1$, $j = \nu, \nu+1, \dots, k-1$ and replacing column ν by column k. The matrix A_k is A. Thus,

$$W_i^\nu = T_i - T_\nu \quad (i = 1, \dots, k; i \neq \nu). \tag{7.6}$$

The random vector $\mathbf{W}^\nu$ is asymptotically distributed as a multivariate normal random vector with mean vector $\mu_\nu = A_\nu\,\mu_T$ and variance-covariance matrix $\Sigma_\nu = A_\nu\,\Sigma_T\,A_\nu'$. Hence, we can state

Theorem 7.1. If Σ_ν is non-singular for $\nu = 1, \dots, k$; then

$$E(S|R_1) \approx \sum_{\nu=1}^{k} K_\nu \int_{-\infty}^{d} \dots \int_{-\infty}^{d} \exp\left[-(\mathbf{w}^\nu - \mu_\nu)'\,\Sigma_\nu^{-1}\,(\mathbf{w}^\nu - \mu_\nu)/2\right] \prod_{\substack{i=1 \\ i \neq \nu}}^{k} dw_i^\nu,$$

$$\tag{7.7}$$

where $K_\nu = [(2\pi)^{k-1}\,|\underset{\nu}{\Sigma}|]^{-\frac{1}{2}}$. For R_3, we have

$$E(S|R_3) \approx \sum_{\nu=1}^{k} \Phi[(\mu_\nu - D)/\sigma_\nu]. \tag{7.8}$$

A similar result can be derived for rule R_2.

Some Monte Carlo results

In order to compare the performance of selection rules R_1 and R_3, some Monte Carlo studies were made. Normal and logistic distributions with variance unity were studied for different configurations of their means. For $k = 3$ and $n = 2, 3, 4$, these configurations were taken to be $(0\cdot1, 0, 0)$,

$(0.2, 0, 0)$, $(0.5, 0, 0)$, $(1.0, 0, 0)$, $(2.0, 0, 0)$, $(0.1, 0.1, 0)$, $(0.2, 0.2, 0)$, $(0.5, 0.5, 0)$, $(1.0, 1.0, 0)$, $(2.0, 2.0, 0)$. The number of simulations were 500 or 1000. The logistic distribution was chosen because equally spaced scores such as ranks yield locally most powerful tests for the location parameter of this distribution. Since there are only a finite number of different possible d and D values for rules R_1 and R_3, there are also only a finite number of values for $\inf_{\Omega} P(\text{CS}|R_1)$ and $\inf_{\Omega} P(\text{CS}|R_3)$; therefore, in general, it is not possible to determine d and D such that both infima yield the same P^*. In such cases there is no definite way of comparing these two rules. For our purposes d and D were determined such that they yielded approximately the same P^* in the case of identical distributions. Then the ratio of $kP(\text{CS}|R)$ and $E(S|R)$ was computed for both rules R_1 and R_3. The bigger ratio for a rule indicates it to be better than the other. For example, for $k = 3$, $n = 2$, then $D = 2$ and $d = 3$ give the probability $14/15$ for the identical case. Using the configuration $(0.1, 0, 0)$, we find that for the normal means the two ratios are 1.012 for R_1 and 1.005 for R_3 so that R_1 seems slightly better than R_3. Using the configuration $(0.5, 0, 0)$, it was found that R_3 was slightly better than R_1; the ratios being 1.045 for R_1 and 1.049 for R_3.

Our Monte Carlo studies showed no significant uniform superiority of either of these procedures. However, R_3 seemed to perform slightly better than R_1 in the cases where the two highest parameters are equal. No difference in the performance of R_1 and R_3 was noticeable when we changed from logistic to normal populations. In all cases the frequency of correct selections for R_1 was higher than the theoretical value exactly calculated for the identical distributions. Thus, there was no indication that the infimum of the probability of a correct selection does not take place when all populations are identically distributed as normal or logistic distributions under shift in location.

Local optimality and monotonicity of R_3

If we use the scores which lead to locally most powerful rank tests of the hypothesis $\boldsymbol{\theta} = (0, 0, ..., 0)$ against $\boldsymbol{\theta} = \alpha(\theta_1, \theta_2, ..., \theta_k)$ (see, Hájek and Šidák [11]), then rules of the type R_3 have the property that the probability of a correct selection increases fastest among all rules based on ranks in the neighborhood of $\boldsymbol{\theta} = (0, 0, ..., 0)$. For example, the locally most powerful rank test for shift in location of the logistic distribution is based on rank sums T_i which for the two-sample case is the Wilcoxon or Mann–Whitney statistic. Hence, the selection rule R_3 based on rank sums is locally optimum in the above sense, provided the underlying distributions are logistic differing only in their location parameters. This result has been shown by K. Nagel.†

A selection rule is called monotone if $\theta_i \geqslant \theta_j$ implies that the population with parameter θ_i is selected with larger probability than that with parameter θ_j. It can be shown that R_3 is monotone if one uses nondecreasing scores and if $F_\theta(x)$ is a stochastically increasing family of distributions.

Asymptotic relative efficiency (ARE) of the rules R_1, R_2 and R_3 relative to a normal means procedure R

We consider here the case of two populations, and so the rules R_1, R_2 and R_3 are equivalent. Hence, we will be concerned with R_1 and R here. Suppose π_1 and π_2 are two independent normal populations with common variance unity. Let the means of $\pi_{(1)}$ and $\pi_{(2)}$ be 0 and $\theta(\geqslant 0)$, respectively. A sample of size n is drawn from each population. Based on

$$X_{ij}, j = 1, \ldots, n \quad (i = 1, 2),$$

let T_i and $\bar{X}_i$ be the rank sum and sample mean, respectively, from π_i, $i = 1, 2$. The procedures to be compared are:

$$R_1: \text{select } \pi_i \text{ iff } T_i \geqslant \max_{j=1,2} T_j - nd \quad (d \geqslant 0), \tag{7.9}$$

$$R: \text{ select } \pi_i \text{ iff } \bar{X}_i \geqslant \max_{j=1,2} \bar{X}_j - b \quad (b \geqslant 0). \tag{7.10}$$

The constants d and b are chosen so that the probability of a correct selection is bounded below by a given number P^*, $\frac{1}{2} < P^* < 1$, for all θ; i.e.

$$\inf_{\theta \geqslant 0} P(\pi_{(2)} \text{ is selected}) \geqslant P^*. \tag{7.11}$$

Procedure R has been investigated by Gupta [10]. Let S^* denote the number of nonbest populations in the selected subset. Since

$$T_{(1)} + T_{(2)} = n(2n+1),$$

$$E(S^* | R_1) = P(T_{(1)} \geqslant T_{(2)} - nd)$$
$$= P\{\sigma^{-1}(T_{(1)} - \mu) \geqslant -\sigma^{-1}[\mu - n(2n+1-d)/2]\}$$
$$\approx \Phi\{\sigma^{-1}[\mu - n(2n+1-d)/2]\}, \tag{7.12}$$

---

† On subset selection rules with certain optimality properties. Department of Statistics, Purdue University. Mimeograph Series No. 222 (1970).

where $\mu = E(T_{(1)}) = n(3n+1)/2 - n^2\Phi(\theta 2^{-\frac{1}{2}}),$ (7.13)

$\sigma^2 = \text{var}(T_{(1)})$

$$= n^2\left[\Phi(\theta 2^{-\frac{1}{2}}) + 2(n-1)\int_{-\infty}^{\infty}\Phi^2(x+\theta)\,\phi(x)\,dx - (2n-1)\,\Phi^2(\theta 2^{-\frac{1}{2}})\right].$$
(7.14)

These moments can be obtained from §5. For $k = 2$ moment expressions are also given in Wilks [24, p. 460]. Now set the right-hand side of (7.12) equal to $\epsilon > 0$ and obtain

$$\mu - n(2n+1-d)/2 = \sigma\Phi^{-1}(\epsilon).$$
(7.15)

From Theorem 3.1, the appropriate value of d is obtained from (7.11) when $\theta = 0$. Equation (6.10) provides a large sample solution for d; namely,

$$d \approx h_1 n^{\frac{1}{2}},$$
(7.16)

where h_1 is independent of n and θ. Actually $h_1 = h(2/3)^{\frac{1}{2}}$, where h is the appropriate value obtained from Gupta [8]. Using (7.13), (7.14) and (7.16) in (7.15) and simplifying yields

$n + h_1 n^{\frac{1}{2}} - 2n\Phi(\theta 2^{-\frac{1}{2}})$

$$= 2\Phi^{-1}(\epsilon)\left[\Phi(\theta 2^{-\frac{1}{2}}) + 2(n-1)\int_{-\infty}^{\infty}\Phi^2(x+\theta)\,\phi(x)\,dx - (2n-1)\,\Phi^2(\theta 2^{-\frac{1}{2}})\right]^{\frac{1}{2}},$$
(7.17)

or upon rearrangement,

$$n(1 - 2\Phi(\theta 2^{-\frac{1}{2}})) + h_1 n^{\frac{1}{2}} = 2\Phi^{-1}(\epsilon)\,(2nB^2(\theta) + R(\theta))^{\frac{1}{2}},$$
(7.18)

where $$B^2(\theta) = \int_{-\infty}^{\infty}\Phi^2(x+\theta)\,\phi(x)\,dx - \Phi^2(\theta 2^{-\frac{1}{2}}),$$
(7.19)

$$R(\theta) = \Phi(\theta 2^{-\frac{1}{2}}) - 2\int_{-\infty}^{\infty}\Phi^2(x+\theta)\,\phi(x)\,dx + \Phi^2(\theta 2^{-\frac{1}{2}}).$$
(7.20)

For large n, the $R(\theta)$ term in (7.18) can be ignored and then that equation simplifies to

$$n^{\frac{1}{2}} \approx [2^{\frac{3}{2}}\Phi^{-1}(\epsilon)\,B(\theta) - h_1][1 - 2\Phi(\theta 2^{-\frac{1}{2}})]^{-1}.$$
(7.21)

Thus, $n \equiv n_{R_1}(\epsilon) \approx [2^{\frac{3}{2}}\Phi^{-1}(\epsilon)\,B(\theta) - h_1]^2[1 - 2\Phi(\theta 2^{-\frac{1}{2}})]^{-2}.$ (7.22)

Now consider rule R:

$$E(S^*|R) = P[\overline{X}_{(1)} \geqslant \overline{X}_{(2)} - b] = \Phi[(b-\theta)\,(n/2)^{\frac{1}{2}}].$$
(7.23)

Again, b is obtained from (7.11) when $\theta = 0$ and is given by

$$b = h_1(3/n)^{\frac{1}{2}}.$$
(7.24)

Setting the right-hand side of (7.23) equal to ϵ and using (7.24) yields

$$n \equiv n_R(\epsilon) = [(3^{\frac{1}{2}}h_1 - 2^{\frac{1}{2}}\Phi^{-1}(\epsilon))/\theta]^2. \qquad (7.25)$$

The asymptotic relative efficiency of R_1 relative to R is defined to be

$$\text{ARE}(R_1, R; \theta) = \lim_{\epsilon \downarrow 0} [n_R(\epsilon)/n_{R_1}(\epsilon)]. \qquad (7.26)$$

From equations (7.22) and (7.25),

$$\text{ARE}(R_1, R; \theta) = \{[2\Phi(\theta 2^{-\frac{1}{2}}) - 1]/2\theta B(\theta)\}^2. \qquad (7.27)$$

If θ is allowed to decrease to 0, then

$$\lim_{\theta \downarrow 0} \text{ARE}(R_1, R; \theta) = 3/\pi = 0 \cdot 9549. \qquad (7.28)$$

Asymptotic relative efficiency of R_2 relative to Gupta's gamma procedure R'

Let π_1, π_2 be two independent exponential populations with independent associated random variables $X_{ij}, j = 1, \ldots, n; i = 1, 2$. The density function of X_{ij} is

$$f_i(x) = \begin{cases} \theta_i^{-1} e^{-x/\theta_i} & (x > 0, i = 1, 2), \\ 0 & (x \leqslant 0), \end{cases} \qquad (7.29)$$

where $1 = \theta_{[1]} \leqslant \theta_{[2]} = \theta$.

Procedure R_2 is given by (2.6) with T_i in the place of H_i and Procedure R' is given by

$$R': \text{ select } \pi_i \text{ iff } \overline{X}_i \geqslant b^{-1} \max_{j=1, 2} \overline{X}_j \quad (b \geqslant 1). \qquad (7.30)$$

The constants c and b are chosen so that

$$\inf_{\theta \geqslant 1} P(\pi_{(2)} \text{ is selected}) \geqslant P^*. \qquad (7.31)$$

Procedure R' has been studied by Gupta [9]. Computing $n_{R_2}(\epsilon)$ and $n_{R'}(\epsilon)$ as before in the case of R_1 and R, we obtain

$$n_{R_2}(\epsilon) \approx 4(\theta + 1)^2 (\theta - 1)^{-2} [6^{-\frac{1}{2}}\Phi^{-1}(P^*) - B(\theta) \Phi^{-1}(\epsilon)]^2 \qquad (7.32)$$

and

$$n_{R'}(\epsilon) = 2(\log \theta)^{-2} [\Phi^{-1}(\epsilon) - \Phi^{-1}(P^*)]^2, \qquad (7.33)$$

where

$$B^2(\theta) = 1 - 2(1 + \theta)^{-1} + (2\theta + 1)^{-1} + \theta(2 + \theta)^{-1} - 2\theta^2(1 + \theta)^{-2}. \qquad (7.34)$$

Hence,

$$\text{ARE}(R_2, R'; \theta) = \lim_{\epsilon \downarrow 0} [n_{R'}(\epsilon)/n_{R_2}(\epsilon)] = [(\theta - 1)/4(\theta + 1) B(\theta) \log \theta]^2. \qquad (7.35)$$

Letting θ decrease to 1 yields

$$\lim_{\theta \downarrow 1} \text{ARE}(R_2, R'; \theta) = \tfrac{3}{4}. \qquad (7.36)$$

ACKNOWLEDGEMENTS
The authors are very thankful to Dr S. Panchapakesan and Mr K. Nagel for valuable assistance during the writing of this paper. Thanks are also due to Mr M. Ward for his assistance in some of the programming work and to Mrs P. McLaughlin for typing earlier drafts as well as the final paper.

REFERENCES

[1] Alam, K. and Rizvi, M. H. (1966). Selection from multivariate normal populations. *Ann. Inst. Statist. Math., Tokyo* **18**, 307–18.

[2] Barlow, R. E. and Gupta, S. S. (1969). Selection procedures for restricted families of probability distributions. *Ann. Math. Statist.* **40**, 905–17.

[3] Barlow, R. E., Gupta, S. S. and Panchapakesan, S. (1969). On the distribution of the maximum and minimum of ratios of order statistics. *Ann. Math. Statist.* **40**, 918–34.

[4] Bartlett, N. S. and Govindarajulu, Z. (1968). Some distribution-free statistics and their application to the selection problem. *Ann. Inst. Statist. Math., Tokyo* **20**, 79–97.

[5] Bechhofer, R. E. (1954). A single sample multiple decision procedure for ranking means of normal populations with known variances. *Ann. Math. Statist.* **25**, 16–39.

[6] Dudewicz, E. J. (1966). The efficiency of a nonparametric selection procedure: largest location parameter case. Department of Industrial Engineering and Operations Research, Cornell University, Technical Report No. 14.

[7] Gupta, S. S. (1956). On a decision rule for a problem in ranking means, Institute of Statistics, University of North Carolina, Chapel Hill, N.C. Mimeograph, Series No. 150.

[8] Gupta, S. S. (1963). Probability integrals of multivariate normal and multivariate *t. Ann. Math. Statist.* **34**, 792–828.

[9] Gupta, S. S. (1963). On a selection and ranking procedure for gamma populations. *Ann. Inst. Statist. Math., Tokyo* **14**, 199–216.

[10] Gupta, S. S. (1965). On some multiple decision rules. *Technometrics* **7**, 225–45.

[11] Hájek, J. and Šidák, Z. (1967). *Theory of Rank Tests.* New York: Academic Press.

[12] Lehmann, E. L. (1959). *Testing Statistical Hypotheses.* New York: John Wiley

[13] Lehmann, E. L. (1963). A class of selection procedures based on ranks. *Math. Annalen* **150**, 268–75.

[14] Mahamunulu, D. M. (1967). Some fixed-sample ranking and selection problems. *Ann. Math. Statist.* **38**, 1079–91.

[15] McDonald, G. C. (1969). On some distribution-free ranking and selection procedures. Department of Statistics, Purdue University. Mimeograph, Series No. 174.

514   SHANTI S. GUPTA & GARY C. MCDONALD

[16] Milton, R. C. (1964). An extended table of critical values for the Mann-Whitney (Wilcoxon) two-sample statistic. *J. Am. Statist. Ass.* **59**, 925–34.
[17] Odeh, R. E. (1967). The distribution of the maximum sum of ranks. *Technometrics* **9**, 271–8.
[18] Patterson, D. W. (1966). Contributions to the theory of ranking and selection problems, Ph.D. Thesis, State University of Rutgers. New Jersey: New Brunswick.
[19] Puri, M. L. (1964). Asymptotic efficiency of a class of *c*-sample tests. *Ann. Math. Statist.* **35**, 102–21.
[20] Puri, P. S. and Puri, M. L. (1968). Selection procedures based on ranks: scale parameter case. *Sankhyā*, Series A **30**, 291–302.
[21] Puri, P. S. and Puri, M. L. (1969). Multiple decision procedures based on ranks for certain problems in analysis of variances. *Ann. Math. Statist.* **40**, 619–32.
[22] Rizvi, M. H. and Sobel, M. (1967). Nonparametric procedures for selecting a subset containing the population with the largest α-quantile. *Ann. Math. Statist.* **38**, 1788–803.
[23] Rizvi, M. H. and Woodworth, G. G. (1968). On selection procedures based on ranks: counterexamples concerning the least favorable configurations. Department of Operations Research and Department of Statistics, Standford University, Technical Report No. 114.
[24] Wilks, S. S. (1962). *Mathematical Statistics*. New York: John Wiley.

INVERSE SAMPLING AND OTHER SELECTION PROCEDURES FOR TOURNAMENTS WITH 2 OR 3 PLAYERS

MILTON SOBEL † AND GEORGE H. WEISS ‡

1 INTRODUCTION

The problem of selecting the best of $k = 2$ or 3 players in a tournament from a ranking and selection point of view is considered when only independent win or lose binary comparisons (or games) are allowed. Some earlier literature on this subject is described in David's monograph [5]. Our main emphasis is on a 'Drop the Loser' (DL) sampling rule: For three players, this rule states that the loser of any game is the one that does not play in the very next game. This DL-sampling rule is regarded as a 3-player binary-comparisons analogue of the Play-the-Winner (PW) rule that has been considered by Robbins [9], [10], Isbell [6] and Smith and Pyke [11] for the 2-arm bandit problem and by Zelen [12] for the same problem in the context of clinical trials. We consider some inverse sampling rules with and without early elimination of non-contenders and attempt to compare our results with those of other new sequential procedures.

Roughly speaking, under our formulation we want a procedure R which has a high probability of a Correct Selection (CS) $P\{CS; R\}$ when one of the players is sufficiently better than the others. Let p_1, p_2, and p_3 denote the single-game probability that A beats B, B beats C, and C beats A, respectively, and let $q_i = 1 - p_i$ ($i = 1, 2, 3$). Let p_0 (which we also write as p below) and P^* denote specified constants with $\frac{1}{2} < p_0 < 1$ and $\frac{1}{3} < P^* < 1$. We define A to be the best player if and only if $\min(p_1, q_3) > \frac{1}{2}$. Although it is possible under this definition that none of the players is best, we assume that there are only three possible decisions available to us, i.e. one of the players has to be selected and declared the best player. We would like to have a procedure R such that for any specified $p_0 > \frac{1}{2}$ and any specified $P^* < 1$

$$P\{CS; R\} = P\{\text{selecting } A; R\} \geqslant P^* \quad \text{whenever } \min(p_1, q_3) \geqslant p_0, \quad (1.1)$$

† This author was supported in part by a National Institutes of Health Special Fellowship and in part by National Science Foundation Grant GP-9018. On leave from University of Minnesota, Minneapolis, Minnesota.
‡ On leave from National Institutes of Health, Bethesda, Maryland.

regardless of the value of p_2; here p_2 is regarded as a nuisance parameter and below it is also denoted by θ.

Some of our procedures (see R_I, R_E, and R_S below) do not satisfy (1.1) and we need a weaker form of (1.1). Suppose we only want (1.1) to hold for specific values of θ and for $p_0 > p > \frac{1}{2}$, where $p = p(\theta)$ may depend on θ and on the procedure R. The function $p(\theta)$ is given explicitly in (3.9) for procedure R_I; the same expression holds for procedures R_E and R_S also.

Taking A as the best player with $\min(p_1, q_3) > p_0$ and θ fixed, we define a least favorable (LF) configuration with θ fixed to be one in which $p_1 = q_3 = p_0$. For procedure R_I a proof is given by N. Elliott and Y. S. Lin in Appendix C that the configuration $(p_1 = q_3 = p_0)$ is least favorable for fixed θ in the usual sense of minimizing the $P\{CS; R_I\}$ over all triples (p_1, p_2, p_3) with $\min(p_1, q_3) \geqslant p_0$ and fixed $p_2 = \theta$.

For the procedures studied we shall also be concerned with the expected number of games required to reach a decision as a function of p, P^*, and θ; we denote this by $E\{N|\mathrm{LF}_\theta\}$.

Since we do not assume the Bradley–Terry Model [3] it is possible to have A preferred to B, B preferred to C, and C preferred to A (i.e. $p_1 > \frac{1}{2}$, $p_2 > \frac{1}{2}$, and $p_3 > \frac{1}{2}$) but such triads will not be in the zone of preference of selection for any of the three players. In the present discussion we will not consider treating the 3-player problem as a succession of 2-player problems; that is, we do not consider testing player A against B until one is found to be better and then testing C against the better of A and B. The present paper is not meant to be an exhaustive study of the 3-player problem or a search for an optimal procedure. Rather we have analyzed several new procedures (the simpler ones being generally less efficient) and made some efficiency comparisons.

In §2 we summarize results for two different 2-player procedures. In §3 we analyze the inverse sampling procedure R_I for 3 players (without early elimination of non-contenders) in which the first player to win r games is selected as the winner of the tournament. Exact recurrence relations are derived for the $P\{CS; R_I\}$ and $E\{N; R_I\}$, conditional on the number of wins already obtained by each player. This Markovian character is then exploited to obtain exact numerical values for the $P\{CS|\mathrm{LF}_\theta\}$ and $E\{N|\mathrm{LF}_\theta\}$ for different values of p, θ, and r0 the Bradley–Terry model is represented by the special case $\theta = \frac{1}{2}$.

2 THE 2-PLAYER PROBLEM

In the case of two contestants all of our procedures can be analyzed exactly and explicit expressions can be derived for the $P\{CS\}$ and $E\{N\}$ functions; we give these results without derivation. In the case of two players, one of our procedures (R_I) occurs in [4] and another procedure (R_S) occurs in the work of Alam [1]; both of these papers deal with the analogous binomial and multinomial ranking problems. Special cases of our formulas and a discussion of these same two rules R_I and R_S for the 2-player problem are also given by Kemeny and Snell ([7], p. 165).

Let $p > \frac{1}{2}$ denote the single-game probability that A beats B, so that A denotes the better player. We consider three procedures for selecting the better player.

(1) The first is a single-stage procedure R_F in which an odd number $n = 2m + 1$ of games are played and the winner of a majority of the games is selected as the better player.

(2) The second is an inverse sampling procedure R_I in which the first player to win r games is selected as the better player.

(3) The third is a sequential procedure R_S in which the first player to win d more games than his opponent is selected as the better player.

For the single-stage procedure R_F, the probability of a correct selection is clearly

$$P\{CS; R_F\} = \sum_{j=m+1}^{2m+1} \binom{2m+1}{j} p^j q^{2m+1-j} = I_p(m+1, m+1), \quad (2.1)$$

where $I_p(a, b)$ is the usual incomplete beta function. These tournaments need only be continued until one player wins $m + 1$ games and then it becomes identical with inverse sampling if we equate r and $m + 1$.

Under the inverse sampling procedure R_I we have

$$P\{CS; R_I\} = p^r \sum_{j=0}^{r-1} \binom{r-1+j}{j} q^j = I_p(r, r) \quad (2.2)$$

since, for a correct selection, A wins r games and B wins $j < r$ games. The expected number of games in a tournament under inverse sampling is

$$E\{N; R_I\} = \sum_{j=0}^{r-1} (j+r) \binom{r+j-1}{j} (p^r q^j + p^j q^r)$$

$$= \frac{r}{p} I_p(r+1, r) + \frac{r}{q} I_q(r+1, r). \quad (2.3)$$

The expected number W_L of games won by the loser of the tournament is

$$E\{W_L; R_I\} = \sum_{j=0}^{r-1} j \binom{r+j-1}{j} (p^r q^j + p^j q^r) = E\{N: R_I\} - r. \quad (2.4)$$

The sequential procedure R_S can be regarded as a gambler's ruin problem and we easily obtain for $\psi = p/q$

$$P\{CS; R_S\} = \frac{\psi^d}{1 + \psi^d}.$$ (2.5)

The expected number of games is given by

$$E\{N; R_S\} = d \left(\frac{\psi^d - 1}{\psi^d + 1}\right) \left(\frac{\psi + 1}{\psi - 1}\right)$$ (2.6)

for $\psi \neq 1$ and equals d^2 for $\psi = 1$. The expected number of games W_L won by the loser is given by

$$E\{W_L; R_S\} = \frac{d}{2} \left\{ \left(\frac{\psi^d - 1}{\psi^d + 1}\right) \left(\frac{\psi + 1}{\psi - 1}\right) - 1 \right\}$$ (2.7)

for $\psi \neq 1$ and equals $d(d-1)/2$ for $\psi = 1$.

Table 1A contains a summary of analytically-obtained results for the 2-player inverse sampling procedure R_I giving the value of $r = m + 1$ as a function of p and P^* for $P^* = 0 \cdot 75$, $0 \cdot 90$, $0 \cdot 95$, and $0 \cdot 99$. Table 1B contains comparable values of d and $E\{N; R_S\}$ for the sequential procedure R_S. The d-value in the table is actually the smallest integer $\geqslant d_0$, where d_0 is the solution obtained by setting (2.5) equal to P^*, i.e.

$$d_0 = \left\{ \ln \left(\frac{P^*}{1 - P^*}\right) \right\} \Big/ \ln \psi;$$ (2.8)

as a result of using the smallest integer $\geqslant d_0$, the tabulated values of $E\{N; R_S\}$ are not smooth.

Table 1A. *Values of* r $(= m + 1)$ *and* $E\{N; R_I\}$ *for the procedure* R_I *with 2 players*

	P^*-values							
	0·75		0·90		0·95		0·99	
p_0	r	$E\{N; R_I\}$	r	$E\{N; R_I\}$	r	$E\{N; R_I\}$	r	$E\{N; R_I\}$
0·55	23	39·8	82	147·7	135	244·8	270	491·3
0·60	6	9·0	21	34·4	34	56·3	67	111·6
0·65	3	4·0	9	13·4	14	21·3	29	44·5
0·70	2	2·4	5	6·9	8	11·3	10	13·8
0·75	1	1·0	3	3·8	5	6·5	9	12·0
0·80	1	1·0	2	2·3	4	4·9	7	8·7
0·85	1	1·0	2	2·2	3	3·5	5	5·9
0·90	1	1·0	1	1·0	2	2·2	3	3·3
0·95	1	1·0	1	1·0	1	1·0	2	2·1

Table 1 B. *Values of d and E{N ; R_S} for the*
procedure R_S with 2 players†

	P*-values							
	0·75		0·90		0·95		0·99	
p_0	d	$E\{N; R_S\}$	d	$E\{N; R_S\}$	d	$E\{N; R_S\}$	d	$E\{N; R_S\}$
0·55	6	32·3	11	88·2	15	135·9	23	225·5
0·60	3	8·1	6	25·2	8	40·0	11	53·7
0·65	2	3·8	4	11·3	5	15·2	7	22·7
0·70	2	3·4	3	6·4	4	9·4	6	12·1
0·75	1	1·0	2	3·2	3	5·6	5	9·1
0·80	1	1·0	2	2·9	3	4·9	4	6·6
0·85	1	1·0	2	2·7	2	2·7	3	4·2
0·90	1	1·0	1	1·0	2	2·4	3	3·7
0·95	1	1·0	1	1·0	1	1·0	2	2·2

† Alam[1] has shown that if we eliminate the discontinuity due to the fact that r and d are integers then

$$E\{N|R_S\} = E\{N|R_I\} \quad \text{for all } p_0 \text{ and } P^*.$$

3 INVERSE SAMPLING PROCEDURE R_I FOR THREE PLAYERS

The procedure R_I consists of a sampling rule, a stopping rule and a decision rule; it is convenient to combine the latter two.

Sampling Rule of Procedure R_I. At the outset randomize between the three possible games ($A\,v.\,B$, $B\,v.\,C$, and $C\,v.\,A$), giving probability 1/3 to each. To determine who plays in succeeding games we use the DL-sampling rule, i.e. the loser of any game sits out the next game.

Stopping and Decision Rule of Procedure R_I. Stop as soon as any one player has a total of r wins and select him as the best player.

Probability of a correct selection $P\{\text{CS}; R_I\}$

Let w_A denote the number of games won by A and let $w'_A = r - w_A$; we define w_B, w'_B, w_C, and w'_C similarly and use W′ to denote the vector (w'_A, w'_B, w'_C). Let 'wt' denote 'wins the tournament' and let 'PNG =' denote 'the players of the next game are'. Let

$$\left. \begin{aligned} R_{k,m,n} &= P\{A\,wt | \mathbf{W}' = (k,m,n) \quad \text{and} \quad \text{PNG} = A\,v.\,B\}, \\ S_{k,m,n} &= P\{A\,wt | \mathbf{W}' = (k,m,n) \quad \text{and} \quad \text{PNG} = B\,v.\,C\}, \\ T_{k,m,n} &= P\{A\,wt | \mathbf{W}' = (k,m,n) \quad \text{and} \quad \text{PNG} = C\,v.\,A\}, \end{aligned} \right\} \quad (3.1)$$

where the procedure R_I is understood. From the DL-sampling rule we obtain the recursive relations

$$\left.\begin{aligned}
R_{k,m,n} &= p_1 T_{k-1,m,n} + q_1 S_{k,m-1,n}, \\
S_{k,m,n} &= p_2 R_{k,m-1,n} + q_2 T_{k,m,n-1}, \\
T_{k,m,n} &= p_3 S_{k,m,n-1} + q_3 R_{k-1,m,n},
\end{aligned}\right\} \tag{3.2}$$

and from the stopping rule we obtain the boundary conditions: for all positive k, m, n

$$\left.\begin{aligned}
R_{0,m,n} &= T_{0,m,n} = 1, \\
R_{k,0,n} &= S_{k,0,n} = S_{k,m,0} = T_{k,m,0} = 0.
\end{aligned}\right\} \tag{3.3}$$

We have solved (3.2) numerically and calculated $A_{r,r,r}$ defined by

$$A_{r,r,r} = \tfrac{1}{3}(R_{r,r\ r} + S_{r,r,r} + T_{r,r,r}), \tag{3.4}$$

which gives the exact $P\{CS; R_I\}$ for procedure R_I. This was done for the $LF(\theta)$ configuration, i.e.

$$\left.\begin{aligned}
p_1 = q_3 &= p \text{ (say)} \quad (\tfrac{1}{2} \leqslant p \leqslant 1), \\
p_2 &= \theta \quad\quad\ (\tfrac{1}{2} \leqslant \theta \leqslant 1),
\end{aligned}\right\} \tag{3.5}$$

for $r = 1(1)\min(40, r^*)$, where r^* is the smallest value of r for which $P\{CS; R_I\} \geqslant 0.99$.

Expected number of comparisons $E\{N; R_I\}$

For $E\{N; R_I\}$ we use a technique similar to the above. Let

$$\left.\begin{aligned}
F_{k,m,n} &= E\{N \,|\, \mathbf{W}' = (k,m,n) \quad \text{and} \quad PNG = A\,v.\,B\}, \\
G_{k,m,n} &= E\{N \,|\, \mathbf{W}' = (k,m,n) \quad \text{and} \quad PNG = B\,v.\,C\}, \\
H_{k,m,n} &= E\{N \,|\, \mathbf{W}' = (k,m,n) \quad \text{and} \quad PNG = C\,v.\,A\}.
\end{aligned}\right\} \tag{3.6}$$

From the DL-sampling rule we have

$$\left.\begin{aligned}
F_{k,m,n} &= p_1 H_{k-1\,m,n} + q_1 G_{k,m-1,n} + 1, \\
G_{k,m,n} &= p_2 F_{k,m-1,n} + q_2 H_{k,m,n-1} + 1, \\
H_{k,m,n} &= p_3 G_{k,m,n-1} + q_3 F_{k-1,m,n} + 1,
\end{aligned}\right\} \tag{3.7}$$

with boundary conditions $F_{k,m,n} = G_{k,m,n} = H_{k,m,n} = 0$ if any one index is zero. Table 4 gives the value of r and $E\{N; R_I\}$ for configurations corresponding to $P^* = 0.75$, 0.90, 0.95, and 0.99, where $E\{N; R_I\}$ is calculated by

$$E\{N; R_I\} = \tfrac{1}{3}(F_{r,r,r} + G_{r,r,r} + P_{r,r,r}). \tag{3.8}$$

Limitations of the procedure R_I

Due to the fact that the procedure R_I depends only on the total number of wins for each player an inefficiency arises which is affected by the sampling rule used. For certain points in the parameter space where A is the best player, the $P\{CS\}$, under the sampling rule of procedure R_I, does not approach 1 as the number of games grows indefinitely. For procedure R_I we delineate these parameter points explicitly. Since our result depends only on the sampling rule and the statistic used (the total number of wins for each player), the same result also holds for procedures R_E and R_S defined below. For the configuration (3.5), our result states that under the DL-sampling rule for the $P\{CS\}$ to approach 1 we need

$$p > \tfrac{1}{2}(-\theta + \sqrt{(\theta^2 + 4\theta)}) \qquad (3.9)$$

and under the vector-at-a-time sampling rule, which gives each player an equal number of games with each of the other players, we need

$$p > \tfrac{1}{3}(1 + \theta). \qquad (3.10)$$

We note that the line in (3.10) is tangent to the curve in (3.9) at $\theta = \tfrac{1}{2}$ and hence any pair (p, θ) satisfying (3.10) also satisfies (3.9); this is an indication of the improvement obtained by using the DL-sampling rule.

To derive (3.9) consider the first m games for large m and let f_{AB} denote the proportion in which we find $A \, v. \, B$ playing; define f_{BC} and f_{CA} similarly. Then under the DL-sampling rule

$$f_{AB} = f_{BC}\theta + f_{AC}q_3, \quad f_{BC} = f_{AB}q_1 + f_{AC}p_3, \qquad (3.11)$$

and hence $\qquad f_{AB}(1 - \theta q_1) = f_{AC}(q_3 + \theta p_3).$ $\qquad (3.12)$

After a large number of games the proportion in which we find A playing is given by the two expressions

$$f_{AB} + f_{AC} = f_{AB}p_1 + f_{AC}q_3 + f_{BC}. \qquad (3.13)$$

From (3.13) and the identity $f_{BC} = 1 - f_{AB} - f_{AC}$, we obtain

$$f_{AB}(2 - p_1) + f_{AC}(2 - q_3) = 1. \qquad (3.14)$$

Solving (3.12) and (3.14) gives

$$f_{AB} = (q_3 + p_3\theta)/D, \quad f_{AC} = (1 - q_1\theta)/D, \qquad (3.15)$$

where $D = 2 + q_1 q_3 + \theta(p_3 - q_1)$. It follows that the long-term frequencies of wins for A, B and C, resp., under the DL- sampling rule are

$$\left.\begin{aligned}
f_A &= \frac{q_3(1+p_1)+\theta(p_3-q_1)}{D} = \frac{p(1+p)}{2+pq}, \\[2mm]
f_B &= \frac{q_1 q_3+\theta(q_1+p_3)}{D} = \frac{q(p+2\theta)}{2+pq}, \\[2mm]
f_C &= \frac{p_1 p_3+(1-\theta)(q_1+p_3)}{D} = \frac{q(p+2-2\theta)}{2+pq},
\end{aligned}\right\} \tag{3.16}$$

where the last expressions in (3.16) are for configuration (3.5). The condition that $f_A > \max(f_B, f_C)$ is that

$$p_1 q_3 > \max\{q_1\theta, p_3(1-\theta)\}. \tag{3.17}$$

For configuration (3.5) this reduces to $p^2 > q\theta$, which easily reduces to (3.9)

A corresponding analysis for vector-at-a-time sampling states that after m vectors for the $P\{CS\}$ to approach 1 as $m \to \infty$ we need

$$f_A = p_1 + q_3 > \max(f_B, f_C) = \max\{\theta + q_1, 1-\theta+p_3\}. \tag{3.18}$$

For the configuration (3.5) this reduces to (3.10).

We notice that the right members of (3.9) and (3.10) are both increasing in θ and equal $(\sqrt{5}-1)/2 = 0{\cdot}618\ldots$ and $\frac{2}{3}$ respectively for $\theta = 1$. This indicates that for $p > (\sqrt{5}-1)/2$ under the DL-sampling rule the $P\{CS\} \to 1$ regardless of how close θ is to 1 and a similar statement holds for $p > \frac{2}{3}$ under the vector-at-a-time sampling rule.

Remark. An alternative definition of best player is to define A as the best player if the average of his probabilities of beating each of the other contestants (in a single game) is larger than the corresponding averages for players B and C, i.e. if and only if

$$\frac{p_1+q_3}{2} > \max\left\{\frac{q_1+\theta}{2}, \frac{p_3+1-\theta}{2}\right\}. \tag{3.19}$$

Since (3.19) is the same as (3.18), it follows that the $P\{CS\}$ will approach 1 for the vector-at-a-time sampling rule. It can also be shown (proof omitted) for $\min(p_1, q_3) > \frac{1}{2}$ that any triple (p_1, p_2, p_3) that satisfies (3.18) also satisfies (3.17). Thus it follows that the $P\{CS\}$ also approaches 1 for the DL-sampling rule. Hence, for this definition of best, the basic condition (1.1) will be satisfied, regardless of the value of θ, under both sampling rules by taking r sufficiently large.

4 BOUNDS AND APPROXIMATIONS FOR INVERSE SAMPLING

In this section an upper bound and approximation for both $P\{CS|LF\}$ and $E\{N|LF\}$ are derived for procedure R_I. Lower bounds for the $P\{CS|LF\}$ can also be found but these do not appear to be as useful, and will not be discussed.

Our derivation of an upper bound is based on a lemma (or three problems) dealing with the number $M = M(r, n)$ of ways of removing r items from an ordered set of n items ($0 \leqslant r \leqslant n$) so that the resulting $r + 1$ groups thus formed by the items remaining all have a preassigned parity; here empty groups are taken into account and their parity is, of course, even.

Lemma. The number M of combinations of r items which (when removed) result in all $r + 1$ groups being even is given by

$$M = \binom{\dfrac{n+r}{2}}{r} \quad if \quad \frac{n+r}{2} \ is \ an \ integer, \qquad (4.1)$$

and $M = 0$ otherwise. In the above, if we specify that the rth item removed must be the last of the n ordered items, then

$$M = \binom{\dfrac{n+r}{2}-1}{r-1} \quad if \quad \frac{n+r}{2} \ is \ an \ integer, \qquad (4.2)$$

and $M = 0$ otherwise. In addition to the latter condition, if we want the first group to be odd and all the others even, then

$$M = \binom{\dfrac{n+r-3}{2}}{r-1} \quad if \quad \frac{n+r-3}{2} \ is \ an \ integer, \qquad (4.3)$$

and $M = 0$ otherwise.

Proof. In the first problem let $x_i \geqslant 0$ ($i = 1, 2, \ldots, r + 1$) denote the even group sizes, so that $x_1 + x_2 + \ldots + x_{r+1} = n - r$. Setting $x_i = 2y_i$, we obtain the equivalent problem of finding non-negative integer solutions of $y_1 + y_2 + \ldots + y_{r+1} = (n - r)/2$. Setting $z_i = y_i + 1$ gives another equivalent problem of finding positive integer solutions of

$$z_1 + z_2 + \ldots + z_{r+1} = r + 1 + (n - r)/2.$$

Hence the answer to our first problem is simply the number of ways of selecting r different spaces from the interior spaces formed by $1 + (n+r)/2$ ordered items, i.e. the combinatorial in (4.1). If $(n+r)/2$ is not an integer there are no solutions.

In the second problem we use the same argument except that there are only r quantities x_i. In the third problem we set $x_1 = 2y_1 + 1$ and $x_i = 2y_i$ for $i > 1$; the remainder of the proof is the same. This proves the lemma.

We now use the symbol A to denote the best of the three players and we let n denote the number of comparisons needed under the DL-sampling rule for A to obtain r wins. Let $P_n(A)$ denote the probability that A obtains r wins on precisely the nth game given that A plays in the first game and the DL-sampling rule is used; let $P_n(\bar{A})$ denote the same except that A does not play in the first game. Of course, waiting for A (the best player) to obtain r wins is not a procedure; we refer to it as a 'pseudo-procedure'.

PCS *bounds.* From the second problem of the lemma (under the LF configuration)

$$P_n(A) = \binom{\dfrac{n+r}{2} - 1}{r-1} p^r q^{(n-r)/2} \tag{4.4}$$

and the sum over n-values $(r \leqslant n < \infty)$ is one for all $p > 0$. Since R_I can yield wrong results, but the above 'pseudo-procedure' does not, we find that

$$P\{\mathrm{CS}|A\} \leqslant \sum_{n=r}^{3r-2} P_n(A) = p^r \sum_{j=0}^{r-1} \binom{j+r-1}{r-1} q^j = I_p(r,r), \tag{4.5}$$

where $P\{\mathrm{CS}|A\}$ is the $P\{\mathrm{CS}|\mathrm{LF}\}$ given that A plays in the first game. Similarly using the third problem of the lemma

$$P\{\mathrm{CS}|\bar{A}\} \leqslant \sum_{n=r}^{3r-2} P_n(\bar{A}) = p^r \sum_{j=0}^{r-2} \binom{j+r-1}{r-1} q^j = I_p(r,r-1). \tag{4.6}$$

From (4.5) and (4.6) we obtain the upper bound UB_1

$$P\{\mathrm{CS}|\mathrm{LF}\} = \frac{2P\{\mathrm{CS}|A\} + P\{\mathrm{CS}|\bar{A}\}}{3} \leqslant \frac{2I_p(r,r) + I_p(r,r-1)}{3} = UB_1. \tag{4.7}$$

It is easy to show that $I_p(r,r)$ is an upper bound for UB_1 and hence also for the $P\{\mathrm{CS}|\mathrm{LF}\}$; we denote it by UB_0.

Suppose we get a wrong decision and (say) B wins the tournament. Consider only the games that B wins, arranged in order of occurrence. If B beats C twice in succession (in this subset of B's wins), then between these wins A must have lost to C, thus providing one factor q for this event. We get another factor q for each game that A loses to B. Since

B wins r games, the minimum power of q is obtained (for example) by alternating the wins of B against C and against A. This gives a minimum power of q equal to $[\frac{1}{2}r]$, i.e. for $q \to 0$

$$P(\text{CS}|\text{LF}) \approx \frac{2I_p(r,r) + I_p(r,r-1)}{3} + \mathcal{O}(q^{[\frac{1}{2}r]}), \qquad (4.8)$$

where $[x]$ denotes the largest integer $\leqq x$.

ASN Bounds. Using the lemma again and the n defined above for the 'pseudo-procedure' we obtain upper bounds for $E\{N|\text{LF}\}$, where N is the number of games required by procedure R_I. Since n is an upper bound on N, we find that if we start with A

$$E\{N|A\} \leqslant p^r \sum_{n=r}^{\infty} n \binom{\frac{n+r}{2}-1}{r-1} q^{(n-r)/2} = \frac{r(1+q)}{p} \qquad (4.9)$$

and if we start without A the result is increased by exactly 1. Hence

$$E\{N|\text{LF}\} \leqslant \frac{r(1+q)}{p} + \frac{1}{3}. \qquad (4.10)$$

A slight improvement on (4.10) is obtained by using the curtailed distribution of n, i.e. by concentrating at $n = 3r-2$ all the probability for $n \geqslant 3r-2$. We give this upper bound

$$E\{N|\text{LF}\} \leqslant 3r - 2 - \frac{4}{3}\left\{(r-1)I_p(r,r) - \frac{qr}{p}I_p(r+1,r-1)\right\}$$

$$- \frac{1}{3}\left\{(2r-3)I_p(r,r-1) - \frac{2rq}{p}I_p(r+1,r-2)\right\} \qquad (4.11)$$

without derivation since the improvement on (4.10) is usually quite small in most cases of interest.

An improved upper bound for the $P\{\text{CS}|\text{LF}\}$. We now use the fact that, even if we condition on the best player (say, A) obtaining his rth win on exactly the jth trial, half of the games that A does not win, i.e. $\left[\dfrac{i-r+1}{2}\right]$ games, are independent binomial contests between B and C with single-game probabilities of winning θ and $1-\theta$, respectively, under the LF configuration (3.5).

An improved upper bound for the $P\{\text{CS}|\text{LF}\}$ is then obtained by distributing A's losses equally between B and C giving the extra one to (say) B when the number is odd. Starting with A as one of the players and using (4.2), we let $i = (j - r)/2$ and obtain for $r \geqslant 3$

$$P\{CS|A\} \leqslant p^r \sum_{i=0}^{r-1} \binom{i+r-1}{r-1} q^i \sum_{\alpha=[\frac{1}{2}(3i+3)]-r}^{r-1-[\frac{1}{2}i]} \binom{i}{\alpha} \theta^\alpha (1-\theta)^{i-\alpha}$$

$$= I_p(r, r) - p^r \sum_{i=[\frac{1}{3}(2r+2)]}^{r-1} \binom{i+r-1}{r-1} q^i$$

$$\times \left\{ I_\theta \left(\left[\frac{2r-i+1}{2}\right], \left[\frac{3i-2r+2}{2}\right] \right) + I_{1-\theta} \left(\left[\frac{2r-i}{2}\right], \left[\frac{3i-2r+3}{2}\right] \right) \right\}.$$

$$(4.12)$$

Similarly, starting without A and using (4.3), we obtain for $r \geqslant 3$

$$P\{CS|\bar{A}\} \leqslant I_p(r, r-1) - p^r \sum_{i=[2r/3]}^{r-2} \binom{i+r-1}{r-1} q^i$$

$$\times \left\{ I_\theta \left(\left[\frac{2r-i+1}{2}\right], \left[\frac{3i-2r+4}{2}\right] \right) + I_{1-\theta} \left(\left[\frac{2r-i}{2}\right], \left[\frac{3i-2r+5}{2}\right] \right) \right\},$$

$$(4.13)$$

where the sum in (4.13) vanishes for $r = 3$. Combining (4.12) and (4.13) we obtain the desired bound

$$P\{CS|LF\} \leqslant UB_1 - \frac{p^r}{3} (2\underset{1}{\textstyle\sum} + \underset{2}{\textstyle\sum}) = UB_2, \qquad (4.14)$$

where $\underset{1}{\sum}$ and $\underset{2}{\sum}$ are the sums in (4.12) and (4.13), respectively. For $\theta = 1$ this reduces to

$$P\{CS|LF, \theta = 1\} \leqslant \frac{1}{3} \left\{ 2I_p\left(r, \left[\frac{2r+2}{3}\right]\right) + I_p\left(r, \left[\frac{2r}{3}\right]\right) \right\}. \qquad (4.15)$$

Approximations for the P{CS} and ASN. From the large-sample analysis of §3 we also obtain another approximation for the PCS which has the same region (see (3.9)) of convergence to one for $r \to \infty$ as the exact $P\{CS\}$ under the DL-rule. We make use of an identity derived in [8] and utilized in [4] between a sum of multinomial probabilities and a Dirichlet integral; the final result is adjusted because successive observations in the limiting multinomial are not independent under the DL-rule. This derivation has some intrinsic interest since it gives a method for applying the results of independent sampling to the corresponding problem with correlated sampling.

    The marginal distribution of the vector (of zeros and a single one) which shows the player that won the jth game for j large is that of a 3-cell multinomial with the 3-cell probabilities given by (3.16). Suppose first that the observations are all mutually independent and let s denote the number of such multinomial observations needed for a corresponding P^*-condition; later we make an adjustment for the lack of independence between successive observations under the DL-rule. If we had independent multinomial vector observations then we can use the identity in Theorem 2.4 of [8] (see also (4.3) of [4]) with $k = 3$ and $N = s$ and the $P\{CS\}$ would then be given by the Dirichlet integral

$$I_{b,c}(s, s; 3s) = \frac{\Gamma(3s)}{\Gamma^3(s)} \int_b^\infty \int_c^\infty \frac{x^{s-1} y^{s-1}}{(1+x+y)^{3s}} \, dx \, dy \qquad (4.16)$$

where $\qquad b = P(W_B)/P(W_A) = q(p+2\theta)/p(1+p)$

and $\qquad c = P(W_C)/P(W_A) = q(p+2-2\theta)/p(1+p).$

Integrating by parts in (4.16), we note that two arguments of the I-function get lowered and by iteration this easily leads to

$$I_{b,c}(s, s; 3s) = \left(\frac{1}{1+b}\right)^s \sum_{j=0}^{s-1} \binom{s-1+j}{s-1} \left(\frac{b}{1+b}\right)^j I_{\frac{1+b}{1+b+c}}(s+j, s). \qquad (4.17)$$

If we divide by the sum of the coefficients of the I functions in (4.17), then we can treat the sum in (4.17) as an average of incomplete beta functions. If we use the identity found in (4.5), let ' $\sim$ ' denote asymptotic equality in the limit $s \to \infty$ and set $j^{(i)} = j(j-1)\ldots(j-i+1)$, it is easy to show that for $i \geqslant 0$

$$\left(\frac{1}{1+b}\right)^s \sum_{j=0}^{s-1} j^{(i)} \binom{s-1+j}{s-1} \left(\frac{b}{1+b}\right)^j = s^{(i)} b^i I_{\frac{1}{1+b}}(s+i, s-i). \qquad (4.18)$$

From this it follows for $s \to \infty$ that the variance of j/s goes to zero, hence we can set j/s equal to b (or j equal to sb) in the I-functions in (4.17). This gives us the desired asymptotic result

$$I_{b,c}(s, s; 3s) \sim I_{\frac{1}{1+b}}(s, s) I_{\frac{1+b}{1+b+c}}(s(1+b), s). \qquad (4.19)$$

The first and second I-function, respectively, on the r.h.s. of (4.19) converge to 1 if and only if $p^2 > q\theta$ as in (3.9) and if and only if $p^2 > q(1-\theta)$, where the latter is implied by the former for $\theta \geqslant \frac{1}{2}$. In fact for θ close to 1 we can neglect the second I-function since it is much closer to 1 than the first.

We set the right side of (4.19) equal to a specified P^* ($\frac{1}{3} < P^* < 1$) and solve for s (not necessarily an integer). This constitutes an approximate solution for the problem in [4] with independent sampling.

We now consider the question of adjusting this s-value to obtain the r-value required by the DL-rule where successive observations are correlated. We associate with the $P\{CS\}$ under procedure R_I the statistic $S = \sum_{i=1}^{r} X_i/r$, where $X_i = 1$ if A wins the ith game and 0 if A loses or does not play in the ith game. We want the statistic S to have the same precision (or variance) under the DL-rule as under independent sampling. Using (3.16) we find that in the asymptotic multinomial

$$\rho_1 = \rho(X_i, X_{i+1}) = \frac{\dfrac{p^2(1+p)}{2+pq} - \left(\dfrac{p(1+p)}{2+pq}\right)^2}{2pq(1+p)^2/(2+pq)^2} = \frac{p}{2}. \tag{4.20}$$

Similarly we can show for each j that in the asymptotic multinomial

$$\rho_j = \rho(X_i, X_{i+j}) = (-q)^{j-1}\frac{p}{2}. \tag{4.21}$$

A general proof of (4.21) can be obtained by using the second result in the lemma proved above, and the identity

$$p^j \sum_{\alpha=0}^{[\frac{1}{2}j]} \binom{j-\alpha}{\alpha} \left(\frac{q}{p^2}\right)^\alpha = \frac{1-(-q)^{j+1}}{1+q} \quad (j = 0, 1, 2, \dots) \tag{4.22}$$

proved in Appendix A. Hence the variance of S (if the common variance of each X_i is σ^2) is asymptotically ($r \to \infty$) given by

$$\sigma^2(S) = \frac{\sigma^2}{r^2}\{r + 2(r-1)\rho_1 + 2(r-2)\rho_2 + \dots + 2\rho_{r-1}\}$$

$$= \frac{\sigma^2}{r^2}\{r + (r-1)p - (r-2)pq + \dots + p(-q)^{r-2}\}$$

$$= \frac{\sigma^2}{r^2}\left(r + \frac{rp}{1+q} - p\left\{\frac{1-(-q)^r}{(1+q)^2}\right\}\right) \sim \frac{2\sigma^2}{r(1+q)}. \tag{4.23}$$

To attain the same precision we set the last expression equal to σ^2/s obtaining the desired adjustment

$$r = \frac{2s}{1+q}, \tag{4.24}$$

where s is the solution obtained by setting the right side of (4.19) equal to P^*. In this case, we would take the nearest integer to the r-value obtained by (4.24); some numerical values based on (4.19), (4.24) are given in Table 2.

Table 2. *Numerical illustrations of approximations, exact values and bounds on r for p = 0·75. (Procedure R_I)*

	$P^* = 0·90$			$P^* = 0·95$		
	$\theta = 0·50$	$\theta = 0·75$	$\theta = 1·00$	$\theta = 0·50$	$\theta = 0·75$	$\theta = 1·00$
Approximation						
Based on (4.19), (4.24)	8	9	13	11	13	22
Exact values	7	9	14	10	13	23
Lower Bounds on r						
Based on (4.14)	7	7	11	8	10	17
Based on (4.7)	5	5	5	6	6	6

Examination of the computer output shows that $E\{N|\mathrm{LF}\}$ is very closely approximated by

$$E\{N|LF\} = Cr + D. \qquad (4.25)$$

Values of C and D are given in Table 3 for $0·60 \leqslant p \leqslant 0·85$ for appropriate values of θ. For $p > 0·85$, C is very close to the upper bound value $(1+q)/p$ and D is very close to $\frac{1}{3}$. In all cases (4.25) differs from the exact value by no more than 1.

Table 3. *Values of C and D for use in (4.25). (Procedure R_I)*

p	θ	C	D
0·60	0·50	2·40	−3·47
0·65	0·50	2·13	−1·62
	0·75	2·11	−2·19
0·70	0·50	1·88	−0·40
	0·75	1·89	−0·92
	1·00	1·89	−1·88
0·75	0·50	1·67	0·16
	0·75	1·68	−0·10
	1·00	1·70	−0·70
0·80	0·50	1·50	0·31
	0·75	1·50	0·24
	1·00	1·51	0·04
0·85	0·50	1·35	0·33
	0·75	1·35	0·33
	1·00	1·34	0·29

5   INVERSE SAMPLING WITH ELIMINATION: PROCEDURE R_E

For three players we develop a Markovian procedure R_E based on inverse sampling and the DL-rule which eliminates one of the three players at or before the tournament is terminated. As in the case of R_I the recurrence formulae provide an algorithm for obtaining exact answers for the $P\{CS|LF\}$ and $E\{N|LF\}$. By building up a table of these functions with $P\{CS|LF\}$-values increasing to one, we can find the specific procedure satisfying any given P^*-condition and the associated $E\{N|LF\}$-value.

At the outset we use randomization and play $A\,v.\,B$, $B\,v.\,C$ or $C\,v.\,A$ each with probability $\frac{1}{3}$. The DL-rule is used for sampling and we eliminate any player as soon as he accumulates a total of r losses. The remaining player is then declared the winner.

When there is no confusion we refer to A as the best player. Let L_A, L_B and L_C denote the number of losses of A, B and C, respectively; let $L'_A = r - L_A$, $L'_B = r - L_B$ and $L'_C = r - L_C$. In analogy with (3.10) we define

$$R_{k, m, n} = P\{A \text{ wins the tournament}|L'_A = k,\ L'_B = \text{m},\ L'_C = n$$

$$\text{and the next game is } A\,v.\,B\} \quad (5.1)$$

and similarly $S_{k, m, n}$ for $B\,v.\,C$ and $T_{k, m, n}$ for $C\,v.\,A$. From the DL-sampling rule we obtain

$$\left.\begin{aligned}
R_{k,\ m,\ n} &= p_1 T_{k,\ m-1,\ n} + q_1 S_{k-1,\ m,\ n},\\
S_{k,\ m,\ n} &= p_2 R_{k,\ m,\ n-1} + q_2 T_{k,\ m-1,\ n},\\
T_{k,\ m,\ n} &= p_3 S_{k-1,\ m,\ n} + q_3 R_{k,\ m,\ n-1},
\end{aligned}\right\} \quad (5.2)$$

and from the elimination rule we obtain the 'boundary' conditions

$$\left.\begin{aligned}
S_{0,\ m,\ n} &= 0,\\
R_{k,\ m,\ 0} &= V_{k,\ m} \quad (\text{say}),\\
T_{k,\ 0,\ n} &= V_{k,\ n} \quad (\text{say}),\\
U_{k,\ 0} &= V_{k,\ 0} = 1,\\
U_{0,\ m} &= V_{0,\ n} = 0.
\end{aligned}\right\} \quad (5.3)$$

After player C is eliminated we use the recurrence formula

$$U_{k,\ m} = p_1 U_{k,\ m-1} + q_1 U_{k-1,\ m} \quad (5.4)$$

and after player B is eliminated we use

$$V_{k,m} = p_3 V_{k,n-1} + q_3 V_{k-1,n}. \tag{5.5}$$

After solving these equations we calculated $PCS(r)$ given by

$$PCS(r) = \tfrac{1}{3}(R_{r,r,r} + S_{r,r,r} + T_{r,r,r}) \tag{5.6}$$

which is the exact $P\{CS\}$ for the procedure R_E. This was done for the configuration (3.5) for $r = 1(1)\min(40, r^*)$, where r^* is the smallest value of r for which $P\{CS; R_E\} \geqslant 0.99$.

For $E\{N\}$ under procedure R_E we define functions $F_{k,m,n}$, $G_{k,m,n}$ and $H_{k,m,n}$ as in (3.6) and write as in (3.7)

$$\left. \begin{aligned} F_{k,m,n} &= p_1 H_{k,m-1,n} + q_1 G_{k-1,m,n} + 1, \\ G_{k,m,n} &= p_2 F_{k,m,n-1} + q_2 H_{k,m-1,n} + 1, \\ H_{k,m,n} &= p_3 G_{k-1,m,n} + q_3 F_{k,m,n-1} + 1. \end{aligned} \right\} \tag{5.7}$$

The boundary conditions assert that $F_{k,m,n}$, $G_{k,m,n}$ and $H_{k,m,n}$ are equal to 0 if two or more subscripts are zero. Letting $X_{k,m}$ denote the common value of

$$F_{k,m,0} = G_{k,m,0} = H_{k,m,0},$$

$Y_{m,n}$ denote the common value of

$$F_{0,m,n} = G_{0,m,n} = H_{0,m,n},$$

and $Z_{k,n}$ denote the common value of

$$F_{k,0,n} = G_{k,0,n} = H_{k,0,n},$$

we then use

$$\left. \begin{aligned} X_{k,m} &= p_1 X_{k,m-1} + q_1 X_{k-1,m} + 1, \\ Y_{m,n} &= p_2 Y_{m,n-1} + q_2 Y_{m-1,n} + 1, \\ Z_{k,n} &= p_3 Z_{k-1,n} + q_3 Z_{k,n-1} + 1, \end{aligned} \right\} \tag{5.8}$$

to complete the algorithm. Solving these equations for the configuration (3.5), we then computed

$$E\{N; R_E\} = \tfrac{1}{3}(F_{r,r,r} + G_{r,r,r} + H_{r,r,r}) \tag{5.9}$$

which is the exact value of $E\{N|LF\}$ for procedure R_E. Table 4 gives the values of r and $E\{N; R_E\}$ for the configuration (3.5) corresponding to selected values of p, θ and P^*.

Table 4. *Three players problem. Comparison of* $E\{N|\mathrm{LF}_\theta\}$ *values (exact and estimated) for six procedures (the r and d values are needed to make the associated procedure explicit;* PS *and* $\overline{N}$ *are Monte Carlo* (MC) *estimates of the* $P\{\mathrm{CS}|\mathrm{LF}_\theta\}$ *and* $E\{N|\mathrm{LF}_\theta\}$ *based on 300 and 600 tournaments respectively)*

$P^* = 0\cdot 75$

		R_I (exact)		R_E (exact)		R_S(MC)		R_{DL}(MC)		R_{DDR}(MC)		R_{DD}(MC)		
p	θ	$E\{N|\mathrm{LF}_\theta\}$	r	$E\{N|\mathrm{LF}_\theta\}$	r	$\overline{N}$§	d§	$\overline{N}$	PS	$\overline{N}$	PS	$\overline{N}$	PS	$\overline{W}$
0·65	0·50	17·4	9	18·5	9	15·7	4	17·0	0·763	15·1	0·760	13·5	0·760	0·792
	0·75	25·3	13	18·4	7	22·5‡	5‡	16·2	0·733	16·3	0·806	12·5	0·817	0·780
	1·00	—	†	23·8	9	—	—	10·6	0·930	8·1	0·883	9·5	0·770	0·813
0·70	0·50	8·4	5	10·0	4	8·8‡	3‡	9·0	0·813	9·7	0·803	8·3	0·803	0·811
	0·75	10·2	6	9·9	4	13·4	4	9·8	0·823	10·4	0·763	7·4	0·777	0·793
	1·00	20·8	12	9·9	4	18·4‡	5‡	7·8	0·840	6·8	0·817	6·4	0·840	0·801
0·75	0·50	6·2	4	7·2	3	7·5‡	3‡	6·1	0·803	6·6	0·820	5·4	0·803	0·816
	0·75	6·1	4	7·2	3	7·6‡	3‡	5·5	0·817	6·6	0·767	5·7	0·850	0·813
	1·00	7·6	5	7·1	3	7·6‡	3‡	5·9	0·870	5·7	0·870	4·9	0·860	0·830
0·80	0·50	4·2	3	4·5	2	3·5‡	3‡	2·9	0·767	4·0	0·820	3·1	0·797	0·786
	0·75	4·2	3	4·5	2	3·7	2	2·8	0·753	4·4	0·787	3·0	0·777	0·774
	1·00	4·1	3	4·5	2	3·6	2	2·8	0·747	3·7	0·830	3·1	0·810	0·808
0·85	0·50	2·5	2	2·0	1	3·5	2	2·8	0·837	3·5	0·893	2·6	0·853	0·833
	0·75	2·5	2	2·0	1	3·4	2	2·7	0·817	3·6	0·807	2·6	0·870	0·828
	1·00	2·5	2	2·0	1	3·4	2	2·7	0·830	3·2	0·810	2·5	0·793	0·821
0·90	0·50	2·5	2	2·0	1	3·1	2	2·0	0·800	2·4	0·823	2·0	0·853	0·845
	0·75	2·5	2	2·0	1	3·3	2	2·0	0·833	2·7	0·810	2·0	0·853	0·839
	1·00	2·5	2	2·0	1	3·1	2	2·0	0·820	2·8	0·830	2·0	0·833	0·840
0·95	0·50	2·4	2	2·0	1	3·0	2	2·0	0·910	2·3	0·887	2·0	0·923	0·919
	0·75	2·4	2	2·0	1	2·9	2	2·0	0·913	2·5	0·907	2·0	0·927	0·917
	1·00	2·4	2	2·0	1	2·9	2	2·0	0·923	2·8	0·930	2·0	0·917	0·919

$P^* = 0\cdot 90$

		R_I (exact)		R_E (exact)		R_S(MC)		R_{DL}(MC)		R_{DDR}(MC)		R_{DD}(MC)		
0·65	0·50	41·1	20	54·3	20	30·8	6	30·6	0·937	30·9	0·890	26·7	0·890	0·915
	0·75	—	†	76·8	28	54·2‡	9	30·2	0·950	30·3	0·923	22·9	0·923	0·913
	1·00	—	†	65·1	24	—	—	12·8	1·000	11·6	0·973	16·1	0·917	0·923
0·70	0·50	20·2	11	23·3	9	17·9‡	5‡	16·5	0·910	16·8	0·927	14·4	0·920	0·921
	0·75	29·3	16	28·6	11	23·9	6	17·9	0·950	19·0	0·943	13·8	0·903	0·918
	1·00	68·0	37	33·9	13	45·3‡	9‡	11·8	0·967	10·2	0·953	11·2	0·943	0·932
0·75	0·50	11·5	7	14·9	6	11·0	4	9·7	0·943	10·8	0·920	9·2	0·937	0·924
	0·75	14·9	9	14·6	6	14·8‡	5‡	11·6	0·933	12·3	0·930	10·2	0·963	0·933
	1·00	23·0	14	20·0	8	19·0	6	9·5	0·970	8·6	0·960	8·0	0·927	0·934
0·80	0·50	7·5	5	9·5	4	9·5	4	7·1	0·943	7·8	0·913	6·5	0·943	0·934
	0·75	9·0	6	9·5	4	9·3	4	7·8	0·923	8·4	0·933	6·7	0·927	0·937
	1·00	11·9	8	11·9	5	9·8	4	6·3	0·953	6·8	0·930	5·5	0·927	0·925
0·85	0·50	5·5	4	6·8	3	5·8‡	3‡	5·0	0·947	5·4	0·947	4·7	0·960	0·933
	0·75	5·5	4	6·8	3	5·6‡	3‡	5·1	0·933	5·9	0·960	4·6	0·917	0·927
	1·00	6·8	5	9·1	4	5·6‡	3‡	5·5	0·963	6·0	0·963	4·7	0·970	0·947
0·90	0·50	3·8	3	4·3	2	3·1	2	4·2	0·957	4·6	0·953	3·7	0·923	0·948
	0·75	3·8	3	4·3	2	3·3	2	4·1	0·937	4·6	0·943	3·7	0·957	0·941
	1·00	3·8	3	6·6	3	3·1	2	4·2	0·953	4·6	0·943	3·7	0·963	0·949
0·95	0·50	2·4	2	2·0	1	3·0	2	2·5	0·940	3·1	0·943	2·5	0·967	0·954
	0·75	2·4	2	2·0	1	2·9	2	2·5	0·933	3·4	0·943	2·6	0·953	0·953
	1·00	2·4	2	4·2	2	2·9	2	2·7	0·957	3·1	0·950	2·4	0·933	0·946

Table 4 (cont.)

p	θ	R_I (exact) $E\{N\|LF_\theta\}$	r	R_E (exact) $E\{N\|LF_\theta\}$	r	R_S(MC) $\bar{N}$§	d§	R_{DL}(MC) $\bar{N}$	PS	R_{DDR}(MC) $\bar{N}$	PS	R_{DD}(MC) $\bar{N}$	PS	$\bar{W}$
							$P^* = 0.95$							
0·65	0·50	58·0	28	65·1	24	43·0	8	40·8	0·957	38·5	0·967	33·8	0·973	0·959
	0·75	—	†	84·5	31	—	—	41·9	0·963	40·9	0·957	30·8	0·973	0·957
	1·00	—	†	—	†	—	—	12·8	1·000	13·2	0·997	21·4	0·977	0·965
0·70	0·50	29·7	16	33·9	13	22·4	6	22·9	0·963	21·5	0·970	18·4	0·957	0·963
	0·75	46·3	25	36·6	14	41·5	10	24·8	0·947	24·9	0·963	18·6	0·953	0·959
	1·00	—	†	44·6	17	—	—	12·5	1·000	12·0	0·990	13·7	0·940	0·965
0·75	0·50	16·7	10	20·0	8	14·4‡	5‡	13·1	0·963	14·2	0·960	11·9	0·970	0·963
	0·75	21·7	13	20·0	8	18·4	6	14·9	0·963	15·9	0·967	12·0	0·970	0·962
	1·00	38·3	23	22·5	9	27·3	8	10·5	0·953	10·0	0·977	9·6	0·973	0·968
0·80	0·50	10·6	7	11·9	5	9·5	4	8·6	0·973	9·5	0·957	8·0	0·980	0·964
	0·75	12·1	8	11·9	5	9·3	4	10·3	0·970	10·6	0·960	8·7	0·953	0·968
	1·00	18·1	12	11·9	5	15·2	6	8·9	0·977	8·6	0·983	7·3	0·970	0·969
0·85	0·50	7·0	5	9·5	4	8·0	4	6·3	0·973	7·3	0·957	5·9	0·973	0·968
	0·75	8·3	6	9·5	4	5·6‡	3‡	6·8	0·977	7·7	0·980	6·1	0·983	0·971
	1·00	9·6	7	9·5	4	7·9	4	6·0	0·967	6·6	0·973	5·2	0·957	0·967
0·90	0·50	5·1	4	6·6	3	4·7‡	3‡	4·4	0·967	5·0	0·957	4·4	0·977	0·969
	0·75	5·1	4	6·6	3	5·0‡	3‡	4·4	0·973	5·3	0·973	4·2	0·967	0·966
	1·00	5·1	4	6·6	3	4·9‡	3‡	5·1	0·990	6·1	0·983	4·6	0·983	0·983
0·95	0·50	3·6	3	4·2	2	3·0	2	3·9	0·980	4·0	0·980	3·3	0·983	0·979
	0·75	3·6	3	4·2	2	2·9	2	3·8	0·997	4·1	0·983	3·3	0·987	0·975
	1·00	3·6	3	4·2	2	2·9	2	3·7	0·990	4·4	0·963	3·5	0·963	0·978
							$P^* = 0.99$							
0·65	0·50	—	†	—	†	—	—	60·6	0·997	57·0	0·997	49·2	0·990	0·992
	0·75	—	†	—	†	—	—	62·9	0·987	59·7	0·983	47·9	1·000	0·991
	1·00	—	†	—	†	—	—	12·5	1·000	13·4	1·000	30·3	0·993	0·992
0·70	0·50	52·3	28	60·0	23	35·3	9	32·8	0·990	33·5	0·997	28·7	1·000	0·992
	0·75	—	†	71·1	27	—	—	39·0	0·997	38·2	0·997	28·4	0·993	0·992
	1·00	—	†	87·0	33	—	—	13·1	1·000	13·6	1·000	18·7	0·990	0·993
0·75	0·50	30·3	18	35·1	14	23·7	8	20·6	0·990	19·4	0·997	17·6	1·000	0·993
	0·75	41·8	25	37·6	15	31·6	10	24·4	0·993	24·8	0·997	18·5	0·994	0·992
	1·00	—	†	42·8	17	—	—	12·7	1·000	13·1	1·000	13·5	0·997	0·994
0·80	0·50	18·3	12	21·6	9	14·2	9	13·2	0·990	13·8	0·990	11·4	0·993	0·993
	0·75	22·8	15	21·6	9	16·8‡	7‡	15·2	0·990	16·1	0·990	12·4	0·990	0·993
	1·00	33·3	22	24·0	10	20·8	8	11·5	0·990	11·4	0·997	9·7	0·997	0·994
0·85	0·50	11·1	8	13·8	6	9·6‡	5‡	9·3	0·993	9·9	0·990	8·0	0·990	0·993
	0·75	13·8	10	13·8	6	9·8‡	5‡	9·7	0·997	10·9	0·997	9·0	1·000	0·994
	1·00	17·9	13	13·7	6	11·8	6	8·4	0·997	9·0	1·000	7·6	0·997	0·993
0·90	0·50	7·6	6	8·8	4	6·4	4	6·0	0·997	7·0	0·993	5·7	0·997	0·993
	0·75	7·6	6	8·8	4	6·7	4	6·2	0·983	7·6	0·993	5·7	0·993	0·992
	1·00	10·1	8	8·8	4	6·7	4	6·9	0·997	8·0	0·997	6·0	0·993	0·994
0·95	0·50	4·7	4	6·3	3	5·5	4	4·6	1·000	5·4	0·997	4·5	0·993	0·997
	0·75	4·7	4	6·3	3	5·5	4	4·6	0·997	5·8	1·000	4·5	0·993	0·997
	1·00	5·8	5	6·3	3	5·6	4	4·5	0·997	5·6	0·997	4·3	1·000	0·996

† Indicates that the value of r is greater than 40.
‡ Indicates values based on linear interpolation in a table of Monte Carlo results.
§ Estimates based on 1000 tournaments.

6   OTHER SEQUENTIAL PROCEDURES

For the purpose of comparison we include a discussion of several sequential procedures with Monte Carlo results. In the first one, called procedure R_S, we used only Monte Carlo methods; in the remaining procedures we make use of some theory in [2] for the P^*-condition and use Monte Carlo methods to estimate the $P\{CS|LF_\theta\}$ and $E\{N|LF_\theta\}$.

Let W_A, W_B and W_C denote the number of games won by A, B and C, respectively; let the ordered values be denoted by $W_1 \leqslant W_2 \leqslant W_3$, where ties are clearly possible. For the procedure R_S we use the DL-sampling rule and for $d > 0$ the following stopping rule.

Stopping rule for procedure R_S. Terminate as soon as $W_3 - W_2 \geqslant d$ and select the player with W_3 wins as best.

Monte Carlo results for the LF_θ configuration (3.5) were obtained for $d = 2(2)10$ and selected values of p and θ. The proportion of successes (PS), i.e. of correct selections, in 1000 tournaments is used as an estimate of the $P\{CS|LF_\theta\}$. The smallest value of d for which this estimate exceeds P^* is taken as the required minimum value needed to satisfy the P^*-condition (1.1); these are given in Table 4 along with the Monte Carlo estimate $\bar{N}$ of $E\{N|LF_\theta\}$ for selected values of p, θ and P^*.

For the remaining three sequential procedures we use a common notation and make use of the remark on p. 14 of [2] that in order for the P^*-condition (see (3.1.11) in [2]) to hold, it is not necessary to take the observations a vector at a time, i.e. to follow the cyclic pattern $A\,v.\,B$, $B\,v.\,C$, $C\,v.\,A$. Thus, for example, we can use the same result for a procedure R_{DL} which uses the DL-sampling rule. This result states that if we use a certain likelihood-ratio statistic W (to be defined explicitly below) and stop at any point when $W \geqslant P^*$, then the P^*-condition will be satisfied. All of the three remaining procedures have the Wald Stopping Structure (see p. 17 of [2]) since they stop as soon as $W \geqslant P^*$ and choose among the decisions (or hypotheses) with maximum likelihood. It should be carefully noted that we have only considered the identification aspect of the problem since no monotonicity properties have been shown with respect to p or θ; in other words, if we knew the values of the common p and θ then we can select the best player with the desired value of P^*. For the ranking aspect of this problem we have to show that if the true $p \geqslant p_0$ (the specified value) and the known value of θ is used, then the achieved $P\{CS\}$ will be at least as large as P^*; this has not been done. Finally we are interested in the robustness of these procedures with respect to θ, i.e. in the values of $P\{CS|LF_\theta\}$ when θ_1 is used in the statistic W defined below and $\theta \ne \theta_1$ is the true value; this also has not been investigated in this paper.

To introduce W, we first define W_{AB} as the number of games in which A beat B and similarly for $W_{BA}, W_{AC}, W_{CA}, W_{BC}$ and W_{CB}. Six likelihoods are defined for the LF_θ configuration (3.5) by

$$
\left.
\begin{aligned}
L_{ABC} &= p^{W_{AB}} q^{W_{BA}} \theta^{W_{BC}} (1-\theta)^{W_{CB}} p^{W_{AC}} q^{W_{CA}}, \\
L_{ACB} &= p^{W_{AC}} q^{W_{CA}} \theta^{W_{CB}} (1-\theta)^{W_{BC}} p^{W_{AB}} q^{W_{BA}}, \\
L_{BAC} &= p^{W_{BA}} q^{W_{AB}} \theta^{W_{AC}} (1-\theta)^{W_{CA}} p^{W_{BC}} q^{W_{CB}}, \\
L_{BCA} &= p^{W_{BC}} q^{W_{CB}} \theta^{W_{CA}} (1-\theta)^{W_{AC}} p^{W_{BA}} q^{W_{AB}}, \\
L_{CAB} &= p^{W_{CA}} q^{W_{AC}} \theta^{W_{AB}} (1-\theta)^{W_{BA}} p^{W_{CB}} q^{W_{BC}}, \\
L_{CBA} &= p^{W_{CB}} q^{W_{BC}} \theta^{W_{BA}} (1-\theta)^{W_{AB}} p^{W_{CA}} q^{W_{AC}}.
\end{aligned}
\right\}
\tag{6.1}
$$

The sum of L_{ABC} and L_{ACB} represents the 'total likelihood' that the common p ($=p_1 = q_3$) is associated with player A, i.e. that A is the best player. Hence we further define

$$
\left.
\begin{aligned}
\mathscr{L}_A &= L_{ABC} + L_{ACB}, \\
\mathscr{L}_B &= L_{BAC} + L_{BCA}, \\
\mathscr{L}_C &= L_{CAB} + L_{CBA}
\end{aligned}
\right\}
\tag{6.2}
$$

and the value of W can then be written as

$$
W = \frac{\max(\mathscr{L}_A, \mathscr{L}_B, \mathscr{L}_C)}{\mathscr{L}_A + \mathscr{L}_B + \mathscr{L}_C}.
\tag{6.3}
$$

For all the three remaining procedures we use the common

Stopping and terminal decision rule. Stop as soon as $W \geqslant P^*$ and select as the best player A if $\mathscr{L}_A$ is the maximum (6.3), B if $\mathscr{L}_B$ is the maximum and C if $\mathscr{L}_C$ is the maximum. For $P^* > \frac{1}{2}$ we cannot have ties for first place at termination. For $\frac{1}{3} < P^* \leqslant \frac{1}{2}$ we may have ties at termination; if t players are tied for first place then we select one by an independent experiment which gives equal probability, i.e. $1/t$, to each of them.

It follows that these three procedures differ only in the sampling rule. Since R_{DL} uses the DL-sampling rule, we need only define the sampling rule for the remaining two.

For the 'Double Duty' (DD) procedure R_{DD} we use the

Sampling Rule. Find the two largest $\mathscr{L}$'s and play the corresponding two players in the next game. In the case of ties and at the outset we use the appropriate randomization, i.e. we use $\frac{1}{2}$ when two are tied for first or second place and $\frac{1}{3}$ when there is a triple tie.

For procedure R_{DDR} we define

$$
\left.
\begin{aligned}
s_1 &= (\mathscr{L}_A + \mathscr{L}_B)/2(\mathscr{L}_A + \mathscr{L}_B + \mathscr{L}_C), \\
s_2 &= (\mathscr{L}_B + \mathscr{L}_C)/2(\mathscr{L}_A + \mathscr{L}_B + \mathscr{L}_C), \\
s_3 &= (\mathscr{L}_C + \mathscr{L}_A)/2(\mathscr{L}_A + \mathscr{L}_B + \mathscr{L}_C),
\end{aligned}
\right\}
\tag{6.4}
$$

which add to one and use the

Sampling Rule. Randomize by taking a uniform deviate U ($0 \leqslant U < 1$) and play $A\,v.\,B$ if $0 \leqslant U < s_1$, play $B\,v.\,C$ if $s_1 \leqslant U < s_1 + s_2$, and play $C\,v.\,A$ if $s_1 + s_2 \leqslant U < 1$.

Procedures R_{DD} and R_{DDR} eliminate noncontenders in a more natural way by gradually lowering the probability of putting them into the next game rather than by a sudden complete withdrawal from the tournament.

Evaluation of Monte Carlo results

The Monte Carlo PS results based on 300 tournaments in Table 4 for the last three sequential procedures R_{DL}, R_{DDR} and R_{DD} indicate that the nominal P^*-value is satisfied since the observed PS value went below P^* in only 8 of the 252 cases (or 3 %) and the maximum error is 0·017. For large values of p (say 0·95), all the procedures have their $E\{N|LF_\theta\}$ (or $\bar{N}$ values) approximately equal and there is not much practical basis for using this as a criterion for choosing one of them. For smaller values of p (say, 0·65 to 0·75) the differences are substantial. If we look at the maximum over the three values of θ studied for each p, then procedure R_{DD} shows a reduction of as much as 50 % when compared to the results of procedure R_I for some values of p and P^*. In general, the results for procedure R_{DD} appear to be better, if we look at the maximum over three values of θ, than any of the other procedures considered.

In the absence of any information about θ, it might be desirable to use a closer net of θ-values and find the value of θ which maximizes $\bar{N}$ and treat the problem as if that were the true value of θ. A more practical approach might be to use the accumulated observations to estimate θ but this has not been considered in this paper.

It is also interesting to note that the value of the statistic W (defined in (6.3) above) at termination is another estimate of the $P\{CS|LF_\theta\}$; the theoretical basis for this is discussed in Section 3.2 of [2]. Monte Carlo values of W averaged over 300 experiments are denoted by $\bar{W}$ and given for procedure R_{DD} in Table 4.

ACKNOWLEDGEMENT

We gratefully acknowledge very capable programming assistance by Mrs Elaine Frankowski and Lt Bar Giora Benyovitch, as well as computer time given us by the Computer Centers of the Israel Defence Forces and the University of Minnesota. We also thank Professors R. E. Bechhofer and J. Kiefer of Cornell University for very helpful comments, several of which were incorporated into the final draft.

APPENDIX A

Proof of the Identity in (4.22). We consider the sum

$$F_j = \sum_{\alpha=0}^{[\frac{1}{2}j]} \binom{j-\alpha}{\alpha} x^\alpha = \sum_{\alpha=0}^{\infty} \binom{j-\alpha}{\alpha} x^\alpha, \tag{A 1}$$

where the sum is extended to ∞, since the binomial coefficients are zero for $\alpha > [\frac{1}{2}j]$. We define the generating function

$$F(s) = \sum_{j=0}^{\infty} F_j s^j = \sum_{\alpha=0}^{\infty} x^\alpha \sum_{j=2\alpha}^{\infty} \binom{j-\alpha}{\alpha} s^j$$

$$= \sum_{\alpha=0}^{\infty} (xs^2)^\alpha \sum_{m=0}^{\infty} \binom{\alpha+m}{\alpha} s^m = \sum_{\alpha=0}^{\infty} \frac{(xs^2)^\alpha}{(1-s)^{\alpha+1}}$$

$$= (1-s-xs^2)^{-1}. \tag{A 2}$$

But this can be expanded by partial fractions and leads to the final result

$$F_j = \frac{1}{2^{j+1}\sqrt{(1+4x)}} \{(1+\sqrt{(1+4x)})^{j+1} - (1-\sqrt{(1+4x)})^{j+1}\}. \tag{A 3}$$

This reduces to (4.22) when x is set equal to q/p^2.

APPENDIX B

Table 5 contains two procedures not previously defined and two that were. The new procedures are defined so that the decrease (or increase) in $E\{N|\mathrm{LF}_\theta\}$ for R_I compared to R_{IC} and for R_{DL} compared to R_{BKS} is due solely to the DL-sampling rule.

Procedure R_{IC} is an inverse sampling procedure with a cyclic sampling pattern, i.e. we start by randomizing (with equal probability) between $A \, v. \, B$, $B \, v. \, C$ and $C \, v. \, A$ and then continue in a cyclic manner with these three types of games (in any prescribed order). Using the same definitions

Table 5. *Comparison of procedures for evaluating the improvement arising only from the DL-sampling rule*

		R_{IC} (exact)		R_I (exact)	R_{BKS}(MC)	R_{DL}(MC)
p	θ	$E\{N\|LF_\theta\}$	r	$E\{N\|LF_\theta\}$	$\bar N$	$\bar N$

$P^* = 0.75$

p	θ	$E\{N\|LF_\theta\}$	r	$E\{N\|LF_\theta\}$	$\bar N$	$\bar N$
0·65	0·50	19·3	9	17·4	18·6	17·0
	0·75	41·9	19	25·3	20·2	16·2
	1·00	—	‡	—	8·7†	10·6†
0·70	0·50	11·8	6	8·4	11·8	9·0
	0·75	13·8	7	10·2	12·9	9·8
	1·00	—	‡	20·8	7·7†	7·8†
0·75	0·50	7·1	4	6·2	10·8	6·1
	0·75	7·1	4	6·1	9·3	5·5
	1·00	18·8	10	7·6	7·5	5·9
0·80	0·50	4·9	3	4·2	3·7	2·9
	0·75	4·9	3	4·2	3·8	2·8
	1·00	6·7	4	4·1	3·6	2·8
0·85	0·50	2·8	2	2·5	3·4	2·8
	0·75	2·8	2	2·5	3·5	2·7
	1·00	2·8	2	2·5	3·5	2·7
0·90	0·50	2·8	2	2·5	3·3	2·0
	0·75	2·8	2	2·5	3·3	2·0
	1·00	2·8	2	2·5	3·3	2·0
0·95	0·50	2·7	2	2·4	3·1	2·0
	0·75	2·7	2	2·4	3·1	2·0
	1·00	2·7	2	2·4	3·2	2·0

$P^* = 0.90$

p	θ	$E\{N\|LF_\theta\}$	r	$E\{N\|LF_\theta\}$	$\bar N$	$\bar N$
0·65	0·50	45·3	20	41·1	39·4	30·6
	0·75	—	‡	—	36·5	30·2
	1·00	—	‡	—	10·8†	12·8†
0·70	0·50	25·1	12	20·2	22·9	16·5
	0·75	46·3	22	29·3	25·0	17·9
	1·00	—	‡	68·0	10·3†	11·8†
0·75	0·50	15·5	8	11·5	15·0	9·7
	0·75	21·4	11	14·9	17·0	11·6
	1·00	69·2	35	23·0	10·9	9·5
0·80	0·50	8·9	5	7·5	9·9	7·1
	0·75	12·6	7	9·0	10·1	7·8
	1·00	23·8	13	11·9	7·4	6·3
0·85	0·50	6·6	4	5·5	9·2	5·0
	0·75	8·4	5	5·5	8·9	5·1
	1·00	11·9	7	6·8	7·4	5·5
0·90	0·50	4·7	3	3·8	7·7	4·2
	0·75	4·6	3	3·8	7·3	4·1
	1·00	6·2	4	3·8	6·9	4·2
0·95	0·50	2·7	2	2·4	3·1	2·5
	0·75	2·7	2	2·4	3·2	2·5
	1·00	2·7	2	2·4	3·1	2·7

† Exceptions to the observed result that the DL-sampling rule reduces $\bar N$ occur only for values of θ near 1.
‡ See Table 4.

Table 5 (*cont.*)

		R_{IC} (exact)		R_I (exact)	R_{BKS}(MC)	R_{DL}(MC)
p	θ	$E\{N|\mathrm{LF}_\theta\}$	r	$E\{N|\mathrm{LF}_\theta\}$	N	N
				$P^* = 0.95$		
0·65	0·50	66·3	29	58·0	50·4	40·8
	0·75	—	‡	—	46·0	41·9
	1·00	—	‡	—	11·7†	12·8†
0·70	0·50	36·0	17	29·7	25·8	22·9
	0·75	74·4	35	46·3	33·0	24·8
	1·00	—	‡	—	11·4†	12·5†
0·75	0·50	21·6	11	16·7	18·6	13·1
	0·75	33·5	17	21·7	20·4	14·9
	1·00	—	‡	38·3	11·1	10·5
0·80	0·50	14·6	8	10·6	13·6	8·6
	0·75	18·4	10	12·1	13·0	10·3
	1·00	39·0	21	18·1	10·6	8·9
0·85	0·50	10·2	6	7·0	8·7	6·3
	0·75	10·2	6	8·3	9·5	6·8
	1·00	19·0	11	9·6	7·2	6·0
0·90	0·50	6·3	4	5·1	7·5	4·4
	0·75	8·1	5	5·1	7·6	4·4
	1·00	9·6	6	5·1	7·0	5·1
0·95	0·50	4·5	3	3·6	6·7	3·9
	0·75	4·5	3	3·6	6·6	3·8
	1·00	6·0	4	3·6	6·7	3·7
				$P^* = 0.99$		
0·65	0·50	—	‡	—	70·7	60·6
	0·75	—	‡	—	67·7	62·9
	1·00	—	‡	—	12·5	12·5
0·70	0·50	61·8	29	52·3	41·2	32·8
	0·75	—	‡	—	48·6	39·0
	1·00	—	‡	—	13·2	13·1
0·75	0·50	37·7	19	30·3	27·2	20·6
	0·75	65·7	33	41·8	32·5	24·4
	1·00	—	‡	—	13·6	12·7
0·80	0·50	24·1	13	18·3	17·4	13·2
	0·75	35·3	19	22·8	20·2	15·2
	1·00	—	‡	33·3	12·3	11·5
0·85	0·50	15·6	9	11·1	13·3	9·3
	0·75	20·9	12	13·8	14·3	9·7
	1·00	36·8	21	17·9	10·0	8·4
0·90	0·50	11·4	7	7·6	9·7	6·0
	0·75	13·1	8	7·6	9·7	6·2
	1·00	18·1	11	10·1	10·5	6·9
0·95	0·50	7·7	5	4·7	6·7	4·6
	0·75	7·7	5	4·7	6·7	4·6
	1·00	9·2	6	5·8	6·5	4·5

as in (3.1), except that we have a new sampling rule, the recursion formulas become

$$R_{k, m, n} = p_1 S_{k-1, m, n} + q_1 S_{k, m-1, n},$$
$$S_{k, m, n} = p_2 T_{k, m-1, n} + q_2 T_{k, m, n-1},$$
$$T_{k, m, n} = p_3 R_{k, m, n-1} + q_3 R_{k-1, m, n}.$$

(B 1)

The boundary conditions are the same as in (3.3). [The two essential changes to make (B 1) the exact equations for $E\{N\}$ namely adding 1 to each equation and replacing 1 by 0 in the boundary conditions, are the same as in (3.7) and need not be repeated.] As in § 3 we are interested in (3.4) and (3.8). Comparable values of $E\{N|\mathrm{LF}_\theta\}$ for procedures R_{IC} and R_I are given in Table 5 for selected values of (p, θ, P^*). The procedure R_I shows an improvement over R_{IC} in all cases computed and in several cases the reduction is as much as 50 %.

The procedure R_{BKS} samples a vector at a time and hence the observed number of games is a multiple of 3 in every tournament. After each set of 3 games the statistic W in (6.3) is calculated and we stop as soon as $W \geqslant P^*$ as in § 6, the only difference between this procedure and R_{DL} being that the latter uses the DL-sampling rule. Except for six cases, in each of which $\theta = 1$, p is small and P^* is moderate (see the entries marked with a dagger in Table 5), the procedure R_{DL} shows an improvement over R_{BKS}; the reduction is less than that observed in the previous comparison but it still surpasses 25 % reduction in many case.

The proportion of successes (PS) in 300 trials went below P^* in 2 out of 64 cases for R_{BKS}, the maximum difference $(P^* - \mathrm{PS})$ being 0·003. For R_{DL} the PS-value went below P^* in 5 out of 64 cases, the maximum difference $(P^* - \mathrm{PS})$ being 0·017. Since these dips below P^* are easily explainable by chance, we have an empirical verification of the basic P^*-condition (1.1) and at the same time a numerical study of the improvement due only to the DL-sampling rule.

APPENDIX C

N. ELLIOTT AND Y. S. LIN

University of Minnesota

Let $R_{k, m, n}$, $S_{k, m, n}$, $T_{k, m, n}$ (with all subscripts non-negative), be as defined in (3.1) with $p_1 = 1 - q_1$, $p_2 = \theta$ and $p_3 = 1 - q_3$ as defined earlier. Using the recursive equations (3.2) and boundary conditions (3.3), we wish to show that the probability of a correct selection given by $A_{r, r, r}$ in (3.4) is a strictly increasing function of p_1.

Let v_N denote the vector (k, m, n) when the sum of the components $k+m+n$ equals N; let w_N denote another vector with component sum N'. We say that v_N majorizes $w_{N'}$ (written $v_N \succ w_{N'}$) if

$$k' \geqslant k, \quad m' \leqslant m, \quad n' \leqslant n \quad \text{and} \quad N' \leqslant N. \tag{C1}$$

Let ψ denote a generic symbol for all three functions, R, S and T.

Lemma 1. If $v_N \succ w_{N'}$ then

$$\psi_{k, m, n} \geqslant \psi_{k', m', n'}. \tag{C2}$$

Proof. We need only consider $N > 3$ and k, m', n' all positive since the proof is trivial otherwise. To complete the induction proof, for example for $\psi = R$, we use (3.2) to write

$$R_{k, m, n} - R_{k', m', n'} = p_1(T_{k-1, m, n} - T_{k'-1, m, n}) + q_1(S_{k, m-1, n} - S_{k', m'-1, n'}) \tag{C3}$$

and the result therefore holds for N if it holds for $N - 1$. A similar proof holds for $\psi = S$ and $\psi = T$.

Lemma 2. For all p_1, θ, p_3

$$T_{k-1, m, n} \geqslant S_{k, m-1, n} \tag{C4}$$

whenever all subscripts in (C4) are non-negative.

Proof. The result is trivial for $k = 1$ or $m = 1$ and we can therefore assume that $k \geqslant 2$ and $m \geqslant 2$.

If we start with the middle equation of (3.2) substitute for R from the first equation, and then use the second equation again to eliminate the new S, we eventually get $S_{k, m-1, n}$ as a function of T's only. This result can be written for all $m \geqslant 2$ as

$$S_{k, m-1, n} = (1 - \theta) \sum_{j=0}^{[\frac{1}{2}(m-2)]} (\theta q_1)^j T_{k, m-1-2j, n-1}$$
$$+ \theta p_1 \sum_{j=0}^{[\frac{1}{2}(m-3)]} (\theta q_1)^j T_{k-1, m-2-2j, n}, \tag{C5}$$

where $[x]$ denotes the largest integer less than or equal to x. The sum of all the coefficients on the right side of (C5) is at most one because the sum is exactly one in (3.2) and, at each step in deriving (C5), we replaced a ψ-function by two terms with coefficients summing to one. Final terms of the form $\psi_{k, 0, n} = 0$ are omitted in (C5) and hence the sum of the coefficients on the right side of (C5) must be at most one. Since

$$v_N = (k - 1, m, n)$$

majorizes every vector on the right side of (C 5), it follows from Lemma 1 that $T_{k-1, m, n}$ is greater than (or equal to) every T on the right side of (C 5). Using the fact that the sum of the coefficients in (C 5) is at most one, it follows from (C 5) that $T_{k-1, m, n} \geqslant S_{k, m-1, n}$; this proves Lemma 2.

Let $\psi_{k, m, n}$ (MF) correspond to the more favorable configuration in which $p_1 = p + \epsilon$, $p_2 = 0$ and $p_3 = p$ for $\epsilon > 0$, and let $\psi_{k, m, n}$ (LF) correspond to the same with $\epsilon = 0$.

Theorem. For all non-negative k, m, n

$$\psi_{k, m, n} (\text{MF}) \geqslant \psi_{k, m, n} (\text{LF}). \tag{C 6}$$

Proof. We can assume k, m, n all positive and $N \geqslant 3$ since the proof is trivial otherwise. Using (3.2) and the inductive hypothesis

$$S_{k, m, n} (\text{MF}) = \theta R_{k, m-1, n} (\text{MF}) + (1 - \theta) T_{k, m, n-1} (\text{MF})$$
$$\geqslant \theta R_{k, m-1, n} (\text{LF}) + (1 - \theta) T_{k, m, n-1} (\text{LF}) = S_{k, m, n} (\text{LF}), \tag{C 7}$$

and a similar proof holds for $T_{k, m, n}$. For $R_{k, m, n}$ we obtain

$$R_{k, m, n} (\text{MF}) = (p + \epsilon) T_{k-1, m, n} (\text{MF}) + (q - \epsilon) S_{k, m-1, n} (\text{MF})$$
$$\geqslant p T_{k-1, m, n} (\text{LF}) + q S_{k, m-1, n} (\text{LF}) + \epsilon \{ T_{k-1, m, n} (\text{MF})$$
$$- S_{k, m-1, n} (\text{MF}) \}$$
$$\geqslant R_{k, m, n} (\text{LF}), \tag{C 8}$$

since Lemma 2 holds for all configurations; this proves the theorem.

Corollary. For procedure R_I with any r

$$P\{\text{CS}|\text{MF}\} \geqslant P\{\text{CS}|\text{LF}\}. \tag{C 9}$$

Proof. Since the $P\{\text{CS}\}$ under procedure R_I is given by (3.4) the result (C 9) is an immediate consequence of the theorem above.

REFERENCES

[1] Alam, K. (1969). On selecting the most probable event from a multinomial population. Submitted for publication.

[2] Bechhofer, R. E., Kiefer, J. and Sobel, M. (1968). *Sequential Identification and Ranking Procedures.* Chicago: The University of Chicago Press.

[3] Bradley, R. A. and Terry, M. E. (1952). The rank analysis of incomplete block designs. I. The method of paired comparisons. *Biometrika* **39**, 324–45.

[4] Cacoullos, T. and Sobel, M. (1966). An inverse sampling procedure for selecting the most probable event in a multinomial distribution. *Multivariate Analysis*. New York: Academic Press Inc., pp. 423–55.

[5] David, H. A. (1963). *The Method of Paired Comparisons*. Griffin's Statistical Monographs and Courses No. 12. New York: Hafner Publ. Co.

[6] Isbell, J. R. (1959). On a problem of Robbins. *Ann. Math. Statist.* 30, 606–10.

[7] Kemeny, J. G. and Snell, J. L. (1960). *Finite Markov Chains*. New York: D. Van Nostrand Company, Inc.

[8] Olkin, I. and Sobel, M. (1965). Integral expressions for tail probabilities of the multinomial and negative multinomial distributions. *Biometrika* 52, 167–79.

[9] Robbins, H. (1952). Some aspects of the sequential design of experiments. *Bull. Am. Math. Soc.* 58, 527–35.

[10] Robbins, H. (1956). A sequential decision problem with a finite memory. *Proc. Natn. Acad. Sci. U.S.A.* 42, 920–23.

[11] Smith, C. V. and Pyke, R. (1965). The Robbins–Isbell two-armed-bandit problem with finite memory. *Ann. Math. Statist.* 36, 1375–86.

[12] Zelen, M. (1969). Play the winner rule and the controlled clinical trial. *J. Am. Statist. Ass.* 64, 131–46.

ON RANKING THE PLAYERS
IN A 3-PLAYER TOURNAMENT†

ROBERT E. BECHHOFER

1 INTRODUCTION

The present paper is an outgrowth of comments made by the writer as
as an official discussant of the very stimulating paper [4] by Milton
Sobel and George Weiss at the Nonparametric Conference. In those
sections of their paper in which 3-player tournaments are considered,
Sobel and Weiss formulate their problem in such a way that only one of
three decisions is permissible after the tournament is completed, i.e.
one must choose one of the decisions 'Player 1 is best,' 'Player 2 is best,'
'Player 3 is best.' As the writer pointed out at the time, this formulation
of the problem, may perhaps be a reasonable one in some situations
(e.g. when the tournament director is told that a prize *must* be awarded
to one of the players); but it would not be appropriate if one permits the
decision that there is no clear-cut 'best' player in the tournament—
to cope with those situations for which the possibility exists that (say)
Player 1 is better than Player 2, Player 2 is better than Player 3, and
Player 3 is better than Player 1. (The failure to provide for this contingency causes some of the Sobel–Weiss sequential procedures (those
described in their § 6, and which are based on the likelihood ratio statistic)
to require an arbitrarily large number of observations to terminate
sampling, when the true state of affairs is such that the probabilities that
i beats j, that j beats k, and that k beats i ($i \neq j, j \neq k, k \neq i, i, j, k = 1, 2, 3$)
all approach one.) The formulation which we propose in the present
paper (and which seems to us to be a very natural one) permits a fourth
decision 'no player is best,' and is thus capable of dealing with such a
possibility.

The present paper has two main parts. In the first part (§ 2) we formulate our *ranking* problem in much the same way as Sobel–Weiss [4]
do theirs, except that we add the possibility of a fourth decision to their
three decisions. We add a probability requirement to their probability
requirement, and propose a single-stage procedure which will guarantee

† This research was supported by the U.S. Army Research Office-Durham under
Contract DA-31-124-ARO-D-474 and the Office of Naval Research under Contract
Nonr-401(53). Reproduction in whole or in part is permitted for any purpose of the
United States Government.

both requirements. A numerical example is provided. The indifference-zone approach employed in [4] and in our §2 is similar in spirit to the approach used earlier by the authors ([1], [2]) for certain classes of ranking problems, although the present problem is quite different in structure from the ones which we considered earlier.

In the second part (§3) we formulate an *identification* problem which also parallels the one formulated by Sobel and Weiss, except that (as with the ranking problem) we add the possibility of a fourth decision, and an additional probability requirement. We propose a sequential identification procedure which will guarantee both requirements. Incorporated into our procedure is a certain prior probability distribution which is assumed to be known. The theory underlying the sequential identification procedures in both the Sobel–Weiss paper and the present paper is developed completely in [2] although some minor modifications are necessary to insure that our procedure guarantees our two requirements.

Our main interest really centers in the formulation of the *ranking* problem of §2, and in *sequential* procedures which guarantee the associated probability requirement. Our *identification* problem of §3, and the sequential procedure which guarantees the associated probability requirement can be regarded as a first step toward finding a sequential procedure for the ranking problem. The relation between such identification and ranking problems is discussed in §4. Finally, a more detailed consideration of certain aspects of the relation between the Sobel–Weiss paper and our paper is given in §5.

Sobel and Weiss proposed several intuitively appealing sampling rules for their sequential identification procedure. The Monte Carlo sampling results which they obtained using these rules are very striking, and demonstrate that this is a very fruitful area for future research. Their sampling rules can be incorporated in our sequential identification procedure, and we would anticipate that the performance of our procedure would thereby be improved.

2   A RANKING PROBLEM

Let $p_{ij} = P\{\text{Player } i \text{ beats player } j\}$ $(i \neq j; i,j = 1, 2, 3)$, $(0 \leq p_{ij} \leq 1)$; the p_{ij} are unknown, and it is also assumed that we have no knowledge as to whether $p_{ij} < \frac{1}{2}$ or $p_{ij} > \frac{1}{2}$. (We assume that the possibility $p_{ij} = \frac{1}{2}$ cannot occur.)† If $p_{ij} > \frac{1}{2}$ we say that Player i is better than Player j. Let $\omega_{ab,cd,ef}$ be the State of Nature in which $p_{ab} > \frac{1}{2}$, $p_{cd} > \frac{1}{2}$, $p_{ef} > \frac{1}{2}$.

† $p_{ij} = \frac{1}{2}$ represents the boundary regions between the Ω_i-regions given in Table 1.

There are then eight possible States of Nature $\omega_{12,23,31}, \ldots, \omega_{21,32,31}$ which we group into sets $\Omega_0, \ldots, \Omega_3$ as in Table 1; the interpretation of each Ω_i is also given in the table.

Table 1

State of Nature		Relation of p_{ij} to $\frac{1}{2}$		Interpretation of Ω_i
		p_{12}	p_{23}	p_{13}	
Ω_0	$\omega_{12,23,31}$	>	>	<	No player is best
	$\omega_{21,32,13}$	<	<	>	
Ω_1	$\omega_{12,23,13}$	>	>	>	Player 1 is best
	$\omega_{12,32,13}$	>	<	>	
Ω_2	$\omega_{21,23,13}$	<	>	>	Player 2 is best
	$\omega_{21,23,31}$	<	>	<	
Ω_3	$\omega_{12,32,31}$	>	<	<	Player 3 is best
	$\omega_{21,32,31}$	<	<	<	

2.1. Goal, and probability requirement

The goal (purpose of the experiment) is stated below along with an associated probability requirement (R).

Goal. We wish to determine which one of the Ω_i $(i = 0, 1, 2, 3)$ represents the true State of Nature.

We permit four† possible terminal decisions. These are given, along with their interpretation, in Table 2.

Table 2

Decision	Meaning of decision
d_0	No player is best: Either $p_{12} > \frac{1}{2}, p_{23} > \frac{1}{2}, p_{13} < \frac{1}{2}$, or $p_{12} < \frac{1}{2}, p_{23} < \frac{1}{2}, p_{13} > \frac{1}{2}$.
d_1	Player 1 is best: Either $p_{12} > \frac{1}{2}, p_{23} > \frac{1}{2}, p_{13} > \frac{1}{2}$, or $p_{12} > \frac{1}{2}, p_{23} < \frac{1}{2}, p_{13} > \frac{1}{2}$.
d_2	Player 2 is best: Either $p_{12} < \frac{1}{2}, p_{23} > \frac{1}{2}, p_{13} > \frac{1}{2}$, or $p_{12} < \frac{1}{2}, p_{23} > \frac{1}{2}, p_{13} < \frac{1}{2}$.
d_3	Player 3 is best: Either $p_{12} > \frac{1}{2}, p_{23} < \frac{1}{2}, p_{13} < \frac{1}{2}$, or $p_{12} < \frac{1}{2}, p_{23} < \frac{1}{2}, p_{13} < \frac{1}{2}$.

Prior to the start of play we specify four constants $\{p_0, P_0^*; p_1, P_1^*\}$ with $\frac{1}{2} < p_0, p_1 < 1$; $\frac{1}{4} < P_0^*, P_1^* < 1$ which are incorporated into the following probability requirement.

† It is also possible to consider an alternative goal which is 'To determine which one of the eight ω's represents the true State of Nature.' There would then be eight possible terminal decisions. Corresponding changes would have to be made in the probability requirement (R).

548   ROBERT E. BECHHOFER

Probability requirement (R):

We require that our procedure guarantee that

$$P_0 = P\{\text{making decision } d_0\} \geqq P_0^*$$

R_0:   when $\min\{p_{12}, p_{23}, p_{31}\} \geqq p_0$ (2.1)

                or

and   when $\min\{p_{21}, p_{32}, p_{13}\} \geqq p_0$,

$$P_i = P\{\text{making decision } d_i\} \geqq P_1^* \quad (i = 1, 2, 3)$$

R_1:   when $\min\{p_{ij}, p_{ik}\} \geqq p_1$ (2.2)

$$(j \neq i;\ k \neq i;\ j \neq k;\ j, k = 1, 2, 3).$$

2.2. Regions of preference and indifference

To assist the reader in visualizing the probability requirement, we give a geometric interpretation of the regions in the parameter space where particular decisions are preferred, and the region of indifference. The true State of Nature can be represented as a point $p = (p_{12}, p_{23}, p_{13})$ in the unit cube $(0 \leqq p_{ij} \leqq 1)$, $(i < j; i, j = 1, 2, 3)$. For simplicity of representation we let $\gamma_{ij} = 2(p_{ij} - \frac{1}{2})$, and transform the parameter space into the cube $(-1 \leqq \gamma_{ij} \leqq 1)$, $(i < j; i, j = 1, 2, 3)$ which is depicted in Fig. 1. Then ⓘ in Figure 1 is the region in the parameter space of the γ_{ij} where decision d_i $(i = 0, 1, 2, 3)$ is preferred. The region not included in the region of preference for any d_i is referred to as the *region of indifference*.

2.3. Single-stage procedure

There may be many statistical procedures (single-stage, two-stage, sequential) which will guarantee the probability requirement (R) of §2.1. We describe here a simple single-stage one.

Single-stage ranking procedure. Conduct a c-cycle tournament, a cycle consisting of 3 games {Player 1 v. Player 2, Player 2 v. Player 3, Player 1 v. Player 3}, where c is a predetermined number, computed as described in §2.4. Calculate $\hat{p}_{12}, \hat{p}_{23}, \hat{p}_{13}$ where $\hat{p}_{ij} = $ (proportion of games in which Player i beat Player j in the c-cycle tournament). Compare each of these $\hat{p}_{ij}$ to $\frac{1}{2}$, and make the obvious terminal decision, e.g. make decision d_0 if $\hat{p}_{12} > \frac{1}{2}$, $\hat{p}_{23} > \frac{1}{2}$, $\hat{p}_{13} < \frac{1}{2}$ or if $\hat{p}_{12} < \frac{1}{2}$, $\hat{p}_{23} < \frac{1}{2}$, $\hat{p}_{13} > \frac{1}{2}$. Randomize between decisions if ties occur, e.g. randomize between decisions d_0 and d_2 with probability one-half if $\hat{p}_{12} = \frac{1}{2}$, $\hat{p}_{23} > \frac{1}{2}$, $\hat{p}_{13} < \frac{1}{2}$; ties cannot occur if c is odd.

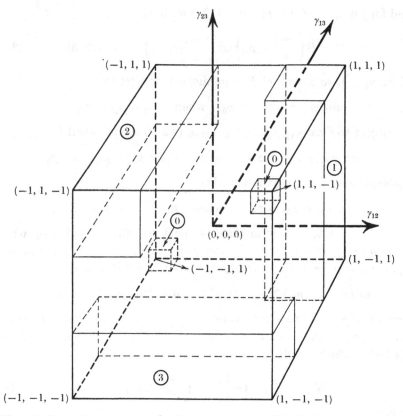

Figure 1. Parameter space: set of points $(\gamma_{12}, \gamma_{23}, \gamma_{13})$ with $-1 \leqq \gamma_{12}, \gamma_{23}, \gamma_{13} \leqq 1$. ⓘ = Region where decision d_i $(i = 0, 1, 2, 3)$ is preferred.

2.4. Determination of number of cycles ($\hat{c}$)

Our problem now is to determine the smallest value of c that will guarantee R. For simplicity we consider only *odd* c (and thereby avoid the necessity of discussing ties). Denoting the cumulative binomial probability

$$\Pr\{X \geqq x \mid n, p\} = \sum_{y=x}^{n} \frac{n!}{y!\,(n-y)!} p^y (1-p)^{n-y} \quad \text{by} \quad B(x \mid n, p),$$

we have that

$$P_0 = B\left(\frac{c+1}{2}\middle| c, p_{12}\right) \cdot B\left(\frac{c+1}{2}\middle| c, p_{23}\right) \cdot B\left(\frac{c+1}{2}\middle| c, p_{31}\right)$$

$$+ \left[1 - B\left(\frac{c+1}{2}\middle| c, p_{12}\right)\right]\left[1 - B\left(\frac{c+1}{2}\middle| c, p_{23}\right)\right]\left[1 - B\left(\frac{c+1}{2}\middle| c, p_{31}\right)\right],$$

$$(2.3)$$

and for $j \neq i$, $k \neq i$, $j \neq k$; $j, k = 1, 2, 3$ we have

$$P_i = B\left(\frac{c+1}{2}\bigg| c, p_{ij}\right) \cdot B\left(\frac{c+1}{2}\bigg| c, p_{ik}\right) \quad (i = 1, 2, 3). \qquad (2.4)$$

Since $p_0 > \frac{1}{2}$ we have that P_0 is minimized, subject to

$$\min\{p_{12}, p_{23}, p_{31}\} \geqq p_0, \quad \text{when} \quad p_{12} = p_{23} = p_{31} = p_0,$$

or subject to $\max\{p_{12}, p_{23}, p_{31}\} \leqq 1 - p_0$ (which is equivalent to

$$\min\{p_{21}, p_{32}, p_{13}\} \geqq p_0) \quad \text{when} \quad p_{12} = p_{23} = p_{31} = 1 - p_0;$$

the same value of P_0 is attained when

$$p_{12} = p_{23} = p_{31} = p_0 \quad \text{as when} \quad p_{12} = p_{23} = p_{31} = 1 - p_0.$$

Also, $P_i\,(i = 1, 2, 3)$ is minimized, subject to $\min\{p_{ij}, p_{ik}\} \geqq p_1$, when $p_{ij} = p_{ik} = p_1$. We refer to the configurations $p_{12} = p_{23} = p_{31} = p_0$ and $p_{12} = p_{23} = p_{31} = 1 - p_0$ as *least favorable* (LF) for R_0 when

$$\min\{p_{12}, p_{23}, p_{31}\} \geqq p_0 \quad \text{or when} \quad \max\{p_{12}, p_{23}, p_{31}\} \leqq 1 - p_0,$$

respectively, and the configurations $p_{ij} = p_{ik} = p_1$ as LF for R_1 when $\min\{p_{ij}, p_{ik}\} \geqq p_1$. Thus if we choose c_0 and c_1 as the smallest (odd) integers c which satisfy

$$B^3\left(\frac{c+1}{2}\bigg| c, p_0\right) + \left[1 - B\left(\frac{c+1}{2}\bigg| c, p_0\right)\right]^3 \geqq P_0^* \qquad (2.5)$$

and

$$B^2\left(\frac{c+1}{2}\bigg| c, p_1\right) \geqq P_1^*, \qquad (2.6)$$

respectively, we see that

$$\hat{c} = \max\{c_0, c_1\} \qquad (2.7)$$

is the smallest (odd) integer which guarantees R.

2.5. Numerical example

Suppose that the experimenter sets up the specification $p_0 = 0 \cdot 60$, $P_0^* = 0 \cdot 75$; $p_1 = 0 \cdot 65$, $P_1^* = 0 \cdot 90$. We find using [3] that

$$[B(22|43, 0 \cdot 60)]^3 + [1 - B(22|43, 0 \cdot 60)]^3 = (0 \cdot 908676)^3 + (0 \cdot 091324)^3$$
$$= 0 \cdot 751048$$

while

$$[B(21|41, 0 \cdot 60)]^3 + [1 - B(21|41, 0 \cdot 60)]^3 = (0 \cdot 903483)^3 + (0 \cdot 096517)^3$$
$$= 0 \cdot 73840;$$

hence $c_0 = 43$. Also, $[B(15|29, 0 \cdot 65)]^2 = (0 \cdot 952363)^2 = 0 \cdot 90700$ while

$$[B(14|27, 0 \cdot 65)]^2 = (0 \cdot 946377)^2 = 0 \cdot 89563;$$

hence $c_1 = 29$. Thus $\hat{c} = \max\{43, 29\} = 43$. But when $c = 43$ we also obtain

$$[B(22|43, 0 \cdot 65)]^2 = (0 \cdot 978598)^2 = 0 \cdot 95765,$$

$$[B(22|43, 0 \cdot 64)]^2 = (0 \cdot 970264)^2 = 0 \cdot 94141,$$

$$[B(22|43, 0 \cdot 63)]^2 = (0 \cdot 959510)^2 = 0 \cdot 92066,$$

$$[B(22|43, 0 \cdot 62)]^2 = (0 \cdot 945923)^2 = 0 \cdot 89477.$$

Thus, for no extra cost in terms of sampling, i.e. using $\hat{c} = 43$, the experimenter could have set up and guaranteed stricter probability requirements R with specifications such as

$$\{p_0, P_0^*; p_1, P_1^*\} = \{0 \cdot 60, 0 \cdot 75; \ 0 \cdot 65, 0 \cdot 95765\} \quad \text{or} \ldots \text{or}$$

$$\{0 \cdot 60, 0 \cdot 75; \ 0 \cdot 622, 0 \cdot 90\}.$$

3 AN IDENTIFICATION PROBLEM

Let p_{ij} be defined as in §2. We assume that p_0, p_1, θ are three *given* probabilities with $\frac{1}{2} < p_0, p_1, \theta \leq 1$, and that there are eight possible States of Nature $\omega_{12,23,31}^{\circ}, \ldots, \omega_{21,32,31}^{\circ}$ which represent *certain* possible pairings of the triple (p_{12}, p_{23}, p_{31}) with the probabilities p_0, p_1, θ. These pairings are given in Table 3, as are the groupings of the ω°'s into sets $\Omega_0^{\circ}, \ldots, \Omega_3^{\circ}$. We further assume a *known* prior probability ξ,

$$\xi(\omega^{\circ}) \geqq 0 \quad (\sum_{\Omega^{\circ}} \xi(\omega^{\circ}) = 1)$$

Table 3

| State of Nature | | Pairings of (p_{12}, p_{23}, p_{31}) with p_0, p_1, θ | | |
		p_{12}	p_{23}	p_{31}
Ω_0°	$\omega_{12,23,31}^{\circ}$	p_0	p_0	p_0
	$\omega_{21,32,13}^{\circ}$	$1 - p_0$	$1 - p_0$	$1 - p_0$
Ω_1°	$\omega_{12,23,13}^{\circ}$	p_1	θ	$1 - p_1$
	$\omega_{12,32,13}^{\circ}$	p_1	$1 - \theta$	$1 - p_1$
Ω_2°	$\omega_{21,23,31}^{\circ}$	$1 - p_1$	p_1	θ
	$\omega_{21,23,13}^{\circ}$	$1 - p_1$	p_1	$1 - \theta$
Ω_3°	$\omega_{12,32,31}^{\circ}$	θ	$1 - p_1$	p_1
	$\omega_{21,32,31}^{\circ}$	$1 - \theta$	$1 - p_1$	p_1

associated with each one of the eight States of Nature in $\Omega^\circ = \bigcup\limits_{i=0}^{3} \Omega_i^\circ$. For

our problem it seems reasonable because of the symmetries, to set

$$\xi(\omega^\circ) = (1 - 3\lambda)/2$$

for each of the two ω°'s in Ω_0°, and $\xi(\omega^\circ) = \lambda/2$ for each of the six ω°'s in $\Omega_1^\circ \cup \Omega_2^\circ \cup \Omega_3^\circ$. Then $\xi_0 = \xi(\Omega_0^\circ) = 1 - 3\lambda$ and $\xi_i = \xi(\Omega_i^\circ) = \lambda$ $(i = 1, 2, 3)$.

3.1. Goal, and probability requirement

The goal is stated below along with an associated probability requirement (R°).

Goal. We wish to determine which one of the Ω_i° $(i = 0, 1, 2, 3)$ represents the true State of Nature.

We permit four possible terminal decisions. These are given, along with their interpretation, in Table 4.

Table 4

Decision	Meaning of decision		
d_0	No player is best: Either	$p_{12} = p_{23} = p_{31} = p_0$,	
	or	$p_{12} = p_{23} = p_{31} = 1 - p_0$.	
d_1	Player 1 is best: Either	$p_{12} = p_{13} = p_1$,	$p_{23} = \theta$,
	or	$p_{12} = p_{13} = p_1$,	$p_{23} = 1 - \theta$.
d_2	Player 2 is best: Either	$p_{21} = p_{23} = p_1$,	$p_{31} = \theta$,
	or	$p_{21} = p_{23} = p_1$,	$p_{31} = 1 - \theta$.
d_3	Player 3 is best: Either	$p_{32} = p_{31} = p_1$,	$p_{12} = \theta$,
	or	$p_{32} = p_{31} = p_1$,	$p_{12} = 1 - \theta$.

Prior to the start of play we specify two constants $\{P_0^*, P_1^*\}$ with $1/4 < P_0^*, P_1^* < 1$ which are incorporated into the following probability requirement.

Probability requirement (R°):

We require that our procedure guarantee that

$$R_0^\circ: \qquad P_0^\circ = P\{\text{making decision } d_0 \text{ when } \Omega_0^\circ \text{ is true}\} \geqq P_0^* \qquad (3.1)$$

and

$$R_1^\circ: \qquad P_i^\circ = P\{\text{making decision } d_i \text{ when } \Omega_i^\circ \text{ is true}\} \geqq P_i^* \qquad (3.2)$$
$$(i = 1, 2, 3).$$

(The notation of (3.1) and (3.2) is justified by the fact that for simplicity we shall consider only procedures whose symmetry insures that for each i

the probability expressions of (3.1) and (3.2) are the same for both members of Ω_i° ($i = 0, 1, 2, 3$).) It should be noted that in this problem we are deciding among four composite hypotheses (each one consisting of two *completely specified* States of Nature) with specified lower bounds on the probability of a correct decision for each of these States of Nature.

3.2. Sequential procedure

In this section we shall describe a sequential identification procedure which will guarantee the probability requirement (R°). The theory underlying this development is derived in [2].

Our sequential procedure is based on eight likelihoods, each one of which is associated with a different one of the eight States of Nature of Table 3. These are defined in (3.3), below. In making these definitions we assume that the tournament is run in cycles (as defined in §2.3). However, this assumption of cyclic play is made only for convenience, and the same results would hold for more general symmetrical playing rules. Let $s_m^{(ij)} = s^{ij}$ (say) denote the number of 'successes' for Player i when he plays Player j, i.e. the number of times that i beats j, in their first m games. We then define the eight likelihoods

$$\left.\begin{aligned} L_m^{(12,23,31)} &= p_0^{s^{12}}(1-p_0)^{s^{21}} p_0^{s^{23}}(1-p_0)^{s^{32}} p_0^{s^{31}}(1-p_0)^{s^{13}} \\ &= p_0^{s^{12}+s^{23}+s^{31}}(1-p_0)^{s^{21}+s^{32}+s^{13}}, \\ L_m^{(21,32,13)} &= (1-p_0)^{s^{12}} p_0^{s^{21}}(1-p_0)^{s^{23}} p_0^{s^{32}}(1-p_0)^{s^{31}} p_0^{s^{13}} \\ &= p_0^{s^{21}+s^{32}+s^{13}}(1-p_0)^{s^{12}+s^{23}+s^{31}}, \\ L_m^{(12,23,13)} &= p_1^{s^{12}}(1-p_1)^{s^{21}} \theta^{s^{23}}(1-\theta)^{s^{32}}(1-p_1)^{s^{31}} p_1^{s^{13}}, \\ L_m^{(12,32,13)} &= p_1^{s^{12}}(1-p_1)^{s^{21}}(1-\theta)^{s^{23}} \theta^{s^{32}}(1-p_1)^{s^{31}} p_1^{s^{13}}, \end{aligned}\right\} \quad (3.3)$$

etc.,

and the four likelihood sums

$$\left.\begin{aligned} \mathscr{L}_m(d_0) &= L_m^{(12,23,31)} + L_m^{(21,32,13)}, \\ \mathscr{L}_m(d_1) &= L_m^{(12,23,13)} + L_m^{(12,32,13)}, \\ \mathscr{L}_m(d_2) &= L_m^{(21,23,31)} + L_m^{(21,23,13)}, \\ \mathscr{L}_m(d_3) &= L_m^{(12,32,31)} + L_m^{(21,32,31)}. \end{aligned}\right\} \quad (3.4)$$

The *a posteriori* probability after cycle m that d_j is the correct decision given ξ and $s_m = (s_m^{(12)}, s_m^{(23)}, s_m^{(31)})$ is

$$P_{jm}^\circ = \frac{\xi_j \mathscr{L}_m(d_j)}{\sum\limits_{k=0}^{3} \xi_k \mathscr{L}_m(d_k)} \quad (j = 0, 1, 2, 3). \quad (3.5)$$

The stopping and terminal decision rules of our sequential procedure are based on the P_{jm}° of (3.5) and on a constant $\hat{P}*$ ($1/4 < \hat{P}* < 1$) which is chosen to guarantee (3.1) and (3.2). As stated in the preliminaries of §3, we shall let $\xi_0 = 1 - 3\lambda$ and $\xi_i = \lambda$ ($i = 1, 2, 3$), where λ is assumed to be known.

SEQUENTIAL IDENTIFICATION PROCEDURE

Sampling rule. Run the tournament in cycles.

Stopping rule. Stop the tournament at the first cycle n for which

$$\max_{0 \leq j \leq 3} P_{jm}^{\circ} \geq \hat{P}*, \qquad (3.6)$$

where

$$\hat{P}* = \begin{cases} P_0^* & \text{for} \quad \lambda = 0, \\ \max\{P_0^* + 3\lambda(1 - P_0^*), 1 - 3\lambda(1 - P_1^*)\} & \text{for} \quad 0 < \lambda < \tfrac{1}{3}, \\ P_1^* & \text{for} \quad \lambda = \tfrac{1}{3}. \end{cases} \qquad (3.7)$$

Terminal decision rule. Make the decision which has associated with it the largest of the P_{jn}°. If, because of equalities among the P_{jn}°, several decisions are associated with the same maximum value, then select one such decision by a random device which assigns the same probability to each.

The terminal decision rule can be stated equivalently as follows. Letting $\bar{d}_j$ be the decision d_j for which $\mathscr{L}_n(\bar{d}_j) = \max_{1 \leq j \leq 3} \mathscr{L}_n(d_j)$, we:

$$\begin{aligned} &\text{Choose} \quad d_0 \quad \text{if} \quad \mathscr{L}_n(d_0) > \frac{\lambda}{1 - 3\lambda} \mathscr{L}_n(\bar{d}_j). \\ &\text{Choose} \quad \bar{d}_j \quad \text{if} \quad \mathscr{L}_n(d_0) < \frac{\lambda}{1 - 3\lambda} \mathscr{L}_n(\bar{d}_j). \\ &\text{Randomize among ties.} \end{aligned} \qquad (3.8)$$

In the next section we show why $\hat{P}*$ of (3.7) guarantees (3.1) and (3.2).

3.3. Determination of stopping constant ($\hat{P}*$)

We first remark that in general it is difficult to determine stopping constants which not only guarantee probability requirements such as R°, but also do so in some optimal sense. This is particularly the case for non-symmetric situations when the number of terminal decisions which the statistician is permitted to make is greater than two (which is our situation). In fact almost all of the theory contained in [2] which deals with sequential identification procedures is concerned with symmetric situations. However, it is possible to make use of certain results and remarks in [2] (specifically those in §3.7(c), pp. 51–2) to obtain solutions for certain particular problems involving nonsymmetric situations. No claim is made that these are the best possible solutions (which they certainly are not), but only that these solutions guarantee (3.1) and (3.2).

It follows from results proved in [2] (specifically from Lemma 3.1.1 and Corollary 3.1.1—see also equation (3.7.2)) that our sequential identification procedure of §3.2 using P^* ($\frac{1}{4} < P^* < 1$) on the r.h.s. of (3.6) guarantees

$$3\lambda P_1^\circ + (1 - 3\lambda) P_0^\circ \geq P^*, \tag{3.9}$$

where P_0° and $P_1^\circ = P_2^\circ = P_3^\circ$ are defined in (3.1) and (3.2), respectively. From (3.8) it follows, by setting first P_1° and then P_0° equal to one, that

$$P_0^\circ \geq (P^* - 3\lambda)/(1 - 3\lambda) \quad \text{for} \quad 0 \leq \lambda < \tfrac{1}{3}, \tag{3.10}$$

$$P_1^\circ \geq (P^* - 1 + 3\lambda)/3\lambda \quad \text{for} \quad 0 < \lambda \leq \tfrac{1}{3}. \tag{3.11}$$

Since for fixed λ the r.h.s. of both (3.10) and (3.11) are increasing functions of P^* which approach one as P^* approaches one, it follows that we can choose P^* sufficiently large that the r.h.s. of (3.10) is $\geq P_0^*$ *and* the r.h.s. of (3.11) is $\geq P_1^*$; if we let $\hat{P}^*$ denote the smallest value of P^* which accomplishes this dual objective, we see that $\hat{P}^*$ is given by (3.7).

3.4. Numerical example

In Table 5 we give the values of $\hat{P}^*$ associated with the specification $P_0^* = 0{\cdot}75$, $P_1^* = 0{\cdot}90$ for selected values of $\xi_i = \lambda$ ($i = 1, 2, 3$).

Table 5

λ	$\hat{P}^*$	$\dfrac{\hat{P}^* - 3\lambda}{1 - 3\lambda}$	$\dfrac{\hat{P}^* - 1 + 3\lambda}{3\lambda}$
$\frac{1}{3}$	0·90	—	0·90
0·33	0·9975	0·75	0·9975⁻
0·30	0·975	0·75	0·9722
$\frac{5}{21}$	$\frac{13}{14}$	0·75	0·90
0·20	0·94	0·85	0·90
0·10	0·97	0·9571	0·90
0·01	0·997	0·9969	0·90
0	0·75	0·75	—

We note that when $\hat{P}^*$ is considered as a function of λ for fixed $\{P_0^*, P_1^*\}$, it is discontinuous at $\lambda = 0$ and $\lambda = \tfrac{1}{3}$, e.g. $\hat{P}^* \to 1$ as $\lambda \to \tfrac{1}{3}$, but $\hat{P}^* = 0{\cdot}90$ when $\lambda = \tfrac{1}{3}$. However, this is to be expected: As $\lambda \to \tfrac{1}{3}$, the *a priori* probability ξ_0 associated with Ω_0° approaches zero. But the procedure is still required to make the decision d_0 at least 75 per cent of the time when Ω_0° occurs (however infrequently), as well as to make the decision d_i at least 90 per cent of the time when Ω_i° occurs ($i = 1, 2, 3$). This will in general require many cycles to terminate the tournament (and hence the expected number of cycles to terminate the experiment will be very large). But when $\lambda = \frac{1}{3}$, Ω_0° cannot occur, and the procedure is only required to decide among $\Omega_1^\circ, \Omega_2^\circ, \Omega_3^\circ$ and make the decision d_i at least 90 per cent of the time when Ω_i° occurs ($i = 1, 2, 3$). A similar analysis holds for $\lambda \to 0$ and $\lambda = 0$.

In Table 5 the minimum of P^*, namely $\frac{13}{14} = 0.9286$, occurs when $\lambda = \frac{5}{21}$. In general this minimum value occurs when

$$\lambda = \bar{\lambda} = (1 - P_0^*)/3(2 - P_0^* - P_1^*)$$

and is associated with $P_0^* = (\hat{P}^* - 3\lambda)/(1 - 3\lambda)$ and $P_1^* = (\hat{P}^* - 1 + 3\lambda)/3\lambda$.

We note that for $\lambda = 0.33$ (say), the experimenter could set up the stricter specification $\{P_0^* = 0.75, P_1^* = 0.9975\}$, and he can guarantee it with the same amount of sampling as is required by the present specification; such a tightening of the specification is possible for all λ-values ($0 < \lambda < \frac{1}{3}$) except $\lambda = \frac{5}{21}$.

A non-Bayesian might regard λ ($0 \leq \lambda \leq \frac{1}{3}$) as a constant, at his disposal, to be set in such a way as to cause the procedure to have certain properties which he deems desirable. Thus he might *set* $\lambda = \bar{\lambda}$, and still be assured that the procedure will guarantee (3.1) and (3.2).

4 RELATION BETWEEN THE RANKING AND IDENTIFICATION PROBLEMS

It was demonstrated in [2] that in many situations of practical interest, appropriate sequential procedures are more efficient than single-stage procedures which guarantee the same probability requirements, i.e. the sequential procedures often guarantee these requirements with a smaller expected number of observations. With this fact in mind, we would like for the problem of §2 to find a *sequential* ranking procedure, to replace the single-stage ranking procedure of that section, which will guarantee the probability requirement (2.1) and (2.2). As a first step in this direction, we set up an artificial identification problem, and displayed in §3.2 a sequential identification procedure which will guarantee the associated probability requirement (3.1) and (3.2) (which corresponds to protecting against the *least* favorable configuration of (2.1) and (2.2) with the addition of the nuisance parameter θ). (Usually, if one cannot solve the identification problem, there is little hope of solving the corresponding ranking problem.) The next logical step in the research process would be to try to produce a sequential ranking procedure which reduces to the sequential identification procedure when the parameters are in the least favorable configuration, and which for an arbitrary configuration of the parameters guarantees the probability requirement (2.1) and (2.2). (This is the same general approach as was used in [2]. See, in particular, §§3.6 and 6.1.) However, we wish to emphasize that *we have made no attempt in the present paper to produce such a sequential ranking procedure*, but rather leave that as a topic for future research.

One would naturally be very much interested in the efficiency (measured in terms of expected number of cycles to terminate the tournament) of our sequential identification procedure of §3.2 relative to that of a single-stage identification procedure which guarantees (3.1) and (3.2). (Note that the determination of the $\hat{c}$ of (2.7) does *not* involve knowledge of the value of the nuisance parameter θ, while the use of the sequential identification procedure of §3.2 does involve such knowledge.) However, we remind the reader that the sequential identification procedure of §3.2 which uses the $\hat{P}^*$ of (3.7) is very conservative for $0 < \lambda < \frac{1}{3}$ because of the crude inequalities (3.10) and (3.11) used in deriving it, i.e. the procedure will actually achieve probabilities of a correct decision under Ω_i° $(i = 0, 1, 2, 3)$ which are substantially higher than the specified values P_0^* and P_1^*. Thus it may be somewhat misleading to compare this procedure with a single-stage identification procedure which is set up to guarantee the same probability requirements (3.1) and (3.2), unless some adjustment is made for 'excess.' (See [2], §3.7 (d) and (e).)

5 RELATION BETWEEN THE SOBEL–WEISS PAPER AND THE PRESENT PAPER

The Sobel–Weiss paper [4] describes selection procedures for tournaments with two or three players. Their formulation of the problem for tournaments with three players is analogous to ours except that they never permit the decision d_0, and hence they consider only the probability requirements R_1 of (2.2) and R_1° of (3.2). Ideally they would like all of their sequential procedures for tournaments with 3 players to guarantee our probability requirement R_1 of (2.2) (their requirement (1.1)). However, at this point in the development of the theory, none of their procedures has been *proved* to do so. They have proved that their procedure R_I does so if the value of the nuisance parameter $\theta = p_{jk}$ in our (2.2) is *known*.

558   ROBERT E. BECHHOFER

Their procedure R_E is shown to guarantee our probability requirement R_1° of (3.2). Their procedure R_S is intended to guarantee R_1° of (3.2); it is studied using Monte Carlo sampling, but no analytic results are obtained.

The Sobel–Weiss procedures R_{DL}, R_{DD}, R_{DDR}, R_{IC}, and R_{BKS} guarantee our R_1° of (3.2); they use the same stopping rule and terminal decision rule as does our sequential identification procedure of §3.2 when we let $\lambda = \frac{1}{3}$ in our procedure, but they use different sampling rules. (Their procedure R_{BKS} is identical to our procedure when we let $\lambda = \frac{1}{3}$.) Their sampling rules could be used with our procedure of §3.2 for *arbitrary* λ ($0 \leqq \lambda \leqq 1/3$) to guarantee (3.1) and (3.2) simply by redefining the L_m's of (3.3) to read $L_{(m_{ab}, m_{cd}, m_{ef})}^{(ab, cd, ef)}$ since in general we will have

$$s_{m_{ab}}^{(ab)}, s_{m_{cd}}^{(cd)}, s_{m_{ef}}^{(ef)} \quad \text{with} \quad m_{ab} \neq m_{cd} \neq m_{ef}.$$

The improvements in $\overline{N}$ values (see the tables in [4] summarizing the results of Monte Carlo sampling experiments) obtained using their sampling rules may very well carry over to our procedure for arbitrary λ ($0 \leqq \lambda \leqq \frac{1}{3}$), and this possibility is certainly worth investigating further.

It should be pointed out that a major drawback of the Sobel–Weiss 3-decision formulation is that $E\{n\}$ will become arbitrarily large for their five procedures R_{DL}, R_{DD}, R_{DDR}, R_{IC}, and R_{BKS} if

$$\min\{p_{12}, p_{23}, p_{31}\} \to 1 \quad \text{or} \quad \min\{p_{21}, p_{13}, p_{31}\} \to 1$$

(since this means that d_0 is the true State of Nature). However, our procedure will tend to terminate early in either of these situations, and make the decision d_0.

6 DIRECTIONS OF FUTURE RESEARCH

It would be desirable to continue the present investigation to find sequential ranking procedures which can be proved analytically to guarantee the probability requirements (2.1) and (2.2), and which employ some of the sampling rules (or variations thereof) studied by Sobel and Weiss. Research should be devoted to finding improved values of the stopping constant $\hat{P}^*$ of (3.7) in order to cut down on the overprotection. It would also be interesting to generalize this approach to tournaments involving more than three players.

7 ACKNOWLEDGEMENT

The writer is happy to acknowledge the assistance of Professor Jack Kiefer with whom he had several very helpful discussions, and who proposed to him the inequalities (3.9) and (3.10). Thanks are also due to Mr Vijay Bawa who called the writer's attention to an error in an earlier version of this paper, and to Professor Milton Sobel who made many constructive comments which were incorporated into the final version.

REFERENCES

[1] Bechhofer, R. E. (1954). A single-sample multiple decision procedure for ranking means of normal populations with known variances. *Ann. Math. Statist.* **30**, 16–39.

[2] Bechhofer, R. E., Kiefer, J. and Sobel, M. (1968). *Sequential identification and ranking procedures.* Chicago: The University of Chicago Press.

[3] National Bureau of Standards (1952). *Tables of the binomial probability distribution*, Applied Mathematics Series 6. Washington, D.C.: U.S. Government Printing Office.

[4] Sobel, M. and Weiss, G. (1970). Inverse sampling and other selection procedures for tournaments with two or three players. In '*Nonparametric Techniques in Statistical Inference*' (M. L. Puri, ed.), pp. 515–43. Cambridge University Press.

PART 6

DECISION THEORETIC AND EMPIRICAL BAYES PROCEDURES

PSEUDO-LIKELIHOOD APPROACH TO SOME TWO-DECISION PROBLEMS UNDER NONPARAMETRIC SET-UPS

SHOUTIR KISHORE CHATTERJEE

1 INTRODUCTION—PSEUDOLIKELIHOOD

Let $\mathbf{X} = (X_1, ..., X_n)$ be a random n-vector whose cumulative distribution function (cdf) is known to belong to the class $\{F_\theta(\mathbf{x}), \theta \in \Omega\}$, $\mathbf{x} = (x_1, ..., x_n)$, of n-variate cdf's. Suppose for each $\theta \in \Omega$, a real-valued measurable function $v_\theta(\mathbf{x})$ satisfying the two following conditions is given.

(a) Whatever θ, for $\mathbf{x} \in R_n$ (the real n-vector-space), $v_\theta(\mathbf{x})$ assumes values within a fixed finite set of real numbers

$$\{y_1, ..., y_N\}, \quad y_1 < y_2 < ... < y_N.$$

(b) When $\mathbf{X}$ follows the distribution $F_\theta(\mathbf{x})$, $v_\theta(\mathbf{X})$ follows a known discrete distribution (same for all $\theta \in \Omega$) with probability mass points $y_i, i = 1, ..., N$ and probability mass function (pmf) $g(y)$.

Then with respect to the class of functions $\{v_\theta(\mathbf{x}), \theta \in \Omega\}$, the pseudo-likelihood of θ is defined to be

$$L(\theta|\mathbf{x}) = g[v_\theta(\mathbf{x})]. \tag{1.1}$$

The expression (1.1) depends on θ through the function $v_\theta(\mathbf{x})$ and not through $g(y)$.

Example. Suppose $n = n_1 + n_2$ and for each $\theta \in \Omega$, there is a continuous univariate cdf $F_\theta^*(x)$ and a real number τ_θ such that

$$F_\theta(\mathbf{x}) = \prod_{i=1}^{n_1} F_\theta^*(x_i) \prod_{j=1}^{n_2} F_\theta^*(x_{n_1+j} - \tau_\theta). \tag{1.2}$$

We take $v_\theta(\mathbf{x}) =$ number of pairs $(x_i, x_{n_1+j}), i = 1, ..., n_1, j = 1, ..., n_2$, for which $x_i \leqslant x_{n_1+j} - \tau_\theta$. Then $v_\theta(\mathbf{x})$ can assume only the values $0, 1, ..., n_1 n_2$, and it is known that when $\mathbf{X}$ follows the distribution $F_\theta(\mathbf{x}), v_\theta(\mathbf{X})$ has the distribution of the Mann–Whitney statistic under the null hypothesis. Writing $g(y)$ for the pmf of this distribution the pseudolikelihood would be given by (1.1).

In this illustration, although $F_\theta(\mathbf{x})$ is determined by both $F_\theta^*(x)$ and $\tau_\theta, v_\theta(\mathbf{x})$, and hence, the pseudolikelihood depends on θ only through τ_θ.

Generally, in any instance, we may choose $v_\theta(\mathbf{x})$ to make the pseudo-likelihood depend only on those aspects of θ which are of interest. Because of this flexibility the pseudolikelihood idea is particularly handy in nonparametric set-ups.

In certain inference problems, pseudolikelihood can be manipulated somewhat, in the same way as ordinary likelihood, to get reasonable procedures. An important instance is the group of estimation procedures suggested by Hodges and Lehmann [2] for parameters of location and shift. A similar procedure was independently suggested by Sen [4] for estimating the relative potency in a problem of bioassay. All these estimates can be interpreted as maximum pseudolikelihood estimates for suitably defined pseudolikelihood functions. The object of this paper is to explore the usefulness of the pseudolikelihood function for finding reasonable solutions to certain nonsequential two-decision problems.

2 DECISION RULES FOR CERTAIN TWO-DECISION PROBLEMS

Suppose, under the set-up of §1, our object is to make one of two given decisions d_1 and d_2 on the basis of an observation on $\mathbf{X}$. Also suppose that Ω is divisible into two disjoint sets Ω_1, Ω_2 such that for $\theta \in \Omega_i$, d_i is the correct decision, $i = 1, 2$. Any decision rule here is representable by a function $\delta(\mathbf{x})\,(0 \leqslant \delta(\mathbf{x}) \leqslant 1)$ giving the probability of choosing d_1 when the observation is $\mathbf{x}$.

Now suppose it is possible to find two functions $v_1(\mathbf{x})$ and $v_2(\mathbf{x})$ such that

(a) Whatever $\mathbf{x} \in R_n$, $v_1(\mathbf{x}), v_2(\mathbf{x})$ assume only the values

$$y_k, \quad k = 1, 2, \ldots, N, \quad y_1 < y_2 < \ldots < y_N,$$

(b) Whatever $\theta \in \Omega_i$, for $\mathbf{X}$ following the distribution $F_\theta(\mathbf{x})$, $v_i(\mathbf{X})$ follows the same discrete distribution with probability mass points y_k, $k = 1, 2, \ldots, N$ and known pmf $g(y)$, $i = 1, 2$.

Here, according to our definition, pseudolikelihood is $g[v_1(\mathbf{x})]$ in Ω_1 and $g[v_2(\mathbf{x})]$ in Ω_2. We propose to consider decision functions $\delta(\mathbf{x})$ which depend on $\mathbf{x}$ only through $v_1(\mathbf{x})$, $v_2(\mathbf{x})$ and satisfy

$$\left. \begin{aligned} \delta(\mathbf{x}) &= 1 && \text{for} \quad g[v_1(\mathbf{x})] > g[v_2(\mathbf{x})], \\ &= 0 && \text{for} \quad g[v_1(\mathbf{x})] < g[v_2(\mathbf{x})], \\ &= p_{k,l} && \text{for} \quad v_1(\mathbf{x}) = y_k, v_2(\mathbf{x}) = y_l, g(y_k) = g(y_l). \end{aligned} \right\} \quad (2.1)$$

A decision function such as (2.1) based on pseudolikelihood is formally similar to an admissible decision function based on ordinary likelihood. In the following, we show that certain decision functions satisfying (2.1) possess various desirable properties under suitable conditions.

Solutions of the form (2.1) to the two-decision problem are appropriate for situations when we are not interested in maintaining any predetermined 'level of significance' either in Ω_1 or Ω_2 but generally want a high probability for correct decision whatever θ. These also provide the basis for generalization to more complex multiple decision problems.

In the following, it will be sometimes convenient to write

$$\delta(\mathbf{x}) = \phi[v_1(\mathbf{x}), v_2(\mathbf{x})], \quad \text{where} \quad \phi(y, y'), y, y' = y_1, y_2, \ldots, y_N$$

is given by

$$\phi(y_k, y_l) = 1 \quad \text{for} \quad g(y_k) > g(y_l),$$
$$\left. \begin{aligned} &= 0 \quad \text{for} \quad g(y_k) < g(y_l), \\ &= p_{k,l} \quad \text{for} \quad g(y_k) = g(y_l). \end{aligned} \right\} \qquad (2.2)$$

Definition. We say that a discrete distribution with pmf $g(y)$ and mass points $y_1, y_2, \ldots, y_N$ is *single humped*, if there are two integers $m_1, m_2, 1 < m_1 \leqslant m_2 < N$, such that

$$g(y_1) < g(y_2) < \ldots < g(y_{m_1}),$$
$$\left. \begin{aligned} &g(y_{m_1}) = g(y_{m_1+1}) = \ldots = g(y_{m_2}), \\ &g(y_{m_2}) > g(y_{m_2+1}) > \ldots > g(y_N). \end{aligned} \right\} \qquad (2.3)$$

y_{m_1} and y_{m_2} are here called the least and largest modes. These, of course, may coincide.

We shall generally assume the following condition.

Condition A. The common distribution of $v_1(\mathbf{X})$ under Ω_1 and $v_2(\mathbf{X})$ under Ω_2, represented by the pmf $g(y)$, $y = y_1, \ldots, y_N$ is single humped.

We state another condition which will be sometimes assumed.

Condition B. Either (i) $v_1(\mathbf{x}) \geqslant v_2(\mathbf{x})$, for all $\mathbf{x}$, or (ii) $v_1(\mathbf{x}) \leqslant v_2(\mathbf{x})$ for all $\mathbf{x}$.

The numbers $p_{k,l}$ $(0 \leqslant p_{k,l} \leqslant 1)$ in (2.1) and (2.2) are to be defined only for $k, l : g(y_k) = g(y_l)$. We shall follow some conventions regarding these. Under condition A, for simplicity, we shall take

$$p_{k,l} = p_0 \quad \text{for} \quad m_1 \leqslant y_k, y_l \leqslant m_2, \qquad (2.4)$$

and also for $y_k \neq y_l$, $g(y_k) = g(y_l)$,

where $0 \leqslant p_0 \leqslant 1$ is a constant. We do not make any general prescription for $p_{k,k}$, $k < m_1$, $k > m_2$. But, whenever condition B holds, we take

$$
\left.
\begin{aligned}
p_{k,k} &= 1 \quad \text{for} \quad k < m_1 \\
&= 0 \quad \text{for} \quad k > m_2 \quad \text{under } B\text{(i).} \\
p_{k,k} &= 0 \quad \text{for} \quad k < m_1 \\
&= 1 \quad \text{for} \quad k > m_2 \quad \text{under } B\text{(ii).}
\end{aligned}
\right\}
\tag{2.5}
$$

Theorem 2.1. Under conditions A and B, a decision function $\delta(\mathbf{x})$ satisfying (2.1), (2.4), *and* (2.5) *is unbiased.*

Proof. We prove the theorem assuming B(i). When B(ii) holds the proof will be same with d_1, d_2 and Ω_1, Ω_2 interchanged. Consider any $\theta_1 \in \Omega_1$, $\theta_2 \in \Omega_2$. Writing $P(d_i|\theta_j, \delta)$ for the probability of choosing d_i according to the decision rule δ when F_{θ_j} is true ($i, j, = 1, 2$), we have

$$
P(d_1|\theta_j, \delta) = E\{\phi[v_1(\mathbf{X}), v_2(\mathbf{X})]|\theta_j\}
\tag{2.6}
$$

$$
= E\{\phi[v_j(\mathbf{X}), v_j(\mathbf{X})]|\theta_j\} + E\{\phi[v_1(\mathbf{X}), v_2(\mathbf{X})]
$$

$$
- \phi[v_j(\mathbf{X}), v_j(\mathbf{X})]|\theta_j\} \quad (j = 1, 2),
\tag{2.7}
$$

whence as

$$
E\{\phi[v_1(\mathbf{X}), v_1(\mathbf{X})]|\theta_1]|\theta_1\} = E\{\phi[v_2(\mathbf{X}), v_2(\mathbf{X})]|\theta_2\}
\tag{2.8}
$$

we have

$$
P(d_1|\theta_1, \delta) = P(d_1|\theta_2, \delta) + E\{\phi[v_1(\mathbf{X}), v_2(\mathbf{X})] - \phi[v_1(\mathbf{X}), v_1(\mathbf{X})]|\theta_1\}
$$

$$
+ E\{\phi[v_2(\mathbf{X}), v_2(\mathbf{X})] - \phi[v_1(\mathbf{X}), v_2(\mathbf{X})]|\theta_2\}.
\tag{2.9}
$$

By condition B(i), the pair $(v_1(\mathbf{x}), v_2(\mathbf{x}))$ can assume only the values (y_k, y_l), $l \leqslant k = 1, \ldots, N$ and by condition A, (2.2), (2.4) and (2.5),

$$
\left.
\begin{aligned}
\min_{l:l \leqslant k} \phi(y_k, y_l) &= \phi(y_k, y_k) \quad (k = 1, \ldots, N), \\
\max_{k:k \geqslant l} \phi(y_k, y_l) &= \phi(y_l, y_l) \quad (l = 1, \ldots, N).
\end{aligned}
\right\}
\tag{2.10}
$$

(For $k < m_1$, $l = 1, 2, \ldots, k$ and for $m_1 \leqslant k \leqslant m_2$, $l = 1, \ldots, m_1 - 1$, $\phi(y_k, y_l) = 1$; for $m \leqslant k \leqslant m_2$, $l = m_1, \ldots, k$, $\phi(y_k, y_l) = p_0 \ (\leqslant 1)$; and for $m_2 < k \leqslant N, \phi(y_k, y_k) = 0$. Hence the first equation follows. The second may similarly be verified.) By (2.10) the mean values appearing as the second and third members on the right of (2.9) are both non-negative. Hence we get

$$
P(d_1|\theta_1, \delta) \geqslant P(d_1|\theta_2, \delta)
\tag{2.11}
$$

which implies

$$
P(d_2|\theta_1, \delta) \leqslant P(d_2|\theta_2, \delta).
\tag{2.12}
$$

These inequalities are true whatever $\theta_1 \in \Omega_i$, $i = 1, 2$.

Note. In the above, the fact that $v_1(\mathbf{X})$ under θ_1 and $v_2(\mathbf{X})$ under θ_2 have the same distribution $g(y)$ has been utilized only in the step (2.8). If instead it were known that under B(i) there are two other functions $v_1'(\mathbf{x}) \geqslant v_1(\mathbf{x})$, and $v_2'(\mathbf{x}) \leqslant v_2(\mathbf{x})$ such that $v_1'(\mathbf{X})$ under θ_1 and $v_2'(\mathbf{X})$ under θ_2 have the same distribution, then also (2.11), (2.12) will hold. This is because $\phi(y_k, y_k) = 1 \ (k < m_1)$, $p_0 \ (m_1 \leqslant k \leqslant m_2)$, $0 \ (m_2 < k)$, being non-increasing in y_k,

$$E\{\phi[v_1(\mathbf{X}), v_1(\mathbf{X})]|\theta_1\} \geqslant E\{\phi[v_1'(\mathbf{X}), v_1'(\mathbf{X})]|\theta_1\}$$

$$= E\{\phi[v_2'(\mathbf{X}), v_2'(\mathbf{X})]|\theta_2\}$$

$$\geqslant E\{\phi[v_2(\mathbf{X}), v_2(\mathbf{X})]|\theta_2\}.$$

A similar observation applies if under B(ii) there are two functions

$$v_1'(\mathbf{x}) \leqslant v_1(\mathbf{x}), \quad v_2'(\mathbf{x}) \geqslant v_2(\mathbf{x}) \quad \text{as above.}$$

In certain applications the distribution $g(y)$ will be symmetric. There we have the following result.

Theorem 2.2. If the distribution $g(y)$ is symmetric, then under conditions A and B, the decision function $\delta(\mathbf{x})$ given by (2.1), (2.4) and (2.5) with $p_0 = \frac{1}{2}$ satisfies

$$P\{d_i|\theta_i, \delta\} \geqslant \tfrac{1}{2} \quad (i = 1, 2), \tag{2.13}$$

whatever $\theta_1 \in \Omega_1$, $\theta_2 \in \Omega_2$.

Proof. Let the center of symmetry of $g(y)$ be μ. Then under condition A the least and largest modes y_{m_1}, y_{m_2} will be located on the two sides of μ at equal distances. Hence, because of symmetry,

$$\sum_{k=1}^{m_1-1} g(y_k) + \frac{1}{2}\sum_{k=m_1}^{m_2} g(y_k) = \sum_{k=m_2+1}^{N} g(y_k) + \frac{1}{2}\sum_{k=m_1}^{m_2} g(y_k) = \tfrac{1}{2}. \tag{2.14}$$

Now assuming B(i) to be true, by (2.2), (2.4), (2.5) and (2.14), for $p_0 = \frac{1}{2}$,

$$P(d_1|\theta_1, \delta) = E\{\phi[v_1(\mathbf{X}), v_1(\mathbf{X})]|\theta_1\} + E\{\phi[v_1(\mathbf{X}), v_2(\mathbf{X})]$$

$$- \phi[v_1(\mathbf{X}), v_1(\mathbf{X})]|\theta_1\}$$

$$= \tfrac{1}{2} + E\{\phi[v_1(\mathbf{X}), v_2(\mathbf{X})] - \phi[v_1(\mathbf{X}), v_1(\mathbf{X})]|\theta_1\} \tag{2.15}$$

and similarly,

$$P(d_2|\theta_2, \delta) = \tfrac{1}{2} + E\{\phi[v_2(\mathbf{X}), v_2(\mathbf{X})] - \phi[v_1(\mathbf{X}), v_2(\mathbf{X})]|\theta_2\}. \tag{2.16}$$

By (2.10), from (2.15) and (2.16), (2.13) follows.

Note. In the usual situations, the second members on the right of both (2.15) and (2.16) would be substantially positive so that the actual values of $P(d_i|\theta_i, \delta), i = 1, 2$ would be considerably larger than $\tfrac{1}{2}$.

In certain situations, we can establish a one-one correspondence between the members of Ω_1 and Ω_2 such that if $\theta_1 \in \Omega_1, \theta_2 \in \Omega_2$ correspond to each other, then either F_{θ_1} or F_{θ_2} is true. Thus for the two-sample location problem considered in the illustration of section 1, if $\Omega_1 = \{\theta : \tau_\theta = \tau^{(1)}\}$, $\Omega_2 = \{\theta : \tau_\theta = \tau^{(2)}\}$, where $\tau^{(1)}, \tau^{(2)}$ are two fixed numbers, $\theta_1 \in \Omega_1$ corresponds to $\theta_2 \in \Omega_2$ if and only if $F^*_{\theta_1}(x) \equiv F^*_{\theta_2}(x)$. In such situations, sometimes it happens that, for a certain number μ, the joint distribution of $v_1(\mathbf{X}) - \mu, v_2(\mathbf{X}) - \mu$ under $\theta_1 \in \Omega_1$ is the same as that of $\mu - v_2(\mathbf{X}), \mu - v_1(\mathbf{X})$ under $\theta_2 \in \Omega_2$, whenever θ_1, θ_2 correspond to each other (the common joint distribution may depend on θ_1). As $v_1(\mathbf{X})$ under θ_1 has the same distribution $g(y)$ as $v_2(\mathbf{X})$ under θ_2, this implies $g(y)$ is symmetric about μ. Here we have the following result even without condition B.

Theorem 2.3. *If for some number μ the joint distribution of $v_1(\mathbf{X}) - \mu$, $v_2(\mathbf{X}) - \mu$ under $\theta_1 \in \Omega_1$ is the same as that of $\mu - v_2(\mathbf{X}), \mu - v_1(\mathbf{X})$ under $\theta_2 \in \Omega_2$, whenever θ_1, θ_2 correspond to each other, then, under condition A, for the decision function $\delta(\mathbf{X})$ given by (2.1) and (2.4) with*

$$p_{k,k} = 1 - p_{l,l} \quad \text{for} \quad y_l = 2\mu - y_k, \quad k = 1, \dots, N. \tag{2.17}$$

we have $\qquad\qquad P(d_1|\theta_1, \delta) = P(d_2|\theta_2, \delta),$ $\qquad\qquad$ (2.18)

for corresponding θ_1, θ_2, $\qquad \theta_i \in \Omega_i \quad (i = 1, 2)$.

Proof. We note that (2.17) implies that in (2.4) $p_0 = \tfrac{1}{2}$. In particular, (2.17) is satisfied, if further, $p_{k,k} = 1$ (or 0 or $\tfrac{1}{2}$) for $k < m_1$ and 0 (or 1, or $\tfrac{1}{2}$) for $k > m_2$.

Consider any corresponding pair $\theta_1, \theta_2, \theta_i \in \Omega_i, i = 1, 2$. Let V_1, V_2 be a pair of random variables following the same joint distribution as $v_1(\mathbf{X}), v_2(\mathbf{X})$ under θ_1. Then, under $\theta_2, v_1(\mathbf{X}), v_2(\mathbf{X})$ are jointly distributed as $2\mu - V_2, 2\mu - V_1$. Therefore,

$$P(d_1|\theta_1, \delta) = E\{\phi(V_1, V_2)\},$$

$$P((d_2|\theta_2, \delta) = 1 - E\{\phi(2\mu - V_2, 2\mu - V_1)\}.$$

As $g(y)$ is symmetric about μ, for any k $(1 \leqslant k \leqslant N)$, $g(y_k) = g(2\mu - y_k)$. Hence from (2.2) and (2.17),

$$\phi(y_k, y_l) = 1 - \phi(2\mu - y_l, 2\mu - y_k), k, l = 1, \dots, N.$$

Thus, whatever the joint distribution of V_1, V_2, (2.18) holds.

The common probability of correct decision given by (2.18) in general depends on the pair θ_1, θ_2.

In the above, condition B was assumed in Theorems 2.1 and 2.2. This condition, however, is rather restrictive. We now drop this and under a weaker set of assumptions prove that any decision rule δ satisfying (2.1) is consistent for every $\theta \in \Omega$.

As now we shall vary n we shall attach n as a subscript to most of the symbols used earlier. Thus we shall write

$$\mathbf{X}_n, F_{n\theta}(\mathbf{x}_n), g_n(y), y_{n1}, \ldots, y_{nN_n}, m_{n1}, m_{n2}, v_{n1}(x_n), v_{n2}(x_n), \delta_n, \text{ and } \phi_n.$$

However, we shall presume that $\Omega, \Omega_1, \Omega_2$ remain the same for all n. This will happen if, for each θ, $F_{n\theta}(\mathbf{x}_n)$ depends on a definite number of parent distributions which, for fixed θ, remain the same as n is varied. We use the symbol Y_n to denote a random variable following the discrete distribution represented by $g_n(y), y = y_{n1}, \ldots, y_{nN_n}$.

Theorem 2.4. Suppose condition A holds for every n and suppose there is a sequence of numbers $\{c_n\}, c_n \to \infty$ as $n \to \infty$, such that, as $n \to \infty$,

$$c_n^{-1} Y_n \overset{P}{\to} \eta_0, \quad c_n^{-1} m_{n1} \to \eta_0, \quad c_n^{-1} m_{n2} \to \eta_0, \qquad (2.19)$$

and for X_n following the distribution $F_\theta(\mathbf{x}_n)$,

$$\left. \begin{aligned} c_n^{-1} v_{n1}(\mathbf{X}_n) &\overset{P}{\to} \eta_1(\theta) \quad \text{if} \quad \theta \in \Omega_2, \\ c_n^{-1} v_{n2}(\mathbf{X}_n) &\overset{P}{\to} \eta_2(\theta) \quad \text{if} \quad \theta \in \Omega_1, \end{aligned} \right\} \qquad (2.20)$$

where η_0 is a fixed number and $\eta_1(\theta) \neq \eta_0, \theta \in \Omega_2, \eta_2(\theta) \neq \eta_0, \theta \in \Omega_1$. Then, if δ_n satisfies (2.1) we have

$$\lim_{n \to \infty} P(d_i | \theta, \delta_n) = 1 \quad \text{for} \quad \theta \in \Omega_i, i = 1, 2. \qquad (2.21)$$

Proof. Consider a fixed $\theta \in \Omega_1$, and suppose $\eta_2(\theta) < \eta_0$. By (2.19) and the second relation in (2.20), for any $\epsilon > 0$, we have

$$\left. \begin{aligned} P\{c_n(\eta_0 - \epsilon) < v_{n1}(\mathbf{X}_n) < c_n(\eta_0 + \epsilon) | \theta\} &\to 1, \\ P\{v_{n2}(\mathbf{X}_n) < c_n(\eta_2(\theta) + \epsilon) | \theta\} &\to 1. \end{aligned} \right\} \qquad (2.22)$$

Now let us choose an ϵ so that $\eta_0 - \epsilon > \eta_2(\theta) + \epsilon$. By (2.19), for sufficiently large n, $m_{n1} > c_n(\eta_0 - \epsilon)$, $m_{n2} < c_n(\eta_0 + \epsilon)$. As condition A holds for every n, this means for y_{nk} lying between $c_n(\eta_0 \pm \epsilon)$ and $y_{nl} < c_n(\eta_2(\theta) + \epsilon)$, $g_n(y_{nk}) > g_n(y_{nl}), k, l = 1, \ldots, N_n$. So, if δ_n satisfies (2.1),

$$P(d_1 | \theta, \delta_n) \geqslant P\{c_n(\eta_0 - \epsilon) < v_{n1}(\mathbf{X}_n) < c_n(\eta_0 + \epsilon),$$

$$v_2(\mathbf{X}_n) < c_n(\eta_2(\theta) + \epsilon) | \theta\}.$$

By (2.22), (2.21) follows for $i = 1$. Modifications required for $\eta_2(\theta) > \eta_0$ are obvious. The proof of (2.21) for $i = 2$ is similar.

3 LARGE SAMPLE APPROXIMATIONS—ROUNDING OFF

To use the decision rule (2.1) we require knowledge of $g(y)$ at $y_1, ..., y_N$. For the rule to possess the properties discussed in §2, $g(y)$ must satisfy condition A and certain other conditions embodied in Theorems 2.1 to 2.4. If $g(y)$ is explicitly known as a function of y these conditions may be verified and the rules (2.1), (2.4), and (2.5) may be simplified. This, unfortunately, is rarely true for the usual nonparametric statistics. However, in a few cases, although the functional form of $g(y)$ is intractable, it is known that $g(y)$ is singlehumped and symmetric about a point μ, with one or two modes at the center. There $g(y_k) >, =,$ or $< g(y_l)$ according as $|y_k - \mu| <, =,$ or $> |y_l - \mu|$ so that with appropriate conventions we may formulate the rule (2.1) in terms of the differences $|v_1(\mathbf{X}) - \mu|$ and $|v_2(\mathbf{X}) - \mu|$. When symmetry of $g(y)$ cannot be presumed, if a table of values of $g(y)$ is available for the sample size considered there is no difficulty. For most nonparametric statistics, however, no exact tables are available for large n and only the limiting form of the distribution is known. We now consider how to draw upon this knowledge to make the proposed decision rule usable for large samples.

Let Y_n stand for a random variable which follows the distribution represented by the pmf $g_n(y)$ over the mass points $y_{n1}, y_{n2}, ..., y_{nN}$ (notation as in the latter part of §2). We consider only situations where normalizing constants $a_n(> 0)$, b_n are available so that as $n \to \infty$,

$$P\{a_n Y_n + b_n < z\} \to H(z), \tag{3.1}$$

$H(z)$ being an absolutely continuous cdf with density, say, $h(z)$. We write

$$g_n^*(z) = g_n[a_n^{-1}(z - b_n)] \quad (z_{nk} = a_n y_{nk} + b_n, k = 1, ..., N_n), \tag{3.2}$$

for the pmf and the mass points of $Z_n = a_n Y_n + b_n$. If

$$v_{ni}^*(\mathbf{x}_n) = a_n v_{ni}(\mathbf{x}_n) + b_n \quad (i = 1, 2),$$

then clearly the decision function $\delta_n(\mathbf{x}_n)$ given by (2.1) may be described as

$$\left. \begin{aligned} \delta_n(\mathbf{x}_n) &= 1 \quad \text{for} \quad g_n^*[v_{n1}^*(\mathbf{x}_n)] > g_n^*[v_{n2}^*(\mathbf{x}_n)], \\ &= 0 \quad \text{for} \quad g_n^*[v_{n1}^*(\mathbf{x}_n)] < g_n^*[v_{n2}^*(\mathbf{x}_n)], \\ &= p_{k,l} \quad \text{for} \quad v_{ni}^*(\mathbf{x}_n) = z_{nk}, \quad v_{n2}^*(\mathbf{x}_n) = z_{nl}, \quad g_n^*(z_{nk}) = g_n^*(z_{nl}). \end{aligned} \right\}$$

$$\tag{3.3}$$

The main difficulty in finding a large sample approximation to (3.3) is that validity of (3.1) is not in itself sufficient ground for approximating $g_n^*(z_{nk})/g_n^*(z_{nl})$ by $h(z_{nk})/h(z_{nl})$. For this, one would require local limit theorems of the type obtained by Gnedenko [1, p. 297]. No such local limit theorems are, however, available for most of the standard nonparametric statistics. To circumvent this difficulty we shall now replace $v_{ni}^*(\mathbf{x}_n)$, $i = 1, 2$, by corresponding *rounded off* versions defined as follows.

For fixed numbers $\epsilon(> 0), \mu$, and integers N_1, N_2 we write

$$\tilde{y}_k = \mu + k\epsilon \quad (k = -N_1, -N_1 + 1, \ldots, N_2 - 1, N_2), \qquad (3.4)$$

and define the rounding off function $R(z)$ by

$$
\begin{aligned}
R(z) &= \tilde{y}_k, && \text{for} \quad \tilde{y}_k - \tfrac{1}{2}\epsilon < z \leqslant \tilde{y}_k + \tfrac{1}{2}\epsilon \quad (k = -N_1 + 1, \ldots, N_2 - 1), \\
&= \tilde{y}_{-N_1}, && \text{for} \quad z \leqslant \tilde{y}_{-N_1} + \tfrac{1}{2}\epsilon, \\
&= \tilde{y}_{N_2}, && \text{for} \quad z > \tilde{y}_{N_2} - \tfrac{1}{2}\epsilon.
\end{aligned}
$$

$$(3.5)$$

We now write

$$\tilde{v}_{n1}(\mathbf{x}_n) = R[v_{n1}^*(\mathbf{x}_n)], \quad \tilde{v}_{n2}(\mathbf{x}_n) = R[v_{n2}^*(\mathbf{x}_n)],$$

and call these the rounded off versions of v_{n1}, v_{n2}. $\tilde{v}_{n1}$ and $\tilde{v}_{n2}$ assume only the values (3.4). Also $\tilde{v}_{n1}(\mathbf{X}_n)$ under Ω_1 and $\tilde{v}_{n2}(\mathbf{X}_n)$ under Ω_2 follow the same discrete distribution over these values with pmf say $\tilde{g}_n(y)$, $y = \tilde{y}_k$, $k = -N_1, \ldots, N_2$.

Clearly, we can define the pseudolikelihood in terms of $\tilde{v}_{n1}, \tilde{v}_{n2}$ and $\tilde{g}(y)$ and formulate the rule (2.1) on the basis of that. Such a decision rule is particularly easy to handle in large samples as known limit theorems apply to them smoothly. From the practical point of view also replacement of v_{ni} by $\tilde{v}_{ni}$, $i = 1, 2$, is not unrealistic as rounding off of statistics is usual in practice for large sample sizes (and in certain situations, e.g. for normal scores tests, is resorted to even in small samples).

If condition A holds for v_{ni} it will hold for $\tilde{v}_{ni}$ as well, under certain additional restrictions. (For instance, if the range

$$(\tilde{y}_{-N_1} - \tfrac{1}{2}\epsilon, \quad \tilde{y}_{N_2} + \tfrac{1}{2}\epsilon)$$

contains all the z_{nk}'s, each interval $\tilde{y}_k \pm \tfrac{1}{2}\epsilon$ contains at least one z_{nk}, and the number of z_{nk}'s contained in any such interval is either constant or progressively decreases on both sides of some central interval located in the modal range of $g_n^*(z)$, $g_n(y)$ being single humped, $\tilde{g}_n(y)$ would also be.) However, for large n, the validity of condition A for $\tilde{g}_n(y)$ usually follows from simpler considerations. As $n \to \infty$, by (3.1) we have

$$
\left.
\begin{aligned}
\tilde{g}_n(\tilde{y}_k) &\to H(\tilde{y}_k + \tfrac{1}{2}\epsilon) - H(\tilde{y}_k - \tfrac{1}{2}\epsilon) = h_k \quad \text{for} \quad k = -N_1+1, \ldots, N_2-1, \\
&\to H(\tilde{y}_{-N_1} + \tfrac{1}{2}\epsilon) = h_{-N_1} \quad \text{for} \quad k = -N_1, \\
&\to 1 - H(\tilde{y}_{N_2} - \tfrac{1}{2}\epsilon) = h_{N_2} \quad \text{for} \quad k = N_2,
\end{aligned}
\right\}
$$

(3.6)

uniformly in $k = -N_1, -N_1+1, \ldots, N_2$. So if $h(z)$ is a unimodal density, $\tilde{g}_n(\tilde{y}_k), k = -N_1, \ldots, N_2$ will represent a single humped distribution for sufficiently large n, irrespective of whether $g_n(y)$ is single humped or not.

If condition B holds for v_{ni} it will hold for $\tilde{v}_{ni}$. Also under mild and obvious restrictions on the choice of a_n, b_n, μ, if the other requirements of Theorems 2.1 to 2.4 hold for v_{ni}, these will hold for $\tilde{v}_{ni}$ as well.

Thus if the original decision rule (2.1) in terms of v_{ni} possesses any of the properties discussed in §1, under essentially the same conditions, the same rule with $\tilde{v}_{ni}$ replacing v_{ni}, $i = 1, 2$ will also possess these properties, at least for large n. In view of (3.6) in defining the rule in terms of $\tilde{v}_{ni}$, for large n, we may replace $\tilde{g}_n(\tilde{y}_k)$ by h_k, $k = -N_1, \ldots, N_2$.

In reality $h(z)$ may be symmetric even though $g_n(y)$ is not so for finite n. Then for suitably chosen $\mu, N_1, N_2, g_n(y), y = \tilde{y}_k, k = -N_1, \ldots, N_2$ will be approximately symmetric and therefore will satisfy the conditions of Theorem 2.2 (and may be 2.3 also) to a close approximation for large n. What is more, there we can formulate the decision rule in terms of

$$
|\tilde{v}_{n1}(\mathbf{x}_n) - \mu| \quad \text{and} \quad |\tilde{v}_{n2}(\mathbf{x}_n) - \mu|
$$

(assuming μ coincides with the center of symmetry of the distribution).

4 APPLICATIONS

4.1. Two-sample location problem

Suppose $n = n_1 + n_2$, and for each $\theta \in \Omega$, there is a univariate cdf $F_\theta^*(x)$ and a real number τ_θ such that

$$
F_\theta(\mathbf{x}) = \prod_1^{n_1} F_\theta^*(x_i) \prod_1^{n_2} F_\theta^*(x_{n_1+j} - \tau_\theta).
$$

Also suppose there are just two possible values $\tau^{(1)}, \tau^{(2)}$ of τ_θ. Our problem is to decide whether $\tau_\theta = \tau^{(1)} (d_1)$ or $\tau^{(2)}(d_2)$. Obviously

$$
\Omega_i = \{\theta : \tau_\theta = \tau^{(i)}\} \quad (i = 1, 2).
$$

For any two sets of numbers $z_1, \ldots, z_{n_1}; z_{n_1+1}, \ldots, z_n$ let

$$
T_q = T_q(z_1, \ldots, z_{n_1}; z_{n_1+1}, \ldots, z_n) \quad (q = 1, 2, 3),
$$

(4.1)

be respectively the Wilcoxon, the normal scores, and the Mood statistic corresponding to first sample observations $z_i, i = 1, \ldots, n_1$ and second sample observations $z_{n_1+j}, j = 1, \ldots, n_2$. If the ordered ranks of z_{n_1+j}, $j = 1, \ldots, n_2$ in the pooled sample are $s_1 < s_2 < \ldots < s_{n_2}$,

$$T_q = \sum_{j=1}^{n_2} u_q(s_j), \qquad (4.2)$$

where, writing ν_{nk} for the expected kth order statistic of a sample of size n from $N(0, 1), k = 1, \ldots, n$

$$\left. \begin{aligned} u_1(k) &= k, \\ u_2(k) &= \nu_{nk}, \\ u_3(k) &= 0, \tfrac{1}{2}, \text{ or } 1 \text{ according as } k < =, \text{ or } > \frac{n+1}{2}. \end{aligned} \right\} \qquad (4.3)$$

We now define

$$v_i(\mathbf{x}) = T_q(x_1, \ldots, x_{n_1}; x_{n_1+1} - \tau^{(i)}, \ldots, x_n - \tau^{(i)}) \quad (i = 1, 2), \qquad (4.4)$$

where q can be taken to be 1, 2, or 3. $v_i(x), i = 1, 2$, clearly satisfy the conditions A, B of §2. For any q let the pmf of the discrete distribution of $v_i(\mathbf{X})$ under Ω_i be $g_q(y)$ with mass points $y_{k(q)}, k = 1, \ldots, N_{(q)}$. As all possible choices of $s_1, \ldots, s_{n_2}$ are equally likely, by (4.2) and (4.3), $g_q(y)$ is symmetric about $n_2 \bar{u}_q$ where $\bar{u}_q = n^{-1} \sum_{k=1}^{n} u_q(k) = \tfrac{1}{2}(n+1), 0, \tfrac{1}{2}$ for $q = 1, 2, 3$ respectively. As for condition A, that $g_3(y)$ is single humped may be seen from its expression. For low values of n the same may be verified for $g_1(y)$ from available tables (see Owen [3], p. 331). (It may be possible to deduce this theoretically.) For moderately large n, from well-known results on the asymptotic normality of T_q, validity of condition A at least for the rounded off versions of $v_i(\mathbf{x})$ is assured for $q = 1, 2, 3$. Condition B, of course, holds for all the three choices. Therefore, here Theorems 2.1 and 2.2 are true for the decision rule (2.1). By well known results on the stochastic convergence of statistics of the form (4.2) consistency as in Theorem 2.4 holds for the decision rule.

For the problem considered the conditions required for the validity of Theorem 2.3 also hold provided either (a) $n_1 = n_2$, or (b) $F_\theta^*(x)$ is symmetric about 0. If $Z_1, \ldots, Z_n$ is a sample from $F_\theta^*(x)$, when $\tau_\theta = \tau^{(1)}$, $v_1(\mathbf{X})$, $v_2(\mathbf{X})$ are distributed as

$$T_q(Z_1, \ldots, Z_{n_1}; Z_{n_1+1}, \ldots, Z_n) \quad \text{and} \quad T_q(Z_1, \ldots, Z_{n_1}; Z_{n_1+1} - \tau, \ldots, Z_n - \tau),$$

$$\tau = \tau^{(2)} - \tau^{(1)}.$$

Under (a), when $\tau_\theta = \tau^{(2)}$, $v_1(\mathbf{X})$, $v_2(\mathbf{X})$ are distributed respectively as

$$T_q(Z_1, ..., Z_{n_1}; Z_{n_1+1} + \tau, ..., Z_n + \tau)$$

$$= n\bar{u}_q - T_q(Z_{n_1+1}, ..., Z_n; Z_1 - \tau, ..., Z_{n_1} - \tau),$$

and

$$T_q(Z_1, ..., Z_{n_1}; Z_{n_1+1}, ..., Z_n) = n\bar{u}_q - T_q(Z_{n_1+1}, ..., Z_n; Z_1, ..., Z_{n_1}).$$

As $n_1 = n_2$, the requisite condition is met. Under (b), when

$$\tau_\theta = \tau^{(2)},$$

$v_1(\mathbf{X})$, $v_2(\mathbf{X})$ are distributed respectively as

$$T_q(-Z_1, ..., -Z_{n_1}; -Z_{n_1+1} + \tau, ..., -Z_n + \tau)$$

$$= 2n_2\bar{u}_q - T_q(Z_1, ..., Z_{n_1}; Z_{n_1+1} - \tau, ..., Z_n - \tau),$$

and

$$T_q(-Z_1, ..., -Z_{n_1}; -Z_{n_1+1}, ..., -Z_n)$$

$$= 2n_2\bar{u}_q - T_q(Z_1, ..., Z_{n_1}; Z_{n_1+1}, ..., Z_n).$$

Thus if either (a) or (b) is true we can ensure the same probability of correct decision for both $\tau_\theta = \tau^{(1)}$, $\tau^{(2)}$ for every $F_\theta^*(x)$. (This common probability depends on $F_\theta^*(x)$ as well as $\tau^{(2)} - \tau^{(1)}$.)

We might use the above procedure even for the extended problem where

$$d_1 : \tau_\theta \leqslant \tau^{(1)}, \quad d_2 : \tau_\theta \geqslant \tau^{(2)} \quad (\text{for } \tau^{(1)} < \tau^{(2)}).$$

By the note under Theorem 2.1 the proposed rule would then be unbiased whatever τ_0 ($\leqslant \tau^{(1)}$ or $\geqslant \tau^{(2)}$).

4.2. Single-sample location problem

In this case for each $\theta \in \Omega$, we have a univariate cdf F_θ^* symmetric about 0 and a location parameter τ_θ such that

$$F_\theta(\mathbf{x}) = \prod_1^n F_\theta^*(x_i - \tau_\theta).$$

The problem is to choose between two possible values $\tau^{(1)}$, $\tau^{(2)}$ of τ_θ. We may take $v_i(\mathbf{x}) = S_q(x_1 - \tau^{(i)}, ..., x_n - \tau^{(i)})$, $i = 1, 2$, where for $q = 1, 2, 3$, $S_q(z_1, ..., z_n)$ are respectively the single sample signed rank, normal scores and the sign test statistic. Almost the same considerations as in the two-sample problem apply here. Here the probability of a correct decision is the same for $\tau_\theta = \tau^{(1)}$, $\tau^{(2)}$.

4.3. Two-sample scale problem for given location

Suppose $n = n_1 + n_2$ and for each θ there is a cdf $F_\theta^*(x)$ with median at 0 and a scale factor β_θ such that

$$F_\theta(\mathbf{x}) = \prod_1^{n_1} F_\theta^*(x_i) \prod_1^{n_2} F_\theta^*(\beta_\theta x_{n_1+j}),$$

and suppose we want to decide whether $\beta_\theta = \beta^{(1)}$ or $\beta^{(2)}$. We may take any standard statistic for scale test to define $v_1(\mathbf{x})$, $v_2(\mathbf{x})$. For instance, if $T(z_1, ..., z_{n_1}; z_{n_1+1}, ..., z_n)$ is the Siegel–Tukey statistic for sample observations $z_1, ..., z_{n_1}$ and $z_{n_1+1}, ..., z_n$, we might take

$$v_i(\mathbf{x}) = T(x_1, ..., x_{n_1}; \beta^{(i)}x_{n_1+1}, ..., \beta^{(i)}x_n) \quad (i = 1, 2).$$

Condition B does not hold here. However, Theorem 2.3 holds if $n_1 = n_2$.

4.4. Three-sample classification problems

Suppose for each θ we have a triplet of continuous cdf's $(F_{1\theta}, F_{2\theta}, F_{3\theta})$ of which $F_{1\theta} \not\equiv F_{2\theta}$, and $F_{3\theta} =$ either $F_{1\theta}$ or $F_{2\theta}$.

$$\Omega_1 = \{\theta : F_{1\theta} \equiv F_{3\theta}\}, \quad \Omega_2 = \{\theta : F_{2\theta} \equiv F_{3\theta}\}.$$

Given independent samples of equal sizes from $F_{1\theta}$ and $F_{2\theta}$, and a third sample from $F_{3\theta}$, say

$$X_1, ..., X_{n_1}; \quad Y_1, ..., Y_{n_1}; \quad Z_1, ..., Z_{n_2}$$

the problem is to identify the third population.

Our approach to the problem will depend on the extent of our knowledge about the mode of divergence of $F_{1\theta}, F_{2\theta}$. Thus if we know $F_{1\theta}, F_{2\theta}$ differ only in location, i.e. $F_{2\theta}(x) = F_{1\theta}(x - \tau_\theta)$, where τ_θ is unknown, we may take for $q = 1, 2,$ or 3,

$$v_1 = T_q(x_1, ..., x_{n_1}; z_1, ..., z_{n_2}),$$

$$v_2 = T_q(y_1, ..., y_{n_1}; z_1, ..., z_{n_2}),$$

where T_q for $q = 1, 2, 3$ has the same interpretation as in Example 4.1. The same v_1, v_2 may be used when it is known that $F_{1\theta} \not\equiv F_{2\theta}$, and either $F_{1\theta}(x) \geqslant F_{2\theta}(x)$ for all x, or $F_{1\theta}(x) \leqslant F_{2\theta}(x)$ for all x. Similarly v_1, v_2 may be constructed from the Siegel–Tukey statistics based on the first and third, and second and third samples if $F_{1\theta}, F_{2\theta}$ are known to have the same location but different scale parameters. In general, when nothing more than $F_{1\theta} \not\equiv F_{2\theta}$ is known, we may use, for instance, the Kolmogorov–Smirnov statistic between the first and third, and second and third samples as v_1, v_2 respectively. Condition A will then hold for the rounded off statistics by virtue of the known unimodal form of the limiting null distribution of this statistic. For none of the solutions for the problem considered here is condition B satisfied.

5 EFFICIENCY INVESTIGATION IN A PARTICULAR CASE

In this section we study the asymptotic efficiency of the decision rule (2.1) for the two-sample location problem of Example 4.1 confining ourselves to the choice of v_1, v_2 corresponding to the Mann–Whitney statistic ($q = 1$). We keep $F_\theta^*(x) = F^*(x)$ fixed and put $\tau^{(1)} = 0$. To keep the probabilities of correct decision bounded away from 1 we take $\tau^{(2)} = n^{-\frac{1}{2}}\eta$, where $\eta \neq 0$ is a fixed number. We suppose $\eta > 0$.

If $u(z_1, z_2) = \pm \frac{1}{2}$ according as $z_1 \leqslant$ or $> z_2$, we can equivalently write

$$\left. \begin{aligned} v_{n1}(\mathbf{X}_n) &= \frac{1}{n_1 n_2} \sum_{i=1}^{n_1} \sum_{j=1}^{n_2} u(x_i, x_{n_1+j}), \\ v_{n2}(\mathbf{X}_n) &= \frac{1}{n_1 n_2} \sum_{i=1}^{n_1} \sum_{j=1}^{n_2} u(x_i, x_{n_1+j} - n^{-\frac{1}{2}}\eta). \end{aligned} \right\} \quad (5.1)$$

Clearly then $v_{n2} \leqslant v_{n1}$ and the distribution $g_n(y)$ (defined in §2) is symmetric about 0. Ignoring possibilities like $v_{n1} = \pm v_{n2}$ as these will have negligible probability for large n, we can describe the decision rule δ_n given by (2.1) as:

$$\left. \begin{aligned} \text{for} \quad v_{n1}+v_{n2} &\leqslant 0 \quad \text{choose } d_1, \\ &> 0 \quad \text{choose } d_2. \end{aligned} \right\} \quad (5.2)$$

By well known results on U-statistics, it may be directly verified from (5.1) that as

$$n \to \infty \quad \text{so that} \quad \frac{n_i}{n} \to \gamma_i \quad (0 < \gamma_i < 1, i = 1, 2, \gamma_1+\gamma_2 = 1), \quad (5.3)$$

for both $\tau_\theta = 0$ and $\tau_\theta = n^{-\frac{1}{2}}\eta$,

$$n^{\frac{1}{2}}v_{n1}(\mathbf{X}_n) - n^{\frac{1}{2}}v_{n2}(\mathbf{X}_n) \xrightarrow{P} \eta\lambda(F^*) \quad (5.4)$$

where $$\lambda = \lambda(F^*) = \int_{-\infty}^{\infty} [f^*(x)]^2 dx \quad (5.5)$$

$f^*(x)$ being the density of $F^*(x)$ which is now assumed to be absolutely continuous. Therefore, for large n, (5.2) reduces to:

$$\left. \begin{aligned} \text{for} \quad 2n^{\frac{1}{2}}v_{n1}(\mathbf{X}_n) - \eta\lambda &\leqslant 0 \quad \text{choose } d_1, \\ &> 0 \quad \text{choose } d_2. \end{aligned} \right\} \quad (5.6)$$

Again, by the properties of U-statistics under (5.3), $n^{\frac{1}{2}}v_{n1}(\mathbf{X}_n)$ is asymptotically normal under both $\tau_\theta = 0$ and $\tau_\theta = n^{-\frac{1}{2}}\eta$ with the same variance

$(12\gamma_1\gamma_2)^{-\frac{1}{2}}$ and means, 0 and $\eta\lambda$ respectively. Hence, writing $H(x)$ for the cdf of the distribution $N(0, 1)$,

$$\lim P(d_1|\tau_\theta = 0, \delta_n) = P(d_2|\tau_\theta = n^{-\frac{1}{2}}\eta, \delta_n) = H(\eta\lambda[3\gamma, \gamma_2]^{\frac{1}{2}}). \quad (5.7)$$

Now let $F^*(x)$ be a cdf for which the second moment exists and let μ^*, σ^* denote its mean and variance. A reasonable decision rule here would be:

$$\text{for} \quad t_n = (\bar{x}_{(1)} - \bar{x}_{(2)})\,\sigma_*^{-1}\left(\frac{1}{n_1} + \frac{1}{n_2}\right)^{-\frac{1}{2}} \begin{cases} \geqslant K, & \text{choose} \quad d_1, \\ < K, & \text{choose} \quad d_2, \end{cases} \quad (5.8)$$

where $\quad \bar{x}_{(1)} = n_1^{-1}\sum_1^{n_1} x_i, \quad \bar{x}_{(2)} = n_2^{-1}\sum_1^{n_2} x_{n_2+j}$

and K is some constant. In the particular case when $F^*(x)$ is the normal cdf a standard admissible decision rule (that corresponding to the uniformly most powerful unbiased t-test) becomes asymptotically equivalent to (5.8). We denote the decision rule (5.8) by Δ_n.

Let $n'(n), n_1'(n), n_2'(n)$ $(n_1'(n) + n_2'(n) = n'(n))$ be sequences of integers such that as $n \to \infty$

$$n'(n), \quad \frac{n_1'(n)}{n'(n)} \to \gamma_1, \quad \frac{n_2'(n)}{n'(n)} \to \gamma_2, \quad \frac{n'(n)}{n} \to \rho, \quad (5.9)$$

where γ_1, γ_2 are as in (5.3). We denote the statistic (5.8) based on samples of sizes $n_1'(n)$, $n_2'(n)$ by $t'_{(n)}$ and the corresponding decision rule by $\Delta'_{(n)}$. Under (5.9), for both $\tau_\theta = 0$ and $\tau_\theta = n^{-\frac{1}{2}}\eta$, $t'_{(n)}$ is asymptotically normal with variance 1 and means 0 and $-\eta(\rho\gamma_1\gamma_2)^{\frac{1}{2}}.\,\sigma^{*-1}$ respectively. Hence, we get

$$\lim P(d_1|\tau_\theta = 0, \quad \Delta'_{(n)}) = 1 - H(K),$$

$$\lim P(d_2|\tau_\theta = n^{-\frac{1}{2}}\eta, \quad \Delta'_{(n)}) = H\left(K + \frac{n}{\sigma^*}(\rho\,\gamma_1\gamma_2)^{\frac{1}{2}}\right). \quad (5.10)$$

As, for δ_n, the limiting probabilities of correct decision given by (5.7) are equal, here also we take $K = -\frac{1}{2}\eta(\rho\gamma_1\gamma_2)^{\frac{1}{2}}\sigma^{*-1}$ so that the two limiting probabilities in (5.10) become equal with common value $H(\frac{1}{2}\eta(\rho\gamma_1\gamma_2)^{\frac{1}{2}}\sigma^{*-1})$. This is equal to (5.7) if

$$\rho = 12\sigma^2\{\lambda(F^*)\}^2, \quad (5.11)$$

which in Pitman's sense is the efficiency of δ_n relative to Δ_n. (5.11) is same as the familiar expression for the Pitman efficiency of the Wilcoxon–Mann-Whitney test relative to the mean test. The comparison between δ_n and Δ_n is particularly relevant when $F^*(x)$ is normal since in this case Δ_n is equivalent to an admissible procedure in large samples. There the value of ρ given by (5.11) and (5.5) is $3/\pi$.

Finally, we note that in small samples, for a particular $F^*(x)$, the rule (5.2) is not admissible even within the class of rules based on v_{n1}, v_{n2}. This is because any such admissible rule should be formulated in terms of the joint pmf's of v_{n1}, v_{n2} under $\tau_\theta = 0$ and $\tau_\theta = \tau$. (Such a rule will of course depend on F^*.) But by (5.4), the rule (5.2), which does not depend on F^* becomes equivalent to an admissible rule in large samples provided as n becomes large τ becomes correspondingly small.

6 CONCLUDING REMARKS

The decision rule proposed in §2 on the basis of pseudolikelihood is not distribution-free. No operating characteristic of the rule is maintained at a fixed level either in Ω_1 or in Ω_2. The decision rule is, however, generally applicable, and as the preceding sections show, it possesses some desirable properties in a broad class of situations.

In the parametric two-decision problem involving a simple hypothesis and a simple alternative, decision is based on the ratio of the two likelihoods. For an admissible rule, one or the other decision is taken according to whether this ratio exceeds or falls short of a positive constant which may be arbitrarily fixed. For the decision rule we have proposed, the counter part of this constant was rigidly taken as unity. However, the situation is not as inflexible as it appears because the functions v_1, v_2 are at our choice. We can replace v_1, v_2 by any transform (not monotonic) $u(v_1), u(v_2)$ and get a different decision rule provided condition A holds for the new set of functions as well. In fact it is even possible to replace $g(y)$ by $g^*(y) = c(y)\, g(y)$, where $c(y) > 0$ for the same v_1, v_2. Provided $g^*(y)$ is single-humped, under the conditions of Theorem 2.1, the decision rule (2.1) with $g^*[v_i(\mathbf{x})]$ for $g[v_i(\mathbf{x})]$, $i = 1, 2$, is still unbiased. For an arbitrary choice of $c(y)$ of course the study of asymptotic properties, etc., would be difficult.

REFERENCES

[1]   Gnedenko, B. V. (1963). *The Theory of Probability* (English Translation), 2nd edition. NewYork: Chelsea.

[2]   Hodges, J. L., Jr. and Lehmann, E. L. (1963). Estimates of location based on rank tests. *Ann. math. Statist.* **34**, 598–611.

[3]   Owen, D. B. (1962). *Handbook of Statistical Tables*. London: Pergamon and Addison-Wesley.

[4]   Sen, P. K. (1963). On the estimation of relative potency in dilution (-direct) assays by distribution-free methods. *Biometrics* **19**, 532–52.

DECISION-THEORETIC EVALUATION OF SOME NONPARAMETRIC METHODS†

HERMAN RUBIN

1 INTRODUCTION

In this paper we shall evaluate for 'large' and for 'moderately large' samples the efficiency of some nonparametric methods, in particular, those of the Kolmogorov–Smirnov type, for both the one-sided and the two-sided testing problems. We also discuss, in general, some of the problems of moderately large samples.

In the two-sided case, the Kolmogorov–Smirnov (K–S) and Kuiper tests are asymptotically as efficient as the median for location parameters in the symmetric unimodal case. We have compared, numerically, the relative efficiency of the K–S test to the median for double-exponential and uniform alternatives for 'reasonable' sample sizes, and we find a slow approach to 1.

For the one-sided case, the situation is different. Here the efficiency depends on the specific test used. The analysis of this situation indicates that a better test statistic than the one-sided K–S test is to use the *difference* of the positive and negative deviations.

2 PRELIMINARIES

For both the one-sided and two-sided tests, we assume that the sample sizes are sufficiently large that the asymptotic distributions are sufficiently good approximations. Also, for the two-sided case we assume that the sample is sufficiently large that the weight measure (loss function times prior probability measure) is approximately proportional to δ_θ for rejection of the null hypothesis and to $|\theta|^q d\theta$ for acceptance, our calculations here are only carried out for $q = 0$. For the one-sided case we assume that the weight function is approximately proportional to $|\theta| \, d\theta$ for either kind of error; the author has not been able to find a practical example of any other kind.

† This research was supported in part by the Office of Naval Research under Contract N00014-67-A-0226-0008. Reproduction in whole or in part is permitted for any purpose of the United States Government.

We assume that the scale has been chosen so that the density at the median is $\frac{1}{2}$, and we define

$$X_+ = 2\sqrt{n}\sup\,(F_n(x) - F(x)),$$

$$X_- = 2\sqrt{n}\sup\,(F(x) - F_n(x)),$$

$$X_0 = 2\sqrt{n}\,(\tfrac{1}{2} - F_n(0)),$$

$$\phi = \sqrt{n}\,\theta,$$

where the null hypothesis is $\theta = 0$.

With this normalization, the median for small θ is approximately $X_0/\sqrt{n}$, and X_0 is approximately normal with mean ϕ and variance 1. Also, under the null hypothesis, by the usual methods the joint cdf of X_+ and X_- is found to be [1] for $a, b \geqslant 0$,

$$P(X_+ \leqslant a, X_- \leqslant b) = \sum_{-\infty}^{\infty} \exp - [\tfrac{1}{2}n^2(a+b)^2]$$

$$- \sum_{-\infty}^{\infty} \exp - [\tfrac{1}{2}(na + (n-1)b)^2]. \quad (1)$$

Under the alternative ϕ,

$$2\sqrt{n}(F_n(x) - F(x)) = 2\sqrt{n}\,(G_n(x-\theta) - F(x-\theta) + F(x-\theta - F(x))$$

$$= Y_n(x-\theta) - 2\phi f(x-\lambda)\theta$$

$$\sim Y_n(x) - 2\phi f(x), \quad (2)$$

and so

$$2\sqrt{n}\,(F_n F^{-1}(t) - (t) \sim X_n(t) - 2\phi f F^{-1}(t). \quad (3)$$

In the double-experimental case $2fF^{-1}(t) = 1 - 2|t - \tfrac{1}{2}|$; in the uniform case $2fF^{-1}(t) = 1$; and in general, $2fF^{-1}(\tfrac{1}{2}) = 1$, $2fF^{-1}(t) \geqslant 0$ and unimodal at $\tfrac{1}{2}$. We will for certain purposes consider an 'extreme' distribution with density $\frac{1}{2}$ 'at θ' and 'θ' elsewhere. In the sequel we shall use h for $2fF^{-1}$.

For the double-exponential case, we can find the limiting distribution of the maximum and minimum of X_+ and X_- given X_0 as

$$P(X_+ \leqslant a, X_- \leqslant b \,|\, X_0 = q) = \left(\sum_{-\infty}^{\infty} \exp - [n^2(a+b)^2 - n(a+b)\,q] \right.$$

$$\left. - \sum_{-\infty}^{\infty} \exp - [(na + (n-1)b)^2 - (na + (n-1)b)\,q] \right)^2 \quad (4)$$

independent of ϕ.

3 THE ONE-SIDED CASE

In this case, the weight density is typically a multiple of $|\theta - \theta_0|$, where θ_0 is that parameter value at which there is indifference as to which action to take. If we use a statistic with variance σ^2/n, we would obtain the Bayes risk as

$$\rho = \frac{C}{n\sigma} \iint_{\phi x > 0} |\phi| \frac{1}{\sqrt{2\pi}} \exp - \left[\frac{1}{2} \frac{(x+\phi)^2}{\sigma^2} \right] dx \, d\phi = \frac{1}{2} \frac{C\sigma^2}{n}. \tag{5}$$

Thus we see that, for regular estimates, the efficiency is proportional to $1/\sigma^2$. We have chosen $\sigma^2 = 1$ for the median by our choice of units. Let us now consider what happens for the one-sided K–S test with uniform alternatives. Here the probability of error is

$$\left. \begin{array}{ll} e^{-\frac{1}{2}(a+\phi)^2} & (\phi > 0), \\[1ex] 1 - e^{-\frac{1}{2}(a+\phi)^2} & (-a < \phi < 0), \\[1ex] 0 & (\phi < -a), \end{array} \right\} \tag{6}$$

so letting $\rho^* = n\rho/C$

$$\rho^* = \tfrac{1}{2}a^2 + \int_{-a}^{\infty} \phi e^{-\frac{1}{2}(a+\phi)^2} d\phi$$

$$= \tfrac{1}{2}a^2 + \int_{0}^{\infty} (x-a) e^{-\frac{1}{2}x^2} dx$$

$$= \tfrac{1}{2}a^2 - a\sqrt{\tfrac{1}{2}\pi} + 1. \tag{7}$$

This is minimized at $a = \sqrt{\tfrac{1}{2}\pi}$ obtaining

$$\hat{\rho}^* = 1 - \tfrac{1}{4}\pi = 0\cdot 2146 \tag{8}$$

for an efficiency of $2\cdot 33$.

If we consider instead the statistic $X_+ - X_-$, we find from (1) that

$$P(X_+ - X_- > \lambda) = \sum_{n=1}^{\infty} \frac{1}{4b^2 - 1} e^{-\frac{1}{2}(n^2\lambda^2)}, \tag{9}$$

yielding

$$\rho^* = \int_{-\infty}^{\infty} |\lambda| \sum_{n=1}^{\infty} \frac{1}{4n^2 - 1} e^{-2n^2\lambda^2} d\lambda^2 = 1 - \frac{\pi^2}{12} = 0\cdot 1776. \tag{10}$$

The efficiency of this test is $2\cdot 82$, so that the test is $1\cdot 21$ times as efficient as the usual one-sided K–S test.

For distributions with very large tails, the one-sided K–S test discriminates well against shifts in one direction, but poorly in the other.

In the 'extreme' case, no one-sided test has finite asymptotic risk. Even in the uniform case, the improved discrimination can be seen from the following table of error probabilities.

$|\phi|$	K–S $(\phi > 0)$	K–S $(\phi < 0)$	DIF
0·0	0·4559	0·5441	0·5000
0·2	0·3478	0·4258	0·3763
0·4	0·2549	0·3052	0·2623
0·6	0·1795	0·1922	0·1664
0·8	0·1215	0·0980	0·0931
1·0	0·0790	0·0276	0·0451
1·2	0·0498	0·0014	0·0187
1·4	0·0296	0	0·0066
1·6	0·0171	0	0·0020
1·8	0·0095	0	0·0005
2·0	0·0050	0	0·0001
2·2	0·0026	0	0·0000
2·4	0·0013	0	0·0000
2·6	0·0006	0	0·0000
2·8	0·0003	0	0·0000
3·0	0·0001	0	0·0000

Computations not yet completed indicate a similar result for the double exponential.

4 THE TWO-SIDED CASE

Here the asymptotic Bayes risk efficiency is easy to obtain—the K–S and Kuiper tests each have asymptotic Bayes risk efficiency *exactly* equal to that of the median (see [2]). However, one must be careful of using the limiting argument here too quickly. For example, let k test statistics be given, so that for a given sample size N the ith test should have a type I error of α_i. If we now decide to reject if any test rejects at level $\min \alpha_i$, the resulting test has at least as good asymptotic relative efficiency as the best one!

We have computed the relative efficiency of the K–S test to that based on the median for various risks (expressed as multiples of the type I risk). The type II loss is here constant. A brief table is appended.

As we can see, there is a very slow approach of the efficiency to 1, as the range of sample sizes in this calculation exceeds 10^{14}.

The author believes that the results will be similar for other distributions and other loss functions, and simulation studies will be made to investigate this.

	Efficiency	
Risk	Uniform	Double-exponential
0·5	1·961	0·798
0·1	1·675	0·847
0·01	1·540	0·880
0·001	1·469	0·898
0·0001	1·422	0·911
0·00001	1·388	0·921
0·000001	1·362	0·928
0·0000001	1·341	0·934

Of course, one really should consider infinite dimensional families of alternatives. Preliminary considerations indicate that the convergence to the asymptotic relative efficiency is slower.

REFERENCES

[1] Doob, J. L. Heuristic approach to the Kolmogorov–Smirnov theorem. *Ann. Math. Statist.* **20** (1949), 393–403.
[2] Rubin, H. and Sethuraman, J. Bayes risk efficiency. *Sankhyā*, Series A, **27**, 347–56.

ON SOME NONPARAMETRIC
EMPIRICAL BAYES MULTIPLE
DECISION PROBLEMS†

J. VAN RYZIN

1 INTRODUCTION

The empirical Bayes approach to statistical decision theory is appropriate when one is confronted repeatedly and independently with the same decision problem. In such instances it is reasonable to formulate the component problems in the sequence as Bayes decision problems with respect to a completely unknown prior distribution on the parameter space and then use the accumulated observations to improve the decision rule at each stage. This approach is due to H. Robbins [7] and is best presented in his paper [8]. Many such empirical Bayes procedures have been shown to be asymptotically optimal in the sense that the risk for the nth decision problem converges to optimal Bayes risk which would have been obtained if the prior distribution were *known* and the Bayes rule with respect to this prior distribution were used.

This paper examines certain nonparametric empirical Bayes decision procedures for some multiple decision problems. In particular, we consider two general problems: (i) a monotone multiple decision problem in §3 and (ii) a selection problem in §4. The selection problem of §4 was treated earlier by Deely [2] and our work here presents certain improvements and extensions of his results.

The sense in which we shall use the word nonparametric in this paper is that, although the (component) decision problems treated depends on a parameter λ, the parametric *form* of the observable random variable whose distribution depends on λ is not assumed known.

It will be seen that this paper is closely related to the paper of Johns [3], in which a nonparametric empirical Bayes estimation and testing problem were first considered. Also, we mention that the methods used herein are similar to some of those introduced in Johns and Van Ryzin [4], [5]. There are many other papers on empirical Bayes procedures which could be mentioned as references and we refer the reader to the bibliographies of the papers cited here.

† Research sponsored by National Science Foundation Grant NSF-GP-9324 at the University of Wisconsin.

2 THE GENERAL PROBLEM AND THE EMPIRICAL BAYES APPROACH

Consider the following multiple decision problem. Let X be an observable random variable with values in a measurable space $(\mathscr{X}, \mathscr{B})$ upon which is defined a σ-finite measure μ. On $(\mathscr{X}, \mathscr{B})$ is defined a family $\mathscr{P} = \{P_\lambda, \lambda \in \Omega\}$ of probability measures dominated by μ and indexed by the parameter λ. Let $f_\lambda(x) = (dP_\lambda/d\mu)\,(x)$ be the μ-density of X when the parameter has value λ. Assume that the statistician is interested in an action space $A = \{a_1, \ldots, a_k\}$ consisting of a finite number of distinct actions. Associated with the problem is a specified loss function

$$L(\lambda, a) \geqslant 0 \quad \text{on} \quad \Omega \times A.$$

Finally, let Λ be a Ω-valued (unobservable) random variable which has *a priori* distribution $G(\lambda)$ on Ω.

In such a statistical decision problem, the statistician's problem is to choose a behavioral (measurable) decision rule

$$t(x) = (t(1|x), \ldots, t(k|x)),$$

where $t(j|x) = \Pr\{\text{taking action } a_j | X = x\}$ and

$$0 \leq t(j|x) \leq 1, \quad \sum_{j=1}^{k} t(j|x) = 1 \quad \text{a.e.}\,\mu. \tag{1}$$

The risk of such a decision rule when λ is the parameter value is

$$R(\lambda, t) = \sum_{j=1}^{k} L(\lambda, a_j)\, E_\lambda[t(j|X)], \tag{2}$$

where E_λ denotes expectation under P_λ. Then, the *Bayes risk* with respect to *a priori* distribution G is

$$r(G, t) = \int R(\lambda, t)\, dG(\lambda)$$

$$= \sum_{j=1}^{k} \int t(j|x) \left[\int L(\lambda, a_j) f_\lambda(x)\, dG(\lambda) \right] d\mu(x) \tag{3}$$

which is clearly minimized by taking $t(j|x) = t_G(j|x), j = 1, \ldots, k$, where

$$t_G(j|x) = 0 \quad \text{if} \quad \int L(\lambda, a_j) f_\lambda(x)\, dG(\lambda) > \min_\nu \int L(\lambda, a_\nu) f_\lambda(x)\, dG(\lambda), \tag{4}$$

subject to (1). Thus, $t_G(x) = (t_G(1|x), \ldots, t_G(k|x))$ defined by (4) or (5)) is a *Bayes rule relative to* G, which we can define equivalently for later use by

$$t_G(j|x) = 0 \quad \text{if} \quad \Delta_G(a_j, x) > \min_\nu \Delta_G(a_\nu, x) \tag{5}$$

for $j = 1, ..., k$, subject to (1), where

$$\Delta_G(a_j, x) = \int \{L(\lambda, a_j) - L(\lambda, a_1)\} f_\lambda(x) \, dG(\lambda). \tag{6}$$

Define
$$r(G) = r(G, t_G) = \min_t r(G, t), \tag{7}$$

which is called the *Bayes risk functional of G*.

When G is fully known to the statistician it is clear that he should choose a Bayes rule relative to G, t_G, defined by (5), to minimize his risk and achieve $r(G)$ in (7) as the minimum attainable risk. This, however, is usually impossible since G is rarely known.

In the empirical Bayes approach of Robbins [8] when one is confronted with a repeated, independent sequence of such problems then one can often find a procedure not knowing G which does almost as well as t_G in the $(n+1)$st problem as the number, n, of problems increases. Specifically let $(X_1, \Lambda_1), (X_2, \Lambda_2), ...$ be a sequence of mutually independent pairs of random variables where each Λ_i is distributed as G on Ω and X_i has conditional density f_λ given $\Lambda_i = \lambda$. The empirical Bayes approach attempts to construct a decision procedure concerning Λ_{n+1} (unobservable) at stage $n+1$ based on $(X_1, ..., X_{n+1})$ which is the data available at stage $n+1$. The $(\Lambda_1, ..., \Lambda_n)$ also remain unobservable. Therefore, we consider decision rules of the form

$$t_n(x) = (t_n(1|x), ..., t_n(k|x)), \quad t_n(j|x) = t_n(j|x_1, ..., x_n; x), \tag{8}$$

$j = 1, ..., n$, subject to (1), and take action a_j with probability $t_n(j|x_{n+1})$ at stage $n+1$. The risk at stage $n+1$ using $t_n(x_{n+1})$ is given by

$$r(G, t_n) = \sum_{j=1}^k E \int t_n(j|x) \left[\int L(\lambda, a_j) f_\lambda(x) \, dG(\lambda) \right] d\mu(x), \tag{9}$$

where E denotes expectation with respect to the n independent random variables $X_1, ..., X_n$ each with common μ-density

$$f_G(x) = \int f_\lambda(x) \, dG(\lambda). \tag{10}$$

Note that since the procedure $t_G(x)$ in (5) achieves the minimum Bayes risk $r(G)$ relative to G, we have

$$r(G, t_n) \geqq r(G) \quad (n = 1, 2, ...). \tag{11}$$

Hence, in empirical Bayes theory the non-negative difference

$$r(G, t_n) - r(G)$$

is used as a measure of goodness of the sequence of procedures $\{t_n\}$ and we say:

Definition 1 (Robbins [8]). The sequence of procedures $\{t_n\}$ is said to be *asymptotically optimal* (a.o.) relative to G if $r(G, t_n) - r(G) = o(1)$ as $n \to \infty$.

Definition 2. The sequence of procedures $\{t_n\}$ is said to be *asymptotically optimal* (a.o.) *of order* a_n relative to G if $r(G, t_n) - r(G) = O(\alpha_n)$ as $n \to \infty$, where $\lim_n \alpha_n = 0$.

In §§3 and 4, we will study some specific nonparametric empirical Bayes multiple decision procedures. In so doing, we will construct functions

$$\Delta_{j,n}(x) = \Delta_{j,n}(x_1, \ldots, x_n; x) \quad \text{such that} \quad \text{a.e. } (\mu)x,$$

$$p \lim_n \Delta_{j,n}(x) = \Delta_G(a_j, x). \tag{12}$$

The procedure $t_n(x) = (t_n(1|x), \ldots, t_n(k|x))$ is then defined by setting $\Delta_{1,n}(x) = 0$ and by taking

$$t_n(j|x) = 0 \quad \text{if} \quad \Delta_{j,n}(x) > \min_\nu \{\Delta_{\nu,n}(x)\} \quad \text{subject to} \quad (1). \tag{13}$$

We shall make extensive use of the following two results concerning the sequence of procedures $\{t_n\}$ defined by (12) and (13).

Lemma 1 (Robbins [8], Corollary 1). *Let* G *be such that*

$$\int L(\lambda, a_j) \, dG(\lambda) < \infty, \quad j = 1, \ldots, k$$

and let $\{t_n(x)\} = \{(t_n(1|x), \ldots, t_n(k|x))\}$

be defined by (13). *Then the sequence* $\{t_n\}$ *of empirical Bayes rules is* a.o. *relative to* G.

Lemma 2. *Let* $\{t_n(x)\} = \{(t_n(1|x), \ldots, t_n(k|x))\}$ *be defined by* (13). *Then,*

$$0 \leq r(G, t_n) - r(G) \leq \sum_{j=2}^{k} \int E|\Delta_{j,n}(x) - \Delta_G(a_j, x)| \, d\mu(x). \tag{14}$$

Proof. By (3) and (9), we have

$$r(G, t_n) - r(G) = \sum_{j=1}^{k} E \int \left[\{t_n(j|x) - t_G(j|x)\} \int L(\lambda, a_j) f_\lambda(x) \, dG(\lambda) \right] d\mu(x)$$

$$= \sum_{j=2}^{k} E \int \{t_n(j|x) - t_G(j|x)\} \Delta_G(a_j, x) \, d\mu(x),$$

where the last equality follows from (1) and (6). Therefore,

$$r(G, t_n) - r(G) = \sum_{j=2}^{k} E \int \{t_n(j|x) - t_G(j|x)\} \{\Delta_G(a_j, x) - \Delta_{j,n}(x)\} \, d\mu(x)$$

$$+ \sum_{j=2}^{k} E \int \{t_n(j|x) - t_G(j|x)\} \Delta_{j,n}(x) \, d\mu(x). \tag{15}$$

But from (13) we have $\{t_n(j|x) - t_G(j|x)\}\,\Delta_{j,n}(x) \leqq 0$, in equality (15) and hence (14) follows from (15) by bounding $|t_n(j|x) - t_G(j|x)| \leqq 1$ in the first term on the right-hand side of (15).

3 A NONPARAMETRIC MONOTONE MULTIPLE DECISION PROBLEM

We consider now the following monotone multiple decision problem. Let $\Omega = (-\infty, +\infty)$ and let $-\infty = \lambda_0^0 < \lambda_1^0 < \lambda_2^0 < \ldots < \lambda_{k-1}^0 < \lambda_k^0 = +\infty$ be given. Let action a_j correspond to 'deciding the value of $\Lambda = \lambda$ is in $(\lambda_{j-1}^0, \lambda_j^0]$,' $j = 1, \ldots, k$, As a loss function we take $L(\lambda, a_j)$ such that

$$L(\lambda, a_{j+1}) - L(\lambda, a_j) = c(\lambda_j^0 - \lambda) \quad (j = 1, \ldots, k-1), \qquad (16)$$

where $c > 0$ is a constant. Then, $L(\lambda, a_{j+1}) - L(\lambda, a_j) \geqq 0$ or $\leqq 0$ according as $\lambda \leqq \lambda_j^0$ or $\lambda \geqq \lambda_j^0$, and the decision problem is seen to be monotone.

Suppose that now one is confronted with such a problem repeatedly and independently in a nonparametric situation where the parametric form of f_λ is not assumed known, and one wishes to apply the empirical Bayes technique. As can be seen from (12) and Lemma 1, we must construct sequences of functions $\{\Delta_{j,n}(x)\}, j = 2, \ldots, k$, which consistently estimate

$$\Delta_G(a_j, x) = \int \{L(\lambda, a_j) - L(\lambda, a_1)\} f_\lambda(x)\, dG(\lambda)$$

$$= \sum_{i=1}^{j-1} \int \{L(\lambda, a_{i+1}) - L(\lambda, a_i)\} f_\lambda(x)\, dG(\lambda)$$

$$= c \sum_{i=1}^{j-1} \int (\lambda_i^0 - \lambda) f_\lambda(x)\, dG(\lambda)$$

$$= c \left\{ \left(\sum_{i=1}^{j-1} \lambda_i^0 \right) f_G(x) - (j-1)\, h_G(x) \right\}, \qquad (17)$$

where $f_G(x)$ is given by (10) and

$$h_G(x) = \int \lambda f_\lambda(x)\, dG(\lambda) = E[\Lambda | x = x] f_G(x). \qquad (18)$$

The solution of the problem of estimating $\Delta_G(a_j, x), j = 2, \ldots, k$ thus reduces to estimating $f_G(x)$ and $h_G(x)$. We can accomplish this estimation problem in two general nonparametric situations wherein we assume: (i) $f_G(x)$ is discrete, and (ii) $f_G(x)$ is continuous. However, throughout the rest of this section we will need to impose the following additional assumptions.

$$E[X | \Lambda = \lambda] = \lambda \qquad (19)$$

and

At each stage i, $i = 1, 2, \ldots$, we observe $Y_i = (Y_{1i}, \ldots Y_{ri})$ in addition to X_i, where conditional on $\Lambda_i = \lambda$, X_i and Y_i are independent and the components of $Y_i, Y_{\nu i}, \nu = 1, \ldots, r$, are independent and identically distributed with density given by f_λ.   (20)

Assumptions (19) and (20) may be used to obtain estimates of $E[\Lambda | x = x]$ involved in the definition of $h_G(x)$ in (18). Let

$$Z_i = \frac{1}{r} \sum_{\nu=1}^{r} Y_{\nu i} \qquad (21)$$

and note that by (19) and (20) we have

$$E[Z_i | X_i] = \frac{1}{r} \sum_{\nu=1}^{r} E[Y_{\nu i} | X_i]$$

$$= \frac{1}{r} \sum_{\nu=1}^{r} E\{E[Y_{\nu i} | X_i, \Lambda_i] | X_i\}$$

$$= \frac{1}{r} \sum_{\nu=1}^{r} E\{E[Y_{\nu i} | \Lambda_i] | X_i\}$$

$$= E[\Lambda_i | X_i]. \qquad (22)$$

Therefore $E[Z_i | X_i = x] = E[\Lambda | X = x]$ a.e. μ.   (23)

This technique of nonparametrically estimating $E[\Lambda | X = x]$ is due to Johns [3].

We now give our results in the two cases mentioned above.

(i) *Discrete case.* Here we assume $f_G(x) = \Pr\{X = x\}$ is a discrete density with respect to counting measure μ. To estimate $f_G(x)$ and $h_G(x)$ define

$$K_i(x) = \begin{cases} 1 & \text{if} \quad X_i = x, \\ 0 & \text{if} \quad X_i \neq x, \end{cases} \qquad (24)$$

$$f_n(x) = \frac{1}{n} \sum_{i=1}^{n} K_i(x) \qquad (25)$$

and (see (21)) $V_i(x) = Z_i K_i(x),$ (26)

$$h_n(x) = \frac{1}{n} \sum_{i=1}^{n} V_i(x). \qquad (27)$$

Then, by the law of large numbers, we have a.e. μ

$$p \lim_{n} f_n(x) = E K_1(x) = f_G(x) \qquad (28)$$

and (see (23))

$$p \lim_{n} h_n(x) = EV_1(x)$$

$$= E[\Lambda | x = x] f_G(x)$$

$$= h_G(x), \tag{29}$$

if $|h_G(x)| < \infty$.

Note that under the assumption that $E_G|\Lambda| = \int |\lambda| \, dG(\lambda) < \infty$, we have $|h_G(x)| < \infty$ a.e. μ. Hence, we have as an immediate consequence of Lemma 1, (17), (18) and (24)–(29),

Theorem 1. In the discrete case monotone multiple decision problem with loss function (16), let G be such that $E_G|\Lambda| < \infty$. Then, the sequence of nonparametric empirical Bayes rules $\{t_n\} = \{(t_n(1|x), ..., t_n(k|x))\}$ defined by (13) with

$$\Delta_{j,n}(x) = c \left\{ \left(\sum_{i=1}^{j-1} \lambda_i^0 \right) f_n(x) - (j-1) h_n(x) \right\}, \tag{30}$$

$j = 1, 2, ..., n$, is a.o. relative to G.

Using Lemma 2, we give a convergence rate result which has some interesting corollaries. In so doing, we will need the following additional assumption.

(A) $E[Y_{11}^2 | \Lambda = \lambda] \leq a + b\lambda^2$ for some positive constants a, b.

Theorem 2. In the discrete case monotone multiple decision problem with loss function (16), let (A) hold and let G be such that

$$\sum_{x \in \mathscr{X}} \{f_G(x)\}^{\frac{1}{2}} < \infty \tag{31}$$

and

$$\sum_{x \in \mathscr{X}} \left\{ \int |\lambda|^2 f_\lambda(x) \, dG(\lambda) \right\}^{\frac{1}{2}} < \infty. \tag{32}$$

Then the sequence of nonparametric empirical Bayes rule $\{t_n\}$ defined by (13) and (30) is a.o. of order $n^{-\frac{1}{2}}$ relative to G.

Proof. Using inequality (14) of Lemma 2, we have by (17), (18), (24)–(27), and (30),

$$r(G, t_n) - r(G)$$

$$\leq \sum_{j=2}^{k} \sum_{x \in \mathscr{X}} cE \left| \left(\sum_{i=1}^{j-1} \lambda_i^0 \right) \{f_G(x) - f_n(x)\} - (j-1) \{h_G(x) - h_n(x)\} \right|$$

$$\leq \sum_{x \in \mathscr{X}} \{c_0 E |f_n(x) - f_G(x)| + c_1 E |h_n(x) - h_G(x)|\}$$

$$\leq n^{-\frac{1}{2}} \{c_0 \sum_{x \in \mathscr{X}} [\mathrm{var}\{K_1(x)\}]^{\frac{1}{2}} + c_1 \sum_{x \in \mathscr{X}} [\mathrm{var}\{V_1(x)\}]^{\frac{1}{2}}\}, \tag{33}$$

where $c_0 = c \sum_{j=2}^{k} \left| \sum_{i=1}^{j-1} \lambda_i^0 \right|$ and $c_1 = \frac{1}{2} ck(k-1)$.

Note that by (20), (21), (24) and (26), we have with $K_1 = K_1(x)$,

$$\operatorname{var}\{V_1(x)\} = \operatorname{var}(Z_1 K_1)$$
$$\leq E Z_1^2 K_1$$
$$= E\{E[Z_1^2 \Lambda_1] E[K_1 | \Lambda_1]\}$$
$$= (1 - r^{-1}) E \Lambda^2 f_\Lambda(x) + r^{-1} E\{E[Y_{11}^2 | \Lambda]] f_\Lambda(x)\}$$
$$\leq (1 + (b-1)r^{-1}) E \Lambda^2 f_\Lambda(x) + ar^{-1} f_G(x),$$

where the last inequality follows by (A). Thus, $\sum\limits_{x \in \mathscr{X}} [\operatorname{var}\{V_1(x)\}]^{\frac{1}{2}} < \infty$
under (31) and (32) and the theorem follows from (33), since also

$$\sum_{x \in \mathscr{X}} [\operatorname{var}\{K_1(x)\}]^{\frac{1}{2}} \leq \sum_{x \in \mathscr{X}} \{f_G(x)\}^{\frac{1}{2}} < \infty \quad \text{under (31)}.$$

Corollary 2.1. *Let P be a discrete lattice-valued family. Let* (A) *hold and G be such that $E_G \Lambda^4 < \infty$. Then, in the monotone multiple decision problem with loss function* (16), *the sequence of nonparametric empirical Bayes rules $\{t_n\}$ defined by* (13) *and* (30) *is a.o. of order $n^{-\frac{1}{2}}$ relative to G.*

Proof. Let $\beta^2 = \sum\limits_{x} (1 + x^2)^{-1}$. Then by Hölder's inequality and assumption
(A), we have

$$\sum_{x \in \mathscr{X}} \left\{ \int \lambda^2 f_\lambda(x) \, dG(\lambda) \right\}^{\frac{1}{2}} \leq \beta \left\{ \int \lambda^2 E_\lambda(1 + X^2) \, dG(\lambda) \right\}^{\frac{1}{2}}$$

$$\leq \beta \{(a+1) E_G \Lambda^2 + b E_G \Lambda^4\}^{\frac{1}{2}}.$$

Thus, with $E_G \Lambda^4 < \infty$ condition (32) of Theorem 2 holds. A similar argument also shows that (31) of Theorem 2 holds under $E_G \Lambda^4 < \infty$.

Corollary 2.2. *Let $\mathscr{P}$ be a discrete lattice-valued and location parameter family, i.e. $f_\lambda(x) = f_0(x - \lambda)$. Let G be such that $E_G |\Lambda|^{3+\epsilon} < \infty$ for some $\epsilon > 0$ and assume $E_0 |X|^{1+\epsilon} < \infty$. Then, in the monotone multiple decision with loss function* (16), *the sequence of nonparametric empirical Bayes rules $\{t_n\}$ defined by* (13) *and* (30) *is a.o. of order $n^{-\frac{1}{2}}$ relative to G.*

Proof. Let $\beta_\epsilon^2 = \sum\limits_{x \in \mathscr{X}} (1 + |x|^{1+\epsilon})^{-1}$. Then by Hölder's inequality, we have

$$\sum_{x \in \mathscr{X}} \left\{ \int \lambda^2 f_\lambda(x) \, dG(\lambda) \right\}^{\frac{1}{2}} \leq \beta_\epsilon \left\{ \int \lambda^2 E_\lambda(1 + |X|^{1+\epsilon}) \, dG(\lambda) \right\}^{\frac{1}{2}}$$

$$\leq \beta_\epsilon \{2^\epsilon + 4^\epsilon E_0 |X|^{1+\epsilon}) E_G \Lambda^2 + 4^\epsilon E_G |\Lambda|^{3+\epsilon}\}^{\frac{1}{2}},$$

where the last inequality follows by

$$(1 + |x|)^{1+\epsilon} \leq 2^\epsilon + 2^\epsilon |x|^{1+\epsilon}$$

and

$$|x|^{1+\epsilon} = |x - \lambda + \lambda|^{1+\epsilon} \leq 2^\epsilon (|x - \lambda|^{1+\epsilon} + |\lambda|^{1+\epsilon}).$$

Hence, we have verified condition (32) of Theorem 2. Verification of (31) is similar.

We point out that Corollaries 2.1 and 2.2 show that the convergence rates of the empirical Bayes procedures are closely related to moment conditions on the prior distribution G. For a more detailed analysis of this phenomenon in some parametric empirical Bayes testing problems see [4] and [5].

(ii) *Continuous case.* Here we assume $\mathscr{X}$ is the real line, μ is Lebesgue measure on $\mathscr{X}$ and that the Lebesgue density $f_G(x)$ is continuous on $\mathscr{X}$. To estimate $f_G(x)$ and $h_G(x)$ define

$$K_{j,n}(x) = \gamma_{n}^{-1} K(\gamma_n^{-1}(x - X_j)), \tag{34}$$

$$f_n(x) = n^{-1} \sum_{j=1}^{n} K_{j,n}(x), \tag{35}$$

$$V_{j,n}(x) = Z_j K_{j,n}(x) \tag{36}$$

and
$$h_n(x) = n^{-1} \sum_{j=1}^{n} V_{j,n}(x), \tag{37}$$

where $K(u)$ is a real-valued Borel measurable function satisfying

$$\int K(u)\, du = 1, \quad \sup_u |K(u)| < \infty, \quad |uK(u)| \to 0 \quad \text{as} \quad |u| \to \infty, \tag{38}$$

and $\{\gamma_n\}$ is a sequence of numbers satisfying

$$\gamma_n \geq 0, \quad \lim_n \gamma_n = 0 \quad \text{and} \quad \lim_n (n\gamma_n) = \infty. \tag{39}$$

Then, by Theorem 1 A of Parzen [6], we have for all $x \in \mathscr{X}$ (we assumed $f_G(x)$ is continuous for all x),

$$(n\gamma_n)\operatorname{var}\{f_n(x)\} \leq \int K^2(u) f_G(x - \gamma_n u)\, du \sim \left(\int K^2(u)\, du\right) f_G(x). \tag{40}$$

Also, by (19), (20), (A) and the same theorem, we have
$(n\gamma_n)\operatorname{var}\{h_n(x)\}$

$$\leq \gamma_n^{-1} E Z_1^2 K^2 (\gamma_n^{-1}(x - X_1))$$

$$= \gamma_n^{-1} E\{E[Z_1^2 | \Lambda_1] E[K^2(\gamma_n^{-1}(x - X_1)) | \Lambda_1]\}$$

$$= \gamma_n^{-1} E\left(\left\{r^{-2} \sum_{i,j=1}^{r} E[Y_{i1} Y_{j1} | \Lambda_1]\right\} E[K^2(\gamma_n^{-1}(x - X_1)) | \Lambda_1]\right)$$

$$= E\left(\{(1 - r^{-1})\Lambda_1^2 + r^{-1} E[Y_{11}^2 | \Lambda_1]\} \int K^2(u) f_{\Lambda_1}(x - \gamma_n u)\, du\right)$$

$$\leq E\{(1 + (b-1) r^{-1})\Lambda_1^2 + a r^{-1}\} \int K^2(u) f_{\Lambda_1}(x - \gamma_n u)\, du$$

$$\leq \int K^2(u) \left\{(1 + |b-1| r^{-1}) \int \lambda^2 f_\lambda(x - \gamma_n u)\, dG(\lambda) + a r^{-1} f_G(x - \gamma_n u)\right\} du$$

$$\sim \left\{\int K^2(u)\, du\right\} \left\{(1 + |b-1| r^{-1}) \int \lambda^2 f_\lambda(x)\, dG(\lambda) + a r^{-1} f_G(x)\right\}. \tag{41}$$

Hence, if $\qquad f_G(x) < \infty$ and $|h_G(x)| < \infty$ for all x (42)

then, by (39), (40) and (41), we have for all x

$$\lim_n \text{var}\{f_n(x)\} = 0, \quad \lim_n \text{var}\{h_n(x)\} = 0.$$

Therefore, by Chebyshev's inequality and Theorem 1 A of Parzen [6], we have under (42), for all x

$$p \lim_n f_n(x) = \lim_n E K_{1,n}(x) = f_G(x) \tag{43}$$

and (see (19), (20) and (38))

$$
\begin{aligned}
p \lim_n h_n(x) &= \lim_n E V_{1,n}(x) \\
&= \lim_n E\{E[V_{1,n}(x)|\Lambda_1]\} \\
&= \lim_n E\{E[Z_1|\Lambda_1] E[\gamma_n^{-1} K(\gamma_n^{-1}(x - X_1))|\Lambda_1]\} \\
&= \lim_n \int \lambda \left\{ \int K(u) f_\lambda(x - \gamma_n u)\, du \right\} dG(\lambda) \\
&= \int \lambda f_\lambda(x)\, dG(\lambda) \\
&= h_G(x).
\end{aligned}
\tag{44}
$$

Finally, noting that under the assumption that $E_G|\Lambda| < \infty$ and

(B) $\qquad\qquad \sup_\lambda f_\lambda(x) < \infty$ for all x

we have (42) holding which implies that $f_n(x)$ and $h_n(x)$ consistently estimate $f_G(x)$ and $h_G(x)$ (see (43) and (44)). Thus, as a consequence of Lemma 1, (17), (18), (43) and (44), we have

Theorem 3. In the continuous case monotone multiple decision problem with loss function (16), let G be such that $E_G|\Lambda| < \infty$. Then, if (A) and (B) hold, the sequence of nonparametric empirical Bayes rules

$$\{t_n\} = \{(t_n(1|x), ..., t_n(k|x))\}$$

defined by (13) and (30), with $f_n(x)$ and $h_n(x)$ as in (35) and (37) where $K(u)$ and $\{\gamma_n\}$ satisfy (38) and (39) respectively, is a.o. relative to G.

Using Lemma 2, we can now prove a convergence rate theorem, which is the continuous case analogy of Theorem 2 in the discrete case.

Theorem 4. *In the continuous case monotone multiple decision problem with loss function* (16), *let* (A) *and* (B) *hold. Furthermore, assume that* $f_\lambda^{(j)}(x)$, *the jth derivative of* f_λ *in x, exist*, $j = 1, ..., q$. *Let G be such that*

(i) $\int |\lambda f_\lambda^{(j)}(x)| dG(\lambda) < \infty$, $j = 1, ..., q-1$, *for all x*;

(ii) $\int \left\{ \int f_{\lambda, \delta}(x) dG(\lambda) \right\}^{\frac{1}{2}} dx < \infty$, $f_{\lambda, \delta}(x) = \sup_{|t| \leq \delta} f_\lambda(x+t)$;

(iii) $\int \left\{ \int \lambda^2 f_{\lambda, \delta}(x) dG(\lambda) \right\}^{\frac{1}{2}} dx < \infty$;

(iv) $\int |\lambda| \alpha_\delta(\lambda) dG(\lambda) < \infty$, $\alpha_\delta(\lambda) = \int \sup_{|t| \leq \delta} |f_\lambda^{(q)}(x+t)| dx$.

Then, choosing $\gamma_n = O(n^{-1/(2q+1)})$ *and* $K(u)$ *such that*

$$\int K(u) du = 1, \quad \int u^j K(u) du = 0, \quad j = 1, ..., q-1,$$

and $K(u) = 0$ *if* $|u| \geq u_0$ *for some* $u_0 < \infty$, *yields a sequence*

$$\{t_n\} = \{(t_n(1|x), ..., t_n(k|x))\}$$

of nonparametric empirical Bayes rules defined by (13) *and* (30) *with* $f_n(x)$ *and* $h_n(x)$ *as in* (35) *and* (37) *which is* a.o. *of order* $n^{-q/(2q+1)}$.

Proof. Note that as in (33), Lemma 2 yields in the continuous case,

$$r(G, t_n) - r(G) \leq c_0 \int E|f_n(x) - f_G(x)| dx + c_1 \int E|h_n(x) - h_G(x)| dx, \quad (45)$$

c_0 and c_1 as in (33).

Furthermore, we have

$$E|h_n(x) - h_G(x)| \leq E|h_n(x) - Eh_n(x)| + |Eh_n(x) - h_G(x)|$$

$$\leq [\text{var}\{h_n(x)\}]^{\frac{1}{2}} + |Eh_n(x) - h_G(x)|. \quad (46)$$

But by the derivation of (41), if $\gamma_n u_0 \leq \delta$, then

$$\text{var}\{h_n(x)\} \leq (n\gamma_n)^{-1} \left\{ a_0 \int \lambda^2 f_{\lambda, \delta}(x) dG(\lambda) + a_1 \int f_{\lambda, \delta}(x) dG(\lambda) \right\}, \quad (47)$$

where

$$a_0 = (1 + |b-1|r^{-1}) \int K^2(u) du \quad \text{and} \quad a_1 = ar^{-1} \int K^2(u) du.$$

By the derivation of (44), condition (i), and our choice of K, we have

$$E|h_n(x) - h_G(x)| = \left| \iint \lambda K(u)\{f_\lambda(x - \gamma_n u) - f_\lambda(u)\}\, du\, dG(\lambda) \right|$$

$$\leqq \left| \sum_{j=1}^{q-1} (j!)^{-1} \gamma_n^j \left(\int u^j K(u)\, du \right) \left(\int \lambda f_\lambda^{(j)}(x)\, dG(\lambda) \right) \right|$$

$$+ (q!)^{-1} \gamma_n^q \left| \iint u^q K(u)\, \lambda f_\lambda^{(q)}(x + \xi_n(\lambda, u))\, du\, dG(\lambda) \right|,$$

where $\qquad |\xi_n(x, u)| < |\gamma_n u| \leqq \gamma_n u_0 \leqq \delta.$

Therefore, if $\gamma_n u_0 \leqq \delta$, we have

$$|E h_n(x) - h_G(x)| \leqq \gamma_n^q a_2 \int |\lambda| \sup_{|t| \leqq \delta} |f_\lambda^{(q)}(x + t)|\, dG(\lambda), \qquad (48)$$

where $\qquad a_2 = (q!)^{-1} \int |u|^q\, |K(u)|\, du.$

Inequalities (46), (47) and (48) together with conditions (ii), (iii) and (iv), thus yield

$$\int E|h_n(x) - h_G(x)|\, dx = O((n\gamma_n)^{-\frac{1}{2}}) + O(\gamma_n^q)$$

$$= O(n^{-q/(2q+1)}) \qquad (49)$$

by our choice of the sequence $\{\gamma_n\}$.

In a similar manner, it can be shown that under the condition of the theorem,

$$\int E|f_n(x) - f_G(x)|\, dx = O(n^{-q/(2q+1)}). \qquad (50)$$

The proof is thus completed by combining (45), (49) and (50).

We must remark that Theorem 4, although an interesting theoretical result, is not very useful in actual application since the choice of K and in particular $\{\gamma_n\}$ depends on the unverifiable assumptions (i)–(iv) and on the differentiability of $f_\lambda(x)$. However, we point out that the choice of $q = 2$, i.e. $\gamma_n = O(n^{-\frac{1}{5}})$ and any K satisfying (38) and $\int u K(u)\, du = 0$ seems reasonable in a nonparametric situation based in the general theory of density estimation in Parzen [6]. We refer the reader to [5] wherein conditions of the type (i)–(iv) are shown to be merely moment conditions on Λ in certain parametric testing problems.

4 A NONPARAMETRIC SELECTION PROBLEM

We consider now a selection problem, which was first studied from a non-parametric empirical Bayes viewpoint by Deely [2]. We shall point out how our results generalize and strengthen his results. Let

$$\Omega = \{\lambda = (\lambda_1, ..., \lambda_k): \lambda_i \in (-\infty, +\infty), i = 1, ..., k\}.$$

Furthermore, let $\mathcal{X} \subset R_k$ and the observed random variable

$$X = (X^{(1)}, ..., X^{(k)})$$

be such that conditional on $\Lambda = \lambda = (\lambda_1, ..., \lambda_k)$, the $X^{(i)}$ are independent with the density of $X^{(i)}$ being given by f_{λ_i}. Define that population (or distribution) among the k to be 'best' if its corresponding parameter value is largest among the k values, that is, the ith population is 'best' if $\lambda_i = \max_{1 \leq \nu \leq k} \{\lambda_\nu\}$. The selection problem is then to choose the 'best' population based on observing $X = (X^{(1)}, ..., X^{(k)})$. Define the action a_j to be 'selecting the jth population as best', or 'selecting λ_j as the largest'.

As a loss function, we take $L(\lambda, a_j)$ such that for $j = 1, ..., k$,

$$L(\lambda, a_j) = c(\lambda_{[k]} - \lambda_j), \tag{51}$$

where $\lambda_{[k]} = \max_\nu \lambda_\nu$ and c a constant, $c > 0$.

Suppose that now one is confronted with such a problem repeatedly and independently in a nonparametric situation and wishes to apply the empirical Bayes technique. As can be seen from (12) what must be constructed are sequences of functions $\{\Delta_{n,j}(x)\}$ which consistently estimate for $j = 2, ..., k$ (see (51)),

$$\Delta_G(a_j, x) = \int \{L(\lambda, a_j) - L(\lambda, a_1)\} f_\lambda(x) \, dG(\lambda)$$

$$= c \int (\lambda_1 - \lambda_j) f_\lambda(x) \, dG(\lambda)$$

$$= c\{h_G^{(1)}(x) - h_G^{(j)}(x)\}, \tag{52}$$

where $\qquad h_G^{(j)}(x) = \int \lambda_j f_\lambda(x) \, dG(\lambda) = E[\Lambda_j | x = x] f_G(x). \tag{53}$

Thus, the solution of the problem of estimating the $\Delta_G(a_j, x)$ is solved by estimating $h_G^{(j)}(x), j = 1, ..., k$. It is seen that our estimation problem is then very similar to that of the previous section where we had to estimate $h_G(x)$ given by (18).

598   J. VAN RYZIN

We make the following additional assumptions for the remainder of this section:

$$E[X^{(j)}|\Lambda = \lambda] = \lambda_j \quad (j = 1, ..., k), \tag{54}$$

and   at each stage i, we observe r.   (55)

k-vectors $Y_i = (Y_{1i}, ..., Y_{ri})$, $Y_{\nu i} = (Y_{\nu i}^{(1)}, ..., Y_{\nu i}^{(k)})$ in addition to the k-vector X_i, where conditional on $\Lambda_i = \lambda$, X_i and Y_i are independent, and the components (k-vectors) of Y_i, the $Y_{\nu i}$, are independent and identically distributed with multivariate μ-density given by f_λ.

Assumption (54) and (55) may be used to obtain estimates of

$$h_G^{(j)}(x), \quad j = 1, ..., k,$$

by defining the random variables

$$Z_i^{(j)} = r^{-1} \sum_{\nu=1}^{r} Y_{\nu i}^{(j)}, \tag{56}$$

and noting that

$$E[Z_i^{(j)}|X_i] = r^{-1} \sum_{\nu=1}^{r} E[Y_{\nu i}^{(j)}|X_i]$$

$$= r^{-1} \sum_{\nu=1}^{r} E\{E[Y_{\nu i}^{(j)}|X_i, \Lambda_i]|X_i\}$$

$$= r^{-1} \sum_{\nu=1}^{r} E\{E[Y_{\nu i}^{(j)}|\Lambda_i]|X_i\}$$

$$= E[\Lambda_j|X_i], \tag{57}$$

therefore, $E[Z_i^{(j)}|X_i = x] = E[\Lambda_j|X = x]$, a.e. μ.

We are now in a position to give our results in two cases: (i) the discrete case and (ii) the continuous case.

(i) *Discrete case.* Here we assume

$$f_G(x) = \Pr\{X = x\} = \Pr\{X^{(1)} = x^{(1)}, ..., X^{(k)} = x^{(k)}\}$$

is a discrete probability mass function. To estimate $h_G^{(j)}(x), j = 1, ..., k$, define (see (56)),

$$V_i^{(j)}(x) = \begin{cases} Z_i^{(j)} & \text{if } X_i = x, \\ 0 & \text{if } X_i \neq x, \end{cases} \tag{58}$$

$$h_n^{(j)}(x) = n^{-1} \sum_{i=1}^{n} V_i^{(j)}(x). \tag{59}$$

Then, by the law of large numbers (54) and (56) we have for $j = 1, ..., k$, a.e. μ

$$p \lim_n h_n^{(j)}(x) = E V_1^{(j)}(x)$$

$$= E[Z_1^{(j)} | X = x] f_G(x)$$

$$= E[\Lambda_j | X = x] f_G(x)$$

$$= h_G^{(j)}(x). \tag{60}$$

Thus, as an immediate consequence of Lemma 1, (52), (53), and (60), we have

Theorem 5. In the discrete case selection problem with loss function (52), *let G be such that $E_G|\Lambda_j| = \int \lambda_j dG_j(\lambda_j) < \infty$ (G_j, the jth marginal of G), $j = 1, ..., k$. Then the sequence of nonparametric empirical Bayes rules $\{t_n\} = \{(t_n(1|x), ..., t_n(k|k))\}$ defined by* (13) *with*

$$\Delta_{j,n}(x) = c\{h_n^{(1)}(x) - h_n^{(j)}(x)\} \quad (j = 1, ..., k), \tag{61}$$

is a.o. relative to G.

Using Lemma 2, we give a convergence rate theorem whose proof, here omitted, is very similar to that of Theorem 2. In so doing, we need the assumption

$$(A^*) \qquad E[Y_{11}^{(j)} | \Lambda] \leqq a + b\lambda_j^2 \quad (j = 1, ..., k)$$

for some positive constants a, b.

Theorem 6. In the discrete case selection problem with loss function (52), *let (A^*) hold and let G be such that*

$$\sum_{x \in \mathcal{X}} \{f_G(x)\}^{\frac{1}{2}} < \infty,$$

and $$\sum_{x \in \mathcal{X}} \left\{ \int |\lambda_j|^2 f_\lambda(x) \, dG(\lambda) \right\}^{\frac{1}{2}} < \infty \quad (j = 1, ..., k).$$

Then the sequence of nonparametric empirical Bayes rules $\{t_n\}$ defined by (13) *and* (61) *is a.o. of order $n^{-\frac{1}{2}}$ relative to G.*

(ii) *Continuous case.* Here we assume $\mathcal{X} = R_k$, μ is k-dimensional Lebesgue measure and $f_G(x)$ is continuous on R_k.

To estimate
$$h_G^{(j)}(x) \quad (j = 1, ..., k)$$

define
$$V_{i,n}^{(j)}(x) = Z_i^{(j)} \gamma_n^{-k} K(\gamma_n^{-1}(x - X_i))$$

and
$$h_n^{(j)}(x) = n^{-1} \sum_{i=1}^{n} V_{i,n}^{(j)}(x), \tag{62}$$

600   J. VAN RYZIN

where $K(u) = K(u_1, \ldots, u_k)$ is a real-valued Borel measurable function satisfying

$$\int K(u)\,du = 1, \quad \sup_{u \in R_k} |K(u)| < \infty, \quad \|u\|^k |K(u)| \to 0 \tag{63}$$

as

$$\|u\| \to \infty, \quad \|u\|^2 = \sum_{i=1}^{k} u_i^2,$$

and $\{\gamma_n\}$ is a sequence satisfying

$$\gamma_n \geq 0, \quad \lim_n \gamma_n = 0, \quad \lim_n (n\gamma_n^k) = \infty. \tag{64}$$

Then, by an analysis similar to that in (40) and (41) using Theorem 2.1 of Cacoullos [1] and Chebyshev's inequality, we have for all $x \in R_k$,

$$p \lim_n h_n^{(j)}(x) = \lim_n E V_{1,n}^{(j)}(x)$$

$$= \lim_n E\{E[V_{i,n}^{(j)}(x)|\Lambda_1]\}$$

$$= \lim_n E\{E[Z_1^{(j)}|\Lambda_1] E[\gamma_n^{-k} K(\gamma_n^{-1}(x - X_1))|\Lambda_1]\}$$

$$= \lim_n \int \lambda_j \left\{ \int K(u) f_\lambda(x - \gamma_n u)\,du \right\} dG(\lambda)$$

$$= \int \lambda_j f_\lambda(x)\,dG(\lambda)$$

$$= h_G^{(j)}(x), \tag{65}$$

provided $\quad |h_G^{(j)}(x)| < \infty \quad$ for all $\quad x \in R_k \quad (j = 1, \ldots, k)$. (66)

Finally, since (66) is implied by (B) if $\max_j E_G|\Lambda_j| < \infty$, we have as a consequence of Lemma 1, (52), (53) and (65).

Theorem 7. In the continuous case selection problem with loss function (52), *let G be such that $\max_j E_G|\Lambda_j| < \infty$. Then, if* (A*) *and* (B) *hold, the sequence of nonparametric empirical Bayes rules* $\{t_n\} = \{(t_n(1|x), \ldots, t_n(k|x))\}$ *defined by* (13) *and* (61) *with $h_n^{(j)}(x)$ as in* (62) *where $K(u)$ and $\{\gamma_n\}$ satisfy* (63) *and* (64) *respectively, is a.o. relative to G.*

We remark that a rate theorem similar to Theorem 4 in the monotone decision problem could also be given in this continuous case selection problem under the assumption that $f_\lambda(x)$ has mixed partials of order q existing plus certain integrability conditions. However, we forego statement of such a theorem here.

Deely [2] has given empirical Bayes rules for the selection problem stated here which requires the assumption that $G(\lambda) = \prod_{j=1}^{k} G_j(\lambda_j)$ to simplify (33) so that its estimation may be done 'componentwise' or 'marginally.' For details see [2]. Under such an assumption he has theorems on asymptotic optimality (similar to our Theorems 5 and 7), but no rate of convergence results.

5 SUMMARY

This paper has extended some of the methods of Johns and Van Ryzin [4], [5] to some nonparametric empirical Bayes multiple decision problems—a monotone multiple decision problem and a selection problem. We note in particular that in the case $k = 2$ (hypothesis testing) for section 3 our Theorems 1 and 3 are essentially due to Johns [3], but the rate results (Theorems 2 and 4) even for $k = 2$ are new. The problem of selection in §4 extends the earlier results of Deely [2] as already noted.

REFERENCES

[1] Cacoullos, T. (1966). Estimation of a multivariate density. *Ann. Inst. Statist. Math. Tokyo* **18**, 179–89.
[2] Deely, J. J. (1966). Non-parametric empirical Bayes procedures for selecting the best of k populations. (Submitted for publication.)
[3] Johns, M. V. Jr. (1957). Non-parametric empirical Bayes procedures. *Ann. Math. Statist.* **28**, 649–69.
[4] Johns, M. V. Jr. and Van Ryzin, J. (1967). Convergence rates for empirical Bayes two-action problems I. Discrete case. Technical report no. 131, Department of Statistics, Stanford University. (Submitted to *Ann. Math. Statist.*)
[5] Johns, M. V. Jr. and Van Ryzin, J. (1967). Convergence rates for empirical Bayes two-action problem II. Continuous case. Technical report no. 132, Department of Statistics, Stanford University. (Submitted to *Ann. Math. Statist.*)
[6] Parzen, E. (1962). On estimation of a probability density. *Ann. Math. Statist.* **33**, 1065–76.
[7] Robbins, H. (1955). An empirical Bayes approach to statistics. *Proc. 3rd Berkeley Symp. Math. Statist. Probab.* **1**, University of California Press, Berkeley, 157–64.
[8] Robbins, H. (1964). The empirical Bayes approach to statistical decision problems. *Ann. Math. Statist.* **35**, 1–20.

DISCUSSION ON VAN RYZIN'S PAPER

J. W. PRATT

Some years ago, Neyman hailed two breakthroughs in statistics. One was the introduction of the ideas relating to empirical Bayes procedures.

Mathematically, this area is very interesting. Then, as I understand it, the interest centered on two questions:

(1) Can you get by without ever learning the values of λ and estimate what you need to by some tricky device?

(2) If nature is malevolent and tries to outwit you, and if her choices of λ are not even required to settle down in a relative frequency sense, can you still get close to the relevant Bayes risk?

In Van Ryzin's paper the restriction to the empirical Bayes problem rather than the compound decision problem excludes the second question. The first question is also pretty directly assumed away through the introduction of extra trials at each value of λ, and the interest centers on:

(3) What is the order of magnitude of the opportunity loss relative to the Bayes risk?

But I don't think Neyman had the mathematical interest of the problem in mind. I think he had in mind either that the method would be of great practical importance or that it would resolve the differences between Bayesian and frequentist views, at least for practical purposes. Perhaps both. There seems to me little evidence that either of these results occurred or is going to, and I would be interested to know what Professor Neyman thinks about the situation now.

I am supposed to be of the Bayesian faith. Actually I am religious only on Sundays, and on working days I am as frequentist as the next statistician. It may nevertheless be worth while to clarify how I would view the problem on Sunday as a fanatical Bayesian.

My first reaction would be that there is no such thing as an unknown prior distribution. G is a parameter (or an infinite vector of parameters if you prefer). But this is a semantic problem only. Since G is unknown, the fanatical Bayesian would put a prior distribution on it. Such a prior distribution is sometimes called a second-story prior distribution— a prior distribution on something one had previously been inclined to consider as itself a prior distribution. This wouldn't be comfortable to assess, but it is what a fully Bayesian analysis would require. That's why

[602]

papers on empirical Bayes procedures aren't really Bayesian and why I am incompetent to discuss them.

There is a possibly interesting mathematical problem here, incidentally: what second-story prior distributions on G would make the Bayesian procedure asymptotically optimal in an empirical Bayes sense?

Now let me revert to my weekday frame of reference, which most of you should be able to share. Given a sequence of decision problems, suppose you subdivide it into sub-sequences and apply the empirical Bayes procedure to each. You will do at least as well asymptotically as if you had applied the procedure to the whole sequence, and if you can find any relevant way to subdivide, you will do better. To be sure, if you restrict yourself to the empirical Bayes problem as stated by Van Ryzin and exclude the possibility of any outside information, there will be no relevant way to subdivide, and you will not be able to do better. In practice, however, one can almost always find some relevant way to subdivide, though it may be highly subjective. A sequence of truly indistinguishable problems almost never arises.

Now if you keep subdividing you will keep doing asymptotically better. Eventually, however, you will subdivide down to the unique problem at hand. Then what? An empirical Bayes procedure is hard to apply to a single problem. In fact, I am not sure it's even defined. At this point, a Bayesian (that is, what I would call a true Bayesian and Bartlett would call a Bayesian Savage), would, of course, resort to using his own prior distribution G. This is the connection I see between empirical Bayes procedures and the Bayesian theory of statistics, and it indicates why I think the empirical Bayes approach does not save orthodox statistical philosophy.

Finally, let me mention another possibly interesting mathematical problem area. Can you say anything useful about the trade-off between the degree of subdivision of the sequence of problems and the order-of-magnitude of the opportunity loss relative to the Bayes risk? How sensitive is this to the degree of relevance of the information you are using to subdivide your sequence of problems?

PART 7

TEACHING OF NONPARAMETRIC
STATISTICS

NONPARAMETRIC METHODS IN
ELEMENTARY STATISTICS

G. E. NOETHER

Recent years have witnessed great changes in what is customarily called nonparametric statistics. Less than twenty years ago in referring to nonparametric procedures Wallis [5] used the adjectives rough-and-ready and Tippet [4] spoke of abbreviated methods. Historically it is presumably true that nonparametric procedures, like the Spearman rank correlation coefficient, were developed to save time. Wilcoxon, after performing a particularly long sequence of t-tests, is supposed to have said: 'There must be an easier way.' Today this easier method is known as Wilcoxon's two-sample or rank sum test, even though a corresponding procedure had been suggested by the German psychologist Gustav Deuchler [2] as early as 1914, thus almost preceding Student's t-test.

Statisticians from many countries met for the First International Symposium on Nonparametric Techniques in Statistical Inference demonstrating dramatically the theoretical advances that have taken place in this rapidly developing branch of statistics. While in general graduate level courses are taking cognizance of these developments, the same can hardly be said of lower level statistics courses. I am primarily interested in discussing introductory statistics courses. But a brief remark about intermediate level courses may not be out of place at this time. What I should like to see are nonparametric methods courses, or even better, courses that integrate nonparametric procedures into the general discussion of statistical methods. However, such an integration should go consideraby beyond the current approach taken by many statistics texts, where one or two chapters are entitled 'Nonparametric Procedures', or possibly 'Distribution-Free Procedures', and where we simply find enumerated five or six nonparametric procedures that happen to strike the author's fancy.

But let me now come to the main topic of my discussion, the introductory statistics course that is being offered each year to tens of thousands, if not hundreds of thousands of nonmathematics and nonstatistics majors all over the country. Students in these courses come from many fields: the life sciences, humanities, education, agriculture, business, but above all from the social sciences. They rarely take statistics voluntarily. They are in the course because of departmental and/or

[607]

graduation requirements. The great majority has minimal preparation in mathematics, not much more than they bring along from high school. They carry over into statistics their prejudices of mathematics. To us as teachers of statistics falls the responsibility to make the course statistically meaningful. But are we doing that?

To answer this question we need only look at the large majority of text-books that have been written for such courses. One recent such book is rather typical. The first half of the book (almost to the page) is entitled *Descriptive Statistics*. After a lengthy discussion of how to organize and graph data, the student is taught successively how to compute the arithmetic mean, the median, the mode, the range, the interquartile range, the mean deviation, the variance, the standard deviation, the correlation coefficient and various regression coefficients. Only then is he told about statistical inference. No wonder the student is bored and decides that statistics is dull and uninteresting.

In recent years another type of text has made its appearance. It contains little or no descriptive statistics. In its place we find something like an axiomatic treatment of probability that takes up at least half the book. This undoubtedly is the right beginning for a student who plans to go deeply into statistics. But I cannot see any justification for spending more than fifty per cent of a student's time set aside for the study of statistics on the study of the theory of probability.

We have to realize that one or even two semesters are not sufficient to produce competent statisticans. All we can hope to achieve is to give the student an appreciation of statistics. I often tell my students that this is going to be a course in statistics appreciation. A single course in music appreciation does not make a musician out of a student. A single course in statistics does not make him a statistician. All it can help him do is to acquire a feel for the statistical way of thinking. Learning how to compute the correlation coefficient from a frequency table hardly achieves this aim. But neither does the axiomatic study of probability, however, intellectually stimulating it may be.

Recently the Committee on the Undergraduate Program in Mathematics of the Mathematical Association of America [1] suggested that a first course in statistics concentrate on basic concepts and then went on: 'Since the main objective of the course is understanding of the basic statistical concepts, proofs and extensive manipulations of formulas should be employed sparingly. While statistics utilizes these, its major focus is on inference from data. By the same token, the course should not dwell upon computational techniques. The amount of computation should be determined by how much it helps the student to understand the principles involved.'

Most standard texts known to me discuss concepts in terms of normal theory methods. But normal theory methods require both computational facility and reasonably sophisticated results from probability, before they can make sense. As a consequence, basic statistical concepts are wrapped up in a great deal of technical detail that detracts from the ideas being discussed. So what else can we do?

In recent years I have been experimenting with a course that is built around the binomial distribution and a relatively small number of nonparametric procedures. For this approach the probabilistic preparations and the computational efforts are minimal, and the results seem rather satisfactory to judge from limited student reaction.

The following are the headings of a fifteen week course with two hours of lectures a week and two additional hours of discussion and testing:

1. What is Statistics?
2. The Meaning of Probability.
3. Basic Concepts of Probability.
4. The Binomial Distribution.
5. The Normal Distribution.
6. Estimation.
7. Tests of Hypotheses.
8. Tests of Hypotheses (*continued*).
9. Chi-Square Tests.
10. Contingency Tables.
11. The One-Sample Problem.
12. The Two-Sample Problem (paired observations).
13. The Two-Sample Problem (independent samples).
14. The k-Sample Problem.
15. Rank Correlation.

Some supplementary remarks are in order. In § 2, I primarily discuss the frequency interpretation of probability. § 3 deals with the addition and multiplication theorems with particular emphasis on the concept and meaning of independence. In § 4, I ignore combinatorial problems, but make extensive use of tables of binomial probabilities. The normal distribution in §5 is introduced primarily as a tool for approximating binomial probabilities. At this stage its use as a mathematical model is mentioned only very lightly. Thus only about 25 per cent of the total course time is taken up with strictly probabilistic discussions.

The ideas of estimation and hypothesis testing have already been introduced intuitively in §1. They are elaborated upon in §§6–8 using the binomial distribution as the underlying model. Discussing tests of hypotheses for binomial situations has two important advantages. Significance levels that can be achieved may differ considerably from such standard levels like 0·05 or 0·01. Thus the student does not get the idea that these customary levels are dictated by statistical considerations. Secondly, the power of a test can be read immediately from tables of binomial probabilities without the need for complicated noncentrality parameters. The chi-square test of §9 is a rather simple extension of the normal approximation for the binomial tests of §§7 and 8. In §10, I discuss the chi-square test both as a test of independence and of homogeneity. In this way I have the possibility to bring up questions about the design of an experiment and to prepare the ground for a possible second semester discussion of normal correlation and regression.

The remaining five sections are concerned with the solution of problems arising from the analysis of continuous data. When discussing the one-sample situation, I use the median, rather than the mean, as the parameter of greatest interest. Conceptually the median is a much simpler quantity than the mean, which in my opinion cannot be satisfactorily 'explained' with the mathematical apparatus available in such a course. An obvious confidence interval for the population median is provided by the smallest and largest observations in the sample. The probability laws developed in §3 are sufficient to compute the confidence coefficient associated with this interval. From this extreme confidence interval it is a simple step to go to a narrower interval bounded by the kth smallest and largest observations. Of course now tables of confidence coefficients have to take over. For symmetric populations it is another 'obvious' step to consider confidence intervals bounded by averages $(x_i + x_j)/2$ rather than simple order statistics. Using the relationship between confidence intervals and tests of hypotheses already brought out in connection with binomial problems, we are then led to the sign and Wilcoxon one-sample tests for a hypothetical median.

The two-sample problem for paired observations in §12 is a useful application of the material of the preceding section. It also provides the opportunity to continue the discussion of design problems. In §13, I discuss the Wilcoxon two-sample test in the Mann–Whitney U-form. The U-form of the test procedure does not only have the advantage of greater symmetry, but it also is more easily converted into a confidence interval for a shift (or scale) parameter.

The k-sample problem of § 14 is handled by the Kruskal–Wallis test in case of independent samples and the Friedman test in case of randomized blocks. I usually also discuss paired comparisons, that is, comparisons in blocks of size 2. There are many practical experiments of this type that students can carry out very simply. Multiple comparisons can be introduced in various ways. Perhaps the simplest method is the one suggested by Dunn [3].

In § 15, I discuss the Kendall rank correlation coefficient, first as a measure of strength of relationship and then in tests of independence against the alternative of a monotone trend.

The preceding discussion emphasizes the greater simplicity of the non-parametric approach compared to the normal theory approach. Let me look at just one example in greater detail. I shall choose the two-sample problem for independent samples, the problem that in elementary statistics texts usually receives the most detailed discussion.

Using the normal theory approach we have to inflict on the poor student the following t-statistic:

$$ t = \frac{(\bar{y} - \bar{x}) - (\mu_y - \mu_z)}{\sqrt{\dfrac{\Sigma(x - \bar{x})^2 + \Sigma(y - \bar{y})^2}{m + n - 2}}} \sqrt{\frac{mn}{m + n}}. $$

To a person with the necessary background in probability this expression makes a great deal of sense. To the great majority of students whom I have in mind this is simply an expression thought up by malevolent statisticians to confuse them. In contrast the corresponding Mann–Whitney U-statistic is simplicity itself:

$$ U = \text{number of } (y_j > x_i). $$

Since presumably we want to find out whether on the average y-observations are greater or smaller than x-observations, what could be more logical than to find out how often these two events occur in the two samples.

For both the t- and U-procedures we have to consult appropriate tables in order to decide whether to accept or reject the given hypothesis or to find confidence intervals. But even here the nonparametric procedure is intuitively more logical. The U-table is entered by using the two sample sizes m and n. The t-table is entered with $m + n - 2$ 'degrees of freedom'. Suppose further that a student should be interested enough to ask how the tabulated values are computed. In the case of t we can only wave our hands and say that mathematical statisticians are able to derive

612 G. E. NOETHER

appropriate formulas. In the case of U, by choosing m and n sufficiently small, say $m = 2$ and $n = 3$, we can derive the complete distribution of U in a minute or two by simple enumeration. The derivation of the null distribution of various nonparametric statistics is a much more useful application of equally likely cases in probability than the perennial games of chance.

Personally I feel that a nonparametric approach is a much preferable way to teach elementary statistics. From my limited experience I can say that students seem to enjoy this kind of course considerably more than the standard normal theory course. On the other hand, I am afraid that there are still a great many statisticians and users of statistics who simply cannot conceive of a first course that does not present Student's t-test. Our job along with the development of new nonparametric theory is to think about ways and means of how this theory can be put to use in achieving a better understanding of statistics.

REFERENCES

[1] CUPM. *A transfer curriculum in mathematics for two year colleges*, Berkeley, 1969.

[2] Deuchler, Gustav. Über die Methoden der Korrelationsrechnung in der Pädagogik und Psychologie. *Z. Pädag. Psychol. Exp. Pädag.* 15 (1914), 114–31, 145–59, and 229–42.

[3] Dunn, O. J. Multiple comparisons using rank sums. *Technometrics*, 6 (1964), 241–52.

[4] Tippet, L. H. C. *The Methods of Statistics*, 4th Edition. New York: Wiley (1952).

[5] Wallis, W. A. Rough-and-ready statistical tests. *Ind. Qual. Control*, 8 (1952), 35–40.

INDEX

Number of cycles 549

Numerical inversion of characteristic function 439–43

Okamoto's inequality 334

Olkin, I., Chebyshev bounds 465

One-sample test 294

One-sided tests 579, 581

Optimal permutation tests 44

Optimal rank tests 42–4

Orbit 217

Order statistic 48, 275, 281, 284, 285, 290, 292, 294, 353

generalized 47

linear combination of 457

methods for studying 349–57

of sample and extended sample 417–23

rth 370

theory of 359, 361

Ordered families, relationship to efficiency robustness problems 99–102

Orthonormal transformation 393, 394

Overall preference 122

p-variate cumulative distribution function 141

p-variate normal random variables 138

p-variate situations 130

Paired comparisons 127–8, 130, 132

Parametric estimator 148

Parametric test 62

Partial orderings of permutations 12–13

Partial sum process 23

Partitioned vectors 142

Partitioning events 422

Percentile-modified tests 101

Permutation analysis 154

Permutation covariance matrix, asymptotic convergence of 137

Permutation distribution 135, 140

asymptotic normality of 137

Permutation groups 216–20

Permutation rank order statistics, asymptotic distribution theory of 135–40

Permutation tests 41, 46, 50, 145, 154, 284

derivation of 44–5

most powerful 50

optimal 44

Permutational covariance matrix 145

Permutationally distribution free test statistic 120, 147

Permutations, partial orderings of 12–13

Pitman efficiency 91, 405–15

Pitman functions 217, 223

Pitman statistic 235

Play-the-Winner (PW) rule 515

Point probability density 429

Poisson processes 215, 216, 226, 228, 249, 251, 252, 263

homogeneous 223–31

nonhomogeneous 223–31

Poisson random variable 230

Power function 44, 48

Power spectrum 128

Pratt, J. W., empirical Bayes procedures 602

Preference, region of 548

Preference equality, test of 111

Preference parameters 112, 113, 126

Preference vector 111

Principal component analysis 128

Probabilistic modelling 154

Probability, bounds on, of correct selection 497

infimum of, of correct selection 495

of error 581, 582

of event 390, 398; determination of 385

that $X_{k,n} = X_{k+j,n+m}$ 417

Probability density 50, 194

bell-shaped 428

Probability density function 345

Probability distribution function 179–81

Probability distributions 113, 428

bell-shaped 432

estimation of 177, 178

Probability function, cumulative 429

for treatment pair 113

Probability measure 243

on measurable spaces 49

Probability model 123

Probability requirements 545–8, 552, 553, 554, 556, 557

Probability space 27, 180, 336

Prohorov's theorem 182, 184

Projection method 26, 75

Prokhorov's metric 28

Pseudolikelihood 563, 564, 571

approach to two-decision problems 563–78

Puri, M. L., incomplete block designs 131

Pyke, R., estimation of regression 196

rank statistics 21

Quantile function 9, 10

Quantile process 356

Quantile statistic 12

see also Sample quantiles

Quantiles in nonstandard cases, asymptotic distributions of 343–8

'Quick' statistical procedures 99

Quick tests 106–8

r-order Kolmogorov condition 179